FLUID MECHANICS

FLUID MECHANICS

Ninth Edition

Victor L. Streeter
Professor Emeritus of Hydraulics
University of Michigan

E. Benjamin Wylie
Professor of Civil and Environmental Engineering
University of Michigan

Keith W. Bedford
Professor of Civil Engineering
Ohio State University

WCB
McGraw-Hill

Boston, Massachusetts • Burr Ridge, Illinois • Dubuque, Iowa
Madison, Wisconsin • New York, New York • San Francisco, California • St. Louis, Missouri

WCB/McGraw-Hill

A Division of The **McGraw·Hill** *Companies*

FLUID MECHANICS

This book is printed on acid-free paper.

1 2 3 4 5 6 7 8 9 0 VNH/VNH 9 0 9 8 7

ISBN 0-07-062537-9

Vice president and editorial director: *Kevin T. Kane*
Publisher: *Tom Casson*
Executive editor: *Eric Munson*
Marketing Manager: *John T. Wannemacher*
Project manager: *Kimberly Schau*
Production supervisor: *Heather D. Burbridge*
Senior designer: *Laurie J. Entringer*
Compositor: *Publication Services, Inc.*
Photo research coordinator: *Sharon Miller*
Typeface: *10/12 Times Roman*
Printer: *Von Hoffmann Press, Inc.*

Library of Congress Cataloging-in-Publication Data

Streeter, Victor L. (Victor Lyle) (date)
    Fluid mechanics / Victor L. Streeter, E. Benjamin Wylie, Keith
  Bedford. – Ninth ed.
        p.   cm.
    Includes index.
    ISBN 0-07-062537-9
    1. Fluid mechanics.   I. Wylie, E. Benjamin.   II. Bedford, Keith
  W.   III. Title.
  TA357.S8   1998
  620.1′06–dc21   97-36324

http://www.mhhe.com

To our families, students, and teachers

PREFACE

In this, the ninth edition, a number of significant changes have been made to the text in scope, organization, focus, and required prerequisite skills.

The scope of the book has been broadened to include heat and mass transport. Therefore, two new chapters on transport fundamentals and transport applications have been added and the chapters concerned with properties, basic equations, dimensional analysis, and measurements have been extended to cover the relevant transport material. With this new material it is now possible to teach two complete junior-senior level introductory courses with this text, one in fluid mechanics and one in transport phenomena.

Other major content changes include the elimination of the chapter on compressible flows, the combining of the steady and unsteady pipe flow chapters into one updated chapter, the splitting of the governing equation chapter into one chapter for the control volume equations and one chapter for the continuum equation methods, a revised chapter on measurement methods, and the incorporation of almost 400 new problems.

Insofar as prerequisite skills, the second series of substantive changes is motivated by the rapid infusion and implementation of computing and information transfer. As mentioned elsewhere in the text, the eighth edition was among the first to fully integrate computing techniques in its presentation. It did so at a time when BASIC and FORTRAN were the computer languages, calculations were computed on a mainframe computer in batch mode, and everyone wrote their own programs from scratch. The present availability of cheap, powerful, networked computing and relevant coursework has resulted in two additional changes to the text. First, we no longer feel that the book needs to teach numerical methods; rather we feel that the existence of required coursework in computer programming and numerical methods in almost all undergraduate technical curricula now allows us to begin concentrating on the routine use of numerical techniques to analyze problems. Further, we have removed the BASIC programs that were present in the eighth edition and present complex problem solutions in Microsoft EXCEL.

The second computer-based change originates in the Internet–World Wide Web availability and the additional fact that new information of use in courses often arrives at a pace that is more frequent than the four- or five-year cycle required to incorporate information in a reissued text. There is also information of relevance to the text that is either too big, too costly, or too transient in its nature to put in a text. For these reasons a Web page is now in place for the book that will serve as a repository for rapidly developing information and communications.

The Web page also will contain substantial information on computing, principally files and information for transfer to the user of large programs used in the text. Tutorials also are included on the use of structured languages such as MATHCAD, MATLAB and MATHEMATICA in analyzing fluid mechanics and transport. Our use of these languages is neither a recommendation nor final evaluation of the relative merits of these languages over any other. These choices are made because of their widespread availability in university student computing facilities.

We strongly urge the students and instructors to scan the Web material before and during the course. We view these tutorials as support information only, and it is not necessary to use any of these tutorials to achieve mastery of the fluids or transport material. However, these are powerful labor-saving tools which make analyzing numbers more enjoyable, and we urge the reader to investigate these languages.

Finally, the authors wish to thank Sean O'Neil and Panagiotis Velissariou for their extremely able help in preparing this book. It has been a pleasure working with them.

K. W. Bedford, Columbus, Ohio
E. B. Wylie, Ann Arbor, Michigan
November 1997

Web Page Address:
http://www.mhhe.com

CONTENTS

FUNDAMENTALS OF FLUID MECHANICS AND TRANSPORT

In the first four chapters of Part 1, the properties of fluids, fluid statics, and the underlying framework of concepts and definitions of fluid dynamics and associated heat and mass transport are discussed. Integral methods leading to control volume equations are presented in Chapter 3, and continuum descriptions leading to nonlinear partial differential equations are presented in Chapter 4. Dimensionless parameters and equation scale analysis are next introduced. Chapters 6 and 7 deal with the effects of friction in real flows either in pipes and channels (internal flows) or over surfaces such as spheres (external flows). The specialized case of frictionless flow is presented next while the final chapter introduces heat and mass transport principles.

1

Fluid Properties

The engineering science of fluid mechanics has developed because of an understanding of fluid properties, the application of the basic laws of mechanics and thermodynamics, and orderly experimentation. The properties of density and viscosity play principal roles in open- and closed-channel flow and in flow around immersed objects. Surface-tension effects are important in the formation of droplets, in the flow of small jets, and in situations where liquid-gas-solid or liquid-liquid-solid interfaces occur, as well as in the formation of capillary waves. The property of vapor pressure, which accounts for changes of phase from liquid to gas, becomes important when reduced pressures are encountered.

In this chapter a fluid is defined, and consistent systems of force, mass, length, time, and temperature units are discussed before the discussion of properties and definition of terms are taken up.

1.1 CONTINUUM

In dealing with fluid-flow relations on a mathematical or analytical basis, it is necessary to consider that the actual molecular structure is replaced by a hypothetical continuous medium, called the *continuum*. For example, velocity at a point in space is indefinite in a molecular medium, as it would be zero at all times except when a molecule occupied this exact point; it then would be the velocity of the molecule and not the mean mass velocity of the particles in the neighborhood. This dilemma is avoided if one considers velocity at a point to be the average or mass velocity of all molecules surrounding the point, say, within a small sphere with a radius that is large compared with the *mean distance between molecules*. With n molecules per cubic centimeter, the mean distance between molecules is of the order $n^{-1/3}$ cm. Molecular theory, however, must be used to calculate fluid properties (e.g., viscosity) which are associated with molecular motions, but continuum equations can be employed with the results of molecular calculations.

In rarefied gases, such as the atmosphere at 50 mi above sea level, the ratio of the mean free path† of the gas to a characteristic length for a body or conduit is used to distinguish the type of flow. The flow regime is called *gas dynamics* for very small values of the ratio; the next regime is called *slip flow;* and for large values of the ratio it is *free molecule flow*. In this text only the gas-dynamics regime is studied.

The quantities density, specific volume, pressure, viscosity, velocity, acceleration, etc., are assumed to vary continuously throughout a fluid (or be constant).

EXERCISE

1.1.1 Under which two of the following flow regimes would the assumption of a continuum be reasonable? (1) free molecule flow, (2) slip flow, (3) gas dynamics, (4) complete vacuum, (5) liquid flow. (*a*) 1, 2; (*b*) 1, 4; (*c*) 2, 3; (*d*) 3, 5; (*e*) 1, 5.

1.2 DEFINITION OF A FLUID

A fluid is a substance that deforms continuously when subjected to a shear stress, no matter how small that shear stress may be. A shear force is the force component tangent to a surface, and this force divided by the area of the surface is the average shear stress over the area. Shear stress at a point is the limiting value of shear force to area as the area is reduced to the point.

In Fig. 1.1 a substance is placed between two closely spaced parallel plates so large that conditions at their edges may be neglected. The lower plate is fixed, and a force F is applied to the upper plate, which exerts a shear stress F/A on any substance between the plates. A is the area of the upper plate. If the force F causes the upper plate to move with a steady (nonzero) velocity, no matter how small the magnitude of F, then the substance between the two plates is a fluid.

The fluid in immediate contact with a solid boundary has the same velocity as the boundary; that is, there is no slip at the boundary [1].‡ This is an experimental observation which has been verified in countless tests with various kinds of fluids

---

†The *mean free path* is the average distance a molecule travels between collisions.
‡Numbered references will be found at the end of this chapter.

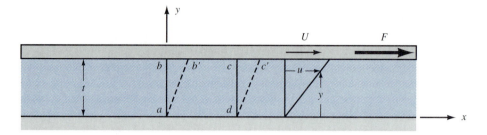

Figure 1.1   Deformation resulting from application of constant shear force.

and boundary materials. The fluid in the area *abcd* flows to the new position *ab'c'd*, each fluid particle moving parallel to the plate and the velocity *u* varying uniformly from zero at the stationary plate to *U* at the upper plate. Experiments show that, other quantities being held constant, *F* is directly proportional to *A* and to *U* and is inversely proportional to thickness *t*. In equation form

$$F = \mu \frac{AU}{t}$$

in which μ is the proportionality factor and includes the effect of the particular fluid. If $\tau = F/A$ for the shear stress,

$$\tau = \mu \frac{U}{t}$$

The ratio *U/t* is the angular velocity of line *ab*, or it is the *rate of angular deformation* of the fluid, that is, the rate of decrease of angle *bad*. The angular velocity can also be written as *du/dy*, as both *U/t* and *du/dy* express the velocity change divided by the distance over which the change occurs. However, *du/dy* is more general, since it holds for situations in which the angular velocity and shear stress change with *y*. The velocity gradient *du/dy* can also be visualized as the rate at which one layer moves relative to an adjacent layer. In differential form,

$$\tau = \mu \frac{du}{dy} \tag{1.2.1}$$

is the relation between shear stress and rate of angular deformation for one-dimensional flow of a fluid. The proportionality factor μ is called the *viscosity* of the fluid, and Eq. (1.2.1) is *Newton's law of viscosity*.

Materials other than fluids cannot satisfy the definition of a fluid. A plastic substance will deform a certain *amount* proportional to the force, but not continuously when the stress applied is below its yield shear stress. A complete vacuum between the plates would cause deformation at an ever-increasing rate. If sand were placed between the two plates, Coulomb friction would require a finite force to cause a continuous motion. Hence, plastics and solids are excluded from the classification of fluids.

Fluids are classified as Newtonian or non-Newtonian. In a Newtonian fluid there is a linear relation between the magnitude of applied shear stress and the resulting rate of deformation [μ in Eq. (1.2.1) is constant], as shown in Fig. 1.2. In a non-Newtonian fluid there is a nonlinear relation between the magnitude of applied shear stress and the rate of angular deformation. An *ideal plastic* has a definite yield stress and a constant linear relation of τ to *du/dy*. A *thixotropic* substance, such as printer's

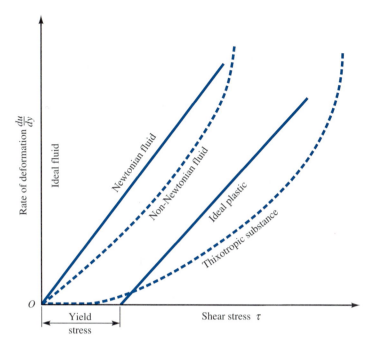

Figure 1.2   Rheological diagram.

ink, has a viscosity that is dependent upon the immediately prior angular deformation of the substance and has a tendency to solidify when at rest. Gases and most common liquids tend to be Newtonian fluids, while thick, long-chained hydrocarbons may be non-Newtonian.

For purposes of analysis, the assumption is frequently made that a fluid is non-viscous. With zero viscosity the shear stress is always zero, regardless of the motion of the fluid. If the fluid is also considered to be incompressible, it is then called an *ideal* fluid and plots as the ordinate in Fig. 1.2.

EXERCISES

1.2.1   A fluid is a substance that (*a*) always expands until it fills any container; (*b*) is practically incompressible; (*c*) cannot be subjected to shear forces; (*d*) cannot remain at rest under action of any shear force; (*e*) has the same shear stress at a point regardless of its motion.

1.2.2   Newton's law of viscosity relates (*a*) pressures, velocity, and viscosity; (*b*) shear stress and rate of angular deformation in a fluid; (*c*) shear stress, temperature, viscosity, and velocity; (*d*) pressure, viscosity, and rate of angular deformation; (*e*) yield shear stress, rate of angular deformation, and viscosity.

1.3   DIMENSIONS AND UNITS

Consistent units of force, mass, length, time, and temperature greatly simplify problem solutions in mechanics; also, derivations can be carried out without reference to any particular consistent system if units are used in a consistent manner. A system of mechanics units is said to be consistent when a unit force causes a unit mass to undergo a unit acceleration. The International System (SI) has been adopted in most countries and is expected to be adopted in the United States soon. This system has the newton (N) as the unit of force, the kilogram (kg) as the unit of mass, the meter (m) as the unit of length, and the second (s) as the unit of time. With the kilogram, meter, and second as defined units, the newton is derived to exactly satisfy Newton's second law of motion

$$1 \text{ N} \equiv 1 \text{ kg} \cdot 1 \ \frac{\text{m}}{\text{s}^2} \qquad (1.3.1)$$

In the United States the consistent set of units at present is the pound (lb) force, the slug mass, the foot (ft) length, and the second (s) time. The slug is the derived unit; it is the unit of mass at which one pound accelerates at one foot per second squared, or

$$1 \text{ lb} \equiv 1 \text{ slug} \cdot 1 \ \frac{\text{ft}}{\text{s}^2} \qquad (1.3.2)$$

The pound-foot-slug-second system will be referred to as the U.S. customary system (USC) in this text.

Some professional engineering groups in the United States use the inconsistent set of units pound (lb) force, pound (lb_m) mass, foot (ft) length, and second (s) time. With inconsistent units a proportionality constant is required in Newton's second law, usually written as

$$F = \frac{m}{g_0} a \qquad (1.3.3)$$

When 1-pound force acts on 1-pound mass at standard gravity in a vacuum, the mass is accelerated at 32.174 ft/s², or

$$1 \text{ lb} = 1 \ \frac{\text{lb}_m}{g_0} \cdot 32.174 \ \frac{\text{ft}}{\text{s}^2}$$

from which g_0 can be determined:

$$g_0 = 32.174 \ \text{lb}_m \cdot \text{lb}/\text{ft} \cdot \text{s}^2 \qquad (1.3.4)$$

g_0 has this fixed value for this set of units, whether applied under standard conditions or on the moon.

The mass M of a body does not change with location, but the weight W of a body is determined by the product of the mass and the local acceleration of gravity g:

$$W = Mg \qquad (1.3.5)$$

For example, where $g = 9.806$ m/s², a body with a gravity force of 10 N has a mass $M = 10/9.806$ kg. At a location where $g = 9.7$ m/s², the weight W is

$$W = \frac{10 \text{ N}}{9.806 \text{ m/s}^2} (9.7 \text{ m/s}^2) = 9.892 \text{ N}$$

Standard gravity, which is acceleration due to gravity at sea level, is 9.806 m/s² in SI. On the inside front cover, many conversions for various units are given. Since they are presented in the form of dimensionless ratios equal to 1, they can be used on one side of an equation, as a multiplier or as a divisor, to convert units.

In Table 1.1 the dimensions and units of the systems as well as values for g_0 are shown. Temperature units, kelvins (K)† and degrees Rankine (°R), are discussed in Section 1.6.

Table 1.1   Common systems of units and values of g_0

System	Mass	Length	Time	Force	Temperature	g_0
SI	kg	m	s	N	K	$1\ \text{kg·m/N·s}^2$
USC	slug	ft	s	lb	°R	$1\ \text{slug·ft/lb·s}^2$
U.S. inconsistent	lb_m	ft	s	lb	°R	$32.174\ \text{lb}_m\text{·ft/lb·s}^2$
Metric, cgs	g	cm	s	dyn	K	$1\ \text{g·cm/dyn·s}^2$
Metric, mks	kg	m	s	kg_f	K	$9.806\ \text{kg·m/kg}_f\text{·s}^2$

Abbreviations of SI units are written in lowercase (small) letters for terms like hours (h), meters (m), and seconds (s). When a unit is named after a person, the abbreviation (but not the spelled form) is capitalized; examples are watt (W), pascal (Pa), and newton (N). The abbreviation L for liter is an exception, made for clarity. Multiples and submultiples in powers of 10^3 are indicated by prefixes, which also are abbreviated. Common prefixes are shown in Table 1.2. Note that prefixes may not be doubled up: the correct form for 10^{-9} is the prefix nano, as in nanometers; combinations of, for example, millimicro, formerly acceptable, are no longer to be used.

Table 1.2   Selected prefixes for powers of 10 SI units

Multiple	SI prefix	Abbreviation	Multiple	SI prefix	Abbreviation
10^9	giga	G	10^{-3}	milli	m
10^6	mega	M	10^{-6}	micro	μ
10^3	kilo	k	10^{-9}	nano	n
10^{-2}	centi	c	10^{-12}	pico	p

EXERCISES

1.3.1   An object has a mass of 2 kg and weighs 19 N on a spring balance. The value of gravity at this location, in meters per second squared, is (a) 0.105; (b) 2; (c) 9.5; (d) 19; (e) none of these answers.

1.3.2   An unbalanced force of 10 N exerted on a 2-kg mass causes an acceleration, in m/s², of (a) 0.2; (b) 2.0; (c) 5.0; (d) 20.0; (e) none of these answers.

1.3.3   The gravity force, in newtons, of a 3-kg mass on a planet where $g = 10$ m/s² is (a) 0.30; (b) 3.33; (c) 29.42; (d) 30; (e) none of these answers.

1.3.4   A pressure intensity of 10^9 Pa can be written as (a) gPa; (b) GPa; (c) kMPa; (d) μPa; (e) none of these answers.

---

†In 1967 the name *degree Kelvin* (°K) was changed to *kelvin* (K).

1.4   VISCOSITY

Absolute Viscosity

A fluid's viscosity is a significant property in the study of fluid flow. The nature and characteristics of viscosity, as well as dimensions and conversion factors for both absolute and kinematic viscosity are discussed in this section. Viscosity is that property of a fluid by virtue of which it offers resistance to shear. Newton's law of viscosity [Eq. (1.2.1)] states that for a given rate of angular deformation of fluid the shear stress is directly proportional to the viscosity. Molasses and tar are examples of highly viscous liquids; water and air have very small viscosities.

The viscosity of a gas increases with temperature, but the viscosity of a liquid decreases with temperature. The variation in temperature trends can be explained by examining the causes of viscosity. The resistance of a fluid to shear depends upon its cohesion and upon its rate of transfer of molecular momentum. A liquid, with molecules much more closely spaced than a gas, has cohesive forces much larger than a gas. Cohesion appears to be the predominant cause of viscosity in a liquid, and since cohesion decreases with temperature, the viscosity does likewise. A gas, on the other hand, has very small cohesive forces. Most of its resistance to shear stress is the result of the transfer of molecular momentum.

As a rough model of how momentum transfer gives rise to an apparent shear stress, consider two idealized railroad cars loaded with sponges and on parallel tracks, as in Fig. 1.3. Assume each car has a water tank and pump so arranged that the water is directed by nozzles at right angles to the track. First, consider A stationary and B moving to the right, with the water from its nozzle striking A and being absorbed by the sponges. Car A will be set in motion owing to the component of the momentum of the jets which is parallel to the tracks, giving rise to an apparent shear stress between A and B. Now if A is pumping water back into B at the same rate, its action tends to slow down B and equal and opposite apparent shear forces result. When both A and B are stationary or have the same velocity, the pumping does not exert an apparent shear stress on either car.

Within fluid there is always a transfer of molecules back and forth across any fictitious surface drawn in it. When one layer moves relative to an adjacent layer, the molecular transfer of momentum brings momentum from one side to the other so that an apparent shear stress is set up that resists the relative motion and tends to equalize the velocities of adjacent layers in a manner analogous to that of Fig. 1.3. The measure of the motion of one layer relative to an adjacent layer is du/dy.

Molecular activity gives rise to an apparent shear stress in gases which is more important than the cohesive forces, and since molecular activity increases with temperature, the viscosity of a gas also increases with temperature.

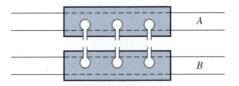

Figure 1.3   Model illustrating transfer of momentum.

For ordinary pressures viscosity is independent of pressure and depends upon temperature only. For very great pressures, gases and most liquids have shown erratic variations of viscosity with pressure.

A fluid at rest or in motion so that no layer moves relative to an adjacent layer will not have apparent shear forces set up, regardless of the viscosity, because du/dy is zero throughout the fluid. Hence, in the study of *fluid statics,* no shear forces can be considered because they do not occur in a static fluid, and the only stresses remaining are normal stresses, or pressures. This greatly simplifies the study of fluid statics, since any free body of fluid can have only gravity forces and normal surface forces acting on it.

The dimensions of viscosity are determined from Newton's law of viscosity [Eq. (1.2.1)]. Solving for the viscosity μ

$$\mu = \frac{\tau}{du/dy}$$

and inserting dimensions F, L, and T for force, length, and time, respectively

$$\tau : FL^{-2} \qquad u : LT^{-1} \qquad y : L$$

shows that μ has the dimensions $FL^{-2}T$. With the force dimension expressed in terms of mass by use of Newton's second law of motion, $F = MLT^{-2}$, the dimensions of viscosity can be expressed as $ML^{-1}T^{-1}$.

The SI unit of viscosity, newton-seconds per square meter ($N \cdot s/m^2$) or kilograms per meter-second ($kg/m \cdot s$), has no name. The USC unit of viscosity (also nameless) is 1 lb·s/ft^2 or 1 slug/ft·s (these are identical). A common unit of viscosity is the cgs unit, called the poise (P); it is 1 dyn·s/cm^2 or 1 g/cm·s. The SI unit is 10 times larger than the poise unit.†

Kinematic Viscosity

The viscosity μ is frequently referred to as the *absolute* viscosity or the *dynamic* viscosity to avoid confusing it with the *kinematic* viscosity ν, which is the ratio of viscosity to mass density:

$$\nu = \frac{\mu}{\rho} \tag{1.4.1}$$

The kinematic viscosity occurs in many applications, for example, in the dimensionless Reynolds number for motion of a body through a fluid, Vl/ν, in which V is the body velocity and l is a representative linear measure of the body size. The dimensions of ν are L^2T^{-1}. The SI unit of kinematic viscosity is 1 m^2/s, and the USC unit is 1 ft^2/s. The cgs unit, called the stoke (St), is 1 cm^2/s.

In SI units, to convert from ν to μ, it is necessary to multiply ν by ρ, the mass density in kilograms per cubic meter. In USC units μ is obtained from ν by multiplying by the mass density in slugs per cubic foot. To change from stoke to the poise, one multiplies by mass density in grams per cubic centimeter, which is numerically equal to specific gravity.

†The conversion from the USC unit of viscosity to the SI unit is

$$\frac{1 \text{ slug}}{\text{ft·s}} \frac{14.594 \text{ kg}}{\text{slug}} \frac{1 \text{ ft}}{0.3048 \text{ m}} = 47.9 \text{ kg/m·s} \quad \text{or} \quad \frac{1 \text{ USC unit viscosity}}{47.9 \text{ SI units viscosity}} = 1$$

| Example 1.1 | **A** liquid has a viscosity of 0.005 kg/m·s and a density of 850 kg/m³. Calculate the kinematic viscosity in (*a*) SI and (*b*) USC units and (*c*) the viscosity in USC units. |

Solution

(*a*) $\quad \nu = \dfrac{\mu}{\rho} = \dfrac{0.005 \text{ kg/m·s}}{850 \text{ kg/m}^3} = 5.882 \ \mu\text{m}^2\text{/s}$

(*b*) $\quad \nu = (5.882 \times 10^{-6} \text{ m}^2\text{/s}) \left(\dfrac{1 \text{ ft}}{0.3048 \text{ m}} \right)^2 = 6.331 \times 10^{-5} \text{ ft}^2\text{/s}$

(*c*) $\quad \mu = (0.005 \text{ kg/m·s}) \dfrac{1 \text{ slug/ft·s}}{47.9 \text{ kg/m·s}} = 0.0001044 \text{ slug/ft·s}$

Viscosity is practically independent of pressure and depends upon temperature only. The kinematic viscosity of liquids, and of gases at a given pressure, is substantially a function of temperature. Charts for the determination of absolute viscosity and kinematic viscosity are given in Appendix C, Figs. C.1 and C.2, respectively.

| Example 1.2 | **I**n Fig. 1.4 a lubricated shaft rotates inside a concentric sleeve bearing at 1200 rpm. The clearance δ is small with respect to the radius R so a linear velocity distribution in the lubricant may be assumed. What are the power requirements to rotate the shaft? $R = 2$ cm, $L = 6$ cm, $\delta = 0.1$ mm, and $\mu = 0.2$ N·s/m². |

Solution

The energy loss, due to viscous shear, per unit time dictates the power requirements. This will be given by the torque required to rotate the shaft at the designated speed.

$$\text{Power} = T\omega$$

The applied torque is given by the shear stress acting over the surface area multiplied by the moment arm R.

$$\tau = \mu \frac{du}{dy} = \mu \frac{\omega R}{\delta} = 0.2(1200) \frac{2\pi}{60} \frac{0.02}{0.0001} = 5026.5 \text{ N/m}^2$$

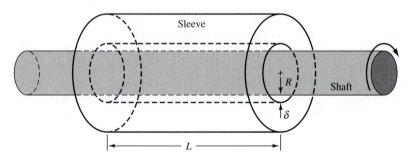

Figure 1.4   Shaft rotating in sleeve.

$$T = \tau(2\pi RL)R = (5026.5)(2\pi)(0.02)(0.06)(0.02) = 0.758 \text{ Nm}$$

$$\text{Power} = T\omega = 0.758(1200)\frac{2\pi}{60} = 95.3 \text{ W}$$

EXERCISES

1.4.1 Viscosity has the dimensions (a) $FL^{-1}T$; (b) $FL^{-1}T^{-1}$; (c)FLT^{-2}; (d) FL^2T; (e) FLT^2.

1.4.2 Select the *incorrect* completion. Apparent shear forces (a) can never occur when the fluid is at rest; (b) may occur owing to cohesion when the liquid is at rest; (c) depend upon molecular interchange of momentum; (d) depend upon cohesive forces; (e) can never occur in a frictionless fluid regardless of its motion.

1.4.3 Correct units for dynamic viscosity are (a) m·s/kg; (b) N·m/s^2; (c) kg·s/N; (d) kg/m·s; (e) N·s/m.

1.4.4 Viscosity, expressed in poises, is converted to the USC unit of viscosity by multiplying by (a) $\frac{1}{479}$; (b) 479; (c) ρ; (d) $1/\rho$; (e) none of these answers.

1.4.5 The dimensions for kinematic viscosity are (a) $FL^{-2}T$; (b) $ML^{-1}T^{-1}$; (c) L^2T^2; (d) L^2T^{-1}; (e) L^2T^{-2}.

1.4.6 Based on Fig. C.1 the viscosity of kerosene at 20°C, in newton-seconds per square meter, is (a) 4×10^{-5}; (b) 4×10^{-4}; (c) 1.93×10^{-3}; (d) 1.93×10^{-2}; (e) 1.8×10^{-2}.

1.4.7 The kinematic viscosity of dry air at 30°C and 760-mm mercury (Hg), in square meters per second, is (a) 1.7×10^{-5}; (b) 1.7×10^{-4}; (c) 1.73×10^{-6}; (d) 1.92×10^{-5}; (e) none of these answers.

1.4.8 For $\mu = 0.06$ kg/m·s and specific gravity = 0.60, ν is, in stokes, (a) 2.78; (b) 1.0; (c) 0.60; (d) 0.36; (e) none of these answers.

1.4.9 For $\mu = 2.0 \times 10^{-4}$ slug/ft·s, the value of μ, in pound-seconds per square foot, is (a) 1.03×10^{-4}; (b) 2.0×10^{-4}; (c) 6.21×10^{-4}; (d) 6.44×10^{-3}; (e) none of these answers.

1.4.10 For $\nu = 3 \times 10^{-8}$ m^2/s and $\rho = 800$ kg/m^3, μ equals (a) 3.75×10^{-11}; (b) 2.4×10^{-5}; (c) 2.4×10^5; (d) 2.4×10^{12}; (e) none of these answers.

1.5 MASS, WEIGHT, AND CONCENTRATION VARIABLES

Engineers are called upon to analyze not only simple fluids of one molecular structure (e.g., water) but also mixtures of two or more components that may be different in molecular structure but similar in phase (e.g., water and oil) or different in both structure and phase (e.g., water and sediment). Measures of mass and weight for simple fluids are straightforward while similar measures for binary (two component) or multicomponent mixtures often require the use of additive measures based on mass fractions or ratios of mass fractions.

Simple Fluids

The *density* ρ of a fluid is defined as its mass per unit volume. To define density at a point, the mass Δm of fluid in a small volume ΔV surrounding the point is divided by ΔV and the limit is taken as ΔV becomes a value ϵ^3 in which ϵ is still large compared with the mean distance between molecules,

$$\rho = \lim_{\Delta V \to \epsilon^3} \frac{\Delta m}{\Delta V} \tag{1.5.1}$$

For water at standard pressure (760-mm Hg) and 4°C (39.2°F), $\rho = 1000 \text{ kg/m}^3$, or 1.94 slugs/ft^3.

The *specific volume* v_s is the reciprocal of the density ρ; that is, it is the volume occupied by unit mass of fluid. Hence,

$$v_s = \frac{1}{\rho} \tag{1.5.2}$$

The *specific weight* γ of a fluid is its weight per unit volume. It changes with location, depending upon gravity. Hence,

$$\gamma = \rho g \tag{1.5.3}$$

It is a convenient property when dealing with fluid statics or with liquids with a free surface.

The *specific gravity* S of a substance is the ratio of its weight to the weight of an equal volume of water at standard conditions. It can also be expressed as a ratio of its density or specific weight to that of water.

Multicomponent Mixtures

Suppose that instead of a clear fluid, several other forms of mass are also in the reservoir of fluid. For example, a lake might contain several classes of sediment, dissolved oxygen, and several molecular forms of nitrogen and phosphorus. The specification of the various mass and weight measures can be done by analyzing the total mixture values or the mass and weight quantities for each class or mass fraction.

The *mixture density* is defined as

$$\rho = \lim_{\Delta V \to \epsilon^3} \frac{\sum_{i}^{n} \Delta m_i}{\Delta V} = \lim_{\Delta V \to \epsilon^3} \frac{\sum_{i}^{n} \rho_i \, \Delta V_i}{\Delta V} \tag{1.5.4}$$

In Eq. (1.5.4) the subscript i refers to all the substances in the mixture, Δm_i and ρ_i are the mass and density of the ith fractions, respectively, and ΔV_i is the total volume of the ith fraction.

The *mass fraction* of the ith component is unitless and defined as

$$\omega_i = \frac{\rho_i \, \Delta V_i}{\rho \, \Delta V} = \frac{\Delta m_i}{m} \tag{1.5.5}$$

and as can be seen, ω_i can be no greater than 1. The *mass concentration* is defined as

$$C_i = \frac{\Delta m_i}{\Delta V} = \frac{\rho_i \, \Delta V_i}{\Delta V} \tag{1.5.6}$$

and has units of (M/L^3). The *volume concentration* is defined as $c_i = \Delta V_i / \Delta V$.

It needs to be noted that whereas the density of each material in the mixture, ρ_i, will be a property of the material, ω_i and C_i are not properties of the mixture but are instead space- and time-varying variables. In reading the literature in many different fields one must also be careful to determine which of the two relative measures is being used since the symbol notation is not universal between fields. The notation used here is commonly utilized by civil and environmental engineers.

C_i units are selected based upon the numerical magnitude of the mass of the component relative to the mixture density. For example, it is common in water quality or waste-water analysis to see concentrations expressed in units of mg/L instead of kg/m^3. This is because the value in the latter is quite small and causes difficulty when used in mathematical analyses (i.e., multiplication of big and small numbers). A 10-mg/L dissolved oxygen content would in turn be equivalent to 0.01 g/L, 0.00001 kg/L, and 0.00000001 kg/m^3.

Occasionally certain mass fraction calculations are used so frequently that they are expressed in a shorthand form. For instance, the labels parts per thousand (ppt) or parts per million (ppm) refer to mass fraction concentration, for example, a salt content of 24 g per kg of mixture is

$$\frac{24 \text{ g}}{1 \text{ kg}} = \frac{24 \text{ g}}{1000 \text{ g}} = 24 \text{ ppt} \tag{1.5.7}$$

While salt or sediment fractions often are high enough to warrant the ppt label, certain water quality variables such as organic or inorganic compound fractions are small enough to require the parts per million (ppm) description. For example, 0.005 g per kg is 0.005 ppt or 5.0 ppm.

Relationship Between Density, Temperature, and Concentration

Adding heat to a volume of most fluids will decrease the density, while as oceanographers are well aware, adding salt to fresh water, that is, adding mass, will cause the density to increase. The detailed relationship between temperature and density for a gas is also dependent upon the state of pressure, and this relation is discussed in the next section.

For liquids the relationship between density and temperature or added mass is also well known. If our attention is confined to fresh water, the relationship between density (kg/m^3) and temperature (°C) is

$$\rho_w = 999.939900 + 4.216485(10^{-2})T - 7.097451(10^{-3})T^2 \tag{1.5.8}$$
$$+ 3.509571(10^{-5})T^3 - 9.9037785(10^{-8})T^4$$

As shown in Fig. 1.5 the density maximum value for fresh water at standard conditions occurs at 4°C while freezing occurs at 0°C. This small difference in density, that is, $\rho(T = 4°C) - \rho(T = 0°C) = (1000.0000 - 999.9399) \text{ kg/m}^3 = 0.06 \text{ kg/m}^3$, is responsible for significant physical differences in the behavior of lakes, rivers, and estuaries during the seasonal heating and cooling cycle. In fact such small differences in density cause considerable dynamical complexity in bodies of water. Therefore, when performing these density calculations it is necessary that accuracies be as high as six significant figures.

The addition of mass in the form of salt or perhaps sediment can also cause changes in the density. From the oceanographic community, an approximate relationship between density, temperature, and salinity, s, (in ppt) for mixtures at low

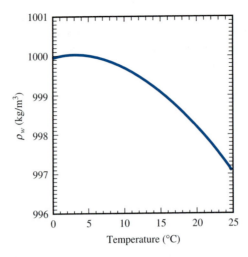

Figure 1.5   Water density versus temperature.

pressure is [2]

$$\rho(s, T) = \rho_w + s\{0.824493 - 4.0899(10^{-3})T + 7.6438(10^{-5})T^2 \qquad \text{(1.5.9)}$$
$$- 8.2467(10^{-7})T^3 + 5.3875(10^{-9})T^4\} + s^{3/2}\{-5.72466(10^{-3})$$
$$+ 1.0227(10^{-4})T - 1.6546(10^{-6})T^2\} + s^2\{4.8314(10^{-4})\}$$

The addition of salt suppresses both the temperature at which the density max-imum value occurs and the freezing point of the water; this discussion however is beyond the scope of this textbook, and an oceanography textbook should be con-sulted.

As seen in Fig. 1.5, a considerable portion of the density versus temperature curve is linear. If analysis is confined to a small variation in density about a fixed or reference density, then a linear form of these density relationships can be used

$$\rho = \rho_r(1 - \beta_t \, \Delta T) \qquad \text{(1.5.10)}$$

$$\rho = \rho_r(1 + \beta_s \, \Delta s) \qquad \text{(1.5.11)}$$

$$\rho = \rho_r(1 + \beta_c \, \Delta C) \qquad \text{(1.5.12)}$$

Here β_t, β_s, and β_c refer to volume "expansion coefficients" for temperature, salt, and sediment concentration, respectively; ρ_r refers to the reference density about which the linear perturbations occur; and ΔT, Δs, and ΔC refer to the differences in temperature, salt, or concentration, respectively, that cause the density to change. It is noted that ρ_r should not be near the density maximum value portion of the temperature-density curve as the relationship is nonlinear in this regime.

1.6   TEMPERATURE AND THERMODYNAMIC VARIABLES

The *heat content* in a volume, ΔV, of fluid equals

$$Q_H = \rho c_p T \, \Delta V \qquad \text{(1.6.1)}$$

where c_p is the *specific heat at constant pressure* and T is the absolute temperature. Heat is measured in dimensions of joules (J) for the SI units, ft-lb in the USC system, and British thermal units (Btu) for the US inconsistent system.

The absolute *temperature* in the SI (and cgs) system has dimensions of kelvins (K) while the dimension of temperature in the British system is degrees Rankine (°R). More common use is made of relative temperature scales, that is, degrees Centigrade (°C) and degrees Fahrenheit (°F). These two systems are related as follows

$$°R = (°F) + 460 \tag{1.6.2}$$

$$K = (°C) + 273 \tag{1.6.3}$$

The Centigrade and Fahrenheit temperatures are related by

$$°F = \frac{9}{5}(°C) + 32 \tag{1.6.4}$$

$$°C = \frac{5}{9}(°F - 32) \tag{1.6.5}$$

The specific heat at constant pressure, c_p, and its companion the *specific heat at constant volume, c_v*, are properties describing the internal energy of the volume, particularly for ideal gases. They are viewed in an elementary way as heat capacities in that c_p represents the amount of heat that must be added to a unit mass of fluid to raise the temperature one unit. At standard conditions 4187 J of heat raises the temperature of 1 kg of water 1 K while the addition of 1 Btu of heat raises the temperature of 1 lb-mass of water 1 degree Fahrenheit. The units are J/kg·K, ft-lb/slug·°R, or Btu/lb$_m$·°R. Values for selected gases are found in Appendix C. Special care should be exercised to ensure that g_o has been used if the mass units are in slugs. For an ideal gas c_v and c_p are related by

$$c_p = c_v + R \tag{1.6.6}$$

and the *ratio of specific heats, c_p/c_v*, is defined as k which for many gases has a value of 1.4 (see Appendix C).

Two other variables used in thermodynamics are *intrinsic energy, u^{**}*, and *enthalpy, h*. Intrinsic energy depends upon p, ρ, and T and is the energy per unit mass due to molecular scale forces and spacings. Enthalpy is defined as

$$h = u^{**} + p/\rho \tag{1.6.7}$$

and has been used in this form due to its frequent occurrence in thermodynamic calculations.

1.7 PRESSURE AND A PERFECT GAS

As will be elaborated upon in Chap. 2 the *pressure* at a point derives from a normal force pushing against a plane defined in the fluid or a plane surface that the fluid is in contact with. The pressure at a point is the ratio of the normal force to the area of the plane as the area approaches a small value enclosing the point. Pressure, p, has units of force per unit area, which may be newtons per square meter, called pascals (Pa), or pounds per square foot (psf), or pounds per square inch (psi).

Liquids normally cannot sustain a tensile (or pulling apart) stress since the liquid would vaporize. Therefore, the absolute pressures used in this book are never negative, since this would imply that the fluid is sustaining a tensile stress. Liquids can often sustain a considerable pressure or compressive force with little or no change in the observed density. However, no universal relationships between pressure and density exist for a liquid.

Gas responds to changes in pressure or compressive force. For ideal gases a relationship between pressure and density can be explicitly stated. The perfect gas, as used herein, is defined as a substance that satisfies the *perfect-gas law*

$$pv_s = RT \tag{1.7.1}$$

and that has constant specific heats; p is the absolute pressure, v_s the specific volume, R the gas constant, and T the absolute temperature. The perfect gas must be carefully distinguished from the ideal fluid. An ideal fluid is frictionless and incompressible. The perfect gas has viscosity and can therefore develop shear stresses, and it is compressible according to Eq. (1.7.1).

Equation (1.7.1) is the equation of state for a perfect gas. It can be written as

$$p = \rho RT \tag{1.7.2}$$

The units of R can be determined from the equation when the other units are known. For p in pascals, ρ in kilograms per cubic meter, and T in kelvins (K)

$$R = \frac{N}{m^2} \frac{m^3}{kg \cdot K} = \frac{m \cdot N}{kg \cdot K} \quad \text{or} \quad m \cdot N/kg \cdot K$$

For USC units, $°R = °F + 459.6$

$$R = \frac{lb}{ft^2} \frac{ft^3}{slug \cdot °R} = \frac{ft \cdot lb}{slug \cdot °R} \quad \text{or} \quad ft \cdot lb/slug \cdot °R$$

For ρ in pounds mass per cubic foot

$$R = \frac{lb}{ft^2} \frac{ft^3}{lb_m \cdot °R} = \frac{ft \cdot lb}{lb_m \cdot °R} \quad \text{or} \quad ft \cdot lb/lb_m \cdot °R$$

The magnitude of R in slugs is 32.174 times greater than in pounds mass. Values of R for several common gases are given in Table C.3 of Appendix C.

Real gases below critical pressure and above the critical temperature tend to obey the perfect-gas law. As the pressure increases, the discrepancy increases and becomes serious near the critical point. The perfect-gas law encompasses both Charles' law and Boyle's law. Charles' law states that for constant pressure the volume of a given mass of gas varies as its absolute temperature. Boyle's law (isothermal law) states that for constant temperature the density varies directly as the absolute pressure. The volume V of m mass units of gas is mv_s; hence,

$$pV = mRT \tag{1.7.3}$$

Certain simplifications result from writing the perfect-gas law on a mole basis. A kilogram mole of gas is the number of kilograms mass of gas equal to the molecular weight; for example, a kilogram mole of oxygen O_2 is 32 kg. With $\bar{v}_s$ being the volume per mole, the perfect-gas law becomes

$$p\bar{v}_s = MRT \tag{1.7.4}$$

if M is the molecular weight. In general, if n is the number of moles of the gas in volume V, then

$$pV = nMRT \tag{1.7.5}$$

since $nM = m$. Now, from Avogadro's law, equal volumes of gases at the same absolute temperature and pressure have the same number of molecules; hence their masses are proportional to the molecular weights. From Eq. (1.7.5) it is seen that MR must be constant, since pV/nT is the same for any perfect gas. The product MR, called the *universal gas constant,* has a value depending only upon the units employed. It is

$$MR = 8312 \text{ m·N/kg·mol·K} \tag{1.7.6}$$

The gas constant R can then be determined from

$$R = \frac{8312}{M} \text{ m·N/kg·K} \tag{1.7.7}$$

In USC units

$$R = \frac{49,709}{M} \text{ ft·lb/slug·°R} \tag{1.7.8}$$

In pounds mass units

$$R = \frac{1545}{M} \text{ ft·lb/lb}_m\text{·°R} \tag{1.7.9}$$

so that knowledge of molecular weight leads to the value of R. In Table C.3 of Appendix C molecular weights of some common gases are listed. Additional relations and definitions used in perfect-gas flow are introduced in Chap. 3.

A gas with molecular weight of 44 is at a pressure of 0.9 MPa and a temperature of 20°C. Determine its density.

Example 1.3

Solution

From Eq. (1.7.7)

$$R = \frac{8312}{44} = 188.91 \text{ m·N/kg·K}$$

Then, from Eq. (1.7.2)

$$\rho = \frac{p}{RT} = \frac{0.9 \times 10^6 \text{ N/m}^2}{(188.91 \text{ m·N/kg·K})(273 + 20 \text{ K})} = 16.26 \text{ kg/m}^3$$

EXERCISES

1.7.1 A perfect gas (*a*) has zero viscosity; (*b*) has constant viscosity; (*c*) is incompressible; (*d*) satisfies $p\rho = RT$; (*e*) fits none of these statements.

1.7.2 The molecular weight of a gas is 28. The value of R in meter-newtons per kilogram-kelvin is (*a*) 29.7; (*b*) 297; (*c*) 2911; (*d*) 8312; (*e*) none of these answers.

1.7.3 The density of air at 10°C and 1-MPa absolute (abs) in SI units is (*a*) 1.231; (*b*) 12.31; (*c*) 65.0; (*d*) 118.4; (*e*) none of these answers.

1.7.4 How many kilograms mass of carbon monoxide gas at 20°C and 200-kPa abs is contained in a volume of 100 L? (*a*) 0.00023; (*b*) 0.23; (*c*) 3.367; (*d*) 3367; (*e*) none of these answers.

1.7.5   A container holds 1-kg air at 30°C and 9-MPa abs. If 1.5-kg air is added and the final temperature is 110°C, the final absolute pressure is (*a*) 7.26 MPa; (*b*) 25.3 MPa; (*c*) 73.4 MPa; (*d*) indeterminable; (*e*) none of these answers.

1.8   BULK MODULUS OF ELASTICITY

In the preceding section the compressibility of a perfect gas is described by the perfect-gas law. For most purposes a liquid can be considered as incompressible, but for situations involving either sudden or great changes in pressure its compressibility becomes important. Liquid (and gas) compressibility also becomes important when temperature changes are involved, for example, free convection. The compressibility of a liquid is expressed by its *bulk modulus of elasticity*. If the pressure of a *unit volume* of liquid is increased by dp, it will cause a volume decrease $-d\mathcal{V}$; the ratio $-dp/d\mathcal{V}$ is the bulk modulus of elasticity K. For any volume $\mathcal{V}$ of liquid

$$K = -\frac{dp}{d\mathcal{V}/\mathcal{V}} \qquad (1.8.1)$$

Since $d\mathcal{V}/\mathcal{V}$ is dimensionless, K is expressed in units of p. For water at 20°C (Table C.1, Appendix C) $K = 2.2$ GPa or, from Table C.2, $K = 311,000$ lb/in^2 for water at 60°F.

To gain some idea about the compressibility of water, consider the application of 0.1 MPa (about 1 atm) to a cubic meter of water.

$$-d\mathcal{V} = \frac{\mathcal{V}\,dp}{K} = \frac{(1.0 \text{ m}^3)(0.1 \text{ MPa})}{2.2 \text{ GPa}} = \frac{1}{22,000} \text{ m}^3$$

or about 45.5 cm^3. As a liquid is compressed, its resistance to further compression increases. At 3000 atm the value of K for water has doubled.

Example 1.4

A liquid compressed in a cylinder has a volume of 1 liter (L = 1000 cm^3) at 1 MN/m^2 and a volume of 995 cm^3 at 2 MN/m^2. What is its bulk modulus of elasticity?

Solution

$$K = -\frac{\Delta p}{\Delta \mathcal{V}/\mathcal{V}} = -\frac{(2-1) \text{ MN/m}^2}{(995-1000)/1000} = 200 \text{ MPa}$$

For most civil and environmnental engineers, water is considered essentially incompressible; however, the oceanographic and limnological communities often work in water depths sufficiently deep to give rise to small but important volume changes due to compression. Two practical consequences are: the density variation with temperature and salt must be modified to include local pressure, and the volume changes can result in anomalous readings of measured concentrations between a deep water site where the data were measured versus shipboard analysis.

The density function correction for compression is beyond the scope of this text and can be found in any modern oceanography textbook. An example of the concentration measurement anomaly follows.

With regard to Example 1.4 suppose the cylinder is a water quality sample bottle used to collect water samples at a predetermined depth. At deep depths the sample bottle has a smaller volume to collect (995 cm^3) due to compression. Suppose that analysis reveals that 15 mg of sediment are collected. What would be the difference in concentration data measured shipboard where the pressure is atmospheric versus the in-situ depths where the sample was collected?

Example 1.5

Solution

Shipboard:

$$C_{ship} = \frac{15 \text{ mg}}{1 \text{ L}} = 1.5(10^{-2}) \text{ kg/m}^3$$

Collection site:

$$C_{site} = \frac{15 \text{ mg}}{0.995 \text{ L}} = 1.507(10^{-2}) \text{ kg/m}^3$$

In other words, though small, the in-situ data for concentration are higher than the shipboard due to the effect of volume compression. This problem becomes more pronounced when very small chemical concentrations are being measured, say for regulatory purposes.

EXERCISES

1.8.1 The bulk modulus of elasticity K for a gas at constant temperature T_0 is given by (a) p/ρ; (b) RT_0; (c) ρp; (d) ρRT_0; (e) none of these answers.

1.8.2 The bulk modulus of elasticity (a) is independent of temperature; (b) increases with the pressure; (c) has the dimensions of $1/p$; (d) is larger when the fluid is more compressible; (e) is independent of pressure and viscosity.

1.8.3 For a 70-atm increase in pressure the density of water has increased, in percent, by about (a) $\frac{1}{300}$; (b) $\frac{1}{30}$; (c) $\frac{1}{3}$; (d) $\frac{1}{2}$; (e) none of these answers.

1.8.4 A pressure of 1 MPa applied to 300-L liquid causes a volume reduction of 0.6 L. The bulk modulus of elasticity in GPa is (a) -0.5; (b) 0.5; (c) 50; (d) 500; (e) none of these answers.

1.9 VAPOR PRESSURE

Liquids evaporate because of molecules escaping from the liquid surface. The vapor molecules exert a partial pressure in the space, known as *vapor pressure*. If the space above the liquid is confined, after a sufficient time the number of vapor molecules striking the liquid surface and condensing is just equal to the number escaping in any interval of time and equilibrium exists. Since this phenomenon depends upon molecular activity, which is a function of temperature, the vapor pressure of a given liquid depends upon temperature and increases with it. When the pressure above a liquid equals the vapor pressure of the liquid, boiling occurs. Boiling of water, for example, can occur at room temperature if the pressure is reduced sufficiently. At

20°C water has a vapor pressure of 2.451-kPa absolute and mercury has a vapor pressure of 0.173-Pa absolute.

In many situations involving the flow of liquids it is possible for very low pressures to be produced at certain locations in the system. Under such circumstances the pressures may be equal to or less than the vapor pressure. When this occurs, the liquid flashes into vapor. This phenomenon is called *cavitation*. A rapidly expanding vapor pocket, or cavity, forms, which is usually swept away from its point of origin and enters regions of the flow where the pressure is greater than vapor pressure. The cavity collapses. This growth and decay of the vapor bubbles affects the operating performance of hydraulic pumps and turbines and can result in erosion of the metal parts in the region of cavitation.

EXERCISE

1.9.1 The vapor pressure of water at 30°C, in pascals, is (*a*) 0.44; (*b*) 7.18; (*c*) 223; (*d*) 4315; (*e*) none of these answers.

1.10   SURFACE TENSION

At the interface between a liquid and a gas, or two immiscible liquids, a film or special layer seems to form on the liquid, apparently owing to attraction of liquid molecules below the surface. It is a simple experiment to place a small needle on a quiet water surface and observe that it is supported there by the film.

The formation of this film can be visualized on the basis of *surface energy* or work per unit area required to bring the molecules to the surface. The surface tension is then the stretching force required to form the film, obtained by dividing the surface-energy term by unit length of the film in equilibrium. The surface tension of water varies from about 0.074 N/m at 20°C to 0.059 N/m at 100°C. Surface tensions, along with other properties, are given for a few common liquids in Table 1.3.

Table 1.3   Approximate properties of common liquids at 20°C and standard atmospheric pressure

Liquid	Specific gravity S	Bulk modulus of elasticity K, GPa	Vapor pressure p_v, kPa	Surface tension† σ, N/m
Alcohol, ethyl	0.79	1.21	5.86	0.0223
Benzene	0.88	1.03	10.0	0.0289
Carbon tetrachloride	1.59	1.10	13.1	0.0267
Kerosene	0.81	...	...	0.023–0.032
Mercury	13.57	26.20	0.00017	0.51
Oil:				
Crude	0.85–0.93	...	...	0.023–0.038
Lubricating	0.85–0.88	...	...	0.023–0.038
Water	1.00	2.2	2.45	0.074

† In contact with air.

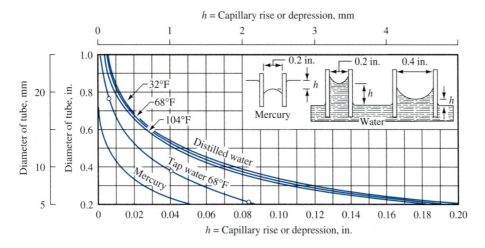

Figure 1.6 Capillarity in circular glass tubes. *(By permission from P.L. Daugherty. "Hydraulics," McGraw-Hill, New York, 1937.)*

The action of surface tension is to increase the pressure within a droplet of liquid or within a small liquid jet. For a small spherical droplet of radius r the internal pressure p necessary to balance the tensile force due to the surface tension σ is calculated in terms of the forces which act on a hemispherical free body (see Sec. 2.6),

$$p\pi r^2 = 2\pi r\sigma \quad \text{or} \quad p = \frac{2\sigma}{r}$$

For the cylindrical liquid jet of radius r, the pipe-tension equation [Eq. 2.6.5] applies:

$$p = \frac{\sigma}{r}$$

Both equations show that the pressure becomes large for a very small radius of droplet or cylinder.

Capillary attraction is caused by surface tension and by the relative value of adhesion between liquid and solid to cohesion of the liquid. A liquid that *wets* the solid has a greater adhesion than cohesion. The action of surface tension in this case is to cause the liquid to rise within a small vertical tube that is partially immersed in it. For liquids that do not wet the solid, surface tension tends to depress the meniscus in a small vertical tube. When the contact angle between liquid and solid is known, the capillary rise can be computed for an assumed shape of the meniscus. Figure 1.6 shows the capillary rise for water and mercury in circular glass tubes in air.

Example 1.6

The conical tube in Fig. 1.7 has its axis horizontal and contains an elongated droplet of liquid, as shown. Find the force tending to move the droplet to the right for angles α between 0 and 12°; $r = 3$ mm, $x = 15$ mm, $\theta = 25°$, and $\sigma = 0.05$ N/m.

Solution

For the left face of the droplet in the conical tube where the radius is r, the force is given by $2\pi r\sigma$. The force component along the longitudinal x-axis is $2\pi r\sigma \cos(\theta - \alpha)$. The force balance in the x direction yields

$$F = 2\pi\sigma[(r + x\tan\alpha)\cos(\theta + \alpha) - r\cos(\theta - \alpha)]$$

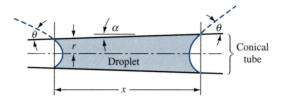

Figure 1.7   Droplet in conical tube.

A calculator or spreadsheet may be used to produce the following results.

α, deg	0	3	6	9	12
$F(10)^6$, N	0	176	341	494	634

PROBLEMS

1.1   If air at standard temperature and pressure has on the order of 10^{18} molecules per cubic millimeter, estimate the mean free path, Δs, by using the information in Sec. 1.1. The flow regime studied in this book, called gas dynamics, is valid for bodies of what characteristic length, ℓ? Assume the appropriate ratio to be greater than 100.

1.2   Classify the substance that has the following rates of deformation and corresponding shear stresses:

du/dy, rad/s	0	1	3	5
τ, kPa	15	20	30	40

1.3   Classify the following substances (maintained at constant temperature):
(*a*)

du/dy, rad/s	0	3	4	6	5	4
τ, lb/ft^2	2	4	6	8	6	4

(*b*)

du/dy, rad/s	0	0.5	1.1	1.8
τ, N/m^2	0	2	4	6

(*c*)

du/dy, rad/s	0	0.3	0.6	0.9	1.2
τ, lb/ft^2	0	2	4	6	8

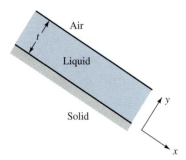

Figure 1.8 Problem 1.4.

1.4 A Newtonian liquid flows down an inclined plane in a thin sheet of thickness t (Fig. 1.8). The upper surface is in contact with air, which offers almost no resistance to the flow. Using Newton's law of viscosity, decide what the value of du/dy, y measured normal to the inclined plane, must be at the upper surface. Would a linear variation of u with y be expected?

1.5 What kinds of rheological materials are paint and grease?

1.6 Determine the weight in pounds of 3-slugs mass at a place where $g = 31.7 \text{ ft/s}^2$.

1.7 When standard scale weights and a balance are used, a body is found to be equivalent in pull of gravity to two of the 1-lb scale weights at a location where $g = 31.5 \text{ ft/s}^2$. What would the body weigh on a correctly calibrated spring balance (for sea level) at this location?

1.8 A mass of gasoline equal to 450 kg is stored in a tank. What is its weight in newtons and in pounds on the earth's surface? What would its mass and weight be if located on the moon's surface where the local acceleration due to gravity is approximately one-sixth that at the earth's surface?

1.9 On another planet, where standard gravity is 3 m/s², what would be the value of the proportionality constant g_0 in terms of the kilogram force, gram, millimeter, and seconds?

1.10 A correctly calibrated spring scale records the weight of a 2-kg body as 17.0 N at a location away from the earth. What is the value of g at this location?

1.11 Does the weight of a 20-N bag of flour at sea level denote a force or the mass of the flour? What is the mass of the flour in kilograms? What are the mass and weight of the flour at a location where the gravitational acceleration is one-seventh that of the earth's standard?

1.12 A Newtonian fluid is in the clearance between a shaft and a concentric sleeve. When a force of 600 N is applied to the sleeve parallel to the shaft, the sleeve attains a speed of 1 m/s. If a 1500-N force is applied, what speed will the sleeve attain? The temperature of the sleeve remains constant.

1.13 Convert 10.4 SI units of kinematic viscosity to USC units of dynamic viscosity if $S = 0.85$.

1.14 A shear stress of 4 N/m² causes a Newtonian fluid to have an angular deformation of 100 rad/s. What is its viscosity?

1.15 A plate 0.5-mm distant from a fixed plate moves at 0.25 m/s and requires a force per unit area of 2 Pa (N/m²) to maintain this speed. Determine the fluid viscosity of the substance between the plates in SI units.

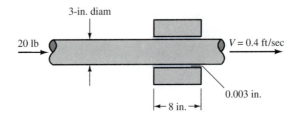

Figure 1.9   Problem 1.16.

1.16   Determine the viscosity of fluid between the shaft and sleeve in Fig. 1.9.

1.17   A flywheel weighing 600 N has a radius of gyration of 300 mm. When it is rotating at 600 rpm, its speed reduces 1 rpm/min owing to fluid viscosity between the sleeve and shaft. The sleeve length is 50 mm; shaft diameter is 20 mm; and radial clearance is 0.05 mm. Determine the fluid viscosity.

1.18   A 25-mm-diameter steel cylinder 300-mm long falls, because of its own weight, at a uniform rate of 0.1 m/s inside a tube of slightly larger diameter. A castor-oil film of constant thickness is between the cylinder and the tube. Determine the clearance between the tube and the cylinder. The temperature is 38°C. The specific gravity of steel = 7.85.

1.19   A piston of diameter 60.00 mm moves inside a cylinder of 60.10 mm. Determine the percent decrease in force necessary to move the piston when the lubricant warms up from 0 to 120°C. Use the crude-oil viscosity from Fig. C.1, Appendix C.

1.20   A 12-kg cube slides down an inclined plane making an angle of 30° with the horizontal. A fluid film 0.1 mm thick separates solid and surface. The fluid viscosity is 0.04 N·s/m². Assuming the velocity distribution in the film is linear, find the terminal velocity of the block. The area of the cube in contact with the film is 0.25 m².

1.21   How much greater is the viscosity of water at 0°C than at 100°C? How much greater is its kinematic viscosity for the same temperature range?

1.22   A fluid has a viscosity of 6 cP and a density of 50 lb_m/ft^3. Determine its kinematic viscosity in USC units and in stokes.

1.23   A fluid has a specific gravity of 0.83 and a kinematic viscosity of $4 (10)^{-4}$ m²/s. What is its viscosity in USC and SI units?

1.24   A body weighing 120 lb with a flat surface area of 2 ft² slides down a lubricated inclined plane making a 30° angle with the horizontal. For viscosity of 0.002 lb·s/ft² and body speed of 3 ft/s, determine the lubricant film thickness.

1.25   What is the viscosity of gasoline at 25°C in poises? Use Fig. C.1, Appendix C.

1.26   A liquid has a specific weight of 48 lb/ft³ and a dynamic viscosity of 3.05 lb·s/ft². What is the kinematic viscosity?

1.27   What is the specific volume in cubic feet per pound mass and cubic feet per slug of a substance of specific gravity 0.75?

1.28   What is the relation between specific volume and specific weight?

1.29   The density of a substance is 2.94 g/cm³. In SI units, what is its (*a*) specific gravity, (*b*) specific volume, and (*c*) specific weight?

1.30 Calculate the value of the gas constant R in SI units if $R = 1545/M$ ft·lb/lb$_m$·°R.

1.31 How much thermal energy is needed to raise the temperature of a liter of water 1°C?

1.32 An airbag is inflated by approximately 0.15 kg of gas in 50 ms. If the temperature change is about 200°C, calculate the average power associated with this deployment.

1.33 A force expressed by $\mathbf{F} = 4\mathbf{i} + 3\mathbf{j} + 9\mathbf{k}$ acts upon a square area 2 by 2 in the xy plane. Resolve this force into a normal- and a shear-force component. What are the pressure and the shear stress? Repeat the calculations for $\mathbf{F} = -4\mathbf{i} + 3\mathbf{j} - 9\mathbf{k}$.

1.34 A "Niskin bottle" is used to sample the water at a point 30 m above the bottom in 100 m of water in the Gulf of Mexico. When the sample is taken, the bottle is immediately brought to the surface and on board a ship. The temperature of the sample is measured to be 11.6°C. Using the typical value of salinity for the Gulf of Mexico water of 33 ppt, what is the density of the water at the sampling point?

1.35 For the previous problem it is determined that there is 1.42 g of sediment in the sample. The size distribution of the suspended sediment is determined using a laser-based particle size analyzer (see Chap. 9) with the results given as:

Component i	Diameter (μm)	Mass fraction†
1	<10	0.1650
2	18	0.2100
3	25	0.2650
4	30	0.1500
5	40	0.0850
6	50	0.0465
7	60	0.0250
8	70	0.0220
9	80	0.0195
10	90	0.0120

† $\Sigma = 1.0000$

Complete the table by determining the mass, volume, and concentration for each component of the sample (including the seawater component), and determine the mixture density. Assume that the sediment is quartz.

1.36 A reservoir contains a two-component mixture of water and sediment. The water density is ρ_w, and the density of the sediment particles is ρ_s. Assuming complete mixing, find the density of the mixture, ρ_m, if the mass fraction of the sediment is ω_s.

1.37 A reservoir contains a multicomponent mixture of water and various species. The water density is ρ_w and the density of each species is ρ_i (where $i = 1, 2, \ldots, n$). Assuming complete mixing, find the density of the mixture, ρ_m (water and species), if the mass fraction of each species is ω_i. This is a generalization of Prob. 1.36.

1.38 The density of fresh water depends upon the temperature as described by Eq. (1.5.8). Prove that the maximum density of the fresh water occurs at $T = 4$°C and find its value.

1.39 For a multicomponent mixture of n species show that

$$\omega_1 + \omega_2 + \cdots + \omega_n = 1$$

$$C_1 + C_2 + \cdots + C_n = \rho_m$$

where ρ_m is the mixture density and ω_i and C_i are the mass fraction and concentration of the species, i, respectively.

1.40 For a binary mixture of species A and B, with $\omega_A/\omega_B = \lambda$, find λ such that the density of the mixture (ρ_m) becomes maximum. What is this maximum value?

1.41 Analyzing a wastewater sample in the laboratory, the following data were obtained

Sample	Volume (mL)	Mass of suspended solids (g)
1	85	85.43

If the specific gravity of the suspended solids is 1.58, determine the volume and the concentration of the suspended solids, and the density of the wastewater sample.

1.42 Three wastewater samples were taken from the same site at the same time, and after analyzing them in the laboratory the following results were obtained

Sample	Volume (mL)	Mass of suspended solids (g)
1	75	23.0
2	83.2	35.6
3	80	Glass container broken

The three samples are of the same density (ρ), and the first one contains solids of specific gravity 1.93. Find the density ρ and the concentration of the suspended solids in the three samples. Take the mass fraction of solids in Sample 3 to be the average of the first two.

1.43 In a juice processing plant, a concentrated juice mixture of orange, pineapple, and kiwi is produced by passing the fresh juice mixture through an evaporator. The mass fractions of the solids contained in the mixture are $\omega_{orange} = 6.7$ percent, $\omega_{pineapple} = 4.35$ percent, and $\omega_{kiwi} = 7.83$ percent. In the evaporator water is removed and the total mass fraction of the solids is increased to $\omega_T = 48.45$ percent. If the fresh juice mixture is delivered to the evaporator at a rate of 850 kg/hr (1.43 m^3/hr), determine:

(*a*) The concentration of the orange, pineapple, and kiwi in the fresh juice.

(*b*) The concentration of the orange, pineapple, and kiwi in the concentrated juice.

(*c*) The density of the fresh juice.

(*d*) The density of the concentrated juice.

1.44 A gas at 20°C and 0.2-MPa abs has a volume of 40 L and a gas constant $R = 210$ m·N/kg·K. Determine the density and mass of the gas.

1.45 What is the specific weight of air at 40-kPa abs and 30°C?

1.46 What is the density of water vapor at 0.4-MPa abs and 15°C in SI units?

1.47 A gas with molecular weight 28 has a volume of 4.0 ft^3 and a pressure and temperature of 2000 psfa (lb/ft^2 abs) and 600°R, respectively. What are its specific volume and specific weight?

1.48 One kilogram of hydrogen is confined in a volume of 150 L at $-40°C$. What is the pressure?

1.49 An automobile tire has a volume of 20 L where the air pressure is 180 kPa and temperature 21°C. What is the air density and weight? After driving two hours, the tire temperature has risen 30 degrees. Estimate the tire pressure, and state any assumptions you make.

1.50 Air is being heated in a heat exchanger at a pressure 45 kPa. If the mass of the air is 4.35 kg, calculate the amount of heat required to heat the air from 45°C to 250°C.

1.51 A gas mixture contains 15-g H_2, 25-g NH_3, and 21-g CO_2. Calculate the molar fractions y_{NH_3}, y_{CO_2}, and y_{H_2}, and the average molecular weight of the gas mixture.

1.52 The volume of the gaseous mixture described in Prob. 1.51 is 250 cm³. What is the density of the mixture? At a temperature 32°C find the pressure of the gas mixture. What are the partial pressures p_{NH_3}, p_{CO_2}, and p_{H_2}?

1.53 An open tank contains water at 22°C with 32.7 percent (by weight) suspended solids of specific gravity 2.32. If the volume of the mixture is 1.2 m³, calculate the concentration of the solids in lb_m/ft^3 and kg/m³.

1.54 Express the bulk modulus of elasticity in terms of density change rather than volume change.

1.55 For constant bulk modulus of elasticity, how does the density of a liquid vary with the pressure?

1.56 What is the bulk modulus of a liquid that has a density increase of 0.02 percent for a pressure increase of 1000 lb/ft²? For a pressure increase of 60 kPa?

1.57 For $K = 2.2$ GPa for bulk modulus of elasticity of water, what pressure is required to reduce its volume by 0.5 percent?

1.58 A steel container expands in volume 1 percent when the pressure within it is increased by 10,000 psi. At standard pressure, 14.7 psi absolute, it holds 0.5-slug of water; $\rho = 1.94$ slug/ft³. For $K = 300,000$ psi, when it is filled, how many slugs of water must be added to increase the pressure to 10,000 psi? What is the weight of the water added?

1.59 For a substance that obeys the perfect gas law, show that $c_p = dh/dT$ and $c_v = du/dT$. (Hint: Differentiate Eq. (1.6.7) with respect to T and compare the result with Eq. (1.6.6)).

1.60 For a perfect gas prove that $c_p = kR/(k-1)$ and $c_v = R/(k-1)$.

1.61 A piston-tank assembly contains nitrogen gas with a mass of 6.73 kg, with initial volume of 0.3 m³, and initial pressure of 450 kPa. It is known that the gas obeys the law $p V^{1.3} = $ constant, in addition to the perfect-gas law. Determine the pressure in the tank when the gas volume is reduced to 0.15 m³. What are the corresponding initial and final temperatures in the tank?

1.62 What is the isothermal bulk modulus for air at 0.4-MPa abs?

1.63 At what pressure can cavitation be expected at the inlet of a pump that is handling water at 20°C?

1.64 In a long horizontal oil pipeline, pumping stations are located every 60 km. If the pressure loss in the pipeline is 100 kPa/km, how much pressure must each pump produce to avoid oil vaporization?

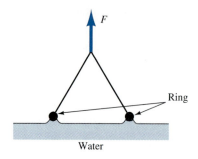

Figure 1.10   Problem 1.70.

1.65   What is the pressure within a droplet of water of 0.002-in. diameter at 68°F if the pressure outside the droplet is standard atmospheric pressure of 14.7 psi?

1.66   A small circular jet of mercury 0.1 mm in diameter issues from an opening. What is the pressure difference between the inside and outside of the jet when at 20°C?

1.67   Determine the capillary rise for distilled water at 104°F in a circular $\frac{1}{4}$-in.-diameter glass tube.

1.68   What diameter of glass tube is required if the capillary effects on the water within are not to exceed 0.5 mm?

1.69   Using the data given in Fig. 1.6, estimate the capillary rise of tap water between two parallel glass plates 0.20 in. apart.

1.70   A method of determining the surface tension of a liquid is to find the force needed to pull a platinum wire ring from the surface (Fig. 1.10). Estimate the force necessary to remove a 20-mm-diameter ring from the surface of water at 20°C.

1.71   Calculate the capillary rise h in the tube of Fig. 1.11 in terms of θ, σ, γ, and r.

1.72   Why would a soap bubble have a relationship

$$p = \frac{4\sigma}{r}$$

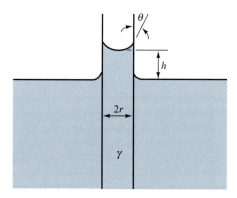

Figure 1.11   Problems 1.71 and 1.73.

when a small spherical droplet has the relationship

$$p = \frac{2\sigma}{r}$$

where p is the internal pressure, σ the surface tension, and r the radius?

1.73 What vertical force due to surface tension would be required to hold the tube of Fig. 1.11? Consider the tube wall thickness to be very small.

1.74 Find the angle at which the surface tension film leaves the glass for a vertical tube immersed in water if the diameter is 0.2 in. and the capillary rise is 0.09 in.; $\sigma = 0.005$ lb/ft.

1.75 Develop a formula for capillary rise h between two concentric glass tubes of radii R and r and contact angle θ.

REFERENCES

1. S. Goldstein, *Modern Developments in Fluid Dynamics,* vol. II, pp. 676–680, Oxford University Press, London, 1938.

2. A. Gill, *Atmosphere-Ocean Dynamics,* App. C, Academic Press, New York, 1982.

ADDITIONAL READING

Cohen, E. and Taylor, B.: "The Fundamental Physical Constants," *Physics Today,* pp. 9–16, August 1994.

Lide, D.: *Handbook of Chemistry and Physics,* 74th ed., CRC Press Inc., Boca Raton, 1993.

W. D. McComb, *The Physics of Fluid Turbulence,* p. 12, Oxford Science Publications, Clarendon Press, Oxford, 1990.

2

Fluid Statics

As in solid mechanics we shall build our knowledge of fluid mechanics by first considering statics followed by the more difficult problem of dynamics. Considering Newton's second law, that is, $d(m\mathbf{v})/dt = \mathbf{F}$, the goal of this chapter is to consider the case where $d(m\mathbf{v})/dt = 0$. This can be achieved either when the fluid velocity is constant or the very special case where the acceleration is constant everywhere in the flow. The first case is the case of fluid statics, while the latter is the special case of solid body acceleration. The overriding assumption necessary to achieve these two conditions is that there is no relative motion of adjacent fluid layers, and consequently the shear stresses (see Chap. 1) are zero. Therefore, only normal or pressure forces are considered to be acting on the fluid surfaces. Before proceeding to the simplified pressure force distribution, however, the definitions and terminology for the fluid forces comprising the force vector, $\mathbf{F}$, are summarized.

2.1 FORCE, STRESS, AND PRESSURE AT A POINT

Two general classes of forces exist on a fluid parcel: *body forces* and *surface forces.* A body force creates its effect on the parcel by action through a distance. Electromagnetic and gravity forces are the only two body forces considered in fluids. The gravity force is responsible for the weight of the parcel and is the only body force considered in this text.

A surface force exists by virtue of direct contact between fluid parcels. Consider the fluid particles all nestled as in Fig. 2.1*a*; each of the particles has a velocity which may be somewhat different in both magnitude and direction than its surrounding neighbors. To proceed with an analysis, we must replace the effect of the particles around the central parcel by an equivalent system of resultant forces acting on the area of contact between the surrounding particle and the central particle (Fig. 2.1*b*). The problem here is to first define a terminology for the force vector distribution in terms of the global coordinate system. Subsequent chapters will detail how this force vector relates to the fluid motion.

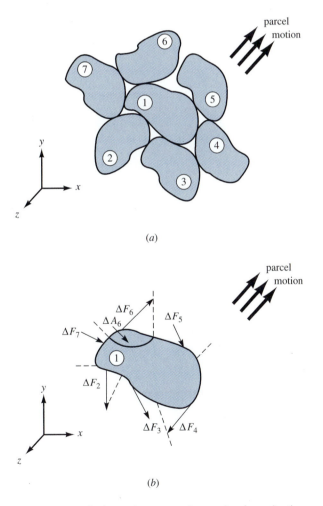

Figure 2.1   Fluid parcel motion and equivalent force distribution.

Consider one of the contact areas, say between parcels 1 and 6, as depicted in Fig. 2.2*a*. The unit normal to the surface area is defined relative to a plane defined by unit direction vectors s_1 and s_2 which are tangent to the point of contact. Therefore, a local coordinate system in the local plane can be defined, and the force vector, $\mathbf{F}_6$, is resolved into its three orthogonal components, ΔF_n, ΔF_{s_1}, and ΔF_{s_2}. ΔF_n is the normal force while ΔF_{s_1} and ΔF_{s_2} are tangential forces.

As area is a vector quantity, and since the surface areas are rotating and changing with time, it is extremely difficult to keep track of the area vector. Therefore, it is convenient to define an intensive representation called stress. Two general classes of stress occur, *normal stress* and *shear stress.* Both are defined in a limit sense as the incremental contact area approaches zero, that is,

$$\sigma_n = \lim_{A \to 0} \frac{\Delta F_n}{\Delta A} \tag{2.1.1}$$

$$\tau_{ss_1} = \lim_{A \to 0} \frac{\Delta F_{s_1}}{\Delta A}$$

$$\tau_{ss_2} = \lim_{A \to 0} \frac{\Delta F_{s_2}}{\Delta A}$$

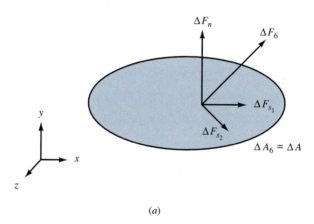

(*a*)

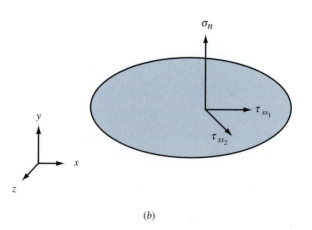

(*b*)

Figure 2.2   Force and stress definition for elemental parcel contact area.

Here the specific area for our example is ΔA_6. σ_n, τ_{ss_1}, and τ_{ss_2} are, respectively, the normal and two shear stresses as defined in local coordinates (Fig. 2.2b). The dimension of stress is force per area.

There are many if not an infinite number of planes (areas) that can arise at a point in the fluid continuum, and as noted earlier it is desirable to do away with the analytical bookkeeping required to keep track of the local area coordinates. Therefore, it is necessary to define the stress in terms of an orthogonal system of forces referenced to global coordinates and an accompanying series of orthogonal planes passing through the origin. In so doing the state of stress can be unambiguously defined with the minimum number of components.

To arrive at this description requires an extensive geometrical derivation which can be found in a number of books on mechanics (see, for example, I. H. Shames, *Mechanics of Fluids*), and only a conceptual description is offered here. With reference to Fig. 2.3, the example stress field in local coordinates is projected to global coordinates by following two steps. First the local plane defined by s_1 and s_2 is projected into its three orthogonal components passing through the origin. Then each of the three local stresses σ_n, τ_{ss_1}, and τ_{ss_2} are projected into three components on each of the orthogonal planes passing through the origin. Since there are three planes with three accompanying stress systems, then a total of nine stress components are formally required to completely describe the state of stress at a point on an arbitrary surface in terms of a global fixed coordinate system. Equation (2.1.2) contains the full *tensor* containing the nine required stresses

$$\begin{pmatrix} \sigma_{xx} & \tau_{yx} & \tau_{zx} \\ \tau_{xy} & \sigma_{yy} & \tau_{zy} \\ \tau_{xz} & \tau_{yz} & \sigma_{zz} \end{pmatrix} = \bar{\bar{\tau}} \qquad (2.1.2)$$

Here the notation is as follows: τ_{yx} is a shear stress acting in the x direction in a plane perpendicular to the y axis (i.e., the xz plane), and σ_{yy} is a normal stress acting in the y direction in a plane perpendicular to the y axis. All of the stress vectors in Fig. 2.3 are defined as positive; the reader will note that positive for the normal

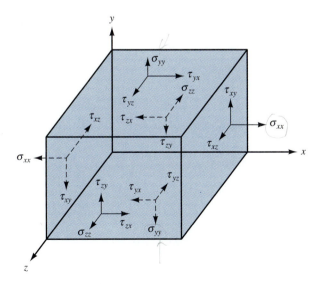

Figure 2.3   Stress distribution definition.

stresses is directed away from the plane while the right-hand screw rule is used to define the positive values for the shear stresses.

The average of the normal stresses is called the *bulk stress, $\bar{\sigma}$*, which in turn is used to define pressure, *p*.

$$p = -\bar{\sigma} = -\frac{1}{3}(\sigma_{xx} + \sigma_{yy} + \sigma_{zz}) \tag{2.1.3}$$

Whereas the positive normal stresses are defined as positive away from the surface, pressure is defined to be positive towards the center of mass of the surface it acts upon.

At a point a fluid at rest has the same pressure in all directions. This means that an element δA of very small area, free to rotate about its center when submerged in a fluid at rest, will have a force of constant magnitude acting on either side of it, regardless of its orientation.

To demonstrate this, a small wedge-shaped free body of unit width is taken at the point (x, y) in a fluid at rest (Fig. 2.4). Since there can be no shear forces, the only forces are the normal surface forces and gravity; so the equations of motion in the *x* and *y* directions are, respectively,

$$\Sigma F_x = p_x\,\delta y - p_s\,\delta s \sin\theta = \frac{\delta x\,\delta y}{2}\rho a_x = 0$$

and

$$\Sigma F_y = p_y\,\delta x - p_s\,\delta s \cos\theta - \left(\gamma\frac{\delta x\,\delta y}{2}\right) = \frac{\delta x\,\delta y}{2}\rho a_y = 0$$

in which p_x, p_y, p_s are the average pressures on the three faces; γ is the specific weight of the fluids; ρ is its density; and a_x and a_y are the accelerations. When the limit is taken as the free body is reduced to zero size by allowing the inclined face to approach (x, y) while maintaining the same angle θ and when the geometric relations

$$\delta s \sin\theta = \delta y \qquad \delta s \cos\theta = \delta x$$

are used, the equations simplify to

$$p_x\,\delta y - p_s\,\delta y = 0 \qquad p_y\,\delta x - p_s\,\delta x - \gamma\frac{\delta x\,\delta y}{2} = 0$$

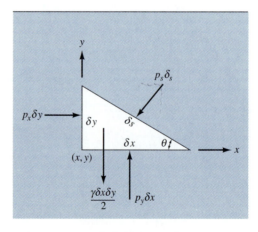

Figure 2.4   Free-body diagram of wedge-shaped particle.

The last term of the second equation is an infinitesimal of higher order of smallness and can be neglected. When divided by δy and δx, respectively, the equations can be combined:

$$p_s = p_x = p_y \tag{2.1.4}$$

Since θ is any arbitrary angle, this equation proves that the pressure is the same in all directions at a point in a static fluid. Although the proof was carried out for a two-dimensional case, it can be demonstrated for the three-dimensional case with the equilibrium equations for a small tetrahedron of fluid with three faces in the coordinate planes and the fourth face inclined arbitrarily.

   If the fluid is in motion so that one layer moves relative to an adjacent layer, shear stresses occur and the normal stresses are, in general, no longer the same in all directions at a point. The pressure is then defined as the average of any three mutually perpendicular normal compressive stresses at a point [Eq. (2.1.3)]. In a fictitious fluid of zero viscosity, that is, a frictionless fluid, no shear stresses can occur for any motion of the fluid, and so at a point the pressure is the same in all directions.

EXERCISE

2.1.1   The normal stress is the same in all directions at a point in a fluid (*a*) *only* when the fluid is frictionless; (*b*) *only* when the fluid is frictionless and incompressible; (*c*) *only* when the fluid has zero viscosity and is at rest; (*d*) when there is no motion of one fluid layer relative to an adjacent layer; (*e*) regardless of the motion of one fluid layer relative to an adjacent layer.

2.2   BASIC EQUATION OF FLUID STATICS

Pressure Variation in a Static Fluid

The forces acting on an element of fluid at rest (Fig. 2.5) consist of surface forces and body forces. With gravity the only body force acting, by taking the y axis vertically upward it is $-\gamma\,\delta x\,\delta y\,\delta z$ in the y direction. With pressure p at its center (x, y, z), the force exerted on the side normal to the y axis closest to the origin is approximately

$$\left(p - \frac{\partial p}{\partial y}\frac{\delta y}{2}\right)\delta x\,\delta z$$

and the force exerted on the opposite side is

$$\left(p + \frac{\partial p}{\partial y}\frac{\delta y}{2}\right)\delta x\,\delta z$$

where $\delta y/2$ is the distance from the center to a face normal to y. Summing the forces acting on the element in the y direction gives

$$\delta F_y = -\frac{\delta p}{\delta y}\,\delta x\,\delta y\,\delta z - \gamma\,\delta x\,\delta y\,\delta z$$

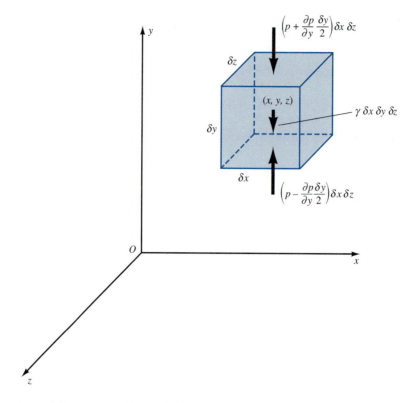

Figure 2.5   Rectangular parallelepiped element of fluid at rest.

For the x and z directions, since no body forces act,

$$\delta F_x = -\frac{\partial p}{\partial x}\,\delta x\,\delta y\,\delta z \qquad\qquad \delta F_z = -\frac{\partial p}{\partial z}\,\delta x\,\delta y\,\delta z$$

The elemental force vector $\delta\mathbf{F}$ is given by

$$\delta\mathbf{F} = \mathbf{i}\,\delta F_x + \mathbf{j}\,\delta F_y + \mathbf{k}\,\delta F_z = -\left(\mathbf{i}\frac{\partial p}{\partial x} + \mathbf{j}\frac{\partial p}{\partial y} + \mathbf{k}\frac{\partial p}{\partial z}\right)\delta x\,\delta y\,\delta z - \mathbf{j}\gamma\,\delta x\,\delta y\,\delta z$$

If the element is reduced to zero size, after dividing through by $\delta x\,\delta y\,\delta z = \delta V$, the expression becomes exact

$$\frac{\delta\mathbf{F}}{\delta V} = -\left(\mathbf{i}\frac{\partial}{\partial x} + \mathbf{j}\frac{\partial}{\partial y} + \mathbf{k}\frac{\partial}{\partial z}\right)p - \mathbf{j}\gamma \quad \lim \delta V \to 0 \tag{2.2.1}$$

This is the resultant force per unit volume at a point, which must be equated to zero for a fluid at rest. The quantity in parentheses is the *gradient,* called ∇ (del),

$$\nabla = \mathbf{i}\frac{\partial}{\partial x} + \mathbf{j}\frac{\partial}{\partial y} + \mathbf{k}\frac{\partial}{\partial z} \tag{2.2.2}$$

and the negative gradient of p, $-\nabla p$, is the vector field $\mathbf{f}$ of the surface pressure force per unit volume,

$$\mathbf{f} = -\nabla p \tag{2.2.3}$$

The fluid static law of variation of pressure is then

$$\mathbf{f} - \mathbf{j}\gamma = 0 \qquad (2.2.4)$$

For an inviscid fluid in motion or a fluid moving so that the shear stress is everywhere zero, Newton's second law takes the form

$$\mathbf{f} - \mathbf{j}\gamma = \rho\mathbf{a} \qquad (2.2.5)$$

where $\mathbf{a}$ is the acceleration of the fluid element and $\mathbf{f} - \mathbf{j}\gamma$ is the resultant fluid force when gravity is the only force acting on the body. Equation (2.2.5) is used to study relative equilibrium in Sec. 2.9 and in the derivation of Euler's equations in Chaps. 4 and 7.

In component form, the combination of Eqs. (2.2.3) and (2.2.4) becomes

$$\frac{\partial p}{\partial x} = 0 \qquad \frac{\partial p}{\partial y} = -\gamma \qquad \frac{\partial p}{\partial z} = 0 \qquad (2.2.6)$$

The partial differential equations, for variation in horizontal directions, are one form of Pascal's law; they state that two points at the same elevation in the same continuous mass of fluid at rest have the same pressure.

Since p is a function of y only,

$$dp = -\gamma\, dy \qquad (2.2.7)$$

This simple differential equation relates the change of pressure to specific weight and change of elevation and holds for both compressible and incompressible fluids.

Pressure Variation in an Incompressible Fluid

For fluids that can be considered homogeneous and incompressible, γ is constant and Eq. (2.2.7), when integrated, becomes

$$p = -\gamma y + c$$

in which c is the constant of integration. The hydrostatic law of variation of pressure is frequently written in the form

$$p = \gamma h \qquad (2.2.8)$$

in which h is measured vertically downward ($h = -y$) from a free-liquid surface and p is the increase in pressure from that found at the free surface. Equation (2.2.8) can be derived by taking as a fluid-free body a vertical column of liquid of finite height h with its upper surface in the free surface. This is left as an exercise for the student.

Example 2.1

An oceanographer needs to design a sea lab 5 m high that will withstand submersion to 100 m, measured from sea level to the top of the sea lab. Find the pressure variation on a side of the container and the pressure on the top if the specific gravity of salt water is 1.020.

Solution

$$\gamma = 1.020(9806 \text{ N/m}^3) = 10 \text{ kN/m}^3$$

At the top $h = 100$ m, and

$$p = \gamma h = 1 \text{ MN/m}^2 = 1 \text{ MPa}$$

If y is measured from the top of the sea lab downward, the pressure variation is

$$p = 10(y + 100) \text{ kPa}$$

Often due to differential heating or the presence of added mass such as salt or sediment, the density in a static homogeneous incompressible fluid may *stratify* or arrange itself in layers where heavier, more dense fluid underlays lighter fluid. The density in each layer remains constant and the pressure varies linearly or hydrostatically with increasing depth into the water column. Figure 2.6 contains an idealized density versus depth diagram of a body of salt water with three constant density regions. The figure also contains the plot of the pressure distribution with depth and it is noted that the pressure is continuous at the interfaces. In practice molecular and turbulent diffusion of salt will marginally "smooth" the discontinuous density interface, but this layered approach to pressure distributions in layered stratified conditions has been a fundamental analysis approach for limnologists and oceanographers for over a hundred years.

From Fig. 2.6 several features are of note. First the pressure in each layer linearly increases with increasing depth. Therefore within layer 1 and subsequent layers the pressure variations are

$$p(0 < h < h_1) = p_0 + \rho_1 g(h) \tag{2.2.9}$$

$$p(h_1 < h < h_2) = p_1 + \rho_2 g(h - h_1) \tag{2.2.10}$$

or for any layer n,

$$p(h_{n-1} < h < h_n) = p_{n-1} + \rho_n g(h - h_{n-1}) \tag{2.2.11}$$

Example 2.2   At a spot in the ocean where the total depth is 450 m, oceanographers measure data at $h_1 = 100$ m, $h_2 = 300$ m, and $h_3 = 450$ m. The specific gravity values for salt

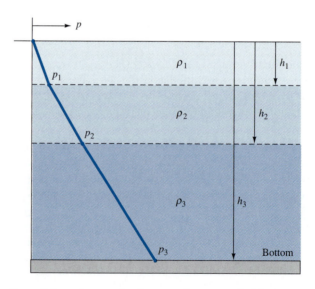

Figure 2.6   Pressure distribution in density stratified fluid at rest.

water in each of the constant density layers are 1.01, 1.02, and 1.025, respectively. Find the pressures at the interfaces. Assume atmospheric pressure at the surface, $p_0 = 0$.

Solution

$$p_1 = p_0 + \rho_1 g(h_1 - 0) = p_0 + S_1 \rho g h_1$$
$$= 0 + 9.904(100) = 990.4 \text{ kPa}$$

$$p_2 = p_1 + \rho_2 g(h_2 - h_1) = p_1 + S_2 \rho g(h_2 - h_1)$$
$$= 990.4 + 10.0(200) = 2990.8 \text{ kPa}$$

$$p_3 = p_2 + \rho_3 g(h_3 - h_2) = p_2 + S_3 \rho g(h_3 - h_2)$$
$$= 2990.8 + 10.05(150) = 4498.5 \text{ kPa}$$

Pressure Variation in a Compressible Fluid

When the fluid is a perfect gas at rest at constant temperature, from Eq. (1.7.2)

$$\frac{p}{\rho} = \frac{p_0}{\rho_0} \tag{2.2.12}$$

in which p is the absolute pressure. When the value of γ in Eq. (2.2.7) is replaced by ρg and ρ is eliminated between Eqs. (2.2.7) and (2.2.9),

$$dy = -\frac{p_0}{g\rho_0}\frac{dp}{p} \tag{2.2.13}$$

It must be remembered that if ρ is in pounds mass per cubic foot, then $\gamma = g\rho/g_0$ with $g_0 = 32.174 \text{ lb}_m\cdot\text{ft/lb}\cdot\text{s}^2$. If $p = p_0$ when $\rho = \rho_0$, integration between limits

$$\int_{y_0}^{y} dy = -\frac{p_0}{g\rho_0}\int_{p_0}^{p}\frac{dp}{p}$$

yields

$$y - y_0 = -\frac{p_0}{g\rho_0}\ln\frac{p}{p_0} \tag{2.2.14}$$

in which ln is the natural logarithm. Then

$$p = p_0 \exp\left(-\frac{y - y_0}{p_0/g\rho_0}\right) \tag{2.2.15}$$

which is the equation for variation of pressure with elevation in an isothermal gas.

The atmosphere frequently is assumed to have a constant temperature gradient which is expressed by

$$T = T_0 + \beta y \tag{2.2.16}$$

For the standard atmosphere, $\beta = -0.00357$ degree Fahrenheit per foot (-0.00651 K/m) up to the stratosphere. The density can be expressed in terms of pressure and elevation from the perfect-gas law:

$$\rho = \frac{p}{RT} = \frac{p}{R(T_o + \beta y)} \tag{2.2.17}$$

Substitution into $dp = -\rho g\, dy$ [Eq. (2.2.7)] permits the variables to be separated and p to be found in terms of y by integration.

Example 2.3

Assuming isothermal conditions to prevail in the atmosphere, compute the pressure and density at 2000-m elevation if $p = 10^5$-Pa abs and $\rho = 1.24$ kg/m^3 at sea level.

Solution

From Eq. (2.2.15)

$$p = (10^5 \text{ N/m}^2)\exp\left\{-\frac{2000 \text{ m}}{(10^5 \text{ N/m}^2)/[(9.806 \text{ m/s}^2)(1.24 \text{ kg/m}^3)]}\right\}$$

$$= 78.4\text{-kPa abs}$$

Then, from Eq. (2.2.12),

$$\rho = \frac{\rho_0}{p_0}p = (1.24 \text{ kg/m}^3)\frac{78,400}{100,000} = 0.972 \text{ kg/m}^3$$

When compressibility of a liquid in static equilibrium is taken into account, Eqs. (2.2.7) and (1.8.1) are used.

EXERCISES

2.2.1   The pressure in the air space above an oil ($S = 0.75$) surface in a tank is 115-kPa abs. The pressure 2.0 m below the surface of the oil, in kPa, is (a) 14.71; (b) 116.5; (c) 129.71; (d) 134.1; (e) none of these answers.

2.2.2   The pressure, in millimeters of mercury gage, equivalent to 80-mm H$_2$O plus 60-mm manometer fluid, sp gr 2.94, is (a) 10.3; (b) 18.8; (c) 20.4; (d) 30.6; (e) none of these answers.

2.2.3   The differential equation for pressure variation in a static fluid (y measured vertically upward) can be written as (a) $dp = -\gamma\, dy$; (b) $d\rho = -\gamma\, dy$; (c) $dy = -\rho\, dp$; (d) $dp = -\rho\, dy$; (e) $dp = -y\, d\rho$.

2.2.4   In an isothermal atmosphere, the pressure (a) remains constant; (b) decreases linearly with elevation; (c) increases exponentially with elevation; (d) varies in the same way as the density; (e) remains constant, as does the density.

2.3   UNITS AND SCALES OF PRESSURE MEASUREMENT

Pressure can be expressed with reference to any arbitrary datum. The usual datums are *absolute zero* and *local atmospheric pressure.* When a pressure is expressed as a difference between its value and a complete vacuum, it is called an *absolute pressure.* When it is expressed as a difference between its value and the local atmospheric pressure, it is called a *gage pressure.*

The *bourdon gage* (Fig. 2.7) is typical of the devices used for measuring gage pressures. The pressure element is a hollow, curved, flat metallic tube closed at one end; the other end is connected to the pressure to be measured. When the internal

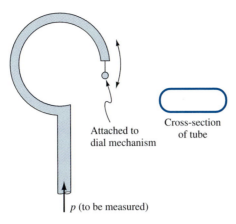

Figure 2.7   Bourdon tube schematic.

pressure is increased, the tube tends to straighten, pulling on a linkage to which is attached a pointer and causing the pointer to move. The dial reads zero when the inside and outside of the tube are at the same pressure, regardless of its particular value. The dial can be graduated to any convenient units, common ones being pascals, pounds per square inch, pounds per square foot, inches of mercury, feet of water, centimeters of mercury, and millimeters of mercury. Owing to its inherent construction, the gage measures pressure relative to the pressure of the medium surrounding the tube, which is the local atmosphere.

Figure 2.8 illustrates the data and the relations of the common units of pressure measurement. Standard atmospheric pressure is the mean pressure at sea level, 29.92-in. Hg. A pressure expressed in terms of *the length of a column* of liquid is equivalent to the force per unit area at the base of the column. The relation for variation of pressure with altitude in a liquid $p = \gamma h$ [Eq. (2.2.8)] shows the relation between the head h, in length of fluid column of specific weight γ, and the pressure p. In consistent units, p is in pascals, γ in newtons per cubic meter, and h in meters, or p is in pounds per square foot, γ in pounds per cubic foot, and h in feet. With the specific weight of any liquid expressed as its specific gravity S times the specific

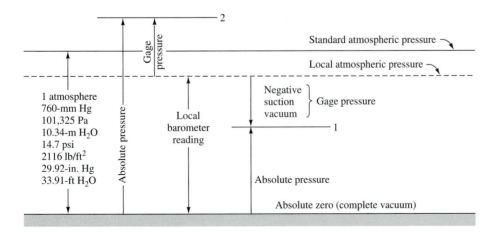

Figure 2.8   Units and scales of pressure measurement.

weight of water, Eq. (2.2.8) becomes

$$p = \gamma_w S\, h \qquad\qquad\qquad (2.3.1)$$

For water γ_w can be taken as 9806 N/m^3 or 62.4 lb/ft^3.

When the pressure is desired in pounds per square inch, both sides of the equation are divided by 144:

$$p_{psi} = \frac{62.4}{144} Sh = 0.433Sh \qquad\qquad (2.3.2)$$

in which h remains in feet.†

Local atmospheric pressure is measured by a mercury barometer (Fig. 2.9) or by an *aneroid* barometer, which measures the difference in pressure between the atmosphere and an evacuated box or tube in a manner analogous to the bourdon gage except that the tube is evacuated and sealed.

A mercury barometer consists of a glass tube closed at one end, filled with mercury, and inverted so that the open end is submerged in mercury. It has a scale so arranged that the height of column R (Fig. 2.9) can be determined. The space above the mercury contains mercury vapor. If the pressure of the mercury vapor h_v is given

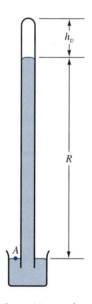

Figure 2.9   Mercury barometer.

---

†In Eq. (2.3.2) the standard atmospheric pressure can be expressed in pounds per square inch,

$$p_{psi} = \frac{62.4}{144}(13.6)\frac{29.92}{12} = 14.7$$

when $S = 13.6$ for mercury. When 14.7 is multiplied by 144, the standard atmosphere becomes 2116 lb/ft^2. Then 2116 divided by 62.4 yields 33.91-ft H$_2$O. Any of these designations is for the standard atmosphere and may be called *one atmosphere* if it is always understood that it is a standard atmosphere and is measured from absolute zero. These various designations of a standard atmosphere (Fig. 2.8) are equivalent and provide a convenient means of converting from one set of units to another. For example, to express 100-ft H$_2$O in pounds per square inch use

$$\frac{100}{33.91}(14.7) = 43.3 \text{ psi}$$

since 100/33.91 is the number of standard atmospheres and each standard atmosphere is 14.7 psi.

in millimeters of mercury and R is measured in the same units, the pressure at A can be expressed as

$$h_v + R = h_A \qquad \text{mm Hg}$$

Although h_v is a function of temperature, it is very small at usual atmospheric temperatures. The barometric pressure varies with location, that is, elevation, and with weather conditions.

In Fig. 2.8 a pressure can be located vertically on the chart, which indicates its relation to absolute zero and to local atmospheric pressure. If the point is below the local-atmospheric-pressure line and is referred to gage datum, it is called *negative, suction,* or *vacuum.* For example, the pressure 460-mm Hg abs, as at 1, with barometer reading 720-mm Hg, can be expressed as -260-mm Hg, 11-in.-Hg suction, or 11-in.-Hg vacuum. It should be noted that

$$p_{\text{abs}} = p_{\text{bar}} + p_{\text{gage}}$$

To avoid any confusion, the convention is adopted throughout this text that a *pressure is gage unless specifically marked absolute,* with the exception of the *atmosphere,* which is an absolute pressure unit.

The rate of temperature change in the atmosphere with change in elevation is called its *lapse rate.* The motion of a parcel of air depends on the density of the parcel relative to the density of the surrounding (ambient) air. However, as the parcel ascends through the atmosphere, the air pressure decreases, the parcel expands, and its temperature decreases at a rate known as the *dry adiabatic lapse rate.* A firm wants to burn a large quantity of refuse. It is estimated that the temperature of the smoke plume at 10 m above the ground will be 11°C greater than that of the ambient air. Determine what will happen to the smoke (*a*) at standard atmospheric lapse rate $\beta = -0.00651°C$ per meter and $t_0 = 20°C$ and (*b*) at an inverted lapse rate $\beta = 0.00365°C$ per meter.

Example 2.4

Solution

Combining Eqs. (2.2.7) and (2.2.17) gives

$$\int_{p_0}^{p} \frac{dp}{p} = -\frac{g}{R} \int_{0}^{y} \frac{dy}{T_0 + \beta y} \qquad \text{or} \qquad \frac{p}{p_0} = \left(1 + \frac{\beta y}{T_0}\right)^{-g/R\beta}$$

The relation between pressure and temperature for a mass of gas expanding without heat transfer is

$$\frac{T}{T_1} = \left(\frac{p}{p_0}\right)^{(k-1)/k}$$

in which T_1 is the initial absolute temperature of the smoke; p_0 is the initial absolute pressure; and k is the specific-heat ratio, 1.4 for air and other diatomic gases. Eliminating p/p_0 in the last two equations gives

$$T = T_1 \left(1 + \frac{\beta y}{T_0}\right)^{-[(k-1)/k](g/R\beta)}$$

Since the gas will rise until its temperature is equal to the ambient temperature,

$$T = T_0 + \beta y$$

the last two equations can be solved for y. Let

$$a = \frac{-1}{(k-1)g/kR\beta + 1}$$

then

$$y = \frac{T_0}{\beta}\left[\left(\frac{T_0}{T_1}\right)^a - 1\right]$$

(a) For $\beta = -0.00651°C$ per meter, $R = 287$ m·N/(kg·K), $a = 2.002$, and $y = 3201$ m.

(b) For the atmospheric temperature inversion $\beta = 0.00365°C$ per meter, $a = -0.2721$, and $y = 809.2$ m.

EXERCISES

2.3.1   Select the correct statement: (a) Local atmospheric pressure is always below standard atmospheric pressure. (b) Local atmospheric pressure depends upon elevation of locality only. (c) Standard atmospheric pressure is the mean local atmospheric pressure at sea level. (d) A barometer reads the difference between local and standard atmospheric pressure. (e) Standard atmospheric pressure is 720-mm Hg abs.

2.3.2   Select the three pressures that are equivalent: (a) 10.0 psi, 23.1-ft H_2O, 4.91-in. Hg; (b) 10.0 psi, 4.33-ft H_2O, 20.3-in. Hg; (c) 10.0 psi, 20.3-ft H_2O, 23.1-in. Hg; (d) 4.33 psi, 10.0-ft H_2O, 20.3-in. Hg; (e) 4.33 psi, 10.0-ft H_2O, 8.83-in. Hg.

2.3.3   When the barometer reads 730-mm Hg, 10-kPa suction is the same as (a) -10.2-m H_2O; (b) 0.075-m Hg; (c) 8.91-m-H_2O abs; (d) 107-kPa abs; (e) none of these answers.

2.3.4   With the barometer reading 29-in. Hg, 7.0 psia is equivalent to (a) 0.476 atm; (b) 0.493 atm; (c) 7.9-psi suction; (d) 7.7 psi; (e) 13.8-in.-Hg abs.

2.4   MANOMETERS

Standard Manometers

Manometers are devices that employ liquid columns for determining differences in pressure. The most elementary manometer, usually called a *piezometer,* is illustrated in Fig. 2.10a; it measures the pressure in a liquid when it is above zero gage. A glass tube is mounted vertically so that it is connected to the space within the container. Liquid rises in the tube until equilibrium is reached. The pressure is then given by the vertical distance h from the meniscus (liquid surface) to the point where the pressure is to be measured, expressed in units of length of the liquid in the container. It is obvious that the piezometer would not work for negative gage pressures, because air would flow into the container through the tube. It is also impractical for measuring large pressures at A, since the vertical tube would need to be very long. If the specific gravity of the liquid is S, the pressure at A is hS units of length of water.

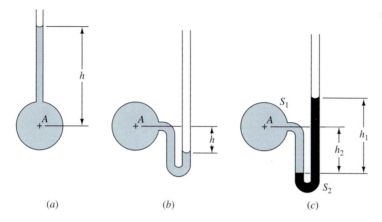

(a)   (b)   (c)

Figure 2.10   Simple manometers.

For measurement of small negative or positive gage pressures in a liquid the tube may take the form shown in Fig. 2.10b. With this arrangement the meniscus can come to rest below A, as shown. Since the pressure at the meniscus is zero gage and since pressure *decreases* with elevation,

$$h_A = -hS \qquad \text{unit of length of } H_2O$$

For greater negative or positive gage pressures a second liquid of greater specific gravity is employed (Fig. 2.10c). It must be immiscible in the first fluid, which may now be a gas. If the specific gravity of the fluid at A is S_1 (based on water) and the specific gravity of the manometer liquid is S_2, the equation for pressure at A can be written, starting at either A or the upper meniscus and proceeding through the manometer, as

$$h_A + h_2 S_1 - h_1 S_2 = 0$$

in which h_A is the unknown pressure, expressed in length units of water, and h_1 and h_2 are in length units. If A contains a gas, S_1 is generally so small that $h_2 S_1$ can be neglected.

A general procedure should be followed in working all manometer problems:

1. Start at one end (or any meniscus if the circuit is continuous) and write the pressure there in an appropriate unit (say pascals) or in an appropriate symbol if it is unknown.

2. Add to this the change in pressure, in the same unit, from one meniscus to the next (plus if the next meniscus is lower and minus if higher).

3. Continue until the other end of the gage (or the starting meniscus) is reached and equate the expression to the pressure at that point, known or unknown.

The expression will contain one unknown for a simple manometer or will give a difference in pressures for the differential manometer. In equation form,

$$p_0 - (y_1 - y_0)\gamma_0 - (y_2 - y_1)\gamma_1 - (y_3 - y_2)\gamma_2$$
$$- (y_4 - y_3)\gamma_3 - \cdots - (y_n - y_{n-1})\gamma_{n-1} = p_n$$

in which $y_0, y_1, \cdots, y_n$ are elevations of each meniscus in length units and $\gamma_0, \gamma_1, \gamma_2, \cdots, \gamma_{n-1}$ are specific weights of the fluid columns. The above expression yields the answer in force per unit area and can be converted to other units by use of Fig. 2.8.

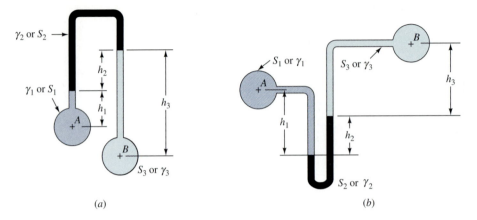

Figure 2.11   Differential manometers.

A differential manometer (Fig. 2.11) determines the difference in pressures at two points A and B when the actual pressure at any point in the system cannot be determined. Application of the procedure outlined above to Fig. 2.11a produces

$$p_A - h_1\gamma_1 - h_2\gamma_2 + h_3\gamma_3 = p_B \quad \text{or} \quad p_A - p_B = h_1\gamma_1 + h_2\gamma_2 - h_3\gamma_3$$

Similarly, for Fig. 2.11b,

$$p_A + h_1\gamma_1 - h_2\gamma_2 - h_3\gamma_3 = p_B \quad \text{or} \quad p_A - p_B = -h_1\gamma_1 + h_2\gamma_2 + h_3\gamma_3$$

No formulas for particular manometers should be memorized. It is much more satisfactory to work them out from the general procedure for each case as needed.

If the pressures at A and B are expressed in length of the water column, the above results can be written, for Fig. 2.11a, as

$$h_A - h_B = h_1 S_1 + h_2 S_2 - h_3 S_3 \qquad \text{units of length } H_2O$$

Similarly, for Fig. 2.11b

$$h_A - h_B = -h_1 S_1 + h_2 S_2 + h_3 S_3$$

in which S_1, S_2, and S_3 are the applicable specific gravities of the liquids in the system.

Example 2.5

In Fig. 2.11a the liquids at A and B are water and the manometer liquid is oil. $S = 0.80$; $h_1 = 300$ mm; $h_2 = 200$ mm; and $h_3 = 600$ mm. (a) Determine $p_A - p_B$, in pascals. (b) If $p_B = 50$ kPa and the barometer reading is 730-mm Hg, find the pressure at A in meters of water absolute.

Solution

(a)
$$h_A(\text{m } H_2O) - h_1 S_{H_2O} - h_2 S_{\text{oil}} + h_3 S_{H_2O} = h_B(\text{m } H_2O)$$

$$h_A - 0.3(1) - 0.2(0.8) + 0.6(1) = h_B$$

$$h_A - h_B = -0.14\text{-m } H_2O$$

$$p_A - p_B = \gamma(h_A - h_B) = (9806 \text{ N/m}^3)(-0.14 \text{ m}) = -1373 \text{ Pa}$$

(b) $h_B = p_B/\gamma = 5 \times 10^4 \text{ N/m}^2/9806 \text{ N/m}^3 = 5.099\text{-m } H_2O$

$$h_B(\text{m } H_2O \text{ abs}) = h_B(\text{m } H_2O \text{ gage}) + (0.73 \text{ m})(13.6)$$

$$= 5.099 + 9.928 = 15.027\text{-m-}H_2O \text{ abs}$$

From (a) $h_{A_{abs}} = h_{B_{abs}} - 0.14 = 15.027 - 0.14 = 14.89\text{-m-}H_2O \text{ abs}$

Micromanometers

Several types of manometers are on the market for determining very small differences in pressure or determining large pressure differences precisely. One type measures the differences in elevation of two menisci of a manometer very accurately. By means of small telescopes with horizontal cross hairs mounted along the tubes on a rack which is raised and lowered by a pinion and slow-motion screw so that the cross hairs can be set accurately, the difference in elevation of menisci (the gage difference) can be read with verniers.

With two immiscible gage liquids a large gage difference R (Fig. 2.12) can be produced in the fluid to be measured for a small pressure difference. The heavier gage liquid fills the lower U tube up to 0-0; then the lighter gage liquid is added to both sides, filling the larger reservoirs up to 1-1. The gas or liquid in the system fills the space above 1-1. When the pressure at C is slightly greater than at D, the menisci move as indicated in Fig. 2.12. The volume of liquid displaced in each reservoir equals the displacement in the U tube; thus,

$$\Delta yA = \frac{R}{2}a$$

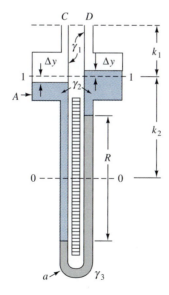

Figure 2.12   Micromanometer using two gage liquids.

in which A and a are the cross-sectional areas of the reservoir and U tube, respectively. The manometer equation can be written, starting at C, in force per unit area, as

$$p_C + (k_1 + \Delta y)\gamma_1 + \left(k_2 - \Delta y + \frac{R}{2}\right)\gamma_2 - R\gamma_3$$

$$- \left(k_2 - \frac{R}{2} + \Delta y\right)\gamma_2 - (k_1 - \Delta y)\gamma_1 = p_D$$

in which $\gamma_1, \gamma_2,$ and γ_3 are the specific weights as indicated in Fig. 2.12. Simplifying and substituting for Δy gives

$$p_C - p_D = R\left[\gamma_3 - \gamma_2\left(1 - \frac{a}{A}\right) - \gamma_1\frac{a}{A}\right] \tag{2.4.1}$$

The quantity in brackets is a constant for specified gage and fluids; hence, the pressure difference is directly proportional to R.

| **Example 2.6** | In the micromanometer of Fig. 2.12 find the pressure difference, in pascals, when air is in the system; $S_2 = 1.0$, $S_3 = 1.10$, $a/A = 0.01$, $R = 5$ mm, $T = 20°C$, and the barometer reads 760-mm Hg. |

Solution

$$\rho_{air} = \frac{p}{RT} = \frac{(0.76\ \text{m})[13.6(9806\ \text{N/m}^3)]}{(287\ \text{N·m/kg·K})(273 + 20\ \text{K})} = 1.205\ \text{kg/m}^3$$

$$\gamma_1\frac{a}{A} = (1.205\ \text{kg/m}^3)(9.806\ \text{m/s}^2)(0.01) = 0.118\ \text{N/m}^3$$

$$\gamma_3 - \gamma_2\left(1 - \frac{a}{A}\right) = (9806\ \text{N/m}^3)(1.10 - 0.99) = 1079\ \text{N/m}^3$$

The term $\gamma_1(a/A)$ can be neglected. Substituting into Eq. (2.4.1) gives

$$p_C - p_D = (0.005\ \text{m})(1079\ \text{N/m}^3) = 5.39\ \text{Pa}$$

The inclined manometer (Fig. 2.13) is frequently used for measuring small differences in gas pressures. It is adjusted to read zero, by moving the inclined scale,

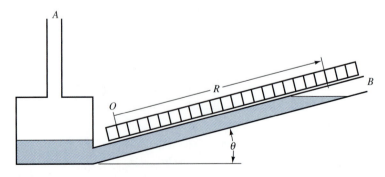

Figure 2.13   Inclined manometer.

when A and B are open. Since the inclined tube requires a greater displacement of the meniscus for a given pressure difference than a vertical tube, it affords greater accuracy in reading the scale.

Surface tension causes a capillary rise in small tubes. If a U tube is used with meniscus in each leg, the surface-tension effects cancel. The capillary rise is negligible in tubes with a diameter of 15 mm or greater.

EXERCISES

2.4.1 In Fig. 2.10*b* the liquid is oil, $S = 0.80$. When $h = 60$ cm, the pressure at A may be expressed as (*a*) -48-cm H_2O abs; (*b*) 48-cm H_2O; (*c*) 48-cm-H_2O suction; (*d*) 52-cm-H_2O vacuum; (*e*) none of these answers.

2.4.2 In Fig. 2.10*c* air is contained in the pipe, water is the manometer liquid, $h_1 = 500$ mm, and $h_2 = 200$ mm. The pressure at A is (*a*) 10.14-m-H_2O abs; (*b*) 0.2-m-H_2O vacuum; (*c*) 0.2-m H_2O; (*d*) 4901 Pa; (*e*) none of these answers.

2.4.3 In Fig. 2.11*a* $h_1 = 2.0$ ft, $h_2 = 1.0$ ft, $h_3 = 4.0$ ft, $S_1 = 0.80$, $S_2 = 0.65$, and $S_3 = 1.0$. Then $h_B - h_A$ in feet of water is (*a*) -3.05; (*b*) -1.75; (*c*) 3.05; (*d*) 6.25; (*e*) none of these answers.

2.4.4 In Fig. 2.11*b* $h_1 = 38$ cm, $h_2 = 33$ cm, $h_3 = 60$ cm, $S_1 = 0.80$, $S_2 = 3.00$, and $S_3 = 1.0$. Then $p_A - p_B$ in kilopascals is (*a*) -7.55; (*b*) 0.098; (*c*) 11.86; (*d*) 19.32; (*e*) none of these answers.

2.4.5 A mercury-water manometer has a gage difference of 500 mm (difference in elevation of menisci). The difference in pressure, measured in meters of water, is (*a*) 0.5; (*b*) 6.3; (*c*) 6.8; (*d*) 7.3; (*e*) none of these answers.

2.4.6 In the inclined manometer of Fig. 2.13 the reservoir is so large that its surface can be assumed to remain at a fixed elevation, $\theta = 30°$. Used as a simple manometer for measuring air pressure, it contains water and $R = 40$ cm. The pressure at A, in centimeters of water, is (*a*) -40; (*b*) 20 vacuum; (*c*) 20; (*d*) 40; (*e*) none of these answers.

2.5 FORCES ON PLANE AREAS

In the preceding sections variations of pressure throughout a fluid have been considered. The distributed forces resulting from the action of fluid on a finite area can be conveniently replaced by a resultant force, insofar as external reactions to the force system are concerned. In this section the magnitude of resultant force and its line of action (pressure center) are determined by integration, by formula, and by using the concept of the pressure prism.

Horizontal Surfaces

A plane surface in a horizontal position in a fluid at rest is subjected to a constant pressure. The magnitude of the force acting on one side of the surface is

$$\int p \, dA = p \int dA = pA$$

The elemental forces $p \, dA$ acting on A are all parallel and in the same sense; therefore, a scalar summation of all such elements yields the magnitude of the resultant

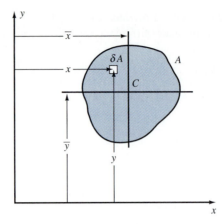

Figure 2.14   Notation for determining the line of action of a force.

force. Its direction is normal to the surface and *towards* the surface if p is *positive*. To find the line of action of the resultant, that is, the point in the area where the moment of the distributed force about any axis through the point is zero, arbitrary xy axes can be selected, as in Fig. 2.14. Then, since the moment of the resultant must equal the moment of the distributed force system about any axis, say the y axis,

$$pAx' = \int_A xp\, dA$$

in which x' is the distance from the y axis to the resultant. Since p is constant,

$$x' = \frac{1}{A} \int_A x\, dA = \overline{x}$$

in which $\overline{x}$ is the distance to the centroid of the area (see Appendix A). Hence, for a horizontal area subjected to static fluid pressure, the resultant passes through the centroid of the area.

Inclined Surfaces

In Fig. 2.15 a plane surface is indicated by its trace $A'B'$. It is inclined $\theta°$ from the horizontal. The intersection of the plane of the area and the free surface is taken as the x axis. The y axis is taken in the plane of the area, with origin O, as shown, in the free surface. The xy plane portrays the arbitrary inclined area. The magnitude, direction, and line of action of the resultant force *due to the liquid,* acting on one side of the area, are sought.

For an element with area δA as a strip with thickness δy with long edges horizontal, the magnitude of force δF acting on it is

$$\delta F = p\,\delta A = \gamma h\,\delta A = \gamma y \sin\theta\,\delta A \tag{2.5.1}$$

Since all such elemental forces are parallel, the integral over the area yields the magnitude of force F, acting on one side of the area.

$$F = \int p\, dA = \gamma\,\sin\theta \int y\, dA = \gamma \sin\theta\, \overline{y}A = \gamma \overline{h}A = p_G A \tag{2.5.2}$$

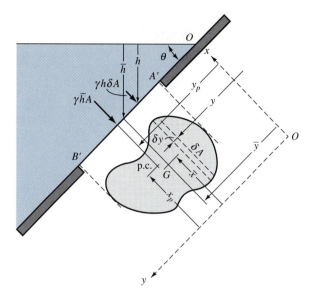

Figure 2.15 Notation for force of liquid on one side of an inclined plane area.

with the relations from Fig. 2.15: $\bar{y}\sin\theta = \bar{h}$ and $p_G = \gamma\bar{h}$ the pressure at the centroid of the area. In words, the magnitude of force exerted on one side of a plane area submerged in a liquid is the product of the area and the pressure at its centroid. In this form, it should be noted, the presence of a free surface is unnecessary. Any means for determining the pressure at the centroid may be used. The sense of the force is to push against the area if p_G is positive. As all force elements are normal to the surface, the line of action of the resultant also is normal to the surface. Any surface can be rotated about any axis through its centroid without changing the magnitude of the resultant if the total area remains submerged in the static liquid.

Center of Pressure

The line of action of the resultant force has its piercing point in the surface at a point called the *pressure center,* with coordinates (x_p, y_p) (Fig. 2.15). Unlike that for the horizontal surface, the center of pressure of an inclined surface is not at the centroid. To find the pressure center, the moments of the resultant $x_p F$ and $y_p F$ are equated to the moment of the distributed forces about the y axis and x axis, respectively; thus,

$$x_p F = \int_A xp\,dA \tag{2.5.3}$$

$$y_p F = \int_A yp\,dA \tag{2.5.4}$$

The area element in Eq. (2.5.3) should be $\delta x\,\delta y$ and not the strip shown in Fig. 2.15. Solving for the coordinates of the pressure center results in

$$x_p = \frac{1}{F}\int_A xp\,dA \tag{2.5.5}$$

$$y_p = \frac{1}{F}\int_A yp\,dA \tag{2.5.6}$$

In many applications Eqs. (2.5.5) and (2.5.6) can be evaluated most conveniently through graphical integration; for simple areas they can be transformed into general formulas as follows (see Appendix A):

$$x_p = \frac{1}{\gamma \bar{y} A \sin \theta} \int_A x \gamma y \sin \theta \, dA = \frac{1}{\bar{y} A} \int_A xy \, dA = \frac{I_{xy}}{\bar{y} A} \tag{2.5.7}$$

In Eqs. (A.10) and (2.5.7)

$$x_p = \frac{\bar{I}_{xy}}{\bar{y} A} + \bar{x} \tag{2.5.8}$$

When either of the centroidal axes $x = \bar{x}$ or $y = \bar{y}$ is an axis of symmetry for the surface, $\bar{I}_{xy}$ vanishes and the pressure center lies on $x = \bar{x}$. Since $\bar{I}_{xy}$ may be either positive or negative, the pressure center may lie on either side of the line $x = \bar{x}$. To determine y_p by formula, with Eqs. (2.5.2) and (2.5.6),

$$y_p = \frac{1}{\gamma \bar{y} A \, \sin \theta} \int_A y \gamma y \, \sin \theta \, dA = \frac{1}{\bar{y} A} \int_A y^2 \, dA = \frac{I_x}{\bar{y} A} \tag{2.5.9}$$

In the parallel-axis theorem for moments of inertia

$$I_x = I_G + \bar{y}^2 A$$

in which I_G is the second moment of the area about its horizontal centroidal axis. If I_x is eliminated from Eq. (2.5.9),

$$y_p = \frac{I_G}{\bar{y} A} + \bar{y} \tag{2.5.10}$$

or

$$y_p - \bar{y} = \frac{I_G}{\bar{y} A} \tag{2.5.11}$$

I_G is always positive; hence, $y_p - \bar{y}$ is always positive and the pressure center is always below the centroid of the surface. It should be emphasized that $\bar{y}$ and $y_p - \bar{y}$ are distances in the plane of the surface.

Example 2.7

The triangular gate CDE (Fig. 2.16) is hinged along CD and is opened by a normal force P applied at E. It holds oil, sp gr 0.80, above it and is open to the atmosphere on its lower side. Neglecting the weight of the gate, find (a) the magnitude of force exerted on the gate by integration and by Eq. (2.5.2); (b) the location of the pressure center; (c) the force P needed to open the gate.

Solution

(a) By integration with reference to Fig. 2.16

$$F = \int_A p \, dA = \gamma \sin \theta \int yx \, dy = \gamma \sin \theta \int_8^{13} xy \, dy + \gamma \, \sin \theta \int_{13}^{18} xy \, dy$$

When $y = 8$, $x = 0$, and when $y = 13$, $x = 6$, with x varying linearly with y; thus,

$$x = ay + b \qquad 0 = 8a + b \qquad 6 = 13a + b$$

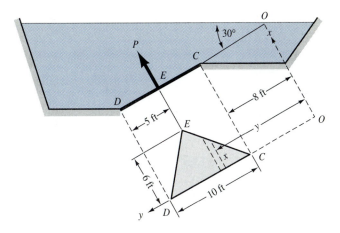

Figure 2.16 Triangular gate.

in which the coordinates have been substituted to find x in terms of y. Solving for a and b gives

$$a = \frac{6}{5} \qquad b = -\frac{48}{5} \qquad x = \frac{6}{5}(y - 8)$$

Similarly, for $y = 13$, $x = 6$, and $y = 18$, $x = 0$, yielding $x = \frac{6}{5}(18 - y)$. Hence,

$$F = \gamma \sin\theta \, \frac{6}{5}\left[\int_{8}^{13}(y - 8)y\,dy + \int_{13}^{18}(18 - y)y\,dy\right]$$

Integrating and substituting for $\gamma \sin\theta$ leads to:

$$F = 62.4(0.8)(.50)\frac{6}{5}\left[\left(\frac{y^3}{3} - 4y^2\right)\Big|_{8}^{13} + \left(9y^2 - \frac{y^3}{3}\right)\Big|_{13}^{18}\right] = 9734.4 \text{ lb}$$

By Eq. (2.5.2)

$$F = p_G A = \gamma\bar{y}\sin\theta\, A = 62.4(0.80)(13)(0.50)(30) = 9734.4 \text{ lb}$$

(b) With the axes as shown, $\bar{x} = 2.0$, $\bar{y} = 13$. In Eq. (2.5.8)

$$x_p = \frac{\bar{I}_{xy}}{\bar{y}A} + \bar{x}$$

$\bar{I}_{xy}$ is zero owing to symmetry about the centroidal axis parallel to the x axis; hence, $\bar{x} = x_p = 2.0$ ft. In Eq. (2.5.11)

$$y_p - \bar{y} = \frac{I_G}{\bar{y}A} = 2\frac{1(6)(5^3)}{12(13)(30)} = 0.32 \text{ ft}$$

that is, the pressure center is 0.32 ft below the centroid, measured in the plane of the area.

(c) When the moments about CD are taken and the action of the oil is replaced by the resultant, then

$$(P)(6) = 9734.4(2) \qquad P = 3244.8 \text{ lb}$$

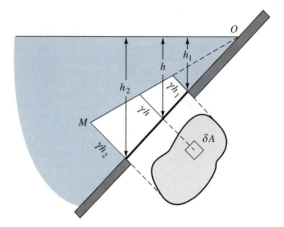

Figure 2.17   Pressure prism.

The Pressure Prism

Another approach to the problem of determining the resultant force and line of action of the force on a plane surface is given by the concept of a pressure prism. It is a prismatic volume with its base the given surface area and with altitude at any point of the base given by $p = \gamma h$. h is the vertical distance to the free surface (Fig. 2.17). (An imaginary free surface can be used to define h if no real free surface exists.) In the figure, γh can be laid off to any convenient scale such that its trace is OM. The force acting on an elemental area δA is

$$\delta F = \gamma h \, \delta A = \delta V \tag{2.5.12}$$

which is an element of volume of the pressure prism. After integrating, $F = V$, the volume of the pressure prism equals the magnitude of the resultant force acting on one side of the surface.

From Eqs. (2.5.5) and (2.5.6)

$$x_p = \frac{1}{V} \int_V x \, dV \qquad y_p = \frac{1}{V} \int_V y \, dV \tag{2.5.13}$$

which shows that x_p and y_p are distances to the *centroid* of the pressure prism [Eq. (A.5)]. Hence, the line of action of the resultant passes through the centroid of the pressure prism. For some simple areas the pressure prism is more convenient than either integration or formula. For example, a rectangular area with one edge in the free surface has a wedge-shaped prism. Its centroid is one-third the altitude from the base; hence, the pressure center is one-third the altitude from its lower edge.

Example 2.8

A structure is so arranged along a channel that it will spill the water out if a certain height y (Fig. 2.18a) is reached. The gate is made of steel plate weighing 2500 N/m². Determine the height of y.

Solution

Using pressure-prism concepts, for unit width normal to the page the force on the horizontal leaf (Fig. 2.18b) is given by the volume of a pressure prism of base 1.2 m² and constant altitude γy N/m², which yields $F_y = 1.2\gamma y$ N acting through the center of the base. The pressure prism for the vertical face (Fig. 2.18c) is a

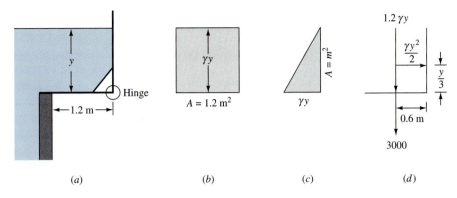

Figure 2.18 Flashboard arrangement on side of channel.

wedge of base y m^2 and altitude varying from 0 to γy N/m^2. The average altitude is $\gamma y/2$, so $F_x = \gamma y^2/2$ N. The centroid of the wedge prism is $y/3$ from the hinge. The weight of the gate floor exerts a force of 3000 N at its center. Figure 2.18d shows all the forces and moment arms. For equilibrium, that is, the value of y for tipping, moments about the hinge must be zero.

$$M = (3000 \text{ N})(0.6 \text{ m}) + (1.2\gamma y \text{ N})(0.6 \text{ m}) - \left(\frac{\gamma y^2}{2} \text{ N}\right)\left(\frac{y}{3} \text{ m}\right) = 0$$

or

$$M = y^3 - 4.32y - 1.1014 = 0$$

This equation has only one positive root, easily seen to be between $y = 2$ and $y = 3$. By use of the Newton-Raphson method (Appendix B.5)

$$y = y - \frac{M(y)}{M'(y)} = y - \frac{y^3 - 4.32y - 1.1014}{3y^2 - 4.32}$$

which is an iterative procedure. A trial pressure of y is assumed, say $y = 2.5$. Substitution into the right-hand side yields an improved value of y. By repeating this procedure three times the root is found to be $y = 2.196$ m. A cubic equation is easily solved with a programmable calculator. Preparing the program requires no more effort than solving the equation once by calculator.

Effects of Atmospheric Pressure on Forces on Plane Areas

In the discussion of pressure forces the pressure datum was not mentioned. The pressures were computed by $p = \gamma h$, in which h is the vertical distance below the free surface. Therefore, the datum taken was gage pressure zero, or local atmospheric pressure. When the opposite side of the surface is open to the atmosphere, a force is exerted on it by the atmosphere equal to the product of the atmospheric pressure p_0 and the area, $p_0 A$, based on absolute zero as datum. On the liquid side the force is:

$$\int (p_0 + \gamma h)\, dA = p_0 A + \gamma \int h\, dA$$

The effect of $p_0 A$ of the atmosphere acts equally on both sides and in no way contributes to the resultant force or its location.

So long as the same pressure datum is selected for all sides of a free body, the resultant and moment can be determined by constructing a free surface at pressure zero on this datum and using the above methods.

Example 2.9

An application of pressure forces on plane areas is given in the design of a gravity dam. The maximum and minimum compressive stresses in the base of the dam are computed from the forces which act on the dam. Figure 2.19 shows a cross section through a concrete dam where the specific weight of concrete has been taken as 2.5γ and γ is the specific weight of water. A 1-ft section of dam is considered as a free body; the forces are due to the concrete, the water, the foundation pressure, and the hydrostatic uplift. Determining the amount of hydrostatic uplift is beyond the scope of this treatment, but it will be assumed to be one-half the hydrostatic head at the upstream edge, decreasing linearly to zero at the downstream edge of the dam. Enough friction or shear stress must be developed at the base of the dam to balance the thrust due to the water; that is, $R_x = 5000\gamma$. The resultant upward force on the base equals the weight of the dam less the hydrostatic uplift, $R_y = 6750\gamma + 2625\gamma - 1750\gamma = 7625\gamma$ lb. The position of R_y is such that the free body is in equilibrium. For moments around O,

$$\Sigma M_0 = 0 = R_y x - 5000\gamma(33.33) - 2625\gamma(5) - 6750\gamma(30) + 1750\gamma(23.33)$$

and

$$x = 44.8 \text{ ft}$$

It is customary to assume that the foundation pressure varies linearly over the base of the dam, that is, that the pressure prism is a trapezoid with a volume equal to R_y; thus,

$$\frac{\sigma_{max} + \sigma_{min}}{2} 70 = 7625\gamma$$

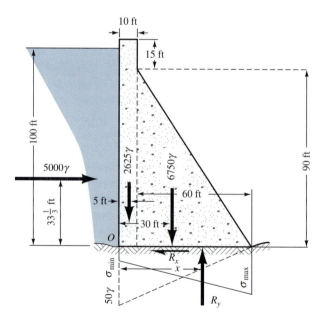

Figure 2.19   Concrete gravity dam.

in which σ_{max} and σ_{min} are, respectively, the maximum and minimum compressive stresses in pounds per square foot. The centroid of the pressure prism is at the point where $x = 44.8$ ft. By taking moments about O to express the position of the centroid in terms of σ_{max} and σ_{min},

$$44.8 = \frac{(\sigma_{min})(70)(\frac{70}{2}) + (\sigma_{max} - \sigma_{min})(\frac{70}{2})\frac{2}{3}(70)}{(\sigma_{max} + \sigma_{min})(\frac{70}{2})}$$

Simplifying gives

$$\sigma_{max} = 11.75\,\sigma_{min}$$

Then

$$\sigma_{max} = 210\gamma = 12{,}500 \text{ lb/ft}^2 \qquad \sigma_{max} = 17.1\gamma = 1067 \text{ lb/ft}^2$$

When the resultant falls within the middle third of the base of the dam, σ_{min} will always be a compressive stress. The poor tensile properties of concrete mean that good design requires the resultant to fall within the middle third of the base.

Example 2.10

Water is held in the pipe of Fig. 2.20 until a depth y is reached that causes sufficient moment to overcome the counterweight. Neglecting the weight of the structure, find y.

Solution

The moment about the hinge of the pressure force on the gate is

$$\text{Moment} = -\gamma \int_{z_o + R}^{z_o + R\cos\theta_o} zhx\,dz = \int_0^{\theta_o} F(\theta)\,d\theta$$

in which

$$z = z_o + R\cos\theta \qquad dz = -R\sin\theta\,d\theta$$
$$h = R(\cos\theta - \cos\theta_o) \qquad x = 2R\sin\theta$$

Then

$$F(\theta) = \gamma(z_o + R\cos\theta)R(\cos\theta - \cos\theta_o)2R^2\sin^2\theta$$

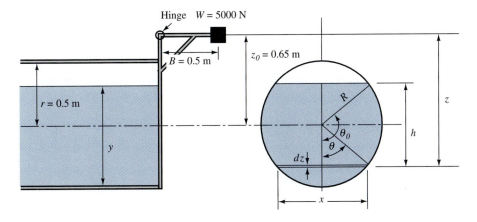

Figure 2.20 Gate at end of pipe.

The bisection method may be used to find θ_o such that the moment due to the liquid equals the moment due to the counterweight. Since we are looking for an answer between $\theta_o = 0$ and $\theta_o = \pi$, we take $\theta_{o_{\max}} = \pi$ and $\theta_{o_{\min}} = 0$. The interval $d\theta$ is θ_o/n, with n the number of intervals. Then the moment contribution from one interval is

$$\text{Moment} = \frac{d\theta}{2}[F(\theta) + F(\theta - d\theta)]$$

which is summed up for the n intervals. If the moment is too great, the next θ_o is reduced according to the bisection method. A spreadsheet, using an optimizer or solver function, may be used to produce the results:

$$\text{Depth} = 0.9027 \text{ m} \qquad \theta_o = 143.64°$$

EXERCISES

2.5.1 The magnitude of force on one side of a circular surface of unit area, with centroid 10 ft below a free-water surface, is (*a*) less than 10γ; (*b*) dependent upon orientation of the area; (*c*) greater than 10γ; (*d*) the product of γ and the vertical distance from free surface to pressure center; (*e*) none of the above.

2.5.2 A rectangular surface 3 by 4 m has the lower 3-m edge horizontal and 6 m below a free-oil surface, $S = 0.80$. The surface is inclined 30° with the horizontal. The force on one side of the surface is (*a*) 38.4γ; (*b*) 48γ; (*c*) 51.2γ; (*d*) 60γ; (*e*) none of these answers.

2.5.3 The pressure center of the surface of Exercise 2.5.2 is vertically below the liquid surface (*a*) 10.133 m; (*b*) 5.133 m; (*c*) 5.067 m; (*d*) 5.00 m; (*e*) none of these answers.

2.5.4 The pressure center is (*a*) at the centroid of the submerged area; (*b*) the centroid of the pressure prism; (*c*) independent of the orientation of the area; (*d*) a point on the line of action of the resultant force; (*e*) always above the centroid of the area.

2.5.5 What is the force exerted on the vertical annular area enclosed by concentric circles of radii 1.0 and 2.0 m? The center is 3.0 m below a free-water surface. (*a*) $3\pi\gamma$; (*b*) $9\pi\gamma$; (*c*) $10.25\pi\gamma$; (*d*) $12\pi\gamma$; (*e*) none of these answers.

2.5.6 The pressure center for the annular area of Exercise 2.5.5 is below the centroid of the area (*a*) 0 m; (*b*) 0.42 m; (*c*) 0.44 m; (*d*) 0.47 m; (*e*) none of these answers.

2.5.7 A vertical triangular area has one side in a free surface, with vertex downward. Its altitude is h. The pressure center is below the free surface (*a*) $h/4$; (*b*) $h/3$; (*c*) $h/2$; (*d*) $2h/3$; (*e*) $3h/4$.

2.5.8 A vertical gate 4 by 4 m holds water with free surface at its top. The moment about the bottom of the gate is (*a*) 42.7γ; (*b*) 57γ; (*c*) 64γ; (*d*) 85.3γ; (*e*) none of these answers.

2.6 FORCE COMPONENTS ON CURVED SURFACES

When the elemental forces $p\,\delta A$ vary in direction, as in the case of a curved surface, they must be added as vector quantities; that is, their components in three mutually perpendicular directions are added as scalars and then the three components are added vectorially. With two horizontal components at right angles and with the vertical component—all easily computed for a curved surface—the resultant can be determined. The lines of action of the components also are readily determined.

Horizontal Component of Force on a Curved Surface

The horizontal component of pressure force on a curved surface is equal to the pressure force exerted on a projection of the curved surface. The vertical plane of projection is normal to the direction of the component. The surface of Fig. 2.21 represents any three-dimensional surface and δA an element of its area, its normal making the angle θ with the negative x direction. Then

$$\delta F_x = p \, \delta A \, \cos \theta$$

is the x component of force exerted on one side of δA. Summing up the x components of force over the surface gives

$$F_x = \int_A p \cos \theta \, dA \tag{2.6.1}$$

in which $\cos \theta \, \delta A$ is the projection δA onto a plane perpendicular to x. The element of force on the projected area is $p \cos \theta \, \delta A$, which is also in the x direction. Projecting each element on a plane perpendicular to x is equivalent to projecting the curved surface as a whole onto the vertical plane. Hence, the force acting on this projection of the curved surface is the horizontal component of force exerted on the curved surface in the direction normal to the plane of projection. To find the horizontal component at right angles to the x direction, the curved surface is projected onto a vertical plane parallel to x and the force on the projection is determined.

When the horizontal component of pressure force on a closed body is to be found, the projection of the curved surface on a vertical plane is always zero, since on opposite sides of the body the area-element projections have opposite signs as indicated in Fig. 2.22. Let a small cylinder of cross section δA with its axis parallel to x intersect the closed body at B and C. If the element of area of the body cut by the prism at B is δA_B and at C is δA_C, then

$$\delta A_B \cos \theta_B = -\delta A_C \cos \theta_C = \delta A$$

as $\cos \theta_C$ is negative. Hence, with the pressure the same at each end of the cylinder

$$p \, \delta A_B \cos \theta_B + p \, \delta A_C \cos \theta_C = 0$$

and similarly for all other area elements.

To find the line of action of a horizontal component of force on a curved surface, the resultant of the parallel force system composed of the force components from each area element is required. This is exactly the resultant of the force on the

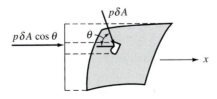

Figure 2.21   Horizontal component of curved surface.

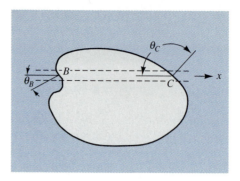

Figure 2.22   Projections of area elements on opposite sides of a body.

projected area, since the two force systems have an identical distribution of elemental horizontal force components. Hence, the pressure center is located on the projected area by the methods of Sec. 2.5.

Example 2.11

The equation of an ellipsoid of revolution submerged in water is $x^2/4 + y^2/4 + z^2/9 = 1$. The center of the body is located 2 m below the free surface. Find the horizontal force components acting on the curved surface that is located in the first octant. Consider the xz plane to be horizontal and y to be positive upward.

Solution

The projection of the surface on the yz plane has an area of $(\pi/4)(2)(3)$ m². Its centroid is located $2 - (4/3\pi)(2)$ m below the free surface. Hence,

$$F_x = -\left[\frac{\pi}{4}(6)\right]\left(2 - \frac{8}{3\pi}\right)\gamma = -(-5.425 \text{ m}^3)(9806 \text{ N/m}^3) = -53.2 \text{ kN}$$

Similarly,

$$F_z = -\left[\frac{\pi}{4}(4)\right]\left(2 - \frac{8}{3\pi}\right)\gamma = (-3.617 \text{ m}^3)(9806 \text{ N/m}^3) = -35.4 \text{ kN}$$

Vertical Component of Force on a Curved Surface

The vertical component of pressure force on a curved surface is equal to the weight of liquid vertically above the curved surface and extending up to the free surface. The vertical component of force on a curved surface can be determined by summing up the vertical components of pressure force on elemental areas δA of the surface. In Fig. 2.23 an area element is shown with the force $p\,\delta A$ acting normal to it. Let θ be the angle the normal to the area element makes with the vertical. Then the vertical component of force acting on the area element is $p\cos\theta\,\delta A$, and the vertical component of force on the curved surface is given by

$$F_v = \int_A p\,\cos\theta\,dA \qquad\qquad \text{(2.6.2)}$$

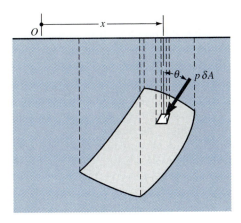

Figure 2.23   Vertical component of force on a curved surface.

Since p is equivalent to γh, in which h is the distance from the area element to the free surface, and since $\cos \theta\, \delta A$ is the projection of δA on a horizontal plane, Eq. (2.6.2) becomes

$$F_v = \gamma \int_A h \cos \theta\, dA = \gamma \int_\forall d\forall \qquad \text{(2.6.3)}$$

in which $\delta\forall$ is the volume of the prism of height h and base $\cos \theta\, \delta A$, or the volume of liquid vertically above the area element. Integrating gives

$$F_v = \gamma \forall \qquad \text{(2.6.4)}$$

When the liquid is below the curved surface (Fig. 2.24) and the pressure magnitude is known at some point, say O, an *imaginary* or equivalent free surface $s\text{-}s$ can be constructed p/γ above O, so that the product of specific weight and vertical distance to any point in the tank is the pressure at the point. The weight of the imaginary volume of liquid vertically above the curved surface is then the vertical component of pressure force on the curved surface. In constructing an imaginary free surface,

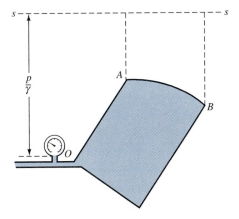

Figure 2.24   Liquid with equivalent free surface.

the imaginary liquid must be of the same specific weight as the liquid in contact with the curved surface; otherwise, the pressure distribution over the surface will not be correctly represented. With an imaginary liquid above a surface, the pressure at a point on the curved surface is equal on both sides, but the elemental force components in the vertical direction are opposite in sign. Hence, the direction of the vertical force component is reversed when an imaginary fluid is above the surface. In some cases a confined liquid can be above the curved surface, and an imaginary liquid must be added (or subtracted) to determine the free surface.

The line of action of the vertical component is determined by equating moments of the elemental vertical components about a convenient axis with the moment of the resultant force. With the axis at O (Fig. 2.23),

$$F_v \bar{x} = \gamma \int_V x \, dV$$

in which $\bar{x}$ is the distance from O to the line of action. Then, since $F_v = \gamma V$,

$$\bar{x} = \frac{1}{V} \int_V x \, dV$$

the distance to the centroid of the volume. Therefore, the line of action of the vertical force passes through the centroid of the volume, real or imaginary, that extends above the curved surface up to the real or imaginary free surface.

| **Example 2.12** | A cylindrical barrier (Fig. 2.25) holds water as shown. The contact between the cylinder and wall is smooth. Consider a 1-m length of cylinder; determine (*a*) its weight and (*b*) the force exerted against the wall. |

Solution

(*a*) For equilibrium the weight of the cylinder must equal the vertical component of force exerted on it by the water. (The imaginary free surface for CD is at elevation A). The vertical force on BCD is

$$F_{v_{BCD}} = \left(\frac{\pi r^2}{2} + 2r^2 \right) \gamma = (2\pi + 8) \gamma$$

The vertical force on AB is

$$F_{v_{AB}} = - \left(r^2 - \frac{\pi r^2}{4} \right) \gamma = -(4 - \pi) \gamma$$

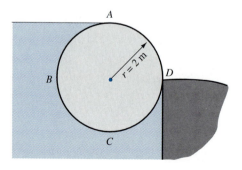

Figure 2.25   Semifloating body.

Hence, the weight per meter of length is

$$F_{v_{BCD}} + F_{v_{AB}} = (3\pi + 4)\gamma = 0.132 \text{ MN}$$

(b) The force exerted against the wall is the horizontal force on ABC minus the horizontal force on CD. The horizontal components of force on BC and CD cancel; the projection of BCD on a vertical plane is zero. Hence,

$$F_H = F_{H_{AB}} = 2\gamma = 19.6 \text{ kN}$$

since the projected area is 2 m² and the pressure at the centroid of the projected area is 9806 Pa.

To find external reactions due to pressure forces, the action of the fluid can be replaced by the two horizontal components and one vertical component acting along their lines of action.

Tensile Stress in a Pipe and Spherical Shell

A circular pipe under the action of an internal pressure is in tension around its periphery. Assuming that no longitudinal stress occurs, the walls are in tension, as shown in Fig. 2.26. A section of pipe of unit length is considered, that is, the ring between two planes normal to the axis and unit length apart. If one-half of this ring is taken as a free body, the tensions per unit length at top and bottom are respectively T_1 and T_2, as shown in the figure. The horizontal component of force acts through the pressure center of the projected area and is $2pr$, in which p is the pressure at the centerline and r is the internal pipe radius.

For high pressures the pressure center can be taken at the pipe center; then $T_1 = T_2$ and

$$T = pr \tag{2.6.5}$$

in which T is the tensile force per unit length. For wall thickness e, the *tensile stress* in the pipe wall, σ, is

$$\sigma = \frac{T}{e} = \frac{pr}{e} \tag{2.6.6}$$

For larger variations in pressure between top and bottom of pipe, the location of pressure center y is computed. Two equations are needed,

$$T_1 + T_2 = 2pr \qquad 2rT_1 - 2pry = 0$$

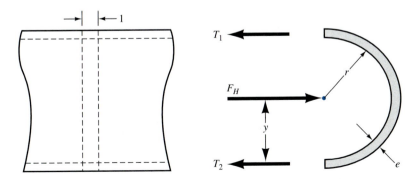

Figure 2.26   Tensile stress in pipe.

The second equation is the moment equation about the lower end of the free body, neglecting the vertical component of force. Solving gives

$$T_1 = py \qquad T_2 = p(2r - y)$$

Example 2.13	**A** 100-m-inner-diameter (ID) steel pipe has a 6-mm wall thickness. For an allowable tensile stress of 70 MPa, what is the maximum pressure?

Solution

From Eq. (2.6.6)

$$p = \frac{\sigma e}{r} = \frac{(70 \text{ MPa})(0.006 \text{ m})}{0.1 \text{ m}} = 4.2 \text{ MPa}$$

If a thin spherical shell is subjected to an internal pressure, neglecting the weight of the fluid within the sphere, the stress in its walls can be found by considering the forces on a free body consisting of a hemisphere cut from the sphere by a vertical plane. The fluid component of force normal to the plane acting on the inside of the hemisphere is $p\pi r^2$, with r being the radius. The stress σ times the cut wall area $2\pi re$, with e being the thickness, must balance the fluid force; hence,

$$\sigma = \frac{pr}{2e}$$

EXERCISES

2.6.1 The horizontal component of force on a curved surface is equal to the (a) weight of liquid vertically above the curved surface; (b) weight of liquid retained by the curved surface; (c) product of pressure at its centroid and area; (d) force on a projection of the curved surface onto a vertical plane; (e) scalar sum of all elemental horizontal components.

2.6.2 A pipe 5 m in diameter is to carry water at 1.4 MPa. For an allowable tensile stress of 55 MPa, the thickness of pipe wall, in millimeters, is (a) 32; (b) 42; (c) 64; (d) 80; (e) none of these answers.

2.6.3 The vertical component of pressure force on a submerged curved surface is equal to (a) its horizontal component; (b) the force on a vertical projection of the curved surface; (c) the product of pressure at its centroid and surface area; (d) the weight of liquid vertically above the curved surface; (e) none of these answers.

2.6.4 The vertical component of force on the upper half of a horizontal right-circular cylinder, 3 ft in diameter and 10-ft long, filled with water and with a pressure of 0.433 psi at the axis, is (a) −458 lb; (b) −333 lb; (c) 124.8 lb; (d) 1872 lb; (e) none of these answers.

2.6.5 A cylindrical wooden barrel is held together by hoops at the top and bottom. When the barrel is filled with liquid, the ratio of tension in the top hoop to tension in the bottom hoop, due to the liquid, is (a) $\frac{1}{2}$; (b) 1; (c) 2; (d) 3; (e) none of these answers.

2.6.6 A 50-mm-ID pipe with 5-mm wall thickness carries water at 0.89 MPa. The tensile stress in the pipe wall, in megapascals, is (a) 4.9; (b) 9.8; (c) 19.6; (d) 39.2; (e) none of these answers.

2.7 BUOYANT FORCE

The resultant force exerted on a body *by a static fluid* in which it is submerged or floating is called the *buoyant force*. The buoyant force always acts vertically upward. There can be no horizontal component of the resultant because the projection of the submerged body or submerged portion of the floating body on a vertical plane is always zero.

The buoyant force on a submerged body is the difference between the vertical component of pressure force on its under side and the vertical component of pressure force on its upper side. In Fig. 2.27 the upward force on the bottom is equal to the weight of liquid, real or imaginary, which is vertically above the surface *ABC*, indicated by the weight of liquid within *ABCEFA*. The downward force on the upper surface equals the weight of liquid *ADCEFA*. The difference between the two forces is a force, vertically upward, due to the weight of liquid *ABCD* that is *displaced* by the solid. In equation form

$$F_B = V\gamma \qquad \text{(2.7.1)}$$

in which F_B is the buoyant force, V is the volume of fluid displaced, and γ is the specific weight of the fluid. The same formula holds for floating bodies when V is taken as the volume of liquid displaced. This is evident from inspection of the floating body in Fig. 2.27.

In Fig. 2.28 the vertical force exerted on an element of the body in the form of a vertical prism of cross section δA is

$$\delta F_B = (p_2 - p_1)\delta A = \gamma h\,\delta A = \gamma\,\delta V$$

in which δV is the volume of the prism. Integrating over the complete body gives

$$F_B = \gamma \int_V dV = \gamma V$$

when γ is considered constant throughout the volume.

To find the line of action of the buoyant force, moments are taken about a convenient axis O and are equated to the moment of the resultant; thus,

$$\gamma \int_V x\,dV = \gamma V \bar{x} \quad \text{or} \quad \bar{x} = \frac{1}{V}\int_V x\,dV$$

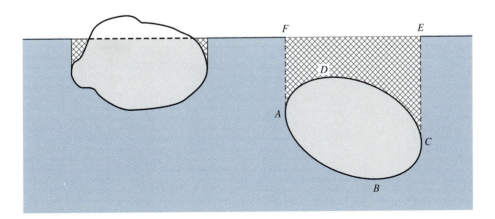

Figure 2.27   Buoyant force on floating and submerged bodies.

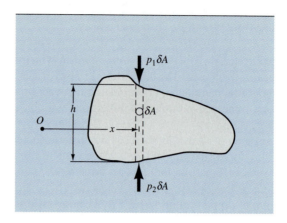

Figure 2.28   Vertical force components on element of body.

in which $\bar{x}$ is the distance from the axis to the line of action. This equation yields the distance to the centroid of the volume; hence, *the buoyant force acts through the centroid of the displaced volume of fluid.* This holds for both submerged and floating bodies. The centroid of the displaced volume of fluid is called the *center of buoyancy.*

In solving a statics problem involving submerged or floating objects, the object is generally taken as a free body and a free-body diagram is drawn. The action of the fluid is replaced by the buoyant force. The weight of the object must be shown (acting through its center of gravity) as well as all other contact forces.

Weighing an odd-shaped object suspended in two different fluids yields sufficient data to determine its weight, volume, specific weight, and specific gravity. Figure 2.29 shows two free-body diagrams for the same object suspended and weighed in two fluids. F_1 and F_2 are the weights when submerged, and γ_1 and γ_2 are the specific weights of the fluids. W and V, the weight and volume of the object, respectively, are to be found.

The equations of equilibrium are written

$$F_1 + V\gamma_1 = W \qquad F_2 + V\gamma_2 = W$$

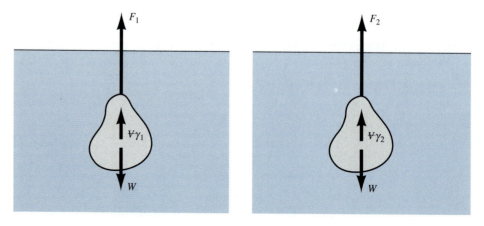

Figure 2.29   Free-body diagrams for body suspended in a fluid.

and solved

$$V = \frac{F_1 - F_2}{\gamma_2 - \gamma_1} \qquad W = \frac{F_1\gamma_2 - F_2\gamma_1}{\gamma_2 - \gamma_1}$$

A *hydrometer* uses the principle of buoyant force to determine specific gravities of liquids. Figure 2.30 shows a hydrometer in two liquids. It has a stem of prismatic cross section a. Considering the liquid on the left to be distilled water, $S = 1.00$, the hydrometer floats in equilibrium when

$$V_0\gamma = W \tag{2.7.2}$$

in which V_0 is the volume submerged, γ is the specific weight of water, and W is the weight of the hydrometer. The position of the liquid surface is marked as 1.00 on the stem to indicate unit specific gravity S. When the hydrometer is floated in another liquid, the equation of equilibrium becomes

$$(V_0 - \Delta V)S\gamma = W \tag{2.7.3}$$

in which $\Delta V = a\Delta h$. Solving for Δh with Eqs. (2.7.2) and (2.7.3) gives

$$\Delta h = \frac{V_0}{a}\frac{S-1}{S} \tag{2.7.4}$$

from which the stem can be marked off to read specific gravities.

A piece of ore weighing 1.5 N in air is found to weigh 1.1 N when submerged in water. What is its volume, in cubic centimeters, and what is its specific gravity?

Example 2.14

Solution

The buoyant force due to air can be neglected. From Fig. 2.29

$$1.5\text{ N} = 1.1\text{ N} + (9806\text{ N/m}^3)V$$

$$V = 0.0000408\text{ m}^3 = 40.8\text{ cm}^3$$

$$S = \frac{W}{\gamma V} = \frac{1.5\text{ N}}{(9806\text{ N/m}^3)(0.0000408\text{ m}^3)} = 3.75$$

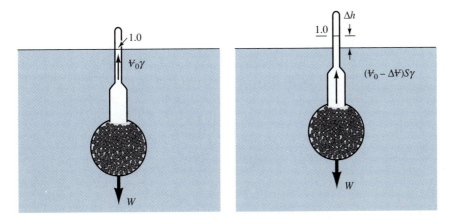

Figure 2.30 Hydrometer in water and in liquid of specific gravity S.

EXERCISES

2.7.1 A slab of wood 1 by 1 by 0.25 m, $S = 0.50$, floats in water with a 400-N load on it. The volume of slab submerged, in cubic meters, is (*a*) 0.043; (*b*) 0.125; (*c*) 0.166; (*d*) 0.293; (*e*) none of these answers.

2.7.2 The line of action of the buoyant force acts through the (*a*) center of gravity of any submerged body; (*b*) centroid of the volume of any floating body; (*c*) centroid of the displaced volume of fluids; (*d*) centroid of the volume of fluid vertically above the body; (*e*) centroid of the horizontal projection of the body.

2.7.3 Buoyant force is (*a*) the resultant force on a body due to the fluid surrounding it; (*b*) the resultant force acting on a floating body; (*c*) the force necessary to maintain equilibrium of a submerged body; (*d*) a nonvertical force for nonsymmerical bodies; (*e*) equal to the volume of liquid displaced.

2.8 STABILITY OF FLOATING AND SUBMERGED BODIES

A body floating in a static liquid has vertical stability. A small upward displacement decreases the volume of liquid displaced, resulting in an unbalanced downward force which tends to return the body to its original position. Similarly, a small downward displacement results in a greater buoyant force, which causes an unbalanced upward force.

A body has linear stability when a small linear displacement in any direction sets up restoring forces tending to return it to its original position. It has rotational stability when a restoring couple is set up by any small angular displacement.

Methods for determining rotational stability are developed in the following discussion. A body may float in stable, unstable, or neutral equilibrium. When a body is in unstable equilibrium, any small angular displacement sets up a couple that tends to increase the angular displacement. With the body in neutral equilibrium any small angular displacement does not set up any couple whatsoever. The three cases of equilibrium are illustrated; in Fig. 2.31*a* a light piece of wood with a metal weight at its bottom is stable; in Fig. 2.31*b* when the metal weight is at the top, the body is in equilibrium but any slight angular displacement causes it to assume

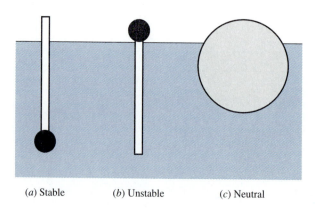

(*a*) Stable   (*b*) Unstable   (*c*) Neutral

Figure 2.31   Examples of stable, unstable, and neutral equilibrium.

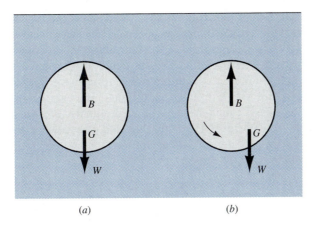

Figure 2.32 Rotationally stable submerged body.

the position in *a;* in Fig. 2.31*c* a homogeneous sphere or right-circular cylinder is in equilibrium for any angular rotation, that is, no couple results from an angular displacement.

A completely submerged object is rotationally stable only when its center of gravity is below the center of buoyancy, as in Fig. 2.32*a*. When the object is rotated counterclockwise, as in Fig. 2.32*b*, the buoyant force and weight produce a couple in the clockwise direction.

Normally, when a body is too heavy to float, it submerges and goes down until it rests on the bottom. Although the specific weight of a liquid increases slightly with depth, the higher pressure tends to cause the liquid to compress the body or to penetrate into pores of solid substances and thus decreases the buoyancy of the body. A ship, for example, is sure to go to the bottom once it is completely submerged, owing to compression of air trapped in its various parts.

Any floating object with center of gravity below its center of buoyancy (centroid of displaced volume) floats in stable equilibrium, as in Fig. 2.31*a*. Certain floating objects, however, are in stable equilibrium when their center of gravity is above the center of buoyancy. The stability of prismatic bodies is first considered, followed by an analysis of general floating bodies for small angles of tip.

Figure 2.33*a* is a cross section of a body with all other parallel cross sections identical. The center of buoyancy is always at the centroid of the displaced volume,

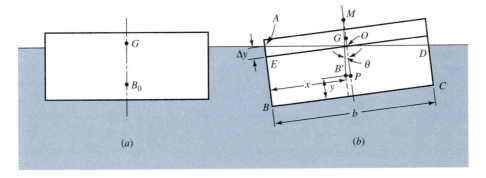

Figure 2.33 Stability of a prismatic body.

which is at the centroid of the cross-sectional area below the liquid surface in this case. Hence, when the body is tipped, as in Fig. 2.33b, the center of buoyancy is at the centroid B' of the trapezoid $ABCD$, the buoyant force acts upward through B', and the weight acts downward through G, the center of gravity of the body. When the vertical through B' intersects the original centerline above G, as at M, a restoring couple is produced; the body is in stable equilibrium. The intersection of the buoyant force and the centerline is called the *metacenter,* designated M. When M is above G, the body is stable; when below G, it is unstable; and when at G, it is in neutral equilibrium. The distance $\overline{MG}$ is called the *metacentric height* and is a direct measure of the stability of the body. The restoring couple is

$$W\overline{MG}\sin\theta$$

in which θ is the angular displacement and W is the weight of the body.

Example 2.15

In Fig. 2.33 a scow 20 ft wide and 60 ft long has a gross weight of 225 short tons (2000 lb). Its center of gravity is 1.0 ft above the water surface. Find the metacentric height and restoring couple when $\Delta y = 1.0$ ft.

Solution

The depth of submergence h in the water is

$$h = \frac{225(2000)}{20(60)(62.4)} = 6.0 \text{ ft}$$

The centroid in the tipped position is located with moments about AB and BC,

$$x = \frac{5(20)(10) + 2(20)(\tfrac{1}{2})(\tfrac{20}{3})}{6(20)} = 9.46 \text{ ft}$$

$$y = \frac{5(20)(\tfrac{5}{2}) + 2(20)(\tfrac{1}{2})(5\tfrac{2}{3})}{6(20)} = 3.03 \text{ ft}$$

By similar triangles AEO and $B'PM$,

$$\frac{\Delta y}{b/2} = \frac{\overline{B'P}}{\overline{MP}}$$

$\Delta y = 1, b/2 = 10, \overline{B'P} = 10 - 9.46 = 0.54$ ft; then

$$\overline{MP} = \frac{0.54(10)}{1} = 5.40 \text{ ft}$$

G is 7.0 ft from the bottom; hence,

$$\overline{GP} = 7.00 - 3.03 = 3.97 \text{ ft}$$

and

$$\overline{MG} = \overline{MP} - \overline{GP} = 5.40 - 3.97 = 1.43 \text{ ft}$$

The scow is stable since $\overline{MG}$ is positive; the righting moment is

$$W\overline{MG}\sin\theta = 225(2000)(1.43)\frac{1}{\sqrt{101}} = 64{,}000 \text{ lb} \cdot \text{ft}$$

EXERCISES

2.8.1 A body floats in stable equilibrium (*a*) when its metacentric height is zero; (*b*) *only* when its center of gravity is below its center of buoyancy; (*c*) when $\overline{GB} - I/\Psi$ is positive and *G* is above *B*; (*d*) when I/Ψ is positive; (*e*) when the metacenter is above the center of gravity.

2.8.2 A closed cubical metal box 1 m on an edge is made of uniform sheet and has a mass of 550 kg. Its metacentric height when placed in oil, $S = 0.90$, with sides vertical, is (*a*) -0.058 m; (*b*) 0.078 m; (*c*) 0.33 m; (*d*) 0.467 m; (*e*) none of these answers.

2.9 RELATIVE EQUILIBRIUM

In fluid statics the variation of pressure is simple to compute, thanks to the absence of shear stresses. For fluid motion such that no layer moves relative to an adjacent layer, the shear stress is also zero throughout the fluid. A fluid with a translation at uniform velocity still follows the laws of static variation of pressure. When a fluid is being accelerated so that no layer moves relative to an adjacent one, that is, when the fluid moves as if it were a solid, no shear stresses occur and variation in pressure can be determined by writing the equation of motion for an appropriate free body. Two cases are of interest, a uniform linear acceleration and a uniform rotation about a vertical axis. When moving thus, the fluid is said to be in *relative equilibrium*.

Although relative equilibrium is not a fluid static phenomenon, it is discussed here because of the similarity of the relations.

Uniform Linear Acceleration

A liquid in an open vessel is given a uniform linear acceleration **a** as in Fig. 2.34. After some time the liquid adjusts to the acceleration so that it moves as a solid; that is, the distance between any two fluid particles remains fixed, and hence no shear stresses occur.

By selecting a cartesian coordinate system with *y* vertical and *x* such that the acceleration vector **a** is in the *xy* plane (Fig. 2.34*a*), the *z* axis is normal to **a** and

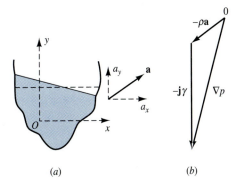

(*a*)   (*b*)

Figure 2.34   Acceleration with free surface.

there is no acceleration component in that direction. Equation (2.2.5) applies to this situation,

$$\mathbf{f} - \mathbf{j}\gamma = -\nabla p - \mathbf{j}\gamma = \rho\mathbf{a} \tag{2.2.5}$$

The pressure gradient ∇p is then the vector sum of $-\rho\mathbf{a}$ and $-\mathbf{j}\gamma$, as shown in Fig. 2.34b. Since ∇p is in the direction of maximum change in p (the gradient), at right angles to ∇p there is no change in p. Surfaces of constant pressure, including the free surface, must therefore be normal to ∇p. To obtain a convenient algebraic expression for variation of p with x, y, and z, that is, $p = p(x, y, z)$, Eq. (2.2.5) is written in component form:

$$\nabla p = \mathbf{i}\frac{\partial p}{\partial x} + \mathbf{j}\frac{\partial p}{\partial y} + \mathbf{k}\frac{\partial p}{\partial z} = -\mathbf{j}\gamma - \frac{\gamma}{g}(\mathbf{i}a_x + \mathbf{j}a_y)$$

or

$$\frac{\partial p}{\partial x} = -\frac{\gamma}{g}a_x \qquad \frac{\partial p}{\partial y} = -\gamma\left(1 + \frac{a_y}{g}\right) \qquad \frac{\partial p}{\partial z} = 0$$

Since p is a function of position (x, y, z), its total differential is

$$dp = \frac{\partial p}{\partial x}dx + \frac{\partial p}{\partial y}dy + \frac{\partial p}{\partial z}dz$$

Substituting for the partial differentials gives

$$dp = -\gamma\frac{a_x}{g}dx - \gamma\left(1 + \frac{a_y}{g}\right)dy \tag{2.9.1}$$

which can be integrated for an incompressible fluid,

$$p = -\gamma\frac{a_x}{g}x - \gamma\left(1 + \frac{a_y}{g}\right)y + c$$

To evaluate the constant of integration c, let $x = 0$, $y = 0$, $p = p_0$; then $c = p_0$ and

$$p = p_0 - \gamma\frac{a_x}{g}x - \gamma\left(1 + \frac{a_y}{g}\right)y \tag{2.9.2}$$

When the accelerated incompressible fluid has a free surface, its equation is given by setting $p = 0$ in Eq. (2.9.2). Solving Eq. (2.9.2) for y gives

$$y = -\frac{a_x}{a_y + g}x + \frac{p_0 - p}{\gamma(1 + a_y/g)} \tag{2.9.3}$$

The lines of constant pressure, $p = $ constant, have a slope

$$-\frac{a_x}{a_y + g}$$

and are parallel to the free surface. The y intercept of the free surface is

$$\frac{p_0}{\gamma(1 + a_y/g)}$$

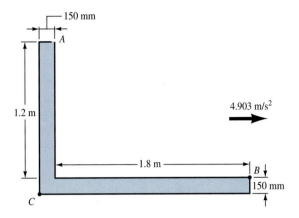

Figure 2.35 Tank completely filled with liquid.

The tank in Fig. 2.35 is filled with oil, sp gr 0.8, and accelerated as shown. There is a small opening in the tank at A. Determine the pressure at B and C, and the acceleration a_x required to make the pressure at B equal to zero.

Example 2.16

Solution

Selecting point A as the origin and applying Eq. (2.9.2) for $a_y = 0$ leads to

$$p = -\gamma \frac{a_x}{g} x - \gamma y = -\frac{0.8(9806 \text{ N/m}^3)(4.903 \text{ m/s}^2)}{9.806 \text{ m/s}^2} x - 0.8(9806 \text{ N/m}^3)y$$

or

$$p = -3922.4x - 7844.8y \quad \text{Pa}$$

At B, $x = 1.8$ m, $y = -1.2$ m, and $p = 2.35$ kPa. At C, $x = -0.15$ m, $y = -1.35$ m, and $p = 11.18$ kPa. For zero pressure at B, from Eq. (2.9.2) with origin at A,

$$0.0 = 0.0 - \frac{0.8(9806 \text{ N/m}^3)}{9.806 \text{ m/s}^2} 1.8a_x - 0.8(9806 \text{ N/m}^3)(-1.2)$$

or

$$a_x = 6.537 \text{ m/s}^2$$

A closed box with horizontal base 6 by 6 units and a height of 2 units is half-filled with liquid (Fig. 2.36). It is given a constant linear acceleration $a_x = g/2$ and $a_y = -g/4$. Develop an equation for variation of pressure along its base.

Example 2.17

Solution

The free surface has the slope

$$\frac{-a_x}{a_y + g} = \frac{-g/2}{-g/4 + g} = -\frac{2}{3}$$

hence, the free surface is located as shown in the figure. When the origin is taken at 0, Eq. (2.9.2) becomes

$$p = p_0 - \frac{\gamma}{2} x - \gamma \left(1 - \frac{1}{4}\right) y = p_0 - \frac{\gamma}{2}\left(x + \frac{3}{2}y\right)$$

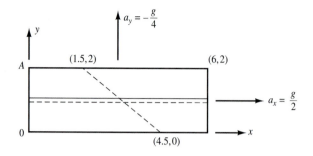

Figure 2.36   Uniform linear acceleration of container.

Since $p = 0$ at the point $y = 0$ and $x = 4.5$, then $p_0 = 2.25\gamma$. Then, for $y = 0$, along the bottom,

$$p = 2.25\gamma - 0.5\gamma x \qquad 0 \le x \le 4.5$$

Uniform Rotation about a Vertical Axis

Rotation of a fluid, moving as a solid, about an axis is called *forced-vortex motion.* Every particle of fluid has the same angular velocity. This motion is to be distinguished from *free-vortex motion,* in which each particle moves in a circular path with a speed varying inversely as the distance from the center. Free-vortex motion is discussed in Chaps. 8 and 11. A liquid in a container, when rotated about a vertical axis at constant angular velocity, moves like a solid after some time interval. No shear stresses exist in the liquid, and the only acceleration that occurs is directed radially inward toward the axis of rotation. If a coordinate system (Fig. 2.37a) with the unit vector **i** in the r direction and **j** in the vertical upward direction with y being the axis of rotation is selected, Eq. (2.2.5) can be applied to determine pressure variation throughout the fluid

$$\nabla p = -\mathbf{j}\gamma - \rho\mathbf{a} \tag{2.2.5}$$

For constant angular velocity, ω, any particle of fluid P has an acceleration $\omega^2 r$ directed radially inward, as $\mathbf{a} = -\mathbf{i}\omega^2 r$. Vector addition of $-\mathbf{j}\gamma$ and $-\rho\mathbf{a}$ (Fig. 2.37b) yields ∇p, the pressure gradient. The pressure does not vary normal to this line at a

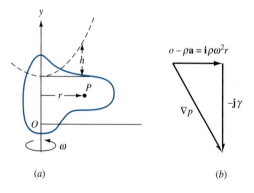

(a)   (b)

Figure 2.37   Rotation of a fluid about a vertical axis.

point; hence, if P is taken at the surface, the free surface is normal to ∇p. Expanding Eq. (2.2.5) gives

$$\mathbf{i}\frac{\partial p}{\partial r} + \mathbf{j}\frac{\partial p}{\partial y} + \mathbf{k}\frac{\partial p}{\partial z} = -\mathbf{j}\gamma + \mathbf{i}\rho\omega^2 r$$

where $\mathbf{k}$ is the unit vector along the z axis (or tangential direction). Then

$$\frac{\partial p}{\partial r} = \frac{\gamma}{g}\omega^2 r \qquad \frac{\partial p}{\partial y} = -\gamma \qquad \frac{\partial p}{\partial z} = 0$$

Since p is a function of y and r only, the total differential dp is

$$dp = \frac{\partial p}{\partial y}dy + \frac{\partial p}{\partial r}dr$$

Substituting for $\partial p/\partial y$ and $\partial p/\partial r$ results in

$$dp = -\gamma\,dy + \frac{\gamma}{g}\omega^2 r\,dr \tag{2.9.4}$$

For a liquid ($\gamma \approx$ constant) integration yields

$$p = \frac{\gamma}{g}\omega^2\frac{r^2}{2} - \gamma y + c$$

in which c is the constant of integration. If the value of pressure at the origin ($r = 0$, $y = 0$) is p_0, then $c = p_0$ and

$$p = p_0 + \frac{\gamma}{g}\omega^2\frac{r^2}{2} - \gamma y \tag{2.9.5}$$

When the particular horizontal plane ($y = 0$) for which $p_0 = 0$ is selected and Eq. (2.9.5) is divided by γ, then

$$h = \frac{p}{\gamma} = \frac{\omega^2 r^2}{2g} \tag{2.9.6}$$

which shows that the head, or vertical depth, varies as the square of the radius. The surfaces of equal pressure are paraboloids of revolution.

When a free surface occurs in a container that is being rotated, the fluid volume underneath the paraboloid of revolution is the original fluid volume. The shape of the paraboloid depends only upon the angular velocity ω.

For a circular cylinder rotating about its axis (Fig. 2.38) the rise of liquid from its vertex to the wall of the cylinder is, from Eq. (2.9.6), $\omega^2 r_0^2/2g$. Since a paraboloid of revolution has a volume equal to one-half its circumscribing cylinder, the volume of the liquid above the horizontal plane through the vertex is

$$\pi r_0^2 \frac{1}{2}\frac{\omega^2 r_0^2}{2g}$$

When the liquid is at rest, this liquid is also above the plane through the vertex to a uniform depth of

$$\frac{1}{2}\frac{\omega^2 r_0^2}{2g}$$

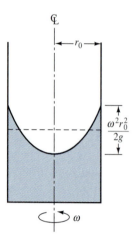

Figure 2.38   Rotation of a circular cylinder about its axis.

Hence, the liquid rises along the walls the same amount as the center drops, thereby permitting the vertex to be located when ω, r_0, and the depth before rotation are given.

Example 2.18

A liquid, $S = 1.2$, is rotated at 200 rpm about a vertical axis. At one point A in the fluid 1 m from the axis, the pressure is 70 kPa. What is the pressure at a point B, which is 2 m higher than A and 1.5 m from the axis?

Solution

When Eq. (2.9.5) is written for the two points,

$$p_A = p_0 + \gamma \frac{\omega^2 r_A^2}{2g} - \gamma y \qquad p_B = p_0 + \gamma \frac{\omega^2 r_B^2}{2g} - \gamma(y + 2)$$

Then $\omega = 200(2\pi/60) = 20.95$ rad/s, $\gamma = 1.2(9806) = 11{,}767$ N/m^3, $r_A = 1$ m, and $r_B = 1.5$ m. When the second equation is subtracted from the first and the values are substituted,

$$70{,}000 - p_B = (2 \text{ m})(11{,}767 \text{ N/m}^3) + \frac{11{,}767 \text{ N/m}^3}{2(9.806 \text{ m/s}^2)}(20.95/\text{s})^2[1 \text{ m}^2 - (1.5 \text{ m})^2]$$

Hence

$$p_B = 375.6 \text{ kPa}$$

If a closed container with no free surface or with a partially exposed free surface is rotated uniformly about some vertical axis, an *imaginary* free surface can be constructed; it consists of a paraboloid of revolution of shape given by Eq. (2.9.6). The vertical distance from any point in the fluid to this free surface is the pressure head at the point.

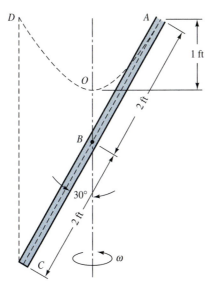

Figure 2.39   Rotation of inclined tube of liquid about a vertical axis.

Example 2.19

A straight tube 4 ft long, closed at the bottom and filled with water, is inclined 30° with the vertical and rotated about a vertical axis through its midpoint 8.02 rad/s. Draw the paraboloid of zero pressure, and determine the pressure at the bottom and midpoint of the tube.

Solution

In Fig. 2.39 the zero-pressure paraboloid passes through point A. If the origin is taken at the vertex, that is, $p_0 = 0$, Eq. (2.9.6) becomes

$$h = \frac{\omega^2 r^2}{2g} = \frac{(8.02)^2}{64.4}(2 \sin 30°)^2 = 1.0 \text{ ft}$$

which locates the vertex at O, 1.0 ft below A. The pressure at the bottom of the tube is $\gamma(\overline{CD})$, or

$$(4 \cos 30°)(62.4) = 216 \text{ lb/ft}^2$$

At the midpoint, $\overline{OB} = 0.732$ ft and

$$p_B = 0.732(62.4) = 45.6 \text{ lb/ft}^2$$

Fluid Pressure Forces in Relative Equilibrium

The magnitude of the force acting on a plane area in contact with a liquid accelerating as a rigid body can be obtained by integration over the surface

$$F = \int p \, dA$$

The nature of the acceleration and orientation of the surface governs the particular variation of p over the surface. When the pressure varies linearly over the plane surface (linear acceleration), the magnitude of force is given by the product of pressure at the centroid and area, since the volume of pressure prism is given by $p_G A$. For nonlinear distributions the magnitude and line of action can be found by integration.

EXERCISES

2.9.1   A closed cubical box, 1 m on each edge, is half filled with water, the other half being filled with oil, $S = 0.75$. When it is accelerated vertically upward 4.903 m/s^2, the pressure difference between the bottom and top, in kilopascals, is (*a*) 4.9; (*b*) 11; (*c*) 12.9; (*d*) 14.7; (*e*) none of these answers.

2.9.2   When the box of Exercise 2.9.1 is accelerated uniformly in a horizontal direction parallel to one side, 16.1 ft/s^2, the slope of the interface is (*a*) 0; (*b*) $-\frac{1}{4}$; (*c*) $-\frac{1}{2}$; (*d*) -1; (*e*) none of these answers.

2.9.3   When the minimum pressure in the box of Exercise 2.9.2 is zero gage, the maximum pressure in meters of water is (*a*) 0.94; (*b*) 1.125; (*c*) 1.31; (*d*) 1.5; (*e*) none of these answers.

2.9.4   Liquid in a cylinder 10 m long is accelerated horizontally $20g$ m/s^2 along the axis of the cylinder. The difference in pressure intensities at the ends of the cylinder, in pascals, is (*a*) 20γ; (*b*) 200γ; (*c*) $20g\gamma$; (*d*) $200\gamma/g$; (*e*) none of these answers.

2.9.5   When a liquid rotates at constant angular velocity about a vertical axis as a rigid body, the pressure (*a*) decreases as the square of the radial distance; (*b*) increases linearly as the radial distance; (*c*) decreases as the square of increase in elevation along any vertical line; (*d*) varies inversely as the elevation along any vertical line; (*e*) varies as the square of the radial distance.

2.9.6   When a liquid rotates about a vertical axis as a rigid body so that points on the axis have the same pressure as points 2 ft higher and 2 ft from the axis, the angular velocity in radians per second is (*a*) 8.02; (*b*) 11.34; (*c*) 64.4; (*d*) not determinable from data given; (*e*) none of these answers.

2.9.7   A right-circular cylinder, open at the top, is filled with liquid, $S = 1.2$, and rotated about its vertical axis at such speed that half the liquid spills out. The pressure at the center of the bottom is (*a*) zero; (*b*) one-fourth its value when the cylinder was full; (*c*) indeterminable for reason of insufficient data; (*d*) greater than a similar case with water as liquid; (*e*) none of these answers.

2.9.8   A forced vortex (*a*) turns in an opposite direction to a free vortex; (*b*) always occurs in conjunction with a free vortex; (*c*) has its velocity decreasing with the radius; (*d*) occurs when fluid rotates as a solid; (*e*) has its velocity decreasing inversely with the radius.

PROBLEMS

2.1   Prove that the pressure is the same in all directions at a point in a static fluid for the three-dimensional case.

2.2   The Empire State Building is 1250 ft high. What is the pressure difference in pounds per square inch of a water column of the same height?

2.3   What is the pressure at a point 10 m below the free surface in a fluid that has a variable density in kilograms per cubic meter given by $\rho = 450 + ah$, in which $a = 12$ kg/m^4 and h is the distance in meters measured from the free surface?

2.4   A vertical gas pipe in a building contains gas, $\rho = 0.002$ slug/ft^3 and $p = 3.0$-in.-H$_2$O gage in the basement. At the top of the building 800 ft high, determine the gas pressure in inches water gage for (*a*) gas assumed incompressible and (*b*) gas assumed isothermal. Barometric pressure is 34-ft H$_2$O and $t = 70°$F.

2.5 Derive the equations that give the pressure and density at any elevation in a static gas when conditions are known at one elevation and the temperature gradient β is known.

2.6 By a limiting process as $\beta \to 0$, derive the isothermal case from the results of Prob. 2.5.

2.7 Use the results of Prob. 2.5 to determine the pressure and density at (a) 3000-m elevation when $p = 100$-kPa abs and $T = 15°C$ and (b) 300-m elevation for air and $\beta = -0.005°C/m$.

2.8 For isothermal air at 0°C determine the pressure and density at 4000 m when the pressure is 0.1-MPa abs at sea level.

2.9 In isothermal air at 80°F what is the vertical distance for reduction of density by 10 percent?

2.10 Express a pressure of 50 kPa in (a) millimeters of mercury, (b) meters of water, (c) meters of acetylene tetrabromide, $S = 2.94$.

2.11 A Bourdon gage reads 2-psi suction, and the barometer is 29.5-in. Hg. Express the pressure in six other customary ways.

2.12 Express 4 atm in meters of water gage with a barometer reading of 750-mm Hg.

2.13 A Bourdon gage A inside a pressure tank (Fig. 2.40) reads 12 psi. Another Bourdon gage B outside the pressure tank and connected with it reads 20 psi, and an aneroid barometer reads 29-in. Hg. What is the absolute pressure measured by A in inches of mercury?

2.14 Determine the heights of columns of water; kerosene, $S = 0.83$; and acetylene tetrabromide, $S = 2.94$, equivalent to 300-mm Hg.

2.15 The container of Fig. 2.41 holds water and air as shown. What is the pressure at A, B, C, and D in pounds per square foot and in pascals.

2.16 The tube in Fig. 2.42 is filled with oil. Determine the pressure at A and B in meters of water.

2.17 Calculate the pressure A, B, C, and D of Fig. 2.43 in pascals.

2.18 In Fig. 2.10a, for a reading $h = 20$ in., determine the pressure at A in pounds per square inch. The liquid has a specific gravity of 1.90.

2.19 Determine the reading h in Fig. 2.10b for $p_A = 30$-kPa suction if the liquid is kerosene, $S = 0.83$.

2.20 In Fig. 2.10b, for $h = 8$ in. and a barometer reading of 29-in. Hg, with water being the liquid, find p_A in feet of water absolute.

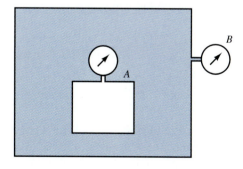

Figure 2.40   Problem 2.13.

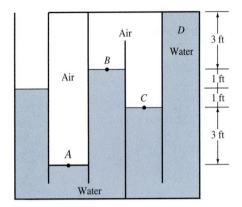

Figure 2.41   Problem 2.15.

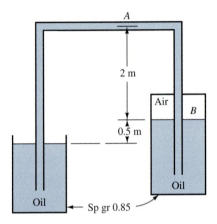

Figure 2.42   Problem 2.16.

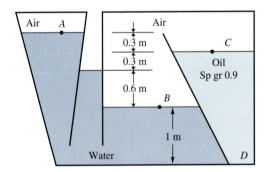

Figure 2.43   Problem 2.17.

2.21 In Fig. 2.10c $S_1 = 0.86$, $S_2 = 1.0$, $h_1 = 150$ mm, and $h_2 = 90$ mm. Find p_A in millimeters of mercury gage. If the barometer reading is 720-mm Hg, what is p_A in meters of water absolute?

2.22 Gas is contained in vessel A of Fig. 2.10c. With water being the manometer fluid and $h_1 = 75$ mm, determine the pressure at A in inches of mercury.

2.23 In Fig. 2.11a $S_1 = 1.0$, $S_2 = 0.95$, $S_3 = 1.0$, $h_1 = h_2 = 280$ mm, and $h_3 = 1$ m. Compute $p_A - p_B$ in millimeters of water.

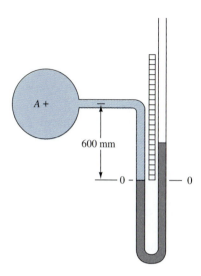

Figure 2.44   Problem 2.27.

2.24   In Prob. 2.23 find the gage difference h_2 for $p_A - p_B = -350$-mm H_2O.

2.25   In Fig. 2.11b $S_1 = S_3 = 0.83$, $S_2 = 13.6$, $h_1 = 150$ mm, $h_2 = 70$ mm, and $h_3 = 120$ mm. (*a*) Find p_A if $p_B = 10$ psi. (*b*) For $p_A = 20$ psia and a barometer reading of 720-mm Hg, find p_B in meters of water gage.

2.26   Find the gage difference h_2 in Prob. 2.25 for $p_A = p_B$.

2.27   In Fig. 2.44 A contains water, and the manometer fluid has a specific gravity of 2.94. When the left meniscus is at zero on the scale, $p_A = 100$-mm H_2O. Find the reading of the right meniscus for $p_A = 8$ kPa with no adjustment of the U tube or scale.

2.28   In Fig. 2.45 find the pressure at A in pascals. What is the pressure of the air in the tube?

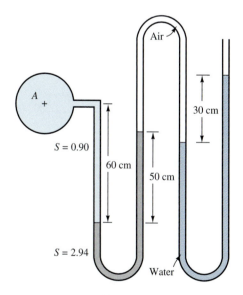

Figure 2.45   Problems 2.28 and 2.29.

2.29 In Fig. 2.45 if the water was replaced with mercury and the measurements remain the same, what would be the pressure at A and what would be the air pressure in the tube?

2.30 In Fig. 2.12 determine R, the gage difference, for a difference in gas pressure of 9-mm H_2O, $\gamma_2 = 9.8$ kN/m³, $\gamma_3 = 10.5$ kN/m³, and $a/A = 0.01$.

2.31 The inclined manometer of Fig. 2.13 reads zero when A and B are the same pressure. The reservoir diameter is 2.0 in. and that of the inclined tube is $\frac{1}{4}$ in. For $\theta = 30°$ and gage fluid $S = 0.832$, find $p_A - p_B$ in pounds per square inch as a function of gage reading R in feet.

2.32 Determine the weight W that can be sustained by the force acting on the piston of Fig. 2.46.

2.33 Neglecting the weight of the container (Fig. 2.47), find (a) the force tending to lift the circular top CD and (b) the compressive load on the pipe wall at A-A.

2.34 Find the force of oil on the top surface CD of Fig. 2.47 if the liquid level in the open pipe is reduced by 1 m.

2.35 The container shown in Fig. 2.48 has a circular cross section. Determine the upward force on the surface of the cone frustum ABCD. What is the downward force on the plane EF? Is this force equal to the weight of the fluid? Explain.

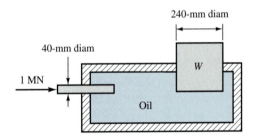

Figure 2.46   Problem 2.32.

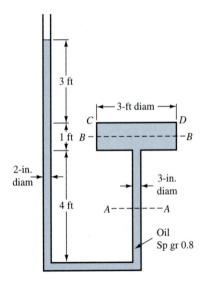

Figure 2.47   Problems 2.33 and
2.34.

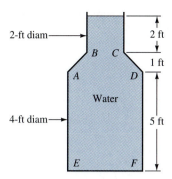

Figure 2.48 Problem 2.35.

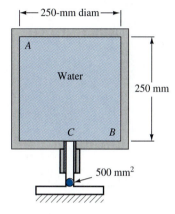

Figure 2.49 Problem 2.36.

2.36 The cylindrical container of Fig. 2.49 weighs 400 N when empty. It is filled with water and supported on the piston. (*a*) What force is exerted on the upper end of the cylinder? (*b*) If an additional 600-N weight is placed on the cylinder, how much will the water force against the top of the cylinder be increased?

2.37 A barrel 600 mm in diameter filled with water has a vertical pipe of 12-mm diameter attached to the top. Neglecting compressibility, how many kilograms of water must be added to the pipe to exert a force of 4 kN on the top of the barrel?

2.38 A vertical right-angled triangular surface has a vertex in the free surface of a liquid (Fig. 2.50). Find the force on one side (*a*) by integration and (*b*) by formula.

2.39 Determine the magnitude of the force acting on one side of the vertical triangle *ABC* of Fig. 2.51 (*a*) by integration and (*b*) by formula.

2.40 Find the moment about *AB* of the force acting on one side of the vertical surface *ABC* of Fig. 2.50. $\gamma = 9000$ N/m^3.

2.41 The triangular gate in the vertical end wall of a tank is hinged along the horizontal axis *AB*, Fig. 2.52. Find the moment necessary to hold the gate in the vertical position.

2.42 Find the moment about *AB* of the force acting on one side of the vertical surface *ABC* of Fig. 2.51.

2.43 Locate a horizontal line below *AB* of Fig. 2.51 such that the magnitude of pressure force on the vertical surface *ABC* is equal above and below the line.

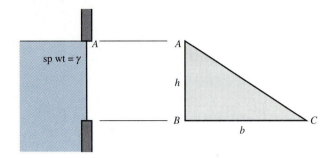

Figure 2.50   Problems 2.38, 2.40, 2.50, and 2.51.

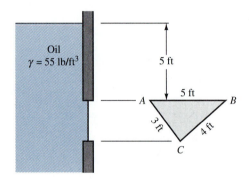

Figure 2.51   Problems 2.39, 2.42, 2.43,
2.48, and 2.49.

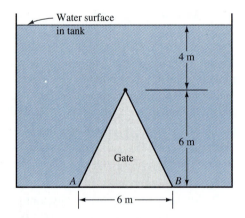

Figure 2.52   Problem 2.41.

2.44 Determine the force acting on one side of the vertical surface $OABCO$ of Fig. 2.53. $\gamma = 9 \text{ kN/m}^3$.

2.45 Calculate the force exerted by water on one side of the vertical annular area shown in Fig. 2.54.

2.46 Determine the moment about A required to hold the gate as shown in Fig. 2.55.

2.47 If there is water on the other side of the gate (Fig. 2.55) up to A, determine the resultant force due to water on both sides of the gate, including its line of action.

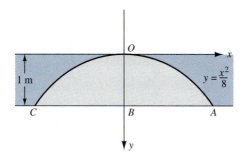

Figure 2.53   Problems 2.44, 2.55, and 2.80.

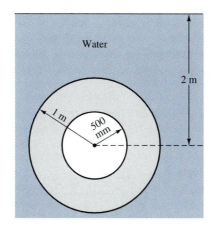

Figure 2.54   Problems 2.45 and 2.52.

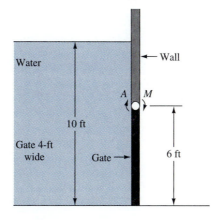

Figure 2.55   Problems 2.46, 2.47, and 2.53.

2.48   Locate the distance of the pressure center below the liquid surface in the triangular area ABC of Fig. 2.51 (a) by integration and (b) by formula.

2.49   By integration locate the pressure center horizontally in the triangular area of ABC of Fig. 2.51.

2.50   Use the pressure prism to determine the resultant force and location for the triangle of Fig. 2.50.

2.51   Determine by integration the pressure center for Fig. 2.50.

2.52   Locate the pressure center for the annular area of Fig. 2.54.

2.53   Locate the pressure center for the gate of Fig. 2.55.

2.54   A vertical square area 6 by 6 ft is submerged in water with upper edge 3 ft below the surface. Locate a horizontal line on the surface of the square such that (a) the force on the upper portion equals the force on the lower portion and (b) the moment of force about the line due to the upper portion equals the moment due to the lower portion.

2.55   Locate the pressure center of the vertical area $OABCO$ of Fig. 2.53.

2.56   Locate the pressure center for the vertical area of Fig. 2.56.

2.57   The gate, OBC, is 4 m wide and is rigid, Fig. 2.57. Neglect the weight of the gate, and assume the hinge friction to be negligible. What force P is necessary to hold the gate closed?

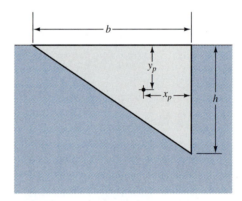

Figure 2.56   Problem 2.56.

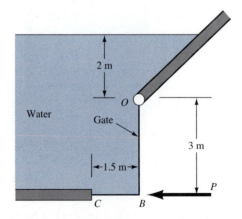

Figure 2.57   Problem 2.57.

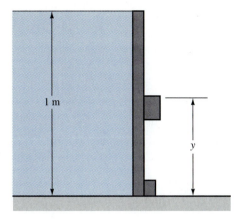

Figure 2.58   Problem 2.58.

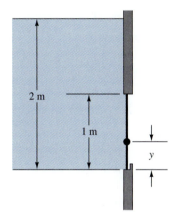

Figure 2.59   Problem 2.59.

2.58 Determine y of Fig. 2.58 so that the flashboards will tumble when water reaches their top.

2.59 Determine the pivot location y of the rectangular gate of Fig. 2.59 so that it will open when the liquid surface is as shown.

2.60 Use the pressure prism to show that the pressure center approaches the centroid of an area as its depth of submergence is increased.

2.61 Demonstrate the fact that the magnitude of the resultant force on a totally submerged plane area is unchanged if the area is rotated about an axis through its centroid.

2.62 An equilateral triangle with one edge in a water surface extends downward at a 45° angle. Locate the pressure center in terms of the length of a side b.

2.63 The shaft of the gate in Fig. 2.60 will fail at a moment of 150 kN·m. Determine the maximum value of liquid depth h.

2.64 In Fig. 2.60 develop the expression for y_p in terms of h.

2.65 The dam of Fig. 2.61 has a strut AB every 6 m. Determine the compressive force in the strut, neglecting the weight of the dam.

2.66 The gate of Fig. 2.62 weighs 300 lb/ft normal to the paper. Its center of gravity is 1.5 ft from the left face and 2.0 ft above the lower face. It is hinged at O. Determine

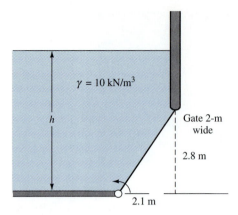

Figure 2.60   Problems 2.63 and 2.64.

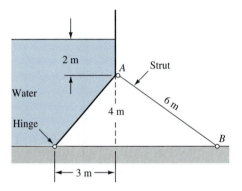

Figure 2.61   Problem 2.65.

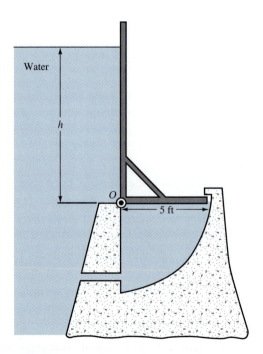

Figure 2.62   Problems 2.66–2.68 and 2.149.

the water-surface position for the gate just to start to come up. (The water surface is below the hinge.)

2.67 Find h of Prob. 2.66 for the gate just to come up to the vertical position shown.

2.68 Determine the value of h and the force against the stop when this force is a maximum for the gate of Prob. 2.66.

2.69 (*a*) Find the magnitude and line of action of force on each side of the gate of Fig. 2.63. (*b*) Find the resultant force due to the liquid on both sides of the gate. (*c*) Determine F to open the gate if it is uniform and has a mass of 2000 kg.

2.70 For linear stress variation over the base of the dam of Fig. 2.64 (*a*) locate where the resultant crosses the base and (*b*) compute the maximum and minimum compressive stresses at the base. Neglect hydrostatic uplift.

2.71 Solve Prob. 2.70 with the addition that the hydrostatic uplift varies linearly from 20 m at A to zero at the toe of the dam.

2.72 Find the moment M at O (Fig. 2.65) to hold the gate closed.

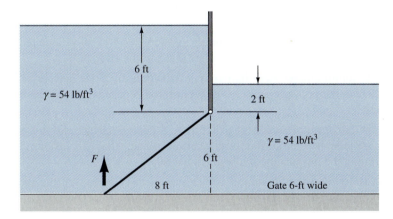

Figure 2.63 Problem 2.69.

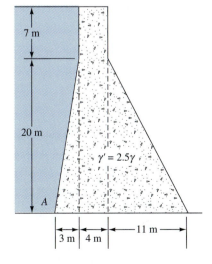

Figure 2.64 Problems 2.70 and 2.71.

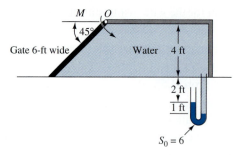

Figure 2.65   Problem 2.72.

2.73   The gate shown in Fig. 2.66 is in equilibrium. Compute W, the weight of counterweight per meter of width, neglecting the weight of the gate. Is the gate in a stable equilibrium?

2.74   To what height h will the water on the right have to rise to open the gate shown in Fig. 2.67? The gate is 5-ft wide and is constructed of material with specific gravity 2.5. Use the pressure-prism method.

2.75   Compute the air pressure required to keep the 700-mm-diameter gate of Fig. 2.68 closed. The gate is a circular plate that weighs 1800 N.

2.76   A 20-mm-diameter steel ball covers a 10-mm-diameter hole in a pressure chamber where the pressure is 30 MPa. What force is required to lift the ball from the opening?

2.77   If the horizontal component of force on a curved surface did *not* equal the force on a projection of the surface onto a vertical plane, what conclusions could you draw regarding the propulsion of a boat (Fig. 2.69)?

2.78   (*a*) Determine the horizontal component of force acting on the radial gate (Fig. 2.70) and its line of action. (*b*) Determine the vertical component of force and

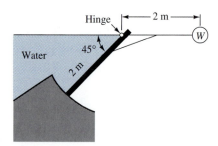

Figure 2.66   Problem 2.73.

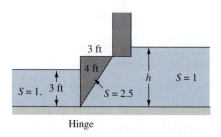

Figure 2.67   Problem 2.74.

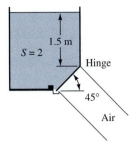

Figure 2.68   Problem 2.75.

Figure 2.69   Problem 2.77.

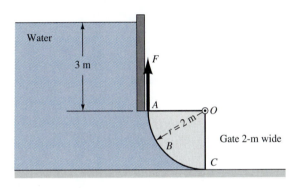

Water

3 m

F

A

$r = 2$ m

B

O

C

Gate 2-m wide

Figure 2.70   Problem 2.78.

its line of action. (*c*) What force *F* is required to open the gate, neglecting its weight? (*d*) What is the moment about an axis normal to the paper and through point *O*?

2.79   Find the vertical component of force on the curved gate of Fig. 2.71, including its line of action.

2.80   What is the force on the surface whose trace is *OA* of Fig. 2.53? The length normal to the paper is 3 m and $\gamma = 9$ kN/m^3.

2.81   Determine the moment needed to hold the gate in Fig. 2.72, neglecting its weight.

2.82   The 2-m-long curved gate in Fig. 2.73 is hinged at *O*. With water the liquid, find the horizontal component of force acting on the gate and its line of action. Find the vertical component of force and its line of action. What force is required to open the gate, neglecting its weight?

2.83   Determine the moment *M* to hold the gate of Fig. 2.71, neglecting its weight.

2.84   Calculate the force *F* required to hold the gate of Fig. 2.74 in a closed position when $R = 2$ ft.

2.85   Calculate the force *F* required to open or hold closed the gate of Fig. 2.74 when $R = 1.5$ ft.

2.86   What is *R* of Fig. 2.74 if no force *F* is required to hold the gate closed or open?

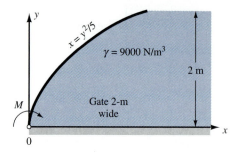

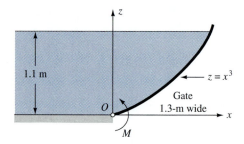

Figure 2.71   Problems 2.79 and 2.83.

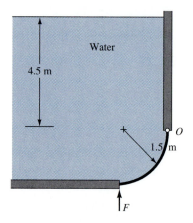

Figure 2.72   Problem 2.81.

Figure 2.73   Problem 2.82.

2.87   A right-circular cylinder is illustrated in Fig. 2.75. The pressure, in pounds per square foot, due to flow around the cylinder varies over the segment ABC as $p = 2\rho(1 - 4\sin^2\theta) + 50$. Calculate the force on ABC.

2.88   If the pressure variation on the cylinder in Fig. 2.75 is $p = 2\rho[1 - 4(1 + \sin\theta)^2] + 50$, determine the force on the cylinder.

2.89   A log holds back water and oil as shown in Fig. 2.76. Determine (*a*) the force per meter pushing it against the dam, (*b*) the weight of the cylinder per meter of length, and (*c*) its specific gravity.

2.90   The cylinder of Fig. 2.77 is filled with liquid as shown. Find (*a*) the horizontal component of force on AB per unit of length, including its line of action, and

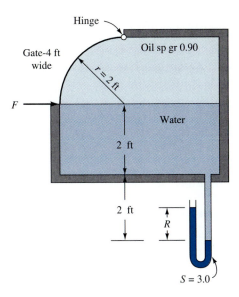

Figure 2.74   Problems 2.84–2.86.

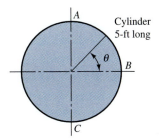

Figure 2.75   Problems 2.87 and 2.88.

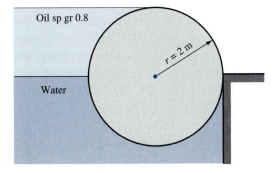

Figure 2.76   Problem 2.89.

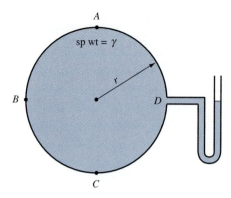

Figure 2.77   Problem 2.90.

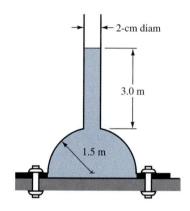

Figure 2.78   Problem 2.91.

(*b*) the vertical component of force on *AB* per unit of length, including its line of action.

2.91   The hemispherical dome in Fig. 2.78 is filled with water. The dome arrangement as shown weighs 28 kN and is fastened to the floor by bolts equally spaced around the circumference at the base. Find the total force required to hold down the dome.

2.92   A cable and semicircular ring suspend a spherical container by the *small* piezometer tube, Fig. 2.79. The top of the tube is open to the atmosphere. Calculate (*a*) the force on the bottom half of the sphere, (*b*) the force on the top half of the sphere, and (*c*) the total tension in the cable. Neglect the weight of the container.

2.93   Find the resultant force, including its line of action, acting on the outer surface of the first quadrant of a spherical shell of radius 600 mm with center at the origin. Its center is 1.2 m below the water surface.

2.94   The volume of the ellipsoid given by $x^2/a^2 + y^2/b^2 + z^2/c^2 = 1$ is $4\pi abc/3$, and the area of the ellipse $x^2/a^2 + z^2/c^2 = 1$ is πac. Determine the vertical force on the surface given in Example 2.10.

2.95   A 16-ft-diameter pressure pipe carries liquid at 200 psi. What pipe-wall thickness is required for maximum stress of 10,000 psi.

2.96   To obtain the same flow area, which pipe system requires the least steel, a single pipe or four pipes having half the diameter? The maximum allowable pipe-wall stress is the same in each case.

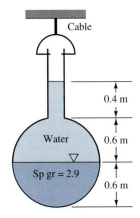

Figure 2.79 Problem 2.92.

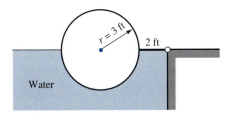

Figure 2.80 Problem 2.99.

2.97 A thin-walled hollow sphere 3 m in diameter holds gas at 1.5 MPa. For allowable stress of 60 MPa determine the minimum wall thickness.

2.98 A cylindrical container 7-ft high and 4 ft in diameter provides for pipe tension with two hoops a foot from each end. When it is filled with water, what is the tension in each hoop due to the water?

2.99 The cylinder gate of Fig. 2.80 is made from a circular cylinder and a plate hinged at the dam. The gate position is controlled by pumping water into or out of the cylinder. The center of gravity of the empty gate is on the line of symmetry 4 ft from the hinge. It is at equilibrium when it is empty in the position shown. How many cubic feet of water must be added per foot of cylinder to hold the gate in its position when the water surface is raised 3 ft?

2.100 A sphere 250 mm in diameter, $S = 1.4$, is submerged in a liquid having a density varying with the depth y below the surface given by $\rho = 1000 + 0.03y$ kg/m³. Determine the equilibrium position of the sphere in the liquid.

2.101 Repeat the calculations for Prob. 2.100 for a horizontal circular cylinder with a specific gravity of 1.4 and a diameter of 250 mm.

2.102 A cube 2 ft on an edge has its lower half of $S = 1.4$ and upper half of $S = 0.6$. It is submerged into a two-layered fluid, the lower $S = 1.2$ and the upper $S = 0.9$. Determine the height of the top of the cube above the interface.

2.103 Determine the density, specific volume, and volume of an object that weighs 3 N in water and 4 N in oil, $S = 0.83$.

2.104 Two cubes of the same size, 1 m³, one of $S = 0.80$ and the other of $S = 1.1$, are connected by a short wire and placed in water. What portion of the lighter cube is above the water surface, and what is the tension in the wire?

2.105 In Fig. 2.81 the hollow triangular prism is in equilibrium as shown when $z = 1$ ft and $y = 0$. Find the weight of prism per foot of length and z in terms of y for equilibrium. Both liquids are water. Determine the value of y for $z = 1.5$ ft.

2.106 How many pounds of concrete, $\gamma = 25$ kN/m^3, must be attached to a beam having a volume of 0.1 m^3 and $S = 0.65$ to cause both to sink in water?

2.107 The timber in Fig. 2.82 is held in a horizontal position by the concrete anchor. The 120-mm by 120-mm by 5-m timber has a specific gravity of 0.6, while that of concrete is 2.5. What must be the minimum total weight of the concrete?

2.108 A spherical balloon 15 m in diameter is open at the bottom and filled with hydrogen. For barometer reading of 28-in. Hg and 20°C, what is the total weight of the balloon and the load to hold it stationary?

2.109 A hot-air balloon weighs 600 lb, including the weight of the basket, one person, and the balloon. The heated air inside the balloon has a temperature of 155°F, and the air temperature is 75°F. Assume standard atmospheric pressure both inside and outside the balloon. What would be the required diameter of the balloon, assuming a spherical shape? If the outside air temperature were 35°F, what would be the balloon size?

2.110 A hydrometer weighs 0.035 N and has a stem 6 mm in diameter. Compute the distance between specific-gravity markings 1.0 and 1.1.

2.111 Design a hydrometer to read specific gravities in the range from 0.80 to 1.10 when the scale is to be 75-mm long.

2.112 The gate of Fig. 2.83 weighs 150 lb/ft normal to the page. It is in equilibrium as shown. Neglecting the weight of the arm and brace supporting the counterweight,

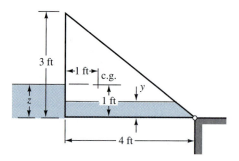

Figure 2.81   Problems 2.105 and 2.150.

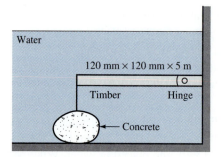

Figure 2.82   Problem 2.107.

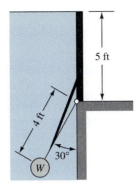

Figure 2.83   Problem 2.112.

(a) find W and (b) determine whether the gate is in stable equilibrium. The weight is made of concrete, $S = 2.50$.

2.113 A wooden cylinder 600-mm in diameter, $S = 0.50$, has a concrete cylinder 600-mm long of the same diameter, $S = 2.50$, attached to one end. Determine the length of the wooden cylinder for the system to float in stable equilibrium with axis vertical.

2.114 Will a beam 4-m long with square cross section, $S = 0.75$, float in stable equilibrium in water with two sides horizontal?

2.115 Determine the metacentric height of the torus shown in Fig. 2.84.

2.116 The plane gate (Fig. 2.85) weighs 2000 N/m normal to the paper, and its center of gravity is 2 m from the hinge at O. (a) Find h as a function of θ for equilibrium of the gate. (b) Is the gate in stable equilibrium for any values of θ?

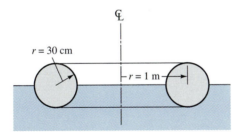

Figure 2.84   Problem 2.115.

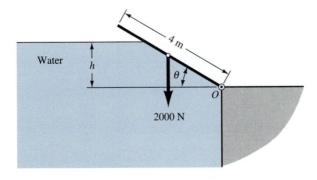

Figure 2.85   Problem 2.116.

2.117   A tank of liquid $S = 0.88$ is accelerated uniformly in a horizontal direction so that the pressure decreases within the liquid 20 kPa/m in the direction of motion. Determine the acceleration.

2.118   The free surface of a liquid makes an angle of 20° with the horizontal when accelerated uniformly in a horizontal direction. What is the acceleration?

2.119   An automobile accelerates uniformly to 60 mph in 5 seconds on a horizontal road. What is the slope of the free surface in a container of gasoline riding with the vehicle?

2.120   In Fig. 2.86, $a_x = 12.88$ ft/s^2 and $a_y = 0$. Find the imaginary free liquid surface and the pressure at B, C, D, and E.

2.121   In Fig. 2.86, $a_x = 0$ and $a_y = -8.05$ ft/s^2. Find the pressure at B, C, D, and E.

2.122   In Fig. 2.86, $a_x = 8.05$ ft/s^2 and $a_y = 16.1$ ft/s^2. Find the imaginary free surface and the pressure at B, C, D, and E.

2.123   In Fig. 2.87, $a_x = 9.806$ m/s^2 and $a_y = 0$. Find the pressure at A, B, and C.

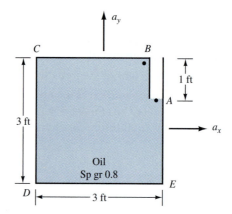

Figure 2.86   Problems 2.120–2.122 and
                     2.126.

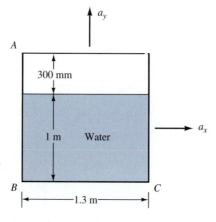

Figure 2.87   Problems 2.123 and
                     2.124.

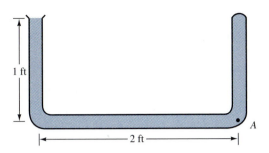

Figure 2.88   Problems 2.127, 2.133, 2.134, and 2.144.

2.124 In Fig. 2.87, $a_x = 4.903$ m/s^2 and $a_y = 9.806$ m/s^2. Find the pressure at A, B, and C.

2.125 A circular cross-sectional tank of 6-ft depth and 4-ft diameter is filled with liquid and accelerated uniformly in a horizontal direction. If one-third of the liquid spills out, determine the acceleration.

2.126 Determine a_x and a_y in Fig. 2.86 if the pressure at A, B, and C is the same.

2.127 The tube of Fig. 2.88 is filled with liquid, $S = 2.40$. When it is accelerated to the right 8.05 ft/s^2, draw the imaginary free surface and determine the pressure at A. For $p_A = 8$-psi vacuum determine a_x.

2.128 A cubical box 1 m on an edge, open at the top and half filled with water, is placed on an inclined plane making a 30° angle with the horizontal. The box alone weighs 500 N and has a coefficient of friction with the plane of 0.30. Determine the acceleration of the box and the angle the free-water surface makes with the horizontal.

2.129 Show that the pressure is the same in all directions at a point in a liquid moving as a solid.

2.130 A closed box contains two immiscible liquids. Prove that when it is accelerated uniformly in the x direction, the interface and zero-pressure surface are parallel.

2.131 Verify the statement made in Sec. 2.9 on uniform rotation about a vertical axis, that is, when a fluid rotates in the manner of a solid body, no shear stresses exist in the fluid.

2.132 A vessel containing liquid, $S = 1.3$, is rotated about a vertical axis. The pressure at one point 0.6 m radially from the axis is the same as at another point 1.2 m from the axis and with elevation 0.6 m higher. Calculate the rotational speed.

2.133 The U tube of Fig. 2.88 is rotated about a vertical axis 6 in. to the right of A at such a speed that the pressure at A is zero gage. What is the rotational speed?

2.134 Locate the vertical axis of rotation and the speed of rotation of the U tube of Fig. 2.88 so that the pressure of liquid at the midpoint of the U tube and at A are both zero.

2.135 An incompressible fluid of density ρ moving as a solid rotates at speed ω about an axis inclined at $\theta°$ with the vertical. Knowing the pressure at one point in the fluid, how do you find the pressure at any other point?

2.136   A right-circular cylinder of radius r_0 and height h_0 with axis vertical is open at the top and filled with liquid. At what speed must it rotate so that half the area of the bottom is exposed?

2.137   A liquid rotating about a *horizontal* axis as a solid has a pressure of 10 psi at the axis. Determine the pressure variation along a vertical line through the axis for density ρ and speed ω.

2.138   Determine the equation for the surfaces of constant pressure for the situation described in Prob. 2.137.

2.139   Prove by integration that a paraboloid of revolution has a volume equal to half its circumscribing cylinder.

2.140   A tank containing two immiscible liquids is rotated about a vertical axis. Prove that the interface has the same shape as the zero-pressure surface.

2.141   A hollow sphere of radius r_0 is filled with liquid and rotated about its vertical axis at speed ω. Locate the circular line of maximum pressure.

2.142   A gas following the law $p\rho^{-n} = $ constant is rotated about a vertical axis as a solid. Derive an expression for pressure in a radial direction for speed ω, pressure p_0, and density ρ_0 at a point on the axis.

2.143   A vessel containing water is rotated about a vertical axis with an angular velocity of 50 rad/s. At the same time the container has a downward acceleration of 16.1 ft/s². What is the equation for a surface of constant pressure?

2.144   The U tube of Fig. 2.88 is rotated about a vertical axis through A at such a speed that the water in the tube begins to vaporize at the closed end above A, which is at 70°F. What is the angular velocity? What would happen if the angular velocity were increased?

2.145   A cubical box 1.3 m on an edge is open at the top and filled with water. When it is accelerated upward at 2.45 m/s², find the magnitude of water force on one side of the box.

2.146   A cube 1 m on an edge is filled with liquid, $S = 0.65$, and is accelerated downward 2.45 m/s². Find the resultant force on one side of the cube due to liquid pressure.

2.147   A cylinder 2 ft in diameter and 6-ft long is accelerated uniformly along its axis in a horizontal direction 16.1 ft/s². It is filled with liquid, $\gamma = 50$ lb/ft³, and it has a pressure along its axis of 10 psi before acceleration starts. Find the horizontal net force exerted against the liquid in the cylinder.

2.148   A closed cube, 300 mm on an edge, has a small opening at the center of its top. When it is filled with water and rotated uniformly about a vertical axis through its center at ω rad/s, find the force on a side due to the water in terms of ω.

2.149   Prepare a program to solve (*a*) Prob. 2.66 and (*b*) Prob. 2.67. (*c*) Determine the head h in Fig. 2.62 for the gate to start to close.

2.150   With reference to Prob. 2.105 use a spreadsheet to find y for increments of z of 0.5 ft from 1.10 to 3.0 ft. The gate weighs 2.722γ lb per ft of length. Also find z for increments of y of 0.5 ft.

2.151   In Example 2.10 a numerical integration of θ was taken from 0 to θ_0 to find the moment on the gate for trial values of θ_0 or y. In this problem integrate the moment equation between 0 and unknown θ_0. Check the results of Example 2.10 by the bisection method.

ADDITIONAL READING

Aris, R.: *Vectors, Tensors, and the Basic Equations of Fluid Mechanics,* Dover Publications, New York, 1989.

Long, R.: *Mechanics of Solids,* Prentice Hall, New Jersey, 1961.

Shames, I.: *Mechanics of Fluids,* 3rd ed., McGraw-Hill, New York, 1992.

chapter
3

Fluid Flow Concepts
and
Basic Control Volume Equations

The statics of fluids, treated in the preceding chapter, is almost an exact science with specific weight (or density) being the only quantity that must be determined experimentally. On the other hand, the nature of *flow* of a real fluid is very complex. This chapter introduces the concepts needed for analysis of fluid motion. The basic equations that enable us to predict fluid behavior are derived; they are equations of continuity and momentum, the first and second laws of thermodynamics, and conservation of mixture mass. In this chapter the control-volume approach is utilized in these derivations. Viscous effects, the experimental determination of losses, and the dimensionless presentation of loss data are presented in Chaps. 6 and 7 after dimensional analysis has been introduced in Chap. 5. In general, one-dimensional-flow theory is developed in this chapter, with applications mainly limited to incompressible cases where viscous effects do not predominate. Chapter 8 deals with two-dimensional flow. Vector notation is used in this chapter as it is in Chap. 4 and reference to Appendix E should be made for review.

3.1 FLOW CONCEPTS AND KINEMATICS

Analysis Approaches

To express the laws of mechanics in a form useful for fluid mechanics and transport requires a different viewpoint than that used to derive these laws in solid mechanics. In solid mechanics the *Lagrangian* approach (named for Joseph-Louis Lagrange 1736–1813) is used wherein the basic equations are derived for a given mass of fluid. This approach is analogous to the "closed system" used in thermodynamics. Energy and momentum may be transferred to and from the system and either fixed or moving coordinate systems can be used to derive the equations. With reference to Fig. 3.1 the solid brick contains a known mass, m_b, at time t_0 and at times t_1 and t_2 later the brick has changed position, linear and angular momentum, and energy, all of which can be described by the simple laws of solid mechanics. Also note that in a solid the relative position of the various particles comprising the mass stays in the same relative position during subsequent motion.

In applying the Lagrangian concept to a fluid, the free-body diagram was used in Chap. 2 as a convenient way to show forces exerted on some arbitrary fixed mass. This is an example of a *system*. A system refers to a definite fixed mass of material and distinguishes it from all other matter, called its *surroundings*. The boundaries of a system form a closed surface. This surface may vary with time, so that it contains the same mass during changes in its condition. For example, a kilogram of gas may be confined in a cylinder and be compressed by motion of a piston; the system boundary coinciding with the end of the piston then moves with the piston. The system may contain an infinitesimal mass or a large finite mass of fluids and solids at the will of the investigator.

From the system point of view the conservation of mass equation states that the mass within the system remains constant with time (disregarding relativity effects). In equation form

$$\frac{dm}{dt} = 0 \tag{3.1.1}$$

where m is the total mass.

Newton's second law of motion is usually expressed for a system as

$$\Sigma\mathbf{F} = \frac{d}{dt}(m\mathbf{v}) \tag{3.1.2}$$

in which it must be remembered that m is the constant mass of the system. $\Sigma\mathbf{F}$ refers to the resultant of all external forces acting on the system, including body forces such as gravity, and $\mathbf{v}$ is the velocity of the center of mass of the system.

However, life is not so neatly ordered for most fluid dynamics cases. Figure 3.1*b* represents the same size "brick" as in Fig. 3.1*a* but is now comprised of fluid clumps, defined as being large enough to have defined properties in the continuum sense but quite small in contrast to the flow geometry. When motion of the fluid clumps occurs, the subsequent motion quickly becomes disordered. The clumps lose contact with each other, change orientation and position, and quickly become *immersed, mixed, or dispersed* in other fluid clumps in the surroundings. Thus, it is particularly difficult if not impossible to use a fixed mass system to derive the fluid dynamics equations.

The *Eulerian* approach (named for Leonhard Euler, 1707–1783) is therefore adopted for most analyses. Here a fixed *control volume* or fixed point in space is adopted, and equations are derived to express changes in mass, momentum, and energy

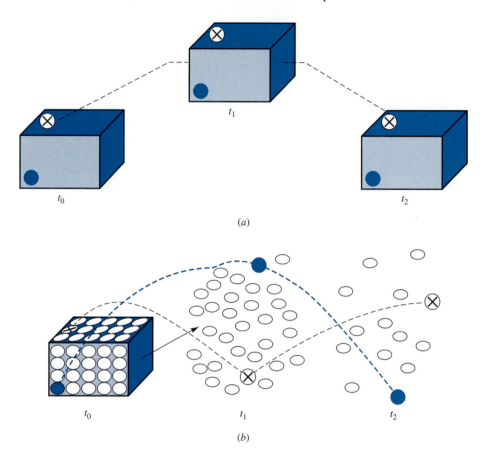

Figure 3.1   Solid and fluid parcel paths.

as the fluid passes through or by the fixed volume or point. The boundary of a control volume is its *control surface*. The size and shape of the control volume are entirely arbitrary, but frequently they are made to coincide with the solid boundaries; otherwise, they are drawn normal to the flow directions as a matter of simplification.

This Eulerian approach leads to an open system description. As suggested above within the Lagrangian or Eulerian approaches there are two possible levels of mathematical abstraction, the *macroscopic level* and the *field level*. At the macroscopic level the laws are derived for finite mass systems or control volumes and a series of integral or average values for the variables are calculated. This level of analysis will yield no information about the point by point variation of the parameters within the system or control volume; therefore, this level of abstraction is useful for initial engineering estimates. More elaborate analyses yielding variations in parameters over the whole domain will quite often be necessary for final problem solution. For example, an analysis of a waste discharger is to be computed from the control volume shown in Fig. 3.2. From a regulatory point of view it is desirable to know the pollutant distribution along the entire shore at all the points along the river's edge. Yet the control volume analysis will only allow average concentration values to be predicted at the control surface inlet *A* and outlet *B*. Furthermore, only the volume average concentration of pollutant can be calculated within the control volume. The point by point variation cannot be calculated.

To calculate the point by point variation of the fluid variable over the whole domain, the *field approach* is used. This approach essentially allows the domain to

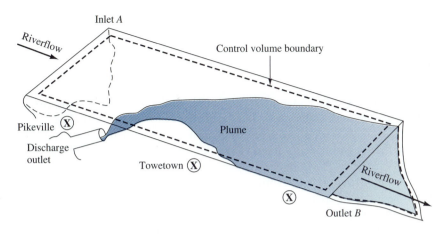

Inlet A

Riverflow

Control volume boundary

Pikeville (X)

Discharge outlet

Plume

Towetown (X)

(X)

Riverflow

Outlet B

Figure 3.2 Control volume for river and plume problem.

consist of a large number of control volumes or systems such that when they shrink to zero a series of nonlinear partial differential equations result. These equations are so difficult that a general solution for the three-dimensional distribution of relevant variables for the types of complex geometries encountered in practice has not yet been achieved, not even on the most powerful parallel processing supercomputer.

All fluid dynamics and accompanying transport essentially vary from point to point in the spatial domain and from one time period to another. These variations can be especially rapid with the smallest scales of variability (above molecular) being approximately 0.1 second and 0.1 cm. A premium is therefore placed on being able to accurately depict or describe all the possible types of physical processes that give rise to this variability, as well as to identify credible procedures for simplifying the mathematical complexity without altering the dominant physical processes.

Descriptions of Fluid Motion

Before proceeding to a description of processes it is useful to explain several geometric concepts which will allow the flow field to be visualized. The concepts of *streamline, pathline,* and *streakline* are all used to both analytically and visually portray the flow patterns.

If the fluid velocity in the Eulerian approach is adopted, then the velocity vector is defined as **v** (or sometimes **q**) where

$$\mathbf{v}(x, y, z, t) = u\mathbf{i} + v\mathbf{j} + w\mathbf{k} \tag{3.1.3}$$

In Equation (3.1.3), u, v, and w are the velocities in the x, y, and z directions, respectively, and each can vary in time or space.

A *streamline* is a continuous line drawn through the fluid so that it has the direction of the velocity vector at every point. There can be no flow across a streamline. Since a particle moves in the direction of the streamline at any instant, its displacement δs, having components δx, δy, δz, has the direction of velocity vector **v** with components u, v, w in the x, y, z directions, respectively. Then

$$\frac{\delta x}{u} = \frac{\delta y}{v} = \frac{\delta z}{w}$$

states that the corresponding components are proportional and hence that δs and **v** have the same direction. Expressing the displacements in differential form

$$\frac{dx}{u} = \frac{dy}{v} = \frac{dz}{w}$$

(3.1.4)

produces the differential equations of a streamline. Equations (3.1.4) are two independent equations. Any continuous line that satisfies them is a streamline.

In steady flow, since there is no change in direction of the velocity vector at any point, the streamline has a fixed inclination at every point and is therefore *fixed in space*. A particle always moves tangent to the streamline; hence, in steady flow the *path of a particle* is a streamline. In unsteady flow, since the direction of the velocity vector at any point may change with time, a streamline may shift in space from instant to instant. A particle then follows one streamline one instant, another one the next instant, and so on, so that the path of the particle may have no resemblance to any given instantaneous streamline.

A dye or smoke is frequently injected into a fluid in order to trace its subsequent motion. The resulting dye or smoke trails are called *streak lines*. In steady flow a streak line is a streamline and the path of a particle.

Streamlines in two-dimensional flow can be obtained by inserting fine, bright particles (aluminum dust) into the fluid, brilliantly lighting one plane, and taking a photograph of the streaks made in a short time interval. Tracing on the picture continuous lines that have the direction of the streaks at every point portrays the streamlines for either steady or unsteady flow.

A *streamtube* is the tube made by all the streamlines passing through a small closed curve. In steady flow it is fixed in space and can have no flow through its walls because the velocity vector has no component normal to the tube surface.

Figure 3.3 contains a schematic of the streamlines for a flow around a cylinder between parallel walls. The streamlines are drawn so that, per unit time, the volume flowing between adjacent streamlines is the same (assuming a unit depth). Thus, when a streamline spacing is closer, the velocity is higher and vice versa.

Flow depiction and visualization are being significantly enhanced by the use of modern computer graphics and animation techniques, thus allowing flow patterns for highly complex geometries to be observed. Heretofore, only simple geometries with mathematically simple physics or simple laboratory renderings were possible.

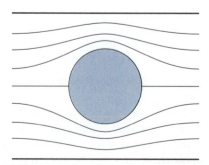

Figure 3.3   Streamlines for steady flow around cylinder between parallel walls.

Descriptors of Fluid Processes and Simplifications

Flow can be classified in many ways such as turbulent versus laminar; real versus ideal; reversible versus irreversible; steady versus unsteady; uniform versus non-uniform; rotational versus irrotational. In this section various types of flow are distinguished.

In *laminar flow,* fluid particles move along smooth paths in laminas, or layers, with one layer gliding smoothly over an adjacent layer. Tendencies to lateral or swirling motion are strongly damped by viscosity. Laminar flow is governed by Newton's law of viscosity [Eq. (1.2.1) or extensions of it to three-dimensional flow], which relates shear stress to rate of angular deformation. Laminar flow is not stable in situations involving combinations of low viscosity, high velocity, or large flow passages, and breaks down into turbulent flow.

Turbulent flow situations are most prevalent in engineering practice. In turbulent flow the fluid particles (small molar masses) move in very irregular, swirling paths, causing an exchange of momentum from one portion of the fluid to another in a manner somewhat similar to the molecular momentum transfer described in Sec. 1.4 but on a much larger scale. The turbulent swirls continuously range in size from very small (say a few thousand fluid molecules) to very large (a large swirl in a river or in an atmospheric gust). In a situation in which the flow could be either turbulent or laminar, the turbulence sets up greater shear stresses throughout the fluid and causes more irreversibilities or losses. Also, in turbulent flow, the losses vary about as the 1.7-to-2 power of the velocity; in laminar flow, they vary as the first power of the velocity.

The calculation of shear stresses introduced by turbulent flow, τ_t, is an extremely challenging problem. Based upon analogies to laminar flow and Newton's law of viscosity as well as concepts from statistical or kinetic theories of particle dynamics the *Boussinesq* approach has been often used with a reasonable degree of success. Here

$$\tau_t = \eta \frac{\partial u}{\partial y} = \tau_{xy} \tag{3.1.5}$$

where η is called the *eddy viscosity*. The eddy viscosity is not a property of the fluid, varies from time to time and point to point, and must be parameterized (often with quite different functions) for each flow field. This subject is elaborated upon in Chap. 6.

With the physical distinction between these flows being so clear a quantitative procedure for classifying flows is reviewed in Chap. 6.

An *ideal fluid* is frictionless and incompressible and should not be confused with a perfect gas (Sec. 1.7). The assumption of an ideal fluid is helpful in analyzing flow situations involving large expanses of fluids, such as the oceans. A frictionless fluid is nonviscous, and its flow processes are reversible or loss-free.

The layer of fluid in the immediate neighborhood of an actual flow boundary that has had its velocity relative to the boundary affected by viscous shear is called the *boundary layer.* Boundary layers may be laminar or turbulent, depending generally upon their length, the viscosity, the velocity of the flow near them, and the boundary roughness.

Rotation of a fluid particle about a given axis, say the z axis, is defined as the average angular velocity of two infinitesimal line elements in the particle that are originally at right angles to each other and to the given axis. If the fluid particles within a region have rotation about any axis, the flow is called *rotational flow,* or *vortex flow.* If the fluid within a region has no rotation, the flow is called *irrotational flow.* If a fluid is at rest and is frictionless, any subsequent motion of this fluid will also be irrotational. However, for most practical circumstances shear stresses introduced via the presence of boundaries, density gradients, or most other interactions between fluid processes will result in rotation.

Adiabatic flow is that flow of a fluid in which no heat is transferred to or from the fluid. *Reversible adiabatic* (frictionless adiabatic) *flow* is called *isentropic flow.*†

With most of the fluids problems being three-dimensional and highly variable in time, it is not often possible to obtain general solutions. Certain spatial and temporal classifications have arisen which both allow and describe procedures for performing simplified analyses. The definition of these classifications concludes this section.

Two temporal classifications exist. *Steady flow* occurs when conditions at any point in the fluid do not change with time. For example, if the velocity at a certain point is 3 m/s in the $+x$ direction in steady flow, it remains exactly that amount and in that direction indefinitely. This can be expressed as $\partial \mathbf{v}/\partial t = 0$, in which space ($x$, y, z coordinates of the point) is held constant. Likewise, in steady flow there is no change in density ρ, pressure p, temperature T, or concentration C with time at any point; thus,

$$\frac{\partial \rho}{\partial t} = 0 \qquad \frac{\partial p}{\partial t} = 0 \qquad \frac{\partial T}{\partial t} = 0 \qquad \frac{\partial C}{\partial t} = 0$$

In turbulent flow, owing to the erratic motion of the fluid particles, there are always small fluctuations occurring at any point. The definition for steady flow must be generalized somewhat to provide for these fluctuations. To illustrate this, a plot of velocity against time, at some point in turbulent flow, is given in Fig. 3.4. When the temporal mean velocity

$$\bar{\mathbf{v}} = \frac{1}{t_p} \int_t^{t+t_p} \mathbf{v}\, dt$$

indicated in the figure by the horizontal line, does not change with time, the flow is said to be steady. Here t_p is called the *averaging period.* The same generalization applies to density, pressure, temperature, etc., when they are substituted for $\mathbf{v}$ in the above formula.

The flow is *unsteady* when conditions at any point change with time, $\partial \mathbf{v}/\partial t \neq 0$, $\partial \mathbf{T}/\partial t \neq 0$, etc. Water being pumped through a fixed system at a constant rate is an example of steady flow. Water being pumped through a fixed system at an increasing rate is an example of unsteady flow. Figure 3.4 also depicts an unsteady turbulent flow where even $\partial \bar{\mathbf{v}}/\partial t \neq 0$.

The description of spatial structure is more complex as flows can be spatially well behaved in one, two, or three dimensions. These definitions follow.

Uniform flow occurs when, at every point, the velocity vector or any of the other fluid variables is identically the same (in magnitude and direction) for any given instant. In equation form, $\partial \mathbf{v}/\partial s = 0$, in which time is held constant and s is a displacement in any direction. The equation states that there is no change in the velocity vector in any direction throughout the fluid at any one instant. It says nothing about the change in velocity at a point with time.

Flow such that the velocity vector varies from place to place at any instant ($\partial \mathbf{v}/\partial s \neq 0$) is *nonuniform flow.* A liquid being pumped through a long straight pipe

---

†An isentropic process, however, can occur in irreversible flow with the proper amount of heat transfer. (Isentropic = constant entropy.)

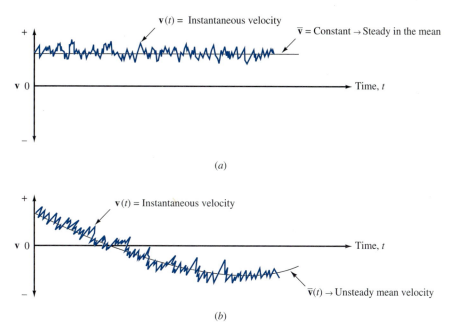

Figure 3.4 Time traces of turbulent and mean velocity data for steady and unsteady mean flow.

has uniform flow. A liquid flowing through a reducing section or through a curved pipe has nonuniform flow.

Examples of steady and unsteady flow and of uniform and nonuniform flow are:

- Liquid flow through a long pipe at a constant rate is *steady uniform flow.*
- Liquid flow through a long pipe at a decreasing rate is *unsteady uniform flow.*
- Flow through an expanding tube at a constant rate is *steady nonuniform flow.*
- Flow through an expanding tube at an increasing rate is *unsteady nonuniform flow.*

One-dimensional (1D) *flow* neglects variations or changes in velocity, pressure, etc., in a plane transverse to the main flow direction. Conditions at a cross section are expressed in terms of average values of velocity, density, and other properties. Flow through a pipe, for example, can usually be characterized as one-dimensional. Many practical problems can be handled by this method of analysis, which is much simpler than two- and three-dimensional methods of analysis. In *two-dimensional* (2D) *flow* and transport all particles are assumed to flow in parallel planes along identical paths in each of these planes; hence, there are no changes in flow normal to these planes. Figure 3.5 is an example of a two-dimensional flow of water wave velocity. The wave is depicted in the vertical plane of the "page" with vertical and horizontal variations of particle path and velocity. These variations are assumed to be identical to those in other parallel planes positioned in front of and behind the figure. *Three-dimensional* (3D) *flow* is the most general and complex flow, and typically only flow and transport in very simple geometries can be handled.

Anyone who has observed a river can easily tell that the flow is three-dimensional yet the description of it in terms of one- and two-dimensional flows suggests that the physics is comprised of processes that are only one- and two-dimensional. The mathematical bridge that many times allows a complex three-dimensional flow to be analyzed in spatially simplified form is *spatial averaging.* As temporal averaging suppresses temporal variability, so will spatial averaging "suppress" spatial

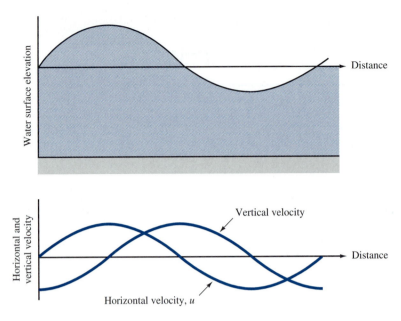

Figure 3.5   Water wave heights and velocities.

complexity. For example, a 3D velocity field defined as $\mathbf{v}(x = x_1, y, z, t)$ can be spatially averaged over the cross section $A_1(x = x_1)$ as

$$v_1 A_1 = \iint \mathbf{v}(\mathbf{x}, t)\, dA$$

Similarly, average velocities can be identified at every x position in the stream. The resulting flow and transport are, therefore, considered to be only one-dimensional not three-dimensional.

The above definition of uniform flow can be extended to spatially averaged flows as well. If parallel cross sections through the conduit or river or estuary channel are identical or prismatic and the average velocity at each cross section is the same at any given instant, then the 1D flow and transport are also said to be uniform.

Summary of Analysis and Solution Approaches

As mentioned previously, three-dimensional time-varying exact or deterministic solutions are not generally available especially for naturally occurring flows governed by very complex geometries. Typically the engineer or analyst employs any one of a number of simplifying analysis and solution strategies. In this book three different simplifying approaches will be presented, all of which in some sense depend upon the construction of a *model* of the process and its use in achieving problem solutions. The three approaches are *exact mathematical solutions*, *laboratory experiments* or *simulations*, and *numerical* or *computational solutions*. Often parts of these approaches or all three are applied to the problem in the hope of achieving credibility for the proposed solution by observing consistent answers between methods. In the case of laboratory data being compared to a model or an exact solution, this is called validation. At other junctures one approach might be used to extend the region of validity of an earlier solution from another approach. This is often the case when exact solutions are elaborated upon by numerical models or a laboratory simulation.

While no one general 3D solution exists, it is also the case that no one of these approaches will provide a general solution to the problem. There are three general considerations in selecting which approach to use. The first is how much is known about the physical processes governing the flow field and are the governing equations well established and accepted; the second consideration is how variable in space or time are the physics; while the third consideration becomes how complex or simple is the flow geometry. In general, if the mathematical functions or equations are yet to be established for a given geometry or flow process, then laboratory experiments are traditionally performed. Procedures for ensuring that the laboratory model and the real world problem or *prototype* are similar are established through *nondimensional analysis*. Chapter 5 introduces these procedures.

Exact solutions or numerical models are used where the physics of the flow and transport can be represented mathematically. Only the processes contributing to the problem are mathematically retained having been identified again through dimensional analysis. In general, the more processes which interact in the problem the more difficult an exact analysis will be, and a numerical solution approach will be optimal. As an example the steady viscous flow of water over a smooth surface involves friction and inertia and is amenable to exact solution. The prediction of a heated effluent plume into a cool river during a storm event involves turbulent diffusion, inertia, stratification, and long wave propagation which is more appropriately handled with a numerical model.

Complex *boundary* or *forcing functions* also suggest numerical modeling. Steady conditions are mathematically simple but multiple or time-varying conditions are quite difficult and often nonlinear. The prediction of erosion off the bottom of a river channel during a storm must be solved by use of knowledge about the time change in the currents as well as the time variation in the surface wind–driven waves and therefore requires a numerical approach.

The type of flow geometry has a strong impact on the selection of an exact or computational solution procedure. Flows may be *external* in that they flow over an object, say a sphere or sediment particle; any confining geometries are therefore so far away they do not impact the local flow. Flows through pipes and channels are called *internal* flows. Often as in the case of wind waves or groundwater the position of the boundary becomes the object of the solution. Such *free surface* problems are highly nonlinear. If there are very high gradients in the flow geometry, then numerical procedures are recommended. Dramatic changes in depth or channel geometry are regularly encountered in natural geometries and it is safe to say that computer models are used almost all the time for natural flow and transport calculations.

As mentioned earlier, spatial simplification can be achieved by spatial averaging of the equations. Such is the case for river flows and pipe flow which are robustly handled with one-dimensional approaches. Flows with significant vertical structure due to density gradients or stratification or counter flow cannot be handled with a one-dimensional approach. Flows in lakes and estuaries, that is, natural flows, are typical of these types of exceptions.

EXERCISES

3.1.1 One-dimensional flow is (*a*) steady uniform flow; (*b*) uniform flow; (*c*) flow which neglects changes in a transverse direction; (*d*) restricted to flow in a straight line: (*e*) none of these answers.

3.1.2 Isentropic flow is (*a*) irreversible adiabatic flow; (*b*) perfect-gas flow; (*c*) ideal-fluid flow; (*d*) reversible adiabatic flow; (*e*) frictionless reversible flow.

3.1.3   In turbulent flow (*a*) the fluid particles move in an orderly manner; (*b*) cohesion is more effective than momentum transfer in causing shear stress; (*c*) momentum transfer is on a molecular scale only; (*d*) one lamina of fluid glides smoothly over another; (*e*) the shear stresses are generally larger than in a similar laminar flow.

3.1.4   The ratio $\eta = \tau/(du/dy)$ for turbulent flow is (*a*) a physical property of the fluid; (*b*) dependent upon the flow and the density; (*c*) the viscosity divided by the density; (*d*) a function of temperature and pressure of fluid; (*e*) independent of the nature of the flow.

3.1.5   Turbulent flow generally occurs for cases involving (*a*) very viscous fluids; (*b*) very narrow passages or capillary tubes; (*c*) very slow motions; (*d*) combinations of (*a*), (*b*), and (*c*); (*e*) none of these answers.

3.1.6   In laminar flow (*a*) experimentation is required for the simplest flow cases; (*b*) Newton's law of viscosity applies; (*c*) the fluid particles move in irregular and haphazard paths; (*d*) the viscosity is unimportant; (*e*) the ratio $\tau/(du/dy)$ depends upon the flow.

3.1.7   An ideal fluid is (*a*) very viscous; (*b*) one which obeys Newton's law of viscosity; (*c*) a useful assumption in problems in conduit flow; (*d*) frictionless and incompressible; (*e*) none of these answers.

3.1.8   Which of the following must be satisfied by the flow of any fluid, real or ideal?

1.   Newton's law of viscosity
2.   Newton's second law of motion
3.   The continuity equation
4. $\tau = (\mu + \eta)\frac{du}{dy}$
5.   The requirement that velocity at the boundary be zero relative to the boundary
6.   The rule that fluid cannot penetrate a boundary

(*a*) 1,2,3; (*b*) 1,3,6; (*c*) 2,3,5; (*d*) 2,3,6; (*e*) 2,4,5.

3.1.9   Steady flow occurs when (*a*) conditions do not change with time at any point; (*b*) conditions are the same at adjacent points at any instant; (*c*) conditions change steadily with the time; (*d*) $\partial v/\partial t$ is constant; (*e*) $\partial v/\partial s$ is constant.

3.1.10   Uniform flow occurs (*a*) whenever the flow is steady; (*b*) when $\partial \mathbf{v}/\partial t$ is everywhere zero; (*c*) only when the velocity vector at any point remains constant; (*d*) when $\partial \mathbf{v}/\partial s = 0$; (*e*) when the discharge through a curved pipe of constant cross-sectional area is constant.

3.1.11   Select the correct practical example of steady nonuniform flow: (*a*) motion of water around a ship in a lake; (*b*) motion of river around bridge piers; (*c*) steadily increasing flow through a pipe; (*d*) steadily decreasing flow through a reducing section; (*e*) constant discharge through a long, straight pipe.

3.1.12   A streamline (*a*) is the line connecting the midpoints of flow cross sections; (*b*) is defined for uniform flow only; (*c*) is drawn normal to the velocity vector at every point; (*d*) is always the path of a particle; (*e*) is fixed in space in steady flow.

3.1.13   In two-dimensional flow around a cylinder the streamlines are 2 in. apart at a great distance from the cylinder, where the velocity is 100 ft/s. At one point near the cylinder the streamlines are 1.5 in. apart. The average velocity there is (*a*) 75 ft/s; (*b*) 133 ft/s; (*c*) 150 ft/s; (*d*) 200 ft/s; (*e*) 300 ft/s.

3.1.14 The head loss in turbulent flow in a pipe (*a*) varies directly as the velocity; (*b*) varies inversely as the square of the velocity; (*c*) varies inversely as the square of the diameter; (*d*) depends upon the orientation of the pipe; (*e*) varies approximately as the square of the velocity.

3.2 THE GENERAL CONTROL VOLUME CONSERVATION EQUATION

Regardless of the nature of the flow, all flow situations are subject to the following relations, which can be expressed in analytic form:

1. Newton's laws of motion, which must hold for every particle at every instant
2. The continuity relation, that is, the law of conservation of mass
3. The conservation of mass applied to mixtures of components within the fluid
4. The first and second laws of thermodynamics
5. Boundary conditions; analytical statements that a real fluid has zero velocity relative to a boundary at a boundary or that frictionless fluids cannot penetrate a boundary

Other relations and equations, such as an equation of state or Newton's law of viscosity, may enter.

In the derivation that follows the *control-volume* concept is related to the *system* in terms of a general property of the system. In subsequent sections it is then applied specifically to obtain continuity, energy, and linear-momentum relations.

To formulate the relation between equations applied to a *system* and those applied to a fixed *control volume*, consider some general flow situations, Fig. 3.6, in which the velocity of a fluid is given relative to an *xyz* coordinate system. At time *t*

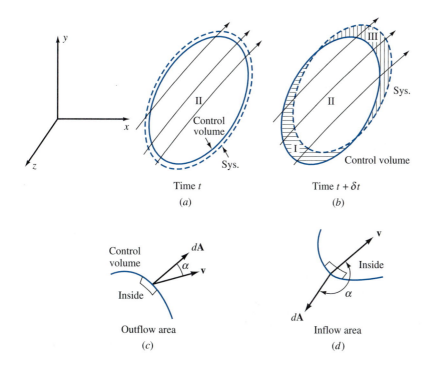

Figure 3.6 System with identical control volume at time *t* in a velocity field.

consider a certain mass of fluid that is contained within a system, having the dotted-line boundaries indicated. Also consider a control volume, fixed relative to the xyz axes, that exactly coincides with the system at time t. At $t + \delta t$ the system has moved somewhat, since each mass particle moves at the velocity associated with its location.

Let N be the total amount of some property (e.g., mass, energy, or momentum) within the system at time t, and let η be the amount of this property, per unit mass, throughout the fluid. The time rate of increase of N for the system is now formulated in terms of the control volume.

At $t + \delta t$ (Fig. 3.6b) the system comprises volumes II and III, while at time t it occupies volume II (Fig. 3.6a). The increase in property N in the system in time δt is given by

$$N_{sys_{t+\delta t}} - N_{sys_t} = \left(\int_{II} \eta \rho \, d\mathrm{V} + \int_{III} \eta \rho \, d\mathrm{V} \right)_{t+\delta t} - \left(\int_{II} \eta \rho \, d\mathrm{V} \right)_t$$

in which $d\mathrm{V}$ is the element of volume. Rearrangement, after adding and subtracting

$$\left(\int_I \eta \rho \, d\mathrm{V} \right)_{t+\delta t}$$

to the right, then dividing through by δt leads to

$$\frac{N_{sys_{t+\delta t}} - N_{sys_t}}{\delta t} = \frac{\left(\int_{II} \eta \rho \, d\mathrm{V} + \int_I \eta \rho \, d\mathrm{V} \right)_{t+\delta t} - \left(\int_{II} \eta \rho \, d\mathrm{V} \right)_t}{\delta t} \qquad \textbf{(3.2.1)}$$
$$+ \frac{\left(\int_{III} \eta \rho \, d\mathrm{V} \right)_{t+\delta t}}{\delta t} - \frac{\left(\int_I \eta \rho \, d\mathrm{V} \right)_{t+\delta t}}{\delta t}$$

The term on the left is the average time rate of increase of N within the system during time δt. In the limit as δt approaches zero, it becomes dN/dt. If the limit is taken as δt approaches zero for the first term on the right-hand side of the equation, the first two integrals are the amount of N in the control volume at $t + \delta t$ and the third integral is the amount of N in the control volume at time t. The limit is

$$\frac{\partial}{\partial t} \int_{cv} \eta \rho \, d\mathrm{V}$$

in which the partial derivative is used since the control volume size is held constant as $\delta t \to 0$.

The next term, which is the time rate of flow of N out of the control volume, in the limit, can be written

$$\lim_{\delta t \to 0} \frac{\left(\int_{III} \eta \rho \, d\mathrm{V} \right)_{t+\delta t}}{\delta t} = \int_{\text{outflow area}} \eta \rho \mathbf{v} \cdot d\mathbf{A} = - \int \eta \rho v \cos \alpha \, dA \qquad \textbf{(3.2.2)}$$

in which $d\mathbf{A}$, Fig. 3.6c, is the vector representing an area element of the outflow area. It has a direction normal to the surface-area element of the control volume, being positive outward, and α is the angle between the velocity vector and the elemental area vector.

Similarly, the last term of Eq. (3.2.1), which is the rate of flow of N into the control volume, is, in the limit, equal to

$$\lim_{\delta t \to 0} \frac{\left(\int_I \eta \rho \, d\mathrm{V} \right)_{t+\delta t}}{\delta t} = - \int_{\text{inflow area}} \eta \rho \mathbf{v} \cdot d\mathbf{A} = - \int \eta \rho v \cos \alpha \, dA \qquad \textbf{(3.2.3)}$$

The minus sign is needed as $\mathbf{v} \cdot d\mathbf{A} = 0$ (or $\cos\alpha$) is negative for inflow, Fig. 3.6d. The last two terms of Eq. (3.2.1), given by Eqs. (3.2.2) and (3.2.3), can be combined into the single term which is an integral over the complete control-volume surface (cs)

$$\lim_{\delta t \to 0} \left(\frac{\left(\int_{III} \eta \rho \, d\mathcal{V}\right)_{t+\delta t}}{\delta t} - \frac{\left(\int_{I} \eta \rho \, d\mathcal{V}\right)_{t+\delta t}}{\delta t} \right) = \int_{cs} \eta \rho \mathbf{v} \cdot d\mathbf{A} = \int_{cs} \eta \rho v \cos\alpha \, dA$$

Where there is no inflow or outflow, $\mathbf{v} \cdot d\mathbf{A} = 0$; hence, the equation can be evaluated over the whole control surface. Collecting the reorganized terms of Eq. (3.2.1) gives

$$\frac{dN}{dt} = \frac{\partial}{\partial t} \int_{cv} \eta \rho \, d\mathcal{V} + \int_{cs} \eta \rho \mathbf{v} \cdot d\mathbf{A} \qquad \text{(3.2.4)}$$

In words, Eq. (3.2.4) states that the time rate of increase of N within a system is just equal to the time rate of increase of the property N within the control volume (fixed relative to xyz) plus the net rate of efflux of N across the control-volume boundary.

Equation (3.2.4) is used to convert from the system form to the control-volume form. The system form, which in effect follows the motion of the parcels, is referred to as the Lagrangian method of analysis; the control-volume approach is called the Eulerian method of analysis, as it observes flow from a reference system fixed relative to the control volume.

Since the xyz frame of reference can be given an arbitrary constant velocity without affecting the dynamics of the system and its surroundings, Eq. (3.2.4) is valid if the control volume, fixed in size and shape, has a uniform velocity of translation.

3.3 THE CONSERVATION OF MASS

The *system* form of the conservation of mass is

$$\frac{dm}{dt} = 0$$

which states that the mass, m, within the system remains constant in time. In Eq. (3.2.4) let $N = m$, then η is the mass per unit mass of $\eta = 1$. Then

$$0 = \frac{\partial}{\partial t} \int_{cv} \rho \, d\mathcal{V} + \int_{cs} \rho \mathbf{v} \cdot d\mathbf{A} \qquad \text{(3.3.1)}$$

The conservation of mass equation states that the time rate of change of mass in the control volume plus the net rate at which mass leaves through the control volume surface equals zero.

Consider the cylindrical tube in Fig. 3.7. Flow enters the tube at station 1 and exits at station 2. No flow is permitted through the solid surface comprising the tube. The application of the conservation of mass proceeds as follows:

1. The control volume is defined to include all the fluid in the tube out to the solid wall and from station 1 to station 2. If at all possible the inlet and outlet stations should be defined or placed in regions where the stream lines (or tubes) are parallel to the boundary such that the entrance and exit fluid velocities are perpendicular to the respective areas.

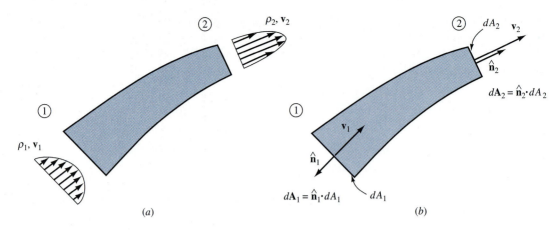

Figure 3.7   Control volume representation of flow through a tube. (a) Flow field. (b) Free body representation.

2. If allowed by the problem statement, it is useful to assume steady flow, in which case Eq. (3.3.1) reduces to

$$0 = \int_{cs} \rho \mathbf{v} \cdot d\mathbf{A} \tag{3.3.2}$$

3. Equation (3.3.2) must be applied to each control surface (cs) where fluid mass (or in later sections, momentum and energy) is entering or leaving; therefore,

$$\int_{cs1} \rho_1 \mathbf{v}_1 \cdot d\mathbf{A}_1 + \int_{cs2} \rho_2 \mathbf{v}_2 \cdot d\mathbf{A}_2 = 0$$

4. If the inlet and outlet velocity vectors are, at each inlet and outlet, perpendicular to their respective areas, then all outflow integral dot products (Fig. 3.6c) are evaluated as $\rho_2 \mathbf{v}_2 \cdot d\mathbf{A}_2 = \rho_2 v_2 \, dA_2$ and inflows are evaluated as $\rho_1 \mathbf{v}_1 \cdot d\mathbf{A}_1 = -\rho_1 v_1 \, dA_1$. Therefore,

$$\int_{cs1} \rho_1 v_1 \, dA_1 = \int_{cs2} \rho_2 v_2 \, dA_2 \tag{3.3.3}$$

It is noted that ρ and v are still functions at A_1 and A_2 and could vary over their respective areas. This is especially the case for the velocities.

5. If ρ_1 and ρ_2 do not vary across the inlet and outlet areas, then

$$\rho_1 \int_{cs1} v_1 \, dA_1 = \rho_2 \int_{cs2} v_2 \, dA_2$$

6. It is more convenient to avoid the spatially varying velocity for initial calculations. Therefore, the spatial average velocity is invoked to reduce the problem to a one-dimensional representation. Hence,

$$V_1 A_1 = \int_{cs1} v_1 \, dA_1 \qquad V_2 A_2 = \int_{cs2} v_2 \, dA_2$$

and

$$\rho_1 V_1 A_1 = \rho_2 V_2 A_2 = \dot{m} \tag{3.3.4}$$

Here $\dot{m}$ is the *mass flow rate* in kg/sec or slugs/sec. For the steady flow problem here the continuity equation says the mass flow rate is a constant.

If the *discharge* Q (also called the *volumetric flow rate* or *flow*) is defined as

$$Q = AV \tag{3.3.5}$$

the continuity equation may take the form

$$\dot{m} = \rho_1 Q_1 = \rho_2 Q_2 \tag{3.3.6}$$

For incompressible, steady flow

$$Q = A_1 V_1 = A_2 V_2 \tag{3.3.7}$$

is a useful form of the equation.

For constant-density flow, steady or unsteady, Eq. (3.3.1) becomes

$$\int_{cs} \mathbf{v} \cdot d\mathbf{A} = 0 \tag{3.3.8}$$

which states that the net volume efflux is zero. (This implies that the control volume is filled with liquid at all times.)

At section 1 of a pipe system carrying water (Fig. 3.8) the velocity is 3.0 ft/s and the diameter is 2.0 ft. At section 2 the diameter is 3.0 ft. Find the discharge and the velocity at section 2.

Example 3.1

Solution

From Eq. (3.3.7)

$$Q = V_1 A_1 = (3.0 \text{ ft/s}) \frac{\pi}{4} (2 \text{ ft})^2 = 9.42 \text{ ft}^3/\text{s}$$

and

$$V_2 = \frac{Q}{A_2} = (9.42 \text{ ft}^3/\text{s}) \frac{4}{\pi(3 \text{ ft})^2} = 1.33 \text{ ft/s}$$

If there are multiple inlets and outlets, the control volume equation must then be extended. Assume a T-intersection as noted in Fig. 3.9; also denoted are the conditions at the inlets (stations 1 and 3) and the outlet (station 2). Further assume that the density at each station is constant (though not necessarily equal); that the velocity vectors are perpendicular to their respective areas; and that the cross-sectional average velocities at each station are therefore defined. Then Eq. (3.3.2) reduces to

$$-\rho_1 V_1 A_1 - \rho_3 V_3 A_3 + \rho_2 V_2 A_2 = 0$$

or

$$\dot{m}_1 + \dot{m}_3 = \dot{m}_2. \tag{3.3.9}$$

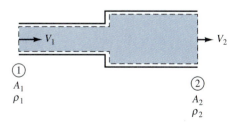

A_1
ρ_1

A_2
ρ_2

Figure 3.8 Control volume for flow through a series of pipes.

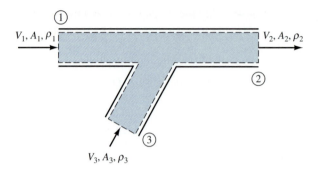

Figure 3.9   Control volume definition for a multiple inlet and outlet geometry.

For the case where the densities are all equal

$$V_1 A_1 + V_3 A_3 = V_2 A_2 \qquad (3.3.10)$$

or the sum of the inflow volume flow rates equals the sum of the volume flow rates out of the control volume.

3.4   THE ENERGY EQUATION

The Basic Equation

The first law of thermodynamics for a system states that the heat Q_H added to a system minus the work W done by the system depends only upon the initial and final states of the system. The difference in states of the system, being independent of the path from initial to final state, must be a property of the system. It is called the internal energy E. The first law in equation form is

$$Q_H - W = E_2 - E_1 \qquad (3.4.1)$$

The internal energy per unit mass is denoted by e; hence, applying Eq. (3.2.4), $N = E$ and $\eta = e$,

$$\frac{dE}{dt} = \frac{\partial}{\partial t} \int_{cv} \rho e \, dV + \int_{cs} \rho e \mathbf{v} \cdot d\mathbf{A} \qquad (3.4.2)$$

or by use of Eq. (3.4.1)

$$\frac{\delta Q_H}{\delta t} - \frac{\delta W}{\delta t} = \frac{dE}{dt} = \frac{\partial}{\partial t} \int_{cv} \rho e \, dV + \int_{cs} \rho e \mathbf{v} \cdot d\mathbf{A} \qquad (3.4.3)$$

The work done by the system on its surroundings can be broken into two parts: the work W_{pr} done by pressure forces on the moving boundaries and the work W_s done by shear forces such as the torque exerted on a rotating shaft. The work done by pressure forces in time δt is

$$\delta W_{pr} = \delta t \int p \mathbf{v} \cdot d\mathbf{A} \qquad (3.4.4)$$

By use of the definitions of the work terms, Eq. (3.4.3) becomes

$$\frac{\delta Q_H}{\delta t} - \frac{\delta W_s}{\delta t} = \frac{\partial}{\partial t} \int_{cv} \rho e \, dV + \int_{cs} \left(\frac{p}{\rho} + e \right) \rho \mathbf{v} \cdot d\mathbf{A} \qquad (3.4.5)$$

In the absence of nuclear, electrical, magnetic, and surface-tension effects, the internal energy e of a pure substance is the sum of potential, kinetic, and "intrinsic" energies. The intrinsic energy u^{**} per unit mass is due to molecular spacing and forces (dependent upon p, ρ, or T). The internal energy then is defined as

$$e = gz + \frac{V^2}{2} + u^{**} \tag{3.4.6}$$

The dimensions of e are in units of work per unit mass or $[FL/M]$, which reduces to dimensions of $[L^2/t^2]$. It should be noticed that the elevation variable, z, in the potential energy term requires the definition of a reference level or datum for each problem. Since relative elevation differences are required, the datum need not be absolute or universal. The velocity in the kinetic energy term is the total velocity magnitude at the point in question in the flow field, that is, $V^2 = \mathbf{v} \cdot \mathbf{v} = u^2 + v^2 + w^2$.

In applying Eq. (3.4.5) to a control volume, consider the mixing chamber defined in Fig. 3.10. Using the procedures for control volume application and simplification from the previous section, the energy equation is developed as follows.

1. Establish the control volume boundaries such that the inlet and outlet areas are in regions of uniform flow where streamlines are, hopefully, parallel to the inlet wall and the velocity vectors are perpendicular to their respective surface areas.

2. Establish a datum for elevation measurement.

3. If the fluid flow is steady, then Eq. (3.4.5) becomes

$$\frac{\delta Q_H}{\delta t} - \frac{\delta W_s}{\delta t} = \int_{cs} \left(\frac{p}{\rho} + e \right) \rho \mathbf{v} \cdot d\mathbf{A} \tag{3.4.7}$$

4. Equation (3.4.7) is then applied to each control surface area

$$\frac{\delta Q_H}{\delta t} - \frac{\delta W_s}{\delta t} = \int_{cs1} \left(\frac{p_1}{\rho_1} + e_1 \right) \rho_1 \mathbf{v_1} \cdot d\mathbf{A_1} + \int_{cs2} \left(\frac{p_2}{\rho_2} + e_2 \right) \rho_2 \mathbf{v_2} \cdot d\mathbf{A_2} \tag{3.4.8}$$

5. With the velocity vectors perpendicular to the areas, the dot products are evaluated as in step number 4 in the previous section (Eq. 3.3.3)

$$\frac{\delta Q_H}{\delta t} - \frac{\delta W_s}{\delta t} = -\int_{cs1} \left(\frac{p_1}{\rho_1} + e_1 \right) \rho_1 v_1 \, dA_1 + \int_{cs2} \left(\frac{p_2}{\rho_2} + e_2 \right) \rho_2 v_2 \, dA_2$$

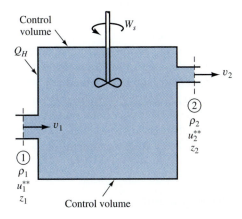

Figure 3.10 Control volume with flow across control surface normal to surface.

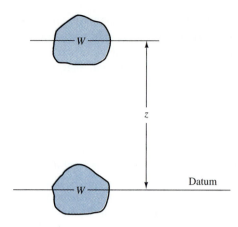

Figure 3.11   Potential energy.

6. Substitution of the definition for e into the above equation gives

$$\frac{\delta Q_H}{\delta t} - \frac{\delta W_s}{\delta t} = -\int_{cs1} \left(\frac{p_1}{\rho_1} + gz_1 + \frac{v_1^2}{2} + u_1^{**} \right) \rho_1 v_1 \, dA_1 \qquad (3.4.9)$$

$$+ \int_{cs2} \left(\frac{p_2}{\rho_2} + gz_2 + \frac{v_2^2}{2} + u_2^{**} \right) \rho_2 v_2 \, dA_2$$

Simplification of the terms in Eq. (3.4.9) will require some mathematical care or some assumptions, but as it stands now, Eq. (3.4.9) is complete and valid for the mixing chamber.

In Eq. (3.4.9), the three energy terms $gz + v^2/2 + p/\rho$ are referred to as available energy. The first term, gz, is potential energy per unit mass. With reference to Fig. 3.11, the work needed to lift W newtons a distance z meters is Wz. The mass of W newtons is W/g kg; hence, the potential energy, in meter-newtons per kilogram, is

$$\frac{Wz}{W/g} = gz$$

The next term, $v^2/2$, is interpreted as follows: The kinetic energy of a particle of mass is $\delta m \, v^2/2$. To place this on a unit mass basis, divide by δm; thus $v^2/2$ is meter-newtons per kilogram kinetic energy.

The term, p/ρ, is the *flow work* or *flow energy* per unit mass. Flow work is net work done by the fluid element on its surroundings while it is flowing. For example, in Fig. 3.12, imagine a turbine consisting of a vaned unit that rotates as fluid passes through it, exerting a torque on its shaft. For a small rotation the pressure drop across a vane times the exposed area of vane is a force on the rotor. When multiplied by the distance from the center of force to the axis of the rotor, a torque is obtained. Elemental work done is $p \, \delta A \, ds$ by $\rho \, \delta A \, ds$ units of mass of flowing fluid; hence, the work per unit mass is p/ρ.

Steady State Energy Equation

Since the distribution of p, ρ, v, and u^{**} are variable across each area, the analysis problem is difficult. Proceeding to a form of Eq. (3.4.9) useful for calculations

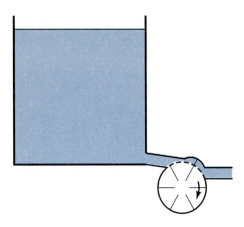

Figure 3.12 Work done by sustained pressure.

requires the following steps or assumptions. The intrinsic energy, u^{**}, is assumed to be constant or uniform across each area; therefore,

$$\int_{cs} u^{**} \rho v \, dA = u^{**} \int_{cs} \rho v \, dA \tag{3.4.10}$$

In order to determine the pressure and potential energy terms a bit more rationale is necessary. Here the potential energy term can be approximated by taking the area-averaged height, z_c, or the height to the centroid of the inlet area. Then

$$\int_{cs} gz \rho v \, dA = gz_c \int_{cs} \rho v \, dA \tag{3.4.11}$$

In order to determine the pressure an arbitrary distribution of pressure can be approximated by determining the average pressure on the area using the methods in Chap. 2. However, simplifying assumptions prevail. First if the diameter of the inlet or outlet is small, say for most typical pipes or conduits, then for most practical purposes the pressure and density are uniformly distributed across each control surface area. In this case

$$\int_{cs} \frac{p}{\rho} \rho v \, dA = \frac{p}{\rho} \int_{cs} \rho v \, dA \tag{3.4.12}$$

For large diameter openings where a significant hydrostatic component prevails then the average pressure may be computed using methods from Chap. 2. If the inlet and outlet areas are selected such that the streamlines are parallel to the pipe wall and perpendicular to gravity $(+z)$, then the sum of $p/\gamma + z$ will be constant everywhere in the vertical coordinate at that station; consequently, we may write

$$\int_{cs} \left(\frac{p}{\rho} + gz \right) \rho v \, dA = \left(\frac{p}{\rho} + gz \right) \int_{cs} \rho v \, dA \tag{3.4.13}$$

The kinetic energy term is not quite so straightforward. The velocity function in the presence of walls varies considerably and cannot, as discussed before, be assumed uniform. However, unlike the continuity example

$$\int_{cs} \frac{v^2}{2} \rho v \, dA = \int_{cs} \rho \frac{v^3}{2} \, dA \neq \rho \frac{V^3}{2} A$$

It is noted that the product of the averages is not equal to the average of the product. The left-hand side of the equation defines the actual kinetic energy passing a cross-section per unit time, in which $\rho v\, dA$ is the mass per unit time passing dA and $v^2/2$ is the kinetic energy per unit mass. It is necessary to compute a correction factor α for $V^2/2$ so that $\alpha V^2/2$ is the average kinetic energy per unit mass passing the section, or

$$\alpha = \frac{1}{A}\int\left(\frac{v^3}{V}\right)dA \tag{3.4.14}$$

If finally $\int \rho v\, dA$ is defined as $\dot{m}$ or as $\rho V A$ (see Eq. 3.3.6), then

$$\frac{\delta Q_H}{\delta t} + \left(\frac{p_1}{\rho_1} + gz_1 + \alpha_1\frac{V_1^2}{2} + u_1^{**}\right)\rho_1 V_1 A_1 \tag{3.4.15}$$

$$= \frac{\delta W_s}{\delta t} + \left(\frac{p_2}{\rho_2} + gz_2 + \alpha_2\frac{V_2^2}{2} + u_2^{**}\right)\rho_2 V_2 A_2$$

It is often convenient to divide by the mass flow rate; therefore, the equation becomes

$$q_H + \frac{p_1}{\rho_1} + gz_1 + \alpha_1\frac{V_1^2}{2} + u_1^{**} = w_s + \frac{p_2}{\rho_2} + gz_2 + \alpha_2\frac{V_2^2}{2} + u_2^{**} \tag{3.4.16}$$

where q_H is the heat added per unit mass of flowing fluid and w_s is the shaft work per unit mass. If work is done on the fluid in the control volume, as with a pump, then w_s is negative, or if work is done by the control volume or extracted from the control volume as in the case of a turbine, the w_s is positive. For the sign convention to be consistently applied, then section 1 is upstream and section 2 is downstream. Equation (3.4.16) is the *energy equation* for steady flow through a control volume.

Equation (3.4.7) can be extended to the case where there are multiple inlets or outlets. If there is a second inlet (Station 3) as in the continuity example, then for Stations 1 and 3 being inlets and Station 2 being the outlet Eq. (3.4.15) is

$$\frac{\delta Q_H}{\delta t} + \left(\frac{p_1}{\rho_1} + gz_1 + \alpha_1\frac{V_1^2}{2} + u_1^{**}\right)\rho_1 V_1 A_1 + \left(\frac{p_3}{\rho_3} + gz_3 + \alpha_3\frac{V_3^2}{2} + u_3^{**}\right)\rho_3 V_3 A_3$$

$$= \frac{\delta W_s}{\delta t} + \left(\frac{p_2}{\rho_2} + gz_2 + \alpha_2\frac{V_2^2}{2} + u_2^{**}\right)\rho_2 V_2 A_2$$

$$\tag{3.4.17}$$

Reversibility, Irreversibility, and Losses

A *process* can be defined as the path of the succession of states through which the system passes, such as the changes in velocity, elevation, pressure, density, or temperature. The expansion of air in a cylinder as the piston moves out and heat is transferred through the walls is an example of a process. Normally, the process causes some change in the surroundings, for example, displacing it or transferring heat to or from its boundaries. When a process can be made to take place in such a manner that it can be *reversed*, that is, made to return to its original state without a final change in either the system or its surroundings, it is said to be *reversible*. In any actual flow of a real fluid or change in a mechanical system, the effect of viscous friction, Coulomb friction, unrestrained expansion, hysteresis, etc., prohibits the process from being reversible. It is, however, an ideal to be strived for in design processes, and their efficiency is usually defined in terms of their nearness to reversibility.

When a certain process has a sole effect upon its surroundings that is equivalent to the raising of a weight, it is said to have done *work* on its surroundings. Any actual process is *irreversible*. The difference between the amount of work a substance can do by changing from one state to another state along a path reversibly and the actual work it produces for the same path is the *irreversibility* of the process. It may be defined in terms of work per unit mass or weight or work per unit time. Under certain conditions the irreversibility of a process is referred to as its *lost work*, that is, the loss of ability to do work because of friction and other causes. In Eq. (3.4.16) terms $gz + V^2/2 + p/\rho$ are *available-energy terms,* or *mechanical-energy terms,* in that they are directly able to do work by virtue of potential energy, kinetic energy, or sustained pressure. In this book when losses are referred to, they mean irreversibility, or lost work, or the transformation of available energy into thermal energy.

In looking at Eq. (3.4.5) there are terms which are not available-energy terms. In control volume form these terms are the shaft work term $\delta W_s/\delta t$, the net heat exchange $\delta Q_H/\delta t$ term, and the net intrinsic energy exchange term

$$\int_{cs} u^{**} \, \rho \mathbf{v} \cdot d\mathbf{A}$$

In cases where u^{**} is uniform across the control surface boundary, $\mathbf{v}$ is perpendicular to the entrance or exit and the area average velocity can be found, then for an inlet outlet control volume the sum

$$-\frac{\delta Q_H}{\delta t} + (u_2^{**} - u_1^{**})\rho V A \qquad \text{(3.4.18)}$$

represents the losses from the system due to irreversibility, lost work, or transformation.

In certain cases we may deal with the intrinsic energy variables directly. For instance thermodynamics or transport phenomena textbooks develop a relationship for an incompressible liquid such that

$$u^{**} = c_v T \simeq c_p T \qquad \text{(3.4.19)}$$

where T is the area averaged temperature at the control volume surface. The loss term, Eq. (3.4.18), therefore becomes

$$-\frac{\delta Q_H}{\delta t} + (T_2 - T_1)c_p \rho V A$$

For an ideal gas which can compress, the situation is a bit more complex in that

$$\frac{p}{\rho} + u^{**} = c_p T$$

with pressure being constant.

For most flows, however, other losses occur which are complex and are not so easily and specifically parameterized. Conversions to heat by friction at the walls, rapid changes in flow geometry, and internal fluid shear stresses arising from rapid acceleration or deceleration all result in losses of energy. While wall friction losses will be explicitly dealt with in Chap. 6, an empirical loss term for the control volume energy equation is introduced here of the form

$$-\frac{\delta Q_H}{\delta t} + (u_2^{**} - u_1^{**})\rho V A = K\frac{V^2}{2}\rho V A$$

or on a per unit mass basis (Eq. 3.4.16)

$$q_H + (u_2^{**} - u_1^{**}) = K\frac{V^2}{2}$$

In this equation K is called the *loss coefficient* and is (with exception of wall friction) specified from correlations with numerous laboratory experiments. Its primary use in control volume analyses is confined to describing a local loss occurring at places in the flow field where the geometry of the flow rapidly changes, for example, diffusers, elbows, and outlets. Occasionally the loss term may also be used to specify total system losses in a bulk or aggregate form. Therefore, Eq. (3.4.16) can be rewritten to include the losses, that is,

$$\frac{p_1}{\rho_1} + gz_1 + \alpha_1 \frac{V_1^2}{2} = w_s + \frac{p_2}{\rho_2} + gz_2 + \alpha_2 \frac{V_2^2}{2} + K\frac{V^2}{2} \tag{3.4.20}$$

Here V in the loss term is the average velocity at the point between station 1 and station 2 where the loss occurred. Equation (3.4.20) is expressed in terms of energy per unit mass. A form of this equation can also be obtained in terms of energy per unit weight by dividing Eq. (3.4.20) by g. This gives

$$\frac{p_1}{\gamma_1} + z_1 + \alpha_1 \frac{V_1^2}{2g} = H_s + \frac{p_2}{\gamma_2} + z_2 + \alpha_2 \frac{V_2^2}{2g} + K\frac{V^2}{2g} \tag{3.4.21}$$

In the above equation the dimensions are now N·m/N or ft·lb/lb leaving the overall dimension of each term as a length. This length is also referred to as "head"; for instance kinetic energy head is $V^2/2g$. The term H_s is the shaftwork, w_s, divided by g and is called the shaft head. This term can represent either pump head, in which case H_s would be negative, or turbine head, where H_s would be positive.

3.5   APPLICATION OF THE ENERGY EQUATION TO STEADY FLUID-FLOW SITUATIONS

The following are a series of examples which use, clarify, or extend the material in Sec. 3.4.

Reservoirs and Large Storage Tanks

The control volume inlet-outlet conditions for which the energy equation was derived automatically included both potential and kinetic energy at the inlets and outlets. Consider the *very* large reservoir in Fig. 3.13 with water issuing from a small (relative to the reservoir) opening. Here, there are two control volume surfaces, one at the outlet (2) and the "inlet" which is considered to be the entire surface of the

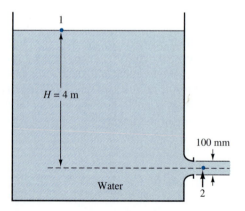

Figure 3.13   Flow through a nozzle from a reservoir.

reservoir. If a datum is drawn through the centerline of the outlet then the piezometer head at the inlet ($p/\gamma + z$) is constant and point 1 can essentially be established anywhere on the reservoir surface. The key assumption is that the volume of material passing through the outlet is extremely small compared to the reservoir volume; so much so that the water surface for all intents and purposes never moves and V_1, the "inlet" velocity, is zero.

Answer the following questions with regard to Fig. 3.13: (*a*) Determine the velocity of efflux from the nozzle in the wall of the reservoir. (*b*) Find the discharge through the nozzle.

Example 3.2

Solution

(*a*) The jet or free jet issues as a cylinder with atmospheric pressure uniformly distributed across the outlet and around its periphery. The energy equation, Eq. (3.4.21) without losses or shaft work, is applied between a point on the water surface and a point downstream from the nozzle.

$$\frac{V_1^2}{2g} + \frac{p_1}{\gamma} + z_1 = \frac{V_2^2}{2g} + \frac{p_2}{\gamma} + z_2$$

With the pressure datum as local atmospheric pressure, $p_1 = p_2 = 0$; with the elevation datum through point 2, $z_2 = 0$ and $z_1 = H$. The velocity on the surface of the reservoir is zero (practically); hence,

$$0 + 0 + H = \frac{V_2^2}{2g} + 0 + 0$$

and

$$V_2 = \sqrt{2gH} = \sqrt{2(9.806)(4)} = 8.86 \text{ m/s}$$

which states that the speed of efflux is equal to the speed of free fall from the surface of the reservoir. This is known as *Torricelli's theorem*.

(*b*) The discharge Q is the product of velocity of efflux and area of stream,

$$Q = A_2 V_2 = \pi(0.05 \text{ m})^2(8.86 \text{ m/s}) = 0.07 \text{ m}^3\text{/s} = 70 \text{ L/s}$$

Area Averaging and the Kinetic Energy Correction Factor

Equation (3.4.14) provides the kinetic energy correction factor so the kinetic energy computed with the average velocity over the flow area is the same as the actual kinetic energy physically present in the true velocity. It is shown later that for laminar flow in a pipe, $\alpha = 2$. For turbulent flow [1]† in a pipe, α varies between about 1.01 and 1.10.

The velocity distribution in turbulent flow in a pipe is given approximately by Prandtl's one-seventh-power law,

Example 3.3

$$\frac{v}{v_{max}} = \left(\frac{y}{r_0}\right)^{1/7}$$

| †Numbered references will be found at the end of this chapter.

with y the distance from the pipe wall and r_0 the pipe radius. Find the kinetic-energy correction factor.

Solution

The average velocity V is expressed by

$$\pi r_0^2 V = 2\pi \int_0^{r_0} rv\,dr$$

in which $r = r_0 - y$. By substituting for r and v,

$$\pi r_0^2 V = 2\pi v_{max} \int_0^{r_0} (r_0 - y)\left(\frac{y}{r_0}\right)^{1/7} dy = \pi r_0^2 v_{max} \frac{98}{120}$$

or

$$V = \frac{98}{120} v_{max} \qquad \frac{v}{V} = \frac{120}{98}\left(\frac{y}{r_0}\right)^{1/7}$$

By substituting into Eq. (3.4.14)

$$\alpha = \frac{1}{\pi r_0^2}\int_0^{r_0} 2\pi r \left(\frac{120}{98}\right)^3 \left(\frac{y}{r_0}\right)^{3/7} dr$$

$$= 2\left(\frac{120}{98}\right)^3 \frac{1}{r_0^2}\int_0^{r_0}(r_0 - y)\left(\frac{y}{r_0}\right)^{3/7} dy = 1.06$$

Losses and Efficiency

When turbines or generators are exchanging energy through the control volume, several measures of system performance exist. In both energy conversion cases shaft work, w_s, represents the energy per unit mass that the blade or impellor puts into the fluid (pump) or extracts from the fluid (turbine). This value is referred to as the pump head, H_p, or the turbine head, H_t and has units of work per weight, that is, [F·L/F] or [L]. The heads, H_p and H_t, are machine-derived performance variables. The total power injected to or extracted from the system is found from $\gamma Q H_p$ and $\gamma Q H_t$, respectively.

From the impellor to the external environment the system efficiency is affected by two conversion steps. With reference to Fig. 3.14 power in the form of electricity,

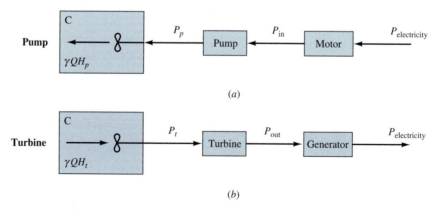

Figure 3.14   Energy flow schematics for pump and turbine configurations.

P_{elec}, is delivered to the pump, P_{in}. The pump in turn delivers power to the impellor, P_p, and the impellor delivers power to the fluid. The power into the pump is greater than the power to the blade, which in turn is greater than the power delivered to the fluid. The reverse is true for the turbine. For the pump $P_{elec} > P_{in} > P_p > \gamma QH_p$, while for the turbine, $\gamma QH_t > P_t > P_{out} > P_{elec}$. System inefficiencies are found by taking the ratios of the power at various parts of the system.

Example 3.4

A hydroelectric plant (Fig. 3.15) has a difference in elevation from head water to tail water of $H = 50$ m and a flow $Q = 5$ m³/s of water through the turbine. The turbine shaft rotates at 180 rpm, and the torque in the shaft is measured to be $T = 1.16 \times 10^5$ N·m. The output of the generator is 2100 kW. Determine (a) the reversible power for the system, (b) the irreversibility, or losses, in the system, and (c) the losses and the efficiency in the turbine and in the generator.

Solution

(a) The potential energy of the water is 50 m·N/N. Hence, for perfect conversion the reversible power is

$$\gamma QH = (9806 \text{ N/m}^3)(5 \text{ m}^3/\text{s})(50 \text{ m·N/N}) = 2,451,500 \text{ N·m/s} = 2451.5 \text{ kW}$$

(b) The irreversibility, or lost power, in the system is the difference between the power into and out of the system, or

$$2451.5 - 2100 = 351.5 \text{ kW}$$

(c) The rate of work by the turbine is the product of the shaft torque and the rotational speed:

$$T\omega = 1.16 \times 10^5 \text{ N·m} \frac{180(2\pi)}{60} \text{ s}^{-1} = 2186.5 \text{ kW}$$

The irreversibility through the turbine is then $2451.5 - 2186.5 = 265.0$ kW, or, when expressed as lost work per unit weight of fluid flowing,

$$(265.0 \text{ kW})\frac{1000 \text{ N·m/s}}{1 \text{ kW}} \frac{1}{9806 \text{ N/m}^3} \frac{1}{5 \text{ m}^3/\text{s}} = 5.4 \text{ m·N/N}$$

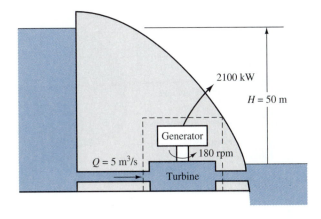

Figure 3.15   Irreversibility in hydroelectric plant.

The generator power loss is $2186.5 - 2100 = 86.5$ kW, or

$$\frac{86.5(1000)}{9806(5)} = 1.76 \text{ m} \cdot \text{N/N}$$

Efficiency of the turbine η_t is

$$\eta_t = 100\frac{50 \text{ m} \cdot \text{N/N} - 5.4 \text{ m} \cdot \text{N/N}}{50 \text{ m} \cdot \text{N/N}} = 89.19 \text{ percent}$$

and efficiency of the generator η_g is

$$\eta_g = 100\frac{50 - 5.4 - 1.76}{50 - 5.4} = 96.05 \text{ percent}$$

Example 3.5

The cooling water plant for a large building is located on a small lake fed by a stream, as shown in Fig. 3.16. The design low-stream flow is 5 cfs, and at this condition the only outflow from the lake is 5 cfs via a gated structure near the discharge channel for the water-cooling system. The temperature of the incoming stream is 80°F. The flow rate of the cooling system is 10 cfs, and the building's heat exchanger raises the cooling water temperature by 10 degrees Fahrenheit. What is the temperature of the cooling water recirculated through the lake, neglecting heat losses to the atmosphere and lake bottom, if these conditions exist for a prolonged period?

Solution

The control volume is shown in Fig. 3.16 with the variables volumetric flow rate Q and temperature T. There is no change in pressure, density, velocity, or elevation from section 1 to 2. Eq. (3.4.18) applied to the control volume is

$$\frac{\delta Q_H}{\delta t} + u_1^{**}\rho Q_1 = u_2^{**}\rho Q_2$$

in which $\delta Q_H/\delta t$ is the time rate of heat addition by the heat exchanger. The intrinsic energy per unit mass at constant pressure and density is a function of temperature only; it is $u_2^{**} - u_1^{**} = c_p(T_2 - T_1)$, in which c_p is the specific heat or heat capacity

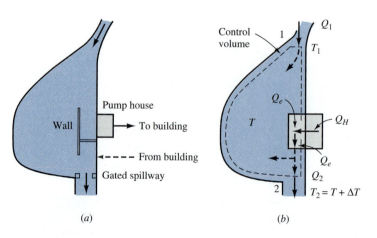

(a)   (b)

Figure 3.16   Water-cooling system.

of water. Hence, the energy equation applied to the control volume is

$$\frac{\delta Q_H}{\delta t} = c_p(T_2 - T_1)\rho Q_1$$

Similarly, the heat added in the heat exchanger is given by

$$\frac{\delta Q_H}{\delta t} = c_p \Delta T \rho Q_e$$

in which $\Delta T = 10$ is the temperature rise and $Q_e = 10$ cfs is the volumetric flow rate through the heat exchanger. Thus,

$$c_p \Delta T \rho Q_e = c_p (T_2 - T_1)\rho Q_1$$

or

$$T_2 = T_1 + \frac{\Delta T Q_e}{Q_1} = 80 + \frac{10(10)}{5} = 100°F$$

Since $T_2 = T + \Delta T$, the lake temperature T is 90°F.

All the terms in the energy equation (Eq. 3.4.21) except the shaft head and loss terms are *available energy*. For real fluids flowing through a system, the available energy decreases in the downstream direction; it is available to do work, as in passing through a water turbine. A plot showing the available energy along a stream tube portrays the *energy grade line* (see Sec. 11.2). A plot of the two terms $z + p/\gamma$ along the control volume portrays the *piezometric head* or *hydraulic grade line*. The energy grade line always slopes downward in real-fluid flow, except at a pump or other source of energy. Reductions in energy grade line are also referred to as *head losses*.

In Fig. 3.17, a pump with a water horsepower (WHP) rating of 10 hp draws water from the reservoir as indicated and delivers water to an outlet 15 feet higher than the reservoir surface for crop irrigation. What is the outlet discharge? Draw and label the hydraulic and energy grade lines. Total system losses from the pump to the outlet are parameterized as $8\ V^2/2g$, but there are no losses from the reservoir inlet to the pump. The delivery pipe diameter is 4.67 in.

Example 3.6

Solution

Assume steady flow and position a datum at the reservoir water surface. The energy equation is applied between the reservoir surface level (1) and outlet (2). At the reservoir surface the pressure is atmospheric, $p_1 = 0$, and the elevation is known ($z_1 = 0$). The volume of the reservoir is considered so big that during the pumping the water volume withdrawn is so small that the water level drop is zero and therefore the velocity of the surface is zero ($V_1 = 0$). At the outlet the only pressure opposing the flow out the nozzle as a free jet is atmospheric; therefore, $p_2 = 0$. The energy equation, Eq. (3.4.21), becomes

$$\frac{V_1^2}{2g} + z_1 + \frac{p_1}{\gamma_1} + H_p = \frac{V_2^2}{2g} + z_2 + \frac{p_2}{\gamma_2} + K\frac{V^2}{2g}$$

in which H_p is the pump head, $H_p = -H_s$, and $K V^2/2g$ represents the losses between 1 and 2.

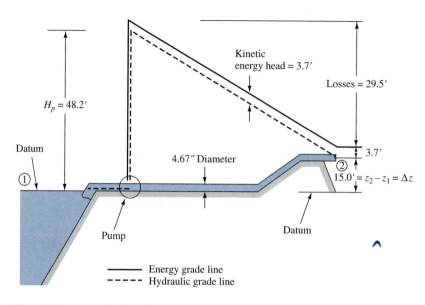

Kinetic
energy head = 3.7′

Losses = 29.5′

$H_p = 48.2'$

Datum

4.67″ Diameter

3.7′

①

②

$15.0' = z_2 - z_1 = \Delta z$

Pump

Datum

——— Energy grade line
‑ ‑ ‑ ‑ Hydraulic grade line

Figure 3.17   Hydraulic and energy grade lines for pumping system.

The flow through the system is constant; therefore, because the pipe diameter is everywhere constant, the system losses can be expressed in terms of the constant outlet velocity, V_2. Hence,

$$0 + 0 + 0 + H_p = \frac{V_2^2}{2g} + 15 + 0 + 0 + 8\frac{V_2^2}{2g}$$

Since

$$WHP = 10 = \frac{\gamma Q H_p \text{ ft·lb/s}}{550 \text{ ft·lb/s/HP}}$$

then

$$H_p = \frac{5500}{\gamma V_2 A} \text{ ft}$$

Therefore,

$$\frac{5500}{\gamma V_2 A} = 9\frac{V_2^2}{2g} + 15$$

Solving for the real root of the cubic equation gives

$$V_2 = 15.4 \text{ ft/s}$$

and

$$Q = V_2 A = 1.83 \text{ ft}^3/\text{s}$$

In units of energy/weight or [F·L/F] or length dimensions the terms in the energy equation can be mapped in units of equivalent head

$$\text{head loss} = 8\frac{V_2^2}{2g} = 29.5 \text{ ft}$$

$$\text{kinetic energy head} = \frac{V_2^2}{2g} = 3.7 \text{ ft}$$

$$\text{pump head} = H_p = 48.2 \text{ ft}$$

Therefore, the hydraulic and energy grade lines can be mapped.

The siphon of Fig. 3.18 is filled with water and discharging at 150 L/s. Find the losses from point 1 to point 3 in terms of velocity head $V^2/2g$. Find the pressure at point 2 if two-thirds of the losses occur between points 1 and 2.

Example 3.7

Solution

The energy equation, Eq. (3.4.21), is first applied to the control volume consisting of all the water in the system upstream from point 3, with elevation datum at point 3 and gage pressure zero for pressure datum:

$$\frac{V_1^2}{2g} + \frac{p_1}{\gamma} + z_1 = \frac{V_3^2}{2g} + \frac{p_3}{\gamma} + z_3 + \text{losses}_{1-3}$$

or

$$0 + 0 + 1.5 = \frac{V_3^2}{2g} + 0 + 0 + K\frac{V_3^2}{2g}$$

in which the losses from 1 to 3 are expressed as $KV_3^2/2g$. From the discharge

$$V_3 = \frac{Q}{A} = 150 \text{ L/s} \frac{1}{\pi(0.1 \text{ m})^2 1000 \text{ L/m}^3} = 4.77 \text{ m/s}$$

and $V_3^2/2g = 1.16$ m. Hence, $K = 0.29$, and the losses are $0.29 \, V_3^2/2g = 0.34$ m·N/N.

The energy equation applied to the control volume between points 1 and 2, with losses $\frac{2}{3}KV_3^2/2g = 0.23$ m, is

$$0 + 0 + 0 = 1.16 + \frac{p_2}{\gamma} + 2 + 0.23$$

The pressure head at 2 is -3.39-m H_2O, or $p_2 = -33.2$ kPa.

The device shown in Fig. 3.19 (called a *pitot tube*) is used to determine the velocity of liquid at point 1. It is a tube with its lower end directed upstream and its other leg vertical and open to the atmosphere. The impact of liquid against opening 2 forces liquid to rise in the vertical leg to the height Δz above the free surface. Determine the velocity at 1.

Example 3.8

Solution

Point 2 is a stagnation point, where the velocity of the flow is reduced to zero. This creates an impact pressure, called the dynamic pressure, which forces the fluid into the vertical leg. Writing the energy equation between points 1 and 2, neglecting losses, which are very small, leads to

$$\frac{V_1^2}{2g} + \frac{p_1}{\gamma} + 0 = 0 + \frac{p_2}{\gamma} + 0$$

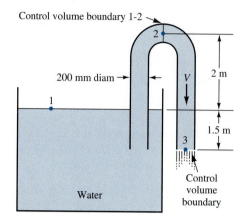

Figure 3.18   Siphon.

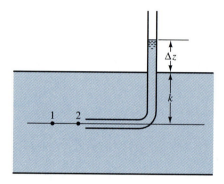

Figure 3.19   Pitot tube.

p_1/γ is given by the height of fluid above point 1 and equals k ft of flowing fluid. p_2/γ is given by the manometer as $k+\Delta z$, neglecting capillary rise. After substituting these values into the equation,

$$\frac{V_1^2}{2g} = \Delta z \quad \text{and} \quad V_1 = \sqrt{2g\,\Delta z}$$

This is the pitot tube in a simple form.

EXERCISES

3.5.1 The kinetic-energy correction factor (*a*) applies to the continuity equation; (*b*) has the units of velocity head; (*c*) is expressed by $\frac{1}{A}\int_A \left(\frac{v}{V}\right) dA$; (*d*) is expressed by $\frac{1}{A}\int_A \left(\frac{v}{V}\right)^2 dA$; (*e*) is expressed by $\frac{1}{A}\int_A \left(\frac{v}{V}\right)^3 dA$.

3.5.2 The kinetic-energy correction factor for the velocity distribution given by Fig. 1.1 is (*a*) 0; (*b*) 1; (*c*) $\frac{4}{3}$; (*d*) 2; (*e*) none of these answers.

3.5.3 A glass tube with a 90° bend is open at both ends. It is inserted into a flowing stream of oil, sp gr 0.90, so that one opening is directed upstream and the other is directed upward. Oil inside the tube is 50 mm higher than the surface of flowing oil. The velocity measured by the tube is, in meters per second, (a) 0.89; (b) 0.99; (c) 1.10; (d) 1.40; (e) none of these answers.

3.5.4 In Fig. 10.9a the gage difference R' for $v_1 = 5$ ft/s, $S = 0.08$, and $S_0 = 1.2$ is, in feet, (a) 0.39; (b) 0.62; (c) 0.78; (d) 1.17; (e) none of these answers.

3.5.5 The theoretical velocity of oil, sp gr 0.75, flowing from an orifice in a reservoir under a head of 4 m is, in meters per second, (a) 6.7; (b) 8.86; (c) 11.8; (d) not determinable from data given; (e) none of these answers.

3.5.6 In which of the following cases is it possible for flow to occur from low pressure to high pressure? (a) Flow through a converging section; (b) adiabatic flow in a horizontal pipe; (c) flow of a liquid upward in a vertical pipe; (d) flow of air downward in a pipe; (e) impossible in a constant-cross-section conduit.

3.5.7 If all losses are neglected, the pressure at the summit of a siphon (a) is a minimum for the siphon; (b) depends upon height of summit above upstream reservoir only; (c) is independent of the length of the downstream leg; (d) is independent of the discharge through the siphon; (e) is independent of the liquid density.

3.6   THE CONTROL VOLUME LINEAR-MOMENTUM EQUATION

Basic Equation

Newton's second law for a system is used as a basis for determining the control volume form of the linear momentum equation. From Eq. (3.2.4) let N be the linear momentum of the system $m\mathbf{v}$ and let η be the linear momentum per unit mass $\rho\mathbf{v}/\rho = \mathbf{v}$. Then Eq. (3.2.4) becomes

$$\mathbf{F} = \frac{d(m\mathbf{v})}{dt} = \frac{\partial}{\partial t}\int_{cv} \rho\mathbf{v}\, d\mathcal{V} + \int_{cs} \mathbf{v}\rho\mathbf{v}\cdot d\mathbf{A} \tag{3.6.1}$$

In words, the vector sum of the real applied external forces acting on the control volume equals the time rate of increase of linear momentum in the control volume plus the net rate at which momentum is leaving through the control surface.

Steady State Equation Analysis

To analyze this equation the approach used in the previous two equations will be used. Consider a section of pipe as in Fig. 3.20 with inlet at 1 and outlet at 2.

1. Define the control volume as before with control surfaces placed where the area is perpendicular to the streamlines and the flow field is well behaved with streamlines parallel to the wall of the flow section.

2. The flow is assumed to be steady.

3. The system of equivalent or resultant forces and flow parameters is established on the control volume and consists of both resultant forces and equivalent momentum exchanges at the inlet and the outlet.

4. The vector sum of the real applied external forces is defined to consist of the following components

$$\mathbf{F} = \mathbf{W} + \mathbf{F}_{p_1} + \mathbf{F}_{p_2} + \mathbf{F}_{\tau_0} + \mathbf{F}_w \tag{3.6.2}$$

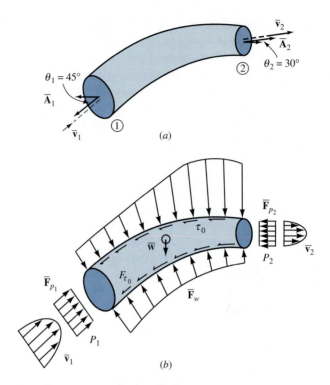

Figure 3.20   Control volume for flow through a pipe.

Each term is defined as follows:

W is the weight force. The control volume fluid has a weight acting in the direction of gravity, and the magnitude equals the volume of the control volume times the specific weight.

$\mathbf{F}_{p_1}$ and $\mathbf{F}_{p_2}$ are end pressure forces. The fluid pressure at the inlet and outlet creates a pressure force on each face. The total vector force equals

$$\mathbf{F}_p = \int_{cs} p \, d\mathbf{A} = \int_{cs} p \, \mathbf{n} \cdot d\mathbf{A} \qquad \text{(3.6.3)}$$

where **n** is the unit normal of the surface area which is always directed positive away from the surface area. As in Chap. 2 the pressure force can be calculated using the methods for finding individual vertical and horizontal forces, or the full vector $\mathbf{F}_p$ can be calculated. In most cases the area average pressure is p or the pressure is constant, then $\mathbf{F}_p = p\mathbf{A}$ and the vector $\mathbf{F}_p$ is perpendicular to the inlet face and directed towards the control volume.

$\mathbf{F}_{\tau_0}$ and $\mathbf{F}_w$ are the wall shear and pressure forces, respectively. The wall exerts both shear and normal stresses on the fluid. The wall shear, τ_0, exerts friction on the fluid and acts in the plane of the surface of the control volume to retard the flow. The wall shear is a function of the type of boundary material and its roughness, the fluid density and speed, and the flow geometry. In general it varies from point to point in the flow. Pipe and certain boundary layer flows are the only flows where an exact calculation of τ_0 and $\mathbf{F}_{\tau_0}$ can occur (see Chaps. 6 and 7). The normal stress, $\mathbf{F}_w$, at the wall or control surface is primarily responsible for maintaining the geometry of the flow field. For example, the flow in Fig. 3.20 bends at outlet 2 because the normal stress is larger on the top resulting in a deflection of the flow. Also, if a normal stress was not provided "underneath" the flow, it would simply "fall" under the effect of gravity.

The normal stress force, along with $\mathbf{F}_{\tau_0}$, is exceedingly difficult to separate; therefore, they are lumped together at this point into a resultant or reaction force vector, $\mathbf{F}$, which will act at the center of gravity of the control volume. Typically the direction and intensity of $\mathbf{F}$ will be determined in the solution.

5. The *momentum exchange,* $\mathbf{M}_1$ and $\mathbf{M}_2$, at the inlet and outlet, respectively, must now be analyzed. For steady flow the right-hand portion of Eq. (3.6.1) is written at the inlet and outlet as

$$\mathbf{M}_1 + \mathbf{M}_2 = \int_{cs_1} \mathbf{v}_1(\rho_1 \mathbf{v}_1 \cdot d\mathbf{A}) + \int_{cs_2} \mathbf{v}_2(\rho_2 \mathbf{v}_2 \cdot d\mathbf{A}) \qquad \text{(3.6.4)}$$

If the velocity at the control surface is perpendicular to the area and the velocity is uniform across the respective area, then the simplest form of Eq. (3.6.4) is

$$\mathbf{M}_1 + \mathbf{M}_2 = -(\rho V_1 A_1)\mathbf{V}_1 + (\rho V_2 A_2)\mathbf{V}_2 \qquad \text{(3.6.5)}$$

Here the minus sign indicates the momentum is entering the control volume. Since the terms have been analyzed only with regard to the local coordinate system of the control surface, the velocity vector must be maintained. Each individual term in Eq. (3.6.5) is defined as the momentum exchange vector

$$\mathbf{M}_1 = -\rho_1 Q \mathbf{V}_1 \qquad \text{(3.6.6)}$$
$$\mathbf{M}_2 = \rho_2 Q \mathbf{V}_2$$

Therefore, at each area $\mathbf{M}$ is perpendicular to the surface, and directed *away* from the control surface regardless of whether it is an inlet or outlet.

6. The final form of the steady control volume form of Newton's second law is

$$\mathbf{W} + \mathbf{F}_{p_1} + \mathbf{F}_{p_2} + \mathbf{F} = \mathbf{M}_1 + \mathbf{M}_2 \qquad \text{(3.6.7)}$$

The Momentum Correction Factor

In the energy equation the product of the area averages is not equal to the average of the products. When the velocity varies over a plane cross section of the control surface, a momentum correction factor β must be introduced before the average velocity can be used.

$$\int_A \rho v^2 \, d\mathbf{A} = \beta \rho V^2 A \qquad \text{(3.6.8)}$$

in which β is dimensionless. Solving for β yields

$$\beta = \frac{1}{A} \int_A \left(\frac{v}{V}\right)^2 dA \qquad \text{(3.6.9)}$$

which is analogous to α, the kinetic-energy correction factor, Eq. (3.4.14). For laminar flow in a straight round tube β is shown to equal $\frac{4}{3}$ in Sec. 6.3. It equals 1 for uniform flow and cannot have a value less than 1. Equation (3.6.7) now becomes

$$\mathbf{W} + \mathbf{F}_{p_1} + \mathbf{F}_{p_2} + \mathbf{F} = \beta_1 \mathbf{M}_1 + \beta_2 \mathbf{M}_2 \qquad \text{(3.6.10)}$$

Multiple Control Surfaces

With just one inlet and outlet the flow through the system is constant and $V_1A_1 = Q_1 = V_2A_2 = Q_2 = Q$. Equation (3.6.10), with $\beta = 1$, can be written as

$$\mathbf{W} + \mathbf{F}_{p_1} + \mathbf{F}_{p_2} + \mathbf{F} = \rho Q(\mathbf{V}_2 - \mathbf{V}_1) \tag{3.6.11}$$

If an additional inlet exists say at station 3, as in Fig. 3.9, then Eq. (3.6.11) requires an additional end pressure force $\mathbf{F}_{p_3}$ (pointed toward the control volume and perpendicular to control surface 3) and an additional momentum exchange vector. The final vector equation therefore becomes

$$\mathbf{W} + \mathbf{F}_{p_1} + \mathbf{F}_{p_2} + \mathbf{F}_{p_3} + \mathbf{F} = \mathbf{M}_1 + \mathbf{M}_2 + \mathbf{M}_3 \tag{3.6.12}$$

or

$$\mathbf{W} + \mathbf{F}_{p_1} + \mathbf{F}_{p_2} + \mathbf{F}_{p_3} + \mathbf{F} = \rho_2 Q_2 \mathbf{V}_2 - (\rho_1 Q_1 \mathbf{V}_1 + \rho_3 Q_3 \mathbf{V}_3) \tag{3.6.13}$$

A similar extension occurs for an additional outlet at station 4

$$\mathbf{W} + \mathbf{F}_{p_1} + \mathbf{F}_{p_2} + \mathbf{F}_{p_3} + \mathbf{F}_{p_4} + \mathbf{F} = \mathbf{M}_1 + \mathbf{M}_2 + \mathbf{M}_3 + \mathbf{M}_4 \tag{3.6.14}$$

or

$$\mathbf{W} + \mathbf{F}_{p_1} + \mathbf{F}_{p_2} + \mathbf{F}_{p_3} + \mathbf{F}_{p_4} + \mathbf{F} = (\rho_1 Q_1 \mathbf{V}_1 + \rho_4 Q_4 \mathbf{V}_4) - (\rho_1 Q_1 \mathbf{V}_1 + \rho_3 Q_3 \mathbf{V}_3)$$

$$\tag{3.6.15}$$

3.7   APPLICATIONS OF THE LINEAR MOMENTUM EQUATION

The following are a series of examples that illustrate several aspects of the use of the linear momentum equation.

A Basic Solution

When Eq. (3.6.7) is written in D'Alembert form where all the terms are collected on the same side of the equation, then

$$\mathbf{W} + \mathbf{F}_{p_1} + \mathbf{F}_{p_2} + \mathbf{F} - \mathbf{M}_1 - \mathbf{M}_2 = 0$$

This has the effect of essentially reversing the direction of the arrows of the momentum exchange vectors such that they both point towards the center of the control volume, just as the pressure force vectors do.

Example 3.9

Suppose in Figs. 3.20 and 3.21 the following pipe conditions apply. The inflow and outflow radii are 25 and 15 cm, respectively; the inflow and outflow angles with respect to the horizontal (θ_1 and θ_2) are 45° and 30°, respectively; Q is 50 L/s; the area average inlet and outlet pressures are 8.5 kPa and 5.83 kPa; and the total fluid weight in the pipe is 2.0 N. Find the horizontal and vertical force required to hold the pipe in place (i.e., the resultant force vector $\mathbf{F}$).

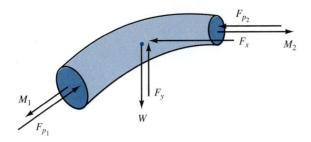

Figure 3.21   Force and momentum vector definitions for the control volume in Fig. 3.20.

Solution

The momentum exchange and force vectors are drawn in Fig. 3.21. The directions of the vectors ($\mathbf{F}_p$ and $\mathbf{M}$) are always known as per the original derivation. The direction of the reaction vectors F_x and F_y are not known, therefore directions for them will be assumed and the calculation will proceed. If the assumption was incorrect the values calculated for F_x and F_y will be negative, suggesting that the original directions need to be reversed. The magnitudes of F_x and F_y remain the same in either case.

The values of the magnitudes of $\mathbf{F}_{p_1}$ and $\mathbf{F}_{p_2}$ are calculated as

$$F_{p_1} = p_1 A_1 = (8500)\pi(.25)^2 = 1669.0 \text{ N}$$
$$F_{p_2} = p_2 A_2 = (5830)\pi(.15)^2 = 412.1 \text{ N}$$

The magnitude of the values of $\mathbf{M}$ are calculated as follows

$$V_1 = Q/A_1 = 0.255 \text{ m/s}$$
$$M_1 = \rho Q V_1 = \rho A_1 V_1 V_1 = (1000)(0.196)(0.255)^2 = 12.75 \text{ N}$$
$$V_2 = Q/A_2 = 0.707 \text{ m/s}$$
$$M_2 = \rho Q V_2 = \rho A_2 V_2 V_2 = (1000)(0.071)(0.707)^2 = 35.48 \text{ N}$$

The components of $\mathbf{F}$ are calculated as

$$-F_x - F_{p_2} \cos 30 + F_{p_1} \cos 45 = -M_1 \cos 45 + M_2 \cos 30$$
$$-W + F_y - F_{p_2} \sin 30 + F_{p_1} \sin 45 = -M_1 \sin 45 + M_2 \sin 30$$

Substitution of the values found above and solving for F_w and F_y gives

$$-F_x - 412.1(0.866) + 1669.0(0.707) = -12.75(0.707) + 35.48(0.866)$$
$$-2.0 + F_y - 412.1(0.500) + 1669.0(0.707) = -12.75(0.707) + 35.48(0.500)$$

or

$$F_x = -801.4 \text{ N}$$
$$F_y = -963.2 \text{ N}$$

The negative signs indicate that the assumed directions for F_x and F_y are incorrect. The actual forces act to the right and down with magnitudes of 801.4 and 963.2 N, respectively.

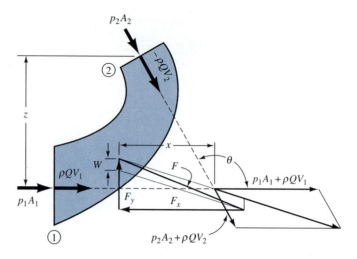

Figure 3.22   Forces on a reducing elbow, including the vector solution.

Example 3.10

The reducing bend of Fig. 3.22 is in a vertical plane. Water is flowing with $D_1 = 6$ ft, $D_2 = 4$ ft, $Q = 300$ cfs, $W = 18{,}000$ lb, $z = 10$ ft, $\theta = 120°$, $p_1 = 40$ psi, $x = 6$ ft, and losses through the bend are $0.5\ V_2^2/2g$ ft·lb/lb. Determine F_x and F_y, and the line of action of the resultant force. $\beta_1 = \beta_2 = 1$.

Solution

The inside surface of the reducing bend comprises the control-volume surface for the portion of the surface with no flow across it. The normal sections 1 and 2 complete the control surface.

$$V_1 = \frac{Q}{A_1} = \frac{300}{\pi(6^2)/4} = 10.61 \text{ ft/s} \qquad V_2 = \frac{Q}{A_2} = \frac{300}{\pi(4^2)/4} = 23.87 \text{ ft/s}$$

By application of the energy equation, Eq. (3.4.21), with $H_s = 0$,

$$\frac{p_1}{\gamma} + \frac{V_1^2}{2g} + z_1 = \frac{p_2}{\gamma} + \frac{V_2^2}{2g} + z_2 + \text{losses}_{1-2}$$

$$\frac{40(144)}{62.4} + \frac{10.61^2}{64.4} + 0 = \frac{p_2}{62.4} + \frac{23.87^2}{64.4} + 10 + 0.5\,\frac{23.87^2}{64.4}$$

in which $p_2 = 4420$ lb/ft^2 = 30.7 psi.

To determine F_x, Eq. (3.6.7) yields

$$p_1 A_1 - p_2 A_2 \cos\theta - F_x = \rho Q(V_2 \cos\theta - V_1)$$

$$40(144)(\pi 6^2)/4 - 4420(\pi 4^2)/4 \cos 120° - F_x$$

$$= 1.935(300)(23.87 \cos 120° - 10.61)$$

since $\cos 120° = -0.5$, then

$$162{,}900 + 27{,}750 - F_x = 580.5(-11.94 - 10.61)$$

$$F_x = 203{,}740 \text{ lb}$$

For the y direction

$$\Sigma F_y = \rho Q(V_{y_2} - V_{y_1})$$

$$F_y - W - p_2 A_2 \sin\theta = \rho Q V_2 \sin\theta$$

$$F_y = 18{,}000 - 4420\pi(4^2)/4 \sin 120°$$

$$= 1.935(300)(23.87)\sin 120°$$

$$F_y = 78{,}100 \text{ lb}$$

To find the line of action of the resultant force, using the momentum flux vectors (Fig. 3.22), $\rho Q V_1 = 6160$ lb, $\rho Q V_2 = 13{,}860$ lb, $p_1 A_1 = 162{,}900$ lb, and $p_2 A_2 = 55{,}560$ lb. Combining these vectors and the weight W in Fig. 3.22 yields the final force of 218,000 lb, which must be opposed by F_x and F_y.

As demonstrated in Example 3.10 a change in direction of a pipeline causes forces to be exerted on the line unless the bend or elbow is anchored in place. These forces are due to both static pressure in the line and dynamic reactions in the turning fluid stream. Expansion joints are placed in large pipelines to avoid stress in the pipe in an axial direction, whether caused by fluid or by temperature change. These expansion joints permit relatively free movement of the line in an axial direction, and hence the static and dynamic forces must be provided for at the bends.

Example 3.11

A jet of water 80 mm in diameter with a velocity of 40 m/s is discharged in a horizontal direction from a nozzle mounted on a boat. What force is required to hold the boat stationary?

Solution

When the control volume is selected as shown in Fig. 3.23, the net efflux of momentum is [Eq. (3.6.6)]

$$\rho Q V_{out} = (1000 \text{ kg/m}^3)\,\frac{\pi}{4}\,(0.08 \text{ m})^2(40 \text{ m/s})^2 = 8.04 \text{ kN}$$

The force exerted against the boat is 8.04 kN in the x direction.

Example 3.12

Find the force exerted by the nozzle on the pipe of Fig. 3.24a. Neglect losses. The fluid is oil, with sp gr $= 0.85$ and $p_1 = 100$ psi.

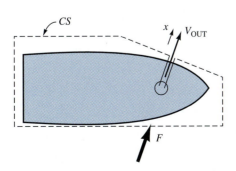

Figure 3.23 Nozzle mounted on a boat.

Solution

To determine the discharge, the energy equation is written for the stream from section 1 to the downstream end of the nozzle, where the pressure is zero.

$$z_1 + \frac{V_1^2}{2g} + \frac{(100 \text{ lb/in}^2)(144 \text{ in}^2/\text{ft}^2)}{0.85(62.4 \text{ lb/ft}^3)} = z_2 + \frac{V_2^2}{2g} + 0$$

Since $z_1 = z_2$ and $V_2 = (D_1/D_2)^2 V_1 = 9V_1$, after substituting,

$$\frac{V_1^2}{2g}(1 - 81) + \frac{(100 \text{ lb/in}^2)(144 \text{ in}^2/\text{ft}^2)}{0.85(62.4 \text{ lb/ft}^3)} = 0$$

and

$$V_1 = 14.78 \text{ ft/s}$$

$$V_2 = 133 \text{ ft/s}$$

$$Q = 14.78\frac{\pi}{4}\left(\frac{1}{4}\right)^2 = 0.725 \text{ ft}^3/\text{s}.$$

Let F_x (Fig. 3.24b) be the force exerted on the liquid control volume by the nozzle; then, with Eq. (3.6.7),

$$(100 \text{ lb/in.}^2)\,\frac{\pi}{4}(3 \text{ in.})^2 - F_x =$$

$$(1.935 \text{ slug/ft}^3)(0.85)(0.725 \text{ ft}^3/\text{s})(133 \text{ ft/s} - 14.78 \text{ ft/s})$$

or $F_x = 565$ lb. The oil exerts a force on the nozzle of 565 lb to the right, and a tension force of 565 lb is exerted by the nozzle on the pipe.

In many situations an unsteady-flow problem can be converted into a steady-flow problem by superposing a constant velocity upon the system and its surroundings, that is, by changing the reference velocity. The dynamics of a system and its surroundings are unchanged by the superposition of a constant velocity; hence, pressures and forces are unchanged. In the next flow situation studied, advantage is taken of this principle.

The Momentum Theory for Propellers

The action of a propeller is to change the momentum of the fluid within which it is submerged and thus to develop a thrust that is used for propulsion. Propellers

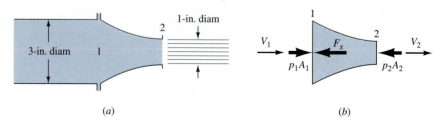

(a)   (b)

Figure 3.24   Nozzle at the end of a pipe.

cannot be designed according to the momentum theory, although some of the relations governing them are made evident by its application. A propeller, with its slipstream and velocity distributions at two sections a fixed distance from it, is shown in Fig. 3.25. The propeller may be either (1) stationary in a flow as indicated or (2) moving to the left with a velocity V_1 through a stationary fluid, since the relative picture is the same. The fluid is assumed to be frictionless and incompressible.

The flow is undisturbed at section 1 upstream from the propeller and is accelerated as it approaches the propeller, owing to the reduced pressure on its upstream side. In passing through the propeller, the fluid has its pressure increased, which further accelerates the flow and reduces the cross section at 4. The velocity V does not change across the propeller, from 2 to 3. The pressure at 1 and 4 is that of the undisturbed fluid, which is also the pressure along the slipstream boundary.

When the momentum Eq. (3.6.7) is applied to the control volume within sections 1 and 4 and the slipstream boundary of Fig. 3.25, the force F exerted by the propeller is the reaction force and is the only external force acting in the axial direction, since the pressure is everywhere the same on the control surface. Therefore,

$$F = \rho Q(V_4 - V_1) = (p_3 - p_2)A \tag{3.7.1}$$

in which A is the area swept over by the propeller blades. The propeller thrust must be equal and opposite to the force on the fluid. After substituting $Q = AV$ and simplifying,

$$\rho V(V_4 - V_1) = p_3 - p_2 \tag{3.7.2}$$

When the energy equation is written for the stream between sections 1 and 2 and between sections 3 and 4,

$$p_1 + \frac{1}{2}\rho V_1^2 = p_2 + \frac{1}{2}\rho V^2 \qquad p_3 + \frac{1}{2}\rho V^2 = p_4 + \frac{1}{2}\rho V_4^2$$

since $z_1 = z_2 = z_3 = z_4$. Solving for $p_3 - p_2$ with $p_1 = p_4$ gives

$$p_3 - p_2 = \frac{1}{2}\rho(V_4^2 - V_1^2) \tag{3.7.3}$$

Eliminating $p_3 - p_2$ in Eqs. (3.7.2) and (3.7.3) gives

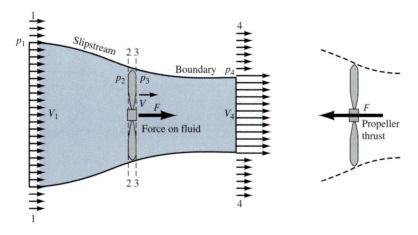

Figure 3.25   Propeller in a fluid stream.

$$V = \frac{V_1 + V_4}{2} \tag{3.7.4}$$

which shows that the velocity through the propeller area is the average of the velocities upstream and downstream from it.

The useful work per unit time done by a propeller moving through still fluid (power transferred) is the product of propeller thrust and velocity, that is,

$$\text{Power} = FV_1 = \rho Q(V_4 - V_1)V_1 \tag{3.7.5}$$

The power input is that required to increase the velocity of fluid from V_1 to V_4. Since Q is the volumetric flow rate,

$$\text{Power input} = \frac{\rho Q(V_4^2 - V_1^2)}{2} \tag{3.7.6}$$

Power input may also be expressed as the useful work (power output) plus the kinetic energy per unit time remaining in the slipstream (power loss)

$$\text{Power input} = \rho Q(V_4 - V_1)V_1 + \frac{\rho Q(V_4 - V_1)^2}{2} \tag{3.7.7}$$

The theoretical mechanical efficiency e_t is given by the ratio of Eqs. (3.7.5) and (3.7.6) or (3.7.7)

$$e_t = \frac{\text{output}}{\text{output} + \text{loss}} = \frac{2V_1}{V_4 + V_1} = \frac{V_1}{V} \tag{3.7.8}$$

If $\Delta V = V_4 - V_1$ is the increase in slipstream velocity, substituting into Eq. (3.7.8) produces

$$e_t = \frac{1}{1 + \Delta V/2V_1} \tag{3.7.9}$$

which shows that maximum efficiency is obtained with a propeller that increases the velocity of slipstream as little as possible, or for which $\Delta V/V_1$ is a minimum.

Owing to compressibility effects, the efficiency of an airplane propeller drops rapidly with speeds above 400 mi/h. Airplane propellers under optimum conditions have actual efficiencies close to the theoretical efficiencies, in the neighborhood of 85 percent. Ship propeller efficiencies are less, around 60 percent, owing to restrictions in diameter.

The windmill can be analyzed by application of the momentum relations. The jet has its speed reduced, and the diameter of the slipstream is increased.

Example 3.13

An airplane traveling at 400 km/h through still air, $\gamma = 12$ N/m³, discharges 1000 m³/s through its two 2.25-m-diameter propellers. Determine (a) the theoretical efficiency, (b) the thrust, (c) the pressure difference across the propellers, and (d) the theoretical power required.

Solution

(a) $V_1 = 400\dfrac{\text{km}}{\text{h}}\ \dfrac{1000\ \text{m}}{1\ \text{km}}\ \dfrac{1\ \text{h}}{3600\ \text{s}} = 111.1$ m/s

$V = \dfrac{500\ \text{m}^3/\text{s}}{(\pi/4)(2.25)^2} = 125.8$ m/s

From Eq. (3.7.8)

$$e_t = \frac{V_1}{V} = \frac{111.1}{125.8} = 88.3 \text{ percent}$$

(b) From Eq. (3.7.4)

$$V_4 = 2V - V_1 = 2(125.8) - 111.1 = 140.5 \text{ m/s}$$

The thrust from the propeller is, from Eq. (3.7.1),

$$F = \frac{12 \text{ N·m}}{9.806 \text{ m/s}^2}(1000 \text{ m}^3/\text{s})(140.5 \text{ m/s} - 111.1 \text{ m/s}) = 36.0 \text{ kN}$$

(c) The pressure difference, from Eq. (3.7.2), is

$$p_3 - p_2 = \frac{12 \text{ N·m}}{9.806 \text{ m/s}^2}(125.8 \text{ m/s})(140.5 \text{ m/s} - 111.1 \text{ m/s}) = 4.52 \text{ kPa}$$

(d) The theoretical power is

$$\frac{FV_1}{e_t} = (36,000 \text{ N})\frac{111.1 \text{ m/s}}{0.883}\frac{1 \text{ kW}}{1000 \text{ N·m/s}} = 4.53 \text{ MW}$$

Fixed and Moving Vanes

The theory of turbomachines, for example, pumps and turbines, is based on the relations between jets and vanes. The mechanics of transfer of work and energy from fluid jets to moving vanes is studied as an application of the momentum principles. When a free jet impinges onto a smooth vane that is curved, as in Fig. 3.26, the jet is deflected, its momentum is changed, and a force is exerted on the vane. The jet is assumed to flow onto the vane in a tangential direction, without shock, and furthermore the frictional resistance between jet and vane is neglected. The velocity is assumed to be uniform throughout the jet upstream and downstream from the vane. Since the jet is open to the air, it has the same pressure at each end of the vane, that is, the control volume. When the small change in elevation between ends, if any, is neglected, application of the energy equation shows that the magnitude of the velocity is unchanged for *fixed* vanes.

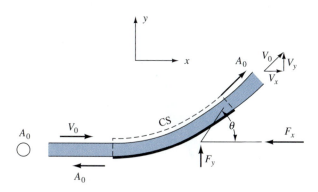

Figure 3.26   Free jet impinging on smooth, fixed vane.

Example 3.14

Find the reaction force exerted on a fixed vane when a jet discharging 60 L/s of water at 50 m/s is deflected through 135°.

Solution

By referring to Fig. 3.26 and applying Eq. (3.6.7) in the x and y directions, it is found that

$$-F_x = \rho V_0 \cos\theta\, V_0 A_0 + \rho V_0(-V_0 A_0) \qquad F_y = \rho V_0 \sin\theta\, V_0 A_0$$

Hence,

$$F_x = -(1000 \text{ kg/m}^3)(0.06 \text{ m}^3/\text{s})(50\cos 135° - 50 \text{ m/s}) = 5.121 \text{ kN}$$
$$F_y = -(1000 \text{ kg/m}^3)(0.06 \text{ m}^3/\text{s})(50\sin 135°) = 2.121 \text{ kN}$$

The force components on the fixed vane are then equal and opposite to F_x and F_y.

Example 3.15

Fluid issues from a long slot and strikes against a smooth inclined flat plate (Fig. 3.27). Determine the division of flow and the force exerted on the plate, neglecting losses due to impact.

Solution

As there are no changes in elevation or pressure before and after impact, the magnitude of the velocity leaving is the same as the initial speed of the jet. The division of flow Q_1 and Q_2 can be computed by applying the momentum equation in the s direction, parallel to the plate. No force is exerted on the fluid by the plate in this direction; hence, the final momentum component must equal the initial momentum component in the s direction. The steady-state momentum equation for the s direction, from Eq. (3.6.1), yields

$$\Sigma F_s = \int_{cs} \rho v_s \mathbf{V} \cdot d\mathbf{A} = 0 = \rho V_0 V_0 A_1 + \rho V_0 \cos\theta(-V_0 A_0) + \rho(-V_0)V_0 A_2$$

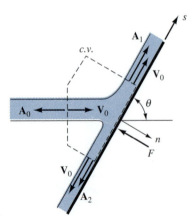

Figure 3.27   Two-dimensional jet impinging on an inclined fixed plane surface.

With the substitution $Q_1 = V_0A_1$, $Q_2 = V_0A_2$, and $Q_0 = V_0A_0$ it reduces to

$$Q_1 - Q_2 = Q_0 \cos\theta$$

and with the continuity equation,

$$Q_1 + Q_2 = Q_0$$

The two equations can be solved for Q_1 and Q_2:

$$Q_1 = \frac{Q_0}{2}(1 + \cos\theta) \qquad Q_2 = \frac{Q_0}{2}(1 - \cos\theta)$$

The force F exerted on the plate must be normal to it. For the momentum equation normal to the plate (Fig. 3.27)

$$\Sigma F_n = \int_{cs} \rho v_n \mathbf{V} \cdot d\mathbf{A} = -F = \rho V_0 \sin\theta(-V_0A_0)$$

$$F = \rho Q_0 V_0 \sin\theta$$

Moving Vanes

Turbomachinery utilizes the forces resulting from motion over moving vanes. No work can be done on or by a fluid that flows over a fixed vane. When vanes can be displaced, work can be done either on the vane or on the fluid. In Fig. 3.28a, a moving vane is shown with fluid flowing onto it tangentially. Forces exerted on the fluid by the vane are indicated by F_x and F_y. To analyze the flow, the problem is reduced to steady state by superposition of vane velocity u to the left (Fig. 3.28b) on

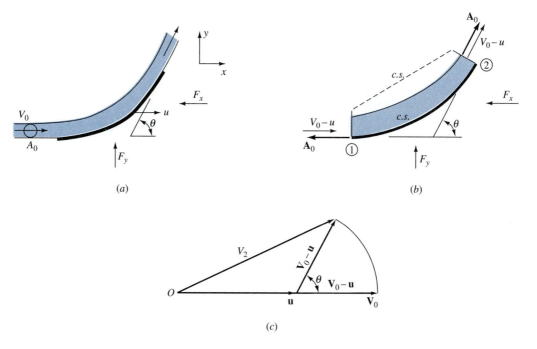

Figure 3.28 (a) Moving vane; (b) vane flow viewed as steady state problem by superposition of velocity u to the left; (c) polar vector diagram.

both vane and fluid. The control volume then encloses the fluid in contact with the vane, with its control surface normal to the flow at sections 1 and 2. Fig. 3.28c shows the *polar vector diagram* for flow through the vane. The absolute-velocity vectors originate at the origin O, and the relative-velocity vector $\mathbf{V}_0 - \mathbf{u}$ is turned through the angle θ of the vane as shown. $\mathbf{V}_2$ is the final absolute velocity leaving the vane. The relative velocity $v_r = V_0 - u$ is unchanged in magnitude as it traverses the vane. The mass per unit time is given by $\rho A_0 v_r$ and is not the mass rate being discharged from the nozzle. If a *series of vanes* is employed, as on the periphery of a wheel, so arranged that one or another of the jets intercepts all flow from the nozzle and the velocity is substantially u, then the mass per second is the total mass per second being discharged. Application of Eq. (3.6.1) to the control volume of Fig. 3.28b yields

$$\Sigma F_x = \int_{cs} \rho v_x \mathbf{V} \cdot d\mathbf{A} = -F_x = \rho(V_0 - u)\cos\theta \; ((V_0 - u)A_0]$$
$$+ \; \rho(V_0 - u)[(-V_0 - u)A_0]$$

or

$$F_x = \rho(V_0 - u)^2 A_0 (1 - \cos\theta)$$
$$\Sigma F_y = \int_{cs} \rho v_y \mathbf{V} \cdot d\mathbf{A} = F_y = \rho(V_0 - u)\sin\theta \; [(V_0 - u)A_0]$$

or

$$F_y = \rho(V_0 - u)^2 A_0 \sin\theta$$

These relations are for the single vane. For a series of vanes they become

$$F_x = \rho Q_0(V_0 - u)(1 - \cos\theta) \qquad F_y = \rho Q_0(V_0 - u)\sin\theta$$

Example 3.16

Determine for the single moving vane of Fig. 3.29a the force components due to the water jet and the rate of work done on the vane.

Solution

Figure 3.29b is the steady-state reduction with a control volume shown. The polar vector diagram is shown in Fig. 3.29c. By applying Eq. (3.6.1) in the x and y directions to the control volume of Fig. 3.29b

$$-F_x = (1000 \text{ kg/m}^3)(60 \text{ m/s})(\cos 170°)(60 \text{ m/s})(0.001 \text{ m}^2)$$
$$+ \; (1000 \text{ kg/m}^3)(60 \text{ m/s})(-60 \text{ m/s})(0.001 \text{ m}^2)$$
$$F_x = 7.145 \text{ kN}$$
$$F_y = (1000 \text{ kg/m}^3)(60 \text{ m/s})(\sin 170°)(60 \text{ m/s})(0.001 \text{ m}^2) = 625 \text{ N}$$

The power exerted on the vane is

$$uF_x = (60 \text{ m/s})(7.145 \text{ kN}) = 428.7 \text{ kW}$$

Example 3.17

Determine the horsepower that can be obtained from a series of vanes (Fig. 3.30a), curved through 150°, moving 60 ft/s away from a 3.0 cfs water jet having a cross section of 0.03 ft². Draw the polar vector diagram and calculate the energy remaining in the jet.

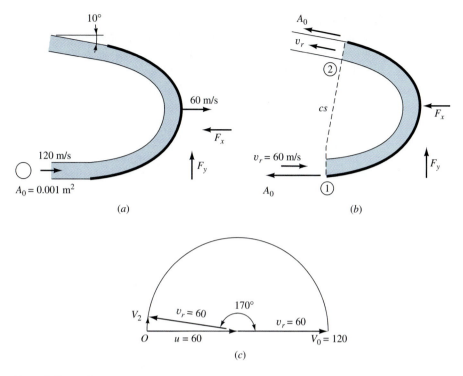

Figure 3.29 Jet acting on a moving vane.

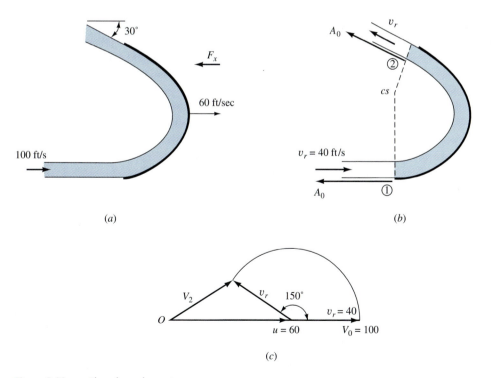

Figure 3.30 Flow through moving vanes.

Solution

The jet velocity is $V_0 = 3/0.03 = 100$ ft/s. The steady-state vane control volume is shown in Fig. 3.30b, and the polar vector diagram is shown in Fig. 3.30c. The force on the series of vanes in the x direction is

$$F_x = (1.94 \text{ slug/ft}^3)(3 \text{ ft}^3/\text{s})(40 \text{ ft/s})(1 - \cos 150°) = 434 \text{ lb}$$

The power is

$$\frac{(434 \text{ lb})(60 \text{ ft/s})}{550 \text{ ft·lb/s/hp}} = 47.4 \text{ hp}$$

The components of absolute velocity leaving the vane, from Fig. 3.30c, are

$$V_{2x} = 60 - 40 \cos 30° = 25.4 \text{ ft/s} \qquad V_{2y} = 40 \sin 30° = 20 \text{ ft/s}$$

and the exit-velocity head is

$$\frac{V_2^2}{2g} = \frac{25.4^2 + 20^2}{64.4} = 16.2 \text{ ft·lb/lb}$$

The kinetic energy remaining in the jet, in foot-pounds per second, is

$$Q\gamma \frac{V_2^2}{2g} = (3 \text{ ft}^3/\text{s})(62.4 \text{ lb/ft}^3)(16.2 \text{ ft}) = 3030 \text{ ft·lb/s}$$

The initial kinetic energy available was

$$(3 \text{ ft}^3/\text{s})(62.4 \text{ lb/ft}^3) \frac{100^2}{64.4} \text{ ft} = 29{,}070 \text{ ft·lb/s}$$

which is the sum of the work done and the energy remaining per second.

When a vane or series of vanes moves toward a jet, work is done by the vane system on the fluid, thereby increasing the energy of the fluid. Figure 3.31 illustrates this situation; the polar vector diagram shows the exit velocity to be greater than the entering velocity.

In turbulent flow, losses generally must be determined from experimental tests on the system or a geometrically similar model of the system. In the following two cases, application of the continuity, energy, and momentum equations permits the losses to be evaluated analytically.

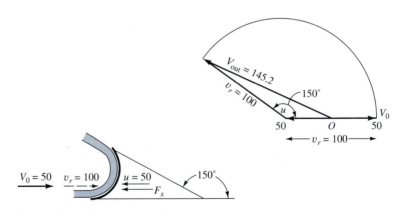

Figure 3.31   Vector diagram for vane doing work on a jet.

Losses Due to Sudden Expansion in a Pipe

The losses due to sudden enlargement in a pipeline can be calculated with both the energy and momentum equations. For steady, incompressible, turbulent flow caused by sudden expansion through the control volume between sections 1 and 2 (Figs. 3.32a and b), the small shear force exerted on the walls between the two sections can be neglected. By assuming uniform velocity over the flow cross sections, which is approached in turbulent flow, application of Eq. (3.6.7) produces

$$p_1 A_2 - p_2 A_2 = \rho V_2 (V_2 A_2) + \rho V_1 (-V_1 A_1)$$

At section 1 the radial acceleration of fluid particles in the eddy along the surface is small, and so generally a hydrostatic pressure variation occurs across the section. The energy equation (Eq. 3.4.21) applied to sections 1 and 2, with the loss term h_l, is (for $\alpha = 1$)

$$\frac{V_1^2}{2g} + \frac{p_1}{\gamma} = \frac{V_2^2}{2g} + \frac{p_2}{\gamma} + h_l$$

Solving for $(p_1 - p_2)/\gamma$ in each equation and equating the results give

$$\frac{V_2^2 - V_2 V_1}{g} = \frac{V_2^2 - V_1^2}{2g} + h_l$$

As $V_1 A_1 = V_2 A_2$

$$h_l = \frac{(V_1 - V_2)^2}{2g} = \frac{V_1^2}{2g}\left(1 - \frac{A_1}{A_2}\right)^2 \qquad \textbf{(3.7.10)}$$

which indicates that the losses in turbulent flow are proportional to the square of the velocity. With respect to the empirical loss term in Sec. 3.6

$$K = \left(1 - A_1/A_2\right)^2$$

Hydraulic Jump

Under proper conditions a rapidly flowing stream of liquid in an open channel suddenly changes to a slowly flowing stream with a large cross-sectional area and a sudden rise in elevation of liquid surface. This phenomenon, known as the *hydraulic jump*, is an example of steady nonuniform flow. The hydraulic jump is the second application of the basic equations to determine losses due to a turbulent flow situa-

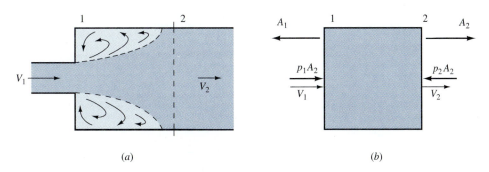

(a) (b)

Figure 3.32   Sudden expansion in a pipe.

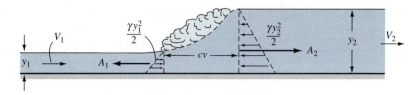

Figure 3.33   Hydraulic jump in a rectangular channel.

tion. In effect, the rapidly flowing liquid jet expands (Fig. 3.33) and converts kinetic energy into potential energy and losses or irreversibilities. A roller develops on the inclined surface of the expanding liquid jet and draws air into the liquid. The surface of the jump is very rough and turbulent, the losses being greater as the jump height is greater. For small heights, the form of the jump changes to a standing wave (Fig. 3.34). The jump is discussed further in Sec. 13.4.

The relations between the variables for the hydraulic jump in a horizontal rectangular channel are obtained by use of the continuity, momentum, and energy equations. For convenience, the width of a channel is taken as unity. The continuity equation (Fig. 3.33) is (with $A_1 = y_1$ and $A_2 = y_2$)

$$V_1 y_1 = V_2 y_2$$

The momentum equation is

$$\frac{\gamma y_1^2}{2} - \frac{\gamma y_2^2}{2} = \rho V_2 (y_2 V_2) + \rho V_1 (-y_1 V_1)$$

and the energy equation is

$$\frac{V_1^2}{2g} + y_1 = \frac{V_2^2}{2g} + y_2 + h_l$$

in which h_l represents losses due to the jump. Eliminating V_2 in the first two equations leads to

$$y_2 = -\frac{y_1}{2} + \sqrt{\left(\frac{y_1}{2}\right)^2 + \frac{2V_1^2 y_1}{g}} \qquad \text{(3.7.11)}$$

in which the plus sign has been taken before the radical (a negative y_2 has no physical significance). The depths y_1 and y_2 are referred to as *conjugate* depths. Solving the energy equation for h_l and eliminating V_1 and V_2 give

$$h_l = \frac{(y_2 - y_1)^3}{4 y_1 y_2} \qquad \text{(3.7.12)}$$

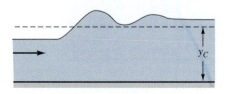

Figure 3.34   Standing wave.

The hydraulic jump, which is very effective in creating irreversibilities, is commonly used at the ends of chutes or the bottoms of spillways to destroy much of the kinetic energy of the flow. It is also an effective mixing chamber, because of the violent agitation that takes place in the roller. Experimental measurements of hydraulic jumps show that the equation yields the correct value of y_2 to within 1 percent.

If 12 m³/s of water per meter of width flows down a spillway onto a horizontal floor and the velocity is 20 m/s, determine the downstream depth required to cause a hydraulic jump and the losses in power by the jump per meter of width.

Example 3.18

$$y_1 = \frac{12 \text{ m}^2/\text{s}}{20 \text{ m/s}} = 0.6 \text{ m}$$

Solution
Substituting into Eq. (3.7.11) gives

$$y_2 = -0.3 + \sqrt{0.3^2 + \frac{2(20^2)(0.6)}{9.806}} = 6.7 \text{ m}$$

With Eq. (3.7.12)

$$\text{losses} = \frac{(6.7 - 0.6)^3}{4(0.6)(6.7)} = 14.1 \text{ m·N/N}$$

$$\text{Power/m} = \gamma Q(\text{losses}) = (9806 \text{ N/m}^3)(12 \text{ m}^3/\text{s})(14.1 \text{ m}) = 1659 \text{ kW}$$

Time Varying Calculations

The inclusion of time varying or *transient* conditions typically requires the application of differential equations.

Find the head H in the reservoir of Fig. 3.35 needed to accelerate the flow of oil, $S = 0.85$, at the rate of 0.5 ft/s² when the flow is 8.02 ft/s. At 8.02 ft/s the steady-state head on the pipe is 20 ft. Neglect entrance loss but do consider the pipe friction losses.

Example 3.19

Solution
The oil can be considered to be incompressible and to be moving uniformly in the pipeline. By application of the energy equation the head loss due to pipe friction can be found from

$$H = \frac{V^2}{2g} + h_l$$

or that

$$h_l = 20 - \frac{(8.02)^2}{2g}$$

If the 1000-ft length of the fluid in the exit pipe is defined as the control volume for the calculation, then the momentum control volume equation for unsteady flow is balanced in the direction of the pipe axis. There is no wall stress contribution.

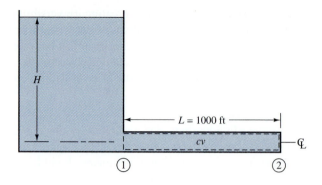

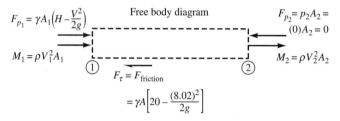

Figure 3.35   Acceleration of liquid in a pipe.

Equation (3.6.1) becomes

$$\frac{d}{dt} \int \rho V d\mathbf{V} + F_{p_1} - F_{p_2} + M_1 - M_2 - F_{\tau_0} = 0$$

By the continuity equation at each instant, $V_1 = V_2 = V$ equals a constant. Therefore, $M_1 = M_2$ and

$$\frac{d}{dt} \int \rho V \, d\mathbf{V} + F_{p_1} - F_{p_2} - F_{\tau_0} = 0$$

Because the outflow is a free jet, $F_{p_2} = 0$ and the energy equation gives

$$p_1 = \gamma \left(H - \frac{V^2}{2g} \right)$$

Therefore,

$$F_{p_1} = p_1 A = \gamma A \left(H - \frac{V^2}{2g} \right)$$

The wall shear force becomes

$$F_{\tau_0} = \gamma A h_l = \gamma A \left(20 - \frac{(8.02)^2}{2g} \right)$$

The volume of the fluid in the pipe does not change with respect to time. Therefore, the time derivative term is written as

$$\frac{d}{dt} \int \rho V \, d\mathbf{V} = \frac{d}{dt} \left(\rho V \int d\mathbf{V} \right) = \frac{d}{dt} (\rho V A L) = (\rho A L) \frac{dV}{dt}$$

The sum of all the contributions results in

$$\gamma A\left(H - \frac{V^2}{2g}\right) - \gamma A\left(20 - \frac{(8.02)^2}{2g}\right) = (\rho AL)\frac{dV}{dt}$$

or that

$$H = 20 + \frac{1000}{32.2}(0.5) = 35.52 \text{ ft}$$

If the reservoir of Ex. 3.19 has an area, A_r, 100 times the pipe area, A, and the head is 35.52 ft when the velocity is 8.02 ft/s, how long does it take for the head in the reservoir to reduce to 20 ft? Assume that the frictional resistance in the pipe varies at the square of the velocity.

Example 3.20

Solution

The resistance to pipe flow is

$$\text{Head loss} = \left(20 - \frac{V_0^2}{2g}\right)\left(\frac{V}{V_0}\right)^2$$

or

$$\text{Head loss} = RV^2$$

The equation of motion reduces to

$$\frac{dV}{dt} = \left(H - \frac{V^2}{2g} - RV^2\right)\frac{g}{L}$$

and the continuity equation is

$$\frac{dH}{dt} = -\frac{AV}{A_r} = -0.01\,V$$

These two equations are solved by the second-order Runge-Kutta method described in the Web page. Table 3.1 contains the spreadsheet algorithm and results. Note that the time steps between 65 and 160 seconds have been deleted. A linear interpolation provides the final time of 165.71 s.

EXERCISES

3.7.1 The equation $\Sigma F_x = \rho Q(V_{x_{out}} - V_{x_{in}})$ requires which two of the following assumptions for its derivation?

1. Velocity constant over the end cross sections
2. Steady flow
3. Uniform flow
4. Compressible fluid
5. Frictionless fluid

(a) 1, 2; (b) 1, 5; (c) 1, 3; (d) 3, 5; (e) 2, 4.

3.7.2 The momentum correction factor is expressed by (a) $\frac{1}{A}\int_A \frac{v}{V}\,dA$; (b) $\frac{1}{A}\int_A \left(\frac{v}{V}\right)^2 dA$; (c) $\frac{1}{A}\int_A \left(\frac{v}{V}\right)^3 dA$; (d) $\frac{1}{A}\int_A \left(\frac{v}{V}\right)^4 dA$; (e) none of these answers.

Table 3.1   Spreadsheet algorithm and results for Ex. 3.20

	A	B	C	D	E	F	G
1	Example 3.20		Reservoir Emptying Time				
2							
3	V0=	8.02	g=	32.2			
4	H0=	35.52	dt=	5			
5	Hf=	20	ApAt=	-0.01			
6	L=	1000	F=	0.295415	=Hf/V0^2-1/(2*g)		
7							
8	at t = 5 s,		dH1= dt*ApAt*G15				
9	at t = 5 s,		dV1= dt*g/L*(F15-G15^2*(F+1/(2*g)))				
10	at t = 5 s,		dH2= dt*ApAt*(G15+C16)				
11	at t = 5 s,		dV2= dt*g/L*(F15+B16-(G15+C16)^2*(F+1/(2*g)))				
12	at t = 5 s,		H= F15+0.5*(B16+D16)				
13	at t = 5 s,		V= G15+0.5*(C16+E16)				
14		time	dH1	dV1	dH2	dV2	H	V
15	0					35.5200	8.0200
16	5	-0.4010	2.4987	-0.5259	0.1151	35.0565	9.3269
17	10	-0.4663	1.2891	-0.5308	-0.0730	34.5580	9.9350
18	15	-0.4967	0.6225	-0.5279	-0.0961	34.0456	10.1982
19	20	-0.5099	0.2747	-0.5236	-0.0917	33.5289	10.2897
20	25	-0.5145	0.0977	-0.5194	-0.0863	33.0119	10.2954
21	30	-0.5148	0.0086	-0.5152	-0.0831	32.4970	10.2581
22	35	-0.5129	-0.0360	-0.5111	-0.0817	31.9849	10.1993
23	40	-0.5100	-0.0582	-0.5071	-0.0810	31.4764	10.1297
24	45	-0.5065	-0.0692	-0.5030	-0.0808	30.9717	10.0547
25	50	-0.5027	-0.0747	-0.4990	-0.0807	30.4708	9.9770
26	55	-0.4989	-0.0774	-0.4950	-0.0807	29.9739	9.8980
27	60	-0.4949	-0.0788	-0.4910	-0.0807	29.4810	9.8182
28	65	-0.4909	-0.0794	-0.4869	-0.0807	28.9920	9.7382
29							
30	160	-0.4146	-0.0801	-0.4106	-0.0807	20.4659	8.2109
31	165	-0.4105	-0.0801	-0.4065	-0.0807	20.0574	8.1305
32	170	-0.4065	-0.0801	-0.4025	-0.0807	19.6529	8.0501
33							
34				time @H=20 ft is		165.71 seconds		

3.7.3   The momentum correction factor for the velocity distribution given by Fig. 1.1 is (a) 0; (b) 1; (c) $\frac{4}{3}$; (d) 2; (e) none of these answers.

3.7.4   The velocity is zero over one-third of a cross section and is uniform over the remaining two-thirds of the area. The momentum correction factor is (a) 1; (b) $\frac{4}{3}$; (c) $\frac{3}{2}$; (d) $\frac{9}{4}$; (e) none of these answers.

3.7.5   The magnitude of the resultant force necessary to hold a 200-mm-diameter 90° elbow under no-flow conditions when the pressure is 0.98 MPa is, in kilonewtons, (a) 61.5; (b) 43.5; (c) 30.8; (d) 0; (e) none of these answers.

3.7.6   A 12-in.-diameter 90° elbow carries water with average velocity of 15 ft/s and pressure of -5 psi. The force component in the direction of the approach velocity necessary to hold the elbow in place is, in pounds, (a) -342; (b) 223; (c) 565; (d) 907; (e) none of these answers.

3.7.7 A 50-mm-diameter 180° bend carries a liquid, $\rho = 1000$ kg/m³ at 6 m/s, at pressure of zero gage. The force tending to push the bend off the pipe is, in newtons, (a) 0; (b) 70.5; (c) 141; (d) 515; (e) none of these answers.

3.8 THE MOMENT-OF-MOMENTUM EQUATION

The general unsteady linear-momentum equation applied to a control volume, Eq. (3.6.1), is

$$\mathbf{F} = \frac{\partial}{\partial t} \int_{cv} \rho \mathbf{v} \, d\mathbb{V} + \int_{cs} \rho \mathbf{v} \mathbf{v} \cdot d\mathbf{A} \qquad (3.8.1)$$

The moment of a force $\mathbf{F}$ about a point O (Fig. 3.36) is given by

$$\mathbf{r} \times \mathbf{F}$$

which is the cross, or vector, product of $\mathbf{F}$ and the position vector $\mathbf{r}$ of a point on the line of action of the vector from O. The cross product of two vectors is a vector at right angles to the plane defined by the first two vectors and with magnitude

$$Fr \sin \theta$$

which is the product of F and the shortest distance from O to the line of action of $\mathbf{F}$. The sense of the final vector follows the right-hand rule. In Fig. 3.36 the force tends to cause a counterclockwise rotation around O. If this were a right-hand screw thread turning in this direction, it would tend to come up, and so the vector is likewise directed up out of the paper. If one curls the fingers of the right hand in the direction the force tends to cause rotation, the thumb yields the direction, or sense, of the vector.

By taking $\mathbf{r} \times \mathbf{F}$, using Eq. (3.8.1),

$$\mathbf{r} \times \mathbf{F} = \frac{\partial}{\partial t} \int_{cv} \rho \, \mathbf{r} \times \mathbf{v} \, d\mathbb{V} + \int_{cs} (\rho \, \mathbf{r} \times \mathbf{v})(\mathbf{v} \cdot d\mathbf{A}) \qquad (3.8.2)$$

The left-hand side of the equation is the torque exerted by any forces on the control volume, and terms on the right-hand side represent the rate of change of *moment of momentum* within the control volume plus the net efflux of moment of momentum

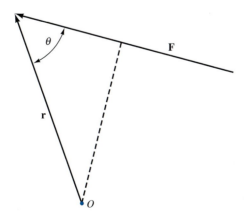

Figure 3.36   Notation for moment of a vector.

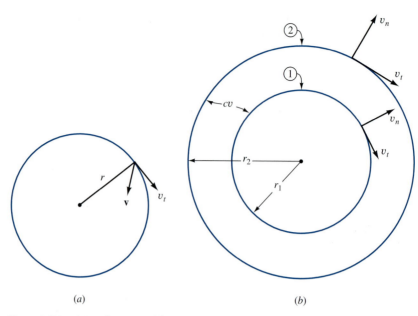

Figure 3.37   Two-dimensional flow in a centrifugal pump impeller.

from the control volume. This is the general moment-of-momentum equation for a control volume. It has great value in the analysis of certain flow problems, for example, in turbomachinery, where torques are more significant than forces.

When Eq. (3.8.2) is applied to a case of flow in the xy plane, with r the shortest distance to the tangential component of the velocity v_t, as in Fig. 3.37a, and v_n is the normal component of velocity

$$F_t r = T_z = \int_{cs} \rho r v_t v_n \, dA + \frac{\partial}{\partial t} \int_{cv} \rho r v_t \, d\forall \tag{3.8.3}$$

in which T_z is the torque. A useful form of Eq. (3.8.3) applied to an annular control volume, in steady flow (Fig. 3.37b), is

$$T_z = \int_{A_2} \rho_2 r_2 v_{t_2} v_{n_2} \, dA_2 - \int_{A_1} \rho_1 r_1 v_{t_1} v_{n_1} \, dA_1 \tag{3.8.4}$$

For complete circular symmetry, where r, ρ, v_t, and v_n are constant over the inlet and outlet control surfaces, it takes the simple form

$$T_z = \rho Q[(r v_t)_2 - (r v_t)_1] \tag{3.8.5}$$

since $\int \rho v_n dA = \rho Q$, being the same at the inlet or outlet.

Example 3.21

The sprinkler shown in Fig. 3.38 discharges water upward and outward from the horizontal plane so that it makes an angle of $\theta°$ with the t axis when the sprinkler arm is at rest. It has a constant cross-sectional flow area of A_0 and discharges q cfs starting with $\omega = 0$ and $t = 0$. The resisting torque due to bearings and seals is the constant T_0, and the moment of inertia of the rotating empty sprinkler head is I_s. Determine the equation for ω as a function of time.

Solution

Equation (3.8.2) can be applied. The control volume is the cylindrical area enclosing the rotating sprinkler head. The inflow is along the axis so that it has no moment of

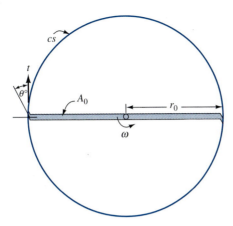

Figure 3.38   Plan view of sprinkler and control surface.

momentum; hence, the torque $-T_0$ due to friction is equal to the time rate of change of moment of momentum of the sprinkler head and fluid within the sprinkler head plus the net efflux of moment of momentum from the control volume. Let $V_r = q/2A_0$

$$-T_0 = 2\frac{d}{dt}\int_0^{r_0} A_0\rho\omega r^2\, dr + I_s\frac{d\omega}{dt} - \frac{2\rho q r_0}{2}(V_r\cos\theta - \omega r_0)$$

The total derivative can be used. Simplifying gives

$$\frac{d\omega}{dt}(I_s + \frac{2}{3}\rho A_0 r_0^3) = \rho q r_0(V_r\cos\theta - \omega r_0) - T_0$$

For rotation to start, $\rho q r_0 V_r\cos\theta$ must be greater than T_0. The equation is easily integrated to find ω as a function of t. The final value of ω is obtained by setting $d\omega/dt = 0$ in the equation.

Example 3.22

A turbine discharging 10 m³/s is to be designed so that a torque of 10 kN·m is to be exerted on an impeller turning at 200 rpm that takes all the moment of momentum out of the fluid. At the outer periphery of the impeller, $r = 1$ m. What must the tangential component of velocity be at this location?

Solution

Equation (3.8.5) is

$$T = \rho Q(rv_t)_{in}$$

in this case, since the outflow has $v_t = 0$. Solving for $v_{t_{in}}$ gives

$$v_{t_{in}} = \frac{T}{\rho Q r} = \frac{10{,}000\ \text{N·m}}{(1000\ \text{kg/m}^3)(10\ \text{m}^3/\text{s})(1\ \text{m})} = 1.000\ \text{m/s}$$

Example 3.23

The sprinkler of Fig. 3.39 discharges 0.01 cfs through each nozzle. Neglecting friction, find its speed of rotation. The area of each nozzle opening is 0.001 ft².

Solution

The fluid entering the sprinkler has no moment of momentum, and no torque is exerted on the system externally; hence, the moment of momentum of fluid leaving

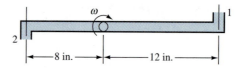

Figure 3.39   Rotating jet system.

must be zero. Let ω be the speed of rotation; then the moment of momentum leaving is

$$\rho Q_1 r_1 v_{t_1} + \rho Q_2 r_2 v_{t_2}$$

in which v_{t_1} and v_{t_2} are absolute velocities. Then

$$v_{t_1} = v_{r_1} - \omega r_1 = \frac{Q_1}{0.001} - \omega r_1 = 10 - \omega$$

and

$$v_{t_2} = v_{r_2} - \omega r_2 = 10 - \frac{2}{3}\omega$$

For the moment of momentum to be zero

$$\rho Q(r_1 v_{t_1} + r_2 v_{t_2}) = 0 \qquad \text{or} \qquad 10 - \omega + \frac{2}{3}(10 - \frac{2}{3}\omega) = 0$$

and $\omega = 11.54$ rad/s, or 110.2 rpm.

3.9   HEAT AND MASS TRANSFER

Before proceeding to the individual control volume equations for heat and mass transfer a few assumptions are invoked. First, and most importantly is that no phase changes in the fluid are permitted. Often the presence of heat or mass in the flowing fluid will result in the necessity of a coupled flow and transport calculation. This computation is often nonlinear and complicated even with no phase changes. Therefore, no phase changes will be considered in this introductory text. Transport phenomena textbooks will provide more elaboration on transfer or transport involving a phase change of the basic fluid. Secondly, one especially prevalent multiphase flow will be considered: sediment and water.

The Basic Equation of Heat Transfer

In advanced transport phenomena textbooks, the derivation of the heat transfer equation proceeds from the conservation of total energy equation discussed in Sec. 3.4. For purposes of this elementary or first-level presentation we shall treat the total heat content per unit mass, $c_p T$, as the conservation variable η in Eq. (3.2.4). Therefore, the control volume equation for heat content becomes

$$\frac{dN}{dt} = \frac{\partial}{\partial t} \int_{cv} \rho c_p T \, d\forall + \int_{cs} c_p T \rho \mathbf{v} \cdot d\mathbf{A} \qquad \text{(3.9.1)}$$

In words Eq. (3.9.1) says that the time rate of change of heat in the system, N, equals the time rate of change of heat in the control volume plus the net rate at which heat leaves through the control surface. The term dN/dt is comprised of three possible contributions: conduction, N_c; radiation, N_r; and dissipation, N_Φ.

1. *Heat Conduction.* Each of the fluid molecules in the flow contains thermal energy by virtue of the fact that the absolute temperature is greater than zero. In an exact analogy to the discussion of viscosity, the collision of fluid particles with different levels of thermal energy will result in a net transport of heat from the higher energy or temperature particles to the lower energy or temperature particles. The flux of heat due to this interparticle transfer mechanism, N_{cx}, is called *heat conduction.* *Fourier* parameterized the flux per unit area of heat in the x direction via conduction as

$$N_{cx} = \frac{q_c}{A} = -k\frac{\partial T}{\partial x} \tag{3.9.2}$$

Here k is the *thermal conductivity* of the fluid, which is a property of the fluid and has dimensions of W/m·K [or Btu/hr·ft·°R]. A watt, W, is defined as a Joule per second. The variable q_c is the heat transfer rate in watts [or Btu/hr]; A is the area perpendicular to the heat flow in m^2 [or ft^2]; and $\partial T/\partial x$ is the temperature gradient in the x direction. The ratio q_c/A is the *heat flux*, N_{cx}, which has dimensions of W/m^2 [or Btu/hr·ft^2]. If k is divided by ρc_p (see Chap. 4), then the *thermal diffusivity*, α, is defined and has dimensions of [L^2/t]. This is an analogous parameter to the kinematic viscosity ν.

   Heat conduction flux is a vector quantity, that is, it can be different in all three coordinate directions; hence, flux expressions similar to Eq (3.9.2) are required for the y and z directions as well.

   Values for k are formed for various fluids in Appendix C, but the values of k for water at 0°C and 100°C are 0.569 W/m·K and 0.680 W/m·K, respectively. Compared to the corresponding change in viscosity over this same range the thermal conductivity change is quite limited.

2. *Radiation.* Thermal radiation is the creation or emission of heat at a point in the volume of a fluid by virtue of the fact that the temperature of the body is greater than zero. Two bodies radiating heat may exchange heat and no medium is required for thermal energy to be exchanged by the radiating bodies. The heat flux from a perfectly radiating blackbody is given by

$$N_r = \frac{q_r}{A} = \sigma T^4 \tag{3.9.3}$$

where N_r is the radiation heat flux with the same dimensions as N_{cx}; A is the area of the body emitting the radiation; and σ is the *Stephan-Boltzmann constant* which is 5.67(10^{-8}) W/m^2·K^4. Absolute temperature must be used in this flux calculation.

3. *Heat Generation by Mechanical Dissipation.* Excessive generation of heat by friction is the third component of dN/dt. For the flows considered here this term will be quite small in contrast to the other terms, and it is neglected.

   The term $c_p T \rho \mathbf{v}$ has units of flux. This flux results from the fluid velocity $\mathbf{v}$ carrying the heat and is called advection or convection. *Advection* or *forced convection* is heat carried by the fluid velocity which in turn has been set into motion by mechanical means, that is, pressure or elevation differentials or shear stress. *Natural* or *free convection* is heat carried by a fluid velocity which originates from instabilities caused by unstable temperature or density differences. For example, when a fluid is heated from below the warm water at the bottom is lighter than the overlaying fluids.

Warmer fluid parcels therefore rise with a finite fluid velocity and carry heat with them.

Mass Transfer and Transport

The Reynold's conservation control volume equation, Eq. (3.2.4), is used to derive the basic equation for a mass species being carried in a moving fluid. If the mass concentration fraction of the species is labeled w_A, then

$$\frac{dN_{ca}}{dt} = \frac{\partial}{\partial t} \int_{cv} \rho w_A \, d\forall + \int_{cs} w_A \rho \mathbf{v_A} \cdot d\mathbf{A} \tag{3.9.4}$$

In this equation $\mathbf{v_A}$ is the velocity vector of the species. In practice the total velocity vector $\mathbf{v}$ at a point is the mass concentration weighted average of the component velocities for all species, for example,

$$\mathbf{v} = \frac{\sum\limits_i^n C_i \mathbf{v}_i}{\sum\limits_i^n C_i} \tag{3.9.5}$$

where $\sum_i^n C_i = \rho$.

For most mixtures the mass fractions of the constituents are quite small compared to the mass of the liquid ($\sim 10^{-3}$), and if the particles are neutrally buoyant or dissolved in the liquid, then the velocity of the mass fraction per species is equal to the bulk velocity of the liquid $\mathbf{v}$. Hence, Eq (3.9.4) becomes

$$\frac{dN_{ca}}{dt} = S_{ca*} = \frac{\partial}{\partial t} \int_{cv} \rho w_A \, d\forall + \int_{cs} w_A \rho \mathbf{v} \cdot d\mathbf{A} \tag{3.9.6}$$

From Chap. 1 the product ρw_a equals C_A, the mass concentration of species A, that is,

$$\frac{dN_{ca}}{dt} = S_{ca} = \frac{\partial}{\partial t} \int_{cv} C_A \, d\forall + \int_{cs} C_A \mathbf{v} \cdot d\mathbf{A} \tag{3.9.7}$$

If the fluid mixture consists of n species, then

$$\sum_{i=1}^n \frac{dN_{ci}}{dt} = \sum_{i=1}^n \left[\frac{\partial}{\partial t} \int_{cv} C_i d\forall + \int_{cs} C_i \, \mathbf{v} \cdot d\mathbf{A} \right] \tag{3.9.8}$$

which sums over all species to

$$0 = \frac{\partial}{\partial t} \int_{cv} \rho \, d\forall + \int_{cs} \rho \mathbf{v} \cdot d\mathbf{A}$$

which is the total continuity equation for the mixture.

Two terms comprise the term dN_{ca}/dt in Eq. (3.9.7):

1. *Molecular Diffusion.* In direct analogy to the viscosity and thermal conductivity, Fick's diffusion law for molecular diffusion in the x direction of a species A in a fluid B is

$$J_{A,x} = -\mathscr{D}_{AB} \frac{\partial C_A}{\partial x} \tag{3.9.9}$$

where $\mathscr{D}_{AB}$ is the binary diffusion coefficient of species A in fluid B. These coefficients are properties of the species and the fluid in which it is immersed. Appendix C contains values for $\mathcal{D}$ for various species in water. If the flux $J_{A,x}$ has dimensions of $[M/(L^2t)]$, that is, mass per unit area perpendicular to x per unit time, then the dimensions of $\mathcal{D}_{AB}$ are $[L^2/t]$.

2. *Chemical and biological transformation.* Due to chemical reaction or biological activity, the mass concentration of species i can be changed within the control volume. Empirical relationships usually are required to describe the biological transformation while direct molar calculations are performed for the chemical transformations. In Eqs. (3.9.6) or (3.9.7) these transformations result in $dN_{ca}/dt \neq 0$ and are empirically described by a *source to sink* term, S_{ca}, with dimensions $[M/L^3t]$.

As in the thermal case the component of the control surface integral also has units of flux, that is, $C_i\mathbf{v}$. This term is also called the *advective* or *forced convective* flux. In this text convection caused by mass density instabilities, that is, *natural* or *free convection*, is much less frequently observed than in the thermal case.

Equation Analysis for Steady Conditions

Based upon control volume application methods explained in the previous sections the steady-state analysis of an inflow (1) and outflow (2) control volume (Fig. 3.40) leads to the following pair of equations for heat and mass transport. A binary mixture is assumed where one component or species, A, is immersed in the fluid

$$\frac{dN_T}{dt} = -\rho_1 c_p T_1 V_1 A_1 + \rho_2 c_p T_2 V_2 A_2 \qquad (3.9.10)$$

$$\frac{dN_{ca}}{dt} = -C_{A_1} V_1 A_1 + C_{A_2} V_2 A_2 \qquad (3.9.11)$$

These equations assume area-averaged uniform values for T, C_A, V, ρ, etc. Along the solid surface in contact with the control volume it can be assumed that no heat or mass flux normal to the wall to control volume interface exists. This is called an *insulated condition* and implies that no heat flux by advection or convection is permitted and no heat or mass flux by molecular diffusion is permitted. Therefore, at the solid boundary the velocity at the wall perpendicular to it must be zero and the temperature or concentration gradient normal to the wall is zero.

Application Examples of Temperature

Two examples of thermal energy transport follow.

Example 3.24

Two streams of water enter the mixing chamber as shown in Fig. 3.41. If the inlet conditions at point 1 are $T_1 = 80°C$ and $\dot{m}_1 = 80$ kg/sec while those at point two are $T_2 = 50°C$ and $\dot{m}_2 = 100$ kg/sec, determine the outlet temperature at point 3.

Solution

From the continuity equation the total mass flow rate into the chamber must equal the total mass outflow rate. For steady flow the continuity equation gives

$$\dot{m}_1 + \dot{m}_2 = \dot{m}_3$$

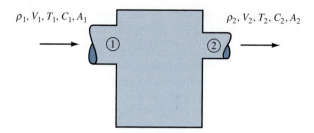

Figure 3.40   Inlet-outlet control volume for mixing chamber.

or

$$\rho_1 V_1 A_1 + \rho_2 V_2 A_2 = \rho_3 V_3 A_3$$

$$80 + 100 = 180 \text{ kg/s}$$

The thermal heat equation gives

$$\rho_1 c_p T_1 V_1 A_1 + \rho_2 c_p T_2 V_2 A_2 = \rho_3 c_p T_3 V_3 A_3$$

Since c_p is constant for this liquid and range of temperatures and since the mass flow rates are already known, then

$$T_1 \dot{m}_1 + T_2 \dot{m}_2 = T_3 \dot{m}_3$$

$$(273 + 80)(80) + (273 + 50)(100) = T_3(180)$$

so

$$T_3 = 336 \text{K} = 63.3°C$$

Since the same answers are obtained by not using absolute temperatures, these calculations may be done using the Centigrade or Fahrenheit scales.

The effect of temperature change on the density may be explicitly determined. From Appendix C the corresponding densities for the inlet-outlet condition above are $\rho_1 = 971.8$, $\rho_2 = 988.1$, and $\rho_3 = 981.5$ kg/m³. This represents a maximum density variation of 1.6 percent, a relatively small variation. Indeed for water and most other liquids such small variations in density are typical and the bulk fluid density is often taken as a constant. If the range of water temperatures begins to

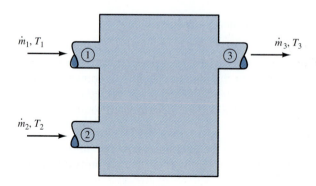

Figure 3.41   Multiple inlet and outlet control volume geometry.

approach the freezing or boiling points, this assumption should be checked by automatic inclusion of temperature-dependent densities before proceeding. Calculations involving any gas or a gas-liquid phase change must *always* account for temperature-dependent density changes.

The above calculation assumed that there was no addition or removal by other agents, that is, heat transfer by conduction at the walls, heat generation by friction, or heat removal via exchanger pipes. The following example includes heat removal by means other than advection, that is, a simple heat exchanger.

A heat exchanger is a device which removes heat from a body of fluid. A pipe containing a very cool fluid is immersed in the tank, and because the coolant is at a very low temperature relative to the fluid in the tank, heat is transfered via conduction from the tank to the fluid in the pipe. The flow rate through the exchanger then removes the heat from the control volume. If the inlet conditions are $\dot{m}_1 = 1.0$ kg/sec and $T_1 = 90°C$, what is the outlet temperature if the exchanger removes heat at the rate of 5 kJ/m^2·s of pipe? Assume the exchanger is 20-m long and 10 cm in diameter.

Example 3.25

Solution

Because the heat exchanger is sealed from the tank, its flows does not affect the calculation for the tank. For steady flows the continuity equation gives

$$\dot{m}_1 = \dot{m}_2 = \dot{m} \qquad \text{or} \qquad \rho_1 V_1 A_1 = \rho_2 V_2 A_2$$

The heat equation becomes

$$c_p T_1 \rho_1 V_1 A_1 = c_p T_2 \rho_2 V_2 A_2 + q_H$$

where q_H is the rate of heat addition (or in this case removal) with dimensions [kJ/s]. The heat removal requires knowledge of the total surface area of the exchanger in contact with the water in the tank:

$$q_H = L P_o (5.0)$$

where L is the pipe length, and P_o is the circumference, or perimeter, of the pipe in contact with the fluid in the control volume. If the pipe diameter is 10 cm, then $P_o = 0.314$ m and $c_p = 4.2$ kJ/kg·K; therefore,

$$T_2 = \frac{c_p T_1 \dot{m} - q_H}{c_p \dot{m}}$$

or

$$T_2 = T_1 - \frac{(5.0) L P_o}{c_p \dot{m}}$$

$$T_2 = 82.5°C$$

It is noticed that for this simple problem the outlet temperature will decrease linearly with increasing length or diameter (circumference) of the heat exchanger and that the temperature will go down proportional to $1/r^2$, where r is the radius of the outlet pipe.

Application Examples of Mass Transfer

Many applications of control volume mass balance equations are applied to stirred tanks or reactors which are marked by an inlet, an outlet, and the possible generation or utilization of the mass species by chemical or biological transformation processes.

The conservation of mass for constituent C is written for unsteady flow as

$$\frac{d}{dt}\int_{cv} C \, d\Psi = V_1 A_1 C_1 - V_2 A_2 C_2 + S \tag{3.9.12}$$

Here, S represents the sources or sinks (generation or utilization) of C due to biological or chemical transformations. If it is assumed that mixing in the tank takes place instantaneously throughout the tank, then C is independent of position, that is, uniformly distributed or mixed in the control volume. The following simplification is invoked

$$\Psi \frac{dC}{dt} = V_1 A_1 C_1 - V_2 A_2 C_2 + S$$

where Ψ is the tank volume. The first order decay or utilization of C per unit volume can be expressed as

$$r_u = -kC$$

where k is the reaction rate constant with dimensions of $[t^{-1}]$. The equation can be written for a system as

$$S = -\Psi k C$$

and

$$\Psi \frac{dC}{dt} = V_1 A_1 C_1 - V_2 A_2 C_2 - \Psi k C \tag{3.9.13}$$

A similar expression for generation (source term) can also be described, that is,

$$r_g = kC$$

For steady flow $dC/dt = 0$ and

$$V_1 A_1 C_1 - V_2 A_2 C_2 - \Psi k C = 0 \tag{3.9.14}$$

the dimensions of this balance equation are $[M/t]$. For a single inlet and outlet reactor tank shown in Fig. 3.42a the following example illustrates the above equations.

Example 3.26

The following data are known for the reactor tank. The tank is 1 m in diameter and 2-m tall. The inlet diameter is 15 cm, the corresponding inlet average velocity (V_1) is 25 cm/sec, and the inlet area average concentration (C_1) is 200 mg/L. What are the outflow velocity and concentration if the discharge pipe is the same size as the inflow pipe, and the utilization of material obeys first order reaction kinetics with a decay rate of 0.03 sec^{-1}?

Solution

For steady conditions, Eq. (3.9.14) is used and can be solved for the outlet concentration, C_2, by making the following observation. It is assumed that the well-mixed tank assumption means that the tank average concentration in Eq. (3.9.13) (or Eq. (3.9.14) for the steady case) must be equal to the outlet concentration C_2. While close inspection will reveal that this is sometimes not realistic, in practice this assumption is fairly standard for elementary stirred tank calculations. Therefore, solving Eq. (3.9.14) for C_2 gives

$$C_2 = \frac{C_1}{1 + k(\Psi/Q)}$$

where $Q = VA = V_1 A_1 = V_2 A_2 = $ the volume flow rate through the system.

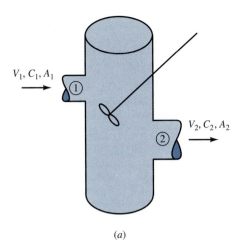

(a)

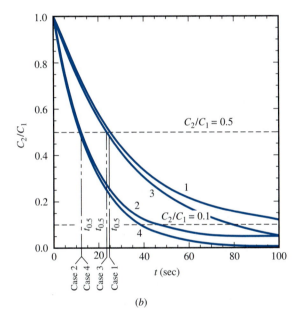

(b)

Figure 3.42 Mixing chamber, Example 3.26

The flow rate through the system is calculated to be

$$Q = V_1 A_1 = (0.25 \text{ m/s}) \, \pi \left(\frac{0.15}{2}\right)^2$$

$$= 0.0044 \text{m}^3/\text{s} = 4.4 \text{ L/s}$$

and from Eq. (3.9.14)

$$C_2 = \frac{200}{1 + 0.03\left(\frac{1570}{4.4}\right)} = 17.09 \text{ mg/L}$$

Both utilization (k_u) and generation (k_g) may coexist in a mass balance and represent different transformation processes. Equation (3.9.14) can be rewritten and solved for steady flow

$$C_2 = \frac{C_1}{1 + (k_u - k_g)(\forall/Q)} \tag{3.9.15}$$

In wastewater or water quality engineering the control volume mass balance equations must often be applied to nonsteady situations. Such would be the case, for example, if a tank originally at uniform concentration conditions suddenly had a new concentration, C_1, steadily introduced at the inlet and one wished to determine the time history of the subsequent concentration in the tank, C_2, and the outlet.

Equation (3.9.13) can be rewritten in the form

$$\frac{dC_2}{dt} + kC_2 + \frac{V_2 A_2}{\forall} C_2 = \frac{V_1 A_1}{\forall} C_1$$

If α is defined as $(k + V_2 A_2/\forall)$, then

$$\frac{dC_2}{dt} + \alpha C_2 = \beta C_1$$

where $\beta = V_1 A_1/\forall$. Both α and β are coefficients. The integration of this equation proceeds in one of several ways (see Chap. 1 in E. Kreyszig, *Advanced Engineering Mathematics*, 7th ed.). The general form of the solution to this initial value problem, which assumes α and β can vary with time, becomes

$$C_2(t) = \exp(\smallint \alpha\, dt)\left[\int \exp(\smallint \alpha\, dt)(\beta C_1)\, dt + \text{const.}\right] \tag{3.9.16}$$

where the constant of integration is to be found from the initial condition, that is,

$$C_2(t = 0) = C_1 = \text{constant}$$

If β and α are constant, Eq. (3.9.16) reduces to

$$C_2 = \beta \frac{C_1}{\alpha}\left(1 - e^{-\alpha t}\right) + C_1 e^{-\alpha t} \tag{3.9.17a}$$

or

$$C_2 = \frac{Q}{\forall}\frac{C_1}{(k + Q/\forall)}\left[1 - \exp\{-(k + Q/\forall)t\}\right] + C_1 \exp\{-(k + Q/\forall)t\} \tag{3.9.17b}$$

As time goes to infinity, the above equation reduces to the steady state form in Eq. (3.9.14).

Example 3.27

For the tank in the previous example find the time it takes after the introduction of C_1 at the inlet for the outlet and tank concentration to reach a value of 10 percent of the original concentration.

Solution

From Eq. (3.9.17a)

$$\frac{C_2}{C_1} = \frac{\beta}{\alpha}\left(1 - e^{-\alpha t}\right) + e^{-\alpha t}$$

Rearranging and solving for the time $t_{0.1}$ when $C_2/C_1 = 0.1$ yields $t_{0.1} = 126.2$ sec or 2.10 min, a relatively quick time. This value can be contrasted with an approximate time to potentially replace or renew the entire volume of fluid in the tank with a "fresh" volume from the inflow. This *residence* or *detention* time can be estimated from

$$t_d = \forall/Q$$

and for the example case (case 1), $t_d = 356.8$ sec (5.95 min). Here new undecayed material is entering the tank quite quickly and consequently the time to equilibrium or the steady state concentration (17.09 mg/L from the previous example) is stretched out.

Fig. 3.42b contains a plot of the time history of the response of case 1. It is seen that the initial period of decay is quite quick with the time to $C_2/C_1 = 0.5$, that is, the *half-life,* being 24.1 sec.

If there were no flow through the tank and just decay processes were taking place, then from Eq. (3.4.14) with $Q = 0$

$$\frac{dC}{dt} = -kC$$

or

$$C_2 = C_1 e^{-kt}$$

and the half-life would be found from

$$\ln\left(\frac{C_2}{C_1}\right) = -kt$$

as $t = 23.1$ sec.

For the tank above, examine the *sensitivity* of the residence time, half-life, and temporal decay of C to changes in k, $\forall$, and Q.

Example 3.28

Solution

From the conditions in case 1 the following conditions will be examined, each case being distinguished by an incremental change in one variable from the previous case. The conditions and results are summarized in Table 3.2 while Figure 3.42b contains the four time traces of C_2/C_1 versus time.

Table 3.2   Reactor conditions and results

Cases	k (sec^{-1})	Q (L/s)	$\forall$ (L)	t_d (sec)	$t_{0.5}$ (sec)	$t_{0.1}$ (sec)
1	0.03	4.4	1570	357	24.1	126.2
2	0.06	4.4	1570	357	11.8	45.3
3	0.03	0.44	15700	$3.57(10^4)$	23.1	77.0
4	0.06	0.44	15700	$3.57(10^4)$	11.6	38.4

In general, case 1 and case 2 show that increasing the decay rate results in a lower final steady state outflow concentration and a much quicker time to equilibrium conditions as shown in the lower values for $t_{0.5}$ and $t_{0.1}$. Cases 1 and 3 demonstrate that a larger tank volume results in much longer tank residence times which, because decay proceeds over a much longer time, also result in a lower equilibrium outflow concentration. The effect of lowering throughflow, Q, is seen in cases 3 and 4 where a further lowering of the residence time and of C_2 occurs. It is important to note that in this formulation the physical steady state variables of the problem, that is, Q and $\forall$, do not have as appreciable an impact on the *speed* with which steady conditions are reached as does the decay rate.

PROBLEMS

3.1   In two-dimensional flow around a circular cylinder (Fig. 3.3), the discharge between streamlines is 0.01 cfs per foot of depth. At a great distance the streamlines are 0.25 in. apart, and at a point near the cylinder they are 0.12 in. apart. Calculate the magnitude of the velocity at these two points.

3.2   A pipeline carries oil, sp gr 0.86, at $V = 2$ m/s through 200-mm-ID pipe. At another section the diameter is 70 mm. Find the velocity at this section and the mass rate of flow in kilograms per second.

3.3   Hydrogen is flowing in a 2.0-in.-diameter pipe at the mass rate of 0.03 lb_m/s. At section 1 the pressure is 30 psia and $t = 80°F$. What is the average velocity?

3.4   A nozzle with a base diameter of 70 mm and with a 30-mm-diameter tip discharges at 10 L/s. Derive an expression for the fluid velocity along the axis of the nozzle. Measure the distance x along the axis from the plane of the larger diameter.

3.5   Consider a cube with 1-m edges parallel to the coordinate axes located in the first quadrant with one corner at the origin. By using the velocity distribution of

$$\mathbf{v} = (5x)\mathbf{i} + (5y)\mathbf{j} + (-10z)\mathbf{k}$$

find the flow through each face and show that no mass is being accumulated within the cube if the fluid is of constant density.

3.6   Find the flow (per foot in the z direction) through each side of the square with the corners at (0,0), (0,1), (1,1), (1,0) due to

$$\mathbf{v} = (16y - 12x)\mathbf{i} + (12y - 9x)\mathbf{j}$$

and show that continuity is satisfied.

3.7   In a flow of liquid through a pipeline the losses are 3 kW for average velocity of 2 m/s and 6 kW for 3 m/s. What is the nature of the flow?

3.8   When tripling the flow in a line causes the losses to increase by 7.64 times, how do the losses vary with velocity and what is the nature of the flow?

3.9   A standpipe 6 m in diameter and 15 m high is filled with water. How much potential energy is in this water if the elevation datum is taken 3 m below the base of the standpipe?

3.10   How much work could be obtained from the water of Prob. 3.9 if the water runs through a 100 percent efficient turbine that discharges into a reservoir with elevation 10 m below the base of the standpipe?

3.11   What is the kinetic-energy flux, in meter-newtons per second, of 0.01 m³/s of oil, sp gr 0.80, discharging through a 30-mm-diameter nozzle?

3.12   Show that the work a liquid can do by virtue of its pressure is $\int p \, d\forall$, in which $\forall$ is the volume of liquid displaced.

3.13   The velocity distribution between two parallel plates separated by a distance a is

$$u = -10\frac{y}{a} + 20\frac{y}{a}\left(1 - \frac{y}{a}\right)$$

in which u is the velocity component parallel to the plate and y is measured from, and normal to, the lower plate. Determine the volume rate of flow and the average velocity. What is the time rate of flow of kinetic energy between the plates? In what direction is the kinetic energy flowing?

3.14 What is the flux of kinetic energy out of the cube given by Prob. 3.5?

3.15 Water is flowing in a channel, as shown in Fig. 3.43. Neglecting all losses, determine the two possible depths of flow y_1 and y_2.

3.16 High-velocity water flows up an inclined plane as shown in Fig. 3.44. Neglecting all losses, calculate the two possible depths of flow at section B.

3.17 Neglecting all losses, in Fig. 3.43 the channel narrows in the drop to 6 ft wide at section B. For uniform flow across section B, determine the two possible depths of flow.

3.18 Some steam locomotives had scoops installed that took water from a tank between the tracks and lifted it into the water reservoir in the tender. To lift the water 4 m with a scoop, neglecting all losses, what speed is required? *Note:* Consider the locomotive stationary and the water moving toward it to reduce to a steady-flow situation.

3.19 Neglecting losses, determine the discharge in Fig. 3.45.

3.20 Neglecting losses and surface-tension effects, derive an equation for the water surface r of the jet of Fig. 3.46 in terms of y/H.

3.21 Neglecting losses, find the discharge through the venturi meter of Fig. 3.47.

3.22 For the venturi meter and manometer installation shown in Fig. 3.48 derive an expression relating the volume rate of flow with the manometer reading.

3.23 In Fig. 3.49 determine V for $R = 12$ in.

3.24 Neglecting losses, calculate H in terms of R for Fig. 3.50.

3.25 A pipeline leads from one water reservoir to another which has its water surface 12 m lower. For a discharge of 0.6 m³/s determine the losses in meter-newtons per kilogram and in kilowatts.

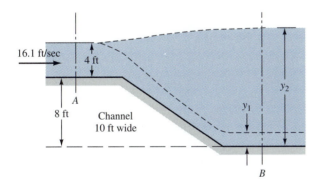

Figure 3.43   Problems 3.15, 3.17, and 3.31.

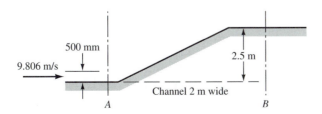

Figure 3.44   Problems 3.16, 3.32, and 3.33.

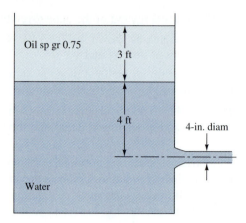

Figure 3.45   Problem 3.19.

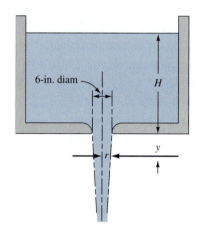

Figure 3.46   Problem 3.20.

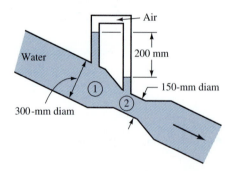

Figure 3.47   Problems 3.21 and 3.50.

3.26   A pump which is located 10 ft above the surface of a lake expels a jet of water vertically upward a distance of 50 ft. If 0.5 cfs is being pumped by a 5-hp electric motor running at rated capacity, what is the efficiency of the motor-pump combination? What is the irreversibility of the pump system when comparing the zenith of the jet and the lake surface? What is the irreversibility after the water falls to the lake surface?

3.27 A blower delivers 2 m³/s of air, $\rho = 1.3$ kg/m³, at an increase in pressure of 150-mm water. It is 72-percent efficient. Determine the irreversibility of the blower in meter-newtons per kilogram and in kilowatts, and determine the torque in the shaft if the blower turns at 1800 rpm.

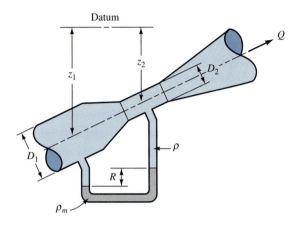

Figure 3.48 Problem 3.22.

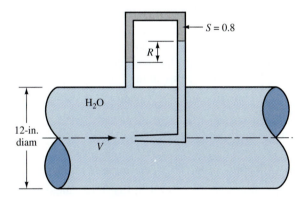

Figure 3.49 Problem 3.23.

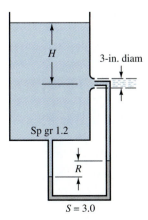

Figure 3.50 Problems 3.24 and 3.52.

3.28 A 6-m-diameter pressure pipe has a velocity of 3 m/s. After passing through a reducing bend, the flow is in a 5-m-diameter pipe. If the losses vary as the square of the velocity, how much greater are they through the 5-m pipe than through the 6-m pipe per 1000 m of pipe?

3.29 The velocity distribution in laminar flow in a pipe is given by

$$v = V_{max}[1 - (r/r_0)^2]$$

Determine the average velocity and the kinetic-energy correction factor.

3.30 For highly turbulent flow the velocity distribution in a pipe is given by

$$\frac{v}{v_{max}} = \left(\frac{y}{r_0}\right)^{1/9}$$

with y the wall distance and r_0 the pipe radius. Determine the kinetic-energy correction factor for this flow.

3.31 If the losses from section A to section B of Fig. 3.43 are 1.9 ft·lb/lb, determine the two possible depths at section B.

3.32 In the situation shown in Fig. 3.44 each kilogram of water increases in temperature 0.0006°C because of losses incurred in flowing between A and B. Determine the lower depth of flow at section B.

3.33 In Fig. 3.44 the channel changes in width from 2 m at section A to 3 m at section B. For losses of 0.3 m·N/N between sections A and B, find the two possible depths at section B.

3.34 At point A in a pipeline carrying water the diameter is 1 m, the pressure 98 kPa, and the velocity 1 m/s. At point B, 2 m higher than A, the diameter is 0.5 m and the pressure 20 kPa. Determine the direction of flow.

3.35 For losses of 0.1 m·N/N, find the velocity at A in Fig. 3.51. The barometer reading is 750-mm Hg.

3.36 The losses in Fig. 3.52 for $H = 25$ ft are $3V^2/2g$ ft·lb/lb. What is the discharge?

3.37 For flow of 750 gpm in Fig. 3.52, determine H for losses of $10V^2/2g$ ft·lb/lb.

3.38 For 1500-gpm flow and $H = 32$ ft in Fig 3.52, calculate the losses through the system in velocity heads, $KV^2/2g$.

3.39 In Fig. 3.53 the losses up to section A are $5\ V_1^2/2g$ and the nozzle losses are $0.05V_2^2/2g$. Determine the discharge and the pressure at A. $H = 8$ m.

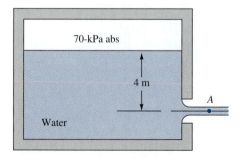

Figure 3.51   Problem 3.35.

3.40 In Fig. 3.53 for a pressure at A of 25 kPa and with the losses found in Prob. 3.39, determine the discharge and the head H.

3.41 The pumping system shown in Fig. 3.54 must have a pressure of 5 psi in the discharge line when cavitation is incipient at the pump inlet. Calculate the length of pipe from the reservoir to the pump for this operating condition if the loss in this pipe can be expressed as $(V_1^2/2g)(0.03\,L/D)$. What horsepower is being supplied to the fluid by the pump? What percent of this power is being used to overcome losses? The barometer reads 30-in. Hg.

3.42 In the siphon of Fig. 3.55, $h_1 = 1$ m, $h_2 = 3$ m, $D_1 = 3$ m, $D_2 = 5$ m, and the losses to section 2 are $2.6V_2^2/2g$, with 10 percent of the losses occurring before section 1. Find the discharge and the pressure at section 1.

3.43 Find the pressure at A of Prob. 3.42 if it is a stagnation point (velocity of zero).

3.44 The siphon of Fig. 3.18 has a nozzle 150-mm long attached at section 3, reducing the diameter to 150 mm. For no losses, compute the discharge and the pressure at sections 2 and 3.

3.45 In Prob. 3.44 with losses from 1 to 2 of $1.7V_2^2/2g$, from 2 to 3 of $0.9V_2^2/2g$, and through the nozzle of $0.06V_E^2/2g$, where V_E is the exit velocity, calculate the discharge and pressure at sections 2 and 3.

3.46 Determine the shaft horsepower for an 80-percent efficient pump to discharge 30 L/s through the system of Fig. 3.56. The system losses, exclusive of pump losses, are $12V^2/2g$ and $H = 16$ m.

3.47 The fluid horsepower produced by the pump of Fig. 3.56 is $Q\gamma H_p/550 = 10$. For $H = 70$ ft and system losses of $8V^2/2g$, determine the discharge and the pump head H_p. Draw the energy grade line.

3.48 If the overall efficiency of the system and turbine in Fig. 3.57 is 80 percent, what horsepower is produced for $H = 200$ ft and $Q = 1000$ cfs?

3.49 Losses through the system of Fig. 3.57 are $4V^2/2g$, exclusive of the turbine. The turbine is 90-percent efficient and runs at 240 rpm. To produce 1000 hp for $H = 300$ ft, determine the discharge and torque in the turbine shaft. Draw the energy grade line.

3.50 With losses of $0.2V_1^2/2g$ between sections 1 and 2 of Fig. 3.47, calculate the flow in gallons per minute.

3.51 In Fig. 3.58 $H = 6$ m and $h = 5.75$ m. Calculate the discharge and the losses in meter-newtons per newton and in watts.

3.52 For losses of $0.1H$ through the nozzle of Fig. 3.50, what is the gage difference R in terms of H?

3.53 A liquid flows through a long pipeline with losses of 6 m·N/N per 30 m of pipe. What is the slope of the hydraulic and energy grade lines?

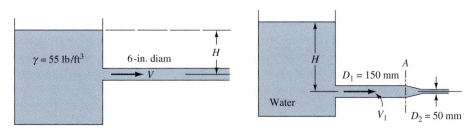

Figure 3.52   Problems 3.36, 3.37, and 3.38.   **Figure 3.53**   Problems 3.39 and 3.40.

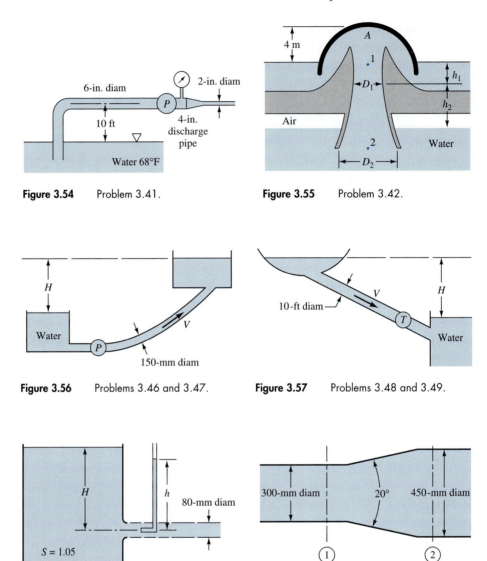

Figure 3.54   Problem 3.41.

Figure 3.55   Problem 3.42.

Figure 3.56   Problems 3.46 and 3.47.

Figure 3.57   Problems 3.48 and 3.49.

Figure 3.58   Problem 3.51.

Figure 3.59   Problem 3.54.

3.54   In Fig. 3.59 100 L/s of water flows from section 1 to section 2 with losses of $0.4(V_1 - V_2)^2/2g$; $p_1 = 80$ kPa. Compute p_2 and plot the energy and hydraulic grade lines through the diffuser.

3.55   In an isothermal, reversible flow at 200°F, 3 Btu/s heat is added to 14 slugs/s flowing through a control volume. Calculate the entropy increase, in foot-pounds per slug and Rankine degree.

3.56   In isothermal flow of a real fluid through a pipe system the losses are 20 m·N/kg per 100 m, and 0.0837 kJ/s per 100 m heat transfer from the fluid is required to hold the temperature at 10°C. What is the entropy change Δs in meter-newtons per kilogram-kelvin of pipe system if the fluid is flowing at 4 kg/s?

3.57   Determine the momentum correction factor for the velocity distribution of Prob. 3.29.

3.58   Calculate the average velocity and momentum correction factor for the velocity distribution in a pipe,

$$\frac{v}{v_{max}} = \left(\frac{y}{r_0}\right)^{1/n}$$

with y the wall distance and r_0 the pipe radius.

3.59   If gravity acts in the negative z direction, determine the z component of the force acting on the fluid within the cube described in Prob. 3.5 for the velocity specified there.

3.60   Find the y component of the force acting on the control volume given in Prob. 3.6 for the velocity given there. Consider gravity to be acting in the negative y direction.

3.61   What force components F_x and F_y are required to hold the black box of Fig. 3.60 stationary? All pressures are zero gage.

3.62   What force F (Fig. 3.61) is required to hold the plate from flowing oil, sp gr 0.83, for $V_0 = 20$ m/s?

3.63   How much is the apparent weight of the tank full of water (Fig. 3.62) increased by the steady jet flow into the tank?

3.64   Does a nozzle on a fire hose place the hose in tension or in compression?

3.65   When a jet from a nozzle is used to aid in maneuvering a fireboat, can more force be obtained by directing the jet against a solid surface such as a wharf than by allowing it to discharge into air?

3.66   Solve Example 3.12 with the flow direction reversed and compare the results.

3.67   In the reducing bend of Fig. 3.22, $D_1 = 4$ m, $D_2 = 3$ m, $\theta = 135°$, $Q = 50$ m^3/s, $W = 392.2$ kN, $z = 2$ m, $p_2 = 1.4$ MPa, $x = 2.2$ m, and losses may be neglected. Find the force components and the line of action of the force which must be resisted by an anchor block.

3.68   If 600 L/s of water flows through an 50-cm-diameter pipeline that contains a horizontal 90° bend and the pressure at the entrance to the bend is 140 kPa, determine the force components, parallel and normal to the approach velocity, required to hold the bend in place. Neglect losses.

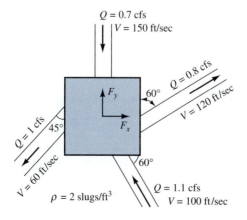

Figure 3.60   Problem 3.61.

3.69   Oil, sp gr 0.83, flows through a 90° expanding pipe bend from a 400- to 600-mm-diameter pipe. The pressure at the bend entrance is 130 kPa, and losses are to be neglected. For a flow of 0.6 m³/s, determine the force components (parallel and normal to the approach velocity) necessary to support the bend.

3.70   Solve Prob. 3.69 with elbow losses of $0.6V_1^2/2g$, and with V_1 being the approach velocity, and compare results.

3.71   A 100-mm-diameter steam line carries saturated steam at 425 m/s velocity. Water is entrained by the steam at the rate of 0.1 kg/s. What force is required to hold a 90° bend in place owing to the entrained water?

3.72   Neglecting losses, determine the x and y components of force needed to hold the Y (Fig. 3.63) in place. The plane of the Y is horizontal.

3.73   Determine the net force on the sluice gate shown in Fig. 3.64. Neglect losses. By noting that the pressure at A and B is atmospheric, sketch the pressure distribution on the surface AB. Is it a hydrostatic distribution? How is it related to the force just calculated?

3.74   The vertical reducing section shown in Fig. 3.65 contains oil, sp gr 0.86, flowing upward at the rate of 0.6 m³/s. The pressure at the larger section is 20 kPa. Neglecting losses but including gravity, determine the force on the contraction.

3.75   Apply the momentum and energy equations to a windmill as if it were a propeller, noting that the slipstream is slowed down and expands as it passes through the blades. Show that the velocity through the plane of the blades is the average of the velocities in the slipstream at the downstream and upstream sections. By defining the theoretical efficiency (neglecting all losses) as the power output divided by the power available in an undisturbed jet having the area at the plane of the blades, determine the maximum theoretical efficiency of a windmill.

3.76   An airplane with propeller diameter of 8.0 ft travels through still air ($\rho = 0.0022$ slug/ft³) at 200 mi/h. The speed of air through the plane of the propeller is 280 mi/h relative to the airplane. Calculate (*a*) the thrust on the plane, (*b*) the kinetic energy per second remaining in the slipstream, (*c*) the theoretical horsepower required to drive the propeller, (*d*) the propeller efficiency, and (*e*) the pressure difference across the blades.

3.77   A boat traveling at 40 km/h has a 500-mm-diameter propeller that discharges 4.5 m³/s through its blades. Determine the thrust on the boat, the theoretical efficiency of the propulsion system, and the power input to the propeller.

3.78   A ship propeller has a theoretical efficiency of 60 percent. If it is 3.2 ft in diameter and the ship travels 20 mi/h, what is the thrust developed and what is the theoretical horsepower required?

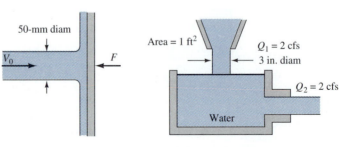

Figure 3.61   Problem 3.62.

Figure 3.62   Problems 3.63 and 3.110.

3.79 In Fig. 3.66, a jet, $\rho = 2$ slugs/ft^3, is deflected by a vane through 180°. Assume that the cart is frictionless and free to move in a horizontal direction. The cart weighs 200 lb. Determine the velocity and the distance traveled by the cart 10 s after the jet is directed against the vane. $A_0 = 0.02$ ft^2 and $V_0 = 100$ ft/s.

3.80 Draw the polar vector diagram for a vane, angle θ, doing work on a jet. Label all the vectors.

3.81 Determine the resultant force exerted on the vane of Fig. 3.26, $A_0 = 0.1$ ft^2; $V_0 = 100$ ft/s; $\theta = 60°$; and $\gamma = 60$ lb/ft^3. How can the line of action be determined?

3.82 In Fig. 3.27 45 percent of the flow is deflected in one direction. What is the plate angle θ?

3.83 A flat plate is moving with velocity u into a jet, as shown in Fig. 3.67. Derive the expression for power required to move the plate.

3.84 At what speed u should the cart of Fig. 3.67 move away from the jet in order to produce maximum power from the jet?

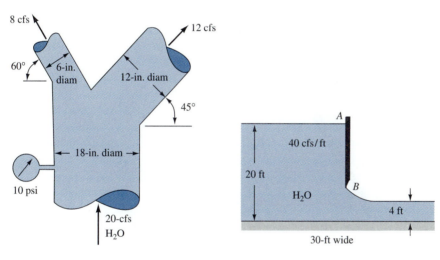

Figure 3.63 Problem 3.72.

Figure 3.64 Problem 3.73.

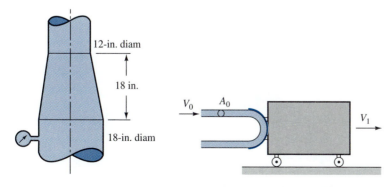

Figure 3.65 Problem 3.74.

Figure 3.66 Problems 3.79 and 3.118.

3.85   Calculate the force components F_x and F_y needed to hold the stationary vane of Fig. 3.68. $Q_0 = 80$ L/s; $\rho = 1000$ kg/m^3; and $V_0 = 120$ m/s.

3.86   If the vane of Fig. 3.68 moves in the x direction at $u = 40$ ft/s, for $Q_0 = 2$ ft^3/s, $\rho = 1.935$ slugs/ft^3, and $V_0 = 120$ ft/s, what are the force components F_x, F_y?

3.87   For the flow divider of Fig. 3.69, find the force components for the following conditions: $Q_0 = 10$ L/s, $Q_1 = 3$ L/s; $\theta_0 = 45°$, $\theta_1 = 30°$, $\theta_2 = 120°$; $V_0 = 10$ m/s; and $\rho = 830$ kg/m^3.

3.88   Solve the preceding problem by graphical vector addition.

3.89   At what speed u should the vane of Fig. 3.28 travel for maximum power from the jet? What should be the angle θ for maximum power?

3.90   Draw the polar vector diagram for the moving vane of Fig. 3.28 for $V_0 = 30$ m/s, $u = 20$ m/s, and $\theta = 160°$.

3.91   Draw the polar vector diagram for the moving vane of Fig. 3.28 for $V_0 = 40$ m/s, $u = -20$ m/s, and $\theta = 150°$.

3.92   What horsepower can be developed from (a) a single vane and (b) a series of vanes (Fig. 3.28), when $A_0 = 10$ in^2, $V_0 = 240$ ft/s, $u = 90$ ft/s, and $\theta = 173°$, for water flowing?

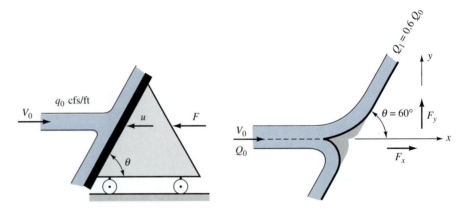

Figure 3.67   Problems 3.83 and 3.84.   **Figure 3.68**   Problems 3.85 and 3.86.

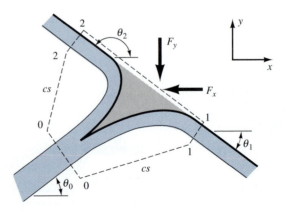

Figure 3.69   Problem 3.87.

3.93 Determine the blade angles θ_1 and θ_2 of Fig. 3.70 so that the flow enters the vane tangent to its leading edge and leaves with no x component of absolute velocity.

3.94 Determine the vane angle required to deflect the absolute velocity of a jet 130° (Fig. 3.71).

3.95 In Prob. 3.18 for a pickup of 40 L/s water at a locomotive speed of 60 km/h, what force is exerted parallel to the tracks?

3.96 Figure 3.72 shows an orifice called a *Borda mouthpiece*. The tube is long enough for the fluid velocity near the bottom of the tank to be nearly zero. Calculate the ratio of the jet area to the tube area.

3.97 Determine the irreversibility in foot-pounds per slug for 5 ft³/s flow of liquid, $\rho = 1.6$ slugs/ft³, through a sudden expansion from a 12- to 24-in.-diameter pipe. $g = 30$ ft/s².

3.98 Air flows through a 650-mm-diameter duct at $p = 70$ kPa, $t = 10°C$, and $V = 60$ m/s. The duct suddenly expands to 800-mm diameter. Considering the gas as incompressible, calculate the losses, in meter-newtons per newton of air, and the pressure difference, in centimeters of water.

3.99 What are the losses when 4 m³/s water discharges from a submerged 1.5-m-diameter pipe into a reservoir?

3.100 Show that, in the limiting case, as $y_1 = y_2$ in Eq. (3.7.11), the relation $V = \sqrt{gy}$ is obtained.

3.101 A jump occurs in a 6-m-wide channel carrying 15 m³/s of water at a depth of 300 mm. Determine y_2, V_2, and the losses in meter-newtons per newton, in kilowatts, and in kilojoules per kilogram.

3.102 Derive an expression for hydraulic jump in a channel having an equilateral triangle as its cross section (symmetric with the vertical).

3.103 Derive Eq. (3.7.12).

3.104 Assuming no losses through the gate of Fig. 3.73 and neglecting $V_0^2/2g$, for $y_0 = 20$ ft and $y_1 = 2$ ft, find y_2 and losses through the jump. What is the basis for neglecting $V_0^2/2g$?

3.105 Under the same assumptions as in Prob. 3.104, for $y_1 = 400$ mm and $y_2 = 2$ m, determine y_0.

3.106 Under the same assumptions as in Prob. 3.104, $y_0 = 20$ ft and $y_2 = 8$ ft. Find the discharge per foot.

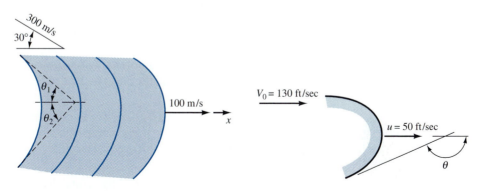

Figure 3.70   Problem 3.93.

Figure 3.71   Problem 3.94.

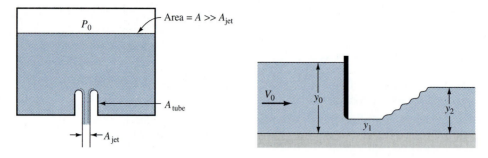

Figure 3.72   Problem 3.96.   **Figure 3.73**   Problem 3.104 to 3.106.

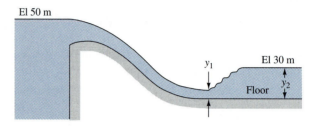

Figure 3.74   Problem 3.107.

3.107   For losses down the spillway of Fig. 3.74 of 2 m·N/N and discharge per meter of 10 m³/s, determine the floor elevation for the jump to occur.

3.108   Water is flowing through the pipe of Fig. 3.75 with velocity $V = 8.02$ ft/s and losses of 10 ft·lb/lb up to section 1. When the obstruction at the end of the pipe is removed, calculate the acceleration of water in the pipe.

3.109   Water fills the piping system of Fig. 3.76. At one instant $p_1 = 10$ psi, $p_2 = 0$, $V_1 = 10$ ft/s, and the flow rate is increasing by 3000 gpm/min. Find the force F_x required to hold the piping system stationary.

3.110   If in Fig. 3.62 Q_2 is 1.0 cfs, what is the vertical force to support the tank? Assume that overflow has not occurred. The tank weighs 20 lb, and water depth is 1 ft.

3.111   In Fig. 3.37b, $r_1 = 120$ mm, $r_2 = 160$ mm, $v_{t_1} = 0$, and $v_{t_2} = 3$ m/s for a centrifugal pump impeller discharging 0.2 m³/s of water. What torque must be exerted on the impeller?

3.112   In a centrifugal pump 25 L/s of water leaves a 200-mm-diameter impeller with a tangential velocity component of 10 m/s. It enters the impeller in a radial direction. For a pump speed of 1200 rpm and neglecting all losses, determine the torque in the pump shaft, the power input, and the energy added to the flow in meter-newtons per newton.

3.113   A water turbine at 240 rpm discharges 40 m³/s. To produce 42 MW, what must be the tangential component of velocity at the entrance to the impeller at $r = 1.6$ m? All whirl is taken from the water when it leaves the turbine. Neglect all losses. What head is required for the turbine?

3.114   The symmetrical water sprinkler of Fig. 3.77 has a total discharge of 14 gpm and is frictionless. Determine its rpm if the diameter of the nozzle tips is $\frac{1}{4}$ in.

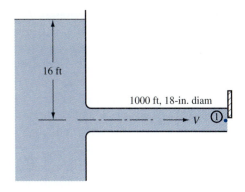

Figure 3.75 Problem 3.108.

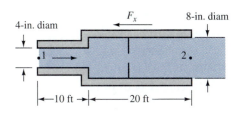

Figure 3.76 Problem 3.109.

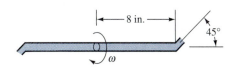

Figure 3.77 Problems 3.114 to 3.117.

3.115 What torque would be required to hold the sprinkler of Prob. 3.114 stationary? Total flow is 2 L/s of water.

3.116 If there is a torque resistance of 0.50 lb·ft in the shaft of Prob. 3.114, what is its speed of rotation?

3.117 For torque resistance of $0.01\omega^2$ in the shaft, determine the speed of rotation of the sprinkler of Prob. 3.114.

3.118 If the resistance to motion of the cart of Prob. 3.79 is $5V_1^2$ lb, determine the velocity and distance traveled in 0.6 s. Use Runge-Kutta solution (second-order) with $H = \frac{1}{64}$ s.

3.119 The thermal conductivity of dry concrete at 68°F is 0.128 W/m·K. What is the value of its thermal conductivity in W/cm·°C, and in Btu/h·ft·°F?

3.120 The thermal conductivity of glass fiber at 20°C is 0.202 Btu/h·ft·°F. What is the value of its thermal conductivity in watts per centimeter per degree centigrade?

3.121 The temperature difference between the inner and outer surface of a concrete wall is 25°C. If the wall is 10-in. thick and its thermal conductivity is 0.98 W/m·K, determine the heat loss per unit area through the wall.

3.122 If the thermal conductivity of a liquid is a linear function of the temperature, derive an expression for the one-dimensional steady temperature distribution as a function of the heat transfer rate q_H, the cross-sectional area A, and the distance x.

3.123 Water of temperature 95.3°F is mixed with water of temperature 43.4°F. The resulting water mass is 10 kg and has temperature of 65°F. Find the mass of the water with temperature 95.3°F.

3.124 Find the conversion factor between 1 cal and 1 Btu.

3.125   A piece of platinum (Pt) with initial temperature T_{Pt} is immersed in a pan containing Hg of temperature T_{Hg}. The same mass of Pt is immersed in another pan that contains twice the amount of Hg with temperature T'_{Hg}. If the final temperatures of the platinum are T_{Pt_1} and T'_{Pt_2} respectively, find the initial temperature of the platinum, T'_{Pt}, in the second experiment.

3.126   Three similar containers contain water of mass m_1, water of mass m_2, and glycerin of mass m_3, respectively. The same amount of heat, Q, is added in all three containers. If the temperature increase in the three containers is ΔT_1, ΔT_2, and ΔT_3 respectively, find the specific heat of the glycerin.

3.127   Water is flowing through a heat exchanger at a constant rate of 6.85 L/min. The temperature of the water at the inlet is 58°C, while the temperature at the outlet is 14°C. What is the amount of heat removal in the heat exchanger per second?

3.128   A piece of stainless steel of mass 0.95 kg, and temperature 65°C is dropped into a heat exchanger which contains 0.4 kg of water with temperature 15°C. If the specific heat of the stainless steel is 460 J/kg·K, what is its final temperature?

3.129   Two containers contain water at 20°C and 92°C, respectively. How much water should be taken out from each container so that the resulting water mixture has a volume 362 L and a temperature of 30°C?

3.130   A heat exchanger made from copper (Cu) of mass 325 g contains 0.4 kg of oil of initial temperature 20°C. A piece of chrome steel with mass 100 g and temperature 85°C is immersed in the tank. The specific heat of Cu is 410 J/kg·K and that of the chrome steel is 460 J/kg·K. Find the specific heat of the oil if the final temperature in the heat exchanger is 28°C.

3.131   Into glycerin of temperature 15°C is dropped a piece of zinc (Zn) of temperature 85°C. The total mass of the glycerin and the Zn is 0.4 kg, and the equilibrium temperature is 22°C. If the specific heat of the glycerin is 2428 J/kg·K and that of the Zn is 0.094 Btu/lb$_m$·°F, find the mass of the glycerin and the Zn.

3.132   The temperature in a container of water drops from 70°C to 60°C in 200 sec. How much time does it take for the temperature to drop from 59°C to 55°C?

3.133   A thermometer initially reading 82°C and made from material with a specific heat of 0.2 Btu/lb$_m$·°F is immersed into the fluid of a heat exchanger with temperature 23°C and a specific heat of 0.098 Btu/lb$_m$·°F. If the thermometer has a mass of 75 g and the mass of the fluid in the heat exchanger is 6.5 kg, find the reading of the thermometer at the equilibrium temperature.

3.134   A bar made from copper (Cu) and an iron bar (Fe) are in contact as shown in Fig. 3.78. The two ends of the system are kept in constant temperatures $T_1 = 0$°C and $T_2 = 100$°C, respectively. The cross-sectional area of the system is 15 cm^2 and the two thermal conductivities are $k_{Cu} = 25.32$ W/m·K and $k_{Fe} = 52$ W/m·K, respectively. What is the heat flux through the system? In what direction does the heat flow?

3.135   The removal efficiency of a reactor tank is 73 percent and its residence time is 28.5 s. For steady flow and first-order kinetics determine the reaction rate coefficient (k) for the tank.

3.136   The efficient operation of a wastewater plant requires, at first stage, that the wastewater should pass through a series of n equally sized, completely mixed reactors. If the concentration of the suspended solids at the inlet is C_o and the flow rate through the system is Q, determine the concentration of the suspended solids at the outlet.

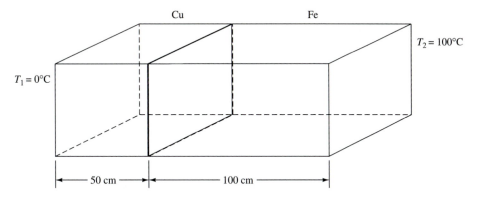

Figure 3.78   Problem 3.134.

3.137   For a reactor tank it is known that at steady state the concentration at the outlet is 22 mg/L. If the concentration at the inlet is 100 mg/L, determine the concentration at the outlet 2.5 min after the process is started. The residence time in the reactor is 8 min.

3.138   In Prob. 3.137 the volume of the tank is 25 m^3. Determine the flow rate at the inlet and the time required to reach steady state.

3.139   A gas mixture of components A and B is flowing in a pipe. The concentrations of the two components are C_A and C_B, respectively, while the average molar velocity of the whole fluid is u_M. Determine the total convective flux, N_A, for component A in the gas mixture as a function of C_A, C, and N_B, where C is the total concentration.

3.140   A gas mixture of species A and B is contained in a pipe. At two sections (1) and (2) the partial pressures of component A are p_1 and p_2, respectively. Find an expression for the diffusion flux J_A, as a function of p_1 and p_2. (Assume steady-state conditions.)

3.141   The following data were collected in a well-mixed reactor tank:

	t_1	t_2	t_3	t_4	t_5	t_6	t_7
C_2 (mg/L)	8.50	7.20	6.50	4.95	3.78	2.50	1.32
dC_2/dt (mg/L/hr)	1.90	2.60	3.75	4.10	5.00	7.00	7.44

where C_2 is the concentration at the outlet. The concentration at the inlet is 23 mg/L. Estimate the residence time t_R and the reaction rate coefficient, k.

3.142   If steady state for the conditions described in Prob. 3.141 is reached at 45-percent efficiency, determine the time required to reach steady state.

REFERENCES

1.   V. L. Streeter, "The Kinetic Energy and Momentum Correction Factors for Pipes and Open Channels of Great Widths," *Civ. Eng.*, N.Y., vol. 12, no. 4, pp. 212–213, 1942.

ADDITIONAL READING

Bird, R., Stewart, W., and Lightfoot, E.: *Transport Phenomena,* John Wiley and Sons, New York, 1968.

Brodkey, R. and Hershey, H.: *Transport Phenomena: A Unified Approach,* McGraw Hill Co., New York, 1988.

Eckart, E. and Drake, R.: *Analysis of Heat and Mass Transfer,* McGraw Hill Co., New York, 1972.

Kreyszig, E.: *Advanced Engineering Mathematics,* 7th ed., Wiley, New York, 1993.

Shames, I.: *Mechanics of Fluids,* 3rd ed., McGraw-Hill, New York, 1992.

chapter

4

Basic Governing Differential Equations

The control volume approach in Chap. 3 is a powerful technique used by the science and engineering community to make accurate calculations. Many of the piping systems used for water supply and wastewater collection are designed with this concept, and initial designs or feasibility studies for most other projects also use this approach. Increasingly however the engineering and scientific community is called upon to provide detailed information about point-by-point variation of velocity, elevation, temperature, or dissolved substances. For this type of analysis the control volume method can no longer be used exclusively.

To arrive at a point-by-point description requires use of the same conservation of mass, momentum, and energy laws used in the control volume analysis as well as the Eulerian or fixed-point approach, but the level of mathematical rigor must now be raised. Two choices are available. The first is the *computational fluid mechanics and transport approach* where the original control volume or system is now considered to be comprised of a large number of small control volumes. Fluxes of the relevant dependent variables are passed between the elements, constrained by the laws of mechanics, and a digital accounting procedure is established to calculate the variables in all the cells or control volumes. This approach will be used in the later stages of this book to handle problems that possess complex geometry or certain nonlinear or steep spatial or temporal gradients.

The *continuum field* approach is developed in this chapter and essentially arrives at a series of nonlinear partial differential equations for each of the laws of mechanics whose solutions give the point-by-point variation in the variables. Here the control volume(s) and accompanying conservation equations are essentially allowed to "shrink" to a point in the limiting sense of the calculus and point-by-point equations result. Before proceeding to the resulting equations, however, some preliminary concepts are reviewed. As in Chap. 3 extensive use of vector notation is made in this chapter and the reader is referred to Appendix E for review.

4.1   KINEMATICS, MOTION, AND DEFORMATION

The use of the Eulerian approach in fluids analysis requires more care in describing the motion of a fluid parcel and the parcel distributions resulting from the motions. In this section methods for describing fluid parcel velocity and acceleration are presented. The methods are extended to general temporal derivatives and are used to quantitatively describe the four types of fluid motion that a fluid parcel is subjected to.

Acceleration at a Point

Consider in Fig. 4.1 a fluid parcel moving from point 1 to point 2 during time dt. The position vector $\mathbf{R}$ is defined as in Eq. (4.1.1)

$$\mathbf{R} = x(t)\mathbf{i} + y(t)\mathbf{j} + z(t)\mathbf{k} \tag{4.1.1}$$

The time rate of change of $\mathbf{R}$ is found from the total differentiation of Eq. (4.1.1) to be

$$\frac{d\mathbf{R}}{dt} = \frac{dx}{dt}\frac{dt}{dt}\mathbf{i} + \frac{dy}{dt}\frac{dt}{dt}\mathbf{j} + \frac{dz}{dt}\frac{dt}{dt}\mathbf{k} \tag{4.1.2}$$

The total velocity vector $\mathbf{v}$ at a point is defined by noting that $dt/dt = 1$ and that, for example, the x direction component fluid parcel velocity is $dx/dt = u$; then

$$\mathbf{v}(x, y, z, t) = \frac{d\mathbf{R}}{dt} = u\mathbf{i} + v\mathbf{j} + w\mathbf{k} \tag{4.1.3}$$

It is noted that the velocity vector is a function of position and time.

The acceleration vector is found from the following approach

$$\mathbf{a}(x, y, z, t) = \frac{d\mathbf{v}(x, y, z, t)}{dt} = \frac{du}{dt}\mathbf{i} + \frac{dv}{dt}\mathbf{j} + \frac{dw}{dt}\mathbf{k}$$

Looking at the x component, chain rule differentiation gives

$$a_x = \frac{du}{dt}(x, y, z, t)$$

$$= \frac{\partial u}{\partial t}\frac{dt}{dt} + \frac{\partial u}{\partial x}\frac{dx}{dt} + \frac{\partial u}{\partial y}\frac{dy}{dt} + \frac{\partial u}{\partial z}\frac{dz}{dt}$$

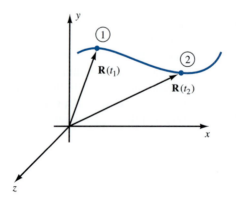

Figure 4.1   Coordinate and velocity definitions.

which can be written as

$$a_x = \frac{du}{dt} = \frac{\partial u}{\partial t} + u\frac{\partial u}{\partial x} + v\frac{\partial u}{\partial y} + w\frac{\partial u}{\partial z} \qquad (4.1.4a)$$

Similar expressions are found for a_y and a_z,

$$a_y = \frac{dv}{dt} = \frac{\partial v}{\partial t} + u\frac{\partial v}{\partial x} + v\frac{\partial v}{\partial y} + w\frac{\partial v}{\partial z} \qquad (4.1.4b)$$

$$a_z = \frac{dw}{dt} = \frac{\partial w}{\partial t} + u\frac{\partial w}{\partial x} + v\frac{\partial w}{\partial y} + w\frac{\partial w}{\partial z}. \qquad (4.1.4c)$$

Using the vector differential operations summarized in Appendix E, Eqs. (4.1.4a–c) can be written in vector form as

$$\mathbf{a} = \frac{\partial \mathbf{v}}{\partial t} + u\frac{\partial \mathbf{v}}{\partial x} + v\frac{\partial \mathbf{v}}{\partial y} + w\frac{\partial \mathbf{v}}{\partial z} = \frac{\partial \mathbf{v}}{\partial t} + (\mathbf{v} \cdot \nabla)\mathbf{v} \qquad (4.1.5)$$

Here $\mathbf{v} \cdot \nabla$ is a vector operator of the form

$$(\mathbf{v} \cdot \nabla)() = u\frac{\partial ()}{\partial x} + v\frac{\partial ()}{\partial y} + w\frac{\partial ()}{\partial z}$$

Example 4.1

For a position vector given by

$$\mathbf{R} = (5xyt^2 + zt)\mathbf{i} + (-2.5y^2t^2 + zt + 3yt)\mathbf{j} + (-3zt + \frac{x}{2}t^2)\mathbf{k}$$

find the velocity and acceleration functions.

Solution

From Eq. (4.1.2)

$$\mathbf{v} = \frac{d\mathbf{R}}{dt} = \left[\frac{d}{dt}(5xyt^2 + zt)\right]\mathbf{i} + \left[\frac{d}{dt}(-2.5y^2t^2 + zt + 3yt)\right]\mathbf{j} + \left[\frac{d}{dt}(-3zt + \frac{x}{2}t^2)\right]\mathbf{k}$$

or

$$\mathbf{v} = (10xyt + z)\mathbf{i} + (-5y^2t + z + 3y)\mathbf{j} + (-3z + xt)\mathbf{k}$$

The acceleration vector is found from Eq. (4.1.5)

$$\mathbf{a} = \frac{\partial \mathbf{v}}{\partial t} + (10xyt + z)\frac{\partial \mathbf{v}}{\partial x} + (-5y^2t + z + 3y)\frac{\partial \mathbf{v}}{\partial y} + (-3z + xt)\frac{\partial \mathbf{v}}{\partial z}$$

Examining the partial derivatives individually

$$\frac{\partial \mathbf{v}}{\partial t} = 10xy\mathbf{i} - 5y^2\mathbf{j} + x\mathbf{k}$$

$$\frac{\partial \mathbf{v}}{\partial x} = 10yt\mathbf{i} + 0\mathbf{j} + t\mathbf{k}$$

$$\frac{\partial \mathbf{v}}{\partial y} = 10xt\mathbf{i} + (-10yt + 3)\mathbf{j} + 0\mathbf{k}$$

$$\frac{\partial \mathbf{v}}{\partial z} = 1\mathbf{i} + 1\mathbf{j} - 3\mathbf{k}$$

and collecting all of the terms gives

$$\mathbf{a} = (50xy^2t^2 + 10xzt + 10yzt + 10xy + 30xyt - 3z + xt)\mathbf{i}$$
$$+ (50y^3t^2 - 30y^2t - 20y^2 - 10yzt + 3xt + 9y)\mathbf{j}$$
$$+ (10xyt^2 + zt - 3xt + x + 9z)\mathbf{k}$$

Inspection of Eqs. (4.1.4) and (4.1.5) reveals that acceleration is composed of two general classes of terms, *temporal acceleration*, that is, the temporal gradients of the velocity vector; and *inertial acceleration* terms comprised of products of the velocity and the spatial gradients of the flow. That terms comprised of spatial gradients should be called acceleration is one of the results of the use of the Eulerian approach and is counterintuitive to the Lagrangian description comprised solely of temporal gradients. A brief qualitative example serves to explain the difference between the two.

Example 4.2

Consider the piston pumping fluid through the nozzle in Fig.4.2a and the two simpler configurations in Figs. 4.2b and c. In Fig. 4.2b the diameter is constant and the change in velocity at point P is solely due to oscillatory or pumping motion of the piston. The time trace of V_p indicates this temporal acceleration. Fig. 4.2c is a steady flow example through the nozzle where, from the continuity equation applied between points 1 and 2 surrounding point P, $V_2 > V_1$, that is, a velocity change corresponding to an inertial acceleration of $V_p(V_2 - V_1)/\Delta x$ has occurred which in the limit as Δx goes to 0 becomes $u\,\partial u/\partial x$. This acceleration results purely from the squeezing of the flow field introduced by the rapid change in the geometry of the flow.

The Total Derivative

The total time rate of change of any variable in the Eulerian system is found in an analogous manner to Eqs. (4.1.2) or (4.1.4). Any variable, be it a scalar or a vector, say α, whose values are a function of position and time is differentiated as follows

$$\frac{d\alpha}{dt}(x, y, z, t) = \frac{\partial \alpha}{\partial t}\frac{dt}{dt} + \frac{\partial \alpha}{\partial x}\frac{dx}{dt} + \frac{\partial \alpha}{\partial y}\frac{dy}{dt} + \frac{\partial \alpha}{\partial z}\frac{dz}{dt}$$

or

$$\frac{d\alpha}{dt} = \frac{\partial \alpha}{\partial t} + u\frac{\partial \alpha}{\partial x} + v\frac{\partial \alpha}{\partial y} + w\frac{\partial \alpha}{\partial z} = (\mathbf{v} \cdot \nabla)\alpha \qquad \text{(4.1.6)}$$

The general operation is called the *total* or *substantial derivative*.

To sort out the various types of temporal derivatives the engineer encounters, the description in Bird, Stewart, and Lightfoot (*Transport Phenomena*, John Wiley Co., 1968) is paraphrased:

An engineer is asked to count and plot the time rate of change of birds, b, which can be done in three ways:

1. The engineer sits at a fixed site and counts the rate of change of b at a fixed observation point in the sky; this derivative is

$$\frac{db}{dt} = \frac{\partial b}{\partial t}$$

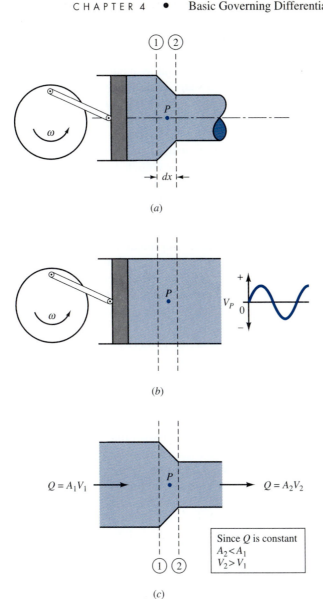

Figure 4.2   (a) Total, (b) temporal, and (c) inertial acceleration.

2. A pilot takes the engineer aloft in a plane that travels with a fixed velocity vector $\mathbf{v}_p = u_p\mathbf{i} + v_p\mathbf{j} + w_p\mathbf{k}$. The engineer counts the rate of change of birds as observed out the window by the plane seat; the time rate of change of b is then

$$\frac{db}{dt} = \frac{\partial b}{\partial t} + u_p\frac{\partial b}{\partial x} + v_p\frac{\partial b}{\partial y} + w_p\frac{\partial b}{\partial z}$$

3. Finally the engineer counts the time rate of change of birds from a balloon floating with the local wind or fluid velocity, $\mathbf{v} = u\mathbf{i} + v\mathbf{j} + w\mathbf{k}$; then the derivative is expressed as

$$\frac{db}{dt} = \frac{\partial b}{\partial t} + u\frac{\partial b}{\partial x} + v\frac{\partial b}{\partial y} + w\frac{\partial b}{\partial z}$$

which is the form used in the Eulerian analysis approach adopted here.

Motion and Deformation

As a fluid parcel moves from one instant to another there are four types of motion and/or deformation of the parcel shape. These include simple *translation* and *rotation* as well as *volume dilation* and *angular deformation*. The first two are examples of linear motion in that the original shape of the parcel is unchanged though its position and orientation may be. The second two represent a deformation of the original shape. Here a brief geometric introduction to these processes will be made by way of relating the velocity fields to resulting motion and deformation. It should be noted that all forms of these motions can occur simultaneously, but for simple understanding each will be considered separately and in two dimensions. Generalization to three dimensions will be assumed to follow.

Translation (Fig. 4.3a) means simply picking up the parcel and moving a distance during a small time period dt. No rotation of the parcel is permitted, nor is any deformation. Deformation will be measured by the degree to which the angle between any pair of lines which were originally orthogonal to each other deforms during the time dt. For the case of translation then the 90° angle between any pair of orthogonal lines defining any plane in the parcel must remain constant. Pure translation, without any other deformation or rotation, can only occur in a very special velocity field; that is, the flow must be spatially uniform and contain no spatial gradients.

Dilation (Fig. 4.3b) refers to a stretching or compressing of the parcel induced by a spatial gradient in the velocity field. No deformation is permitted; rather, only a linear extension or compression of the orthogonal axes defining the plane is permitted. The velocity field accompanying this change is again restricted. For example, in Fig. 4.3b the change of shape to the dashed line configuration preserves the 90° angle between all the orthogonal axes but the velocity field is restricted to changes only in the direction of the axes. Therefore for the figure in the x direction, only u can vary, not v; while in the y direction only v can change and not u. Whether the shape change results in a volume change is an extremely important question. From the figure the original volume V is found to be $dx\,dy$. In the rearranged form the incremental length changes during time period dt are found from a Taylor's series expansion (correct to first order) as labeled in Fig. 4.3; therefore the volume at time dt later is found as

$$V_{t+dt} = \left(dx + \frac{\partial u}{\partial x}\,dx\,dt\right)\left(dy + \frac{\partial v}{\partial y}\,dy\,dt\right)$$

After multiplying the terms and dropping second and higher order terms, the time rate of change of relative volume V_R can be found in terms of the velocity field as

$$\frac{d\left(\frac{V_{t+dt}-V_t}{V_t}\right)}{dt} = \frac{dV_R}{dt} = \frac{\partial u}{\partial x} + \frac{\partial v}{\partial y}$$

In the full three dimensions,

$$\frac{dV_R}{dt} = \frac{\partial u}{\partial x} + \frac{\partial v}{\partial y} + \frac{\partial w}{\partial z} = \nabla \cdot \mathbf{v} \tag{4.1.7}$$

Therefore, volume dilation can be directly related to the spatial structure of the velocity gradients, and this relationship will take on significant physical meaning as regards the continuity equation in Sec. 4.3.

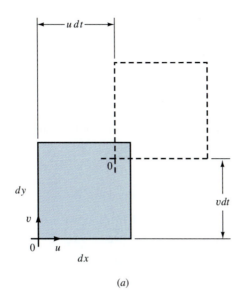

(a)

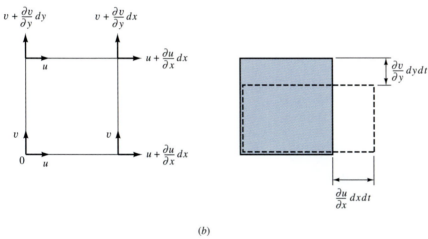

(b)

Figure 4.3 (a) Translation and (b) dilation of a fluid parcel and their relationship to the velocity field gradients. (*continues next page*)

Rotation is defined as the average angular velocity of two elements originally at right angles to each other. As can be seen in Fig. 4.3c spatial gradients in the velocity field or shear must be present to sustain rotation over the period dt. With regard to the element dx, and for small angles,

$$\tan d\theta_1 \simeq d\theta_1 \simeq \left(\frac{\partial v}{\partial x}\, dx\, dt\right)/dx$$

therefore,

$$\dot{\theta}_1 = \frac{d\theta_1}{dt} = \frac{\partial v}{\partial x}$$

For the vertical element dy

$$\tan d\theta_2 \approx d\theta_2 \approx -\frac{\partial u}{\partial y}\,dy\,dt$$

therefore,

$$\dot\theta_2 = \frac{d\theta_2}{dt} = -\frac{\partial u}{\partial y}$$

The average of the two is the parcel's angular velocity about the z axis

$$\omega_z = \frac{1}{2}\left(\frac{\partial v}{\partial x} - \frac{\partial u}{\partial y}\right) = \frac{1}{2}\left(\dot\theta_1 + \dot\theta_2\right) \tag{4.1.8a}$$

Rotation about the other two axes is defined as

$$\omega_y = \frac{1}{2}\left(\frac{\partial u}{\partial z} - \frac{\partial w}{\partial x}\right) \tag{4.1.8b}$$

$$\omega_x = \frac{1}{2}\left(\frac{\partial w}{\partial y} - \frac{\partial v}{\partial z}\right) \tag{4.1.8c}$$

Angular velocity is a vector quantity

$$\boldsymbol{\Omega} = \omega_x \mathbf{i} + \omega_y \mathbf{j} + \omega_z \mathbf{k}$$

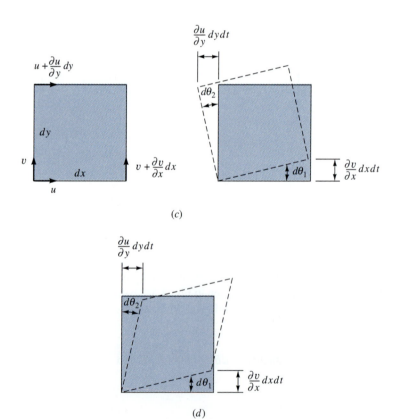

(c)

(d)

Figure 4.3   (continued from previous page) (c) Rotation and (d) deformation of a fluid parcel and their relationship to the velocity field gradients.

Vorticity is defined as twice the angular velocity; therefore,

$$\Gamma = 2\Omega = \left(\frac{\partial w}{\partial y} - \frac{\partial v}{\partial z}\right)\mathbf{i} + \left(\frac{\partial u}{\partial z} - \frac{\partial w}{\partial x}\right)\mathbf{j} + \left(\frac{\partial v}{\partial x} - \frac{\partial u}{\partial y}\right)\mathbf{k} \tag{4.1.9}$$

It is left to the reader to show that

$$\Gamma = \nabla \times \mathbf{v} \tag{4.1.10}$$

Irrotational flow occurs when the cross gradients of the velocity (or shear) are zero or (in the quite unlikely case) cancel each other out. Figure 4.4 contains a schematic of a fluid parcel traveling on a streamline down a gradual spillway. Parcels in Fig. 4.4a are moving in an irrotational flow as there is no rotation of a pair of orthogonal axes embedded in the parcel. Figure 4.4b shows the rotational analogy.

Angular deformation or *strain rate* is defined as the average of the difference in angular velocities of two originally perpendicular elements. Again spatial velocity gradients or shear must be present. From Figure 4.3d the previous deformations hold and for the velocity field indicated in the sketch,

$$\tan d\theta_1 \simeq d\theta_1 = \left(\frac{\partial u}{\partial y} dy\,dt\right)/dy = \frac{\partial u}{\partial y} dt$$

$$\dot{\theta}_1 = \frac{d\theta_1}{dt} = -\frac{\partial u}{\partial y}$$

The minus sign occurs as a result of the clockwise rotation which is negative. A similar derivation for $\dot{\theta}_2$ yields

$$\dot{\theta}_2 = \frac{d\theta_2}{dt} = \frac{\partial v}{\partial y}$$

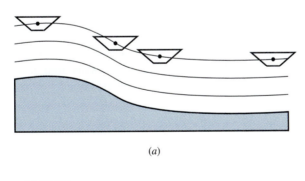

(a)

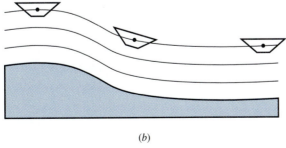

(b)

Figure 4.4   (a) Irrotational and (b) rotational flow.

therefore,

$$\varepsilon_z = \frac{1}{2}\left(\dot{\theta}_1 - \dot{\theta}_2\right) = \frac{1}{2}\left(\frac{\partial v}{\partial x} + \frac{\partial u}{\partial y}\right)$$

(4.1.11a)

and in the other two dimensions

$$\varepsilon_x = \frac{1}{2}\left(\frac{\partial v}{\partial z} + \frac{\partial w}{\partial y}\right)$$

(4.1.11b)

$$\varepsilon_y = \frac{1}{2}\left(\frac{\partial u}{\partial z} + \frac{\partial w}{\partial x}\right)$$

(4.1.11c)

Example 4.3

In a two-dimensional vertical plane many wind-driven water waves can be described by a linearized velocity field of the form

$$u = A(z)\cos\left(\frac{2\pi x}{L} - \frac{2\pi t}{T}\right) = A(z)\cos\Theta$$

$$w = B(z)\sin\left(\frac{2\pi x}{L} - \frac{2\pi t}{T}\right) = B(z)\sin\Theta$$

where L is the wavelength, T is the wave period, and from elementary physics the ratio of L/T is the wave phase speed C. $A(z)$ and $B(z)$ are amplitude functions which depend on the depth, z, measured from the still water level or average water surface. They are defined as

$$A(z) = \frac{h}{2}\frac{gT}{L}\frac{\cosh\left[2\pi(z + d)/L\right]}{\cosh(2\pi d/L)}$$

$$B(z) = \frac{h}{2}\frac{gT}{L}\frac{\sinh\left[2\pi(z + d)/L\right]}{\cosh(2\pi d/L)}$$

The problem is to find the vorticity and strain rate at the water surface ($z = 0$) and at the depth $z = -5$ m, when the horizontal velocity is a maximum. Assume the wave height $H = 2$ m, the wavelength $L = 50$ m, and the period $T = 6$ s.

Solution

In the xy plane the angular velocity is found from Eq (4.1.8b) while Eq. (4.1.9) states the vorticity equals $2\omega_y$, that is,

$$\Omega_y = \left(\frac{\partial u}{\partial z} - \frac{\partial w}{\partial x}\right)$$

From the above formula for the wave velocities

$$\Omega_y = \frac{\partial}{\partial z}\left[A(z)\cos\Theta\right] - \frac{\partial}{\partial x}\left[B(z)\sin\Theta\right]$$

or

$$\Omega_y = \left[\cos\Theta\frac{\partial}{\partial z}A(z)\right] - \left[B(z)\frac{\partial}{\partial x}\sin\Theta\right]$$

After some algebra the final expression for vorticity (or angular velocity) is

$$\Omega_y = \left(\frac{\pi}{L^2}gTH\right) \cdot \frac{\sinh\left[2\pi(z+d)/L\right]}{\cosh(2\pi d/L)} \cdot \cos\left(\frac{2\pi x}{L} - \frac{2\pi t}{T}\right)$$
$$- \left(\frac{\pi}{L^2}gTH\right) \cdot \frac{\sinh\left[2\pi(z+d)/L\right]}{\cosh(2\pi d/L)} \cdot \cos\left(\frac{2\pi x}{L} - \frac{2\pi t}{T}\right)$$

or $\Omega_y = 0$. In other words, there is no rotation of fluid parcels in the flow field. In fact the irrotational flow assumption is central to the derivation of the linear wave field in Chap. 8.

The strain rate ε_y is considerable

$$\varepsilon_y = \frac{1}{2}\left(\frac{\partial u}{\partial z} + \frac{\partial w}{\partial x}\right)$$

$$\varepsilon_y = \left(\frac{2\pi gTH}{L^2}\right) \cdot \frac{\sinh\left[2\pi(z+d)/L\right]}{\cosh(2\pi d/L)} \cdot \cos\left(\frac{2\pi x}{L} - \frac{2\pi t}{T}\right)$$

If this formula is evaluated for the condition when the horizontal velocity is a maximum, then $\cos\Theta = 1$ and for $z = 0$

$$\varepsilon_y = [0.295]\left[\frac{1.6}{2.36}\right][1] = 0.2 \text{ s}^{-1}$$

For $z = -5$ m

$$\varepsilon_y = [0.295]\left[\frac{0.67}{2.36}\right][1] = 0.084 \text{ s}^{-1}$$

4.2 THE GENERAL REYNOLDS TRANSPORT EQUATION

In order to derive the general governing differential equations valid at a point, two choices are available. The first is the simple expansion of relevant variables by a Taylor's series expansion about an elemental volume followed by a limit analysis as the volume shrinks to a point. A second approach, yielding the same results, derives a systematic representation from the Reynolds control volume equation [Eq. (3.2.4)] valid for any conservation property.

The Reynolds control volume equation is

$$\frac{dN}{dt} = \frac{\partial}{\partial t}\int_{cv} \rho\eta\,d\forall + \int_{cs} \eta\rho\mathbf{v}\cdot d\mathbf{A}$$

where η is the mass of the constituent per unit mass of the mixture. It is desirable to have the surface integral be a volume integral, and to do so the Gauss divergence theorem is employed. If a vector $\mathbf{B}$ and its spatial derivatives exist, that is, no discontinuities in either are permitted, then the divergence theorem says that

$$\int_{cv} \nabla\cdot\mathbf{B}\,d\forall = \int_{cs} \mathbf{B}\cdot\hat{n}\,dA$$

In the theorem $\hat{n}$ is the unit normal vector. Proof of this theorem can be found in many texts, for example E. Kreyszig, *Advanced Engineering Mathematics,* 7th ed., or I. H. Shames, *Mechanics of Fluids*. In nonvector form this equation says that

$$\int_{cv}\left(\frac{\partial B_x}{\partial x} + \frac{\partial B_y}{\partial y} + \frac{\partial B_z}{\partial z}\right)d\Psi = \int_{cs}\left(B_x \cos\alpha_x + B_y \cos\alpha_y + B_z \cos\alpha_z\right)dA$$

where α_x, α_y, and α_z are the angles between the unit normal, $\hat{n}$, and the positive x, y, and z axes, respectively.

Using this theorem in the control volume formulation gives

$$\frac{dN}{dt} = \int_{cv}\left[\frac{\partial}{\partial t}(\rho\eta) + \nabla\cdot(\eta\rho\mathbf{v})\right]d\Psi \qquad (4.2.1)$$

N is defined as the volume integral of the intensive (per unit volume) quantity n, that is,

$$N = \int_{cv} n\,d\Psi$$

Then Eq. (4.2.1) can be rewritten as

$$\frac{d}{dt}\int_{cv} n\,d\Psi = \int_{cv}\left[\frac{\partial}{\partial t}(\rho\eta) + \nabla\cdot(\eta\rho\mathbf{v})\right]d\Psi$$

By letting the parcel shrink to an infinitesimally small (but not molecular in size) parcel, the mean value theorem prevails and the particle volume $\int d\Psi$ divides out leaving

$$\frac{dn}{dt} = \frac{\partial}{\partial t}(\rho\eta) + \nabla\cdot(\eta\rho\mathbf{v}) \qquad (4.2.2)$$

Equation (4.2.2) is the Reynolds differential transport theorem.

4.3   THE CONTINUITY EQUATION

The conservation of mass or continuity equation is of fundamental importance as it must hold in every flow field no matter what type of simplifying assumptions have been made. The total time rate of change of mass per unit volume must be equal to zero; therefore, $dn/dt = 0$, and η, the mass per unit mass, equals 1. Equation (4.2.2) becomes

$$\frac{\partial\rho}{\partial t} + \nabla\cdot(\rho\mathbf{v}) = 0 \qquad (4.3.1)$$

or

$$\frac{\partial\rho}{\partial t} + \frac{\partial(\rho u)}{\partial x} + \frac{\partial(\rho v)}{\partial y} + \frac{\partial(\rho w)}{\partial z} = 0 \qquad (4.3.2)$$

This equation holds for all flow fields.

By using the chain rule, the differentiation terms may be reclustered to give

$$\frac{\partial\rho}{\partial t} + u\frac{\partial\rho}{\partial x} + v\frac{\partial\rho}{\partial y} + w\frac{\partial\rho}{\partial z} + \rho\left(\frac{\partial u}{\partial x} + \frac{\partial v}{\partial y} + \frac{\partial w}{\partial z}\right) = 0 \qquad (4.3.3a)$$

and from Eq. (4.1.5) it is seen that this can be written in vector form as

$$\frac{\partial \rho}{\partial t} + (\mathbf{v} \cdot \nabla)\rho + \rho(\nabla \cdot \mathbf{v}) = 0 \qquad\qquad \textbf{(4.3.3b)}$$

or

$$\frac{1}{\rho}\frac{D\rho}{Dt} + (\nabla \cdot \mathbf{v}) = 0 \qquad\qquad \textbf{(4.3.3c)}$$

where D/Dt stands for the substantial or total derivative.

For an incompressible fluid ρ is constant; therefore,

$$\frac{1}{\rho}\frac{D\rho}{Dt} = 0$$

which gives

$$(\nabla \cdot \mathbf{v}) = \frac{\partial u}{\partial x} + \frac{\partial v}{\partial y} + \frac{\partial w}{\partial z} = 0 \qquad\qquad \textbf{(4.3.4)}$$

as the continuity equation for incompressible flow.

Physical insight to the restriction that the continuity equation places on the flow can be seen from Eq. (4.1.7), where it was shown that the volume dilation or time rate of relative volume change was related to

$$\frac{d\mathcal{V}_r}{dt} = \nabla \cdot \mathbf{v}$$

Contrasting Eq. (4.1.7) to the continuity equation for an incompressible fluid in Eq. (4.3.4), it is seen that the time rate of change of relative volume for an incompressible fluid must *always* equal zero. A velocity vector is not proper if it does not satisfy Eq. (4.3.4) for an incompressible fluid or Eqs. (4.3.3a–c) for a compressible fluid.

Does the velocity field from the acceleration example (Ex. 4.1) satisfy the continuity equation?

Example 4.4

Solution

In Ex. 4.1 the velocity vector was

$$\mathbf{v} = (10xyt + z)\mathbf{i} + \left(-5y^2t + z + 3y\right)\mathbf{j} + \left(-3z + xt\right)\mathbf{k}$$

For continuity to apply, $\nabla \cdot \mathbf{v} = 0$, so

$$\frac{\partial u}{\partial x} = \frac{\partial}{\partial x}(10xyt + z) = 10yt$$

$$\frac{\partial v}{\partial y} = \frac{\partial}{\partial y}\left(-5y^2t + z + 3y\right) = -10yt + 3$$

$$\frac{\partial w}{\partial z} = \frac{\partial}{\partial z}\left(-3z + xt\right) = -3$$

Summing

$$10yt - 10yt + 3 - 3 = 0$$

and continuity is satisfied.

For further practice the reader should verify that the two-dimensional continuity equation

$$\frac{\partial u}{\partial x} + \frac{\partial w}{\partial z} = 0$$

is satisfied by the linear wave velocity field in Ex. 4.3.

4.4   THE MOMENTUM EQUATION

The Basic Vector Equation

The force-momentum balance results in a vector differential equation consisting of three nonlinear partial differential equations. These are the most difficult equations to solve in fluid mechanics. With reference to Eq. (4.2.2) η is now the momentum per unit mass or $m\mathbf{v}/m = \mathbf{v}$, while dn/dt equals the force vector per unit volume $\mathbf{f}$. From Eq. (4.2.2) then

$$\mathbf{f} = \frac{\partial}{\partial t}(\rho\mathbf{v}) + \nabla \cdot (\rho\mathbf{v}\mathbf{v}) \tag{4.4.1}$$

the product $\rho\mathbf{v}\mathbf{v}$ is called the dyadic product which can be quickly simplified by chain rule differentiation and expansion of Eq. (4.2.1) to give

$$\mathbf{f} = \mathbf{v}\left(\frac{\partial \rho}{\partial t} + \nabla \cdot (\rho\mathbf{v})\right) + \rho\left(\frac{\partial \mathbf{v}}{\partial t} + (\mathbf{v} \cdot \nabla)\mathbf{v}\right) \tag{4.4.2}$$

The first term in parentheses is the continuity equation which from Eq. (4.3.1) equals zero. The second term in parentheses is the acceleration vector $\mathbf{a}$ from Eq. (4.1.5); therefore Eq. (4.4.2) is recognized as the Eulerian form of Newton's second law expressed in a "per unit volume" basis. The component form of the vector equation is

$$f_x = \rho\left[\frac{\partial u}{\partial t} + u\frac{\partial u}{\partial x} + v\frac{\partial u}{\partial y} + w\frac{\partial u}{\partial z}\right] \tag{4.4.3a}$$

$$f_y = \rho\left[\frac{\partial v}{\partial t} + u\frac{\partial v}{\partial x} + v\frac{\partial v}{\partial y} + w\frac{\partial v}{\partial z}\right] \tag{4.4.3b}$$

$$f_z = \rho\left[\frac{\partial w}{\partial t} + u\frac{\partial w}{\partial x} + v\frac{\partial w}{\partial y} + w\frac{\partial w}{\partial z}\right] \tag{4.4.3c}$$

Equations (4.3.1), (4.4.2), and (4.4.3a–c) should form a complete solvable set of equations. However, even if we assume ρ to be known, there are still four equations and six unknowns (u, v, w, f_x, f_y, and f_z). Therefore, $\mathbf{f}$ must be specified in terms of u, v, and w or aspects of the flow geometry (which is known).

The Force Description

The force vector is broken into a surface force and body force per unit volume

$$\mathbf{f} = \mathbf{f}_s + \mathbf{f}_g$$

where the body force vector solely results from gravity and the surface force arises from direct fluid particle contact. An elemental volume δV (Fig. 4.5) is configured

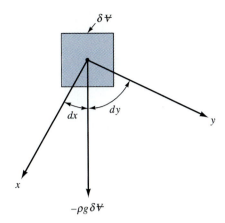

Figure 4.5 An elemental fluid volume.

in a coordinate system x, y, z such that it is rotated, locally, relative to a coordinate system where the vertical coordinate, h, aligns with gravity. The weight component per unit volume of the parcel (ρg) in each coordinate direction is given by

$$\mathbf{f}_g = \rho g \left(\cos \alpha_x \mathbf{i} + \cos \alpha_y \mathbf{j} + \cos \alpha_z \mathbf{k} \right)$$

Here δV has been divided out to place the expression on the per unit volume basis required by Eq. (4.4.3). The angles $(\alpha_x, \alpha_y, \text{ and } \alpha_z)$ are the angles between the vertical (h) coordinate and the local coordinate system. In vector notation an equivalent form is

$$\mathbf{f}_g = -\rho g \, \nabla h \qquad (4.4.4)$$

A popular equivalent way to express the gravity force in local coordinates is to create a new acceleration field due to gravity which coincides with the local coordinates

$$\mathbf{f}_g = -\rho \left(g_x \mathbf{i} + g_y \mathbf{j} + g_z \mathbf{k} \right)$$

where, for example,

$$g_x = g \cos \alpha_x$$

Finally it is noticed that when the local and global coordinate systems coincide,

$$\alpha_x = 90° \qquad \alpha_y = 90° \qquad \alpha_z = 0°$$

then

$$\cos \alpha_x = 0 \qquad \cos \alpha_y = 0 \qquad \cos \alpha_z = 1$$

and

$$\mathbf{f}_g = -\rho g \mathbf{k} \qquad (4.4.5)$$

The vector stress force per unit volume is found via a Taylor's series expansion. Using the definition of stress from Chap. 2, a stress distribution is configured on the parallelopiped in Fig. 4.6. Here only the x-component force per unit volume is used as an example. A Taylor's series expansion is used to allow for differentially different values between the two planes in the figure.

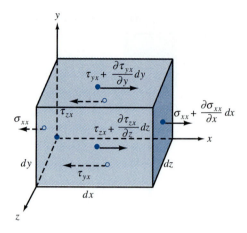

Figure 4.6   Stress force components in the x direction.

$$F_x = \left[\left(\sigma_{xx} + \frac{\partial \sigma_{xx}}{\partial x}dx\right)dy\,dz + \left(\tau_{yx} + \frac{\partial \tau_{yx}}{\partial y}dy\right)dx\,dz + \left(\tau_{zx} + \frac{\partial \tau_{zx}}{\partial z}dz\right)dx\,dy\right]$$
$$- \left[\sigma_{xx}dy\,dz + \tau_{yx}dx\,dz + \tau_{zx}dx\,dy\right]$$

Expanding out terms, eliminating the lowest order terms, and dividing by the volume $dx\,dy\,dz$ give

$$f_{sx} = \frac{\partial \sigma_{xx}}{\partial x} + \frac{\partial \tau_{yx}}{\partial y} + \frac{\partial \tau_{zx}}{\partial z} \tag{4.4.6a}$$

Similar expressions for the y and z components result in

$$f_{sy} = \frac{\partial \tau_{xy}}{\partial x} + \frac{\partial \sigma_{yy}}{\partial y} + \frac{\partial \tau_{zy}}{\partial z} \tag{4.4.6b}$$

$$f_{sz} = \frac{\partial \tau_{xz}}{\partial x} + \frac{\partial \tau_{yz}}{\partial y} + \frac{\partial \sigma_{zz}}{\partial z} \tag{4.4.6c}$$

It is customary to separate the pressure from the normal shear stress term. This is accomplished by subtracting the bulk stress term from each of the three normal stress terms and employing the relationship between bulk stress and pressure, that is, $p = -\sigma$; therefore, for example, $\tau_{yy} + \sigma = \tau_{yy} - p = \sigma_{yy}$ and Eqs. (4.4.6a–c) are rewritten as

$$f_{sx} = -\frac{\partial p}{\partial x} + \left(\frac{\partial \tau_{xx}}{\partial x} + \frac{\partial \tau_{yx}}{\partial y} + \frac{\partial \tau_{zx}}{\partial z}\right) \tag{4.4.7a}$$

$$f_{sy} = -\frac{\partial p}{\partial y} + \left(\frac{\partial \tau_{xy}}{\partial x} + \frac{\partial \tau_{yy}}{\partial y} + \frac{\partial \tau_{zy}}{\partial z}\right) \tag{4.4.7b}$$

$$f_{sz} = -\frac{\partial p}{\partial z} + \left(\frac{\partial \tau_{xz}}{\partial x} + \frac{\partial \tau_{yz}}{\partial y} + \frac{\partial \tau_{zz}}{\partial z}\right) \tag{4.4.7c}$$

These terms can be combined with Eqs. (4.4.3) and (4.4.4) and written in vector form as

$$\rho\left(\frac{\partial \mathbf{v}}{\partial t} + (\mathbf{v} \cdot \nabla)\mathbf{v}\right) = -\rho g \nabla h - \nabla p + \nabla \cdot \overline{\overline{\tau}}^{*} \tag{4.4.8}$$

The notation $\overline{\overline{\tau}}^*$ stands for the stress tensor in Eqs. (4.4.7). It is noted however that the analysis so far has only made the problem worse in that there are now 13 unknowns (u, v, w, p, and the stress field) while there still are only four equations.

The Navier-Stokes Equations

It remained for Navier (Louis Marie Henri, 1785–1836) and Stokes (George Gabriel, 1819–1903) to finish the derivation by relating the stress field to the deformation of the field resulting from the time and space varying velocity field. Here, Stokes' viscosity law, a generalization of Newton's viscosity law, is invoked (see I. H. Shames, *Mechanics of Fluids,* for a theoretical discussion). If it is assumed that the fluid is incompressible, then the following relations can be shown to hold

$$\tau_{xx} = 2\mu\frac{\partial u}{\partial x} \tag{4.4.9}$$

$$\tau_{yy} = 2\mu\frac{\partial v}{\partial y}$$

$$\tau_{zz} = 2\mu\frac{\partial w}{\partial z}$$

and

$$\tau_{xy} = \tau_{yx} = 2\mu\dot{\varepsilon}_z = \mu\left(\frac{\partial u}{\partial y} + \frac{\partial v}{\partial x}\right) \tag{4.4.10}$$

$$\tau_{xz} = \tau_{zx} = 2\mu\dot{\varepsilon}_y = \mu\left(\frac{\partial w}{\partial x} + \frac{\partial u}{\partial z}\right)$$

$$\tau_{yz} = \tau_{zy} = 2\mu\dot{\varepsilon}_x = \mu\left(\frac{\partial v}{\partial z} + \frac{\partial w}{\partial y}\right)$$

Therefore, the core of the shear stress relationship is the linear dependence of shear stress to strain or deformation rate [Eqs. (4.1.11a–c)] with the coefficient of proportionality being Newton's viscosity coefficient.

Assembling Eqs. (4.4.9) and (4.4.10) into Eq. (4.4.8) gives the *Navier-Stokes equations*

$$\rho\frac{D\mathbf{v}}{Dt} = \rho\left(\frac{\partial\mathbf{v}}{\partial t} + (\mathbf{v}\cdot\nabla)\mathbf{v}\right) = -\rho g\nabla h - \nabla p + \mu\nabla^2\mathbf{v} \tag{4.4.11}$$

or in component form

$$\rho\frac{Du}{Dt} = \rho\left(\frac{\partial u}{\partial t} + u\frac{\partial u}{\partial x} + v\frac{\partial u}{\partial y} + w\frac{\partial u}{\partial z}\right) = -\rho g\frac{\partial h}{\partial x} - \frac{\partial p}{\partial x} + \mu\left(\frac{\partial^2 u}{\partial x^2} + \frac{\partial^2 u}{\partial y^2} + \frac{\partial^2 u}{\partial z^2}\right)$$

$$\tag{4.4.12}$$

$$\rho\frac{Dv}{Dt} = \rho\left(\frac{\partial v}{\partial t} + u\frac{\partial v}{\partial x} + v\frac{\partial v}{\partial y} + w\frac{\partial v}{\partial z}\right) = -\rho g\frac{\partial h}{\partial y} - \frac{\partial p}{\partial y} + \mu\left(\frac{\partial^2 v}{\partial x^2} + \frac{\partial^2 v}{\partial y^2} + \frac{\partial^2 v}{\partial z^2}\right)$$

$$\rho\frac{Dw}{Dt} = \rho\left(\frac{\partial w}{\partial t} + u\frac{\partial w}{\partial x} + v\frac{\partial w}{\partial y} + w\frac{\partial w}{\partial z}\right)$$

$$= -\rho g\frac{\partial h}{\partial z} - \frac{\partial p}{\partial z} + \mu\left(\frac{\partial^2 w}{\partial x^2} + \frac{\partial^2 w}{\partial y^2} + \frac{\partial^2 w}{\partial z^2}\right)$$

While a number of subsequent chapters will be devoted to special cases and solutions of these equations, two points should be noted. First, if all fluid motion is stopped ($\mathbf{v} = 0$ and $\mathbf{a} = 0$) and z is selected as vertical up in the line of action of gravity, the hydrostatic equation emerges as a special case of the general case of full fluid motion. Indeed the study of the Navier-Stokes equations must concentrate on collections of specialized solutions such as this for specific flow conditions for the following reason. Due to the nonlinearities in these equations, there is an almost bewildering variety of possible results, so much so that these equations have never been solved completely in a general analytical fashion. In fact the U.S. National Academy of Sciences has listed their full solution as a top priority in its list of Grand Challenges, a list of the most pressing problems to be tackled by the science community.

4.5   THE CONSERVATION OF MECHANICAL ENERGY AND THE BERNOULLI EQUATION

Euler's Equation along a Streamline

One of the earliest approaches to solving the Navier-Stokes equations is to simply argue that the friction terms are quite small relative to the other terms and to eliminate them. This assumption results in Euler's Equations of Motion for an inviscid flow field. From Eq. (4.4.11) the equations become

$$\rho\frac{D\mathbf{v}}{Dt} = \rho\left[\frac{\partial \mathbf{v}}{\partial t} + (\mathbf{v} \cdot \nabla)\mathbf{v}\right] = -\rho g\nabla h - \nabla p \qquad \textbf{(4.5.1)}$$

In full three-dimensional form these are still challenging equations. However by concentrating on motion along a streamline, Euler's equations along a streamline may be developed. When integrated it yields Bernoulli's equation, which is a statement of conservation of mechanical energy between any two points along a streamline.

In Fig. 4.7 a prismatic control volume of very small size, with cross-sectional area δA and length δs, is selected. By virtue of the definition of a streamline, the only velocity permitted is along streamline s. By assuming that the viscosity is zero,

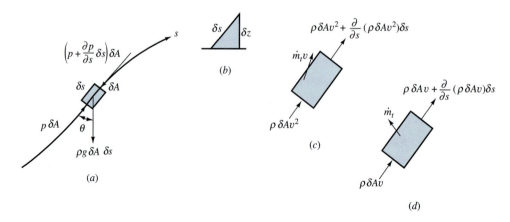

Figure 4.7   Application of continuity and momentum to flow through a control volume in the streamline, s, direction.

or that the flow is frictionless, the only forces acting on the control volume in the s direction are the end forces and the weight. The momentum equation is applied to the control volume for the s component.

$$\Sigma F_s = \frac{\partial}{\partial t}(\rho v)\,\delta s \delta A + \sum_{cs} \rho v \mathbf{v} \cdot d\mathbf{A} \tag{4.5.2}$$

where δs and δA are not functions of time. The acting forces are

$$\Sigma F_s = p\,\delta A - \left(p\,\delta A + \frac{\partial p}{\partial s}\,\delta s\,\delta A\right) - \rho g\,\delta s\,\delta A\,\cos\theta$$

$$= -\frac{\partial p}{\partial s}\,\delta s\,\delta A - \rho g\,\frac{\partial z}{\partial s}\,\delta s\,\delta A \tag{4.5.3}$$

since, as s increases, the vertical coordinate increases in such a manner that $\cos\theta = \partial z/\partial s$.

The net efflux of s momentum must consider flow through the cylindrical surface $\dot{m}_t$, as well as flow through the end faces (Fig. 4.7c).

$$\sum_{cs} \rho v \mathbf{v} \cdot d\mathbf{A} = \dot{m}_t v - \rho\,\delta A v^2 + \left[\rho\,\delta A\,v^2 + \frac{\partial}{\partial s}(\rho\,\delta A\,v^2)\,\delta s\right] \tag{4.5.4}$$

To determine the value of $\dot{m}_t$, the continuity equation [Eq. (4.3.2)] is applied to the control volume (Fig. 4.7d)

$$0 = \frac{\partial \rho}{\partial t}\,\delta A\,\delta s + \dot{m}_t + \frac{\partial}{\partial s}(\rho v)\,\delta A\,\delta s \tag{4.5.5}$$

Now, eliminating $\dot{m}_t$ in Eqs. (4.5.4) and (4.5.5) and simplifying give

$$\sum_{cs} \rho v \mathbf{v} \cdot d\mathbf{A} = \left(\rho v \frac{\partial v}{\partial s} - v\frac{\partial \rho}{\partial t}\right)\delta A\,\delta s \tag{4.5.6}$$

Next, substituting Eqs. (4.5.3) and (4.5.6) into Eq. (4.5.2) results in

$$\left(\frac{\partial p}{\partial s} + \rho g \frac{\partial z}{\partial s} + \rho v \frac{\partial v}{\partial s} + \rho \frac{\partial v}{\partial t}\right)\delta A\,\delta s = 0$$

After dividing through by $\rho\,\delta A\,\delta s$ and by taking the limit as δA and δs approach zero, the equation reduces to

$$\frac{1}{\rho}\frac{\partial p}{\partial s} + g\frac{\partial z}{\partial s} + v\frac{\partial v}{\partial s} + \frac{\partial v}{\partial t} = 0 \tag{4.5.7}$$

Two assumptions have been made: (1) that the flow is along a streamline and (2) that the flow is frictionless. If the flow is also steady, Eq. (4.5.7) reduces to

$$\frac{1}{\rho}\frac{\partial p}{\partial s} + g\frac{\partial z}{\partial s} + v\frac{\partial v}{\partial s} = 0 \tag{4.5.8}$$

Now s is the only independent variable, and total differentials can replace the partials.

$$\frac{dp}{\rho} + g\,dz + v\,dv = 0 \tag{4.5.9}$$

Equation (4.5.9) is Euler's equation along a streamline.

An alternative derivation of Euler's equation is as follows: At a point in the fluid construct an element δs of the streamline, with δz taken in the vertical upward direction, then the component of Eq. (2.2.5) in the s direction is

$$\frac{\partial p}{\partial s} = -\gamma \frac{\partial z}{\partial s} - \rho a_s \tag{4.5.10}$$

Since the acceleration component a_s of the particle along the streamline is a function of distance s along the streamline and time, then a total differential of a_s gives

$$a_s = \frac{dv}{dt} = \frac{\partial v}{\partial s}\frac{ds}{dt} + \frac{\partial v}{\partial t} = v\frac{\partial v}{\partial s} + \frac{\partial v}{\partial t}$$

as ds/dt is the time rate of displacement of the particle, which is its velocity v. It is readily noticed that this is a simplified form of the acceleration definition found in Sec. 4.1. After rearranging Eq. (4.5.10) with substitution of a_s, Eq. (4.5.7) is obtained. A frictionless fluid was assumed in deriving Eq. (2.2.5), and the component along s, the streamline, was taken in Eq. (4.5.10); hence, the same assumptions are made in obtaining Eq. (4.5.7).

Equation (4.5.9) is one form of Euler's equation, which required the three assumptions (1) frictionless flow, (2) motion along a streamline, and (3) steady flow. It can be integrated if ρ is a function of p or is constant. When ρ is constant, the Bernoulli equation is obtained.

The Bernoulli Equation

Integration of Eq. (4.5.9) for constant density yields the Bernoulli equation

$$gz + \frac{v^2}{2} + \frac{p}{\rho} = \text{constant} \tag{4.5.11}$$

The constant of integration (called the *Bernoulli constant*) in general varies from one streamline to another but remains constant along a *streamline* in *steady, frictionless, incompressible* flow. These four assumptions are needed and must be kept in mind when applying this equation. Each term has the dimensions $(L/T)^2$ or the units meter-newtons per kilogram

$$\frac{\text{m·N}}{\text{kg}} = \frac{\text{m·kg·m/s}^2}{\text{kg}} = \frac{\text{m}^2}{\text{s}^2}$$

because $1\ \text{N} = 1\ \text{kg·m/s}^2$. Therefore, Eq. (4.5.11) is interpreted as energy per unit mass. When it is divided by g,

$$z + \frac{v^2}{2g} + \frac{p}{\gamma} = \text{const} \tag{4.5.12}$$

it can be interpreted as energy per unit weight, meter-newtons per newton (or foot-pounds per pound). This form is particularly convenient for dealing with liquid problems with a free surface.

Each of the terms of Bernoulli's equation [Eq. (4.5.12)] may be interpreted as a form of available energy. This equation also is referred to as a *conservation of mechanical energy* equation. It is particularly important to note that this energy equation was derived from the conservation of momentum equation. Incorporation of energy losses due to friction and heat transfer can only be achieved with the full differential energy equation which is discussed in the next section.

By applying Eq. (4.5.12) to two points on a streamline,

$$z_1 + \frac{p_1}{\gamma} + \frac{v_1^2}{2g} = z_2 + \frac{p_2}{\gamma} + \frac{v_2^2}{2g} \qquad \text{(4.5.13)}$$

or

$$z_1 - z_2 + \frac{p_1 - p_2}{\gamma} + \frac{v_1^2 - v_2^2}{2g} = 0$$

This equation shows that it is the difference in potential energy, flow energy, and kinetic energy that actually has significance in the equation. Thus $z_1 - z_2$ is independent of the particular elevation datum as is the difference in elevation of the two points. Similarly, $p_1/\gamma - p_2/\gamma$ is the difference in pressure heads expressed in units of length of the fluid flowing and is not altered by the particular pressure datum selected. Since the velocity terms are not linear, their datum is fixed.

Water is flowing in an open channel (Fig. 4.8) at a depth of 2 m and a velocity of 3 m/s. It then flows down a contracting chute into another channel where the depth is 1 m and the velocity is 10 m/s. Assuming frictionless flow, determine the difference in elevation of the channel floors.

Example 4.5

Solution

The velocities are assumed to be uniform over the cross sections and the pressures hydrostatic. The points 1 and 2 can be selected on the free surface, as shown, or they can be selected at other depths. If the difference in elevation of floors is y, Bernoulli's equation is

$$\frac{V_1^2}{2g} + \frac{p_1}{\gamma} + z_1 = \frac{V_2^2}{2g} + \frac{p_2}{\gamma} + z_2$$

then $z_1 = y + 2$, $z_2 = 1$, $V_1 = 3$ m/s, $V_2 = 10$ m/s, and $p_1 = p_2 = 0$.

$$\frac{3^2}{2(9.806)} + 0 + y + 2 = \frac{10^2}{2(9.806)} + 0 + 1$$

and $y = 3.64$ m.

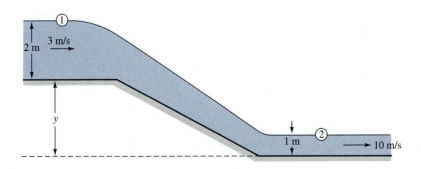

Figure 4.8   Open channel.

Figure 4.9   Venturi meter.

Example 4.6

A venturi meter, consisting of a converging portion followed by a throat portion of constant diameter and then a gradually diverging portion, is used to determine rate of flow in a pipe (Fig. 4.9). The diameter at section 1 is 6.0 in., and at section 2 it is 4.0 in. Find the discharge through the pipe when $p_1 - p_2 = 3$ psi and oil, sp gr 0.90, is flowing.

Solution

From the continuity equation

$$Q = A_1 V_1 = A_2 V_2 = \frac{\pi}{16} V_1 = \frac{\pi}{36} V_2$$

in which Q is the discharge (volume per unit time flowing). By applying Eq. (4.5.13) for $z_1 = z_2$

$$p_1 - p_2 = 3(144) = 432 \text{ lb/ft}^2 \qquad \gamma = 0.90(62.4) = 56.16 \text{ lb/ft}^3$$

$$\frac{p_1 - p_2}{\gamma} = \frac{V_2^2}{2g} - \frac{V_1^2}{2g} \quad \text{or} \quad \frac{432}{56.16} = \frac{Q^2}{\pi^2} \frac{1}{2g} (36^2 - 16^2)$$

Solving for discharge gives $Q = 2.20$ cfs.

With care each of the four assumptions underlying Bernoulli's equation can be relaxed and the equations used in the following four conditions.

1. When all streamlines originate from a reservoir, where the energy content is everywhere the same, the constant of integration does not change from one streamline to another and points 1 and 2 for application of Bernoulli's equation can be selected arbitrarily, that is, not necessarily on the same streamline.

2. In the flow of a gas, as in a ventilation system, where the change in pressure is only a small fraction (a few percent) of the absolute pressure, the gas can be considered incompressible.

3. For unsteady flow with gradually changing conditions, for example, emptying a reservoir, Bernoulli's equation can be applied without appreciable error.

4. Bernoulli's equation is of use in preliminary analysis of real-fluid cases by first neglecting viscous shear to obtain the results. Design results can then be obtained by use of the energy equation developed in Secs. 3.4 and 4.6.

Example 4.7

The water supply reservoir shown in Fig. 4.10 has an average depth of 20 m, a surface area of 20 km^2, and an outlet whose centerline is 15 m below the water surface. If the outflow diameter is 1 m, what is the outflow and its associated velocity? What would be the draw downs (drop in water surface elevation) during one-week and one-day periods?

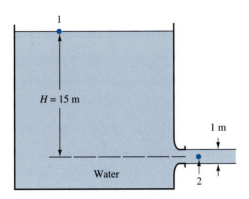

H = 15 m

1 m

Water

Figure 4.10   Flow through a nozzle from a reservoir.

Solution

In Fig. 4.10 the datum is placed at the centerline ($z_2 = 0$) of the outlet and the steady state form of Bernoulli's equation is applied

$$\frac{V_1^2}{2g} + z_1 = \frac{V_2^2}{2g} + z_2$$

Since the pressures at points 1 and 2 are atmospheric, the pressure of flow work terms cancel.

The velocity of the water surface (point 1) is assumed to be zero. If the outlet velocity, V_2, is assumed uniform across the outlet area, then

$$V_2 = \sqrt{2gz_1}$$

and from the given data

$$V_2 = 17.15 \text{ m/s}$$

The corresponding outflow is then

$$Q = V_2 A_2 = 13.46 \text{ m}^3/\text{s}$$

Conditions 1 and 3 above have been used in order to achieve this solution. The impact or applicability of them is checked as follows. Insofar as condition 1 is concerned it is a simple matter to move point 1 in Fig. 4.10 to other water surface locations and compute the same result. Condition 3 essentially says if the flow is weakly unsteady it can still be treated as a steady flow. To check this the volume of fluid lost out of the reservoir can be computed for one day and one week. The one-day volume removed is

$$V_1 = Q(24)(3600) = 1.163(10^6) \text{ m}^3$$

The seven-day volume removed is

$$V_7 = Q(7)(24)(3600) = 8.141(10^6) \text{ m}^3$$

As a percentage of the original reservoir volume these removals represent 0.3 and 2.0 percent of the total volume, respectively; these represent very small percentages.

These removals result in a steady drop of the water surface: the one-day drop being

$$0.003(20 \text{ m}) = 0.06 \text{ m} = 6 \text{ cm}$$

and the seven-day drop being 42 cm. The seven-day drop represents 2.1 percent of the total elevation resulting in an outlet velocity decrease from 17.15 m/sec to 16.9 m/sec, a change of -1.4 percent. Even the seven-day drawdown with its time-varying water surface can be well approximated by a steady flow assumption.

Finally the steady drop of 6 cm/day translates into a "velocity" of $0.06/(24 \times 3600) = 6.9(10^{-7})$ m/sec. Certainly the steady assumption works quite well for large reservoir volumes relative to daily discharge volumes.

4.6   THE ENERGY EQUATION

The conservation of energy equation must account for the sources, exchanges, and dissipation of energy in all its various forms. A formal development will proceed as it did in the continuity and momentum sections by defining η and n and proceeding to the full three-dimensional nonlinear equation. However, the incorporation of the second law of thermodynamics and the explanation of loss terms can more easily be handled by a simple derivation in which energy between any two points in the flow field is conserved. Therefore, this section begins with the two-point energy equation. In contrast to the Bernoulli equation the inviscid approximation and the assumption of motion along a streamline are both eliminated and a generalized energy equation between two points is derived.

The Two-Point Energy Equation

Figure 4.11 is a streamtube in the flow field with differentially small cross section inlet and outlet areas δA_1 and δA_2, respectively. The streamtube length is δs. Assuming only steady flow, the control volume energy equation [Eq. (3.4.15)] is applied to this elemental control volume

$$\frac{\delta Q_H}{\delta t} + \left(\frac{p_1}{\rho_1} + gz_1 + \frac{v_1^2}{2} + u_1^{**}\right)\rho_1 v_1\, \delta A_1 = \frac{\delta W_s}{\delta t} + \left(\frac{p_2}{\rho_2} + gz_2 + \frac{v_2^2}{2} + u_2^{**}\right)\rho_2 v_2\, \delta A_2$$

Dividing by the constant mass flow through the elemental volume, $\rho_1 v_1\, \delta A_1 = \rho_2 v_2\, \delta A_2$, gives

$$q_H + \frac{p_1}{\rho_1} + gz_1 + \frac{v_1^2}{2} + u_1^{**} = w_s + \frac{p_2}{\rho_2} + gz_2 + \frac{v_2^2}{2} + u_2^{**}$$

To take this control volume equation and find a differential equation valid at a point in the flow field is conceptually simple. First the equation is rearranged into a difference equation (for now shaft work is ignored without loss of rigor)

$$\left(\frac{p_2}{\rho_2} - \frac{p_1}{\rho_1}\right) + \left(\frac{v_2^2}{2} - \frac{v_1^2}{2}\right) + g(z_2 - z_1) + (u_2^{**} - u_1^{**}) - q_H = 0$$

When the control volume is allowed to shrink to a point, the equation becomes

$$d\left(\frac{p}{\rho}\right) + g\, dz + v\, dv + du^{**} - dq_H = 0 \tag{4.6.1}$$

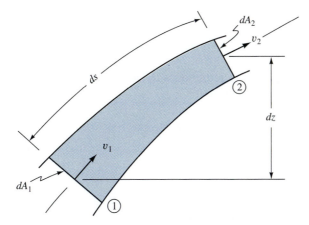

Figure 4.11   Steady stream tube.

This is a form of the *First Law of Thermodynamics*. Rearranging gives

$$\frac{dp}{\rho} + g\,dz + v\,dv + du^{**} + p\,d\left(\frac{1}{\rho}\right) - dq_H = 0 \qquad \text{(4.6.2)}$$

For *frictionless* flow the sum of the first three terms equals zero from the Euler equation [Eq. (4.5.8)] and the last three terms equal

$$dq_H = p\,d\left(\frac{1}{\rho}\right) + du^{**} \qquad \text{(4.6.3)}$$

Now, for reversible flow, *entropy, s* per unit mass is defined by

$$ds = \left(\frac{dq_H}{T}\right)_{rev} \qquad \text{(4.6.4)}$$

in which T is the absolute temperature. Entropy is shown to be a fluid property in texts on thermodynamics. In this equation it may have the units Btu per slug and Rankine degree or foot-pounds per slug and Rankine degree, as heat can be expressed in foot-pounds (1 Btu = 778 ft·lb). In SI units s is in joules per kilogram-kelvin. Since Eq. (4.6.3) is for a frictionless fluid (reversible), dq_H can be eliminated from Eqs. (4.6.3) and (4.6.4),

$$T\,ds = du^{**} + p\,d\left(\frac{1}{\rho}\right) \qquad \text{(4.6.5)}$$

which is a very important thermodynamic relation. Although it was derived for a reversible process, since all terms are thermodynamic properties, it must also hold for irreversible flow cases as well. The following is a brief discussion of the specification of losses by use of the Second Law of Thermodynamics.

Losses and the Second Law of Thermodynamics

The First Law of Thermodynamics must be "closed" in the sense that the internal energy and heat transfer terms must be expressed in variables related to the flow field. The Second Law of Thermodynamics is utilized. Substituting for $du^{**} + p\,d(1/\rho)$ in

Eq. (4.6.2) gives

$$dw_s + \frac{dp}{\rho} + v\,dv + g\,dz + T\,ds - dq_H = 0 \qquad \textbf{(4.6.6)}$$

Here the shaft work term is reintroduced. The Clausius inequality, or the Second Law of Thermodynamics, states that

$$ds \geq \frac{dq_H}{T}$$

or

$$T\,ds \geq dq_H \qquad \textbf{(4.6.7)}$$

Thus, $T\,ds - dq_H \geq 0$. The equals sign applies to a reversible process. If the quantity called losses or irreversibilities is identified as

$$d(\text{losses}) \equiv T\,ds - dq_H \qquad \textbf{(4.6.8)}$$

it is seen that $d(\text{losses})$ is positive in irreversible flow, is zero in reversible flow, and can never be negative. Substituting Eq. (4.6.8) into Eq. (4.6.6) yields

$$dw_s + \frac{dp}{\rho} + v\,dv + g\,dz + d(\text{losses}) = 0 \qquad \textbf{(4.6.9)}$$

This is a most important form of the energy equation. In general, the losses must be determined by experimentation. It implies that some of the available energy is converted into intrinsic energy during an irreversible process. Losses occur when some of the available energy in fluid flow is converted into thermal energy through viscous shear or turbulence. This equation, in the absence of the shaft work, differs from Euler's equation by the loss term only. In integrated form,

$$\frac{v_1^2}{2} + gz_1 = \int_1^2 \frac{dp}{\rho} + \frac{v_2^2}{2} + gz_2 + w_s + \text{losses}_{1-2} \qquad \textbf{(4.6.10)}$$

If work is done on the fluid in the streamtube, say for example by a pump, then w_s is negative. Station 1 is upstream, and station 2 is downstream.

If the flow is incompressible then $\rho \neq f(p)$, $\rho_1 = \rho_2 = \rho$, and Eq. (4.6.10) becomes

$$\frac{v_1^2}{2} + gz_1 + \frac{p_1}{\rho} = \frac{v_2^2}{2} + gz_2 + \frac{p_2}{\rho} + w_s + \text{losses}_{1-2} \qquad \textbf{(4.6.10a)}$$

or upon dividing by g

$$\frac{v_1^2}{2g} + z_1 + \frac{p_1}{\gamma} = \frac{v_2^2}{2g} + z_2 + \frac{p_2}{\gamma} + H_s + h_{l_{1-2}} \qquad \textbf{(4.6.10b)}$$

where H_s is the shaft work head and $h_{l_{1-2}}$ is the loss term in units of head, or head loss.

Comparison with the one-inlet–one-outlet control volume reveals considerable similarity in the mathematical form of the two equations. Formally the control volume form is comprised of bulk or average quantities such as area average inlet or outlet velocities or pressures, while due to its differential origins, Eqs. (4.6.10) are comprised of quantities valid at a point in the flow field. However, the streamtube origin of Eqs. (4.6.10) suggests that there will be considerable overlap or merging of these approaches particularly as the control volume geometry becomes small and the streamtube becomes larger. The analysis of conduit, piping, and pumping systems falls prey to this merging or blending of equations. What is important is that care must be exercised in remembering the mathematical requirements of the differential-point form and the control volume form. Because the latter requires averages while the former requires point data, the type of data supplied from field observations, for example, must be well known. It has been the case that in lieu of area average data, point data are substituted under the guise of uniform flow. This procedure is not recommended as it has been shown through the kinetic energy correction factor [Ex. 3.3] and it will be shown in Chap. 6 that significant error can occur.

4.7 THE DIFFERENTIAL HEAT EQUATION

Two different strategies are possible in deriving the differential equation. From Reynolds' equation, η is defined as the heat content per unit mass, $c_p T$, where T is the absolute temperature. The differential heat transport equation then becomes

$$\frac{\partial}{\partial t}\left(\rho c_p T\right) + \nabla \cdot \left(\rho c_p T \mathbf{v}\right) = \frac{dn}{dt}$$

The continuity equation can be eliminated by chain rule differentiation to give

$$\rho\left[\frac{\partial T}{\partial t} + \mathbf{v} \cdot \nabla\left(c_p T\right)\right] = \frac{dn}{dt}$$

In keeping with the emphasis on incompressible fluids, c_p can be assumed constant and after division by ρc_p, the equation becomes

$$\frac{\partial T}{\partial t} + \mathbf{v} \cdot \nabla T = \frac{1}{\rho c_p}\frac{dn}{dt} \qquad (4.7.1)$$

The source-sink term on the right side is comprised of four components, *conduction, radiation, heat generation* via shear, and *heat exchange* due to chemical reactions; therefore, Eq. (4.7.1) can be rewritten as

$$\frac{\partial T}{\partial t} + \mathbf{v} \cdot \nabla_T = \frac{1}{\rho c_p}\left[\frac{dq}{dt} + R^* + S_\tau^* + S_c^*\right] \qquad (4.7.2)$$

The *heat conduction* term uses Fourier's heat conduction law [Eq. (3.9.2)] by first recognizing that $\mathbf{N}_c$ is a vector of the form

$$\mathbf{N}_c = N_{cx}\mathbf{i} + N_{cy}\mathbf{j} + N_{cz}\mathbf{k} = -k\frac{\partial T}{\partial x}\mathbf{i} - k\frac{\partial T}{\partial y}\mathbf{j} - k\frac{\partial T}{\partial z}\mathbf{k}$$

This equation can be written in vector form as

$$\mathbf{N}_c = -k\nabla T$$

which after substitution in Eq. (4.7.2) gives

$$\frac{dq}{dt} = \nabla \cdot \mathbf{N}_c = \nabla \cdot \left(k\nabla T\right) = k\nabla^2 T \qquad (4.7.3)$$

$$= k\left(\frac{\partial^2 T}{\partial x^2} + \frac{\partial^2 T}{\partial y^2} + \frac{\partial^2 T}{\partial z^2}\right)$$

The rate of heat generation via *viscous dissipation* is found as follows. For the moment the radiation and reaction terms, R^* and S_c^*, are ignored. If the incompressibility condition is assumed, then the internal energy, u^{**}, equals $c_p T$ and after some algebra it can be shown that $S_\tau^* = -\Phi$. Therefore, Eq. (4.7.2) becomes

$$\frac{\partial T}{\partial t} + \mathbf{v} \cdot \nabla T = \left(\frac{k}{\rho c_p} \nabla^2 T + \frac{\Phi}{\rho c_p} + \frac{R^*}{\rho c_p} + \frac{S_c^*}{\rho c_p} \right) \tag{4.7.4a}$$

or in full form

$$\frac{\partial T}{\partial t} + u \frac{\partial T}{\partial x} + v \frac{\partial T}{\partial y} + w \frac{\partial T}{\partial z} = \alpha \left(\frac{\partial^2 T}{\partial x^2} + \frac{\partial^2 T}{\partial y^2} + \frac{\partial^2 T}{\partial z^2} \right) + \frac{\Phi}{\rho c_p} + R + S_c \tag{4.7.4b}$$

In Eq. (4.7.4) the dissipation term is not written out in its full form and is rarely if ever used for incompressible flows. In fact for most geophysical flows the term is dropped altogether due to its small size relative to the other transport terms of conduction and advection.

Insofar as terminology is concerned, the term $\alpha \nabla^2 T$, as noted before, is the heat conduction or diffusion term. The term $(\mathbf{v} \cdot \nabla)T$ is the heat transport due to *convection* or *advection*. *Natural* or *free convection* is the heat being transported by the fluid velocity originating from temperature instabilities, for example, a hot fluid underlying a cold fluid. *Advection* or *forced convection* is heat transport by a fluid velocity originating from shear, elevation, or pressure gradients imposed on the flow field. It is considered to be poor practice to use the word convection without the associated descriptor.

4.8   DIFFERENTIAL MASS BALANCE FOR A SPECIES

The final field equation to be presented is the transport equation for each mass component of a mixture. The species may be dissolved or particulate in form and chemical, biological, and physical in origin.

Proceeding from Eq. (4.2.2) an equation for each fraction, i, of a mixture is found as follows. First, η is defined as the mass of the ith fraction per mass, ω_i. Therefore the Reynolds transport equation becomes

$$\frac{\partial}{\partial t} (\rho \omega_i) + \nabla \cdot (\rho \mathbf{v} \omega_i) = \frac{dn}{dt} \tag{4.8.1}$$

Here $\mathbf{v}_i$ is the velocity vector of the ith species. Remembering from Eq. (1.5.5) that the mass fraction

$$\omega_i = C_i / \rho$$

where C_i is the mass concentration of the ith species and ρ is the mixture density, then Eq. (4.8.1) becomes

$$\frac{\partial C_i}{\partial t} + \nabla \cdot (\mathbf{v} C_i) = \frac{dn}{dt} \tag{4.8.2}$$

where dn/dt is the net rate of production ($+$ or $-$) of C_i by nontransport agents. This term is called the source-sink term for the ith component and will be labeled as S_i. The terms in these equations have units of [$M/L^3/t$].

The vector product $\mathbf{v}_i C_i$ has units of [$M/L^2/t$] and is the flux vector, $\mathbf{N}_i$, of species i. Substituting $\mathbf{N}$ into Eq. (4.8.2), the following constraint applies to an

N-component mixture

$$\sum_{i=1}^{N}\left(\frac{\partial C_i}{\partial t} + \nabla \cdot \mathbf{N} - S_i\right) = \frac{\partial \rho}{\partial t} + \nabla \cdot (\rho \mathbf{v}) = 0 \qquad \textbf{(4.8.3)}$$

As noted in Sec. 3.9 there are two perspectives which one can adopt in further analysis of the mass transport flux term. The first is to simply adopt the individual flux specification as defined, and the equation becomes

$$\frac{\partial C_i}{\partial t} + \nabla \cdot \mathbf{N}_i = \frac{\partial C_i}{\partial t} + \nabla \cdot (\mathbf{v} C_i) = S_i \qquad \textbf{(4.8.4)}$$

This form has a decided disadvantage in that it is *extremely* difficult to measure and therefore know the velocity vector, $\mathbf{v}_i$, of each individual species. At best one can measure only the mixture average velocity.

The universally accepted approach is to use Fick's diffusion law (Sec. 3.9) which is based upon defining the total flux to be comprised of two components: advective flux which is based upon the mixture average velocity, $\mathbf{v}$, and a diffusion flux which is based upon individual species motion relative to the average velocity. Therefore, the relative flux is defined as

$$\mathbf{J}_i = C_i(\mathbf{v}_i - \mathbf{v}) \qquad \textbf{(4.8.5)}$$

where $\mathbf{v}$ is the mixture average velocity vector. Fick's diffusion term is defined as

$$\mathbf{J}_i = -\rho \mathcal{D}_{iw} \nabla \omega_i \qquad \textbf{(4.8.6)}$$

As in Sec. 3.9, $\mathcal{D}_{iw}$ is Fick's diffusion coefficient for species i. Since most mixtures are dominated by one particular liquid, diffusion coefficients are determined relative to the dominant liquid type. Since water is a primary liquid in this text, $\mathcal{D}_{iw}$ should be read as "the diffusion coefficient of the ith species in water."

If Eq. (4.8.6) is substituted into Eq. (4.8.5) and solved for the total flux, then

$$\mathbf{N}_i = C_i \mathbf{v}_i = \mathbf{J}_i + C_i \mathbf{v} = -\rho \mathcal{D}_{iw} \nabla \omega_i + C_i \mathbf{v} \qquad \textbf{(4.8.7)}$$

If Eq. (4.8.7) is substituted into Eq. (4.8.4), then an equation in terms of the mixture average velocity is found

$$\frac{\partial C_i}{\partial t} + \nabla \cdot (\mathbf{v} C_i) = \nabla \cdot (\rho \mathcal{D}_{iw} \nabla \omega_i) + S_i$$

After cancellation of the ρ's the final equation becomes

$$\frac{\partial C_i}{\partial t} + \nabla \cdot (\mathbf{v} C_i) = \nabla \cdot (\mathcal{D}_{iw} \nabla C_i) + S_i \qquad \textbf{(4.8.8)}$$

Some special cases occur. If the mixture density distribution is incompressible, then $\nabla \cdot \mathbf{v} = 0$, and after performing chain rule differentiation, Eq. (4.8.8) becomes

$$\frac{\partial C_i}{\partial t} + (\mathbf{v} \cdot \nabla) C_i = \nabla \cdot (\mathcal{D}_{iw} \nabla C_i) + S_i \qquad \textbf{(4.8.9)}$$

Finally $\mathcal{D}_{iw}$ is often constant. This isn't nearly as universal a condition as the heat diffusion assumption, but if it can be ascertained from the data that it is true, then

$$\frac{\partial C_i}{\partial t} + (\mathbf{v} \cdot \nabla) C_i = \mathcal{D}_{iw} \nabla^2 C_i + S_i \qquad \textbf{(4.8.10a)}$$

which can be written in full form as

$$\frac{\partial C_i}{\partial t} + u\frac{\partial C_i}{\partial x} + v\frac{\partial C_i}{\partial y} + w\frac{\partial C_i}{\partial z} = \mathcal{D}_{iw}\left(\frac{\partial^2 C}{\partial x^2} + \frac{\partial^2 C}{\partial y^2} + \frac{\partial^2 C}{\partial z^2}\right) + S_i \quad \text{(4.8.10}b\text{)}$$

As with the heat transfer case the terms $\nabla \cdot (C_i\mathbf{v}_i)$ or $(\mathbf{v} \cdot \nabla)C_i$ are referred to as the *advection* or *convection* terms while the term $\mathcal{D}_{iw}\nabla^2 C_i$ is referred to as the *diffusion* term.

PROBLEMS

4.1   Prove, for any two vectors **a** and **b**, that

$$|\mathbf{a} - \mathbf{b}|^2 + |\mathbf{a} + \mathbf{b}|^2 = 2(|\mathbf{a}|^2 + |\mathbf{b}|^2).$$

4.2   Find the angle between the two vectors $\mathbf{a} = 10\mathbf{i}+3\mathbf{j}+2\mathbf{k}$ and $\mathbf{b} = -\mathbf{i}+2\mathbf{j}+3\mathbf{k}$.

4.3   Given $\mathbf{u} = -2\mathbf{i}+5\mathbf{j}$, $\mathbf{v} = \mathbf{i}+2\mathbf{j}+3\mathbf{k}$, and $\mathbf{w} = 2\mathbf{i}+3\mathbf{k}$, find the products ($a$) $\mathbf{u} \cdot (\mathbf{v} \times \mathbf{w})$ and (b) $(\mathbf{u} \times \mathbf{v}) \cdot \mathbf{w}$.

4.4   Given $\mathbf{a} = 3\mathbf{i}+4\mathbf{j}-\mathbf{k}$ and $\mathbf{b} = 2\mathbf{i}+5\mathbf{k}$, find the value of α so that $\mathbf{a} + \alpha\mathbf{b}$ is orthogonal to **b**.

4.5   Find the products (a) $\mathbf{i} \cdot \mathbf{i}$, $\mathbf{j} \cdot \mathbf{j}$, $\mathbf{i} \cdot \mathbf{j}$, and $\mathbf{j} \cdot \mathbf{k}$ and (b) $\mathbf{j} \times \mathbf{k}$, $\mathbf{k} \times \mathbf{i}$, and $\mathbf{i} \times \mathbf{j}$.

4.6   Given $\mathbf{u} = x^2 y z^{1/2}$, find the gradient of **u**.

4.7   Given $\mathbf{u} = x\mathbf{i} + y^2\mathbf{j} + 3z\mathbf{k}$, find the divergence of **u**.

4.8   Given the velocity $\mathbf{v} = u\mathbf{i}+v\mathbf{j}+w\mathbf{k}$, ($a$) Is the product $\mathbf{v} \cdot \nabla$ a vector or a scalar? (b) How can $\mathbf{v} \cdot \nabla$ be expressed in terms of the velocity components in Cartesian coordinates? (c) What is the physical meaning of the term $\mathbf{v} \cdot \nabla$?

4.9   Given the velocity field $\mathbf{v} = 2x^2 y\mathbf{i} - 3y\mathbf{j} + 8t\mathbf{k}$, determine the acceleration field of the flow. What is its value at $\mathbf{x} = 8\mathbf{i} + 12\mathbf{j}$ and $t = 6$ sec?

4.10   If a velocity field is given as $\mathbf{v} = 10\mathbf{i} + (x^2 + y^2)\mathbf{j} - 2xyz\mathbf{k}$, find the acceleration field of the flow at $\mathbf{x} = 2\mathbf{i} - 3\mathbf{j} + 2\mathbf{k}$.

4.11   Given the velocity $\mathbf{v} = u\mathbf{i}+v\mathbf{j}+w\mathbf{k}$, determine that the acceleration of a fluid particle is given by

$$\mathbf{a} = \frac{\partial \mathbf{v}}{\partial t} + \nabla\left(\frac{|\mathbf{v}|^2}{2}\right) + \Gamma \times \mathbf{v}$$

where Γ is the vorticity.

4.12   Given the vector $\mathbf{v} = -1.5\mathbf{i}+3\mathbf{j}-4.5\mathbf{k}$, what is a unit vector in the direction of **v**?

4.13   A two-dimensional flow can be described by $u = -y/b^2$ and $v = x/a^2$. Verify that this is the flow of an incompressible fluid.

4.14   Given an arbitrary function, ψ, such that

$$u(x, y) = \frac{\partial \psi}{\partial y} \qquad \text{and} \qquad v(x, y) = -\frac{\partial \psi}{\partial x}$$

find the vorticity, Γ, in terms of ψ, where ψ is called the *stream function*.

4.15   Prove that the vorticity in any flow satisfies $\nabla \cdot \Gamma = 0$.

4.16 Does the velocity distribution

$$\mathbf{v} = 5x\mathbf{i} + 5y\mathbf{j} + (-10z)\mathbf{k}$$

satisfy the law of mass conservation for incompressible flow?

4.17 Consider a cube with 1-m edges parallel to the coordinate axes located in the first quadrant with one corner at the origin. By using the velocity distribution of Prob. 4.16, find the flow through each face and show that no mass is being accumulated within the cube if the fluid is of constant density.

4.18 Find the flow (per foot in the z direction) through each side of the square with the corners at (0,0), (0,1), (1,1), (1,0) due to

$$\mathbf{v} = (16y - 12x)\mathbf{i} + (12y - 9x)\mathbf{j}$$

and show that continuity is satisfied.

4.19 Show that the velocity

$$\mathbf{v} = \frac{4x}{x^2 + y^2}\mathbf{i} + \frac{4y}{x^2 + y^2}\mathbf{j}$$

satisfies continuity at every point except the origin.

4.20 Problem 4.19 describes a velocity distribution that is everywhere radial from the origin with magnitude $v_r = 4/r$. Show that the flow through each circle concentric with the origin (per unit length in the z direction) is the same.

4.21 By introducing the following relations between cartesian coordinates and plane polar coordinates, obtain a form of the continuity equation in plane polar coordinates:

$$x^2 + y^2 = r^2 \qquad \frac{y}{x} = \tan\theta \qquad \frac{\partial}{\partial x} = \frac{\partial}{\partial r}\frac{\partial r}{\partial x} + \frac{\partial}{\partial \theta}\frac{\partial \theta}{\partial x}$$

$$u = v_r \cos\theta - v_\theta \sin\theta \qquad v = v_r \sin\theta + v_\theta \cos\theta.$$

4.22 If in a one-dimensional flow $\mathbf{v} = u(x, t)\mathbf{i} + 0\mathbf{j} + 0\mathbf{k}$ and the density is not constant, but is given as $\rho = \rho_o(1.5 + \cos wt)$, find the velocity, $\mathbf{v}$, such that $u(0, t) = U$, a constant.

4.23 Given an arbitrary function, $\phi(x, y)$, such that

$$u(x, y) = \frac{\partial \phi}{\partial x} \qquad \text{and} \qquad v(x, y) = \frac{\partial \phi}{\partial y}$$

show for an incompressible flow that the *potential function* satisfies Laplace's equation $\nabla^2 \phi = 0$.

4.24 If the potential of a flow is $\phi(x, y) = x$, find the velocities $u(x, y)$ and $v(x, y)$ and plot the velocity field of the flow.

4.25 Repeat Prob. 4.24 for $\phi(x, y) = \ln(x^2 + y^2)/4\pi$.

4.26 For an incompressible fluid the velocity components $u(x, y, z)$ and $w(x, y, z)$ satisfy

$$u = (1 + xy)(a_o + a_1 x + a_2 x^2) \quad \text{and} \quad w = 0 \qquad 0 \le x, y, z \le 1.$$

Find the velocity component $v(x, y, z)$. Is this flow field reasonable?

4.27 For an incompressible flow, show that the rate of volume change is zero.

4.28 Given the following stress field

$$\tau_{xx} = 16x^2 - 8xy \quad \tau_{yy} = 16y^2 + 8xy \quad \tau_{xy} = -5x^2 \quad \tau_{zz} = \tau_{xz} = \tau_{yz} = 0$$

find a function for the bulk stress as a scalar field. What is the bulk stress at the point $2\mathbf{i} + 4\mathbf{j} + 3\mathbf{k}$?

4.29   Given that $p = xy + (x + z^2) + 10$, find the force per unit volume in the direction $\mathbf{n} = 2\mathbf{i} + 3\mathbf{j}$.

4.30   Label the stresses acting upon the material element as shown in Figure 4.12.

4.31   Given the following stress distribution

$$\sigma_{xx} = 2x^2 + 4xy - 3y^2 \qquad \tau_{xy} = 3x^2 - 6xy - 2y^2$$
$$\sigma_{yy} = 2y^2 - 4xy + 3x^2 \qquad \sigma_{zz} = \tau_{xz} = \tau_{yz} = 0$$

determine if equilibrium exists in the absence of the body force.

4.32   For a two-dimensional flow field it is known that the body force is nonexistent and the stress field satisfies $\sigma_{zz} = \tau_{zx} = \tau_{zy} = 0$. Show that there is an arbitrary function $\phi(x, y)$ such that

$$\sigma_{xx} = \frac{\partial^2 \phi}{\partial y^2} \qquad \sigma_{yy} = \frac{\partial^2 \phi}{\partial x^2} \qquad \tau_{xy} = -\frac{\partial^2 \phi}{\partial x \partial y}$$

4.33   Given that the arbitrary function $\phi(x, y)$ in Prob. 4.32 can be expressed as $\phi = x^2 y^3 - e^{-xy}$, find the stress tensor at $(x,y) = (2,7)$ in the absence of the body force.

4.34   In a fluid the pressure distribution is given as $p = \alpha(x^2 + y^2)$ where α is a constant. (*a*) What are the dimensions of the constant α? (*b*) What is the magnitude of the pressure gradient?

4.35   Perform the operation $\nabla \cdot \mathbf{v}$ on the velocity vector of Prob. 4.19.

4.36   In a velocity field where $\mathbf{v} = 112.5(y^2\mathbf{i} + x^2\mathbf{j})$ (in m/s), determine the pressure gradient at a point (1.25, 2). The specific gravity of the fluid is 1.4 and the viscous effects are negligible.

4.37   Determine the time rate of x momentum passing out of the cube of Prob. 4.17. *Hint:* Consider all six faces of the cube.

4.38   Calculate the y-momentum efflux from the figure described in Prob. 4.18 for the velocity given there.

4.39   If gravity acts in the negative z direction, determine the z component of the force acting on the fluid within the cube described in Prob. 4.17 for the velocity specified there.

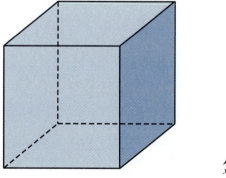

Figure 4.12   Problem 4.30.

4.40 Find the y component of the force acting on the control volume given in Prob. 4.18 for the velocity given there. Consider gravity to be acting in the negative y direction.

4.41 Show that in a two-dimensional flow field the vorticity satisfies the equation

$$\frac{D\Gamma}{Dt} = \nu\nabla^2\Gamma$$

4.42 For a steady two-dimensional, incompressible flow derive the continuity equation in polar coordinates.

4.43 For a steady two-dimensional, incompressible flow derive the heat transfer equation in polar coordinates.

4.44 By analogy write the mass transfer equation in polar coordinates for the same conditions as in Prob. 4.43.

4.45 For the fluid described in Prob. 4.26 the temperature distribution is given as

$$T = T_o\,e^{-kt}\sin ax \cdot \cos by$$

where k, a, and b are constants. Find an expression for the material rate of change of the temperature (DT/Dt).

4.46 Using the Navier-Stokes equations, derive the equations for the one-dimensional steady, viscous, incompressible flow.

4.47 From the heat transfer differential equation derive the differential equation that governs the one-dimensional, steady conduction through a wall without internal heat generation. If temperature at one face of the wall is kept at T_1 and at the other at T_2, and the wall thickness is d, find the temperature distribution in the wall.

4.48 A liquid flows over a thin, flat sheet of a slightly soluble soil. Over the region in which diffusion is occurring, the liquid velocity may be assumed to be parallel to the plate and to be given by $u = y^2/2$, where y is the distance from the plate and D_{Aw} is a constant. Show that the equation governing the mass transfer, with certain simplifying assumptions, is

$$D_{Aw}\left(\frac{\partial^2 C_A}{\partial x^2} + \frac{\partial^2 C_A}{\partial y^2}\right) = \frac{y^2}{2}\frac{\partial C_A}{\partial x}$$

List the simplifying assumptions.

4.49 Hydrogen is flowing in a 2.0-in.-diameter pipe at the mass rate of 0.03 lb_m/s. At section 1 the pressure is 30 psia and $t = 80°$F. What is the average velocity?

4.50 Fluid is flowing in a long horizontal circular cylindrical tube of radius R. The fluid has viscosity μ. Show that $u = C(R^2 - r^2)/4\mu$ where r is the radial distance from the centerline of the tube and C is a constant.

4.51 Given the conditions in Prob. 4.50, determine the mean flow velocity in the tube. What is the rate of mass flow through the tube?

4.52 An incompressible fluid is confined between two parallel vertical plates as shown in Fig. 4.13. The left is stationary, while the other is moving upwards with a velocity v_w. Assuming laminar flow, determine the velocity profile within the fluid.

4.53 From the Navier-Stokes equations and the continuity equation, derive the velocity profile for a flow of a viscous incompressible fluid between two flat, parallel plates.

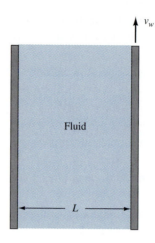

Figure 4.13   Problem 4.52.

4.54 Repeat Prob. 4.46 (under the same conditions) for a two-dimensional flow field in the xy plane.

4.55 Calculate the average velocity and momentum correction factor for the velocity distribution in a pipe,

$$\frac{v}{v_{max}} = \left(\frac{y}{r_0}\right)^{1/n}$$

with y the wall distance and r_0 the pipe radius.

4.56 Newtonian liquid flows down an inclined plane in a thin sheet of thickness t (Fig. 4.14). Assuming that there are no end effects on the velocity profile, derive the equation for the velocity profile as a function of x.

4.57 From the velocity profile derived in Prob. 4.56 find the mean velocity of the flow. What is the flowrate Q?

4.58 Determine the distribution of τ_{xz} from the velocity profile derived in Prob. 4.56.

4.59 By using Bernoulli's equation, resolve Prob. 3.15.

4.60 By using Bernoulli's equation, resolve Prob. 3.16.

4.61 By using Bernoulli's equation, resolve Prob. 3.17.

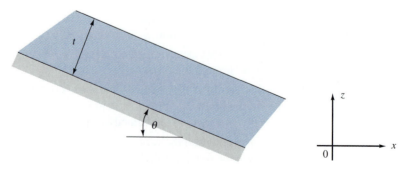

Figure 4.14   Problem 4.56.

4.62 By neglecting air resistance, determine the height a vertical jet of water will rise with a velocity of 60 ft/s.

4.63 If the water jet of Prob. 4.62 is directed upward 45° with the horizontal and air resistance is neglected, how high will it rise and what is the velocity at its high point?

4.64 A submarine travels at a depth of 70 ft in the Atlantic Ocean, with a speed of 10 mph. What is the pressure on the bow stagnation point of the submarine?

4.65 Calculate the kinetic energy correction factor α for a two-dimensional laminar flow between two parallel flat plates (refer to Prob. 4.52).

4.66 Calculate the kinetic energy correction factor α for a laminar flow in a circular tube (refer to Prob. 4.50).

4.67 What angle α of jet is required to reach the roof of the building of Fig. 4.15 with minimum jet velocity V_0 at the nozzle? What is the value of V_0?

4.68 A standpipe 30 ft in diameter and 40-ft high is filled with water. How much potential energy is in this water if the elevation datum is taken 10 ft below the base of the standpipe?

4.69 How much work could be obtained from the water of Prob. 4.68 if it runs through a 100-percent efficient turbine that discharged into a reservoir with elevation 30 ft below the base of the standpipe?

4.70 What is the flux of kinetic energy out of the cube given by Prob. 4.17 for the velocity prescribed in Prob. 4.16?

4.71 In Fig. 4.16 oil discharges from a "two-dimensional" slot as indicated at A into the air. At B oil discharges from under a gate onto a floor. Neglecting all losses, determine the discharges at A and at B per foot of width. Why do they differ?

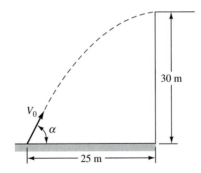

Figure 4.15   Problem 4.67.

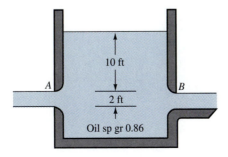

Figure 4.16   Problem 4.71.

4.72   At point A in a pipeline carrying water the diameter is 1 m, the pressure 98 kPa, and the velocity 1 m/s. At point B, 2 m higher than A, the diameter is 0.5 m and the pressure 20 kPa. Determine the direction of flow.

4.73   The losses in Fig. 4.17 for $H = 25$ ft are $3V^2/2g$ ft·lb/lb. What is the discharge?

4.74   For a flow of 750 gpm in Fig. 4.17, determine H for losses of $10V^2/2g$ ft·lb/lb.

4.75   For 1500-gpm flow and $H = 32$ ft in Fig. 4.17, calculate the losses through the system in velocity head, $KV^2/2g$.

4.76   The pumping system shown in Fig. 3.54 must have pressure of 5 psi in the discharge line when cavitation is incipient at the pump inlet. Calculate the length of pipe from the reservoir to the pump for this operating condition if the loss in this pipe can be expressed as $(V_1^2/2g)(0.03\ L/D)$. What horsepower is being supplied to the fluid by the pump? What percent of this power is being used to overcome losses? The barometer reads 30-in. Hg.

4.77   In the siphon of Fig. 3.55, $h_1 = 1$ m, $h_2 = 3$ m, $D_1 = 3$ m, $D_2 = 5$ m, and the losses to section 2 are $2.6V_2^2/2g$, with 10 percent of the losses occurring before section 1. Find the discharge and the pressure at section 1.

4.78   Find the pressure at A of Prob. 4.77 if it is a stagnation point (velocity equals zero).

4.79   A water jet has an area of 0.10 ft^2 and a velocity of 85 ft/s. The jet entrains a secondary stream of water having a velocity of 5 ft/s (see Fig. 4.18). The diameter

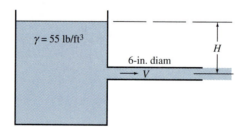

$\gamma = 55$ lb/ft^3

6-in. diam

$\longrightarrow V$

H

Figure 4.17   Problems 4.73, 4.74, and 4.75.

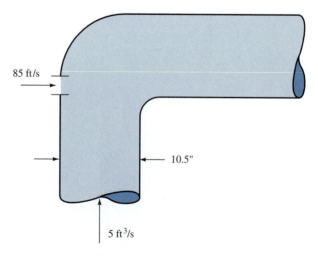

85 ft/s

10.5"

5 ft^3/s

Figure 4.18   Problem 4.79.

of the pipe is 10.5 in. What is the average velocity at section 2? Assume that at section 2, the mixing is complete.

4.80 The piston with cross-sectional area A forces the oil of density ρ out of the tank and into the atmosphere through a small pipe of cross-sectional area α (Fig. 4.19). The oil jet has a cross-sectional area of $C_c\alpha$. Show that the coefficient, C_c, is given as $C_c = 1/(2 - \alpha/A)$.

4.81 Water is flowing in a rectangular open channel with width of 8 ft. At a contracted section downstream the width is reduced to 7 ft while the bottom is raised 1 ft (Fig. 4.20). If the water depth far upstream is 5 ft and at the contracted section it is 3.5 ft, determine the flow rate.

4.82 Water flows in a rectangular, horizontal open channel 2-m wide at a depth of 10 cm. The channel bottom is gradually raised by $\Delta z = 5$ cm. As the water passes over the raised bottom of the channel the water depth increases by 10 cm (Figure 4.21). What is the flow rate?

4.83 Solve the equation

$$\frac{\partial T}{\partial t} = \alpha \nabla^2 T + R$$

for the case of one-dimensional distribution with $T = T_0$ at $x = 0$ and $T = T_L$ at $x = L$. The internal heat generation per unit volume (R) varies according to $R = R_0 e^{-bx/L}$.

4.84 Repeat Prob. 4.83 with boundary conditions $T = T_0$ at $x = 0$ and $dT/dx = 0$ at $x = L$.

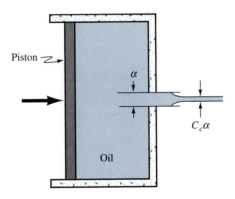

Figure 4.19   Problem 4.80.

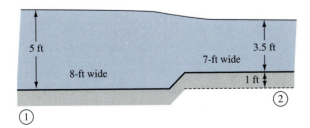

Figure 4.20   Problem 4.81.

4.85   Repeat Prob. 4.83 with boundary conditions $T = T_0$ at $x = 0$ and $dT/dx = c$ at $x = L$ (c = constant).

4.86   Repeat Prob. 4.47 considering a uniform internal heat generation q_H.

4.87   For the conditions described in Prob. 4.86 find the maximum temperature in the wall and the location where $T = T_{max}$.

4.88   A large truck is carrying sludge (sp gr = 1.8) in a 50-ft cylindrical tank with diameter 10 ft. The truck travels with constant speed of 55 mph. The rate of generation of bacteria in the sludge is assumed to be proportional to the bacteria concentration, C_b. Assuming a steady state process, derive the governing equation that describes the concentration distribution of the bacteria in the tank.

4.89   A small diameter tube is immersed in a pan full of a liquid with sp gr = 1.32 (Fig. 4.22). A jet of a gas flows across the mouth of the tube carrying away the gas vapors of the liquid in the pan. Assuming that the evaporation of the liquid is a steady state process, find the governing differential equation which describes this phenomenon. State the assumptions (if any) necessary to derive the governing equation.

4.90   The cell division of a microorganism placed in a stagnant fluid follows the first-order reaction

$$M \longrightarrow 2M$$

Write the differential equation which describes the concentration profile for the microorganism.

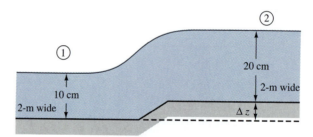

Figure 4.21   Problem 4.82.

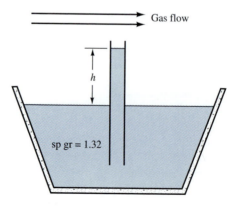

Figure 4.22   Problem 4.89.

4.91 On interstate I-270 an accident occurs and a large truck which carries an herbicide overturns and spills the herbicide over an adjacent field. The fluid starts evaporating into the atmosphere in about an hour after the occurrence of the accident. Assuming that the evaporation of the herbicide into the atmosphere is a steady state process, derive the governing equation which describes the phenomenon.

ADDITIONAL READING

Aris, R.: *Vectors, Tensors, and the Basic Equations of Fluid Mechanics,* Dover Publications, New York, 1989.

Bird, R., Stewart, W., and Lightfoot, E.: *Transport Phenomena,* John Wiley and Sons, New York, 1968.

Brodkey, R. and Hershey, H.: *Transport Phenomena: A Unified Approach,* McGraw Hill Co., New York, 1988.

Eckart, E. and Drake, R.: *Analysis of Heat and Mass Transfer,* McGraw Hill Co., New York, 1972.

Kreyszig, E. *Advanced Engineering Mathematics,* 7th ed., Wiley, New York, 1993.

Long, R.: *Mechanics of Solids,* Prentice Hall, New Jersey, 1961.

Shames, I.H.: *Mechanics of Fluids,* 3rd ed., McGraw-Hill, New York, 1992.

Dimensional Analysis and Dynamic Similitude

Dimensionless parameters significantly deepen our understanding of fluid-flow phenomena in a way which is analogous to the case of a hydraulic jack, where the ratio of piston diameters, a dimensionless number which is independent of the overall size of the jack, determines the mechanical advantage. They permit limited experimental results to be applied to situations involving different physical dimensions and often different fluid properties. The concepts of dimensional analysis introduced in this chapter plus an understanding of the mechanics of the type of flow under study make this generalization of experimental data possible. The consequence of such generalization is manifold, since one is now able to describe the phenomenon in its entirety and is not restricted to discussing the specialized experiment that was performed. Thus, it is possible to conduct fewer, although highly selective, experiments to uncover the hidden facets of the problem and thereby achieve important savings in time and money. The results of an investigation can also be presented to other engineers and scientists in a more compact and meaningful way to facilitate their use. Equally important is the fact that, through such incisive and uncluttered presentations of information, researchers are able to discover new features and missing areas of knowledge of the problem at hand. This directed advancement of our understanding of a phenomenon would be impaired if the tools of dimensional analysis were not available. In the following chapter, dealing primarily with viscous effects, one parameter is highly significant, that is, the Reynolds number. In Chap. 12, dealing with open channels, the Froude number has the greatest significance.

Many of the dimensionless parameters can be viewed as a ratio of a pair of fluid forces, the relative magnitude indicating the relative importance of one of the forces with respect to the other. If some forces in a particular flow situation are very much larger than a few others, it is often possible to neglect the effect of the smaller forces and treat the phenomenon as though it were completely determined by the major forces. This means that simpler, although not necessarily easy, mathematical and experimental procedures can be used to solve the problem. For situations with several forces of the same magnitude, such as inertial, viscous, and gravitational forces, special techniques are required. After a discussion of dimensions, dimensional analysis and dimensionless parameters, dynamic similitude and model studies are presented.

5.1 DIMENSIONAL HOMOGENEITY AND DIMENSIONLESS RATIOS

Solving practical design problems in fluid mechanics usually requires both theoretical developments and experimental results. By grouping significant quantities into dimensionless parameters it is possible to reduce the number of variables and to make this compact result (equations or data plots) applicable to all similar situations.

If one were to write the equation of motion $\mathbf{F} = m\mathbf{a}$ for a fluid parcel, including all types of force terms that could act on the parcel, such as gravity, pressure, viscous, elastic, and surface-tension forces, an equation of the sum of these forces equated to $m\mathbf{a}$, the inertial force, would result. As with all physical equations, each term must have the same dimensions, in this case, force. The division of each term of the equation by any one of the terms would make the equation dimensionless. For example, dividing through by the inertial force term would yield a sum of dimensionless parameters equated to unity. The relative size of any one parameter, compared with unity, would indicate its importance. If one were to divide the force equation through by a different term, say the viscous force term, another set of dimensionless parameters would result. Without experience in the flow case it is difficult to determine which parameters will be most useful.

An example of the use of dimensional analysis and its advantages is given by considering the hydraulic jump, treated in Sec. 3.8. The momentum equation for this case

$$\frac{\gamma y_1^2}{2} - \frac{\gamma y_2^2}{2} = \frac{V_1 y_1 \gamma}{g}(V_2 - V_1) \tag{5.1.1}$$

can be rewritten as

$$\frac{\gamma}{2} y_1^2 \left[1 - \left(\frac{y_2}{y_1}\right)^2 \right] = V_1^2 \frac{\gamma}{g} y_1 \left(1 - \frac{y_2}{y_1}\right)\frac{y_1}{y_2}$$

Clearly, the right-hand side represents the inertial forces, and the left-hand side represents the pressure forces due to gravity. These two forces are of equal magnitude, since one determines the other in this equation. Furthermore, the term $\gamma y_1^2/2$ has the dimensions of force per unit width, and it multiplies a dimensionless number which is specified by the geometry of the hydraulic jump.

If one divides this equation by the geometric term $1 - y_2/y_1$ and a number representative of the gravity forces, one has

$$\frac{V_1^2}{g y_1} = \frac{1}{2}\frac{y_2}{y_1}\left(1 + \frac{y_2}{y_1}\right) \tag{5.1.2}$$

It is now clear that the left-hand side is the ratio of the inertia and gravity forces, even though the explicit representation of the forces has been obscured through the cancellation of terms that are common in both the numerator and denominator. This ratio is equivalent to a dimensionless parameter, actually the square of the *Froude number*, which will be discussed in further detail later in this chapter. It is also interesting to note that this ratio of forces is known once the ratio y_2/y_1 is given, regardless of the values of y_2 and y_1. From this observation one can obtain an appreciation of the increased scope that Eq. (5.1.2) affords over Eq. (5.1.1) even though one is only a rearrangement of the other.

In writing the momentum equation which led to Eq. (5.1.2), only inertia and gravity forces were included in the original problem statement. But other forces, such as surface tension and viscosity, are present. These were neglected as being small in comparison with gravity and inertia forces; however, only experience with the phenomenon or with phenomena similar to it would justify such an initial simplification. For example, if viscosity had been included because one was not sure of the magnitude of its effect, the momentum equation would become

$$\frac{\gamma y_1^2}{2} - \frac{\gamma y_2^2}{2} - F_{\text{viscous}} = V_1\, y_1 \frac{\gamma}{g}\, (V_2 - V_1)$$

with the result that

$$\frac{V_1^2}{g y_1} + \frac{F_{\text{viscous}}\, y_2}{\gamma\, y_1^2 (y_1 - y_2)} = \frac{1}{2} \frac{y_2}{y_1} \left(1 + \frac{y_2}{y_1}\right)$$

This statement is more complete than that given by Eq. (5.1.2). However, experiments would show that the second term on the left-hand side is usually a small fraction of the first term which can be neglected in making initial tests on a hydraulic jump.

In the last equation the ratio y_2/y_1 can be considered to be a dependent variable which is determined for each of the various values of the force ratios, $V_1^2/g y_1$ and $F_{\text{viscous}}/\gamma y_1^2$, which are the independent variables. For the previous discussion it appears that the latter variable plays only a minor role in determining the values of y_2/y_1. Nevertheless, if one observed that the ratio of the forces, $V_1^2/g y_1$ and $F_{\text{viscous}}/\gamma y_1^2$, had the same values in two different tests, one would expect, on the basis of the last equation, that the values of y_2/y_1 would be the same in the two situations. If the ratio for $V_1^2/g y_1$ was the same in the two tests but the ratio $F_{\text{viscous}}/\gamma y_1$, which has only a minor influence for this case, was not, one would conclude that the values of y_2/y_1 for the two cases would be almost the same.

This is the key to much of what follows. For if one can create in a model experiment the same geometric and force ratios that occur on the full-scale unit, then the dimensionless solution for the model is valid for the prototype also. Often, as will be seen, it is not possible to have all the ratios equal in the model and prototype. Then one attempts to plan the experimentation in such a way that the dominant force ratios are as nearly equal as possible. The results obtained with such incomplete modeling are often sufficient to describe the phenomenon in the detail that is desired.

Writing a force equation for a complex situation may not be feasible, and another process, *dimensional analysis*, is then used if one knows the pertinent quantities that enter into the problem.

In a given situation several of the forces may be of little significance, leaving perhaps two or three forces of the same order of magnitude. With three forces of the same order of magnitude, two dimensionless parameters are obtained; one set of experimental data on a geometrically similar model provides the relations between parameters for all other similar flow cases.

EXERCISE

5.1.1   Select a common dimensionless parameter in fluid mechanics from the following: (*a*) angular velocity; (*b*) kinematic viscosity; (*c*) specific gravity; (*d*) specific weight; (*e*) none of these answers.

5.2   DIMENSIONS AND UNITS

The dimensions of mechanics are force, mass, length, and time; they are related by Newton's second law of motion,

$$\mathbf{F} = m\mathbf{a} \tag{5.2.1}$$

Force and mass units are discussed in Sec. 1.2. For all physical systems, it would probably be necessary to introduce two more dimensions, one dealing with electromagnetics and the other with thermal effects. For the compressible work in this text, it is unnecessary to include a thermal unit, because the equations of state link pressure, density, and temperature.

Newton's second law of motion in dimensional form is

$$F = MLT^{-2} \tag{5.2.2}$$

which shows that only three of the dimensions are independent. F is the force dimension, M the mass dimension, L the length dimension, and T the time dimension. One common system employed in dimensional analysis is the $MLT\Theta$ system, where Θ is the dimension of temperature.

Table 5.1 lists some of the quantities used in fluid flow, together with their symbols and dimensions. To avoid confusion the temperature has been labeled as T' for this chapter only.

Table 5.1   Dimensions of physical quantities used in fluid mechanics

Quantity	Symbol	Dimensions
Length	l	L
Time	t	T
Mass	m	M
Force	F	MLT^{-2}
Velocity	V	LT^{-1}
Acceleration	a	LT^{-2}
Area	A	L^2
Discharge	Q	L^3T^{-1}
Pressure	p	$ML^{-1}T^{-2}$
Gravity	g	LT^{-2}
Density	ρ	ML^{-3}
Specific weight	γ	$ML^{-2}T^{-2}$
Dynamic viscosity	μ	$ML^{-1}T^{-1}$
Kinematic viscosity	ν	L^2T^{-1}
Surface tension	σ	MT^{-2}
Bulk modulus of elasticity	K	$ML^{-1}T^{-2}$
Temperature	T'	Θ
Mass concentration	C	ML^{-3}
Thermal conductivity	k	$MLT^{-3}\Theta^{-1}$
Thermal diffusivity	α	L^2T^{-1}
Mass diffusivity	$\mathcal{D}$	L^2T^{-1}
Heat capacity	c_p	$L^2T^{-2}\Theta^{-1}$
Reaction rate	k_1	T^{-1}

EXERCISE

5.2.1   A dimensionless combination of Δp, ρ, l, and Q is

$(a)\ \sqrt{\dfrac{\Delta p}{\rho}\,\dfrac{Q}{l^2}}\,;$ $(b)\ \dfrac{\rho Q}{\Delta p l^2}\,;$ $(c)\ \dfrac{\rho l}{\Delta p Q^2}\,;$ $(d)\ \dfrac{\Delta p l Q}{\rho}\,;$ $(e)\ \sqrt{\dfrac{\rho}{\Delta p}}\,\dfrac{Q}{l^2}\,.$

5.3   THE Π THEOREM: MOMENTUM AND ENERGY

The Buckingham Π theorem [1]† proves that, in a physical problem including n quantities in which there are m dimensions, the quantities can be arranged into $n - m$ independent dimensionless parameters. Let $A_1, A_2, A_3, \ldots, A_n$ be the quantities involved, such as pressure, viscosity, velocity, etc. All the quantities are known to be essential to the solution, and hence some functional relation must exist

$$F(A_1, A_2, A_3, \ldots, A_n) = 0 \tag{5.3.1}$$

If $\Pi_1, \Pi_2, \ldots,$ represent dimensionless groupings of the quantities $A_1, A_2, A_3, \ldots,$ then with m dimensions involved, an equation of the form

$$f(\Pi_1, \Pi_2, \Pi_3, \ldots, \Pi_{n-m}) = 0 \tag{5.3.2}$$

exists.

Proof of the Π theorem can be found in [1, 2]. The method of determining the Π parameters is to select m of the A quantities, with different dimensions, that contain among them the m dimensions, and to use them as repeating variables* together with one of the other A quantities for each Π. For example, let A_1, A_2, A_3 contain $M, L,$ and T, not necessarily in each one, but collectively. Then the first Π parameter is defined as

$$\Pi_1 = A_1^{x_1} A_2^{y_1} A_3^{z_1} A_4 \tag{5.3.3}$$

the second one as

$$\Pi_2 = A_1^{x_2} A_2^{y_2} A_3^{z_2} A_5$$

and so on, until

$$\Pi_{n-m} = A_1^{x_{n-m}} A_2^{y_{n-m}} A_3^{z_{n-m}} A_n$$

In these equations the exponents are to be determined so that each Π is dimensionless. The dimensions of the A quantities are substituted, and the exponents of $M, L,$ and T are set equal to zero respectively. These produce three equations in three unknowns for each Π parameter, so the x, y, z exponents can be determined, and hence the Π parameter.

If only two dimensions are involved, two of the A quantities are selected as repeating variables and two equations in the two unknown exponents are obtained for each Π term.

In many cases the grouping of A terms is such that the dimensionless arrangement is evident by inspection. The simplest case is that when two quantities have

---

†Numbered references will be found at the end of this chapter.

*It is essential that none of the m selected quantities used as repeating variables be derivable from the other repeating variables.

the same dimensions, for example, length, the ratio of these two terms is the Π parameter. The procedure is best illustrated by several examples.

Example 5.1

The discharge through a horizontal capillary tube is thought to depend upon the pressure drop per unit length, the diameter, and the viscosity. Find the form of the equation.

Solution

The quantities are listed with their dimensions:

Quantity	Symbol	Dimensions
Discharge	Q	L^3T^{-1}
Pressure drop per length	$\Delta p/l$	$ML^{-2}T^{-2}$
Diameter	D	L
Viscosity	μ	$ML^{-1}T^{-1}$

Then

$$F\left(Q, \frac{\Delta p}{l}, D, \mu\right) = 0$$

Three dimensions are used, and with four quantities there will be one Π parameter:

$$\Pi = Q^{x_1}\left(\frac{\Delta p}{l}\right)^{y_1} D^{z_1} \mu$$

Substituting in the dimensions gives

$$\Pi = (L^3T^{-1})^{x_1}(ML^{-2}T^{-2})^{y_1}L^{z_1}ML^{-1}T^{-1} = M^0L^0T^0$$

The exponents of each dimension must be the same on both sides of the equation. With L first

$$3x_1 - 2y_1 + z_1 - 1 = 0$$

and similarly for M and T

$$y_1 + 1 = 0$$
$$-x_1 - 2y_1 - 1 = 0$$

from which $x_1 = 1, y_1 = -1, z_1 = -4$, and

$$\Pi = \frac{Q\mu}{D^4 \Delta p/l}$$

After solving for Q,

$$Q = C\frac{\Delta p}{l}\frac{D^4}{\mu}$$

from which dimensional analysis yields no information about the numerical value of the dimensionless constant C; experiment (or analysis) shows that it is $\pi/128$ [Eq. (6.3.10a)].

When the dimensional analysis is used, the variables in a problem must be known. In the last example if kinematic viscosity had been used in place of dynamic viscosity, an incorrect formula would have resulted.

Example 5.2

A V-notch weir is a vertical plate with a notch of angle ϕ cut into the top of it and placed across an open channel. The liquid in the channel is backed up and forced to flow through the notch. The discharge Q is a function of the elevation H of upstream liquid surface above the bottom of the notch. In addition, the discharge depends upon gravity and upon the velocity of approach V_0 to the weir. Determine the form of discharge equation.

Solution

A functional relation

$$F(Q, H, g, V_0, \phi) = 0$$

is to be grouped into dimensionless parameters. Since ϕ is dimensionless, it is one Π parameter. Only two dimensions are used, L and T. If g and H are the repeating variables,

$$\Pi_1 = H^{x_1} g^{y_1} Q = L^{x_1} (LT^{-2})^{y_1} L^3 T^{-1}$$
$$\Pi_2 = H^{x_2} g^{y_2} V_0 = L^{x_2} (LT^{-2})^{y_2} LT^{-1}$$

Then

$$x_1 + y_1 + 3 = 0 \qquad x_2 + y_2 + 1 = 0$$
$$-2y_1 - 1 = 0 \qquad -2y_2 - 1 = 0$$

from which $x_1 = -\frac{5}{2}$, $y_1 = -\frac{1}{2}$, $x_2 = -\frac{1}{2}$, $y_2 = -\frac{1}{2}$, and

$$\Pi_1 = \frac{Q}{\sqrt{g} H^{5/2}} \qquad \Pi_2 = \frac{V_0}{\sqrt{gH}} \qquad \Pi_3 = \phi$$

or

$$f\left(\frac{Q}{\sqrt{g} H^{5/2}}, \frac{V_0}{\sqrt{gH}}, \phi \right) = 0$$

This can be written as

$$\frac{Q}{\sqrt{g} H^{5/2}} = f_1 \left(\frac{V_0}{\sqrt{gH}}, \phi \right)$$

in which both f and f_1 are unknown functions. After solving for Q,

$$Q = \sqrt{g} H^{5/2} f_1 \left(\frac{V_0}{\sqrt{gH}}, \phi \right)$$

Either experiment or analysis is required to yield additional information about the function f_1. If H and V_0 were selected as repeating variables in place of g and H,

$$\Pi_1 = H^{x_1} V_0^{y_1} Q = L^{x_1} (LT^{-1})^{y_1} L^3 T^{-1}$$
$$\Pi_2 = H^{x_2} V_0^{y_2} g = L^{x_2} (LT^{-1})^{y_2} LT^{-2}$$

Then

$$x_1 + y_1 + 3 = 0 \qquad x_2 + y_2 + 1 = 0$$
$$-y_1 - 1 = 0 \qquad -y_2 - 2 = 0$$

from which $x_1 = -2$, $y_1 = -1$, $x_2 = 1$, $y_2 = -2$, and

$$\Pi_1 = \frac{Q}{H^2 V_0} \qquad \Pi_2 = \frac{gH}{V_0^2} \qquad \Pi_3 = \phi$$

or

$$f\left(\frac{Q}{H^2 V_0}, \frac{gH}{V_0^2}, \phi\right) = 0$$

Since any of the Π parameters can be inverted or raised to any power without affecting their dimensionless status, then

$$Q = V_0 H^2 f_2\left(\frac{V_0}{\sqrt{gH}}, \phi\right)$$

The unknown function f_2 has the same parameters as f_1, but it could not be the same function. The last form is not very useful, in general, because frequently V_0 may be neglected with V-notch weirs. This shows that a term of minor importance should not be selected as a repeating variable.

Another method of determining alternative sets of Π parameters would be the arbitrary recombination of the first set. If four independent Π parameters Π_1, Π_2, Π_3, and Π_4 are known, the term

$$\Pi_a = \Pi_1^{a_1} \Pi_2^{a_2} \Pi_3^{a_3} \Pi_4^{a_4}$$

with the exponents chosen at will, would yield a new parameter. Then Π_1, Π_2, Π_3, and Π_4 would constitute a new set. This procedure can be continued to find all possible sets.

Example 5.3

The losses $\Delta p/l$ in turbulent flow through a smooth horizontal pipe depend upon velocity V, diameter D, dynamic viscosity μ, and density ρ. Use dimensional analysis to determine the general form of the equation

$$F\left(\frac{\Delta p}{l}, V, D, \rho, \mu\right) = 0$$

Solution

If V, D, and ρ are repeating variables, then

$$\Pi_1 = V^{x_1} D^{y_1} \rho^{z_1} \mu = (LT^{-1})^{x_1} L^{y_1} (ML^{-3})^{z_1} ML^{-1} T^{-1}$$
$$x_1 + y_1 - 3z_1 - 1 = 0$$
$$-x_1 \qquad\qquad - 1 = 0$$
$$z_1 + 1 = 0$$

from which $x_1 = -1$, $y_1 = -1$, $z_1 = -1$, and

$$\Pi_2 = V^{x_2} D^{y_2} \rho^{z_2} \frac{\Delta p}{l} = (LT^{-1})^{x_2} L^{y_2} (ML^{-3})^{z_2} ML^{-2} T^{-2}$$

$$x_2 + y_2 - 3z_2 - 2 = 0$$
$$-x_2 \qquad\qquad - 2 = 0$$
$$z_2 + 1 = 0$$

from which $x_2 = -2$, $y_2 = 1$, and $z_2 = -1$. Then

$$\Pi_1 = \frac{\mu}{VD\rho} \qquad \Pi_2 = \frac{\Delta p/l}{\rho V^2/D}$$

$$F\left(\frac{VD\rho}{\mu}, \frac{\Delta p/l}{\rho V^2/D}\right) = 0$$

since the Π quantities can be inverted if desired. The first parameter, $VD\rho/\mu$, is the *Reynolds number* **R**, one of the most important dimensionless parameters in fluid mechanics. Its size determines the nature of the flow. The Reynolds number is discussed in Sec. 6.1. After solving for $\Delta p/l$, we have

$$\frac{\Delta p}{l} = f_1\left(\mathbf{R}, \frac{\rho V^2}{D}\right)$$

The usual formula is

$$\frac{\Delta p}{l} = f(\mathbf{R})\frac{\rho V^2}{2D}$$

or, in terms of head loss,

$$\frac{\Delta h}{l} = f(\mathbf{R})\frac{1}{D}\frac{V^2}{2g}$$

Example 5.4   **A** fluid-flow situation depends on the velocity V; the density ρ; several linear dimensions l, l_1, and l_2; pressure drop Δp; gravity g; viscosity μ; surface tension σ; and bulk modulus of elasticity K. Apply dimensional analysis to these variables to find a set of Π parameters

$$F(V, \rho, l, l_1, l_2, \Delta p, g, \mu, \sigma, K) = 0$$

Solution

As three dimensions are involved, three repeating variables are selected. For complex situations, V, ρ, and l are generally helpful. There are seven Π parameters:

$$\Pi_1 = V^{x_1} \rho^{y_1} l^{z_1} \Delta p$$
$$\Pi_2 = V^{x_2} \rho^{y_2} l^{z_2} g$$
$$\Pi_3 = V^{x_3} \rho^{y_3} l^{z_3} \mu$$
$$\Pi_4 = V^{x_4} \rho^{y_4} l^{z_4} \sigma$$
$$\Pi_5 = V^{x_5} \rho^{y_5} l^{z_5} K$$
$$\Pi_6 = \frac{l}{l_1}$$
$$\Pi_7 = \frac{l}{l_2}$$

By expanding the Π quantities into dimensions,

$$\Pi_1 = (LT^{-1})^{x_1}(ML^{-3})^{y_1}L^{z_1}ML^{-1}T^{-2}$$

$$x_1 - 3y_1 + z_1 - 1 = 0$$
$$- x_1 \qquad\qquad - 2 = 0$$
$$y_1 \qquad\quad + 1 = 0$$

from which $x_1 = -2$, $y_1 = -1$, and $z_1 = 0$.

$$\Pi_2 = (LT^{-1})^{x_2}(ML^{-3})^{y_2}L^{z_2}LT^{-2}$$

$$x_2 - 3y_2 + z_2 + 1 = 0$$
$$- x_2 \qquad\qquad - 2 = 0$$
$$y_2 \qquad\qquad = 0$$

from which $x_2 = -2$, $y_2 = 0$, and $z_2 = 1$.

$$\Pi_3 = (LT^{-1})^{x_3}(ML^{-3})^{y_3}L^{z_3}ML^{-1}T^{-1}$$

$$x_3 - 3y_3 + z_3 - 1 = 0$$
$$- x_3 \qquad\qquad - 1 = 0$$
$$y_3 \qquad\quad + 1 = 0$$

from which $x_3 = -1$, $y_3 = -1$, and $z_3 = -1$.

$$\Pi_4 = (LT^{-1})^{x_4}(ML^{-3})^{y_4}L^{z_4}MT^{-2}$$

$$x_4 - 3y_4 + z_4 \qquad = 0$$
$$- x_4 \qquad\qquad - 2 = 0$$
$$y_4 \qquad\quad + 1 = 0$$

from which $x_4 = -2$, $y_4 = -1$, and $z_4 = -1$.

$$\Pi_5 = (LT^{-1})^{x_5}(ML^{-3})^{y_5}L^{z_5}ML^{-1}T^{-2}$$

$$x_5 - 3y_5 + z_5 - 1 = 0$$
$$- x_5 \qquad\qquad - 2 = 0$$
$$y_5 \qquad\quad + 1 = 0$$

$$\Pi_5 = (LT^{-1})^{x_5}(ML^{-3})^{y_5}L^{z_5}ML^{-1}T^{-2}$$

from which $x_5 = -2$, $y_5 = -1$, and $z_5 = 0$.

$$\Pi_1 = \frac{\Delta p}{\rho V^2} \quad \Pi_2 = \frac{gl}{V^2} \quad \Pi_3 = \frac{\mu}{Vl\rho} \quad \Pi_4 = \frac{\sigma}{V^2\rho l}$$

$$\Pi_5 = \frac{K}{\rho V^2} \quad \Pi_6 = \frac{l}{l_1} \quad \Pi_7 = \frac{l}{l_2}$$

and

$$f\left(\frac{\Delta p}{\rho V^2}, \frac{gl}{V^2}, \frac{\mu}{Vl\rho}, \frac{\sigma}{V^2\rho l}, \frac{K}{\rho V^2}, \frac{l}{l_1}, \frac{l}{l_2}\right) = 0$$

It is convenient to invert some of the parameters and to take some square roots,

$$f_1\left(\frac{\Delta p}{\rho V^2}, \frac{V}{\sqrt{gl}}, \frac{Vl\rho}{\mu}, \frac{V^2l\rho}{\sigma}, \frac{V}{\sqrt{K/\rho}}, \frac{l}{l_1}, \frac{l}{l_2}\right) = 0$$

The first parameter, usually written $\Delta p/(\rho V^2/2)$, is the *pressure coefficient*, the second parameter is the *Froude* number $\mathbf{F}$; the third is the *Reynolds* number $\mathbf{R}$; the fourth is the *Weber* number $\mathbf{W}$; and the fifth is the *Mach* number $\mathbf{M}$. Hence,

$$f_1\left(\frac{\Delta p}{\rho V^2}, \mathbf{F}, \mathbf{R}, \mathbf{W}, \mathbf{M}, \frac{l}{l_1}, \frac{l}{l_2}\right) = 0$$

After solving for pressure drop,

$$\Delta p = \rho V^2 f_2\left(\mathbf{F}, \mathbf{R}, \mathbf{W}, \mathbf{M}, \frac{l}{l_1}, \frac{l}{l_2}\right)$$

in which f_1 and f_2 must be determined from analysis or experiment. By selecting other repeating variables, a different set of Π parameters could be obtained.

Figure 6.20 is a representation of a functional relation of the type just given as it applies to the flow in pipes. Here the parameters $\mathbf{F}$, $\mathbf{W}$, and $\mathbf{M}$ are neglected as being unimportant; l is the pipe diameter D, l_1 is the length of the pipe L, and l_2 is a dimension which is representative of the effective height of the surface roughness of the pipe and is given by ϵ. Thus,

$$\frac{\Delta p}{\rho V^2} = f_3\left(\mathbf{R}, \frac{L}{D}, \frac{\epsilon}{D}\right)$$

The fact that the pressure drop in the pipeline varies linearly with the length (i.e., doubling the length of pipe doubles the loss in pressure) appears reasonable, so that

$$\frac{\Delta p}{\rho V^2} = \frac{L}{D} f_4\left(\mathbf{R}, \frac{\epsilon}{D}\right) \quad \text{or} \quad \frac{\Delta p}{\rho V^2(L/D)} = f_4\left(\mathbf{R}, \frac{\epsilon}{D}\right)$$

The term on the left-hand side is commonly given the notation $f/2$, as in Fig. 6.21. The curves shown in this figure have f and $\mathbf{R}$ as ordinate and abscissa, respectively, with ϵ/D as a parameter which assumes a given value for each curve. The nature of these curves was determined through experiment. Such experiments show that when the parameter $\mathbf{R}$ is below the value of 2000, all the curves for the various values of ϵ/D coalesce into one. Hence, f is independent of ϵ/D, and the result is

$$f = f_5(\mathbf{R})$$

This relation will be predicted in Chap. 6 on the basis of theoretical considerations, but it remained for an experimental verification of these predictions to indicate the power of the theoretical methods.

| Example 5.5 | The thrust due to any one of the family of geometrically similar airplane propellers is to be determined experimentally from a wind-tunnel test on a model. Use dimensional analysis to find suitable parameters for plotting test results. |

Solution

The thrust F_T depends upon speed of rotation ω, speed of advance V_0, diameter D, air viscosity μ, density ρ, and speed of sound c. The function

$$F(F_T, V_0, D, \omega, \mu, \rho, c) = 0$$

is to be arranged into four dimensionless parameters, since there are seven quantities and three dimensions. Starting first by selecting ρ, ω, and D as repeating variables

$$\Pi_1 = \rho^{x_1}\omega^{y_1}D^{z_1}F_T = (ML^{-3})^{x_1}(T^{-1})^{y_1}L^{z_1}MLT^{-2}$$
$$\Pi_2 = \rho^{x_2}\omega^{y_2}D^{z_2}V_0 = (ML^{-3})^{x_2}(T^{-1})^{y_2}L^{z_2}LT^{-1}$$
$$\Pi_3 = \rho^{x_3}\omega^{y_3}D^{z_3}\mu = (ML^{-3})^{x_3}(T^{-1})^{y_3}L^{z_3}ML^{-1}T^{-1}$$
$$\Pi_4 = \rho^{x_4}\omega^{y_4}D^{z_4}c = (ML^{-3})^{x_4}(T^{-1})^{y_4}L^{z_4}LT^{-1}$$

Writing the simultaneous equations in x_1, y_1, z_1, etc., as before and solving them gives

$$\Pi_1 = \frac{F_T}{\rho\omega^2 D^4} \qquad \Pi_2 = \frac{V_0}{\omega D} \qquad \Pi_3 = \frac{\mu}{\rho\omega D^2} \qquad \Pi_4 = \frac{c}{\omega D}$$

Solving for the thrust parameter leads to

$$\frac{F_T}{\rho\omega^2 D^4} = f_1\left(\frac{V_0}{\omega D}, \frac{\rho\omega D^2}{\mu}, \frac{c}{\omega D}\right)$$

Since the parameters can be recombined to obtain other forms, the second term is replaced by the product of the first and second terms, $VD\rho/\mu$, and the third term is replaced by the first term divided by the third term, V_0/c; thus,

$$\frac{F_T}{\rho\omega^2 D^4} = f_2\left(\frac{V_0}{\omega D}, \frac{V_0 D\rho}{\mu}, \frac{V_0}{c}\right)$$

Of the dimensionless parameters, the first is probably of the most importance since it relates speed of advance to speed of rotation. The second parameter is a Reynolds number and accounts for viscous effects. The last parameter, speed of advance divided by speed of sound, is a Mach number, which would be important for speeds near or higher than the speed of sound. Reynolds effects are usually small, so that a plot of $F_T/\rho\omega^2 D^4$ against $V_0/\omega D$ should be most informative.

The steps in a dimensional analysis can be summarized as follows:

1. Select the pertinent variables. This requires some knowledge of the process.
2. Write the functional relations, for example,

$$F(V, D, \rho, \mu, c, H) = 0$$

3. Select the repeating variables. (Do not make the dependent quantity a repeating variable.) These variables should contain all the m dimensions of the problem. Often one variable is chosen because it specifies the scale, another the kinematic conditions. In the cases of major interest in this chapter one variable which is related to the forces or mass of the system, for example, D, V, or ρ, is chosen.
4. Write the Π parameters in terms of unknown exponents, for example,

$$\Pi_1 = V^{x_1}D^{y_1}\rho^{z_1}\mu = (LT^{-1})^{x_1}L^{y_1}(ML^{-3})^{z_1}ML^{-1}T^{-1}$$

5. For each of the Π expressions write the equations of the exponents, so that the sum of the exponents of each dimension will be zero.
6. Solve the equations simultaneously.
7. Substitute back into the Π expressions of step 5 the exponents to obtain the dimensionless Π parameters.

8. Establish the functional relation

$$f_1(\Pi_1, \Pi_2, \Pi_3, \ldots, \Pi_{n-m}) = 0$$

or solve for one of the Π's explicitly:

$$\Pi_2 = f(\Pi_1, \Pi_3, \ldots, \Pi_{n-m})$$

9. Recombine, if desired, to alter the forms of the Π parameters, keeping the same number of independent parameters.

Alternative Formulation of Π Parameters

A rapid method for obtaining Π parameters, developed by Hunsaker and Rightmire [3], uses the repeating variables as primary quantities and solves for M, L, and T in terms of them. In Example 5.3 the repeating variables are V, D, and ρ; therefore,

$$V = LT^{-1} \qquad D = L \qquad \rho = ML^{-3} \qquad \text{(5.3.4)}$$
$$L = D \qquad T = DV^{-1} \qquad M = \rho D^3$$

Now, by use of Eqs. (5.3.4),

$$\mu = ML^{-1}T^{-1} = \rho D^3 D^{-1} D^{-1} V = \rho DV$$

hence, the Π parameter is

$$\Pi_1 = \frac{\mu}{\rho DV}$$

Equations (5.3.4) can be used directly to find the other Π parameters. For Π_2

$$g = LT^{-2} = DD^{-2}V^2 = V^2 D^{-1}$$

and

$$\Pi_2 = \frac{g}{V^2 D^{-1}} = \frac{gD}{V^2}$$

This method does not require the repeated solution of three equations in three unknowns for each Π parameter determination.

Pressure Coefficient

The pressure coefficient $\Delta p/(\rho V^2/2)$ is the ratio of pressure to dynamic pressure. When the pressure coefficient is multiplied by the area, the product is the ratio of pressure force to inertial force, as $(\rho V^2/2)A$ would be the force needed to reduce the velocity to zero. It can also be written as $\Delta h/(V^2/2g)$ by dividing by γ. For pipe flow the Darcy-Weisbach equation relates losses h_l to length of pipe L, diameter D, and velocity V by a dimensionless friction factor f†

$$h_l = f\frac{L}{D}\frac{V^2}{2g} \qquad \text{or} \qquad \frac{fL}{D} = \frac{h_l}{V^2/2g} = f_2\left(\mathbf{R}, \mathbf{F}, \mathbf{W}, \mathbf{M}, \frac{l}{l_1}, \frac{l}{l_2}\right)$$

---

†There are several friction factors in general use. This is the Darcy-Weisbach friction factor, which is 4 times the size of the *Fanning friction factor*, also called *f*.

as fL/D is shown to be equal to the pressure coefficient (see Ex. 5.4). In pipe flow, gravity has no influence on losses; therefore, **F** can be dropped. Similarly, surface tension has no effect, and **W** drops out. For steady liquid flow compressibility is not important, and **M** is dropped; l may refer to D; l_1 refers to roughness height projection ϵ in the pipe wall; and l_2 refers to spacing ϵ'. Hence,

$$\frac{fL}{D} = f_2\left(\mathbf{R}, \frac{\epsilon}{D}, \frac{\epsilon'}{D}\right) \tag{5.3.5}$$

Pipe-flow problems are discussed in Chaps. 6 and 12. If compressibility is important,

$$\frac{fL}{D} = f_2\left(\mathbf{R}, \mathbf{M}, \frac{\epsilon}{D}, \frac{\epsilon'}{D}\right) \tag{5.3.6}$$

With orifice flow, studied in Chap. 10, $V = C_v\sqrt{2gH}$ and

$$\frac{H}{V^2/2g} = \frac{1}{C_v^2} = f_2\left(\mathbf{R}, \mathbf{W}, \mathbf{M}, \frac{l}{l_1}, \frac{l}{l_2}\right) \tag{5.3.7}$$

in which l may refer to orifice diameter and l_1 and l_2 to upstream dimensions. Viscosity and surface tension are unimportant for large orifices and low-viscosity fluids since the numerators of the Reynolds and Weber numbers are very large compared with their denominators. Compressibility effects are not important when the Mach number is small as compared with 1. They become very important as the Mach number approaches or is greater than unity.

In steady, uniform open-channel flow, discussed in Chap. 6, the Chézy formula relates average velocity V, slope of channel S, and hydraulic radius of cross section R (the area of section divided by the wetted perimeter) by

$$V = C\sqrt{RS} = C\sqrt{R\frac{\Delta h}{L}} \tag{5.3.8}$$

where C is a coefficient depending upon size, shape, and roughness of the channel. Then

$$\frac{\Delta h}{V^2/2g} = \frac{2gL}{R}\frac{1}{C^2} = f_2\left(\mathbf{F}, \mathbf{R}, \frac{l}{l_1}, \frac{l}{l_2}\right) \tag{5.3.9}$$

since surface tension and compressible effects are usually unimportant.

The drag F on a body is expressed by $F = C_D A\rho V^2/2$, in which A is a typical area of body, usually the projection of the body onto a plane normal to the flow. Then F/A is equivalent to Δp, and

$$\frac{F}{A\rho V^2/2} = C_D = f_2\left(\mathbf{R}, \mathbf{F}, \mathbf{M}, \frac{l}{l_1}, \frac{l}{l_2}\right) \tag{5.3.10}$$

The term **R** is related to *skin friction* drag due to viscous shear as well as to *form*, or *profile*, drag resulting from *separation* of the flow streamlines from the body; **F** is related to wave drag if there is a free surface; for large Mach numbers C_D may vary more markedly with **M** than with the other parameters; and the length ratios may refer to shape or roughness of the surface.

The Reynolds Number

The Reynolds number $VD\rho/\mu$ is the ratio of inertial forces to viscous forces. A *critical* Reynolds number distinguishes among flow regimes, such as laminar or turbulent flow in pipes, in the boundary layer, or around immersed objects. The particular value depends upon the situation. In compressible flow, the Mach number is generally more significant than the Reynolds number.

The Froude Number

The Froude number $V/\sqrt{gl}$, when squared and then multiplied and divided by ρA, is a ratio of dynamic (or inertial) force to gravity force. With free liquid-surface flow (where l is replaced by y, the depth) the nature of the flow (rapid† or tranquil) depends upon whether the Froude number is greater or less than unity. It is useful in calculations of hydraulic jump, in design of hydraulic structures, and in ship design.

The Weber Number

The Weber number $V^2 l\rho/\sigma$ is the ratio of inertial forces to surface-tension forces (evident when the numerator and denominator are multiplied by l). It is important at gas-liquid or liquid-liquid interfaces and also where these interfaces are in contact with a boundary. Surface tension causes small (capillary) waves and droplet formation and has an effect on discharge of orifices and weirs at very small heads. The effect of surface tension on wave propagation is shown in Fig. 5.1. To the left of the curve's minimum the wave speed is controlled by surface tension (the waves are called ripples), and to the right of the curve's minimum gravity effects are dominant.

The Mach Number

The speed of sound in a liquid is written $\sqrt{K/\rho}$ if K is the bulk modulus of elasticity (Sec. 1.8) or $c = \sqrt{kRT}$ where k is the specific heat ratio and T is the absolute temperature for a perfect gas. V/c or $V/\sqrt{K/\rho}$ is the Mach number. It is a measure of the

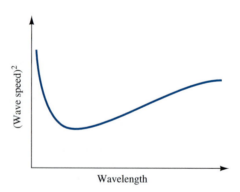

Figure 5.1   Wave speed versus wavelength for surface waves.

---

†Open-channel flow at depth y is *rapid* when the flow velocity is greater than the speed $\sqrt{gy}$ of an elementary wave in quiet liquid. *Tranquil* flow occurs when the flow velocity is less than $\sqrt{gy}$.

ratio of inertial to elastic forces. When V/c is squared and multiplied by $\rho A/2$ in the numerator and denominator, the numerator is dynamic force and the denominator is the dynamic force at sonic flow. It can also be shown to be a measure of the ratio of kinetic energy of the flow to internal energy of the fluid. It is the most important correlating parameter when velocities are near or above local sonic velocities.

EXERCISES

5.3.1 An *incorrect* arbitrary recombination of the Π parameters

$$F\left(\frac{V_0}{\omega D}, \frac{\rho\omega D^2}{\mu}, \frac{c}{\omega D}\right) = 0$$

is given by (a) $F\left(\dfrac{c}{V_0}, \dfrac{\rho c D}{\mu}, \dfrac{c}{\omega D}\right) = 0$ (b) $F\left(\dfrac{V_0}{\omega D}, \dfrac{\rho c D^2}{\mu}, \dfrac{c}{\omega D}\right) = 0$

(c) $F\left(\dfrac{V_0}{\omega D}, \dfrac{V_0 c \rho}{\omega \mu}, \dfrac{\rho c D}{\mu}\right) = 0$ (d) $F\left(\dfrac{V_0 \mu}{\omega^2 D^3 \rho}, \dfrac{V_0 \rho D}{\mu}, \dfrac{c}{\omega D}\right) = 0$ (e) none of these answers.

5.3.2 The repeating variables in a dimensional analysis should (a) include the dependent variable; (b) have two variables with the same dimensions if possible; (c) exclude one of the dimensions from each variable if possible; (d) include the variables not considered very important factors; (e) satisfy none of these answers.

5.3.3 Select the quantity in the following that is *not* a dimensionless parameter: (a) pressure coefficient; (b) Froude number; (c) Darcy-Weisbach friction factor; (d) kinematic viscosity; (e) Weber number.

5.3.4 How many Π parameters are needed to express the function $F(a, V, t, \nu, L) = 0$? (a) 5, (b) 4; (c) 3; (d) 2; (e) 1.

5.3.5 Which of the following could be a Π parameter of the function $F(Q, H, g, V_0, \phi) = 0$ when Q and g are taken as repeating variables?   (a) Q^2/gH^4 (b) $V_0^2/g^2 Q$ (c) $Q/g\phi^2$ (d) $Q/\sqrt{gH}$ (e) none of these answers.

5.3.6 Which of the following has the form of a Reynolds number? (a) $\dfrac{ul}{\nu}$ (b) $\dfrac{VD\mu}{\rho}$ (c) $\dfrac{u\nu}{l}$ (d) $\dfrac{V}{gD}$ (e) $\dfrac{\Delta p}{\rho V^2}$

5.3.7 The Reynolds number can be defined as the ratio of (a) viscous forces to inertial forces; (b) viscous forces to gravity forces; (c) gravity forces to inertial forces; (d) elastic forces to pressure forces; (e) none of these answers.

5.3.8 The pressure coefficient can take the form (a) $\dfrac{\Delta p}{\gamma H}$ (b) $\dfrac{\Delta p}{\rho V^2/2}$ (c) $\dfrac{\Delta p}{l\mu V}$ (d) $\Delta p \dfrac{\rho}{\mu^2 l^4}$ (e) none of these answers.

5.3.9 The pressure coefficient is a ratio of pressure forces to (a) viscous forces; (b) inertial forces; (c) gravity forces; (d) surface-tension forces; (e) elastic-energy forces.

5.3.10 Select the situation in which inertial forces would be *unimportant:* (a) flow over a spillway crest; (b) flow through an open-channel transition; (c) waves breaking against a sea wall; (d) flow through a long capillary tube; (e) flow through a half-opened valve.

5.3.11   Which two forces are most important in laminar flow between closely spaced parallel plates? (*a*) inertial and viscous; (*b*) pressure and inertial; (*c*) gravity and pressure; (*d*) viscous and pressure; (*e*) none of these answers.

5.3.12   If the capillary rise Δh of a liquid in a circular tube of diameter D depends upon surface tension σ and specific weight γ, the formula for capillary rise could take the form (*a*) $\Delta h = \sqrt{\frac{\sigma}{\gamma}}F\left(\frac{\sigma}{\gamma D^2}\right)$ (*b*) $\Delta h = c\left(\frac{\sigma}{\gamma D^2}\right)^n$ (*c*) $\Delta h = cD\left(\frac{\sigma}{\gamma}\right)^n$

(*d*) $\Delta h = \sqrt{\frac{\gamma}{\sigma}}F\left(\frac{\gamma D^2}{\sigma}\right)$ (*e*) none of these answers.

5.4   THE Π THEOREM: HEAT AND MASS TRANSPORT

The Buckingham Π procedure can be extended to the case of heat and mass transport as well; several illustrative examples follow. Example 5.6 considers the heat exchanger apparatus.

Example 5.6

It is desirable to find a series of dimensionless groups that would allow the heat carried away via the average velocity in an exchanger pipe to be related to other relevant variables.

Solution

The relevant variables are assumed to be the pipe diameter D, the fluid density ρ, the viscosity μ, the heat capacity c_p, the velocity V, the heat transfer coefficient h, and the heat conductivity k. Essentially the heat transfer coefficient is the unknown variable and has dimensions of $[MT^{-3}\Theta]$. Since there are seven variables and four independent dimensions, M, L, T, and Θ, it is necessary to find three dimensionless groups. If D, k, μ, and V are chosen as the repeaters, then the following system of equations is formed

$$\Pi_1 = D^{x_1}\mu^{y_1}V^{z_1}k^{w_1}c_p$$
$$\Pi_2 = D^{x_2}\mu^{y_2}V^{z_2}k^{w_2}\rho$$
$$\Pi_3 = D^{x_3}\mu^{y_3}V^{z_3}k^{w_3}h$$

Looking at Π_3 in detail by expanding powers on the dimensions gives

$$
\begin{aligned}
L &: x_3 - y_3 + z_3 + w_3 & = 0 \\
M &: -y_3 + z_3 + w_3 + 1 = 0 \\
T &: -y_3 - z_3 - 3w_3 - 3 = 0 \\
\Theta &: -w_3 - 1 = 0
\end{aligned}
$$

which is easily solved to give

$$x_3 = 1, y_3 = z_3 = 0, w_3 = -1$$

or

$$\Pi_3 = \frac{hD}{k} \tag{5.4.1}$$

The two other numbers can be similarly found

$$\Pi_1 = \frac{\mu c_p}{k} \tag{5.4.2}$$

$$\Pi_2 = \frac{\rho DV}{\mu} = \frac{DV}{\nu} \tag{5.4.3}$$

While Π_2 is recognized as the Reynolds number, the new Π_3 grouping is called the *Nusselt number,* $\mathbf{N}_u$, and is a measure of the intensity of convection to conduction in heat transport mechanisms. The Π_1 grouping is called the *Prandtl number,* $\mathbf{P}_r$, which represents the ratio of heat diffusion to momentum diffusion. If both the top and bottom of Π_1 are multiplied by ρ then Π_1 becomes ν/k, the Prandtl number formed from the ratio of the diffusivities of momentum and heat. For the example problem laboratory experiments are now required to correlate

$$\mathbf{N}_u = f(\mathbf{R}, \mathbf{P}_r)$$

Cooling ponds, lakes, lagoons, etc., are subjected to intensive radiative heating during the spring and summer seasons. Heat rapidly builds up at the surface and wind-induced velocity and shear stress mix the heat down into the deeper waters. Wind mixing is not strong enough to fully mix the water column to a uniform temperature; consequently a *stratification* results marked by a strong vertical temperature gradient. If the linear relationship between density and temperature [Eq. (1.5.10)] is assumed, then the relevant variables are depth d, viscosity μ, velocity v, expansion coefficient β, surface to bottom temperature difference $\Delta T'$, acceleration due to gravity g, density ρ, and modified thermal conductivity $k^* = k/c_p$. Find the dimensionless groups relating the stratification strength to the relevant variables.

Example 5.7

Solution

There are eight variables and four dimensions yielding four nondimensional groups. If d, μ, β, and g are selected as the repeating variables, then the four nondimensional groups are defined as

$$\Pi_1 = d^{x_1} \mu^{y_1} \beta^{z_1} g^{w_1} \rho$$
$$\Pi_2 = d^{x_2} \mu^{y_2} \beta^{z_2} g^{w_2} v$$
$$\Pi_3 = d^{x_3} \mu^{y_3} \beta^{z_3} g^{w_3} \Delta T'$$
$$\Pi_4 = d^{x_4} \mu^{y_4} \beta^{z_4} g^{w_4} k^*$$

After some algebra the four groups become

$$\Pi_1 = \frac{d^{3/2} g^{1/2} \rho}{\mu} \qquad \Pi_2 = \frac{v}{d^{1/2} g^{1/2}} = \mathbf{F}$$

$$\Pi_3 = \beta \Delta T' \qquad \Pi_4 = \frac{\mu}{k^*} = \mathbf{P}_r$$

It is immediately seen that Π_2 is the Froude number while Π_4 is the Prandtl number. Without loss of rigor Π_1 can be squared to eliminate the fractional powers, and since Π_3 is already dimensionless, it can be multiplied with Π_1 to give a new number called the *Grashof number,* $\mathbf{G}_r$,

$$\Pi_5 = \mathbf{G}_r = \Pi_1^2 \Pi_3 = \frac{\beta d^3 \rho^2 g \Delta T'}{\mu^2} \tag{5.4.4}$$

The Grashof number is a common dimensionless group that is used when analyzing the potential effect of convection introduced by large temperature differences or a large density gradient. Therefore, if it was desired to establish a relationship between the stratification (i.e., $\Delta T'$) and the problem variables, then the functional relationship between $\mathbf{G}_r$, $\mathbf{F}$, and $\mathbf{P}_r$ would need to be established in the laboratory, that is,

$$\mathbf{G}_r = f(\mathbf{F}, \mathbf{P}_r)$$

Example 5.8	The concentration of dissolved oxygen at the bottom of a water supply reservoir is a function of both physical and chemical processes. The amount of dissolved oxygen exchanged between the near surface sediments, C_b, and the overlying water column, C_{wc}, is a function of the concentration difference between the two, that is, $C_b - C_{wc}$. The concentration C_{wc} is also affected because decaying, dead phytoplankton and zooplankton consume dissolved oxygen according to a first-order reaction rate $-k_1 C_{wc}$. The following variables are considered essential to the problem: the depth d, density ρ, velocity v, kinematic viscosity ν, mass diffusivity $\mathscr{D}$, mass transfer coefficient h, and the decay rate k_1. Find a relationship between the variables which demonstrates the importance of the source-sink term.

Solution

With there being seven variables and three dimensions, four dimensionless groups are required. If $\mathscr{D}$, ρ, and d are used as repeaters, then

$$\Pi_1 = d^{x_1} \rho^{y_1} \mathscr{D}^{z_1} v$$
$$\Pi_2 = d^{x_2} \rho^{y_2} \mathscr{D}^{z_2} h$$
$$\Pi_3 = d^{x_3} \rho^{y_3} \mathscr{D}^{z_3} \nu$$
$$\Pi_4 = d^{x_4} \rho^{y_4} \mathscr{D}^{z_4} k_1$$

The algebraic solution yields

$$\Pi_1 = \frac{dv}{\mathscr{D}} \qquad \Pi_2 = \frac{dh}{\mathscr{D}}$$
$$\Pi_3 = \frac{\nu}{\mathscr{D}} \qquad \Pi_4 = \frac{d^2 k_1}{\mathscr{D}}$$

The Π_2 group is called the *Sherwood number*, $\mathbf{S}_h$, or *mass Nusselt number* and, as in the thermal case, contrasts the relative strengths of convection versus diffusive transport. Π_3 is the *Schmidt number*, $\mathbf{S}_c$, which contrasts momentum and mass diffusivities, and if Π_1 is divided by Π_3 the result is the Reynolds number, $\mathbf{R}$. Finally, the Π_4 group has no name by itself; however, if Π_4 is divided by Π_1 then the *Damkohler number*, $\mathbf{D}_N$, is defined. The four final groups then become

$$\mathbf{D}_N = \frac{k_1 d}{v} \tag{5.4.5}$$

$$\mathbf{S}_h = \frac{dh}{\mathscr{D}} \tag{5.4.6}$$

$$\mathbf{S}_c = \frac{\nu}{\mathscr{D}} \tag{5.4.7}$$

$$\mathbf{R} = \frac{dv}{\nu} \tag{5.4.8}$$

Nusselt Number and Sherwood Number

$\mathbf{N}_u$ and $\mathbf{S}_h$ compare the relative strength of convection or advection processes to molecular diffusion of heat or mass. They are similar to the Reynolds number in that they are fundamental design variables when designing a flow and transport geometry to achieve specified rates of heat exchange or mass flux.

Prandtl Number and Schmidt Number

$\mathbf{P}_r$ and $\mathbf{S}_h$ compare fluid properties. $\mathbf{P}_r$ compares momentum to heat diffusivity and $\mathbf{S}_h$ compares momentum to mass diffusivity. While important numbers in flow and transport process design, their importance in naturally occurring flows is diminished in contrast to other transport agents.

All the above numbers are fundamental to *any* heat or mass transport problem. The following two numbers describe processes that are not necessarily universal.

Grashof Number

G_r compares the relative strengths of convection in the transport field. Originally applied to natural convection from unstable temperature or density fields, its use has become more widespread in the analysis of all flows with large spatial density gradients.

Damkohler Number

D_N simply contrasts the strength of a chemical or biological transformation to a change in mass concentration brought about by advection. Its use is widespread in industrial process design but has achieved little recognition in environmental transport analyses.

5.5 NONDIMENSIONAL ANALYSIS OF GOVERNING EQUATIONS

Single Scale Normalization

As noted in the introduction and the previous section, dimensional analysis can be approached with two objectives in mind: to discover the initial form of previously unknown correlations between variables, and to compare the relative size or importance of one fluid mechanics or transport process to another. The second goal is approached in this subsection and presumes full knowledge of the physics of the problem as expressed through the governing differential equation derived in the previous chapter.

The procedure is based upon a normalization of all the dependent and independent variables. Normalization in this context means to reference all the quantities in the equation to constant values which are presumed to be the largest values encountered in the problem. In this fashion new dependent and independent variables are created all of which range in value from ± 1 to 0. By inserting these new variable definitions into the governing equations and striving to achieve full consistency or similarity between the newly transformed equations and the original equations, quantitative measures in the form of nondimensional parameters result. These dimensionless groups or numbers allow direct evaluation of the importance of the various processes in the problem.

For example, it is desirable to determine the heat, dissolved oxygen, and flow field characteristics of a water supply reservoir. The length of the reservoir is 20 km, the maximum temperature is T_m, and the maximum dissolved oxygen concentration is C_m. The dimensional analysis begins by defining the scaling parameters from the given dimensions and defining new nondimensional variables. The dependent variables are u, v, w, p, T, and C, which are nondimensionalized in the following fashion: $u_* = u/u_m$, $v_* = v/u_m$, $w_* = w/u_m$, $p_* = p/p_{ref}$, $T_* = T/T_m$, and $C_* = C/C_m$. Here, u_m is the reference velocity, which has yet to be specified, and p_{ref} is the reference pressure which also has yet to be specified. The independent variables are x, y, z, and t which are normalized as follows: $x_* = x/L$, $y_* = y/L$, $z_* = z/L$, and $t_* = t/t_{ref}$.

The insertion of the new definitions for the variables in the following three types of operations requires some care. The derivatives are transformed by first redefining the independent variable. Therefore, from the calculus the temporal derivative is found as

$$\frac{\partial}{\partial t} = \frac{\partial}{\partial t_*}\frac{\partial t_*}{\partial t} = \frac{1}{t_{ref}}\frac{\partial}{\partial t_*} \tag{5.5.1}$$

In this equation $\partial t_*/\partial t = 1/t_{ref}$. The spatial derivatives are found in a similar fashion, that is,

$$\frac{\partial}{\partial x} = \frac{\partial}{\partial x_*}\frac{\partial x_*}{\partial x} = \frac{1}{L}\frac{\partial}{\partial x_*} \tag{5.5.2}$$

and

$$\frac{\partial^2}{\partial x^2} = \frac{\partial}{\partial x}\frac{\partial}{\partial x} = \frac{1}{L}\frac{\partial}{\partial x_*}\frac{\partial}{\partial x} = \frac{1}{L^2}\frac{\partial^2}{\partial x_*^2} \tag{5.5.3}$$

The dependent variables are then inserted in these three classes of differentiations. If, for example, concentration is used, then the time derivative is

$$\frac{\partial C}{\partial t} = \frac{1}{t_{ref}}\frac{\partial}{\partial t_*}(C_* C_m) = \frac{C_m}{t_{ref}}\frac{\partial C_*}{\partial t_*} \tag{5.5.4}$$

while the spatial derivatives transform as

$$\frac{\partial C}{\partial x} = \frac{1}{L}\frac{\partial}{\partial x_*}(C_* C_m) = \frac{C_m}{L}\frac{\partial C_*}{\partial x_*} \tag{5.5.5}$$

and

$$\frac{\partial^2 C}{\partial x^2} = \frac{1}{L^2}\frac{\partial^2}{\partial x_*^2}(C_* C_m) = \frac{C_m}{L^2}\frac{\partial^2 C_*}{\partial x_*^2} \tag{5.5.6}$$

If the original governing differential equation [Eq. (4.8.10)] is

$$\frac{\partial C}{\partial t} + u\frac{\partial C}{\partial x} + v\frac{\partial C}{\partial y} + w\frac{\partial C}{\partial z} = \mathscr{D}\left[\frac{\partial^2 C}{\partial x^2} + \frac{\partial^2 C}{\partial y^2} + \frac{\partial^2 C}{\partial z^2}\right] + S_c$$

then using the transformation operations in Eqs. (5.5.1) to (5.5.6) yields

$$\left(\frac{C_m}{t_{ref}}\right)\frac{\partial C_*}{\partial t} + \left(\frac{u_m C_m}{L}\right)u_*\frac{\partial C_*}{\partial x_*} + \left(\frac{u_m C_m}{L}\right)v_*\frac{\partial C_*}{\partial y_*} + \left(\frac{u_m C_m}{L}\right)w_*\frac{\partial C_*}{\partial z_*}$$

$$= \mathscr{D}\left[\left(\frac{C_m}{L^2}\right)\frac{\partial^2 C_*}{\partial x_*^2} + \left(\frac{C_m}{L^2}\right)\frac{\partial^2 C_*}{\partial y_*^2} + \left(\frac{C_m}{L^2}\right)\frac{\partial^2 C_*}{\partial z_*^2}\right] + (S_{mc})S_{c*} \tag{5.5.7}$$

Collecting terms gives

$$\left(\frac{C_m}{t_{ref}}\right)\frac{\partial C_*}{\partial t} + \left(\frac{u_m C_m}{L}\right)(\mathbf{v}_* \cdot \nabla_*)C_* = \frac{\mathscr{D}C_m}{L^2}\nabla_*^2 C_* + (S_{mc})S_{c*} \tag{5.5.8}$$

In Eq. (5.5.8), ∇_* and ∇_*^2 are the nondimensional versions of the operators, for example,

$$\nabla_* = \mathbf{i}\frac{\partial}{\partial x_*} + \mathbf{j}\frac{\partial}{\partial y_*} + \mathbf{k}\frac{\partial}{\partial z_*} \tag{5.5.9}$$

The continuity, momentum, and heat transport equations for an incompressible fluid can be similarly transformed. The continuity equation is

$$\left(\frac{u_m}{L}\right)\nabla_* \cdot \mathbf{v}_* = 0 \qquad (5.5.10)$$

The momentum equation is

$$\left(\frac{u_m}{t_{ref}}\right)\frac{\partial \mathbf{v}_*}{\partial t_*} + \left(\frac{u_m^2}{L}\right)(\mathbf{v}_* \cdot \nabla_*)\mathbf{v}_* = -\left(\frac{p_{ref}}{\rho L}\right)\nabla_* p_* - g\nabla_* h_* + \left(\frac{\nu u_m}{L^2}\right)\nabla_*^2 \mathbf{v}_* \qquad (5.5.11)$$

and the heat transport equation is

$$\left(\frac{T_m}{t_{ref}}\right)\frac{\partial T_*}{\partial t} + \left(\frac{u_m T_m}{L}\right)\mathbf{v}_* \cdot \nabla_* T_* = \left(\frac{\alpha T_m}{L^2}\right)\nabla_*^2 T_* + (S_{mT})S_{T*} \qquad (5.5.12)$$

At this stage, each term in parentheses is considered to be a scale factor serving to expand or contract each differential term to equal its original value in the untransformed case. The virtue of this form is that the differential terms now range in size from 0 to 1.

Two other comments before proceeding. First, dividing the continuity equation by u_m/L leaves a transformed continuity equation that is exactly identical in form to the untransformed equation. This is as it should be because volume and hence the mass must be identically preserved regardless of the transformation.

Second, the gravitational term in Eq. (5.5.11) was formally transformed, but since it was already nondimensional (a cosine term), it was not required.

In order to compare the size or importance of one set of differential terms to another, the scale factors may be compared. A common comparison is to ratio all scale factor terms in parentheses to the scale factor for the inertial acceleration scale factor. Therefore, dividing the momentum equation through by (u_m^2/L), Eq. (5.5.11) then becomes

$$\left(\frac{L}{u_m t_{ref}}\right)\frac{\partial \mathbf{v}_*}{\partial t_*} + (\mathbf{v}_* \cdot \nabla_*)\mathbf{v}_* = -\left(\frac{p_{ref}}{\rho u_m^2}\right)\nabla_* p_* - \left(\frac{gL}{u_m^2}\right)\nabla_* h_* + \left(\frac{\nu}{u_m L}\right)\nabla_*^2 \mathbf{v}_* \qquad (5.5.13)$$

Several items are immediately apparent. First and most importantly is that all the terms in parentheses are nondimensional or dimensionless. In fact comparison with Sec. 5.4 reveals that these numbers are identical to those derived empirically through Buckingham Π procedures. Eq. (5.5.13) is rewritten as

$$[\mathbf{S}_t]\frac{\partial \mathbf{v}_*}{\partial t_*} + (\mathbf{v}_* \cdot \nabla_*)\mathbf{v}_* = -[\mathbf{E}]\nabla_* p_* - \frac{1}{[\mathbf{F}]^2}\nabla_* h_* + \frac{1}{[\mathbf{R}]}\nabla_*^2 \mathbf{v}_* \qquad (5.5.14)$$

where

$\mathbf{S}_t = L/u_m t_{ref}$ is a form of the Strouhal number

$\mathbf{E} = p_{ref}/\rho u_m^2$ is the Euler number

$\mathbf{F} = \sqrt{u_m^2/gL}$ is the Froude number

$\mathbf{R} = u_m L/\nu$ is the Reynolds number

Essentially then these numbers allow each term's "strength" to be compared to the inertia terms. For instance the Euler number is the ratio of the pressure term to the inertia term, the Reynolds number is the ratio of the inertia to viscous terms, and the Froude number is the ratio of the inertia to gravity force.

Retaining the Strouhal number allows a comparison of a periodic phenomenon with a strong temporal acceleration, say for example a wind-driven gravity wave, to be compared to the inertial acceleration term which may derive from wholly different processes, say for example wind-driven circulation. It is often the case, however, that the time base in the Strouhal number, t_{ref}, is selected to be L/u_m in which case $\mathbf{S}_t = 1$ and the nondimensional acceleration is completely identical or similar to the original Eulerian acceleration.

It is highly desirable to have full similarity between the original and the normalized equations. However, a severe conflict occurs if we attempt to judiciously select values for L and u_m that allow the remaining nondimensional numbers to collapse to 1. Looking at the definitions

$$\mathbf{R} = \frac{u_m L}{\nu} \qquad \mathbf{E} = \frac{p_{ref}}{\rho u_m^2} \qquad \mathbf{F} = \frac{u_m}{\sqrt{gL}}$$

it is readily seen that $\mathbf{R}$ and $\mathbf{F}$ can never be simultaneously reduced because $\mathbf{R}$ is proportional to L while $\mathbf{F} \propto \sqrt{L}$. Consequently, the maximum reduction possible is to leave the final equation in terms of either $\mathbf{R}$ or $\mathbf{F}$.

Returning now to the concentration and heat transport equations, the scale factor equations, Eqs. (5.5.8) and (5.5.12), are also divided by the advection scale factor, resulting in a concentration equation of

$$\left[\mathbf{S}_t\right]\frac{\partial C_*}{\partial t_*} + (\mathbf{v}_* \cdot \nabla_*)T_* = \frac{1}{\left[\mathbf{P}_{em}\right]}\nabla_*^2 C_* + \left[\Phi_c\right]S_{c*} \tag{5.5.15}$$

and heat transport of

$$\left[\mathbf{S}_t\right]\frac{\partial T_*}{\partial t_*} + (\mathbf{v}_* \cdot \nabla_*)T_* = \frac{1}{\left[\mathbf{P}_e\right]}\nabla_*^2 C_* + \left[\Phi_T\right]S_{T*} \tag{5.5.16}$$

Once again the Strouhal number appears on the temporal gradient term, and as before if t_{ref} is selected as L/U_m, then the Strouhal number equals 1.

The second bracketed term in Eq. (5.5.15) is the inverse of the *mass Peclèt number*, $\mathbf{P}_{em}$, which is defined as $u_m L/\mathcal{D}$. Similarly, in Eq. (5.5.16) $u_m L/\alpha$ is called the *Peclèt number*, $\mathbf{P}_e$. Both numbers assess the diffusion transfer. The Peclèt numbers are often rewritten in terms of the Reynolds number as follows:

$$\mathbf{P}_e = \mathbf{R}\mathbf{P}_r \tag{5.5.17}$$

where

$$\mathbf{P}_r = \nu/\alpha$$
$$\mathbf{P}_{em} = \mathbf{R}\mathbf{S}_c \tag{5.5.18}$$

where

$$\mathbf{S}_c = \nu/\mathcal{D}$$

The *Prandtl number*, $\mathbf{P}_r$, and the *Schmidt number*, $\mathbf{S}_c$, assess the relative strength of the "momentum diffusivity" as parameterized by kinematic viscosity to molecular diffusion of heat and mass, respectively.

Finally, the source-sink dimensionless groups Φ_c and Φ_T are defined as

$$\Phi_c = \left[\frac{S_{mc} L}{u_m C_m}\right] = \left[\frac{S_{mc} t_{ref}}{C_m}\right] \tag{5.5.19}$$

$$\Phi_T = \left[\frac{S_{mT} L}{u_m T_m}\right] = \left[\frac{S_{mT} t_{ref}}{T_m}\right] \tag{5.5.20}$$

However, they are nameless groups at this point because there are numerous forms of Φ_c and Φ_T depending on the functional form of the source-sink term. Hence there are as many named groups as there are functional forms. For instance, if the source-sink term in dimensional form was a first-order reaction rate parameterized by a constant k_1, then $S_c = k_1 C$, S_{mc} in Eq. (5.5.19) becomes $k_1 C_m$, and the number $\left[k_1 L/u_m\right]$ is a form of the *Damkohler number* which assesses the importance of chemical generation of C to advective transport.

Two-Scale Normalization: The Boundary Layer

A boundary layer is defined when there are two length scales in the problem. Typically the horizontal scale is several orders of magnitude greater than the vertical scale. Examples include most large scale, naturally occurring flows in lakes, rivers, estuaries, and the atmosphere. All engineered flows near solid boundaries such as solid walls, wing surfaces, or ship hulls exhibit boundary layers. Using the normalization methods above with the two length scales yields the ubiquitous boundary layer equations to be analyzed in Chap. 7.

5.6   MODEL STUDIES AND SIMILITUDE

Model studies of proposed hydraulic structures and machines are frequently undertaken as an aid to the designer. They permit visual observation of the flow and make it possible to obtain certain numerical data, for example, calibrations of weirs and gates, depths of flow, velocity distributions, forces on gates, efficiencies and capacities of pumps and turbines, pressure distributions, and losses.

If accurate quantitative data are to be obtained from a model study, there must be dynamic similitude between model and prototype. This similitude requires (1) that there be exact geometric similitude and (2) that the ratio of dynamic pressures at corresponding points be a constant. The second requirement can also be expressed as a kinematic similitude, that is, the streamlines must be geometrically similar.

Geometric similitude extends to the actual surface roughness of model and prototype. If the model is one-tenth the size of the prototype in every linear dimension, the height of roughness projections must be in the same ratio. For dynamic pressures to be in the same ratio at corresponding points in model and prototype, the ratios of the various types of forces must be the same at corresponding points. Hence, for strict dynamic similitude, the Mach, Reynolds, Froude, and Weber numbers must be the same in both model and prototype.

Strict fulfillment of these requirements is generally impossible to achieve, except with a 1:1 scale ratio. Fortunately, in many situations only two of the forces are of the same magnitude. Discussion of a few cases will make this clear.

As an aid to understanding the requirements of similarity consider the laboratory analysis of flow over a sphere; the prototype (or real world) and model spheres are depicted in Fig. 5.2. Geometric similarity is of course ensured if the model is also a sphere. Furthermore every linear dimension must be in the ratio of D_m/D_p. This includes even small-scale roughness projections.

Dynamic similarity is ensured by requiring that the "force" polygons on the model and prototype be similar. Each sphere has acting upon it three net forces, the

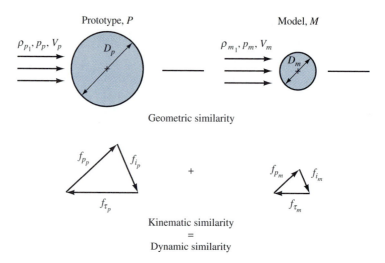

Figure 5.2   Geometric and dynamic similitude for flow over a sphere.

pressure force, f_p; the viscous or shear force, f_τ; and the inertia force due to acceleration, f_i. These forces must form a closed polygon as depicted for the prototype in Fig. 5.2. The model force polygon must be similar to the prototype in that it must be closed and scale linearly. To ensure such similarity, ratios of sides must be maintained, that is,

$$\left[\frac{f_p}{f_i}\right]_{\text{prototype}} = \left[\frac{f_p}{f_i}\right]_{\text{model}} \tag{5.6.1}$$

and

$$\left[\frac{f_\tau}{f_i}\right]_{\text{prototype}} = \left[\frac{f_\tau}{f_i}\right]_{\text{model}} \tag{5.6.2}$$

Note that these ratios are formed from the nondimensional groupings in the previous section. The force polygons are considered similar if

$$\mathbf{E}_p = \mathbf{E}_m \tag{5.6.3}$$

$$\mathbf{R}_p = \mathbf{R}_m \tag{5.6.4}$$

In other words ensuring equality between model and prototype force polygons is attained by seeking the equality of the nondimensional numbers between model and prototype. Strict fulfillment of these requirements is generally impossible to achieve unless the scale ratio is 1:1. Following are several example cases to illustrate the requirements.

Wind- and Water-Tunnel Tests

This equipment is used to examine the streamlines and the forces that are induced as the fluid flows past a fully submerged body. The type of test being conducted and the availability of the equipment determine which kind of tunnel will be used. Because the kinematic viscosity of water is about one-tenth that of air, a water tunnel can be used for model studies at relatively high Reynolds numbers. The drag effect of various parachutes was studied in a water tunnel! At very high air velocities the effects of compressibility, and consequently Mach number, must be taken into consideration and indeed may be the chief reason for undertaking an investigation. Figure 5.3 shows a model of an aircraft carrier being tested in a low-speed tunnel to study the flow pattern around the ship's superstructure. The model has been inverted and suspended from the ceiling so that the wool tufts can be used to give an indication of the flow direction. Behind the model is an apparatus for sensing the air speed and direction at various locations along an aircraft's glide path.

Pipe Flow

In steady flow in a pipe viscous and inertial forces are the only ones of consequence; hence, when geometric similitude is observed, the same Reynolds number in model and prototype provides dynamic similitude. The various corresponding pressure coefficients are the same. For testing with fluids having the same kinematic viscosity in model and prototype, the product, VD, must be the same. Frequently this requires very high velocities in small models.

Open Hydraulic Structures

Structures such as spillways, stilling pools, channel transitions, and weirs generally have forces due to gravity (from changes in elevation of liquid surfaces) and inertial forces that are greater than viscous and turbulent shear forces. In these cases geometric similitude and the same value for Froude's number in the model and prototype

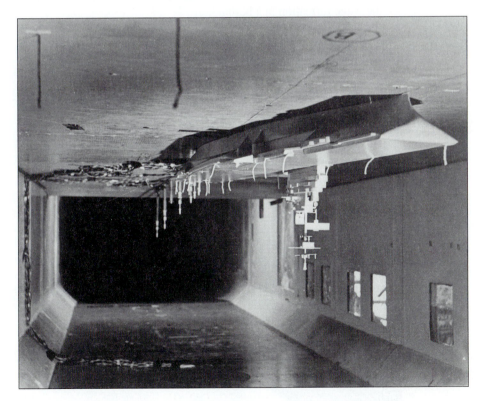

Figure 5.3   Wind tunnel tests on an aircraft carrier superstructure. Model is inverted and suspended from ceiling. (Photograph is taken in Aerospace Engineering Laboratories of the University of Michigan for the Dyna Sciences Corp.)

produce a good approximation to dynamic similitude, that is,

$$\frac{V_m^2}{g_m l_m} = \frac{V_p^2}{g_p l_p}$$

Since gravity is the same, the velocity ratio varies as the square root of the scale ratio $\lambda = l_p/l_m$,

$$V_p = V_m \sqrt{\lambda}$$

The corresponding times for events to take place (as time for passage of a particle through a transition) are related; thus,

$$t_m = \frac{l_m}{V_m} \quad t_p = \frac{l_p}{V_p} \quad t_p = t_m \frac{l_p}{l_m} \frac{V_m}{V_p} = t_m \sqrt{\lambda}$$

The discharge ratio Q_p/Q_m is

$$\frac{Q_p}{Q_m} = \frac{l_p^3/t_p}{l_m^3/t_m} = \lambda^{5/2}$$

Force ratios, for example, on gates, F_p/F_m, are

$$\frac{F_p}{F_m} = \frac{\gamma h_p l_p^2}{\gamma h_m l_m^2} = \lambda^3$$

where h is the head. In a similar fashion other pertinent ratios can be derived so that model results can be interpreted as prototype performance.

Figure 5.4 shows a model test conducted to determine the effect of a breakwater on the wave formation in a harbor.

Figure 5.4　　Model test of a harbor. (Department of Civil Engineering, University of Michigan.)

Ship's Resistance

The resistance to motion of a ship through water is composed of pressure drag, skin friction, and wave resistance. Model studies are complicated by the three types of forces that are important: inertia, viscosity, and gravity. Skin-friction studies should be based on equal Reynolds numbers in model and prototype, but wave resistance depends upon the Froude number. To satisfy both requirements, the model and prototype must be the same size.

The difficulty is surmounted by using a small model and measuring the total drag on it when towed. The skin friction is then computed for the model and subtracted from the total drag. The remainder is stepped up to the prototype size by using Froude modeling, and the prototype skin friction is computed and added to yield total resistance due to the water. Figure 5.5 shows the dramatic change in the wave profile which resulted from a redesigned bow. From such tests it is possible to predict using Froude modeling the wave formation and drag that would occur on the prototype.

Hydraulic Machinery

The rotational speed of hydraulic machinery introduces an extra variable. The moving parts in a hydraulic machine require an extra parameter to ensure that the streamline patterns will be similar in model and prototype. This parameter must relate the throughflow (discharge) to the speed of moving parts. For geometrically similar machines, if the vector diagrams of velocity entering or leaving the moving parts are similar, the units are *homologous,* that is, for practical purposes dynamic similitude exists. The Froude number is unimportant, but the Reynolds number effects (called *scale effects* because it is impossible to maintain the same Reynolds number in homologous units) may cause a discrepancy of 2 or 3 percent in efficiency between model and prototype. The Mach number is also of importance in axial-flow compressors and gas turbines.

Example 5.9

The valve coefficients $K = \Delta p/(\rho V^2/2)$ for a 600-mm-diameter valve are to be determined from tests on a geometrically similar 300-mm-diameter valve using atmospheric air at 80°F. The ranges of tests should be for flow of water at 70°F at 1 to 2.5 m/s. What ranges of airflows are needed?

Solution

The range for Reynolds number for the prototype valve is

$$\left(\frac{VD}{\nu}\right)_{min} = \frac{(1 \text{ m/s})(0.6\text{m})}{(1.059 \times 10^{-5} \text{ ft}^2/\text{s})(0.3048 \text{ m/ft})^2} = 610{,}000$$

$$\left(\frac{VD}{\nu}\right)_{max} = 610{,}000(2.5) = 1{,}525{,}000$$

For testing with air at 80°F

$$\nu = (1.8 \times 10^{-4} \text{ ft}^2/\text{s})(0.3048 \text{ m/ft})^2 = 1.672 \times 10^{-5} \text{ m}^2/\text{s}$$

Figure 5.5   Model tests showing the influence of a bulbous bow-wave formation. (Department of Naval Architecture and Marine Engineering, University of Michigan.)

Then the ranges of air velocities are

$$\frac{V_{min}(0.3 \text{ m})}{1.672 \times 10^{-5} \text{ m}^2/\text{s}} = 610,000 \qquad V_{min} = 30.6 \text{ m/s}$$

$$\frac{V_{max}(0.3 \text{ m})}{1.672 \times 10^{-5} \text{ m}^2/\text{s}} = 1,525,000 \qquad V_{max} = 85 \text{ m/s}$$

$$Q_{min} = \frac{\pi}{4}(0.3 \text{ m})^2(30.6 \text{ m/s}) = 2.16 \text{ m}^3/\text{s}$$

$$Q_{max} = \frac{\pi}{4}(0.3 \text{ m})^2(85 \text{ m/s}) = 6.0 \text{ m}^3/\text{s}$$

EXERCISES

5.6.1 What velocity of oil, $\rho = 1.6$ slugs/ft^3 and $\mu = 0.20$ P, must occur in a 1-in.-diameter pipe to be dynamically similar to 10-ft/s water velocity at 68°F in a $\frac{1}{4}$-in.-diameter tube? (*a*) 0.60 ft/s; (*b*) 9.6 ft/s; (*c*) 4.0 ft/s; (*d*) 60 ft/s; (*e*) none of these answers.

5.6.2 The velocity at a point on a model dam crest was measured to be 1 m/s. The corresponding prototype velocity for $\lambda = 25$ is, in meters per second, (*a*) 25; (*b*) 5; (*c*) 0.2; (*d*) 0.04; (*e*) none of these answers.

5.6.3 The height of a hydraulic jump in a stilling pool was found to be 4.0 in. in a model, $\lambda = 36$. The prototype jump height is (*a*) 12 ft; (*b*) 2 ft; (*c*) not determinable from data given; (*d*) less than 4 in.; (*e*) none of these answers.

5.6.4 A ship's model, scale 1:100, had a wave resistance of 10 N at its design speed. The corresponding prototype wave resistance is, in kilonewtons, (*a*) 10; (*b*) 100; (*c*) 1000; (*d*) 10,000; (*e*) none of these answers.

5.6.5 A 1:5 scale model of a projectile has a drag coefficient of 3.5 at **M** = 2.0. How many times greater would the prototype resistance be when fired at the same Mach number in air of the same temperature and half the density? (*a*) 0.312; (*b*) 3.12; (*c*) 12.5; (*d*) 25; (*e*) none of these answers.

PROBLEMS

5.1 Show that Eqs. (4.5.11), (4.6.5), and (3.7.1) are dimensionally homogeneous.

5.2 Arrange the following group into dimensionless parameters: (*a*) Δp, ρ, and V; (*b*) ρ, g, V, and F; (*c*) μ, F, Δp, and t.

5.3 By inspection, arrange the following groups into dimensionless parameters (*a*) a, l, and t; (*b*) v, l, and t; (*c*) A, Q, and ω; (*d*) K, σ, and A.

5.4 Derive the unit of mass consistent with the units inches, minutes, and tons.

5.5 In terms of M, L, and T, determine the dimensions of radians, angular velocity, power, work, torque, and moment of momentum.

5.6 Find the dimensions of the quantities in Prob. 5.5 in the FLT system.

5.7 Solve Ex. 5.2 using Q and H as repeating variables.

5.8 Using the variables Q, D, $\Delta H/l$, ρ, μ, and g as pertinent to smooth-pipe flow, arrange them into dimensionless parameters with Q, ρ, and μ as repeating variables.

5.9   If the shear stress τ is known to depend upon viscosity and rate of angular deformation du/dy in one-dimensional laminar flow, determine the form of Newton's law of viscosity by dimensional reasoning.

5.10   The variation Δp of pressure in static liquids is known to depend upon specific weight γ and elevation difference Δz. By dimensional reasoning determine the form of the hydrostatic law of variation of pressure.

5.11   When viscous and surface-tension effects are neglected, the velocity V of efflux of liquid from a reservoir is thought to depend upon the pressure drop Δp of the liquid and its density ρ. Determine the form of expression for V.

5.12   The buoyant force F_B on a body is thought to depend upon its submerged volume $\forall$ and the gravitational body force acting on the fluid. Determine the form of the buoyant-force equation.

5.13   In a fluid rotated as a solid about a vertical axis with angular velocity ω, the pressure rise p in a radial direction depends upon speed ω, radius r, and fluid density ρ. Obtain the form of equation for p.

5.14   In Ex. 5.3 find two other sets of dimensionless parameters by recombining the dimensionless parameters given.

5.15   Find the dimensionless parameter of Ex. 4.4 using Δp, ρ, and l as repeating variables.

5.16   The Mach number **M** for flow of a perfect gas in a pipe depends upon the specific-heat ratio k (dimensionless), the pressure p, the density ρ, and the velocity V. Obtain by dimensional reasoning the form of the Mach number expression.

5.17   Work out the scaling ratio for torque T on a disk of radius r that rotates in fluid of viscosity μ with angular velocity ω and clearance y between disk and fixed plate.

5.18   The velocity at a point in a model of a spillway for a dam is 1 m/s. For a ratio of prototype to model of 10:1 what is the velocity at the corresponding point in the prototype under similar conditions?

5.19   The power input to a pump depends upon the discharge Q, the pressure rise Δp, the fluid density ρ, size D, and efficiency e. Find the expression for power by the use of dimensional analysis.

5.20   The torque delivered by a water turbine depends upon discharge Q, head H, specific weight γ, angular velocity ω, and the efficiency e. Determine the form of equation for torque.

5.21   Extended experimental studies for the problem of convective heat transfer on cylindrical rods has revealed that the heat transfer coefficient, h_c, depends upon the set of variables listed in the following table:

Symbol	Name	Units
u	Velocity	m/s
ρ	Density	kg/m³
μ	Viscosity	kg/m·s
d	Diameter	m
k	Thermal conductivity	kg·m/s³·K
c_p	Specific heat	m²/s²·K
h_c	Heat transfer coefficient	kg/s³·K

Using the above variables find all the necessary dimensionless numbers that could be used to describe such physical conditions.

5.22 It is required to establish the functional relationship among the dimensionless numbers found in Prob. 5.21. Describe how it is possible to quantitatively determine the functional relationship and what data are essential for the conditions given in Prob. 5.21.

5.23 The set of variables that describe the transient heat transfer in infinite slabs, without heat generation, are the heat transfer coefficient h_c, the thermal diffusivity α, the distance x, the thermal conductivity k, the temperature T, the reference temperature T_{ref}, and the time t. Determine all the suitable, for this problem, dimensionless parameters.

5.24 Reformulate Prob. 5.23 considering a heat source, and therefore heat generation, q_H.

5.25 The cooling of a small billet (a time-varying process), which is removed from a furnace at a uniform temperature, T_f, and suddenly immersed in cool water of uniform temperature, T_w, is described by the average heat transfer coefficient $\overline{h_c}$, the thermal diffusivity α, the thermal conductivity k, the density of the billet ρ, the surface area of the billet A, and finally the length L. Determine all the dimensionless parameters for this problem.

5.26 It is known that the mass transfer coefficient, k_c, depends upon the following variables:

Symbol	Name	Units
u	Velocity	m/s
ρ	Density	kg/m^3
μ	Viscosity	kg/m·s
L_{ref}	Reference length	m
$\mathcal{D}$	Diffusion coefficient	m^2/s
k_c	Mass transfer coefficient	kg/m^2·s

Find all the dimensionless parameters.

5.27 The sedimentation in filter separators depends upon particle and medium characteristics such as the particle diameter d_s, the particle density ρ_s, the particle terminal velocity w_s, the particle diffusion coefficient $\mathcal{D}_s$, the gas velocity u, the density ρ, the viscosity μ, the filter fiber diameter d_f, and the acceleration due to gravity g. What are the dimensionless parameters for the sedimentation in filter separators?

5.28 Derive the nondimensional form of the two-dimensional (in the xy plane) heat transport equation using the single scale normalization. Show all the details.

5.29 In flows past solid bodies, the force exerted on the body by the fluid motion depends upon the fluid velocity u, the density ρ, the viscosity μ, and the principal dimension of the body L. What are the dimensionless parameters for the given conditions?

5.30 The drag coefficient, C_D, represents the ratio of the surface shear stress to the free stream kinetic energy. Water flows over a flat surface, with principal dimension L, and the local shear stress given by

$$\tau(x) = 0.3 \, (\rho\mu/x)^{1/2} \, U^{3/2}$$

where x is the distance from the leading edge and U is the free stream velocity. From the above equation obtain nondimensional relationships for the local and average drag coefficient, C_D.

5.31 Plot the dimensionless parameters, **R** and **Pr**, versus temperature for water, using the characteristic length of $L = 4.5$ m and the characteristic velocity $U = 3.2$ m/s.

5.32 For flows over horizontal plates it is known that the following correlation exists: $\mathbf{Nu} = \alpha \mathbf{R}^\beta \mathbf{Pr}^\gamma$ where the dimensionless numbers are evaluated over the length, L, of the plate. Future evaluations of such flows require the knowledge of the constants, α, β, and γ. Determine the necessary experimental data required to efficiently evaluate these coefficients.

5.33 Suppose that in Prob. 5.32, a set of experimental data is available and an engineer needs to find the values of the coefficients, α, β, and γ. Can you suggest a procedure for the determination of these coefficients?

5.34 The following relationship for the concentration distribution of species A was suggested for use in large water bodies

$$C_A(z, t) = \alpha\, C_{Ao}\, e^{-z^2/D_z t}$$

where α is a correction factor accounting for temperature changes. Is the above equation in nondimensional form? If not, what is its nondimensional representation?

5.35 Evaluate the dimensionless parameters, $\mathbf{R}_x^{1/2}\mathbf{Pr}^{1/3}$, $\mathbf{Nu}_x$, using the following data, for water at 45°C flowing over a thin plate: $x = 25$ ft, $c_p = 4176$ J/kg·K, $h_c = 3.0$ Btu/hr·ft^2·°F, $k = 0.369$ Btu/hr·ft·°F, and $u = 3.31$ ft/s.

5.36 Evaluate the dimensionless parameters, **R**, **Pr**, and $(\mathbf{R} \cdot \mathbf{Pr})^{1/2}$ for water flowing over a thin plate of length 3 m, using the following data: $T = 80°F$, $c_p = 0.997$ Btu/lbm·°F, $h_c = 3.0$ Btu/hr·ft^2·°F, $k = 0.615$ W/m·K, and $u = 2.18$ m/s.

5.37 A model of a venturi meter has linear dimensions one-fifth those of the prototype. The prototype operates on water at 20°C, and the model on water at 95°C. For a throat diameter of 600 mm and a velocity at the throat of 6 m/s in the prototype, what discharge is needed through the model for similitude?

5.38 The drag F on a high-velocity projectile depends upon the speed V of the projectile, density of fluid ρ, acoustic velocity c, diameter of projectile D, and viscosity μ. Develop an expression for the drag.

5.39 The wave drag on a model of a ship is 16 N at a speed of 3 m/s. For a prototype fifteen times as long, what will the corresponding speed and wave drag be if the liquid is the same in each case?

5.40 Determine the specific gravity of spherical particles, $D = 0.13$ mm, which drop through air at 0°C at a speed U of 0.1 m/s. The drag force on a small sphere in laminar motion is given by $3\pi\mu DU$.

5.41 A small liquid sphere of radius r_0 and density ρ_0 settles at velocity U in a second liquid of density ρ and viscosity μ. The tests are conducted inside vertical tubes of radius r. By dimensional analysis determine a set of dimensionless parameters to be used in determining the influence of the tube wall on the settling velocity.

5.42 The losses in a Y in a 1.2-m-diameter pipe system carrying gas ($\rho = 40$ kg/m^3, $\mu = 0.002$ P, and $V = 25$ m/s) are to be determined by testing a model with water at 20°C. The laboratory has a water capacity of 75 L/s. What model scale should be used, and how are the results converted into prototype losses?

5.43 Ripples have a velocity of propagation that is dependent upon the surface tension and density of the fluid as well as the wavelength. By dimensional analysis justify the shape of Fig. 5.1 for small wavelengths.

5.44 In very deep water the velocity of propagation of waves depends upon the wavelength, but in shallow water it is independent of this dimension. Upon what variables does the speed of advance depend for shallow-water waves? Is Fig. 5.1 in agreement with this problem?

5.45 If a vertical circular conduit which is not flowing full is rotated at high speed, the fluid will attach itself uniformly to the inside wall as it flows downward (see Sec. 2.9). Under these conditions the radial acceleration of the fluid yields a radial force field which is similar to gravitational attraction and a hydraulic jump can occur on the inside of the tube, whereby the fluid thickness suddenly changes. Determine a set of dimensionless parameters for studying this rotating hydraulic jump.

5.46 A nearly spherical fluid drop oscillates as it falls. Surface tension plays a dominant role. Determine a meaningful dimensionless parameter for this natural frequency.

5.47 The lift and drag coefficients for a wing are shown in Fig. 7.17. If the wing has a chord of 10 ft, determine the lift and drag per foot of length when the wing is operating at zero angle of attack at a Reynolds number, based on the chord length, of 4.5×10^7 in air at 50°F. What force would be on a 1:20 scale model if the tests were conducted in water at 70°F? What would be the speed of the water? Comment on the desirability of conducting the model tests in water.

5.48 A 1:5 scale model of a water pumping station piping system is to be tested to determine overall head losses. Air at 25°C and 1 atm is available. For a proto-type velocity of 500 mm/s in a 4-m-diameter section with water at 15°C determine the air velocity and quantity needed and how losses determined from the model are converted into prototype losses.

5.49 Full-scale wind tunnel tests of the lift and drag on hydrofoils for a boat are to be made. The boat will travel at 55 km/h through water at 15°C. What velocity of air ($p = 200$-kPA abs and $T = 32$°C) is required to determine the lift and drag? *Note:* The lift coefficient C_L is dimensionless. Lift $= C_L A \rho V^2 / 2$.

5.50 The resistance to ascent of a balloon is to be determined by studying the ascent of a 1:50 scale model in water. How would such a model study be conducted and the results converted to prototype behavior?

5.51 The moment exerted on a submarine by its rudder is to be studied with a 1:20 scale model in a water tunnel. If the torque measured on the model is 5N · m for a tunnel velocity of 15 m/s, what are the corresponding torque and speed for the prototype?

5.52 For two hydraulic machines to be homologous they must (*a*) be geometrically similar, (*b*) have the same discharge coefficient when viewed as an orifice, $Q_1/(A_1\sqrt{2gH_1}) = Q_2/(A_2\sqrt{2gH_2})$, and (*c*) have the same ratio of peripheral speed to fluid velocity, $\omega D/(Q/A)$. Show that the scaling ratios can be expressed as $Q/ND^3 = $ constant and $H/(ND)^2 = $ constant. N is the rotational speed.

5.53 Use the scaling ratios of Prob. 5.52 to determine the head and discharge of a 1:4 model of a centrifugal pump that produces 600 L/s at a 30-m head when turning 240 rpm. The model operates at 1200 rpm.

REFERENCES

1. E. Buckingham, "Model Experiments and the Form of Empirical Equations," *Trans. ASME,* vol. 37, pp. 263–296, 1915.

2. L. I. Sedov, *Similarity and Dimensional Methods in Mechanics* (translated by M. Holt), Academic Press, New York, 1959.

3. J. C. Hunsaker and B. G. Rightmire, *Engineering Applications of Fluid Mechanics,* McGraw-Hill, New York, 1961.

ADDITIONAL READING

Bridgeman, P. W.: *Dimensional Analysis,* Yale University Press, New Haven, Conn., 1931. (The paperback Y-82, 1963.)

Hansen, A.: *Similarity Analyses of Boundary Value Problems in Engineering,* Prentice Hall, New Jersey, 1964.

Holt, M.: "Dimensional Analysis," sec. 15 in V. L. Streeter (ed.), *Handbook of Fluid Dynamics,* McGraw-Hill, New York, 1961.

Hydraulic Models, *ASCE Man. Eng. Pract.* 25, 1942.

Ipsen, D. C.: *Units, Dimensions, and Dimensionless Numbers,* McGraw-Hill, New York, 1960.

Kline, S. J.: *Similitude and Approximation Theory,* McGraw-Hill, New York, 1965.

Langhaar, H. L.: *Dimensional Analysis and Theory of Models,* John Wiley, New York, 1951.

Seshadri, R. and Na, T. Y.: *Group Invariance in Engineering Boundary Value Problems,* Springer Verlag, New York, 1985.

chapter

6

Viscous Flow: Pipes and Channels

In Chaps. 3 and 4 the basic equations used in the analysis of fluid-flow situations were discussed. This chapter deals with real fluids, that is, situations in which irreversibilities are important. Viscosity is the fluid property that causes shear stresses in a moving fluid; viscosity is also one means by which losses are developed. In turbulent flows random fluid motions, superposed on the average, create apparent shear stresses that are more important than those due to viscous shear. These topics are the central theme in the chapter. First, the concept of the Reynolds number, introduced in Chap. 5, is developed. The characteristics that distinguish laminar from turbulent flow are presented, and the categorization of flows into internal versus external is established. This chapter concentrates on internal-flow cases. Steady, laminar, incompressible flows are first developed since the losses can be computed analytically. Resistance due to steady, uniform, incompressible, turbulent flow is then examined for open and closed conduits. Free-surface flow in open channels is introduced, followed by a more detailed treatment of pipe flow.

6.1 LAMINAR AND TURBULENT FLOWS: INTERNAL AND EXTERNAL FLOWS

The Reynolds Number

Laminar flow is defined as flow in which the fluid moves in layers, or laminas, one layer gliding smoothly over an adjacent layer with only a molecular interchange of momentum. Any tendencies toward instability and turbulence are damped out by viscous shear forces that resist relative motion of adjacent fluid layers. Turbulent flow, however, has very erratic motion of fluid particles, with a violent transverse interchange of momentum. The nature of the flow, that is, whether laminar or turbulent, and its relative position along a scale indicating the relative importance of turbulent to laminar tendencies are indicated by the *Reynolds number.* The concept of the Reynolds number and its interpretation are discussed in this section. In Chap. 4 the equations of motion were developed, and these equations are complicated nonlinear partial differential equations for which no general solution has been obtained. In the last century Osborne Reynolds studied them to try to determine when two different flow situations would be similar [1].†

As reviewed in Chap. 5, two flow cases are said to be *dynamically similar* when

1. They are geometrically similar, that is, corresponding linear dimensions have a constant ratio.

2. The corresponding force polygons are geometrically similar, or pressures at corresponding points have a constant ratio.

In considering two geometrically similar flow situations, Reynolds deduced that they would be dynamically similar if the general differential equations describing their flow were identical. By changing the units of mass, length, and time in one set of equations and determining the condition that must be satisfied to make them identical to the original equations, Reynolds found that the dimensionless group $ul\rho/\mu$ must be the same for both cases. The quantity u is a characteristic velocity, l a characteristic length, ρ the mass density, and μ the viscosity. This group, or parameter, is now called the Reynolds number $\mathbf{R}$ and equals

$$\mathbf{R} = \frac{ul\rho}{\mu} \tag{6.1.1}$$

To determine the significance of the dimensionless group, Reynolds conducted his experiments on flow of water through glass tubes, illustrated in Fig. 6.1. A glass tube was mounted horizontally with one end in a tank and a valve on the opposite end. A smooth bellmouth entrance was attached to the upstream end, with a dye jet so arranged that a fine stream of dye could be ejected at any point in front of the bellmouth. Reynolds took the average velocity V as the characteristic velocity and the diameter of the tube D as the characteristic length, so that $\mathbf{R} = VD\rho/\mu$.

For small flows the dye stream moved as a straight line through the tube, showing that the flow was laminar. As the flow rate increased, the Reynolds number increased, since D, ρ, and μ were constant and V was directly proportional to the rate of flow. With increasing discharge a condition was reached at which the dye stream

---

† Numbered references will be found at the end of this chapter.

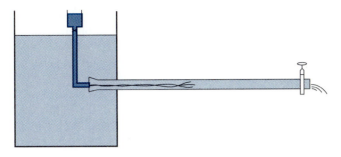

Figure 6.1   Reynolds apparatus.

wavered and then suddenly broke up and was diffused or dispersed throughout the tube. The flow had changed to turbulent flow with its violent interchange of momentum that had completely disrupted the orderly movement of laminar flow. By careful manipulation Reynolds was able to obtain a value $\mathbf{R} = 12{,}000$ before turbulence set in. A later investigator, using Reynolds' original equipment, obtained a value of 40,000 by allowing the water to stand in the tank for several days before the experiment and by taking precautions to avoid vibrations of the water or equipment. These numbers, referred to as the *Reynolds upper critical numbers,* have no practical significance in that an ordinary pipe installation has irregularities that cause turbulent flow at a much smaller value of the Reynolds number.

Starting with turbulent flow in the glass tube, Reynolds found that it always becomes laminar when the velocity is reduced to make $\mathbf{R}$ less than 2000. This is the *Reynolds lower critical number* for pipe flow and is of practical importance. With the usual piping installation, the flow will change from laminar to turbulent in the range of Reynolds numbers from 2000 to 4000. For the purpose of this treatment it is assumed that the change occurs at $\mathbf{R} = 2000$. In laminar flow the losses are directly proportional to the average velocity, while in turbulent flow the losses are proportional to the velocity to a power varying from 1.7 to 2.0.

There are many Reynolds numbers in use today in addition to the one for straight round tubes. For example, the motion of a sphere through a fluid can be characterized by $UD\rho/\mu$, in which U is the velocity of the sphere, D is the diameter of the sphere, and ρ and μ are the fluid density and viscosity, respectively.

The *nature* of a given flow of an incompressible fluid is characterized by its Reynolds number. For large values of $\mathbf{R}$ one or all of the terms in the numerator are large compared with the denominator. This implies a large expanse of fluid, high velocity, great density, extremely small viscosity, or combinations of these extremes. The numerator terms are related to *inertial forces* or to forces set up by acceleration or deceleration of the fluid. The denominator term is the cause of viscous shear forces. Thus, the Reynolds number parameter can also be considered as a ratio of inertial to viscous forces. A large $\mathbf{R}$ indicates a highly turbulent flow with losses proportional to the square of the velocity. The turbulence may be *fine-scale,* composed of a great many small eddies that rapidly convert mechanical energy into irreversibilities through viscous action, or it may be *large-scale,* like the huge vortices and swirls in a river or gusts in the atmosphere. The large eddies generate smaller eddies, which in turn create fine-scale turbulence. Turbulent flow may be thought of as a smooth, possibly uniform flow, with a secondary flow superposed on it. A fine-scale turbulent flow has small fluctuations in velocity that occur with high frequency.

The root-mean-square value of the fluctuations and the frequency of change of sign of the fluctuations are quantitative measures of turbulence. In general, the intensity of turbulence increases as the Reynolds number increases. For intermediate values of **R** both viscous and inertial effects are important, and changes in viscosity change the velocity distribution and the resistance to flow.

For the same **R** two geometrically similar closed-conduit systems (one, say, twice the size of the other) will have the same ratio of losses to velocity head. The Reynolds number provides a means of using experimental results with one fluid to predict results in a similar case with another fluid.

Internal and External Flows

Another method of categorizing flows is by examining the geometry of the flow field. Internal flow involves flow in a bounded region, as the name implies. External flow involves fluid in an unbounded region in which the focus of attention is on the flow pattern around a body immersed in the fluid.

The motion of a real fluid is influenced significantly by the presence of the boundary. Fluid particles at the wall remain at rest in contact with the wall. In the flow field a strong velocity gradient exists in the vicinity of the wall, a region referred to as the boundary layer. A retarding shear force is applied to the fluid at the wall, the boundary layer being a region of significant shear stresses.

This chapter deals with flows constrained by walls in which the boundary effect is likely to extend through the entire flow. The boundary influence is easily visualized at the entrance to a pipe from a reservoir (Fig. 6.2). At section *A-A*, near a well-rounded entrance, the velocity profile is almost uniform over the cross section. The action of the wall shearing stress is to slow down the fluid near the wall. As a consequence of continuity, the velocity must increase in the central region. Beyond a transitional length *L'* the velocity profile is fixed since the boundary influence has extended to the pipe center line. The transition length is a function of the Reynolds number; for laminar flow Langhaar [2] developed the theoretical formula

$$\frac{L'}{D} = 0.058\mathbf{R} \qquad (6.1.2)$$

which agrees well with observation. In turbulent flow the boundary layer grows more rapidly and the transition length is considerably shorter than given by Eq. (6.1.2).

In external flows, with an object in an unbounded fluid, the frictional effects are confined to the boundary layer next to the body. Examples include a golf ball in flight, a sediment particle, and a boat. The fully developed velocity profile, presented in Fig. 6.2 for an internal flow, is unlikely to exist in external flows. Typically interest

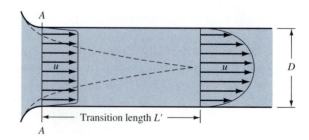

Figure 6.2   Entrance zone of pipeline.

is focused on drag forces on the object or the lift characteristics developed on the body by the particular flow pattern. These flow situations will be addressed in the next chapter.

The Navier-Stokes Equations

To describe the variation of the flow field parameters at every point in the continuum the Navier-Stokes equations, developed in Chap. 4, apply. Equations (4.4.12) are listed for completeness.

$$\rho\frac{Du}{Dt} = \rho\left(\frac{\partial u}{\partial t} + u\frac{\partial u}{\partial x} + v\frac{\partial u}{\partial y} + w\frac{\partial u}{\partial z}\right) = -\rho g\frac{\partial h}{\partial x} - \frac{\partial p}{\partial x} + \mu\left(\frac{\partial^2 u}{\partial x^2} + \frac{\partial^2 u}{\partial y^2} + \frac{\partial^2 u}{\partial z^2}\right)$$

$$\rho\frac{Dv}{Dt} = \rho\left(\frac{\partial v}{\partial t} + u\frac{\partial v}{\partial x} + v\frac{\partial v}{\partial y} + w\frac{\partial v}{\partial z}\right) = -\rho g\frac{\partial h}{\partial y} - \frac{\partial p}{\partial y} + \mu\left(\frac{\partial^2 v}{\partial x^2} + \frac{\partial^2 v}{\partial y^2} + \frac{\partial^2 v}{\partial z^2}\right)$$

$$\rho\frac{Dw}{Dt} = \rho\left(\frac{\partial w}{\partial t} + u\frac{\partial w}{\partial x} + v\frac{\partial w}{\partial y} + w\frac{\partial w}{\partial z}\right) = -\rho g\frac{\partial h}{\partial z} - \frac{\partial p}{\partial z} + \mu\left(\frac{\partial^2 w}{\partial x^2} + \frac{\partial^2 w}{\partial y^2} + \frac{\partial^2 w}{\partial z^2}\right)$$

What follows in this and several succeeding chapters are analysis techniques for simplifying these equations and solving them for practical problems. Simple laminar flow cases are introduced in the following two sections.

EXERCISES

6.1.1 The upper critical Reynolds number is (*a*) important from a design viewpoint; (*b*) the number at which turbulent flow changes to laminar flow; (*c*) about 2000; (*d*) not more than 2000; (*e*) of no practical importance in pipe-flow problems.

6.1.2 The Reynolds number for pipe flow is given by (*a*) VD/ν; (*b*) $VD\mu/\rho$; (*c*) $VD\rho/\nu$; (*d*) VD/μ; (*e*) none of these answers.

6.1.3 The lower critical Reynolds number for pipe flow has a value of (*a*) 200; (*b*) 1200; (*c*) 12,000; (*d*) 40,000; (*e*) none of these answers.

6.1.4 The Reynolds number for a 30-mm-diameter sphere moving 3 m/s through oil, $S = 0.90$ and $\mu = 0.10$ kg/m·s, is (*a*) 404; (*b*) 808; (*c*) 900; (*d*) 8080; (*e*) none of these answers.

6.1.5 The Reynolds number for 10 cfs discharge of water at 68°F through a 12-in.-diameter pipe is (*a*) 2460; (*b*) 980,000; (*c*) 1,178,000; (*d*) 14,120,000; (*e*) none of these answers.

6.2 LAMINAR, INCOMPRESSIBLE, STEADY FLOW BETWEEN PARALLEL PLATES

Velocity Distribution

The general case of steady flow between parallel inclined plates is first developed for laminar flow, the upper plate having a constant velocity U (Fig. 6.3). Flow between fixed plates is a special case obtained by setting $U = 0$. In Fig. 6.3 the upper plate

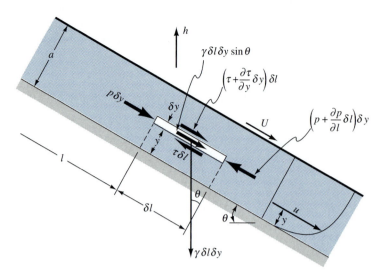

Figure 6.3   Flow between inclined parallel plates with the upper plate in motion.

moves parallel to the flow direction, and there is a pressure variation in the l direction. The flow is analyzed by taking a thin lamina of unit width as a free body. In steady flow the lamina moves at constant velocity u. The equation of motion yields

$$p\,\delta y - \left(p\,\delta y + \frac{\partial p}{\partial l}\,\delta l\,\delta y\right) - \tau\,\delta l + \left(\tau\,\delta l + \frac{\partial \tau}{\partial y}\,\delta y\,\delta l\right) + \gamma\,\delta l\,\delta y\,\sin\theta = 0$$

Dividing through by the volume of the element, using $\sin\theta = -\partial h/\partial l$, and simplifying yield

$$\frac{\partial \tau}{\partial y} = \frac{\partial}{\partial l}(p + \gamma h)$$

Since u is a function of y only, $\partial\tau/\partial y = d\tau/dy$; and since $p + \gamma h$ does not change value in the y direction (no acceleration), $p + \gamma h$ is a function of l only. Hence, $\partial(p + \gamma h)/\partial l = d(p + \gamma h)/dl$, and

$$\frac{d\tau}{dy} = \mu\frac{d^2u}{dy^2} = \frac{d}{dl}(p + \gamma h) \tag{6.2.1}$$

Integrating Eq. (6.2.1) with respect to y yields

$$\mu\frac{du}{dy} = y\frac{d}{dl}(p + \gamma h) + A$$

Integrating again with respect to y leads to

$$u = \frac{1}{2\mu}\frac{d}{dl}(p + \gamma h)y^2 + \frac{A}{\mu}y + B$$

in which A and B are constants of integration. To evaluate them, take $y = 0$, $u = 0$ and $y = a$, $u = U$ and obtain

$$B = 0 \qquad U = \frac{1}{2\mu}\frac{d}{dl}(p + \gamma h)a^2 + \frac{Aa}{\mu} + B$$

Eliminating A and B results in

$$u = \frac{Uy}{a} - \frac{1}{2\mu}\frac{d}{dl}(p + \gamma h)(ay - y^2) \qquad \text{(6.2.2)}$$

For horizontal plates, $h = C$, a constant; if there is no gradient due to pressure or elevation, that is, a hydrostatic pressure distribution, $p + \gamma h = C$ and the velocity has a straight-line distribution. For fixed plates, $U = 0$, and the velocity distribution is parabolic.

The discharge past a fixed cross section is obtained by integration of Eq. (6.2.2) with respect to y, yielding

$$Q = \int_0^a u\, dy = \frac{Ua}{2} - \frac{1}{12\mu}\frac{d}{dl}(p + \gamma h)a^3 \qquad \text{(6.2.3)}$$

In general, the maximum velocity is not at the midplane.

Example 6.1

In Fig. 6.4 one plate moves relative to the other as shown; $\mu = 0.08$ N·s/m² and $\rho = 850$ kg/m³. Determine the velocity distribution, the discharge, and the shear stress exerted on the upper plate.

Solution

At the upper point

$$p + \gamma h = 1400\ \text{Pa} + (850\ \text{kg/m}^3)(9.806\ \text{m/s}^2)(3\ \text{m}) = 26{,}405\ \text{Pa}$$

and at the lower point

$$p + \gamma h = 800\ \text{Pa}$$

to the same datum. Hence,

$$\frac{d}{dl}(p + \gamma h) = \frac{800\ \text{Pa} - 26{,}405\ \text{Pa}}{3\sqrt{2}\ \text{m}} = -6035\ \text{N/m}^3$$

From the figure, $a = 0.006$ m and $U = -1$ m/s; and from Eq. (6.2.2)

$$u = \frac{(-1\ \text{m/s})(y\ \text{m})}{0.006\ \text{m}} + \frac{6035\ \text{N/m}^3}{2(0.08\ \text{N·s/m}^2)}(0.006y - y^2\ \text{m}^2)$$
$$= 59.646y - 37{,}718y^2\ \text{m/s}$$

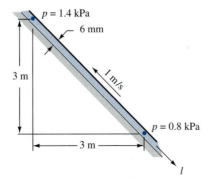

p = 1.4 kPa

6 mm

3 m

1 m/s

p = 0.8 kPa

3 m

l

Figure 6.4   Flow between inclined flat plates.

The maximum positive velocity occurs where $du/dy = 0$, or $y = 0.00079$ m, and it is $u_{max} = 0.0236$ m/s. The absolute maximum velocity occurs at the moving plate, $y = 0.006$ m, where the velocity is -1.0 m/s. The discharge per meter of width is

$$Q = \int_0^{0.006} u \, dy = \left[29.823y^2 - 12,573y^3 \right]_0^{0.006} = -0.00164 \text{ m}^2/\text{s}$$

which is upward. To find the shear stress on the upper plate,

$$\left. \frac{du}{dy} \right|_{y=0.006} = 59.646 - 75,436y \left. \right|_{y=0.006} = -392.97 \text{ s}^{-1}$$

and

$$\tau = \mu \frac{du}{dy} = 0.08(-392.97) = -31.44 \text{ Pa}$$

This is the fluid shear at the upper plate; hence, the shear force per unit area on the plate is 31.44 N resisting the motion of the plate.

Losses in Laminar Flow

Expressions for irreversibilities are developed for one-dimensional, incompressible, steady, laminar flow. For steady flow in a tube, between parallel plates, or in a film flow at constant depth, the kinetic energy does not change and the reduction in $p + \gamma h$ represents the work done on the fluid per unit volume. The work done is converted into irreversibilities by the action of viscous shear. Here, it is shown that the losses in length L are $Q\Delta(p + \gamma h)$ per unit time.

If u is a function of y, the transverse direction, and the change in $p + \gamma h$ is a function of distance x in the direction of flow, total derivatives can be used throughout the development. First, from Eq. (6.2.1)

$$\frac{d(p + \gamma h)}{dx} = \frac{d\tau}{dy} \tag{6.2.4}$$

With reference to Fig. 6.5, a particle of fluid of rectangular shape of unit width has its center at (x, y), where the shear is τ, the pressure p, the velocity u, and the elevation h. This particle of fluid moves in the x direction. In unit time it has work done on it by the surface boundaries as shown, and it gives up potential energy $\gamma \, \delta x \, \delta y \, u \sin \theta$. As there is no change in kinetic energy of the particle, the net work done and the loss of potential energy represent the losses per unit time due to irreversibilities. Collecting the terms from Fig. 6.5, dividing through by the volume $\delta x \, \delta y$, and taking the limit as $\delta x \, \delta y$ goes to zero yield

$$-u\frac{dp}{dx} - \gamma u\frac{dh}{dx} + \tau\frac{du}{dy} + u\frac{d\tau}{dy} = \frac{\text{net power input}}{\text{unit volume}} \tag{6.2.5}$$

By combining with Eq. (6.2.4)

$$\frac{\text{Net power input}}{\text{Unit volume}} = \tau\frac{du}{dy} = \mu\left(\frac{du}{dy}\right)^2 = \frac{\tau^2}{\mu} \tag{6.2.6}$$

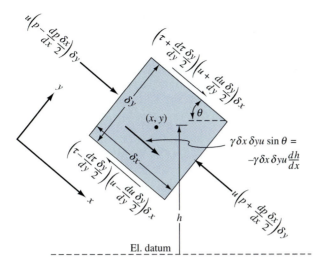

Figure 6.5   Work done and loss of potential energy for a fluid particle in one-dimensional flow.

Integrating this expression over a length L between two parallel plates, with Eq. (6.2.2) for $U = 0$, gives

$$\text{Net power input} = \int_0^a \mu \left(\frac{du}{dy}\right)^2 L \, dy$$

$$= \mu L \int_0^a \left[\frac{1}{2\mu} \frac{d(p + \gamma h)}{dx} (2y - a)\right]^2 dy$$

$$= \left[\frac{d(p + \gamma h)}{dx}\right]^2 \frac{a^3 L}{12\mu}$$

Substituting for Q from Eq. (6.2.3) for $U = 0$ yields

$$\text{Losses} = \text{net power input} = -Q\frac{d(p + \gamma h)}{dx} L = Q\Delta(p + \gamma h)$$

in which $\Delta(p + \gamma h)$ is the drop in $p + \gamma h$ in the length L. The expression for power input per unit volume [Eq. (6.2.6)] is also applicable to laminar flow in a tube. The irreversibilities are greatest when du/dy is greatest.

EXERCISES

6.2.1   The shear stress in a fluid flowing between two fixed parallel plates (*a*) is constant over the cross section; (*b*) is zero at the plates and increases linearly to the midpoint; (*c*) varies parabolically across the section; (*d*) is zero at the midplane and varies linearly with distance from the midplane; (*e*) is none of these answers.

6.2.2   The velocity distribution for flow between two fixed parallel plates (*a*) is constant over the cross section; (*b*) is zero at the plates and increases linearly to the midpoint; (*c*) varies parabolically across the section; (*d*) varies as the three-halves power of the distance from the midpoint; (*e*) is none of these answers.

6.2.3   The discharge between two parallel plates, distance a apart, when one has the velocity U and the shear stress is zero at the fixed plate, is (a) $Ua/3$; (b) $Ua/2$; (c) $2Ua/3$; (d) Ua; (e) none of these answers.

6.2.4   Fluid is in laminar motion between two parallel plates, distance a apart, with one plate in motion, and is under the action of a pressure gradient so that the discharge through any fixed cross section is zero. The minimum velocity occurs at a point which is distant from the fixed plate (a) $a/6$; (b) $a/3$; (c) $a/2$; (d) $2a/3$; (e) none of these answers.

6.2.5   In Exercise 6.2.4 the value of the minimum velocity is (a) $-3U/4$; (b)$-2U/3$; (c) $-U/2$; (d) $-U/3$; (e) $-U/6$.

6.2.6   The relation between pressure and shear stress in one-dimensional laminar flow in the x direction is given by

(a) $\dfrac{dp}{dx} = \mu\dfrac{d\tau}{dy}$ (b) $\dfrac{dp}{dy} = \dfrac{d\tau}{dx}$ (c) $\dfrac{dp}{dy} = \mu\dfrac{d\tau}{dx}$ (d) $\dfrac{dp}{dx} = \dfrac{d\tau}{dy}$

(e) none of these answers.

6.2.7   The expression for power input per unit volume to a fluid in one-dimensional laminar motion in the x direction is

(a) $\tau\dfrac{du}{dy}$ (b) $\dfrac{\tau}{\mu^2}$ (c) $\mu\dfrac{du}{dy}$ (d) $\tau\left(\dfrac{du}{dy}\right)^2$

(e) none of these answers

6.2.8   When liquid is in laminar motion at constant depth and flowing down an inclined plate (with y being measured normal to the surface), (a) the shear is zero throughout the liquid; (b) $d\tau/dy = 0$ at the plate; (c) $\tau = 0$ at the surface of the liquid; (d) the velocity is constant throughout the liquid; (e) there are no losses.

6.2.9   A 4-in.-diameter shaft rotates at 240 rpm in a bearing with a radial clearance of 0.006 in. The shear stress in an oil film, $\mu = 0.1$ P, is, in pounds per square foot, (a) 0.15; (b) 1.75; (c) 3.50; (d) 16.70; (e) none of these answers.

6.3   LAMINAR FLOW THROUGH CIRCULAR TUBES AND CIRCULAR ANNULI

Annular-Shaped Tubes

For steady, incompressible, laminar flow through a circular tube or an *annulus,* a cylindrical infinitesimal sleeve (Fig. 6.6) is taken as a free body. The equation of motion is applied in the l direction, with acceleration equal to zero. From the figure,

$$2\pi r\,\delta r p - \left(2\pi r\,\delta r p + 2\pi r\,\delta r \frac{dp}{dl}\,\delta l\right) + 2\pi r\,\delta l\tau$$

$$-\left[2\pi r\,\delta l\tau + \frac{d}{dr}(2\pi r\,\delta l\tau)\delta r\right] + \gamma 2\pi r\,\delta r\,\delta l\,\sin\theta = 0$$

Replacing $\sin\theta$ by $-dh/dl$ and dividing by the volume of the free body, $2\pi r\,\delta r\,\delta l$, give

$$\frac{d}{dl}(p + \gamma h) + \frac{1}{r}\frac{d}{dr}(\tau r) = 0 \tag{6.3.1}$$

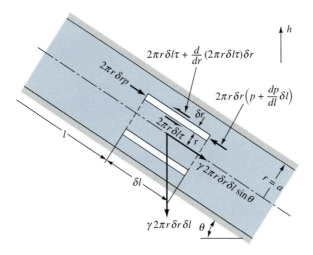

Figure 6.6 Free-body diagram of cylindrical sleeve element for laminar flow in an inclined circular tube.

Since $d(p + \gamma h)/dl$ is not a function of r, the equation can be multiplied by $r \, \delta r$ and integrated with respect to r, yielding

$$\frac{r^2}{2} \frac{d}{dl}(p + \gamma h) + \tau r = A \tag{6.3.2}$$

in which A is the constant of integration. For a circular tube this equation must be satisfied when $r = 0$; hence, $A = 0$ for this case. Substituting

$$\tau = -\mu \frac{du}{dr}$$

yields

$$du = \frac{1}{2\mu} \frac{d}{dl}(p + \gamma h)r \, dr - \frac{A}{\mu} \frac{dr}{r}$$

Note that the minus sign is required to obtain the sign of the τ term in Fig. 6.6. The velocity u is considered to decrease with r; hence, du/dr is negative. Another integration gives

$$u = \frac{r^2}{4\mu} \frac{d}{dl}(p + \gamma h) - \frac{A}{\mu} \ln r + B \tag{6.3.3}$$

For the annular case, to evaluate A and B, $u = 0$ when $r = b$, the inner tube radius, and $u = 0$ when $r = a$ (Fig. 6.7). When A and B are eliminated,

$$u = -\frac{1}{4\mu} \frac{d}{dl}(p + \gamma h)\left[a^2 - r^2 + \frac{a^2 - b^2}{\ln(b/a)} \ln\left(\frac{a}{r}\right)\right] \tag{6.3.4}$$

and for discharge through an annulus (Fig. 6.7),

$$Q = \int_b^a 2\pi r u \, dr = -\frac{\pi}{8\mu} \frac{d}{dl}(p + \gamma h)\left[a^4 - b^4 - \frac{(a^2 - b^2)^2}{\ln(a/b)}\right] \tag{6.3.5}$$

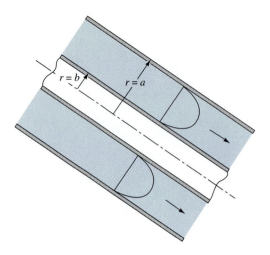

Figure 6.7   Flow through an annulus.

Circular Tube: The Hagen-Poiseuille Equation

For the circular tube, $A = 0$ in Eq. (6.3.3) and $u = 0$ for $r = a$,

$$u = -\frac{a^2 - r^2}{4\mu}\frac{d}{dl}(p + \gamma h) \tag{6.3.6}$$

The maximum velocity u_{max} is given for $r = 0$ as

$$u_{max} = -\frac{a^2}{4\mu}\frac{d}{dl}(p + \gamma h) \tag{6.3.7}$$

Since the velocity distribution is a paraboloid of revolution (Fig. 6.8), its volume is one-half that of its circumscribing cylinder; therefore, the average velocity is one-half of the maximum velocity,

$$V = -\frac{a^2}{8\mu}\frac{d}{dl}(p + \gamma h) \tag{6.3.8}$$

The discharge Q is equal to $V\pi a^2$; therefore,

$$Q = -\frac{\pi a^4}{8\mu}\frac{d}{dl}(p + \gamma h) \tag{6.3.9}$$

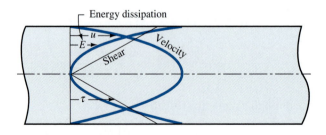

Figure 6.8   Distribution of velocity, shear, and losses per unit volume for a round tube.

The discharge can also be obtained by integration of the velocity u over the area, that is,

$$Q = \int_0^a 2\pi r u \, dr$$

For a horizontal tube, $h = $ constant; writing the pressure drop Δp in the length L gives

$$\frac{\Delta p}{L} = -\frac{dp}{dl}$$

and substituting diameter D leads to

$$Q = \frac{\Delta p \pi D^4}{128 \mu L} \qquad \text{(horizontal only)} \qquad \textbf{(6.3.10a)}$$

In terms of average velocity,

$$V = \frac{\Delta p D^2}{32 \mu L} \qquad \text{(horizontal only)} \qquad \textbf{(6.3.10b)}$$

Equation (6.3.10a) can then be solved for pressure drop, which represents losses per unit volume, as

$$\Delta p = \frac{128 \mu L Q}{\pi D^4} \qquad \text{(horizontal only)} \qquad \textbf{(6.3.11)}$$

The losses are seen to vary directly as the viscosity, the length, and the discharge, and to vary inversely as the fourth power of the diameter. It should be noted that tube roughness does not enter into the equations. Equation (6.3.10a) is known as the *Hagen-Poiseuille equation;* it was determined experimentally by Hagen in 1839 and independently by Poiseuille in 1840. The analytical derivation was made by Wiedemann in 1856.

Determine the direction of flow through the tube shown in Fig. 6.9, in which $\gamma = 8000 \text{ N/m}^3$ and $\mu = 0.04 \text{ kg/m·s}$. Find the quantity flowing in liters per second and calculate the Reynolds number for the flow.

Example 6.2

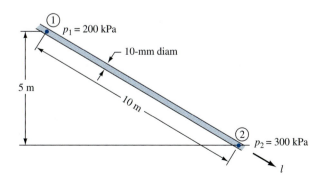

Figure 6.9   Flow through an inclined tube.

Solution

At section 1

$$p + \gamma h = 200{,}000 \text{ N/m}^2 + (8000 \text{ N/m}^3)(5 \text{ m}) = 240 \text{ kPa}$$

and at section 2

$$p + \gamma h = 300 \text{ kPa}$$

if the elevation datum is taken through section 2. The flow is from section 2 to section 1, since the energy is greater at section 2 (kinetic energy must be the same at each section) than at section 1. To determine the quantity flowing, the expression is written

$$\frac{d}{dl}(p + \gamma h) = \frac{300{,}000 - 240{,}000 \text{ N/m}^2}{10 \text{ m}} = 6000 \text{ N/m}^3$$

with l positive from section 1 to section 2. Substituting into Eq. (6.3.9) gives

$$Q = -\frac{\pi(0.005 \text{ m})^4}{8(0.04 \text{ N·s/m}^2)} 6000 \text{ N/m}^3 = -0.0000368 \text{ m}^3/\text{s} = -0.0368 \text{ L/s}$$

The average velocity is

$$V = \frac{0.0000368 \text{ m}^3/\text{s}}{\pi(0.005 \text{ m})^2} = 0.4686 \text{ m/s}$$

and the Reynolds number (Sec. 5.4) is

$$\mathbf{R} = \frac{VD\rho}{\mu} = \frac{(0.4686 \text{ m/s})(0.01 \text{ m})}{(0.04 \text{ N·s/m}^2)} \frac{8000 \text{ N/m}^3}{9.806 \text{ m/s}^2} = 95.6$$

If the Reynolds number had been above 2000, the Hagen-Poiseuille equation would no longer apply, as discussed in Sec. 6.1.

The kinetic-energy correction factor α [Eq. (3.4.14)] can be determined for laminar flow in a tube by use of Eqs. (6.3.6) and (6.3.7)

$$\frac{u}{V} = 2\frac{u}{u_{\max}} = 2\left[1 - \left(\frac{r}{a}\right)^2\right] \tag{6.3.12}$$

Substituting into the expression for α gives

$$\alpha = \frac{1}{A}\int\left(\frac{u}{V}\right)^3 dA = \frac{1}{\pi a^2}\int_0^a \left\{2\left[1 - \left(\frac{r}{a}\right)^2\right]\right\}^3 2\pi r\, dr = 2 \tag{6.3.13}$$

There is twice as much energy in the flow as in uniform flow at the same average velocity. The momentum correction factor is obtained by replacing the exponent 3 with exponent 2, yielding $\beta = \frac{4}{3}$. The distribution of shear stress, velocity, and losses per unit volume is shown in Fig. 6.8 for a round tube.

EXERCISES

6.3.1   The shear stress in a fluid flowing in a round pipe (*a*) is constant over the cross section; (*b*) is zero at the wall and increases linearly to the center; (*c*) varies parabolically across the section; (*d*) is zero at the center and varies linearly with the radius; (*e*) is none of these answers.

6.3.2 When the pressure drop in a 24-in.-diameter pipeline is 10 psi in 100 ft, the wall shear stress, in pounds per square foot, is (*a*) 0; (*b*) 7.2; (*c*) 14.4; (*d*) 720; (*e*) none of these answers.

6.3.3 In laminar flow through a round tube the discharge varies (*a*) linearly as the viscosity; (*b*) as the square of the radius; (*c*) inversely as the pressure drop; (*d*) inversely as the viscosity; (*e*) as the cube of the diameter.

6.3.4 When a tube is inclined, the term $-dp/dl$ is replaced by

$$(a) -\frac{dh}{dl} \qquad (b) -\gamma\frac{dh}{dl} \qquad (c) -\frac{d(p+h)}{dl} \qquad (d) -\frac{d(p+\rho h)}{dl}$$

$$(e) -\frac{d(p+\gamma h)}{dl}$$

6.4 TURBULENT FLOW RELATIONS

Temporal Averaging

In turbulent flow the random fluctuations of each velocity component and pressure term in Eqs. (4.4.12) make exact analysis difficult if not impossible, even with numerical methods. It becomes more convenient to separate the quantities into mean or time-averaged values and fluctuating parts [3]. The *x* component of velocity *u*, for example, is represented by

$$u = \bar{u} + u' \tag{6.4.1}$$

as shown in Fig. 6.10, in which the mean value is the time-averaged quantity defined by

$$\bar{u} = \frac{1}{T_o} \int_t^{t+T_o} u \, dt \tag{6.4.2}$$

The limit T_o on the integration is an averaging time period, suitable to the particular problem, that is greater than any period of the fine-scale turbulent variations. We

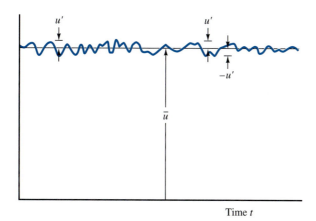

Time *t*

Figure 6.10   Turbulent fluctuations in direction of flow.

note in Fig. 6.10 and from the definition that the fluctuation has a mean value of zero

$$\overline{u'} = \frac{1}{T_o} \int_t^{t+T_o} (u - \bar{u}) \, dt = \bar{u} - \bar{u} = 0 \tag{6.4.3}$$

However, the mean square of each fluctuation is not zero

$$\overline{u'^2} = \frac{1}{T_o} \int_t^{t+T_o} (u - \bar{u})^2 dt \neq 0 \tag{6.4.4}$$

The square root of this quantity, the root mean square of measured values of the fluctuations, is a measure of the intensity of the turbulence. Reynolds [4] split each property into mean and fluctuating variables

$$v = \bar{v} + v' \qquad w = \bar{w} + w' \qquad p = \bar{p} + p'$$

In each case the mean value of the fluctuation is zero and the mean square is not. Nor are mean products of the fluctuations as $\overline{u'v'}$, $\overline{u'w'}$, etc., zero.

Substitution of the mean and fluctuating parts of the variables into the continuity equation [Eq. (4.3.4)] for incompressible flow yields

$$\frac{\partial \bar{u}}{\partial x} + \frac{\partial \bar{v}}{\partial y} + \frac{\partial \bar{w}}{\partial z} = 0 \tag{6.4.5}$$

The time-averaged momentum equations contain the product of the fluctuating x, y, and z component velocities. In the x direction the equation becomes

$$\rho \left[\frac{\partial \bar{u}}{\partial t} + \frac{\partial}{\partial x}(\overline{uu}) + \frac{\partial}{\partial x}(\overline{uv}) + \frac{\partial}{\partial x}(\overline{uw}) \right] = -\frac{\partial}{\partial x}(\bar{p} + \gamma h) + \mu \left(\frac{\partial^2 \bar{u}}{\partial x^2} + \frac{\partial^2 \bar{u}}{\partial y^2} + \frac{\partial^2 \bar{u}}{\partial z^2} \right)$$

As they stand the temporal average of the inertial acceleration terms $\overline{uu}$, $\overline{uv}$, and $\overline{uw}$ is not in tractable form. It is necessary to have these terms in the form of products of the averages but not as averages of products; as noted in Chap. 3 these two types of averages are not equal. Therefore, further analysis is required. Instead of pursuing a formulation such as the kinetic energy or momentum correction factor approach, Reynolds used the averaging decomposition on the acceleration terms which after some algebra results in the following

$$\overline{(\bar{u} + u')(\bar{u} + u')} = \overline{\bar{u}\bar{u}} + \overline{u'u'} \simeq \bar{u}\,\bar{u} + \overline{u'u'}$$

similarly

$$\overline{uv} \simeq \bar{u}\,\bar{v} + \overline{u'v'}$$

$$\overline{uw} \simeq \bar{u}\,\bar{w} + \overline{u'w'}$$

Assembling these terms into the previous equation yields

$$\rho \left[\frac{\partial \bar{u}}{\partial t} + \frac{\partial}{\partial x}(\bar{u}\,\bar{u} + \overline{u'u'}) + \frac{\partial}{\partial x}(\bar{u}\,\bar{v} + \overline{u'v'}) + \frac{\partial}{\partial x}(\bar{u}\,\bar{w} + \overline{u'w'}) \right]$$

$$= -\frac{\partial}{\partial x}(\bar{p} + \gamma h) + \mu \nabla^2 \bar{u}$$

The terms $\overline{\rho u'u'}$, $\overline{\rho u'v'}$, and $\overline{\rho u'w'}$ are complex terms physically but for the simple turbulent flows analyzed by Reynolds it was shown (and verified by others since) that these terms provide a stress-like effect. Consequently these terms as well as those in the y and z equations are called *Reynolds stresses*. These terms are responsible for considerable momentum exchange and mixing in turbulent flow and their magnitude completely dominates the viscous stress terms in turbulent flows. It should be remembered, however, that these terms come from the inertial acceleration terms and for many geophysical flows have a more complex physics than simple "shear-like" behavior. The label "Reynolds stress," therefore, does not refer to a robust physical description so much as it pays homage to the originator of the analysis concept and viewpoint.

The complex suite of equations is written for the x direction as

$$\rho \frac{D\bar{u}}{Dt} = \rho \left\{ \frac{\partial \bar{u}}{\partial t} + \bar{u}\frac{\partial \bar{u}}{\partial x} + \bar{v}\frac{\partial \bar{u}}{\partial y} + \bar{w}\frac{\partial \bar{u}}{\partial z} \right\} = -\frac{\partial}{\partial x}(\bar{p} + \gamma h) \qquad \text{(6.4.6a)}$$

$$+ \frac{\partial}{\partial x}\left\{ -\rho\overline{u'u'} + \mu\frac{\partial \bar{u}}{\partial x} \right\} + \left\{ -\rho\overline{u'v'} + \mu\frac{\partial \bar{u}}{\partial y} \right\} + \left\{ -\rho\overline{u'w'} + \mu\frac{\partial \bar{u}}{\partial z} \right\}$$

or

$$\rho \frac{D\bar{u}}{Dt} = -\frac{\partial}{\partial x}(\bar{p} + \gamma h) + \frac{\partial \tau_{xx}}{\partial x} + \frac{\partial \tau_{yx}}{\partial y} + \frac{\partial \tau_{zx}}{\partial z}$$

where

$$\tau_{xx} = -\rho\overline{u'u'} + \mu\frac{\partial \bar{u}}{\partial x}, \quad \text{etc.}$$

for the y direction as

$$\rho \frac{D\bar{v}}{Dt} = \rho \left\{ \frac{\partial \bar{v}}{\partial t} + \bar{u}\frac{\partial \bar{v}}{\partial x} + \bar{v}\frac{\partial \bar{v}}{\partial y} + \bar{w}\frac{\partial \bar{v}}{\partial z} \right\} = -\frac{\partial}{\partial y}(\bar{p} + \gamma h) \qquad \text{(6.4.6b)}$$

$$+ \frac{\partial}{\partial x}\left\{ -\rho\overline{u'v'} + \mu\frac{\partial \bar{v}}{\partial x} \right\} + \left\{ -\rho\overline{v'v'} + \mu\frac{\partial \bar{v}}{\partial y} \right\} + \left\{ -\rho\overline{v'w'} + \mu\frac{\partial \bar{v}}{\partial z} \right\}$$

or

$$\rho \frac{D\bar{v}}{Dt} = -\frac{\partial}{\partial x}(\bar{p} + \gamma h) + \frac{\partial \tau_{xy}}{\partial x} + \frac{\partial \tau_{yy}}{\partial y} + \frac{\partial \tau_{zy}}{\partial z}$$

and for the z direction as

$$\rho \frac{D\bar{w}}{Dt} = \rho \left\{ \frac{\partial \bar{w}}{\partial t} + \bar{u}\frac{\partial \bar{w}}{\partial x} + \bar{v}\frac{\partial \bar{w}}{\partial y} + \bar{w}\frac{\partial \bar{w}}{\partial z} \right\} = -\frac{\partial}{\partial z}(\bar{p} + \gamma h) \qquad \text{(6.4.6c)}$$

$$+ \frac{\partial}{\partial x}\left\{ -\rho\overline{u'w'} + \mu\frac{\partial \bar{w}}{\partial x} \right\} + \left\{ -\rho\overline{v'w'} + \mu\frac{\partial \bar{w}}{\partial y} \right\} + \left\{ -\rho\overline{w'w'} + \mu\frac{\partial \bar{w}}{\partial z} \right\}$$

or

$$\rho \frac{D\bar{w}}{Dt} = -\frac{\partial}{\partial z}(\bar{p} + \gamma h) + \frac{\partial \tau_{xz}}{\partial x} + \frac{\partial \tau_{yz}}{\partial y} + \frac{\partial \tau_{zz}}{\partial z}$$

Since in general the Reynolds stresses are unknown, empirical methods based upon intuitive reasoning, dimensional analysis, or physical experiments are used to assist in their analyses. In one-dimensional flow in the x direction the turbulent stress $-\rho\overline{u'v'}$ is the most important, and the linear-momentum equation can be approximated by

$$-\frac{\partial}{\partial x}(\bar{p} + \gamma h) + \frac{\partial \tau}{\partial y} \approx \rho\frac{\partial \bar{u}}{\partial t} \qquad \text{(6.4.7)}$$

in which

$$\tau_{xy} = \tau = \mu \frac{\partial \overline{u}}{\partial y} - \overline{\rho u'v'} \tag{6.4.8}$$

is a total shear made up of laminar τ_l and turbulent τ_t components. In view of the difficulty of evaluating $-\overline{\rho u'v'}$, Prandtl [5] introduced the mixing-length theory, which relates the apparent shear stress to the temporal mean velocity distribution.

The apparent shear stress in turbulent flow is expressed in a form similar to Newton's viscosity law, that is,

$$-\overline{\rho u'v'} = \tau_t = \eta \frac{\partial \overline{u}}{\partial y} \tag{6.4.9}$$

where η is an empirical coefficient called the *eddy viscosity*. The following is a brief explanation of a procedure to *close* the turbulent flow equations by finding a way to relate the average velocity to the fluctuating quantities. *To avoid confusion with vector notation all overbars denoting time averaging will be dropped.*

Prandtl's Mixing Length

In Prandtl's theory [6] expressions for u' and v' are obtained in terms of a mixing-length distance l and the velocity gradient du/dy, in which u is the temporal mean velocity at a point (the bar over u has been dropped) and y is the distance normal to u, usually measured from the boundary. In a gas, one molecule, before striking another, travels an average distance known as the *mean free path* of the gas. Using this as an analogy (Fig. 6.11a), Prandtl assumed that a particle of fluid is displaced a distance l before its momentum is changed by the new environment. The fluctuation u' is then related to l by

$$u' \sim l \frac{du}{dy}$$

which means that the amount of the change in velocity depends upon the changes in temporal mean velocity at two points a distance l apart in the y direction. From the continuity equation he reasoned that there must be a correlation between u' and v' (Fig. 6.11b), so that v' is proportional to u'

$$v' \sim u' \sim l \frac{du}{dy}$$

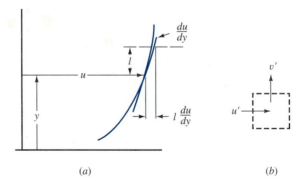

(a)   (b)

Figure 6.11   Notation for mixing-length theory.

By substituting for u' and v' and letting l absorb the proportionality factor, the defining equation for mixing length is obtained as

$$\tau_{xy_t} = \tau_t = -\rho\overline{u'v'} = \rho l^2 \left(\frac{du}{dy}\right)^2 \tag{6.4.10}$$

τ always acts in the sense that causes the velocity distribution to become more uniform. When Eq. (6.4.9) is compared with Eq. (6.4.10), it is found that the eddy viscosity is

$$\eta = \rho l^2 \frac{du}{dy} \tag{6.4.11}$$

But η is not a fluid property like dynamic viscosity; instead η depends upon the density, the velocity gradient, and mixing length l, and in general varies from point to point in the flow field. In turbulent flow there is a violent interchange of parcels of fluid except at a boundary, or very near to it, where this interchange is reduced to zero; hence, l must approach zero at a fluid boundary. The particular relationship of l to wall distance y is not given by Prandtl's derivation. Von Kármán [7] suggested, after considering similitude relations in a turbulent fluid, that

$$l = \kappa \frac{du/dy}{d^2u/dy^2} \tag{6.4.12}$$

in which κ is a universal constant in turbulent flow, regardless of the boundary configuration or value of the Reynolds number. Von Kármán's coefficient, κ, has a value of 0.4.

   In turbulent flows η, the eddy viscosity, is generally much larger than μ. It can be considered as a coefficient of momentum transfer, expressing the transfer of momentum from points where the concentration is high to points where it is lower. It is convenient to use a *kinematic eddy viscosity* $\epsilon = \eta/\rho$, which depends upon the flow parameters alone and is analogous to kinematic viscosity.

Velocity Distributions

In turbulent flows, conditions next to a surface are considerably more complex than in laminar flows. It is convenient to visualize the turbulent-shear layer near a smooth wall to be divided into three layers (Fig. 6.12). In the viscous wall layer or laminar sublayer the shear stress in the fluid is essentially constant and equal to the shear stress at the wall τ_0. The velocity distribution is related to the shear stress and absolute viscosity within the region $y \le \delta'$ by Newton's viscosity law

$$\frac{\tau_0}{\rho} = \frac{\mu}{\rho}\frac{u}{y} = \nu\frac{u}{y} \qquad y \le \delta' \tag{6.4.13}$$

Here δ' is the laminar sublayer height and the term $\sqrt{\tau_0/\rho}$ has the dimensions of a velocity and is called the *shear-stress* or *friction velocity* u_*. Hence,

$$\frac{u}{u_*} = \frac{u_* y}{\nu} \qquad y \le \delta' \tag{6.4.14}$$

shows a linear relation between u and y in the laminar film. It extends to a value of $u_* y/\nu \approx 5$, that is,

$$\delta' = 5\frac{\nu}{u_*} \tag{6.4.15}$$

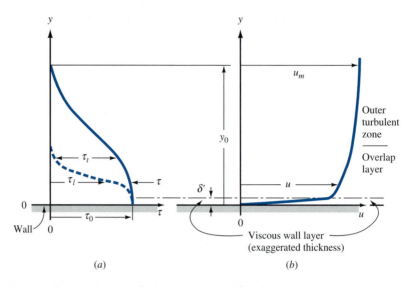

Figure 6.12   Schematic of (a) shear stress and (b) velocity distribution near wall in turbulent flow.

In the overlap layer it is assumed that the shear stress is approximately equal to the shear stress at the wall (Fig. 6.12), but turbulence dominates and the viscous shear stress expressed in Eq. (6.4.13) is unimportant. Hence, Eq. (6.4.9) produces

$$\tau_0 = \rho l^2 \left(\frac{du}{dy}\right)^2 \tag{6.4.16}$$

Since l has the dimensions of a length and from dimensional considerations would be proportional to y (the only significant linear dimension), assume $l = \kappa y$. Substituting into Eq. (6.4.16) and rearranging give

$$\frac{du}{u_*} = \frac{1}{\kappa}\frac{dy}{y} \tag{6.4.17}$$

and integration leads to

$$\frac{u}{u_*} = \frac{1}{\kappa}\ln y + \text{constant} \tag{6.4.18}$$

It is to be noted that this value of u substituted in Eq. (6.4.12) also determines l proportional to y (d^2u/dy^2 is negative, since the velocity gradient decreases as y increases). Equation (6.4.18) agrees well with experiment and, in fact, is also useful when τ is a function of y, because most of the velocity change occurs near the wall, where τ is substantially constant. It is quite satisfactory to apply the equation to turbulent flow in pipes.

Example 6.3

By integrating Eq. (6.4.18), find the relation between the average velocity V and the maximum velocity u_m in turbulent flow in a pipe.

Solution

When $y = r_0$, then $u = u_m$, so that

$$\frac{u}{u_*} = \frac{u_m}{u_*} + \frac{1}{\kappa}\ln\frac{y}{r_0}$$

The discharge $V \pi r_0^2$ is obtained by integrating the velocity over the area as

$$V \pi r_0^2 = 2\pi \int_0^{r_0 - \delta'} ur \, dr = 2\pi \int_{\delta'}^{r_0} \left(u_m + \frac{u_*}{\kappa} \ln \frac{y}{r_0} \right) (r_0 - y) \, dy$$

The integration cannot be carried out to $y = 0$ since the equation holds in the turbulent zone only. The volume per second flowing in the laminar zone is so small that it can be neglected. Then

$$V = 2 \int_{\delta'/r_0}^{1} \left(u_m + \frac{u_*}{\kappa} \ln \frac{y}{r_0} \right) \left(1 - \frac{y}{r_0} \right) d\frac{y}{r_0}$$

in which the variable of integration is y/r_0. Integrating gives

$$V = 2 \left\{ u_m \left[\frac{y}{r_0} - \frac{1}{2} \left(\frac{y}{r_0} \right)^2 \right] + \frac{u_*}{\kappa} \left[\frac{y}{r_0} \ln \frac{y}{r_0} - \frac{y}{r_0} - \frac{1}{2} \left(\frac{y}{r_0} \right)^2 \ln \frac{y}{r_0} + \frac{1}{4} \left(\frac{y}{r_0} \right)^2 \right] \right\}_{\delta'/r_0}^{1}$$

Since δ'/r_0 is very small, such terms as δ'/r_0 and $\delta'/r_0 \ln (\delta'/r_0)$ become negligible ($\lim_{x \to 0} x \ln x = 0$); thus,

$$V = u_m - \frac{3}{2} \frac{u_*}{\kappa} \qquad \text{or} \qquad \frac{u_m - V}{u_*} = \frac{3}{2\kappa}$$

In an open channel flow of depth d find a relationship between the average velocity V and measured point values for velocity in the boundary layer. Find the depth at which the point velocity is equal to the average velocity.

Example 6.4

Solution

From Eq. (6.4.18) applied to free surface flow

$$\frac{u}{u_*} = \frac{1}{\kappa} \ln \frac{y}{y_0}$$

where y_0 is an integration factor derived from the condition that is very near the rough channel bottom where $y = y_0$. As a result $u = 0$. The average velocity is found from

$$V = \frac{1}{d - y_0} \int_{y_0}^{d} \left(\frac{u_*}{\kappa} \ln \frac{y}{y_0} \right) dy$$

or, with $\dfrac{y_0}{d} \ll 1$, as

$$V = \frac{u_*}{\kappa} \ln \frac{d}{y_0} - \frac{u_*}{\kappa}$$

A velocity-deficit law is found as

$$\frac{V - u}{u_*} = \frac{1}{\kappa} \ln \frac{d}{y} - \frac{1}{\kappa}$$

or

$$u = -\frac{u_*}{\kappa} \ln \frac{d}{y} + \frac{u_*}{\kappa} + V$$

By equating $u = V$, the depth at which the point velocity is equal to the average velocity is found to be

$$\ln \frac{y}{y_0} = \ln \frac{d}{y_0} - 1$$

$$y = 0.3679d$$

In practice, the point velocity is sometimes measured at $y = 0.4d$, and the value is used as the average velocity, rather than measuring the velocity distribution to find the average. Alternatively, the mean of two point measurements, at $y = 0.8d$ and $y = 0.2d$, is often used as the average velocity.

To evaluate the constant in Eq. (6.4.18), the methods of Bakhmeteff [8] are used, that is, $u = u_w$, the *wall velocity*, when $y = \delta'$. According to Eq. (6.4.14)

$$\frac{u_w}{u_*} = \frac{u_* \delta'}{\nu} = N \qquad\qquad (6.4.19)$$

from which it is reasoned that $u_* \delta' / \nu$ should have a critical value N at which flow changes from laminar to turbulent since it is a Reynolds number in form. Substituting $u = u_w$ when $y = \delta'$ into Eq. (6.4.18) and using Eq. (6.4.19) yield

$$\frac{u_w}{u_*} = N = \frac{1}{\kappa} \ln \delta' + \text{const} = \frac{1}{\kappa} \ln \frac{N\nu}{u_*} + \text{constant}$$

Eliminating the constant gives

$$\frac{u}{u_*} = \frac{1}{\kappa} \ln \frac{y u_*}{\nu} + N - \frac{1}{\kappa} \ln N$$

or

$$\frac{u}{u_*} = \frac{1}{\kappa} \ln \frac{y u_*}{\nu} + A \qquad\qquad (6.4.20)$$

The coefficient $A = N - (1/\kappa) \ln N$ has been found experimentally by plotting u/u_* against $\ln (y u_*/\nu)$. For smooth-wall pipes Nikuradse's [9] experiments yielded $\kappa = 0.40$ and $A = 5.5$. Figure 6.13 shows the logarithmic velocity distribution, Eq. (6.4.20), for turbulent flow in smooth pipes, together with an indication of experimental error provided from a number of sources.

In the outer turbulent zone (Fig. 6.12) a velocity-deficit law applies. The dimensionless velocity deficiency $(u_m - u)/u_*$ is a function of the ratio of y to the thickness y_0. For pipe flow $y_0 = r_0$, the pipe radius. The velocity-deficit law applies to rough as well as smooth pipes. From Eq. (6.4.18) evaluating the constant for $u = u_m$ when $y = r_0$ gives

$$\frac{u_m - u}{u_*} = \frac{1}{\kappa} \ln \frac{r_0}{y} \qquad\qquad (6.4.21)$$

For rough wall flows, the velocity can be assumed to be u_w at the wall distance $y_w = m\epsilon'$, in which ϵ' is a typical height of the roughness projections and m is a form coefficient depending upon the nature of the roughness. Substituting into Eq. (6.4.21)

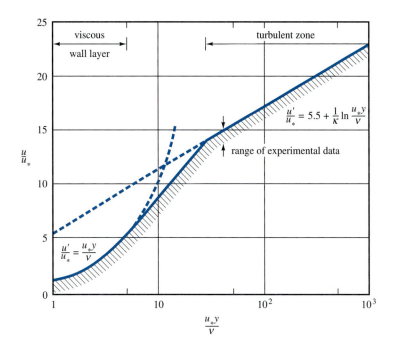

Figure 6.13 Turbulent flow in smooth pipes.

and eliminating u_m/u_* between the two equations lead to

$$\frac{u}{u_*} = \frac{1}{\kappa} \ln \frac{y}{\epsilon'} + \frac{u_w}{u_*} - \frac{1}{\kappa} \ln m \qquad \textbf{(6.4.22)}$$

in which the last two terms on the right-hand side are constant for a given type of roughness,

$$\frac{u}{u_*} = \frac{1}{\kappa} \ln \frac{y}{\epsilon'} + B \qquad \textbf{(6.4.23)}$$

In Nikuradse's experiments with sand-roughened pipes constant-size sand particles (those passing a given screen and being retained on a slightly finer screen) were glued to the inside pipe walls. If ϵ' represents the diameter of sand grains, experiment shows that $\kappa = 0.40$ and $B = 8.48$.

The logarithmic law, Eq. (6.4.20), has the broadest applicability since it overlaps and matches quite well the defect law in most flows. The viscous wall layer physically covers only a very small portion of the flow in turbulent flows and is of course virtually nonexistent in rough wall flows as ϵ' is typically the same size or larger than δ' for equal Reynolds number smooth flows.

Prandtl has developed a convenient exponential velocity-distribution formula for turbulent pipe flow,

$$\frac{u}{u_m} = \left(\frac{y}{r_0}\right)^n \qquad \textbf{(6.4.24)}$$

in which n varies with the Reynolds number. This empirical equation is valid only at some distance from the wall. For $\mathbf{R}$ less than 100,000, $n = \frac{1}{7}$, and for greater values of $\mathbf{R}$, n decreases. The velocity-distribution Eqs. (6.4.20) and (6.4.24) both have the fault of a nonzero value of du/dy at the center of the pipe.

Example 6.5

Find an approximate expression for mixing-length distribution in turbulent flow in a pipe from Prandtl's one-seventh-power law.

Solution

Writing a force balance for steady flow in a round tube (Fig. 6.14) gives

$$\tau = -\frac{dp}{dl}\frac{r}{2}$$

At the wall

$$\tau_0 = \frac{dp}{dl}\frac{r_0}{2}$$

Hence,

$$\tau = \tau_0\frac{r}{r_0} = \tau_0\left(1 - \frac{y}{r_0}\right) = \rho l^2\left(\frac{du}{dy}\right)^2$$

Solving for l gives

$$l = \frac{u_*\sqrt{1 - y/r_0}}{du/dy}$$

From Eq. (6.4.24)

$$\frac{u}{u_m} = \left(\frac{y}{r_0}\right)^{1/7}$$

the approximate velocity gradient is obtained as

$$\frac{du}{dy} = \frac{u_m}{r_0}\frac{1}{7}\left(\frac{y}{r_0}\right)^{-6/7} \qquad \text{and} \qquad \frac{l}{r_0} = \frac{u_*}{u_m}7\left(\frac{y}{r_0}\right)^{6/7}\sqrt{1 - \frac{y}{r_0}}$$

EXERCISES

6.4.1   The Prandtl mixing length is (a) independent of radial distance from pipe axis; (b) independent of the shear stress; (c) zero at the pipe wall; (d) a universal constant; (e) useful for computing laminar flow problems.

6.4.2   The average velocity divided by the maximum velocity, as given by the one-seventh-power law, is (a) $\frac{49}{120}$; (b) $\frac{1}{2}$; (c) $\frac{6}{7}$; (d) $\frac{98}{120}$; (e) none of these answers.

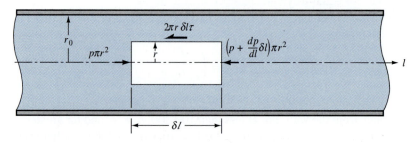

Figure 6.14   Free-body diagram for steady flow through a round tube.

6.5 TURBULENT FLOW LOSSES IN OPEN AND CLOSED CONDUITS

In steady, uniform turbulent incompressible flow in conduits of constant cross section the wall shear stress varies approximately in proportion to the square of the average velocity

$$\tau_0 = \lambda \frac{\rho}{2} V^2 \qquad (6.5.1)$$

in which λ is a dimensionless coefficient. In open channels and noncircular closed conduits the shear stress is not constant over the surface. In these cases τ_0 is used as the average wall shear stress. Secondary flows† occurring in noncircular conduits act to equalize the wall shear stress.

In Fig. 6.15 a steady uniform flow is indicated in either an open or a closed conduit. For an open channel p_1 and p_2 are equal, and flow occurs as a result of reduction in potential energy, $z_1 - z_2$ m·N/N. For closed-conduit flow, energy for flow could be supplied by the potential energy drop as well as by a drop in pressure $p_1 - p_2$. With flow vertically downward in a pipe, p_2 could increase in the flow direction, but potential energy drop $z_1 - z_2$ would have to be greater than $(p_2 - p_1)/\gamma$ to supply energy to overcome the wall shear stress.

We can write the energy equation [Eq. (3.4.21)] to relate losses to available energy reduction

$$\frac{p_1}{\gamma} + \frac{V_1^2}{2g} + z_1 = \frac{p_2}{\gamma} + \frac{V_2^2}{2g} + z_2 + \text{losses}_{1-2}$$

Since the velocity head $V^2/2g$ is the same,

$$\text{Losses}_{1-2} = \frac{p_1 - p_2}{\gamma} + z_1 - z_2 \qquad (6.5.2)$$

Owing to the uniform flow assumption, the linear-momentum equation [Eq. (3.6.7)] applied in the l direction yields

$$\Sigma F_l = 0 = (p_1 - p_2)A + \gamma AL \sin \theta - \tau_0 LP$$

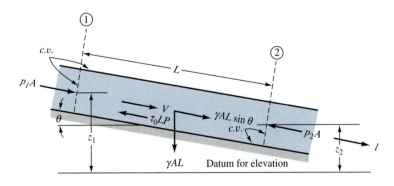

Figure 6.15   Axial forces on a control volume in a conduit.

---

†Secondary flows are transverse components that cause the main central flow to spread out into corners or near walls.

in which P is the *wetted perimeter* of the conduit, that is, the portion of the perimeter where the wall is in contact with the fluid (free-liquid surface excluded). Since $L \sin \theta = z_1 - z_2$,

$$\frac{p_1 - p_2}{\gamma} + z_1 - z_2 = \frac{\tau_0 LP}{\gamma A} \tag{6.5.3}$$

From Eqs. (6.5.2) and (6.5.3), using Eq. (6.5.1),

$$\text{Losses}_{1-2} = \frac{\tau_0 LP}{\gamma A} = \lambda \frac{\rho}{2} V^2 \frac{LP}{\gamma A} = \lambda \frac{L}{R} \frac{V^2}{2g} \tag{6.5.4}$$

in which $R = A/P$ has been substituted. R, called the *hydraulic radius* of the conduit, is most useful in dealing with open channels. For a pipe $R = D/4$.

The loss term in Eq. (6.5.4) is in units of meter-newtons per newton or foot-pounds per pound. It is given the name h_f, *head loss due to friction*. Defining S as the losses per unit weight per unit length of the channel leads to

$$S = \frac{h_f}{L} = \frac{\lambda}{R} \frac{V^2}{2g} \tag{6.5.5}$$

After solving for V,

$$V = \sqrt{\frac{2g}{\lambda}} \sqrt{RS} = C\sqrt{RS} \tag{6.5.6}$$

The coefficient λ, or coefficient C, must be found by experiment. This is the Chézy formula, in which originally the Chézy coefficient C was thought to be a constant for any size conduit or wall-surface condition. Various formulas for C are now generally used.

For pipes, when $\lambda = f/4$ and $R = D/4$, the Darcy-Weisbach equation is obtained as

$$h_f = f \frac{L}{D} \frac{V^2}{2g} \tag{6.5.7}$$

in which D is the pipe inside diameter. This equation can be applied to open channels in the form

$$V = \sqrt{\frac{8g}{f}} \sqrt{RS} \tag{6.5.8}$$

with values of f determined from pipe experiments.

EXERCISES

6.5.1 The hydraulic radius is given by (a) the wetted perimeter divided by the area; (b) area divided by the square of the wetted perimeter; (c) the square root of the areas; (d) the area divided by the wetted perimeter; (e) none of these answers.

6.5.2 The hydraulic radius of an open channel 60 mm wide by 120 mm deep is, in millimeters, (a) 20; (b) 24; (c) 40; (d) 60; (e) none of these answers.

6.6 STEADY UNIFORM FLOW IN OPEN CHANNELS

For incompressible, steady flow at constant depth in a prismatic open channel with bed slope, S_o, the *Manning* formula is widely used. It can be obtained from the Chézy formula (6.5.6) by setting

$$C = \frac{C_m}{n} R^{1/6} \qquad (6.6.1)$$

so that

$$V = \frac{C_m}{n} R^{2/3} S^{1/2} \qquad (6.6.2)$$

which is the Manning formula.

The value of C_m is 1.49 and 1.0 for USC and SI units, respectively; V is the average velocity at a cross section; R is the hydraulic radius (Sec. 6.5); and S is the losses per unit weight and unit length of the channel. For steady uniform flow it can readily be shown that S equals S_o, the slope of the bottom of the channel. It is also the slope of the water surface, which for steady uniform flow is parallel to the channel bottom. The coefficient n was thought to be an absolute roughness coefficient, that is, dependent upon surface roughness only, but it actually depends upon the size and shape of the channel cross section in some unknown manner. Values of the coefficient n determined by many tests on actual canals are given in Table 6.1. Equation (6.6.2) must have consistent USC or SI units as indicated for use with the values in Table 6.1.†

Table 6.1 Average values of the Manning roughness factor for various boundary materials‡

Boundary material	Manning n
Planed wood	0.012
Unplaned wood	0.013
Finished concrete	0.012
Unfinished concrete	0.014
Cast iron	0.015
Brick	0.016
Riveted steel	0.018
Corrugated metal	0.022
Rubble	0.025
Earth	0.025
Earth, with stones or weeds	0.035
Gravel	0.029

‡ Work by the U.S. Bureau of Reclamation and other government agencies indicates that the Manning roughness factor should be increased (say, 10 to 15 percent) for hydraulic radii greater than about 10 ft. The loss in capacity of large channels is due to the roughening of the surfaces with age, marine and plant growths, deposits, and the addition of bridge piers or other channel constrictions.

†To convert the empirical equation in USC units to SI units, n is taken to be dimensionless. Then the constant has dimensions and $(1.49 \text{ ft}^{1/3}/\text{s})(0.3048 \text{ m/ft})^{1/3} = 1.0 \text{ m}^{1/3}/\text{s}$.

When Eq. (6.6.2) is multiplied by the cross-section area A, the Manning formula takes the form

$$Q = \frac{C_m}{n} A R^{2/3} S^{1/2}$$ (6.6.3)

When the cross-sectional area is known, any one of the other quantities can be obtained from Eq. (6.6.3) by direct solution.

Example 6.6

Determine the discharge for a trapezoidal channel (Fig. 6.16) with a bottom width $b = 8$ ft and side slopes 1 on 1. The depth is 6 ft, and the slope of the bottom is 0.0009. The channel has a finished concrete lining.

Solution

From Table 6.1, $n = 0.012$. The area is

$$A = 8(6) + 6(6) = 84 \text{ ft}^2$$

and the wetted perimeter is

$$P = 8 + 2(6\sqrt{2}) = 24.96 \text{ ft}$$

By substituting into Eq. (6.6.3),

$$Q = \frac{1.49}{0.0012} (84) \left(\frac{84}{24.96} \right)^{2/3} (0.0009^{1/2}) = 703 \text{ cfs}$$

Trial solutions are required in some instances when the cross-sectional area is unknown. Expressions for both the hydraulic radius and the area contain the depth in a form that cannot be solved explicitly.

Example 6.7

What depth is required for 4 m³/s flow in a rectangular planed-wood channel 2 m wide with a bottom slope of 0.002?

Solution

If the depth is y, $A = 2y$, $P = 2 + 2y$, and $n = 0.012$. By substituting in Eq. (6.6.3),

$$4 \text{ m}^3/\text{s} = \frac{1.00}{0.012} (2y) \left(\frac{2y}{2 + 2y} \right)^{2/3} 0.002^{1/2}$$

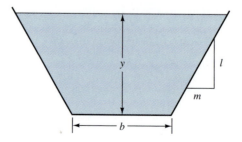

Figure 6.16   Notation for trapezoidal cross section.

Simplifying gives

$$f(y) = y\left(\frac{y}{1+y}\right)^{2/3} = 0.536$$

Assuming $y = 1$ m, then $f(y) = 0.63$. Assuming $y = 0.89$ m, then $f(y) = 0.538$. The correct depth is about 0.89 m. A spreadsheet could also be used to solve for y.

Example 6.8

A developer has been required by environmental regulatory authorities to line an open channel to prevent erosion. The channel is trapezoidal in cross section and has a slope of 0.0009. The bottom width is 10 ft and side slopes are 2:1 (horizontal to vertical). If a roughly spherical rubble ($\gamma_s = 135$ lb/ft^3) is used for the lining, find the minimum D_{50} of the rubble that can be used. The design flow is 1000 cfs. Assume the shear that rubble can withstand is described by

$$\tau = 0.040(\gamma_s - \gamma)D_{50} \qquad \text{lb/ft}^2$$

in which γ_s is the unit weight of rock and D_{50} is the average rock diameter in feet.

Solution

A Manning n of 0.03 is appropriate for the rubble. To find the size of the channel, from Eq. (6.6.3)

$$1000 = \frac{1.49}{0.03}\frac{[y(10 + 2y)]^{5/3}}{(10 + 2\sqrt{5}y)^{2/3}} 0.03$$

By trial solution the depth is $y = 8.62$ ft and the hydraulic radius $R = 4.84$ ft. From Eqs. (6.5.4) and (6.5.5)

$$\tau_0 = \gamma R S_0 = 62.4(4.84)(0.0009) = 0.272 \text{ lb/ft}^2$$

To find the D_{50} size for incipient movement, $\tau = \tau_0$ and

$$0.040(135 - 62.4)D_{50} = 0.272$$

Hence, $D_{50} = 0.0936$ ft.

More general cases of open-channel flow are considered in Chap. 13.

EXERCISES

6.6.1 The losses in open-channel flow generally vary as the (a) first power of the roughness; (b) inverse of the roughness; (c) square of the velocity; (d) inverse square of the hydraulic radius; (e) velocity.

6.6.2 The most simple form of open-channel-flow computation is (a) steady uniform; (b) steady nonuniform; (c) unsteady uniform; (d) unsteady nonuniform; (e) gradually varied.

6.6.3 In an open channel of great width the hydraulic radius equals (a) $y/3$; (b) $y/2$; (c) $2y/3$; (d) y; (e) none of these answers.

6.6.4 The Manning roughness coefficient for finished concrete is (a) 0.002; (b) 0.020; (c) 0.20; (d) dependent upon hydraulic radius; (e) none of these answers.

6.7   STEADY INCOMPRESSIBLE FLOW THROUGH SIMPLE PIPES

Colebrook Formula

A force balance for steady flow (no acceleration) in a horizontal pipe (Fig. 6.17) yields

$$\Delta p \pi r_0^2 = \tau_0 2\pi r_0 \, \Delta L$$

This simplifies to

$$\tau_0 = \frac{\Delta p}{\Delta L} \frac{r_0}{2} \qquad (6.7.1)$$

which holds for laminar or turbulent flow. The Darcy-Weisbach equation (6.5.7) can be written as

$$\Delta p = \gamma h_f = f \frac{\Delta L}{2r_0} \rho \frac{V^2}{2}$$

Eliminating Δp in the two equations and simplifying gives

$$\sqrt{\frac{\tau_0}{\rho}} = \sqrt{\frac{f}{8}} \, V \qquad (6.7.2)$$

which relates the wall shear stress, friction factor, and average velocity. The average velocity V can be obtained from Eq. (6.4.20) by integrating over the cross section. Substituting for V in Eq. (6.7.2) and simplifying produces the equation for friction factor in smooth-pipe flow,

$$\frac{1}{\sqrt{f}} = A_s + B_s \ln(\mathbf{R}\sqrt{f}) \qquad (6.7.3)$$

With the Nikuradse [9] data for smooth pipes the equation becomes

$$\frac{1}{\sqrt{f}} = 0.869 \, \ln (\mathbf{R}\sqrt{f}) - 0.8 \qquad (6.7.4)$$

For rough pipes in the complete turbulence zone,

$$\frac{1}{\sqrt{f}} = F_2 \left(m, \frac{\epsilon'}{D} \right) + B_r \ln \frac{\epsilon}{D} \qquad (6.7.5)$$

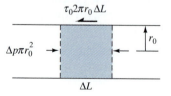

Figure 6.17   Equilibrium conditions for steady flow in a pipe.

in which F_2 is, in general, a constant for a given form and spacing of the roughness elements. For the Nikuradse sand-grain roughness discussed later in the section (Fig. 6.20), Eq. (6.7.5) becomes

$$\frac{1}{\sqrt{f}} = 1.14 - 0.869 \ln \frac{\epsilon}{D} \qquad (6.7.6)$$

The roughness height ϵ for sand-roughened pipes can be used as a measure of the roughness of commercial pipes. If the value of f is known for a commercial pipe in the fully developed wall-turbulence zone, that is, large Reynolds numbers and the loss proportional to the square of the velocity, the value of ϵ can be computed by Eq. (6.7.6). In the transition region, where f depends upon both ϵ/D and $\mathbf{R}$, sand-roughened pipes produce different results from commercial pipes. This is made evident by a graph based on Eqs. (6.7.4) and (6.7.6) with both sand-roughened and commercial-pipe-test results shown. Rearranging Eq. (6.7.6) gives

$$\frac{1}{\sqrt{f}} + 0.869 \ln \frac{\epsilon}{D} = 1.14$$

and adding $0.869 \ln (\epsilon/D)$ to each side of Eq. (6.7.4) leads to

$$\frac{1}{\sqrt{f}} + 0.869 \ln \frac{\epsilon}{D} = 0.869 \ln \left(\mathbf{R}\sqrt{f}\frac{\epsilon}{D} \right) - 0.8$$

By selecting $1/\sqrt{f} + 0.869 \ln(\epsilon/D)$ as ordinate and $\ln(\mathbf{R}\sqrt{f}\,\epsilon/D)$ as abscissa (Fig. 6.18), smooth-pipe-test results plot as a straight line with slope $+0.869$ and rough-pipe-test results in the complete turbulence zone plot as the horizontal line. Nikuradse sand-roughness-test results plot along the dashed line in the transition region, and commercial-pipe-test results plot along the lower curved line.

The reason for the difference in shape of the artificial roughness curve of Nikuradse and the commercial roughness curve is that the laminar sublayer, or laminar film, covers all the artificial roughness or allows it to protrude uniformly as the film thickness decreases. With commercial roughness, which varies greatly in uniformity, small portions extend beyond the film first, as the film decreases in thickness with increasing Reynolds number. An empirical transition function for commercial pipes for the region between smooth pipes and the complete turbulence zone has been

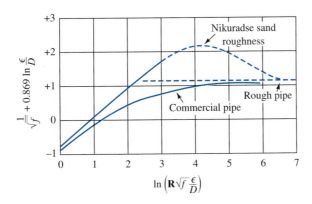

Figure 6.18 Colebrook transition function.

developed by Colebrook [10] as

$$\frac{1}{\sqrt{f}} = -0.869 \ln\left(\frac{\epsilon/D}{3.7} + \frac{2.523}{\mathbf{R}\sqrt{f}}\right) \qquad (6.7.7)$$

which is the basis for the Moody diagram (Fig. 6.21) discussed in detail in the following section.

Pipe Flow

In steady incompressible flow in a pipe the irreversibilities are expressed in terms of a head loss or drop in *hydraulic grade line* (Sec. 12.2). The hydraulic grade line is p/γ above the center of the pipe, and if z is the elevation of the center of the pipe, then $z + p/\gamma$ is the elevation of a point on the hydraulic grade line. The locus of values of $z + p/\gamma$ along the pipeline gives the hydraulic grade line. Losses, or irreversibilities, cause this line to drop in the direction of flow. The Darcy-Weisbach equation [Eq. (6.5.7)]

$$h_f = f \frac{L}{D} \frac{V^2}{2g}$$

is generally adopted for pipe-flow calculations. h_f is the head loss, or drop in hydraulic grade line, in the pipe length L, having an inside diameter D and an average velocity V. h_f has the dimension length and is expressed in terms of foot-pounds per pound or meter-newtons per newton. The friction factor f is a dimensionless factor that is required to make the equation produce the correct value for losses. All quantities in Eq. (6.5.7) except f can be measured experimentally. A typical setup is shown in Fig. 6.19. By measuring the discharge and inside diameter, the average velocity can be computed. The head loss h_f is measured by a differential manometer attached to piezometer openings at sections 1 and 2, distance L apart.

Experimentation shows the following to be true in turbulent flow:

1. The head loss varies directly as the length of the pipe.
2. The head loss varies almost as the square of the velocity.
3. The head loss varies almost inversely as the diameter.
4. The head loss depends upon the surface roughness of the interior pipe wall.
5. The head loss depends upon the fluid properties of density and viscosity.
6. The head loss is independent of the pressure.

The friction factor f must be so selected that Eq. (6.5.7) correctly yields the head loss; hence, f cannot be a constant but must depend upon velocity V, diameter D, density ρ, viscosity μ, and certain characteristics of the wall roughness signified

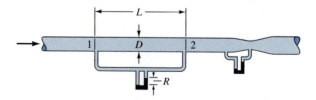

Figure 6.19   Experimental arrangement for determining head loss in a pipe.

by ϵ, ϵ', and m, where ϵ is a measure of the *size* of the roughness projections and has the dimensions of a length, ϵ' is a measure of the *arrangement* or *spacing* of the roughness elements and also has the dimensions of a length, and m is a form factor that is dependent upon the *shape* of the individual roughness elements and is dimensionless. The term f, instead of being a simple constant, turns out to depend upon seven quantities:

$$f = f(V, D, \rho, \mu, \epsilon, \epsilon', m) \tag{6.7.8}$$

Since f is a dimensionless factor, it must depend upon the grouping of these quantities into dimensionless parameters. For a *smooth* pipe $\epsilon = \epsilon' = m = 0$, leaving f dependent upon the first four quantities. They can be arranged in only one way to make them dimensionless, namely, $VD\rho/\mu$, which is the Reynolds number. For *rough* pipes the terms ϵ and ϵ' can be made dimensionless by dividing by D. Therefore, in general,

$$f = f\left(\frac{VD\rho}{\mu}, \frac{\epsilon}{D}, \frac{\epsilon'}{D}, m\right) \tag{6.7.9}$$

The proof of this relation is left to experimentation. For smooth pipes a plot of all experimental results shows the functional relation, subject to a scattering of ± 5 percent. The plot of friction factor against the Reynolds number on a log-log chart is called a *Stanton diagram*. Blasius [11], the first to correlate the smooth-pipe experiments in turbulent flow, presented the results by an empirical formula that is valid up to about $\mathbf{R} = 100,000$. The Blasius formula is

$$f = \frac{0.316}{\mathbf{R}^{1/4}} \tag{6.7.10}$$

In rough pipes the term ϵ/D is called the *relative roughness*. Nikuradse [12] proved the validity of the relative-roughness concept by his tests on sand-roughened pipes. He used three sizes of pipes and glued sand grains ($\epsilon =$ diameter of the sand grains) of practically constant size to the interior walls so that he had the same values of ϵ/D for different pipes. These experiments (Fig. 6.20) show that for one value

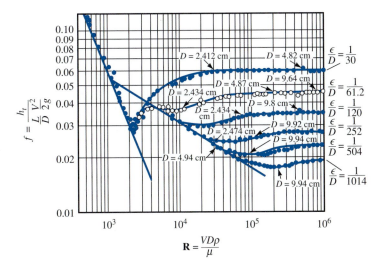

Figure 6.20   Nikuradse's sand-roughened-pipe tests.

of ϵ/D the f versus $\mathbf{R}$ curve is smoothly connected regardless of the actual pipe diameter. These tests did not permit variation of ϵ'/D or m but proved the validity of the equation

$$f = f\left(\mathbf{R}, \frac{\epsilon}{D}\right)$$

for one type of roughness.

Because of the extreme complexity of naturally rough surfaces, most of the advances in understanding the basic relations have been developed from experiments on artificially roughened pipes. Moody [13] has constructed one of the most convenient charts for determining friction factors in clean, commercial pipes. The chart (Fig. 6.21) is the basis for pipe-flow calculations in this chapter. The chart is a Stanton diagram that expresses f as a function of relative roughness and the Reynolds number. The values of absolute roughness of the commercial pipes are determined by experiments in which f and $\mathbf{R}$ are found and substituted into the Colebrook formula, Eq. (6.7.7), which closely represents natural pipe trends. These are listed in the table in the lower left-hand corner of Fig. 6.21. The Colebrook formula provides the shape of the $\epsilon/D =$ constant curves in the transition region.

The straight line marked "laminar flow" in Fig. 6.21 is the Hagen-Poiseuille equation. Equation (6.3.10b),

$$V = \frac{\Delta p r_0^2}{8\mu L}$$

can be transformed into Eq. (6.5.7) with $\Delta p = \gamma h_f$ and by solving for h_f

$$h_f = \frac{V 8\mu L}{\gamma r_0^2} = \frac{64\mu}{\rho D}\frac{L}{D}\frac{V}{2g} = \frac{64}{\rho D V/\mu}\frac{L}{D}\frac{V^2}{2g}$$

or

$$h_f = f\frac{L}{D}\frac{V^2}{2g} = \frac{64}{\mathbf{R}}\frac{L}{D}\frac{V^2}{2g} \tag{6.7.11}$$

from which

$$f = \frac{64}{\mathbf{R}} \tag{6.7.12}$$

This equation, which plots a straight line with slope -1 on a log-log chart, can be used for the solution of laminar flow problems in pipes. It applies to all roughnesses, as the head loss in laminar flow is independent of wall roughness. The Reynolds critical number is about 2000, and the *critical zone,* where the flow may be either laminar or turbulent, is about 2000 to 4000.

It should be noted that the relative-roughness curves $\epsilon/D \leq 0.001$ approach the smooth-pipe curve for decreasing Reynolds numbers. This can be explained by the presence of a laminar film at the wall of the pipe that decreases in thickness as the Reynolds number increases. For certain ranges of Reynolds numbers in the transition zone the film completely covers small roughness projections, and the pipe has a friction factor the same as that of a smooth pipe. For larger Reynolds numbers projections protrude through the laminar film, and each projection causes extra turbulence that increases the head loss. For the zone marked "complete turbulence, rough pipes," the film thickness is negligible compared with the height of roughness

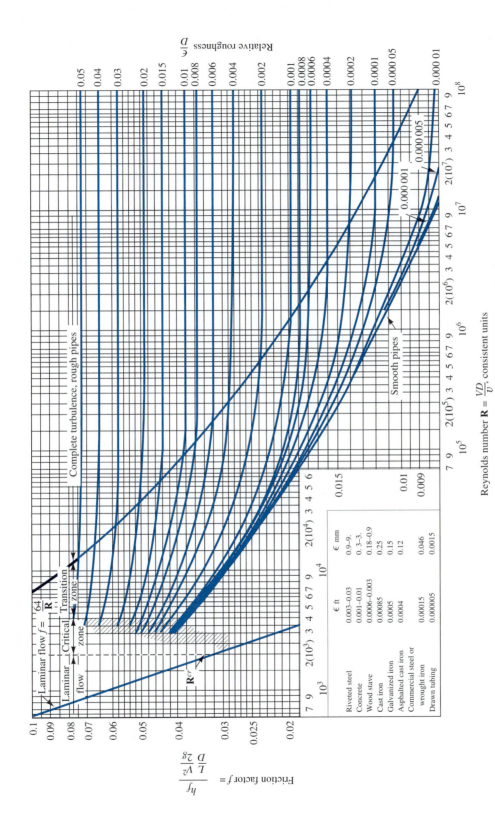

Figure 6.21 Moody diagram.

293

projections, and each projection contributes fully to the turbulence. Viscosity does not affect the head loss in this zone, as evidenced by the fact that the friction factor does not change with the Reynolds number. In this zone the loss follows the V^2 law; that is, it varies directly as the square of the velocity.

Simple Pipe Problems

By "simple pipe problems" we refer to pipes or pipelines in which pipe friction is the only loss. The pipe may be laid at any angle with the horizontal. Six variables enter into the problems (the fluid is treated as incompressible): Q, L, D, h_f, ν, and ϵ. In general, L, ν, and ϵ, the length, kinematic viscosity of fluid, and absolute roughness, are given or may be determined. Simple pipe problems can then be treated as three types:

Type	Given	Unknown
I	Q, L, D, ν, ϵ	h_f
II	h_f, L, D, ν, ϵ	Q
III	h_f, Q, L, ν, ϵ	D

In each case the Darcy-Weisbach equation, the continuity equation, and the Moody diagram are used to find the unknown quantity. In place of the Moody diagram, the following explicit formula [14, 15] for f can be utilized with the restrictions placed on it:

$$f = \frac{1.325}{[\ln(\epsilon/3.7D + 5.74/\mathbf{R}^{0.9})]^2} \qquad \begin{array}{c} 10^{-6} \leq \dfrac{\epsilon}{D} \leq 10^{-2} \\[2mm] 5000 \leq \mathbf{R} \leq 10^8 \end{array} \qquad (6.7.13)$$

This equation yields an f within about 1 percent of the Colebrook equation (6.7.7) and can be conveniently used with a hand calculator. Solutions to each of the three types of problems follow.

Type I: Solution for h_f.   With Q, ϵ, and D known, $\mathbf{R} = VD/\nu = 4Q/\pi D\nu$ and f can be looked up in Fig. 6.21 or calculated from Eq. (6.7.13). Substitution into Eq. (6.5.7) yields h_f, the energy loss due to flow through the pipe per unit weight of fluid.

Example 6.9

Determine the head (energy) loss for a flow of 140 L/s of oil, $\nu = 0.00001$ m²/s, through 400 m of 200-mm-diameter cast-iron pipe.

Solution

$$\mathbf{R} = \frac{4Q}{\pi D\nu} = \frac{4(0.140 \text{ m}^3/\text{s})}{\pi(0.2 \text{ m})(0.00001 \text{ m}^2/\text{s})} = 89,127$$

The relative roughness is $\epsilon/D = 0.25$ mm/200 mm $= 0.00125$. From Fig. 6.21, by interpolation, $f = 0.023$. Solving Eq. (6.7.13), $f = 0.0234$; hence,

$$h_f = f \frac{L}{D} \frac{V^2}{2g} = 0.023 \frac{400 \text{ m}}{0.2 \text{ m}} \left[\frac{0.14}{(\pi/4)(0.2 \text{ m})^2} \right]^2 \frac{1}{2(9.806 \text{ m/s}^2)} = 46.58 \text{m·N/N}$$

Type II: Solution for Discharge Q. In the second case, V and f are unknowns, and the Darcy-Weisbach equation and Moody diagram must be used simultaneously to find their values. Since ϵ/D is known, a value of f can be assumed by inspection of the Moody diagram. Substitution of this trial f into the Darcy-Weisbach equation produces a trial value of V, from which a trial Reynolds number is computed. With the Reynolds number an improved value of f is found from the Moody diagram. When f has been found correct to two significant figures, the corresponding V is the value sought and Q is determined by multiplying by the area.

Example 6.10

Water at 15°C flows through a 300-mm-diameter riveted steel pipe, $\epsilon = 3$ mm, with a head loss of 6 m in 300 m. Determine the flow.

Solution

The relative roughness is $\epsilon/D = 0.003/0.3 = 0.01$, and from Fig. 6.21, a trial f is taken as 0.04. By substituting into Eq. (6.5.7),

$$6 \text{ m} = 0.04 \frac{300 \text{ m}}{0.3 \text{ m}} \frac{(V \text{ m/s})^2}{2(9.806 \text{ m/s}^2)}$$

from which $V = 1.715$ m/s. From Appendix C, $\nu = 1.13 \times 10^{-6}$ m²/s, and so

$$\mathbf{R} = \frac{VD}{\nu} = \frac{(1.715 \text{ m/s})(0.30 \text{ m})}{1.13 \times 10^{-6} \text{ m}^2/\text{s}} = 455{,}000$$

From the Moody diagram $f = 0.038$, and

$$Q = AV = \pi(0.15 \text{ m})^2 \sqrt{\frac{(6 \text{ m})(0.3 \text{ m})(2)(9.806 \text{ m/s}^2)}{0.038(300 \text{ m})}} = 0.1245 \text{ m}^3/\text{s}$$

An explicit solution for discharge Q can be obtained from the Colebrook equation (6.7.7) and from the Darcy-Weisbach equation (6.5.7). From Eq. (6.5.7)

$$h_f = f \frac{L}{D} \frac{Q^2}{2g[(\pi/4)D^2]^2} \tag{6.7.14}$$

Solving for $1/\sqrt{f}$

$$\frac{1}{\sqrt{f}} = \frac{\sqrt{8}Q}{\pi \sqrt{gh_f D^5/L}}$$

Substitution of $1/\sqrt{f}$ into Eq. (6.7.7) and simplifying give

$$Q = -0.965D^2 \sqrt{\frac{gDh_f}{L}} \ln\left(\frac{\epsilon}{3.7D} + \frac{1.784\nu}{D\sqrt{gDh_f/L}}\right) \tag{6.7.15}$$

This equation, first derived by Swamee and Jain [15] is as accurate as the Colebrook equation and holds for the same range of values of ϵ/D and $\mathbf{R}$. Substitution of the variables from Ex. 6.10, $D = 0.3$ m, $g = 9.806$ m/s², $h_f/L = 0.02$, $\epsilon/D = 0.01$, and $\nu = 1.13 \times 10^{-6}$ m²/s, yields $Q = 0.1231$ m³/s.

Type III: Solution for Diameter D.   In the third case, with D unknown, there are three unknowns in Eq. (6.5.7), f, V, and D; two in the continuity equation, V and D; and three in the Reynolds number equation, V, D, and $\mathbf{R}$. The relative roughness also is unknown. Using the continuity equation to eliminate the velocity in Eq. (6.5.7) and in the expression for $\mathbf{R}$ simplifies the problem. Equation (6.5.7) becomes

$$h_f = f\frac{L}{D}\frac{Q^2}{2g(D^2\pi/4)^2}$$

or

$$D^5 = \frac{8LQ^2}{h_f g\pi^2}f = C_1 f \tag{6.7.16}$$

in which C_1 is the known quantity $8LQ^2/h_f g\pi^2$. As $VD^2 = 4Q/\pi$ from continuity,

$$\mathbf{R} = \frac{VD}{\nu} = \frac{4Q}{\pi\nu}\frac{1}{D} = \frac{C_2}{D} \tag{6.7.17}$$

in which C_2 is the known quantity $4Q/\pi\nu$. The solution is now effected by the following procedure:

1. Assume a value of f.
2. Solve Eq. (6.7.16) for D.
3. Solve Eq. (6.7.17) for $\mathbf{R}$.
4. Find the relative roughness ϵ/D.
5. With $\mathbf{R}$ and ϵ/D, look up a new f from Fig. 6.21.
6. Use the new f, and repeat the procedure.
7. When the value of f does not change in the first two significant figures, all equations are satisfied and the problem is solved.

Normally only one or two trials are required. Since standard pipe sizes are usually selected, the next larger size of pipe than that given by the computation is taken.

| Example 6.11 | Determine the size of clean wrought-iron pipe required to convey 4000-gpm (gallons per minute) of oil, $\nu = 0.0001$ ft^2/s; for 10,000 ft with a head loss of 75 ft·lb/lb. |

Solution

The discharge is

$$Q = \frac{4000 \text{ gpm}}{448.8 \text{ gpm/cfs}} = 8.93 \text{ cfs}$$

From Eq. (6.7.16)

$$D^5 = \frac{8(10,000)(8.93^2)}{75(32.2)(\pi^2)}f = 267.0f$$

and from Eq. (6.7.17)

$$\mathbf{R} = \frac{4(8.93)}{\pi(0.0001)}\frac{1}{D} = \frac{113,800}{D}$$

and from Fig. 6.21, $\epsilon = 0.00015$ ft.

If $f = 0.02$, then $D = 1.398$ ft, $\mathbf{R} = 81,400$, and $\epsilon/D = 0.00011$. From Fig. 6.21 $f = 0.019$. In repeating the procedure, $D = 1.382$, $\mathbf{R} = 82,300$, and $f = 0.019$. Therefore, $D = 1.382(12) = 16.6$ in. (1.38 ft).

Following Swamee and Jain [15], an empirical equation to determine diameter directly by using dimensionless relations and an approach similar to development of the Colebrook equation yields

$$D = 0.66 \left[\epsilon^{1.25} \left(\frac{LQ^2}{gh_f} \right)^{4.75} + \nu Q^{9.4} \left(\frac{L}{gh_f} \right)^{5.2} \right]^{0.04} \qquad \text{(6.7.18)}$$

Solution of Ex. 6.11 by use of Eq. (6.7.18) for

$$Q = 8.93 \text{ cfs} \qquad \epsilon = 0.00015 \text{ ft} \qquad L = 10,000 \text{ ft}$$
$$h_f = 75 \text{ ft·lb/lb} \qquad \nu = 0.0001 \text{ ft}^2/\text{s} \qquad g = 32.2 \text{ ft/s}^2$$

yields $D = 1.404$ ft.

Equation (6.7.18) is valid for the ranges

$$3 \times 10^3 \le \mathbf{R} \le 3 \times 10^8 \qquad 10^{-6} \le \frac{\epsilon}{D} \le 2 \times 10^{-2}$$

and will yield a D within 2 percent of the value obtained by the method using the Colebrook equation.

In each of the cases considered, the loss has been expressed in units of energy per unit weight. For horizontal pipes, this loss shows up as a gradual reduction in pressure along the line. For nonhorizontal cases, the energy equation (3.4.20) is applied to the two end sections of the pipe and the loss term is included; thus,

$$\frac{V_1^2}{2g} + \frac{p_1}{\gamma} + z_1 = \frac{V_2^2}{2g} + \frac{p_2}{\gamma} + z_2 + h_f \qquad \text{(3.4.20)}$$

in which the kinetic-energy correction factors have been taken as unity. The upstream section is given the subscript 1 and the downstream section the subscript 2. The total head at section 1 is equal to the sum of the total head at section 2 and all the head losses between the two sections.

In the preceding example, for $D = 16.6$ in., if the specific gravity is 0.85, $p_1 = 40$ psi, $z_1 = 200$ ft, and $z_2 = 50$ ft, determine the pressure at section 2.   **Example 6.12**

Solution

In Eq. (3.4.20) $V_1 = V_2$; hence,

$$\frac{40 \text{ psi}}{0.85(0.433 \text{ psi/ft})} + 200 \text{ ft} = \frac{p_2 \text{ psi}}{0.85(0.433 \text{ psi/ft})} + 50 \text{ ft} + 75 \text{ ft}$$

so $p_2 = 67.6$ psi.

EXERCISES

6.7.1   In turbulent flow a rough pipe has the same friction factor as a smooth pipe (*a*) in the zone of complete turbulence and rough pipes; (*b*) when the friction factor is independent of the Reynolds number; (*c*) when the roughness projections are much smaller than the thickness of the laminar film; (*d*) everywhere in the transition zone; (*e*) when the friction factor is constant.

6.7.2   The friction factor in turbulent flow in smooth pipes depends upon (*a*) V, D, ρ, L, and μ; (*b*) Q, L, μ, and ρ; (*c*) V, D, ρ, p, and μ; (*d*) V, D, μ, and ρ; (*e*) p, L, D, Q, and V.

6.7.3   In a *given* rough pipe, the losses depend upon (*a*) f and V; (*b*) μ and ρ; (*c*) $\mathbf{R}$; (*d*) Q only; (*e*) none of these answers.

6.7.4   In the complete-turbulence zone and rough pipes, (*a*) rough and smooth pipes have the same friction factor; (*b*) the laminar film covers the roughness projections; (*c*) the friction factor depends upon the Reynolds number only; (*d*) the head loss varies as the square of the velocity; (*e*) the friction factor is independent of the relative roughness.

6.7.5   The friction factor for flow of water at 60°F through a 2-ft-diameter cast-iron pipe with a velocity of 5 ft/s is (*a*) 0.013; (*b*) 0.017; (*c*) 0.019 (*d*) 0.021; (*e*) none of these answers.

6.7.6   The procedure to follow in solving for losses when Q, L, D, ν, and ϵ are given is to (*a*) assume an f, look up $\mathbf{R}$ on a Moody diagram, etc.; (*b*) assume an h_f, solve for f, check against $\mathbf{R}$ on a Moody diagram; (*c*) assume an f, solve for h_f, compute $\mathbf{R}$, etc.; (*d*) compute $\mathbf{R}$, look up f for ϵ/D, solve for h_f; (*e*) assume an $\mathbf{R}$, compute V, look up f, solve for h_f.

6.7.7   The procedure to follow in solving for discharge when h_f, L, D, ν, and ϵ are given is to (*a*) assume an f, compute V, $\mathbf{R}$, and ϵ/D, look up f, and repeat if necessary; (*b*) assume an $\mathbf{R}$, compute f, check ϵ/D, etc.; (*c*) assume a V, compute $\mathbf{R}$, look up f, compute V again, etc.; (*d*) solve Darcy-Weisbach for V and compute Q; (*e*) assume a Q, compute V and $\mathbf{R}$, look up f, etc.

6.7.8   The procedure to follow in solving for pipe diameter when h_f, Q, L, ν, and ϵ are given is to (*a*) assume a D, compute V, $\mathbf{R}$, ϵ/D, look up f, and repeat; (*b*) compute V from continuity, assume an f, and solve for D; (*c*) eliminate V in $\mathbf{R}$ and the Darcy-Weisbach equation, using continuity, assume an f, solve for D and $\mathbf{R}$, look up f, and repeat; (*d*) assume an $\mathbf{R}$ and an ϵ/D, look up f, solve the Darcy-Weisbach equation for V^2/D, and solve simultaneously with continuity for V and D, compute new $\mathbf{R}$, etc.; (*e*) assume a V, solve for D, $\mathbf{R}$, and ϵ/D, look up f, and repeat.

6.8   MINOR LOSSES

Losses which occur in pipelines because of bends, elbows, joints, valves, etc., are called *local* or *minor losses*. This is a misnomer because in many situations they are more important than the losses due to pipe friction considered so far in this chapter, but the name is conventional. In almost all cases the minor loss is determined by experiment. However, one important exception is the head loss due to a *sudden expansion* in a pipeline (Sec. 3.7).

Equation (3.7.10) can also be written as

$$h_e = K \frac{V_1^2}{2g} = \left[1 - \left(\frac{D_1}{D_2}\right)^2\right]^2 \frac{V_1^2}{2g} \qquad (6.8.1)$$

in which

$$K = \left[1 - \left(\frac{D_1}{D_2}\right)^2\right]^2 \qquad (6.8.2)$$

From Eq. (6.8.1) it is obvious that the head loss varies as the square of the velocity. This is substantially true for all minor losses in turbulent flow. A convenient method of expressing the minor losses in flow is by means of the coefficient K, usually determined by experiment.

If the sudden expansion is from a pipe to a reservoir, $D_1/D_2 = 0$ and the loss becomes $V_1^2/2g$, that is, the complete kinetic energy in the flow is converted into thermal energy.

The head loss due to gradual expansions (including pipe friction over the length of the expansion) has been investigated experimentally by Gibson [16], whose results are given in Fig. 6.22. Diffuser pipes similar to the one shown in Fig. 6.22 are commonly used for pressure recovery in fluid systems. In addition to being a function of the diameter ratio and expansion angle, as illustrated, the actual loss coefficient and the amount of pressure rise in the direction of flow depend on a number of other parameters [17]. Items of importance at a given diffuser section include velocity distribution, flow symmetry, boundary-layer thickness at the entrance, and free discharge or attached outlet pipe at the exit.

The head loss h_c due to a *sudden contraction* in the pipe cross section, illustrated in Fig. 6.23, is subject to the same analysis as the sudden expansion, provided that the amount of contraction of the jet is known. The process of converting pressure head into velocity head is very efficient; hence, the head loss from section 1 to the *vena contracta*† is small compared with the loss from section 0 to section 2, where velocity head is being reconverted into pressure head. By applying Eq. (3.7.10) to

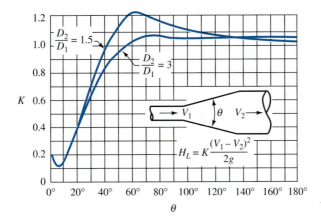

Figure 6.22   Loss coefficients for conical expansions.

| †The *vena contracta* is the section of greatest contraction of the jet.

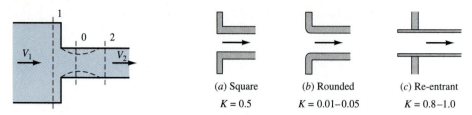

(a) Square
K = 0.5

(b) Rounded
K = 0.01–0.05

(c) Re-entrant
K = 0.8–1.0

Figure 6.23   Sudden contraction in a pipeline.

Figure 6.24   Head-loss coefficient K, in a number of velocity heads, $V^2/2g$, for a pipe entrance.

this expansion, the head loss is computed to be

$$h_c = \frac{(V_0 - V_2)^2}{2g}$$

With the continuity equation $V_0 C_c A_2 = V_2 A_2$, a contraction coefficient C_c can be defined, that is, the area of jet at section 0 divided by the area of section 2. The head loss is computed to be

$$h_c = \left(\frac{1}{C_c} - 1\right)^2 \frac{V_2^2}{2g} \tag{6.8.3}$$

The contraction coefficient C_c for water was determined by Weisbach [18] as follows:

A_2/A_1	0.1	0.2	0.3	0.4	0.5	0.6	0.7	0.8	0.9	1.0
C_c	0.624	0.632	0.643	0.659	0.681	0.712	0.755	0.813	0.892	1.00

The head loss at the entrance to a pipeline from a reservoir is usually taken as $0.5V^2/2g$ if the opening is square-edged. For well-rounded entrances, the loss is between $0.01V^2/2g$ and $0.05V^2/2g$ and can usually be neglected. For re-entrant openings, as with the pipe extending into the reservoir beyond the wall, the loss is taken as $1.0V^2/2g$ for thin pipe walls (Fig. 6.24).

Experimental data show wide variations in coefficients for special fittings. For example, values of K for a wide-open globe valve vary from 4 to 25, depending on the size and manufacturer. Representative values are given in Table 6.2.

Table 6.2   Representative head-loss coefficients, K, for various fittings [19,20]

Fittings	K
Globe valve (fully open)	10.0
Angle valve (fully open)	5.0
Swing check valve (fully open)	2.5
Gate valve (fully open)	0.19
Close return bend	2.2
Standard tee	1.8
Standard elbow	0.9
Medium sweep elbow	0.75
Long sweep elbow	0.60

Minor losses can be expressed in terms of the equivalent length L_e of pipe that has the same head loss in meter-newtons per newton (foot-pounds per pound) for the same discharge; thus,

$$f \frac{L_e}{D} \frac{V^2}{2g} = K \frac{V^2}{2g}$$

in which K may refer to one minor head loss or to the sum of several losses. Solving for L_e gives

$$L_e = \frac{KD}{f} \qquad\qquad (6.8.4)$$

For example, if the minor losses in a 12-in. pipeline add to $K = 20$, and if $f = 0.020$ for the line, then to the actual length of line can be added $20(1/0.020) = 1000$ ft, and this additional or equivalent length causes the same resistance to flow as the minor losses.

Find the discharge through the pipeline in Fig. 6.25 for $H = 10$ m and determine the head loss H for $Q = 60$ L/s.

Example 6.13

Solution

The energy equation applied between points 1 and 2, including all the losses, can be written as

$$H_1 + 0 + 0 = \frac{V_2^2}{2g} + 0 + 0 + \frac{1}{2}\frac{V_2^2}{2g} + f\frac{102 \text{ m}}{0.15 \text{ m}}\frac{V_2^2}{2g} + 2(0.9)\frac{V_2^2}{2g} + 10\frac{V_2^2}{2g}$$

in which the entrance loss coefficient is $\frac{1}{2}$, each elbow is 0.9, and the globe valve is 10. Then

$$H_1 = \frac{V_2^2}{2g}(13.3 + 680f)$$

When the head is given, this problem is solved as the second type of simple pipe problem. If $f = 0.022$,

$$10 = \frac{V_2^2}{2g}[13.3 + 680(0.022)]$$

and $V_2 = 2.63$ m/s. From Appendix C,

$$\nu = 1.01 \ \mu \text{ m}^2\text{/s} \qquad \frac{\epsilon}{D} = 0.0017 \qquad \mathbf{R} = \frac{(2.63 \text{ m/s})(0.15 \text{ m})}{1.01 \times 10^{-6} \text{ m}^2\text{/s}} = 391{,}000$$

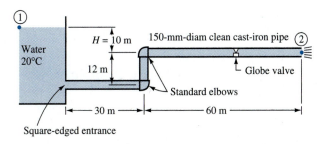

Figure 6.25   Pipeline with minor losses.

From Fig. 6.21, $f = 0.023$. Repeating the procedure gives $V_2 = 2.60$ m/s, $\mathbf{R} = 380,000$, and $f = 0.023$. The discharge is

$$Q = V_2 A_2 = (2.60 \text{ m/s})\frac{\pi}{4}(0.15 \text{ m})^2 = 45.9 \text{ L/s}$$

For the second part, with Q known, the solution is straightforward:

$$V_2 = \frac{Q}{A} = \frac{0.06 \text{ m}^3/\text{s}}{(\pi/4)(0.15 \text{ m})^2} = 3.40 \text{ m/s} \qquad \mathbf{R} = 505,000 \qquad f = 0.023$$

and

$$H_1 = \frac{(3.4 \text{ m/s})^2}{2(9.806 \text{ m/s}^2)}[13.3 + 680(0.023)] = 17.06 \text{ m}$$

With equivalent lengths [Eq. (6.8.4)] the value of f is approximated, say $f = 0.022$. The sum of minor losses is $K = 13.3$, in which the kinetic energy at 2 is considered a minor loss, yielding

$$L_e = \frac{13.3(0.15)}{0.022} = 90.7 \text{ m}$$

Hence, the total length of pipe is $90.7 + 102 = 192.7$ m. The first part of the problem is solved by

$$10 \text{ m} = f\frac{L + L_e}{D}\frac{V_2^2}{2g} = f\frac{192.7 \text{ m}}{0.15 \text{ m}}\frac{(V_2 \text{ m/s})^2}{2g \text{ m/s}^2}$$

If $f = 0.022$, $V_2 = 2.63$ m/s, $\mathbf{R} = 391,000$, and $f = 0.023$, then $V_2 = 2.58$ m/s and $Q = 45.6$ L/s. Normally, it is not necessary to use the new f to improve L_e.

Minor losses can be neglected when they constitute only 5 percent or less of the head losses due to pipe friction. The friction factor, at best, is subject to about 5 percent error, and it is meaningless to select values to more than three significant figures. In general, minor losses can be neglected when, on the average, there is a length of 1000 diameters between each minor loss. Complex pipe-flow situations are treated in Chap. 12.

The solution for head loss is direct, much as it is for a single pipe, since D, Q, ν, ϵ, L, and K are known; $\mathbf{R}$, ϵ/D, A, and f can be calculated; and

$$h_f = \frac{f}{D}\left(L + \frac{KD}{f}\right)\frac{Q^2}{2gA^2} \tag{6.8.5}$$

An iterative solution for the discharge proceeds as follows: Replacing L in Eq. (6.7.15) by $L + KD/f$ gives an equation for Q in terms of an unknown f. Let

$$Y = \sqrt{\frac{gDh_f}{L + KD/f}} = \sqrt{\frac{R_3}{1 + R_4/f}} \tag{6.8.6}$$

with $R_3 = gDh_f/L$ and $R_4 = KD/L$. Then Eq. (6.7.15) becomes

$$Q = -0.965D^2Y\left[\ln\left(\frac{\epsilon}{3.7D} + \frac{1.784\nu}{DY}\right)\right] = R_2Y\ln\left(R_1 + \frac{R_0}{Y}\right) \tag{6.8.7}$$

with $R_0 = 1.784\nu/D$, $R_1 = \epsilon/3.7D$, and $R_2 = -0.965D^2$. The Reynolds number is given by

$$\mathbf{R} = \frac{4Q}{\pi D\nu} = R_5 Q \qquad (6.8.8)$$

with $R_5 = 4/\pi D\nu$. The friction factor Eq. (6.7.13) becomes

$$f = \frac{R_7}{[\ln(R_1 + R_6/\mathbf{R}^{0.9})]^2} \qquad (6.8.9)$$

with $R_7 = 1.325$ and $R_6 = 5.74$. A value of f, say $f = 0.022$, is assumed, with R_0 to R_7 being the stored constants, and Eqs. (6.8.6) to (6.8.9) are solved in sequence. This procedure is continued until f and Q do not change (to four significant figures), and f and Q have been determined. Example 6.13 yields $f = 0.0231$ and $Q = 45.6$ L/s after three iterations. The procedure converges satisfactorily when KD/f is much larger than L.

An iterative solution for the diameter may proceed in the following manner: In Eq. (6.7.18) L can be replaced by $L + KD/f$, yielding

$$\mathbf{R} = \frac{R_5}{D} \qquad (6.8.10)$$

$$f = \frac{R_7}{[\ln(R_3/D + R_2/\mathbf{R}^{0.9})]^2} \qquad (6.8.11)$$

$$x = R_6 + \frac{R_4 D}{f} \qquad (6.8.12)$$

$$D = R_0(x^{4.75} + R_1 x^{5.2})^{0.04} \qquad (6.8.13)$$

where

$$R_0 = 0.66(\epsilon^{1.25}Q^{9.5})^{0.04} \qquad R_1 = \frac{\nu}{\epsilon^{1.25}Q^{0.1}} \qquad R_2 = 5.74$$

$$R_3 = \frac{\epsilon}{3.7} \qquad R_4 = \frac{K}{gh_f} \qquad R_5 = \frac{4Q}{\pi\nu}$$

$$R_6 = \frac{L}{gh_f} \qquad R_7 = 1.325$$

The procedure solves Eqs. (6.8.10) to (6.8.13) in sequence after a trial D has been selected.

Example 6.14

Assume that water at 10°C is to be conveyed at 300 L/s through 500 m of commercial steel pipe with a total head drop of 6 m. Minor losses are $12V^2/2g$. Determine the required diameter.

Solution

With $\nu = 1.308 \ \mu \ m^2/s$ and $\epsilon = 46 \ \mu \ m$, the constants in Eqs. (6.8.10) to (6.8.13) become $R_0 = 0.25351$, $R_1 = 0.38945$, $R_3 = 1.2432 \times 10^{-5}$, $R_4 = 0.20396$, $R_5 = 292{,}027$, and $R_6 = 8.4982$. Assume $D = 1$ m and then solve the equations in order, yielding $D = 438$ mm and $f = 0.0141$ after four iterations.

EXERCISES

6.8.1   The losses due to a sudden contraction are given by

(a) $\left(\dfrac{1}{C_c^2} - 1\right)\dfrac{V_2^2}{2g}$ (b) $(1 - C_c^2)\dfrac{V_2^2}{2g}$ (c) $\left(\dfrac{1}{C_c} - 1\right)^2\dfrac{V_2^2}{2g}$ (d) $(C_c - 1)^2\dfrac{V_2^2}{2g}$

(e) none of these answers.

6.8.2   The losses at the exit of a submerged pipe in a reservoir are (a) negligible; (b) $0.05(V^2/2g)$; (c) $0.5(V^2/2g)$; (d) $V^2/2g$; (e) none of these answers.

6.8.3   Minor losses usually may be neglected when (a) there is 100 ft of pipe between special fittings; (b) their loss is 5 percent or less of the friction loss; (c) there are 500 diameters of pipe between minor losses; (d) there are no globe valves in the line; (e) rough pipe is used.

6.8.4   The length of pipe ($f = 0.025$) in diameters, equivalent to a globe valve, is (a) 40; (b) 200; (c) 300; (d) 400; (e) not determinable, insufficient data.

PROBLEMS

6.1   Determine the formulas for shear stress on each plate and for the velocity distribution for flow in Fig. 6.3 when an adverse pressure gradient exists such that $Q = 0$.

6.2   In Fig. 6.3, with U positive as shown, find the expression for $d(p+\gamma h)/dl$ such that the shear is zero at the fixed plate. What is the discharge for this case?

6.3   In Fig. 6.26a $U = 0.7$ m/s. Find the rate at which oil is carried into the pressure chamber by the piston, the shear force, and the total force F acting on the piston.

6.4   Determine the force on the piston of Fig. 6.26a due to shear and the leakage from the pressure chamber for $U = 0$.

6.5   Find F and U in Fig. 6.26a such that no oil is lost through the clearance from the pressure chamber.

6.6   Derive an expression for the flow past a fixed cross section of Fig. 6.26b for laminar flow between the two moving plates.

6.7   In Fig. 6.26b, for $p_1 = p_2 = 0.1$ MPa, $U = 2V = 2$ m/s, $a = 1.5$ mm, and $\mu = 0.5$ P, find the shear stress at each plate.

6.8   Compute the kinetic-energy and momentum correction factors for laminar flow between fixed parallel plates.

6.9   Determine the formula for angle θ for fixed parallel plates so that laminar flow at constant pressure takes place.

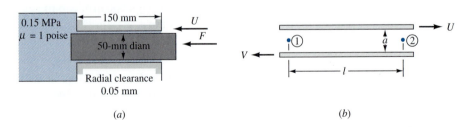

(a)   (b)

Figure 6.26   Problems 6.3 to 6.7.

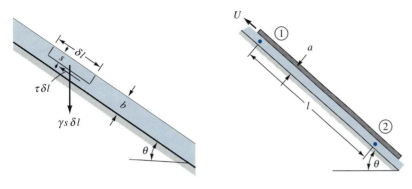

Figure 6.27   Problems 6.10 and 6.11.   **Figure 6.28**   Problems 6.13 and 6.14.

6.10   With a free body, as in Fig. 6.27, for uniform flow of a thin lamina of liquid down an inclined plane, show that the velocity distribution is

$$u = \frac{\gamma}{2\mu}(b^2 - s^2)\sin\theta$$

and that the discharge per unit width is

$$Q = \frac{\gamma}{3\mu}b^3 \sin\theta$$

6.11   Derive the velocity distribution of Prob. 6.10 by inserting into the appropriate equation prior to Eq. (6.2.2) the condition that the shear at the free surface must be zero.

6.12   A thin film of water flows over a parking lot of bottom slope 0.003. Find the depth if the flow is 0.08 L/s per meter of width and $\nu = 10^{-6}$ m^2/s.

6.13   In Fig. 6.28, $p_1 = 6$ psi, $p_2 = 8$ psi, $l = 4$ ft, $a = 0.006$ ft, $\theta = 30°$, $U = 4$ ft/s, $\gamma = 50$ lb/ft^3, and $\mu = 0.8$ P. Determine the tangential force per square foot exerted on the upper plate and its direction.

6.14   For $\theta = 90°$ in Fig. 6.28, what speed U is required for no discharge? $S = 0.87$, $a = 3$ mm, $p_1 = p_2$, and $\mu = 0.2$ kg/m·s.

6.15   Find the pressure gradient that results in zero shear stress at the lower wall, where $y = 0$, for flow between two parallel plates. The plates are spaced a apart and are horizontal. The top plate speed is U with respect to the stationary bottom plate.

6.16   What is the time rate of momentum and kinetic energy passing though a cross section that is normal to the flow if in Eq. (6.2.3) $Q = 0$?

6.17   A film of fluid 0.005 ft thick flows down a fixed vertical surface with a surface velocity of 2 ft/s. Determine the fluid viscosity. $\gamma = 55$ lb/ft^3.

6.18   Determine the momentum correction factor for laminar flow in a round tube.

6.19   Water at standard conditions is in laminar flow in a tube at pressure p_1 and diameter d_1. This tube expands to a diameter of $2d_1$ and pressure p_2, and the flow is again described by Eq. (6.3.6) some distance downstream of the expansion. Determine the force on the tube which results from the expansion.

6.20   At what distance r from the center of a tube of radius r_0 does the average velocity occur in laminar flow?

6.21   Determine the maximum wall shear stress for laminar flow in a tube of diameter D with fluid properties μ and ρ given.

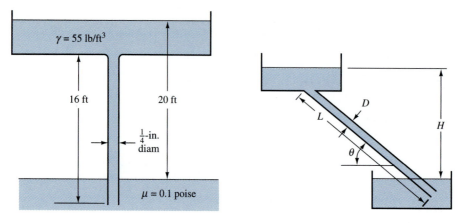

Figure 6.29   Problems 6.28 and 6.95.   **Figure 6.30**   Problems 6.29, 6.30, and 6.101.

6.22   Show that laminar flow between parallel plates can be used in place of flow through an annulus for 2 percent accuracy if the clearance is no more than 4 percent of the inner radius.

6.23   What are the losses per kilogram per meter of tubing for the flow of mercury at 35°C through a 0.6-mm-diameter tube at $\mathbf{R} = 1600$?

6.24   Determine the shear stress at the wall of a $\frac{1}{16}$-in.-diameter tube when water at 80°F flows through it with a velocity of 1.5 ft/s.

6.25   Determine the pressure drop per meter of 3-mm-ID horizontal tubing for flow of liquid, with $\mu = 60$ cP and $S = 0.83$ at $\mathbf{R} = 150$.

6.26   Glycerin at 100°F flows through a horizontal $\frac{3}{8}$-in.-diameter pipe with a pressure drop of 5 psi/ft. Find the discharge and the Reynolds number.

6.27   Calculate the diameter of vertical pipe needed for flow of liquid at $\mathbf{R} = 1400$ when the pressure remains constant, $\nu = 1.5\ \mu\text{m}^2/\text{s}$.

6.28   Calculate the discharge of the system in Fig. 6.29, neglecting all losses except through the pipe.

6.29   In Fig. 6.30, $H = 24$ m, $L = 40$ m, $\theta = 30°$, $D = 8$ mm, $\gamma = 10$ kN/m³, and $\mu = 0.08$ kg/m·s. Find the head loss per unit length of pipe and the discharge, in liters per minute.

6.30   In Fig. 6.30 and Prob. 6.29, find H if the velocity is 0.1 m/s.

6.31   Oil, $S = 0.85$ and $\mu = 0.06$ N·s/m², flows through an annulus with $a = 15$ mm and $b = 7$ mm. When the shear stress at the outer wall is 12 Pa, calculate (a) the pressure drop per meter for a horizontal system, (b) the discharge, in liters per hour, and (c) the axial force exerted on the inner tube per meter of length.

6.32   An annular pipe system is arranged so that flow occurs through the inner pipe then back through the annulus, having the same pressure drop per unit length. The flow is laminar, radius of inside pipe is 5 cm, and pipe thickness is 3 mm. Find the radius of the outside pipe.

6.33   What is the Reynolds number of flow of 0.3-m³/s oil, with $S = 0.86$ and $\mu = 0.025$ N·s/m², through a 450-mm-diameter pipe?

6.34   A small-diameter horizontal tube, $D = 3.0$ mm and $L = 40$ m, is connected to a supply reservoir as shown in Fig. 6.31. If $3(10)^{-5}$ m³ is captured at the outlet in 10 s, calculate the viscosity of the water.

Figure 6.31   Problem 6.34.

6.35 Show that the power input for laminar flow in a round tube is $Q \, \Delta p$ by integration of Eq. (6.2.6).

6.36 Use the one-seventh-power law of velocity distribution, $u/u_{max} = (y/r_0)^{1/7}$, to determine the mixing-length distribution l/r_0 in terms of y/r_0 from Eq. (6.4.12).

6.37 Plot a curve of $\epsilon/u_* r_0$ as a function of y/r_0 using Eq. (6.4.18) for velocity distribution in a pipe.

6.38 Find the value of y/r_0 in a pipe where the velocity equals the average velocity for turbulent flow.

6.39 A 4-cm-smooth, horizontal pipe transports 0.004 m³/s of water. $\nu = 1(10)^{-6}$ m²/s and $\tau_0 = 25.3$ Pa. Calculate (a) the shear velocity, u_*, (b) the maximum velocity, and (c) the pressure drop over 10 m of length.

6.40 Plot the velocity profiles for Prandtl's exponential velocity formula for values of n of $\frac{1}{7}$, $\frac{1}{8}$, and $\frac{1}{9}$.

6.41 The Chézy coefficient is 127 for flow in a rectangular channel 6 ft wide, 3 ft deep, and with a bottom slope of 0.0016. What is the discharge?

6.42 A rectangular channel 1 m wide, $\lambda = 0.005$, and $S = 0.0064$, carries 1 m³/s. Determine the velocity.

6.43 What is the value of the Manning roughness factor n in Prob. 6.42?

6.44 A rectangular brick-lined channel 6 ft wide and 4 ft deep carries 210 cfs. What slope is required for the channel?

6.45 The channel cross section shown in Fig. 6.32 is made of unplaned wood and has a slope of 0.001. What is the discharge?

6.46 A trapezoidal unfinished concrete channel carries water at a depth of 2 m. Its bottom width is 3 m, with side slopes of 1 horizontal to $1\frac{1}{2}$ vertical. For a bottom slope of 0.004 what is the discharge?

6.47 A trapezoidal channel with a bottom slope of 0.003, bottom width of 1.2 m, and side slopes of 2 horizontal to 1 vertical carries 6 m³/s at a depth of 1.2 m. What is the Manning roughness factor?

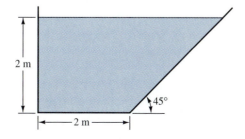

Figure 6.32   Problem 6.45.

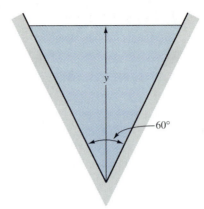

Figure 6.33   Problems 6.54 to 6.56

6.48   A trapezoidal earth canal, with a bottom width of 8 ft and side slopes of 2 horizontal to 1 vertical, must carry 280 cfs. The best velocity for nonscouring is 2.8 ft/s with this material. What bottom slope is required?

6.49   What diameter is required of a semicircular corrugated-metal channel that will carry 2 m³/s when its slope is 0.006?

6.50   A semicircular corrugated-metal channel 9 ft in diameter has a bottom slope of 0.004. What is its capacity when flowing full?

6.51   Calculate the depth of flow of 60 m³/s in a gravel trapezoidal channel with a bottom width of 4 m, side slopes of 3 horizontal to 1 vertical, and a bottom slope of 0.0009.

6.52   What is the velocity of flow of 260 cfs in a rectangular channel 12 ft wide? $S = 0.0049$ and $n = 0.014$.

6.53   A trapezoidal brick-lined channel must carry 35 m³/s a distance of 8 km with a head loss of 5 m. The bottom width is 4 m and has side slopes of 1 horizontal to 1 vertical. What is the velocity?

6.54   How does the discharge vary with depth in Fig. 6.33?

6.55   How does the velocity vary with depth in Fig. 6.33?

6.56   Determine the depth of flow in Fig. 6.33 for discharge of 12 cfs. The channel is made of riveted steel with a bottom slope of 0.01.

6.57   Determine the depth y (Fig. 6.34) for a maximum velocity and given n and S.

6.58   Determine the depth y (Fig 6.34) for a maximum discharge and given n and S.

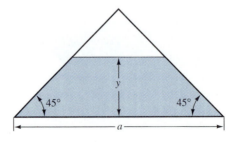

Figure 6.34   Problems 6.57 and 6.58.

6.59 A trapezoidal channel (Fig. 6.16) has $b = 4$ m, $m = 2$, $n = 0.014$, and $S = 0.0006$. Use a spreadsheet to determine the uniform flow depth for discharges of 60, 90, 120, and 150 m³/s.

6.60 A test on 300-mm-diameter pipe with water showed a gage difference of 280 mm on a mercury-water manometer connected to two piezometer rings 120 m apart. The flow was 0.23 m³/s. What is the friction factor?

6.61 Use the Blasius equation (6.7.10) to determine the friction factor in order to find the horsepower per mile required to pump 3.0-cfs liquid, $\nu = 3.3 \times 10^{-4}$ ft²/s and $\gamma = 55$ lb/ft³, through an 18-in. pipeline.

6.62 Determine the head loss per kilometer required to maintain a velocity of 3 m/s in a 10-mm-diameter pipe. $\nu = 4 \times 10^{-5}$ m²/s.

6.63 Fluid flows through a 10-mm-diameter tube at a Reynolds number of 1800. The head loss is 30 m in 100 m of tubing. Calculate the discharge in liters per minute.

6.64 What size galvanized-iron pipe is needed to be "hydraulically smooth" at $\mathbf{R} = 3.5 \times 10^5$? (A pipe is said to be hydraulically smooth when it has the same losses as a smoother pipe under the same conditions.)

6.65 Above what Reynolds number is the flow through a 3-m-diameter riveted steel pipe, $\epsilon = 3$ mm, independent of the viscosity of the fluid?

6.66 Determine the absolute roughness of a 1-ft-diameter pipe that has a friction factor $f = 0.03$ for $\mathbf{R} = 10^6$.

6.67 What diameter clean galvanized-iron pipe has the same friction factor for $\mathbf{R} = 100{,}000$ as a 300-mm-diameter cast-iron pipe?

6.68 Under what conditions do the losses in an artificially roughened pipe vary as the velocity raised to a power greater than two?

6.69 Why does the friction factor increase as the velocity decreases in laminar flow in a pipe?

6.70 Use Eq. (6.7.13) to calculate the friction factor for atmospheric air at 80°F, $V = 50$ ft/s, in a 3-ft-diameter galvanized pipe.

6.71 Water at 20°C is to be pumped through 1 km of 200-mm-diameter wrought-iron pipe at the rate of 60 L/s. Compute the head loss and power required.

6.72 If 16,000-ft³/min atmospheric air at 90°F is conveyed 1000 ft through a 4-ft-diameter wrought-iron pipe, what is the head loss in inches of water?

6.73 What power is required of a fan motor to circulate standard air in a wind tunnel at 500 km/h? The tunnel is a closed loop 60 m long, and it can be assumed to have a constant circular cross section with a 2-m diameter. Assume a smooth pipe.

6.74 Must provision be made to cool the air at some section of the tunnel described in Prob. 6.73? To what extent?

6.75 Assume that 2.0-cfs oil, $\mu = 0.16$ P and $\gamma = 54$ lb/ft³, is pumped through a 12-in. pipeline of cast iron. If each pump produces 80 psi, how far apart can they be placed?

6.76 A 60-mm-diameter smooth pipe 150 m long conveys 10-L/s water at 25°C from a water main, $p = 1.6$ MN/m², to the top of a building 25 m above the main. What pressure can be maintained at the top of the building?

6.77 For water at 150°F calculate the discharge for the pipe of Fig. 6.35.

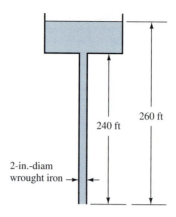

Figure 6.35   Problems 6.77 and 6.78.

6.78   In Fig. 6.35, how much power would be required to pump 160 gpm of water at 60°F from a reservoir at the bottom of the pipe to the reservoir shown?

6.79   Determine the discharge in the horizontal smooth pipe shown in Fig. 6.36. The liquid has $S = 0.83$ and $\nu = 6(10)^{-6}$ m²/s. The pressure in the air cap in the closed tank A is 162.8 kPa, and reservoir B is open to the atmosphere.

6.80   A 12-mm-diameter commercial steel pipe 15 m long is used to drain an oil tank. Determine the discharge when the oil level in the tank is 2 m above the exit end of the pipe. $\mu = 0.10$ P and $\gamma = 8$ kN/m³.

6.81   Two liquid reservoirs are connected by 200 ft of 2-in.-diameter smooth tubing. What is the flow rate when the difference in elevation is 50 ft? $\nu = 0.001$ ft²/s. Use the Moody diagram and Eq. (6.7.15).

6.82   For a head loss of 80-mm H_2O in a length of 200 m for flow of atmospheric air at 15°C through a 1.25-m-diameter duct, $\epsilon = 1$ mm, calculate the flow in cubic meters per minute. Use the Moody diagram and Eq. (6.7.15).

6.83   A gas of molecular weight 37 flows through a 24-in.-diameter galvanized duct at 90 psia and 100°F. The head loss per 100 ft of duct is 2-in. H_2O. What is the flow in slugs per hour? $\mu = 0.0194$ cP.

6.84   What is the power per kilometer required for a 70-percent efficient blower to maintain the flow of Prob. 6.82?

6.85   The 100-lb$_m$/min air required to ventilate a mine is admitted through 3000 ft of 12-in.-diameter galvanized pipe. Neglecting minor losses, what head, in inches of water, does a blower have to produce to furnish this flow? $p = 14$ psia and $t = 90°F$.

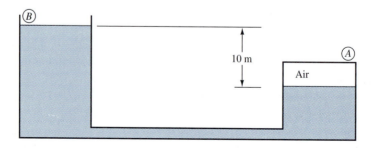

Figure 6.36   Problem 6.79.

6.86 In Fig. 6.30 $H = 20$ m, $L = 150$ m, $D = 50$ mm, $S = 0.85$, $\mu = 4$ cP, and $\epsilon = 1$ mm. Determine the discharge in newtons per second.

6.87 In a process, 10,000 lb/h of distilled water at 70°F is conducted through a smooth tube between two reservoirs 30 ft apart and having a difference in elevation of 4 ft. What size tubing is needed?

6.88 What size of new cast-iron pipe is needed to transport 400-L/s water at 25°C for 1 km with a head loss of 2 m? Use the Moody diagram and Eq. (6.7.18).

6.89 Two types of steel plates have surface roughnesses of $\epsilon_1 = 0.0003$ ft and $\epsilon_2 = 0.0001$ ft. The smooth plate costs 10 percent more. With an allowable stress in each of 10,000 psi, which plate should be selected to convey 100-cfs water at 200 psi with a head loss of 6 ft/mi?

6.90 An old pipe 2 m in diameter has a roughness of $\epsilon = 30$ mm. A 12-mm-thick lining would reduce the roughness to $\epsilon = 1$ mm. How much in annual pumping costs would be saved per kilometer of pipe for water at 20°C with discharge of 6 m³/s? The pumps and motors are 80-percent efficient, and power costs 4 cents per kilowatt-hour.

6.91 Calculate the diameter of a new wood-stave pipe in excellent condition needed to convey 300-cfs water at 60°F with a head loss of 1 ft per 1000 ft of pipe. Use the Moody diagram and Eq. (6.7.18).

6.92 Two oil reservoirs with a difference in elevation of 5 m are connected by 300 m of commercial steel pipe. The pipe must be what size to convey 50 L/s? $\mu = 0.05$ kg/m·s and $\gamma = 8$ kN/m³.

6.93 If 300-cfs air, $p = 16$ psia and $t = 70$°F, is to be delivered to a mine with a head loss of 3-in. H_2O per 1000 ft, what size galvanized pipe is needed?

6.94 Compare the smooth-pipe curve on the Moody diagram with Eq. (6.7.4) for $\mathbf{R} = 10^5$, 10^6, and 10^7.

6.95 Check the location of line $\epsilon/D = 0.0002$ on the Moody diagram with Eq. (6.7.7).

6.96 Show that Eq. (6.7.7) reduces to Eq. (6.7.4) when $\epsilon = 0$ and that it reduces to Eq. (6.7.6) when $\mathbf{R}$ is very large.

6.97 Equation (6.7.13) is an approximation of the Colebrook equation (6.7.7). Use a spreadsheet to determine the error in Eq. (6.7.13) for given ϵ/D and Reynolds number. Prepare a table with $\epsilon/D = 10^{-6}$, 10^{-5}, 10^{-4}, 10^{-3}, and 10^{-2} and with Reynolds number of 0.5×10^4, 10^4, 10^5, 10^6, 10^7, and 10^8.

6.98 Solve Eq. (6.7.15) by computer for Ex. 6.10.

6.99 Compute the losses, in joules per newton, caused by the flow of 25-m³/min air, $p = 1$ atm and $t = 20$°C, through a sudden expansion from 300- to 900-mm pipe. How much head would be saved by using a 10° conical diffuser?

6.100 Calculate the value of H in Fig. 6.37 for 125-L/s water at 15°C through commercial steel pipe. Include minor losses.

6.101 In Prob. 6.28 what would be the discharge if a globe valve were inserted into the line? Assume a smooth pipe and a well-rounded inlet, with $\mu = 1$ cP. Use the Moody diagram and the iterative method with Eqs. (6.8.6) to (6.8.9).

6.102 In Fig. 6.37 for $H = 3$ m, calculate the discharge of oil, $S = 0.8$ and $\mu = 7$ cP, through a smooth pipe. Include minor losses.

6.103 If a valve is placed in the line in Prob. 6.102 and adjusted to reduce the discharge by one-half, what is K for the valve and what is its equivalent length of pipe at this setting?

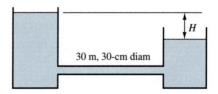

Figure 6.37   Problems 6.100, 6.102,
                       and 6.103.

6.104   A water line connecting two reservoirs at 70°F has 5000 ft of 24-in.-diameter
steel pipe, three standard elbows, a globe valve, and a re-entrant pipe entrance. What
is the difference in reservoir elevations for 20 cfs?

6.105   Determine the discharge in Prob. 6.104 if the difference in elevation is
40 ft.

6.106   What size commercial steel pipe is needed to convey 200-L/s water at 20°C
5 km with a head drop of 4 m? The line connects two reservoirs, has a re-entrant
entrance, a submerged outlet, four standard elbows, and a globe valve.

6.107   What is the equivalent length of 2-in.-diameter pipe, $f = 0.022$, for (*a*) a
re-entrant pipe entrance, (*b*) a sudden expansion from 2- to 4-in. diameter, (*c*) a globe
valve and a standard tee?

6.108   Find H in Fig. 6.38 for 200-gpm oil flow, $\mu = 0.1$ P and $\gamma = 60$ lb/ft^3, for
the angle valve wide open.

6.109   Find K for the angle valve in Prob. 6.108 for a flow of 10 L/s at the same H.

6.110   What is the discharge through the system of Fig. 6.38 for water at 25°C when
$H = 8$ m?

6.111   The pumping system shown in Fig. 6.39 has a pump head–discharge curve
$H = 40 - 24Q^2$ with head in meters and discharge in cubic meters per second. The pipe lengths include a correction for minor losses. Determine the flow
through the system in liters per second. If the efficiency of the pumping system is 72
percent, determine the power required. The pump requires a suction head of at least
$\frac{1}{2}$ atm to avoid cavitation. What is the maximum discharge and the power required
to reach this maximum rate of discharge?

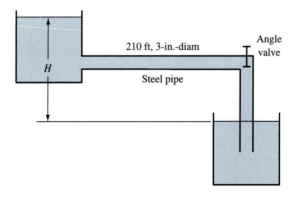

Figure 6.38   Problems 6.108 to 6.110.

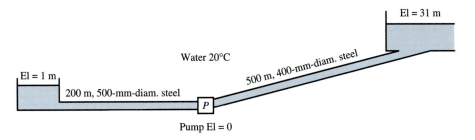

Figure 6.39   Problem 6.111.

REFERENCES

1.   O. Reynolds, "An Experimental Investigation of the Circumstances Which Determine Whether the Motion of Water Shall Be Direct or Sinuous, and of the Laws of Resistance in Parallel Channels," *Trans. R. Soc. Lond.,* vol. 174, 1883.

2.   H. L. Langhaar, "Steady Flow in the Transition Length of a Straight Tube," *J. Appl. Mech.,* vol. 9, pp. 55–58, 1942.

3.   H. Tennekes and J. L. Lumley, *A First Course in Turbulence,* The MIT Press, Cambridge, MA, 1972.

4.   O. Reynolds, "On the Dynamical Theory of Incompressible Viscous Fluids and the Determination of the Criterion," *Phil. Trans. R. Soc. Lond.,* Ser. A, 186, 123, 1895.

5.   L. Prandtl, *Essentials of Fluid Dynamics,* pp. 105–145, Hafner, New York, 1952.

6.   L. Prandtl, "Bericht über Untersuchungen zur ausgebildeten Turbulenz," *Z. Angew. Math. Mech.,* vol. 5, no. 2, p. 136, 1925.

7.   T. von Kármán, "Turbulence and Skin Friction," *J. Aeronaut. Sci.,* vol. 1, no. 1, p. 1, 1934.

8.   B. A. Bakhmeteff, *The Mechanics of Turbulent Flow,* Princeton University Press, Princeton, 1941.

9.   J. Nikuradse, "Gesetzmassigkeiten der turbulenten Strömung in glatten Rohren," *Ver. Dtsch. Ing. Forschungsh.,* vol. 356, 1932.

10.   C. F. Colebrook, "Turbulent Flow in Pipes, with Particular Reference to the Transition Region between the Smooth and Rough Pipe Laws," *J. Inst. Civ. Eng. Lond.,* vol. 11, pp. 133–156, 1938–1939.

11.   H. Blasius, "Das Ähnlichkeitsgesetz bei Reibungsvorgängen in Flüssigkeiten," *Ver. Dtsch. Ing. Forschungsh.,* vol. 131, 1913.

12.   J. Nikuradse, "Strömungsgesetze in rauhen Rohren," *Ver. Dtsch. Ing. Forschungsh.,* vol. 361, 1933.

13.   L. F. Moody, "Friction Factors for Pipe Flow," *Trans. ASME,* November 1944.

14.   S. W. Churchill, "Empirical Expressions for the Shear Stress in Turbulent Flow in Commercial Pipe," *A. I. Ch. E. J.,* vol. 19, no. 2, pp. 375–376, 1973.

15.   P. K. Swamee and A. K. Jain, "Explicit Equations for Pipe-Flow Problems," *J. Hydr. Div., Proc. ASCE,* pp. 657–664, May 1976.

16.   A. H. Gibson, "The Conversion of Kinetic to Pressure Energy in the Flow of Water through Passages Having Divergent Boundaries," *Engineering,* vol. 93, p. 205, 1912.

17.   D. S. Miller, *Internal Flow Systems,* Second Edition, BHRA Fluid Engineering, Cranfield, Bedford, UK, 1990.

18.   Julius Weisbach, *Die Experimental-Hydraulik,* p. 133, Englehardt, Freiburg, 1855.

19.   Crane Company, "Flow of Fluids through Valves, Fittings, and Pipe," *Tech. Pap.* 410, 1979.

20.   *Engineering Data Book,* Hydraulic Institute, 2nd Edition, Cleveland, 1990.

7

External Flows

Chapter 6 focused on confined flows, and one of the primary concerns involved energy dissipation associated with through-flow. This chapter shifts the focus to forces generated when a body moves through the fluid in which it is immersed. The same categorizations of laminar and turbulent flows are important, as are the boundary-layer and velocity distribution generated in the vicinity of the no-slip condition at the body surface. Drag and lift forces are the main concern. They are important in many applications—airfoils, automobiles, buildings, ships, to name a few—but they are also relevant in less obvious but equally important situations such as particle transport, erosion mechanisms, and pump impeller design.

The chapter begins with a qualitative description of drag and lift, followed by a more detailed discussion of boundary layers, both laminar and turbulent. The concept of flow separation is also needed as an integral part of drag and lift. The chapter ends with a section on unsteady or acceleration effects and introduces the concept of added mass.

7.1 SHEAR AND PRESSURE FORCES

Drag and lift are defined as the force components exerted on a body by the moving fluid parallel and normal, respectively, to the relative approach velocity. Both pressure and viscous stresses act on an immersed body and either or both contribute to the resultant forces. It is the dynamic action of the moving fluid that develops drag and lift; other forces such as the gravitational body force and buoyant force are not included in either drag or lift.

Flow over an airfoil provides an introductory example. Shear stresses can be visualized as acting along the surface of the foil (Fig. 7.1). The velocity of flow over the top of the foil is greater than the free-stream velocity; thus, by application of the Bernoulli equation, the pressure over the top is less than the free-stream pressure. The underside velocity is less than the free-stream velocity, resulting in a pressure greater than the free-stream velocity. This pressure gradient is largely responsible for the lift force on the airfoil, while the drag force is the result of both the pressure differences and the shear stresses.

Conceptually the drag and lift forces can be computed directly from the pressure and viscous stresses. Two-dimensional flow is visualized in Fig. 7.1, with the flow in the plane of the page. Our attention is directed to a slice of airfoil of unit thickness. By focusing on a differential surface area dA (Fig. 7.1), the drag force is given by

$$d(\text{Drag}) = p\,dA\sin\theta + \tau_0\,dA\cos\theta \qquad (7.1.1)$$

Integrating over the surface area, with positive pressure beneath the foil and negative above, the total drag force is obtained as

$$\text{Drag} = \int (p\sin\theta + \tau_0\cos\theta)\,dA \qquad (7.1.2)$$

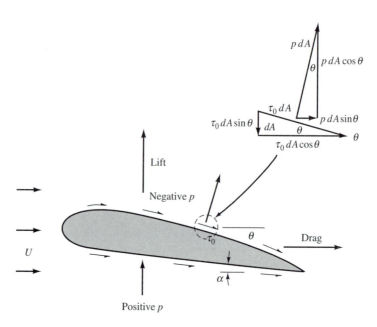

Figure 7.1   Viscous and pressure forces on an airfoil.

Similarly the elemental lift force

$$d(\text{Lift}) = p\,dA\cos\theta - \tau_0\,dA\sin\theta \tag{7.1.3}$$

yields the total lift after integration over the surface area

$$\text{Lift} = \int (p\cos\theta - \tau_0\sin\theta)\,dA \tag{7.1.4}$$

In the airfoil the shear stress contributes a very small portion of the total lift and can generally be neglected. The flow pattern around the immersed body controls the magnitude of the drag and lift forces, and the development of the boundary layer plays an important role in defining the forces. Unfortunately in most bodies the complete flow and pressure pattern cannot be calculated exactly, and Eqs. (7.1.2) and (7.1.4), although very formal, become of limited practical value. More commonly the forces are computed with empirically defined drag and lift coefficients.

A thin plate of unit width into the page is used to illustrate this. When the plate is in the direction of flow (Fig. 7.2a), the drag force can be computed with Eq. (7.1.2). With symmetrical flow over the plate the boundary layer develops as shown, and balanced pressures exist above and below. The pressure term drops out of both Eqs. (7.1.2) and (7.1.4). There is no lift on the plate since the flow is totally symmetrical. When the plate is placed at right angles to the flow (Fig. 7.2b), a positive pressure develops on the front side of the plate while a much lower pressure exists on the leeward side, as a result of the separation that occurs at the edges of the plate. In this case the first term in Eq. (7.1.2) is the sole contributor to the drag force on the plate. Again, due to symmetry the lift force is zero. Experiments are necessary to identify the drag force on the plate oriented as shown in Fig. 7.2b. In rounded objects the point at which the boundary layer separates from the object is not easy to predict, making the direct application of Eq. (7.1.2) difficult. The next sections illustrate cases for which computations are feasible and provide coefficients for empirical determinations for many other practical body shapes.

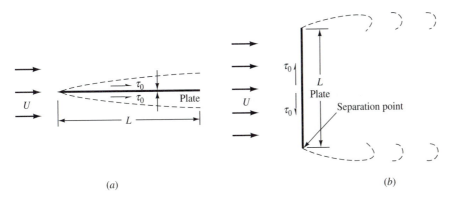

(a)   (b)

Figure 7.2   Flow past a flat plate.

7.2   BOUNDARY LAYER CONCEPTS: FLAT PLATES

In 1904 Prandtl [1]† developed the concept of the boundary layer. It provides an important link between ideal-fluid flow and real-fluid flow. *For fluids having relatively small viscosity, the effect of internal friction in a fluid is appreciable only in a narrow region surrounding the fluid boundaries.* From this hypothesis, the flow outside the narrow region near the solid boundaries can be considered as ideal flow or potential flow. Relations within the boundary-layer region can be computed from the general equations for viscous fluids, but use of the momentum equation permits the developing of approximate equations for boundary-layer growth and drag. In this section the boundary layer is described and the momentum equation is applied to it. Two-dimensional flow along a flat plate is studied by means of the momentum relations for both the laminar and the turbulent boundary layer. The phenomenon of separation of the boundary layer and formation of the wake is described.

Description of the Boundary Layer

When motion is started in a fluid having very small viscosity, the flow is essentially irrotational (Secs. 3.1 and 4.1) in the first instants. Since the fluid at the boundaries has zero velocity relative to the boundaries, there is a steep velocity gradient from the boundary into the flow. This velocity gradient in a real fluid sets up shear forces near the boundary that reduce the flow speed to that of the boundary. That fluid layer which has had its velocity affected by the boundary shear is called the *boundary layer.* The velocity in the boundary layer approaches the velocity in the main flow asymptotically. The boundary layer is very thin at the upstream end of a streamlined body at rest in an otherwise uniform flow. As this layer moves along the body, the continual action of shear stress tends to slow additional fluid particles down, causing the thickness of the boundary layer to increase with distance from the upstream point. The fluid in the layer is also subjected to a pressure gradient, imposed by and determined from the potential flow, that increases the momentum of the layer if the pressure decreases downstream and decreases its momentum if the pressure increases downstream (*adverse* pressure gradient). The flow outside the boundary layer may also *inject* momentum into the layer.

For smooth upstream boundaries the boundary layer starts out as a *laminar boundary layer* in which the fluid particles move in smooth layers. As the laminar boundary layer increases in thickness, it becomes unstable and finally transforms into a *turbulent boundary layer* in which the fluid particles move in haphazard paths, although their velocity has been reduced by the action of viscosity at the boundary. When the boundary layer has become turbulent, there is still a very thin layer next to the boundary that has laminar motion. It is called the *laminar sublayer.*

Various definitions of boundary-layer thickness δ have been suggested. The most basic definition refers to the displacement of the main flow due to slowing down of fluid particles in the boundary zone. This thickness δ_1, called the *displacement thickness,* is expressed by

$$U\delta_1 = \int_0^\delta (U - u)\,dy \qquad \text{(7.2.1)}$$

---

†Numbered references will be found at the end of this chapter.

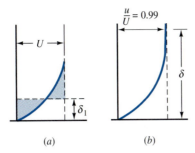

Figure 7.3 Definitions of boundary-layer thickness.

in which δ is that value of y at which $u = U$ in the undisturbed flow. In Fig. 7.3a the line $y = \delta_1$ is so drawn that the shaded areas are equal. This distance is, in itself, not the distance that is strongly affected by the boundary but is the distance the main flow must be shifted away from the boundary. In fact, that region is frequently taken as $3\delta_1$. Another definition, expressed by Fig. 7.3b, is the distance to the point where $u/U = 0.99$.

Momentum Equation Applied to the Boundary Layer

By utilizing von Kármán's method [2], the principle of momentum conservation can be applied directly to the boundary layer in steady flow along a flat plate. In Fig. 7.4 a control volume is taken enclosing the fluid above the plate, as shown, extending the distance x along the plate. In the y direction it extends a distance h so great that the velocity is undisturbed in the x direction, although some flow occurs along the upper surface, leaving the control volume.

The momentum equation for the x direction is

$$\Sigma F_x = \frac{\partial}{\partial t} \int_{cv} \rho u \, d\mathbb{V} + \int_{cs} \rho u \mathbf{v} \cdot d\mathbf{A}$$

and will be applied to the case of incompressible steady flow. The only force acting on the control volume is due to drag or shear at the plate, since the pressure is constant

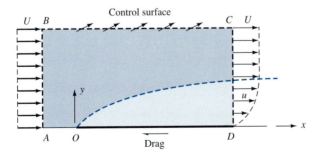

Figure 7.4 Control volume applied to fluid flowing over one side of a flat plate.

around the periphery of the control volume. For unit widths of plate normal to the paper

$$-\text{Drag} = \rho \int_0^h u^2 \, dy - \rho U^2 h + U\rho \int_0^h (U - u) \, dy$$

The first term on the right-hand side of the equation is the efflux of x momentum from CD, and the second term is the x-momentum influx through AB. The integral in the third term is the net volume influx through AB and CD, which, by continuity, must just equal the volume efflux through BC. It is multiplied by $U\rho$ to yield x-momentum efflux through BC. Combining the integrals gives

$$\text{Drag} = \rho \int_0^h u(U - u) \, dy \qquad \text{(7.2.2)}$$

The drag $D(x)$ on the plate is in the reverse direction, so that

$$D(x) = -\rho \int_0^h u(U - u) \, dy \qquad \text{(7.2.3)}$$

The drag on the plate can also be expressed as an integral of the shear stress along the plate as

$$D(x) = -\int_0^x \tau_0 \, dx \qquad \text{(7.2.4)}$$

Equating the last two expressions and then differentiating with respect to x leads to

$$\tau_0 = \rho \frac{\partial}{\partial x} \int_0^h u(U - u) \, dy \qquad \text{(7.2.5)}$$

which is the momentum equation for two-dimensional flow along a flat plate.

Calculations of boundary-layer growth, in general, are complex and require advanced mathematical treatment. The parallel-flow cases, laminar or turbulent, along a flat plate can be worked out approximately by using momentum methods that do not give any detail regarding the velocity distribution. In fact, a velocity distribution must be assumed. The results can be shown to agree closely with the more exact results obtained from general viscous flow differential equations.

For an assumed distribution which satisfies the boundary conditions $u = 0$, $y = 0$ and $u = U$, $y = \delta$, the boundary-layer thickness as well as the shear at the boundary can be determined. The velocity distribution is assumed to have the same form at each value of x,

$$\frac{u}{U} = F\left(\frac{y}{\delta}\right) = F(\eta) \qquad \eta = \frac{y}{\delta}$$

when δ is unknown.

Laminar Boundary Layer

For the laminar boundary layer Prandtl assumed that

$$\frac{u}{U} = F = \frac{3}{2}\eta - \frac{\eta^3}{2} \qquad 0 \leq y \leq \delta \qquad \text{and} \qquad F = 1 \qquad y \geq \delta$$

which satisfy the boundary conditions. Equation (7.2.5) can be rewritten

$$\tau_0 = \rho U^2 \frac{\partial \delta}{\partial x} \int_0^1 \left(1 - \frac{u}{U}\right)\frac{u}{U}\,d\eta$$

$$\tau_0 = \rho U^2 \frac{\partial \delta}{\partial x} \int_0^1 \left(1 - \frac{3}{2}\eta + \frac{\eta^3}{2}\right)\left(\frac{3}{2}\eta - \frac{\eta^3}{2}\right)d\eta = 0.139\rho U^2 \frac{\partial \delta}{\partial x}$$

At the boundary

$$\tau_0 = \mu \frac{\partial u}{\partial y}\Big|_{y=0} = \mu \frac{U}{\delta}\frac{\partial F}{\partial \eta}\Big|_{\eta=0} = \mu \frac{U}{\delta}\frac{\partial}{\partial \eta}\left(\frac{3}{2}\eta - \frac{\eta^3}{2}\right)\Big|_{\eta=0} = \frac{3\mu}{2}\frac{U}{\delta} \quad (7.2.6)$$

Equating the two expressions for τ_0 yields

$$\frac{3\mu}{2}\frac{U}{\delta} = 0.139\rho U^2 \frac{\partial \delta}{\partial x}$$

and rearranging gives

$$\delta\,d\delta = 10.78\,\frac{\mu\,dx}{\rho U}$$

since δ is a function of x only in this equation. Integrating gives

$$\frac{\delta^2}{2} = 10.78\,\frac{\nu}{U}\,x + \text{constant}$$

If $\delta = 0$ for $x = 0$, the constant of integration is zero. Solving for δ/x leads to

$$\frac{\delta}{x} = 4.65\sqrt{\frac{\nu}{Ux}} = \frac{4.65}{\sqrt{\mathbf{R}_x}} \quad (7.2.7)$$

in which $\mathbf{R}_x = Ux/\nu$ is a Reynolds number based on the distance x from the leading edge of the plate. This equation for boundary-layer thickness in laminar flow shows that δ increases as the square root of the distance from the leading edge.

Substituting the value of δ into Eq. (7.2.6) yields

$$\tau_0 = 0.322\sqrt{\frac{\mu\rho U^3}{x}} \quad (7.2.8)$$

The shear stress varies inversely as the square root of x and directly as the three-halves power of the velocity. The drag on one side of the plate, of unit width, is

$$\text{Drag} = \int_0^1 \tau_0\,dx = 0.644\sqrt{\mu\rho U^3 l} \quad (7.2.9)$$

Selecting other velocity distributions does not radically alter these results. The exact solution, worked out by Blasius [11] from the general equations of viscous motion, yields the coefficients 0.332 and 0.664 for Eqs. (7.2.8) and (7.2.9), respectively.

The drag can be expressed in terms of a drag coefficient C_D times the stagnation pressure $\rho U^2/2$ and the area of plate l (per unit breath),

$$\text{Drag} = C_D \frac{\rho U^2}{2}\,l$$

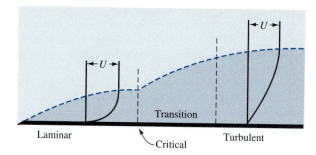

Figure 7.5   Boundary-layer growth; the vertical scale is greatly enlarged.

in which, for the laminar boundary layer,

$$C_D = \frac{1.328}{\sqrt{\mathbf{R}_l}}$$

(7.2.10)

and $\mathbf{R}_l = Ul/\nu$.

The boundary layer becomes turbulent when the Reynolds number for the plate has values between 500,000 and 1,000,000. Figure 7.5 indicates the growth and transition from the laminar to the turbulent boundary layer. The critical Reynolds number depends upon the initial turbulence of the fluid stream, the upstream edge of the plate, and the plate roughness.

Turbulent Boundary Layer

The momentum equation can be used to determine turbulent boundary-layer growth and shear stress along a smooth plate in a manner analogous to the treatment of the laminar boundary layer. The universal velocity-distribution law for smooth pipes, Eq. (6.4.20), provides the best basis, but the calculations are involved. A simpler approach is to use Prandtl's *one-seventh-power* law. It is $u/u_{max} = (y/r_o)^{1/7}$, in which y is measured from the wall of the pipe and r_o is the pipe radius. Applying it to flat plates produces

$$F = \frac{u}{U} = \left(\frac{y}{\delta}\right)^{1/7} = \eta^{1/7}$$

and

$$\tau_0 = 0.0228\rho U^2 \left(\frac{\nu}{U\delta}\right)^{1/4}$$

(7.2.11)

in which the latter expression is the shear stress at the wall of a smooth plate with a turbulent boundary layer.† The method used to calculate the laminar boundary layer gives

$$\tau_0 = \rho U^2 \frac{d\delta}{dx} \int_0^1 (1 - \eta^{1/7})\eta^{1/7}\, d\eta = \frac{7}{72}\rho U^2 \frac{d\delta}{dx}$$

(7.2.12)

---

†Equation (7.2.11) is obtained from the following pipe equations: $\tau_0 = \rho f V^2/8$, $f = 0.316\mathbf{R}^{1/4}$ (Blasius equation), $\mathbf{R} = V 2 r_0 \rho/\mu$, and $V = u_m/1.235$. To transfer to the flat plate, assume $r_0 \sim \delta$ and $u_m \sim U$.

By equating the expressions for shear stress, the differential equation for boundary-layer thickness δ is obtained as

$$\delta^{1/4} d\delta = 0.234 \left(\frac{\nu}{U}\right)^{1/4} dx$$

After integrating and then assuming that the boundary layer is turbulent over the whole length of the plate so that the initial conditions $x = 0$, $\delta = 0$ can be used,

$$\delta^{5/4} = 0.292 \left(\frac{\nu}{U}\right)^{1/4} x$$

Solving for δ gives

$$\delta = 0.37 \left(\frac{\nu}{U}\right)^{1/5} x^{4/5} = \frac{0.37x}{(Ux/\nu)^{1/5}} = \frac{0.37x}{\mathbf{R}_x^{1/5}} \tag{7.2.13}$$

The thickness increases more rapidly in the turbulent boundary layer. In it the thickness increases as $x^{4/5}$, but in the laminar boundary layer δ varies as $x^{1/2}$.

To determine the drag on a smooth, flat plate, δ is eliminated in Eqs. (7.2.11) and (7.2.13), and

$$\tau_0 = 0.029\rho U^2 \left(\frac{\nu}{Ux}\right)^{1/5} \tag{7.2.14}$$

The drag per unit width on one side of the plate is

$$\text{Drag} = \int_0^l \tau_0 \, dx = 0.036\rho U^2 l \left(\frac{\nu}{Ul}\right)^{1/5} = \frac{0.036\rho U^2 l}{\mathbf{R}_l^{1/5}} \tag{7.2.15}$$

In terms of the drag coefficient,

$$C_D = 0.072 \mathbf{R}_l^{-1/5} \tag{7.2.16}$$

in which $\mathbf{R}_l$ is the Reynolds number based on the length of plate.

The above equations are valid only for the range in which the Blasius resistance equation holds. For larger Reynolds numbers in smooth-pipe flow the exponent in the velocity-distribution law is reduced. For $\mathbf{R} = 400{,}000$, $n = \frac{1}{8}$, and for $\mathbf{R} = 4{,}000{,}000$, $n = \frac{1}{10}$. The drag law, Eq. (7.2.15), is valid for a range

$$5 \times 10^5 < \mathbf{R}_l < 10^7$$

Experiment shows that the drag is slightly higher than is predicted by Eq. (7.2.16),

$$C_D = 0.074 \mathbf{R}_l^{-1/5} \tag{7.2.17}$$

The boundary layer is actually laminar along the upstream part of the plate. Prandtl [3] has subtracted the drag from the equation for the upstream end of the plate up to the critical Reynolds number and then added the drag as given by the laminar equation for this portion of the plate, producing

$$C_D = 0.074 \mathbf{R}_l^{-1/5} - \frac{1700}{\mathbf{R}_l} \qquad 5 \times 10^5 < \mathbf{R}_l < 10^7 \tag{7.2.18}$$

In Fig. 7.6 a log-log plot of C_D versus $\mathbf{R}_l$ shows the trend of the drag coefficients.

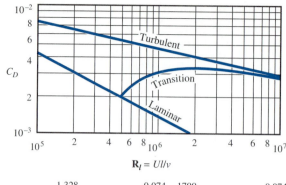

$$\text{Laminar } C_D = \frac{1.328}{\sqrt{\mathbf{R}_l}}, \text{ transition } C_D = \frac{0.074}{\mathbf{R}_l^{1/5}} - \frac{1700}{\mathbf{R}_l}, \text{ turbulent } C_D = \frac{0.074}{\mathbf{R}_l^{1/5}}$$

Figure 7.6   The drag law for smooth plates.

Use of the logarithmic velocity distribution, Eq. (6.4.18), produces

$$C_D = \frac{0.455}{(\log \mathbf{R}_l)^{2.58}} \qquad 10^6 < \mathbf{R} < 10^9 \qquad \text{(7.2.19)}$$

in which the constant term has been selected for best agreement with experimental results.

Example 7.1   **A** smooth, flat plate 3 m wide and 30 m long is towed through still water at 20°C with a speed of 6 m/s. Determine the drag on one side of the plate and the drag on the first 3 m of the plate.

Solution

For the whole plate

$$\mathbf{R} = \frac{Vl}{\nu} = \frac{(6 \text{ m/s})(30 \text{ m})}{1.007 \times 10^{-6} \text{ m}^2/\text{s}} = 1.787 \times 10^8$$

From Eq. (7.2.19)

$$C_D = \frac{0.455}{[\log(1.787 \times 10^8)]^{2.58}} = 0.00196$$

The drag on one side is

$$\text{Drag} = C_D b l \rho \frac{U^2}{2} = 0.00196(3 \text{ m})(30 \text{ m})(998.2 \text{ kg/m}^3)\frac{(6 \text{ m/s})^2}{2} = 3169 \text{ N}$$

in which b is the plate width and ν and ρ are from Table C.1. If the critical Reynolds number occurs at 5×10^5, the length l_0 to the transition is

$$\frac{(6 \text{ m/s})(l_0 \text{ m})}{1.007 \times 10^{-6} \text{ m}^2/\text{s}} = 5 \times 10^5 \qquad l_0 = 0.084 \text{ m}$$

For the first 3 m of the plate $\mathbf{R}_l = 1.787 \times 10^7$ and, using Eq. (7.2.19) again,

$$\text{Drag} = C_D b l \rho \frac{U^2}{2} = \frac{0.455(3 \text{ m})}{[\log(1.787 \times 10^7)]^{2.58}}(3 \text{ m})(998.2 \text{ kg/m}^3)\frac{(6 \text{ m/s})^2}{2} = 443 \text{ N}$$

Calculation of the turbulent boundary layer over rough plates proceeds in similar fashion, starting with the rough-pipe tests using sand roughnesses. At the upstream end of the flat plate the flow may be laminar; then, in the turbulent boundary layer, where the boundary layer is still thin and the ratio of roughness height to boundary-layer thickness ϵ/δ is significant, the region of fully developed roughness occurs and the drag is proportional to the square of the velocity. For long plates this region is followed by a transition region where ϵ/δ becomes increasingly smaller, and eventually the plate becomes hydraulically smooth, that is, the loss would not be reduced by reducing the roughness. Prandtl and Schlichting [4] have carried through these calculations, which are too complicated for reproduction here.

EXERCISES

7.2.1 The displacement thickness of the boundary layer is (a) the distance from the boundary affected by boundary shear; (b) one-half the actual thickness of the boundary layer; (c) the distance to the point where $u/U = 0.99$; (d) the distance the main flow is shifted; (e) none of these answers.

7.2.2 The shear stress at the boundary of a flat plate is

(a) $\dfrac{\partial p}{\partial x}$ (b) $\mu\dfrac{\partial u}{\partial y}\Big|_{y=0}$ (c) $\rho\dfrac{\partial u}{\partial y}\Big|_{y=0}$ (d) $\mu\dfrac{\partial u}{\partial y}\Big|_{y=\delta}$ (e) none of these answers.

7.2.3 Which of the following velocity distributions u/U satisfies the boundary conditions for flow along a flat plate if $\eta = y/\delta$? (a) e^{η}; (b) $\cos(\pi\eta/2)$; (c) $\eta - \eta^2$; (d) $2\eta - \eta^3$; (e) none of these answers.

7.2.4 The drag coefficient for a flat plate ($D =$ drag) is (a) $2D/\rho U^2 l$; (b) $\rho U l/D$; (c) $\rho U l/2D$; (d) $\rho U^2 l/2D$; (e) none of these answers.

7.2.5 The laminar-boundary-layer thickness varies as (a) $1/x^{1/2}$; (b) $x^{1/7}$; (c) $x^{1/2}$; (d) $x^{6/7}$; (e) none of these answers.

7.2.6 The turbulent-boundary-layer thickness varies as (a) $1/x^{1/5}$; (b) $x^{1/5}$; (c) $x^{1/2}$; (d) $x^{4/5}$; (e) none of these answers.

7.2.7 In flow along a rough plate the order of flow type from upstream to downstream is (a) laminar, fully developed wall roughness, transition region, hydraulically smooth; (b) laminar, transition region, hydraulically smooth, fully developed wall roughness; (c) laminar, hydraulically smooth, transition region, fully developed wall roughness; (d) laminar, hydraulically smooth, fully developed wall roughness, transition region; (e) laminar, fully developed wall roughness, hydraulically smooth, transition region.

7.3 FLOW AND DRAG: SPHERES

Horizontal Flow

Another external flow of considerable importance is flow over spheres. Spheres are used as surrogates for the irregularly shaped particles which comprise (among many examples) sediment transport, fluidized bed reactors, airborne dust and smog pollutants, and wastewater treatment plant processes. Flow and corresponding drag calculations were originally done by Stokes in 1851 [5] with further fluid mechanics

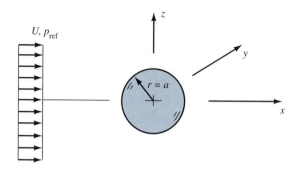

Figure 7.7   Low velocity flow around a sphere.

elaboration reported upon in Schlichting [11]. The most elementary exact solution is achieved by assuming steady, very slow flow with Reynolds number (UD/ν) based upon an equivalent spherical diameter, D (based upon volume conservation), of less than 1. This assumption ensures that the streamlines very near the surface of the sphere stay *attached* or follow the shape of the sphere. If the flow over the sphere is horizontal (Fig. 7.7), then gravity forces are considered unimportant and the Navier-Stokes equations [Eqs. (4.4.11)] reduce to

$$\nabla p = \mu \nabla^2 \mathbf{v} \tag{7.3.1}$$

The boundary conditions are the no-slip condition on the wall ($r = a$) and no flow normal to the wall ($r = a$). The velocity and pressure fields corresponding to these conditions are

$$u = U\left[\frac{3}{4}\frac{ax^2}{r^3}\left(\frac{a^2}{r^2} - 1\right) - \frac{1}{4}\frac{a}{r}\left(3 + \frac{a^2}{r^2}\right) + 1\right] \tag{7.3.2a}$$

$$v = U\left[\frac{3}{4}\frac{axy}{r^3}\left(\frac{a^2}{r^2} - 1\right)\right] \tag{7.3.2b}$$

$$w = U\left[\frac{3}{4}\frac{axz}{r^3}\left(\frac{a^2}{r^2} - 1\right)\right] \tag{7.3.2c}$$

$$p = p_{ref} - \frac{3}{2}\mu U\frac{ax}{r^3} \tag{7.3.2d}$$

Here the origin of the axes is the center of the sphere with the x axis coinciding with the direction of the free-stream velocity, U. By the use of Eqs. (7.3.2a–d) above and the definitions for drag [Eqs. (7.1.1) and (7.1.2)] and lift [Eqs. (7.1.3) and (7.1.4)], it can be shown that the net lift on the sphere is zero because the flow and pressure fields are symmetric above and below the centerline. The drag on the sphere is given by

$$\text{Drag} = \text{Form drag} + \text{Skin friction drag} = \mu \pi DU + 2\mu \pi DU = 3\pi \mu DU \tag{7.3.3}$$

It is seen in Eq. (7.3.3) that there are two components to the total drag: *form drag* and *skin friction drag*. The form drag is associated with the total pressure drop or

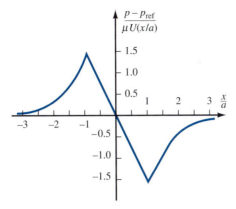

Figure 7.8 Pressure distribution for horizontal flow around a sphere.

pressure gradient between the front (upstream) and back (downstream) of the sphere. Figure 7.8 contains a plot of the normalized pressure field down the centerline of the sphere. The large pressure difference between the stagnation point at $x = r = a$ and the back of the sphere is the origin of the form drag. The skin friction component results from the viscous shear acting on the wall of the sphere as the flow passes by. While the ratio of the form to skin friction drag for the sphere is 1:2, many flows are such that the form drag dominates the skin friction. The geometry of the object and the Reynolds number determine the relative ratio.

Using the drag coefficient concept introduced in the previous section, it is easy to show that

$$\text{Drag} = 3\mu\pi DU = C_D\rho A \frac{U^2}{2} \tag{7.3.4}$$

or that for the slow flow over a sphere,

$$C_D = 24/\mathbf{R} \tag{7.3.5}$$

where the Reynolds number is defined based upon the particle-equivalent spherical diameter, D.

Settling Velocity

When a particle is composed of material with specific gravity greater than that of the displaced fluid, its absolute weight will be greater than the weight of displaced fluid and the particle will sink or settle due to gravity. The *settling velocity* is useful in determining the viscosity of a fluid, the design of settling basins for separating solid particles from the fluid, and desilting river flow. If the settling velocity, or terminal velocity, w_t, is used in the Reynolds number definition and if the resulting Reynolds number is still less than one, then a momentum force balance gives

$$\frac{4}{3}\pi\left(\frac{D}{2}\right)^3\gamma_f + 3\mu\pi Dw_t = \frac{4}{3}\pi\left(\frac{D}{2}\right)^3\gamma_s \tag{7.3.6}$$

where γ_f is the specific weight of the fluid and γ_s is the specific weight of the particle.

The primary assumption in Eq. (7.3.6) is that all the acceleration required for the particle to attain speed w_t has already occurred and the particle is traveling at a steady velocity. Solving for w_t gives

$$w_t = \frac{D^2}{18\mu}(\gamma_s - \gamma_f) \qquad (7.3.7)$$

In terms of the drag coefficient formulation

$$w_t^2 = \frac{4}{3}\frac{D}{\rho C_D}(\gamma_s - \gamma_f) \qquad (7.3.8)$$

For Stokes flow, defined as $\mathbf{R} < 1$, $C_D = 24/\mathbf{R}$, Eq. (7.3.7) is recovered exactly.

7.4   THE EFFECT OF PRESSURE GRADIENTS: SEPARATION AND WAKES

Along a flat plate the boundary layer continues to grow in the downstream direction, regardless of the length of the plate, when the pressure gradient remains zero. With the pressure decreasing in the downstream direction, as in the conical reducing section, the boundary layer tends to be reduced in thickness.

For *adverse* pressure gradients, that is, with pressure increasing in the downstream direction, the boundary layer thickens rapidly. The adverse gradient and the boundary shear decrease the momentum in the boundary layer, and if both act over a sufficient distance, they cause the boundary layer to come to rest. This phenomenon is called *separation*. Figure 7.9a illustrates this case. The boundary streamline must leave the boundary at the separation point, and downstream from this point the adverse pressure gradient causes backflow near the wall. This region downstream from the streamline that separates from the boundary is known as the *wake*. The effect of separation is to decrease the net amount of flow work that can be done by a fluid element on the surrounding fluid at the expense of its kinetic energy, with the net result that pressure recovery is incomplete and flow losses (drag) increase. Figures 7.9b and c illustrate actual flow cases, the first with a very small adverse pressure gradient, which causes thickening of the boundary layer, and the second with a large diffuser angle, which causes separation and backflow near the boundaries.

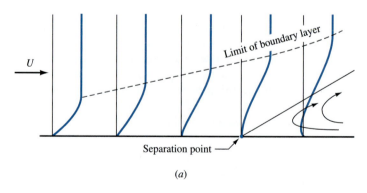

(*a*)

Figure 7.9   (a) Effect of adverse pressure gradient on boundary-layer separation.

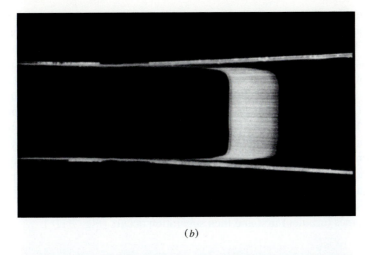

(b)

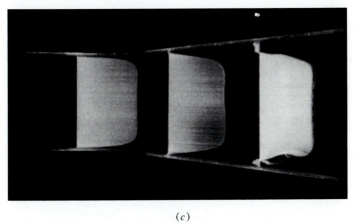

(c)

Figure 7.9
(continued)

(b) Boundary-layer growth in a small-angle diffuser. (c) Boundary-layer separation in a large-angle diffuser. *[Parts (b) and (c) from the film "Fundamental of Boundary Layers," by the National Committee for Fluid Mechanics Films and the Educational Development Center.]*

As discussed in Sec. 7.1 drag and lift have two components, form drag and skin friction or *viscous drag.* Separation and the wake that accompanies the phenomenon have a profound influence on the form drag on bodies. If separation in flow over a body can be avoided, the boundary layer remains thin and the pressure reduction in the wake is avoided, thereby minimizing pressure drag. Rounding the forward face of bodies to reduce the opportunity for flow separation at sharp edges is effective. Even more important is streamlining the trailing portion of the body (Fig. 7.10) to assure that the separation point will occur as far downstream along the body as possible.

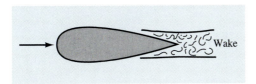

Figure 7.10   Streamlined body.

The laminar-versus-turbulent nature of the boundary layer is also of importance in influencing the position of separation point. Greater momentum transfer within a turbulent boundary layer requires a larger adverse pressure gradient to cause separation than within a more orderly laminar flow. Flow around a sphere can be used to illustrate this (Fig. 7.11). For very small Reynolds numbers, $UD/\nu < 1$, the flow is everywhere nonturbulent, and the drag is referred to as *deformation drag*. Stokes law [Eq. (7.3.4)] gives the drag force for this case. For large Reynolds numbers the flow can be considered potential flow except in the boundary layer and the wake. The boundary layer forms at the forward stagnation point and is generally laminar. In the laminar boundary layer an adverse pressure gradient causes separation more readily than in a turbulent boundary layer because of the small amount of momentum brought into the laminar layer. If separation occurs in the laminar boundary layer, the location is farther upstream on the sphere (Fig. 7.11*b*) than it is when the boundary layer becomes turbulent first and then separation occurs (Fig. 7.11*c*).

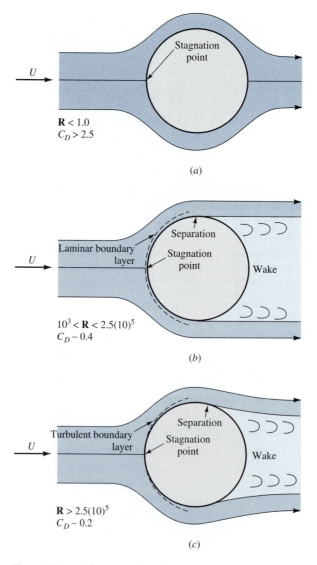

Figure 7.11   Flow around a sphere.

In Fig. 7.12 this is graphically portrayed by the photographs of the two spheres dropped into water at 25 ft/s. In Fig. 7.12a separation occurs in the laminar boundary layer that forms along the smooth surface and causes a very large wake with a resulting large pressure drag. In Fig. 7.12b the nose of the sphere, roughened by sand glued to it, induced an early transition to turbulent boundary layer before separation occurred. The high-momentum transfer in the turbulent boundary layer delayed the separation so that the wake is substantially reduced, resulting in a total drag on the sphere less than half that occurring in Fig. 7.12a. The importance of the roughened surface of a golf ball or a tennis ball or the seam on a baseball becomes clear in the light of this discussion. A graph of the drag coefficient versus Reynolds number is presented in the next section for spheres.

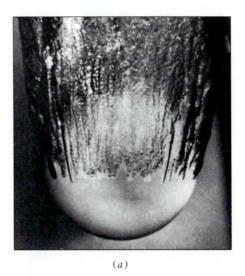

(a)

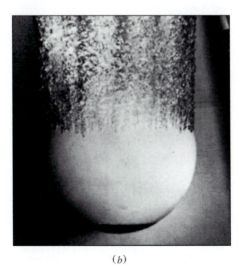

(b)

Figure 7.12   Shift in separation point due to induced turbulence: (a) 8.5-in. bowling ball with a smooth surface and 25-ft/s entry velocity into water; (b) the same ball with a 4-in.-diameter patch of sand on the nose. *(Official U.S. Navy photograph made at Naval Ordnance Test Station, Pasadena Annex.)*

EXERCISES

7.4.1 Separation is caused by (*a*) reduction of the pressure to vapor pressure; (*b*) reduction of the pressure gradient to zero; (*c*) an adverse pressure gradient; (*d*) the boundary-layer thickness reducing to zero; (*e*) none of these answers.

7.4.2 Separation occurs when (*a*) the cross section of a channel is reduced; (*b*) the boundary layer comes to rest; (*c*) the velocity of sound is reached; (*d*) the pressure reaches a minimum; (*e*) a valve is closed.

7.4.3 The wake (*a*) is a region of high pressure; (*b*) is the principal cause of skin friction; (*c*) always occurs when deformation drag predominates; (*d*) always occurs after a separation point; (*e*) is none of these answers.

7.5   DRAG ON IMMERSED BODIES

The drag formulae for flat plates and spheres were found by use of some rather strict assumptions or limitations. As seen in the previous section when an increasingly higher Reynolds number combines with adverse pressure gradients, separation occurs and the exact solutions are no longer linear or well behaved. Empirical relationships must therefore be established by laboratory or numerical experiments to relate the drag to flow field variables for these more difficult flows, and the relationships should be robust enough to predict the simpler flows as well. The drag coefficient approach will be used for this relationship. As defined in Sec. 7.1, drag is the force component parallel to the relative approach velocity exerted on the body by the moving fluid. The drag coefficient is defined by

$$\text{Drag} = C_D A \rho \frac{U^2}{2} \tag{7.5.1}$$

in which A is the projected area of the body on a plane normal to the flow.

The drag-coefficient curves for spheres and circular disks are shown in Fig. 7.13 (three-dimensional cases). The drag coefficient in Eq. (7.5.1) for Stokes flow is

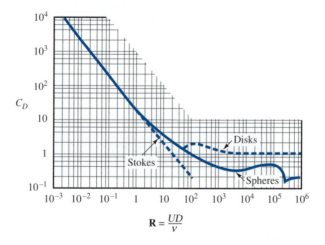

Figure 7.13   Drag coefficients for spheres and circular disks.

24/**R**. Figure 7.13 shows a plot of the coefficient for Stokes law, along with the drag coefficient versus Reynolds number for smooth spheres for separated laminar and turbulent flows. It shows the shift from laminar to turbulent boundary-layer flow, as evidenced by the sudden drop in the drag coefficient. The exact Reynolds number for the sudden shift depends upon the smoothness of the sphere and the turbulence in the fluid stream. In fact, the sphere is frequently used as a turbulence meter by determining the Reynolds number at which the drag coefficient is 0.30, a point located in the center of the sudden drop (Fig. 7.13). Using a hot-wire anemometer, Dryden [6] correlated the turbulence level of the fluid stream to the Reynolds number for the sphere at $C_D = 0.30$. The greater the turbulence of the fluid stream, the smaller the Reynolds number for a shift in the separation point. Drag-coefficient values for other three-dimensional bodies are listed in Table 7.1.

In Fig. 7.14 the drag coefficient for an infinitely long cylinder (two-dimensional case) is plotted against the Reynolds number. Like the sphere, this case also has the sudden shift in separation point. In Table 7.2 typical drag coefficients are shown for several cylinders. In general, the values given are for the range of Reynolds numbers in which the coefficient changes little with Reynolds number.

Table 7.1   Approximate C_D values for three-dimensional bodies at $\mathbf{R} > 10^4$ [7, 8]

Body shape*		C_D
Cube		1.1
60° cone		0.5
Open hemisphere		1.4 ——— 0.4
Rectangular plate		1.18 (1)† 1.2 (5) 1.3 (10) 1.5 (20) 2.0 (∞)
Cylinder flow along axis		1.15 (0.5)‡ 0.90 (1) 0.85 (2) 0.87 (4) 0.99 (8)

* Arrow indicates the direction of flow.
† The number in the parentheses indicates the value for b/h.
‡ The number in the parentheses indicates the value for l/D.

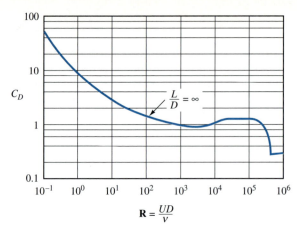

Figure 7.14   Drag coefficients for circular cylinders.

Table 7.2   Typical drag coefficients for various cylinders in two-dimensional flow [8]

Body shape*			C_D	Reynolds number
Circular cylinder	⟶ ◯		1.2	10^4 to 1.5×10^5
Elliptical cylinder	⟶ ⬭	2:1	0.6	4×10^4
			0.46	10^5
	⟶ ⬭	4:1	0.32	2.5×10^4 to 10^5
	⟶ ⬭	8:1	0.29	2.5×10^4
			0.20	2×10^5
Square cylinder	⟶ ▢		2.0	3.5×10^4
	⟶ ◇		1.6	10^4 to 10^5
Triangular cylinders	⟶ ▷ 120°		2.0	10^4
	⟶ ◁ 120°		1.72	10^4
	⟶ ▷ 90°		2.15	10^4
	⟶ ◁ 90°		1.60	10^4
	⟶ ▷ 60°		2.20	10^4
	⟶ ◁ 60°		1.39	10^4
	⟶ ▷⊦ 30°		1.8	10^5
	⟶ ⊦◁ 30°		1.0	10^5
Semitubular	⟶)		2.3	4×10^4
	⟶ (		1.12	4×10^4

* Arrow indicates the direction of flow.

Example 7.2

Dredging operations in a river yield large volumes of sediments whose smallest particle is measured to be coarse clay with a diameter of 4 microns (4×10^{-6} m) and whose largest particle is coarse sand with a diameter of 1000 microns or 1 mm. Determine the settling velocity for each size class. $\gamma_w = 9764$ N/m³; $S_{sand} = 2.65$; $S_{clay} = 1.6$; and the viscosity at 30°C is 0.8×10^{-3} N·s/m².

Solution

1. *Clay particles.* We will first assume that the particle Reynolds number is less than 1 and solve Eq. (7.3.7) for the settling velocity as

$$
\begin{aligned}
W_{t_{clay}} &= \frac{[4(10^{-6})\text{ m}]^2[9764(1.6 - 1.0)\text{ N/m}^3]}{18(0.8)(10^{-3})\text{ N·s/m}^2} \\
&= \frac{16(10^{-12})(5858.4)}{14.4(10^{-3})} = 6.510(10^{-6})\text{ m/s} \\
&= 6.51(10^{-4})\text{ cm/s}
\end{aligned}
$$

Next, check to see if the Reynolds number is less than one.

$$
\mathbf{R}_D = \frac{Dw_t}{\nu} = \frac{4(10^{-6})(6.51(10^{-6}))}{0.8(10^{-3})} = 3.25(10^{-8}) < 1
$$

Therefore, the Stokes assumption is correct. The settling velocities of clay particles are extremely small; in fact this particle will settle only 0.56 m in one day.

2. *Sand particles.* Anticipating that the larger particle will have a Reynolds number greater than one, Eq. (7.3.8) and Fig. 7.13 will be directly employed in an iterative fashion. To begin the calculation, assume $\mathbf{R}_D = 100$, then the drag coefficient from Fig. 7.13 is estimated as 1 and from Eq. (7.3.8)

$$
w_t^2 = \frac{4}{3}\frac{1(10^{-3})}{(995.7)(1)}[9764(2.65 - 1)] = \frac{0.02157}{C_D}
$$

or

$$
w_t = \frac{0.147}{\sqrt{C_D}}
$$

For the first iteration then

$$
w_{t_1} = 0.147\text{ m/s}
$$

Checking the Reynolds number

$$
\mathbf{R}_D = \frac{Dw_t}{\nu} = 1.24(10^3)w_t
$$

or

$$
\mathbf{R}_{D_1} = 1.24(10^3)(0.147) = 182
$$

Therefore, there is a mismatch between our assumed and calculated Reynolds numbers. After selecting another Reynolds number and continuing the iteration several times, the Reynolds number is found to be approximately 220, $C_D \approx 0.7$, and

$$w_{t_{sand}} = 0.175 \text{ m/s}$$

The sand particle will settle approximately 10.5 m per minute.

Example 7.3

The body in Fig. 7.15 with $S = 6.0$, cross-sectional area of 0.25 m², and volume of 0.1 m³ has a drag coefficient of 1.1. It is propelled horizontally into water at 13 m/s. Use a spreadsheet to determine its trajectory.

Solution

S_V = Vertical distance; S_H = Horizontal distance
V_V = Vertical velocity; V_H = Horizontal velocity

$$V_V = \frac{dS_V}{dt} \qquad\qquad V_H = \frac{dS_H}{dt}$$

$$F_V = W - F_B - 0.5C_D A\rho V^2 \sin\theta = \frac{W}{g}\frac{dV_V}{dt}$$

$$F_H = -0.5C_D A\rho V^2 \cos\theta = \frac{W}{g}\frac{dV_H}{dt}$$

$$V = \sqrt{V_{H^2} + V_{V^2}} \qquad \cos\theta = V_H/V \qquad \sin\theta = V_V/V$$

A second-order Runge-Kutta method (described at the Web site) may be used to solve the four differential equations simultaneously.

$$dV_{V_1} = \frac{dt}{W}g(W - F_B - 0.5C_d A_R\rho V_V V) \qquad dS_{V_1} = dtV_V$$

$$dV_{H_1} = -\frac{dt}{W}g(0.5C_d A_R\rho V_H V) \qquad\qquad dS_{H_1} = dtV_H$$

Spreadsheet results are shown in Table 7.3. In computing these results, a time step, dt, of 0.05 s was used through 0.6 s, where the time step was changed to 0.1 s.

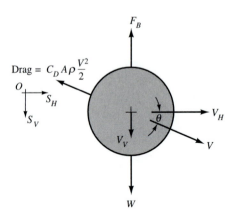

Figure 7.15   Forces on a body.

Table 7.3   Results for Ex. 7.3.

Time	V_V	S_V	V_H	S_H	V
0.0	0.00	0.00	13.00	0.00	13.00
0.2	1.33	0.14	8.16	2.03	8.26
0.4	2.36	0.51	5.88	3.41	6.34
0.6	3.24	1.08	4.49	4.43	5.54
0.8	3.98	1.80	3.51	5.23	5.31
1.0	4.57	2.66	2.76	5.85	5.34
1.2	5.01	3.62	2.15	6.34	5.46
1.4	5.33	4.66	1.67	6.72	5.59
1.6	5.55	5.75	1.29	7.01	5.70
1.8	5.70	6.88	0.99	7.24	5.79
2.0	5.80	8.03	0.76	7.41	5.85
2.2	5.86	9.20	0.58	7.54	5.89
2.4	5.90	10.38	0.44	7.65	5.92
2.6	5.93	11,56	0.34	7.72	5.94
2.8	5.95	12.75	0.26	7.78	5.95
3.0	5.96	13.94	0.20	7.83	5.96
3.2	5.96	15.13	0.15	7.86	5.96
3.4	5.97	16.32	0.11	7.89	5.97
3.6	5.97	17.52	0.09	7.91	5.97
3.8	5.97	18.71	0.07	7.92	5.97
4.0	5.97	19.90	0.05	7.93	5.97

EXERCISES

7.5.1   Pressure drag results from (a) skin friction; (b) deformation drag; (c) occurrence of a wake; (d) none of these answers.

7.5.2   The terminal velocity of a small sphere settling in a viscous fluid varies as the (a) first power of its diameter; (b) inverse of the fluid viscosity; (c) inverse square of the diameter; (d) inverse of the diameter; (e) square of the difference in specific weights of solid and fluid.

7.5.3   A sudden change in position of the separation point in flow around a sphere occurs at a Reynolds number of about (a) 1; (b) 300; (c) 30,000; (d) 3,000,000; (e) none of these answers.

7.6   LIFT

As defined in Sec. 7.1, lift is the fluid force component on a body at right angles to the relative approach velocity. If the lift force does not coincide with gravity but is at right angles to the approach velocity, it is often called a *transverse force*. The lift coefficient C_L is defined by

$$\text{Lift} = C_L A \frac{\rho U^2}{2} \tag{7.6.1}$$

In the design of lifting bodies, such as hydrofoils, airfoils, or vanes, the objective is to create a large force normal to the free-stream flow and minimize drag at the same

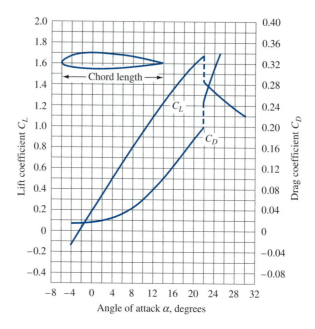

Figure 7.16   Typical lift and drag coefficients for an airfoil; C_L and C_D are based on maximum projected wing area.

time. Figure 7.16 provides lift and drag coefficients for an airfoil section. In the calculation of drag and lift in Eqs. (7.3.4) and (7.6.1) the area is defined as the chord length times the wing length (maximum projected area of the wing). This convention is adopted since the cross-sectional area changes with the angle of attack, both in the flow direction and at right angles to it. The angle of attack α is the angle between the chord of the surface section and the free-stream velocity vector.

At small angles of attack the boundary layer adheres to the foil and even though the trailing surfaces have an adverse pressure gradient, there is little separation. The lack of symmetry produces a lift at a 0° angle of attack. As the angle is increased, the adverse gradient on the upper surface strengthens and the separation point moves forward. At approximately 20°, depending on the foil design, a maximum lift is achieved with the airfoil. Further increase in the angle of attack causes a sudden decrease in the lift coefficient and increase in the drag coefficient. This condition is called *stall*.

Various techniques are available to improve lift and drag characteristics of foils for special purposes such as takeoff and landing. They generally include airfoil-section variations through the use of flaps or boundary-layer control methods by the addition of slots.

Moving surfaces that influence the boundary layer and separation points on bodies also appear in a number of common physical situations. Spinning spheres play an important role in many sporting events including curve balls or screw balls in baseball or hooks and slices in soccer or golf. Figure 7.17*a* shows velocities developed in the boundary layer of a spinning body in quiescent fluid. If this is superimposed on a moving fluid, the condition shown in Fig. 7.17*b* develops, which shows a shift in the separation points on the body, with the wake positioned nonsymmetrically. A lift force is created in the direction shown since the pressure is reduced on the upper

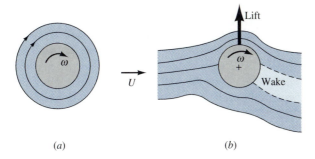

Figure 7.17 Spinning sphere.

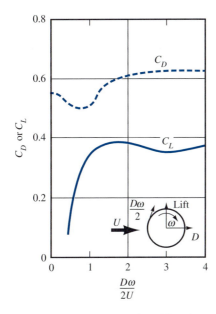

Figure 7.18 Rotating sphere lift and drag coefficients, $\mathbf{R} \approx 10^5$.

surface and increased on the lower surface. Figure 7.18 provides the lift and drag coefficient [9, 10] for various dimensionless spin ratios for a spinning sphere.

EXERCISES

7.6.1 The lift on a body immersed in a fluid stream is (a) due to buoyant force; (b) always opposite in direction to gravity; (c) the resultant fluid force on the body; (d) the dynamic fluid force component exerted on the body normal to the approach velocity; (e) the dynamic fluid force component exerted on the body parallel to the approach velocity.

7.6.2 A ball moving forward with top spin through a quiescent fluid will (a) drift to the right; (b) drift to the left; (c) travel in a straight line; (d) only drop due to gravity; (e) drop due to gravity and spin.

7.7   ACCELERATION AND INERTIAL FORCES

The drag and lift calculations done in the previous sections all assumed a steady uniform flow. However, many flows contain temporal accelerations which require an additional or *inertial force* to be considered. Temporal accelerations may be *rectilinear* if the flow proceeds in the same direction but the velocity magnitude changes with time, or flows may be *oscillatory* as marked by repetitive reversals of direction and continuously time-varying velocity and acceleration. A riverine flood wave caused by storm water runoff is an example of the former while a wind-driven gravity wave is an example of the latter. Instead of accelerating the fluid through the flow geometry the opposite frame of reference may also be chosen where an object such as a submarine, soccer ball, or airplane is to be accelerated from rest through a quiescent fluid.

In either acceleration problem the force distribution on the object in the fluid consists of the drag and the lift forces discussed previously plus the force required to accelerate the fluid that would occupy the space now occupied by the object. This extra force or inertial force is calculated by use of the *added mass* concept and will require knowledge of the geometry of the object in the fluid and the specification of the fluid (or solid) acceleration in the vicinity of the object. These requirements are contrasted to the drag and lift calculation which also required knowledge of the geometry but only the fluid (or solid) velocity.

Consider the vertical circular cylinder of diameter D in Figure 7.19 subjected to an oscillatory flow introduced by a train of waves. In the ocean and coastal engineering field these *piles* are used to support docks, boardwalks, or shallow oil and gas drilling rigs. Were the flow steady the total force, F_t, introduced on the incremental height Δz of the pile would consist only of that due to form and skin friction drag, that is,

$$f_t = f_D = \frac{1}{2}\rho C_D A u^2 = \frac{1}{2}\rho C_D (dz\, D) u^2$$

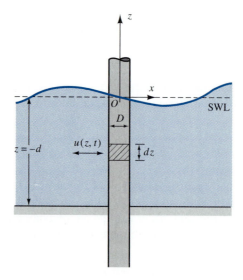

Figure 7.19   Inertial forces on vertical circular cylinder.

where A is the area, $dz \cdot D$, perpendicular to the flow and C_D is found from Table 7.2 to be approximately 1.2 for a Reynolds number of 10^4. In the accelerated, oscillatory flow the total force consists of f_D plus an inertial force f_i as

$$f_t = f_D + f_i = \frac{1}{2}\rho C_D(dz\,D)u^2 + \rho\frac{\pi D^2}{4}\,dz\frac{du}{dt} \tag{7.7.1}$$

Here, the inertial force has been calculated as the mass of the fluid that *would* occupy the space occupied by the pile $(\rho\,dz\,\pi D^2/4)$ times the temporal acceleration. The *added mass* then is $(\rho\,dz\,\pi D^2/4)$. The total force on the pile is found by integrating over the length of the pile exposed to the fluid

$$F_t = \int_{-d}^{\eta} f_D\,dz + \int_{-d}^{\eta} f_i\,dz \tag{7.7.2}$$

or for the case of the pile

$$F_t = \frac{1}{2}\rho C_D D\int_{-d}^{\eta} u^2\,dz + \rho\frac{\pi D^2}{4}\int_{-d}^{\eta}\frac{du}{dt}\,dz \tag{7.7.3}$$

In Eq. (7.7.3) it has been assumed that C_D, though a function of the Reynolds number (Table 7.2), remains constant in the vertical direction. Experiments first done by Morison (1950) [12] show that for a limited range of Reynolds numbers this expression for added mass is suitable if not exact. However, when flow separation and vortex shedding occur, this expression is more generally augmented by addition of an inertia coefficient C_i; therefore, in Eq. (7.7.1)

$$f_t = \frac{1}{2}\rho C_D(dz\,D)u^2 + \rho\,C_i\frac{\pi D^2}{4}\,dz\frac{du}{dt} \tag{7.7.4}$$

Therefore, the added mass per unit length is $\rho C_i\pi D^2/4$. Laboratory experience reported by the U.S. Army Corps of Engineers [13] reveals that the range of C_i for piles is 1.0–2.5 and is Reynolds number dependent.

For shapes other than those depicted in Table 7.4, Table 2.3 in Sarpkaya and Isaacson [14] is an excellent compendium of added mass functions for a variety of shapes.

PROBLEMS

7.1 Discuss the origin of the drag on a disk when its plane is parallel to the flow and when it is normal to it.

7.2 The velocity distribution in a boundary layer is given by $u/U = 3(y/\delta) - 2(y/\delta)^2$. Show that the displacement thickness of the boundary layer is $\delta_1 = \delta/6$.

7.3 Using the velocity distribution $u/U = \sin(\pi y/2\delta)$, determine the equation for growth of the laminar boundary layer and for shear stress along a smooth flat plate in two-dimensional flow.

7.4 Compare the drag coefficients that are obtained with the velocity distributions given in Probs. 7.2 and 7.3.

7.5 Solve the equations for growth of the turbulent boundary layer, based on the exponential law $u/U = (y/\delta)^{1/9}$ and $f = 0.185\mathbf{R}^{1/5}$. $(\tau_0 = \rho f V^2/8.)$

Table 7.4   Added mass formula for various shapes

Body shape	Flow orientation ($\longleftrightarrow$ motion)*	Added mass per unit length
Circular cylinder		$\rho \frac{\pi D^2}{4}$
Ellipse		$\rho \pi b^2$
Plate		$\rho \pi b^2$
Rectangular plate		1.51 $\rho \pi b^2$ (1.0)† 1.98 $\rho \pi b^2$ (1.5) 1.21 $\rho \pi b^2$ (5.0) 1.14 $\rho \pi b^2$ (10)

* All the flows in the table are assumed to be horizontal.
† The number in the parentheses indicates the value for b/a.

7.6   Air at 20°C, 100-kPa abs flows along a smooth plate with a velocity of 150 km/h. How long does the plate have to be to obtain a boundary-layer thickness of 8 mm?

7.7   Estimate the skin-friction drag on an airship 100 m long, average diameter 20 m, with velocity of 130 km/h traveling through air at 90-kPa abs and 25°C.

7.8   The walls of a wind tunnel are sometimes made divergent to offset the effect of the boundary layer in reducing the portion of the cross section in which the flow is of constant speed. At what angle must plane walls be set so that the displacement thickness does not encroach upon the tunnel's constant-speed cross section at distances greater than 0.8 ft from the leading edge of the wall? Use the data of Prob. 7.6.

7.9   An advertising sign is towed by a small plane at a velocity of 35 m/s. The dimensions of the sign are 1.4 by 38 m, $p = 1$ atm, and $t = 15$°C. Assuming the sign to be a flat plate, calculate the power required to tow the sign.

7.10 A high-speed train travels at 160 mi/h. A train 400 ft long can be visualized as having a surface area of 400 by 28 ft. Estimate the skin-friction drag and the power required to overcome this resistance only. Assume standard pressure and 60°F.

7.11 Determine the settling velocity of small metal spheres, sp gr 4.5 and diameter 0.1 mm, in crude oil at 25°C, sp gr 0.86.

7.12 A spherical dust particle at an altitude of 80 km is radioactive as a result of an atomic explosion. Determine the time it will take to settle to earth if it falls in accordance with Stokes law. Its size and sp gr are 25 μm and 2.5. Neglect wind effects. Use isothermal atmosphere at $-18°C$.

7.13 What is the upper limit for a spherical particle of dust, sp gr 2.5, that settles in atmospheric air at 20°C in obedience to Stokes law? What is the settling velocity?

7.14 At what speed must a 120-mm sphere travel through water at 10°C to have a drag of 5 N?

7.15 How many 30-m-diameter parachutes (C_D = 1.2) should be used to drop a bulldozer weighing 45 kN at a terminal speed not more than 10 m/s through air at 100-kPa abs at 20°C?

7.16 An object weighing 400 lb is attached to a circular disk and dropped from a plane. What diameter should the disk have so that the object strikes the ground at 72 ft/s? The disk is so attached that it is normal to the direction of motion. p = 14.7 psia and t = 70°F.

7.17 A circular disk 3 m in diameter is held normal to a 100-km/h airstream (ρ = 1.1 kg/m^3). What force is required to hold it at rest?

7.18 A parachute jumper with associated gear weighs 250 lb. The vertical component of the landing speed should not be more than 20 ft/s. By assuming the parachute to be an open hemisphere, determine the required diameter of the parachute to be used at standard pressure and 80°F.

7.19 A 0.8-m cubical box is placed on the luggage carrier on top of a station wagon. Estimate the additional power requirements for the vehicle to travel at (*a*) 80 km/h and (*b*) 110 km/h.

7.20 Two circular cups are attached to the ends of circular rods (Fig. 7.20). The device is rotated about the vertical axis to mix additives in a container filled with liquid. ρ = 1075 kg/m^3 and ν = 10^{-6} m^2/s. The rotational speed is 40 rpm, and the rods are 8 mm in diameter. Determine the power requirements to drive the mixer. The cups are open in the direction of rotation.

7.21 A semitubular cylinder of 6-in. radius with concave side upstream is submerged in water flowing 3 ft/s. Calculate the drag for a cylinder 24 ft long.

7.22 A 1.8-m-diameter smokestack, 55 m high, is designed to resist a 35 m/s wind. C_D = 0.7. What is the total force on the smokestack, and what is the moment at the base?

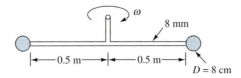

Figure 7.20   Problem 7.20.

7.23   Find the bending moment at the base of a cylindrical antenna 8 mm in diameter, extended 2 m, on an automobile when traveling at 100 km/hr.

7.24   What is the terminal velocity of a 2-in.-diameter metal ball, sp gr 3.5, dropped in oil, sp gr 0.80 and $\mu = 1$ P? What would be the terminal velocity for the same-size ball but with a 7.0 sp gr? How do these results agree with the experiments attributed to Galileo at the Leaning Tower of Pisa?

7.25   A spherical balloon that contains helium ascends through air at 14 psia and 40°F. The balloon and payload weigh 300 lb. What diameter permits ascension at 10 ft/s? $C_D = 0.21$. If the balloon is tethered to the ground in a 10 mi/h wind, what is the angle of inclination of the retaining cable?

7.26   A steel sphere of diameter 6.5 mm ($S = 7.8$) is released in a tank of oil ($S = 0.83$). What is the viscosity of the oil if the terminal velocity of the sphere is 0.1 m/s?

7.27   Determine the settling velocity of a particle of sand ($S = 2.55$) in water at 20°C if the particles may be assumed spherical in shape, for diameters, (a) 0.1 mm, (b) 1.0 mm, and (c) 10 mm.

7.28   A harpoon with a sharp tip is 2 cm in diameter and 1.5 m in length. If the harpoon is launched in water at a velocity of 8 m/s, find the drag force. What is the maximum thickness of the boundary layer? Assume the water temperature to be 20°C.

7.29   Give some reason for the discontinuity in the curves of Fig. 7.16 at the angle of attack of 22°.

7.30   What is the ratio of lift to drag for the airfoil section of Fig. 7.16 for an angle of attack of 2°?

7.31   A boat fitted with a hydrofoil weighs 5000 lb. At a velocity of 50 ft/s, what size hydrofoil is needed to support the boat? Use the lift characteristics in Fig. 7.16 at an angle of attack of 4°.

7.32   A tennis player hitting from the baseline develops a forward velocity of 70 ft/s and a backspin of 5000 rpm. The ball weighs 0.125 lb and has a diameter of 2.56 in. Assume standard pressure and 70°F and neglect the drag force. Including the lift provided by the backspin, how much will the tennis ball have dropped by the time it reaches the net 39 ft away?

7.33   A pitcher delivers a curve ball at 50 mi/h with a spin of 2500 rpm about a vertical axis. The baseball weighs 0.3 lb and has a diameter of 2.9 in. Assume a constant velocity toward home plate, 60 ft away; $p = 1$ atm and $t = 80°F$. How far will the ball curve in traveling to the plate?

7.34   A pitcher delivers a baseball at 85 mi/h with backspin of ω rad/s about a horizontal axis. With the same assumptions as in Prob. 7.33, what is the approximate value of ω if the ball travels on a horizontal path, not dropping due to gravity?

7.35   As a result of alternate shedding of vortices, a periodic pressure (lift) pulsation may develop on a stationary cylinder in a flow field. The process is described by the Strouhal number, $S_t = fD/V$, where f is the frequency in hertz. Over a wide range of Reynolds numbers the critical Strouhal number is approximately 0.2. What frequency of oscillation is produced by a 100 km/hr wind blowing over a 2-mm wire?

7.36   A golf ball leaves the tee with a velocity of 55 m/s. A golf ball weighs 0.418 N and has a diameter of 4.2 cm. Assume the ball is rising at an angle of 30° and assume $C_D = 0.42$. Use a spreadsheet to calculate the distance the drive will travel (*a*) with no spin, (*b*) with back spin such that $C_L = 0.2$, and (*c*) with top spin such that $C_L = -0.2$.

7.37 In Prob. 7.36, for case (*a*), find the influence of a head wind of 20 km/hr and of a tail wind of 20 km/hr.

REFERENCES

1. L. Prandtl, "Über Flussigkeitsbewegung bei sehr kleiner Reibung," *Verh. III Int. Math.-Kongr., Heidelb*, 1904.

2. T. von Kármán, "On Laminar and Turbulent Friction," *Z. Angew. Math. Mech.,* vol. 1, pp. 235–236, 1921.

3. L. Prandtl, "Über den Reibungswiderstand strömender Luft," *Result. Aerodyn. Test Inst. Goett.,* III Lieferung, 1927.

4. L. Prandtl and H. Schlichting, "Das Widerstandsgesetz rauher Platten," *Werft, Reederei, Hafen,* p. 1, 1934.

5. G. Stokes, *Trans. Camb. Phil. Soc.,* vol. 8, 1845; vol. 9, 1851.

6. H. Dryden, "Reduction of Turbulence in Wind Tunnels," *NACA Tech. Rep.* 392, 1931.

7. S. F. Hoerner, *Fluid Dynamic Drag,* 2d ed., the author, Midland Park, NJ, 1965.

8. W. F. Lindsey, *NACA Tech. Rep.* 619, 1938.

9. S. Goldstein (ed.), *Modern Developments in Fluid Dynamics,* vol. II, Clarendon Press, Oxford, 1938.

10. H. M. Barkla and L. J. Auchterlonie, "The Magnus Effect on Rotating Spheres," *J. Fluid Mech.,* vol. 47, p. 3, 1971.

11. H. Schlichting, *Boundary Layer Theory,* McGraw-Hill, New York, 1979.

12. J. R. Morison, M. P. O'Brien, J. W. Johnson, and S. A. Schaaf, "The Forces Exerted by Surface Winds in Piles," *AIME Petroleum Transactions,* vol. 189, pp. 149–157, 1950.

13. U.S. Army Corps of Engineers, *Shore Protection Manual,* vol.II, pp. 7-144–7-145, Superintendent of Documents, U.S. Government Printing Office, Washington, D. C., 1984.

14. T. Sarpkaya and M. Isaacson, *Mechanics of Wave Forces on Offshore Structures,* pp. 47–51, Van Nostrand Reinhold Company, New York, 1981.

8

Ideal-Fluid Flow

In the preceding chapters most of the relations have been developed for one-dimensional flow, that is, flow in which the average velocity at each cross section is used and variations across the section are neglected. Many design problems in fluid flow, however, require more exact knowledge of velocity and pressure distributions, such as flow over curved boundaries along an airplane wing, through the passages of a pump or compressor, or over the crest of a dam. An understanding of two- and three-dimensional flow of a nonviscous, incompressible fluid provides the student with a much broader approach to many real fluid-flow situations. There are also analogies that permit the same methods to apply to flow through porous media.

In this chapter the principles of irrotational flow of an ideal fluid are developed and applied to elementary flow cases. After the flow requirements are established, Euler's equation is derived and the velocity potential is defined. Euler's equation is integrated to obtain Bernoulli's equation, and stream functions and boundary conditions are developed. Flow cases are then studied in two dimensions.

8.1 REQUIREMENTS FOR IDEAL-FLUID FLOW

The Prandtl hypothesis (Sec. 7.2) states that for fluids of low viscosity the effects of viscosity are appreciable only in the narrow boundary-layer region surrounding the fluid boundaries or at fluid interfaces with large density gradients. For incompressible-flow situations in which the boundary layer remains thin, ideal-fluid results can be applied to flow of a real fluid as an initial approximation. Converging or accelerating flow situations generally have thin boundary layers, but decelerating flow may have separation of the boundary layer and development of a large wake that is difficult to predict analytically.

An ideal fluid must satisfy the following requirements:

1. The continuity equation (Sec. 4.3) $\nabla \cdot \mathbf{v} = 0$, or

$$\frac{\partial u}{\partial x} + \frac{\partial v}{\partial y} + \frac{\partial w}{\partial z} = 0$$

2. Newton's second law of motion at every point at every instant

3. Neither penetration of fluid into, nor gaps between, the fluid and boundary at any solid boundary

If, in addition to requirements 1, 2, and 3, the assumption of irrotational flow is made, the resulting fluid motion closely resembles real-fluid motion for fluids of low viscosity outside boundary layers.

Using the above conditions, the application of Newton's second law to a fluid parcel leads to Euler's equation, which, together with the assumption of irrotational flow, can be integrated to obtain Bernoulli's equation. The unknowns in a fluid-flow situation with given boundaries are velocity and pressure at every point. Unfortunately, in most cases it is impossible to proceed directly to equations for velocity and pressure distribution from the boundary conditions.

8.2 EULER'S EQUATION OF MOTION

Cartesian Coordinate System

Euler's equations of motion were developed in Sec. 4.5 [Eq. (4.5.1)] by use of the momentum and continuity equations. The component equations of Eq. (4.5.1) are

$$-\frac{1}{\rho}\frac{\partial}{\partial x}(p + \gamma h) = u\frac{\partial u}{\partial x} + v\frac{\partial u}{\partial y} + w\frac{\partial u}{\partial z} + \frac{\partial u}{\partial t} \tag{8.2.1}$$

$$-\frac{1}{\rho}\frac{\partial}{\partial y}(p + \gamma h) = u\frac{\partial v}{\partial x} + v\frac{\partial v}{\partial y} + w\frac{\partial v}{\partial z} + \frac{\partial v}{\partial t} \tag{8.2.2}$$

$$-\frac{1}{\rho}\frac{\partial}{\partial z}(p + \gamma h) = u\frac{\partial w}{\partial x} + v\frac{\partial w}{\partial y} + w\frac{\partial w}{\partial z} + \frac{\partial w}{\partial t} \tag{8.2.3}$$

As noted previously the first three terms on the right-hand side of the equations are *inertial-acceleration* terms, depending upon changes of velocity with space. The last term is the *local* or *temporal acceleration,* depending upon velocity change with time at a point.

Natural Coordinates in Two-Dimensional Flow

Euler's equations in two dimensions are obtained from the general-component equations by setting $w = 0$ and $\partial/\partial z = 0$; thus,

$$-\frac{1}{\rho}\frac{\partial}{\partial x}(p + \gamma h) = u\frac{\partial u}{\partial x} + v\frac{\partial u}{\partial y} + \frac{\partial u}{\partial t} \tag{8.2.4}$$

$$-\frac{1}{\rho}\frac{\partial}{\partial y}(p + \gamma h) = u\frac{\partial v}{\partial x} + v\frac{\partial v}{\partial y} + \frac{\partial v}{\partial t} \tag{8.2.5}$$

By taking particular directions for the x and y axes, they can be reduced to a form that makes them easier to understand. If the x axis, called the s axis, is taken parallel to the velocity vector at a point (Fig. 8.1), it is then tangent to the streamline through the point. The y axis, called the n axis, is directed normal to the streamline coordinate s, which for the case in Figure 8.1 is toward the center of curvature of the streamline. The velocity component u is v_s, and the component v is v_n. As v_n is zero at the point, Eq. (8.2.4) becomes

$$-\frac{1}{\rho}\frac{\partial}{\partial s}(p + \gamma h) = v_s\frac{\partial v_s}{\partial s} + \frac{\partial v_s}{\partial t} \tag{8.2.6}$$

Although v_n is zero at the point (s, n), its rates of change with respect to s and t are not necessarily zero. Equation (8.2.5) becomes

$$-\frac{1}{\rho}\frac{\partial}{\partial n}(p + \gamma h) = v_s\frac{\partial v_n}{\partial s} + \frac{\partial v_n}{\partial t} \tag{8.2.7}$$

When the velocity at s and at $s + \delta s$ along the streamline is considered, v_n changes from zero to δv_n. With r the radius of curvature of the streamline at s, from similar triangles (Fig. 8.1),

$$\frac{\delta s}{r} = \frac{\delta v_n}{v_s} \qquad \text{or} \qquad \frac{\partial v_n}{\partial s} = \frac{v_s}{r}$$

Substituting into Eq. (8.2.7) gives

$$-\frac{1}{\rho}\frac{\partial}{\partial n}(p + \gamma 8) = \frac{v_s^2}{r} + \frac{\partial v_n}{\partial t} \tag{8.2.8}$$

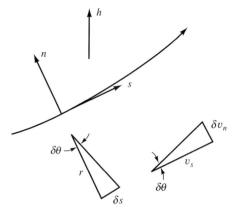

Figure 8.1   Notation for natural coordinates.

For steady flow of an incompressible fluid Eqs. (8.2.4) and (8.2.8) can be written

$$-\frac{1}{\rho}\frac{\partial}{\partial s}(p + \gamma h) = \frac{\partial}{\partial s}\left(\frac{v_s^2}{2}\right) \qquad (8.2.9)$$

and

$$-\frac{1}{\rho}\frac{\partial}{\partial n}(p + \gamma h) = \frac{v_s^2}{r} \qquad (8.2.10)$$

Equation (8.2.9) can be integrated with respect to s to produce Eq. (4.5.10), with the constant of integration varying with n, that is, from one streamline to another. Equation (8.2.10) shows how the pressure head varies across streamlines. With v_s and r known functions of n, Eq. (8.2.10) can be integrated.

A container of liquid is rotated with angular velocity ω about a vertical axis as a solid. Determine the variation of pressure in the liquid.

Example 8.1

Solution

n is the radial distance, measured inwardly; $dn = -dr$; and $v_s = \omega r$. Integrating Eq. (8.2.10) gives

$$-\frac{1}{\rho}(p + \gamma h) = -\int \frac{\omega^2 r^2\, dr}{r}$$

or

$$\frac{1}{\rho}(p + \gamma h) = \frac{\omega^2 r^2}{2} + \text{constant}$$

To evaluate the constant, if $p = p_0$ when $r = 0$ and $h = 0$,

$$p = p_0 - \gamma h + \rho\frac{\omega^2 r^2}{2}$$

which shows that the pressure is hydrostatic along a vertical line and increases as the square of the radius. Integration of Eq. (8.2.9) shows that the pressure is constant for a given h and v_s, that is, along a streamline. These results are the same as for rotation in relative equilibrium determined in Sec. 2.9.

EXERCISES

8.2.1   The units for Euler's equations of motion are given by (*a*) force per unit mass; (*b*) velocity; (*c*) energy per unit weight; (*d*) force per unit weight; (*e*) none of these answers.

8.2.2   Euler's equations of motion are a mathematical statement that at every point (*a*) the rate of mass inflow equals the rate of mass outflow; (*b*) the force per unit mass equals acceleration; (*c*) the energy does not change with the time; (*d*) Newton's third law of motion holds; (*e*) the fluid momentum is constant.

8.3   IRROTATIONAL FLOW: VELOCITY POTENTIAL

In this section it is shown that the assumption of irrotational flow leads to the existence of a velocity potential. By using these relations and assuming a conservative body force, the Euler equations can be integrated.

The individual parcels of a frictionless incompressible fluid initially at rest cannot be caused to rotate. This can be visualized by considering a small free body of fluid in the shape of a sphere. Surface forces act normal to its surface since the fluid is frictionless and, therefore, act through the center of the sphere. Similarly, the body force acts at the mass center. Hence, no torque can be exerted on the sphere, and it remains without rotation. Likewise, once an ideal fluid has rotation, there is no way of altering it, as no torque can be exerted on an elementary sphere of the fluid.

An analytical expression for fluid rotation of a parcel about an axis parallel to the z axis was developed in Sec. 4.1 [Eqs. (4.1.8a–c)]. By assuming that the fluid has no rotation, that is, it is irrotational, $\nabla \times \mathbf{v} = 0$, or from Eqs. (4.1.8$a$–$c$)

$$\frac{\partial v}{\partial x} = \frac{\partial u}{\partial y} \qquad \frac{\partial w}{\partial y} = \frac{\partial v}{\partial z} \qquad \frac{\partial u}{\partial z} = \frac{\partial w}{\partial x} \tag{8.3.1}$$

These restrictions on the velocity must hold at every point (except special singular points or lines). The first equation is the irrotational condition for two-dimensional flow in the xy plane. It is the condition that the differential expression

$$u\,dx + v\,dy$$

is exact, say

$$u\,dx + v\,dy = -d\phi = -\frac{\partial \phi}{\partial x}\,dx - \frac{\partial \phi}{\partial y}\,dy \tag{8.3.2}$$

The minus sign is arbitrary; it is a convention that causes the value of ϕ to decrease in the direction of the velocity. By comparing terms in Eq. (8.3.2), $u = -\partial \phi/\partial x$ and $v = -\partial \phi/\partial y$. This proves the existence, in two-dimensional flow, of a function ϕ such that its negative derivative with respect to any direction is the velocity component in that direction. It can also be demonstrated for three-dimensional flow. In vector form,

$$\mathbf{v} = -\nabla \phi \tag{8.3.3}$$

is equivalent to

$$u = -\frac{\partial \phi}{\partial x} \qquad v = -\frac{\partial \phi}{\partial y} \qquad w = -\frac{\partial \phi}{\partial z} \tag{8.3.4}$$

The assumption of a velocity potential is equivalent to the assumption of irrotational flow, as

$$\mathrm{curl}(-\mathrm{grad}\,\phi) = \nabla \times (-\nabla \phi) = 0 \tag{8.3.5}$$

because $\nabla \times \nabla = 0$. This is shown from Eq. (8.3.4) by cross-differentiation

$$\frac{\partial u}{\partial y} = -\frac{\partial^2 \phi}{\partial x\,\partial y} \qquad \frac{\partial v}{\partial x} = -\frac{\partial^2 \phi}{\partial y\,\partial x}$$

proving $\partial v/\partial x = \partial u/\partial y$, etc. Substitution of Eqs. (8.3.4) into the continuity equation

$$\frac{\partial u}{\partial x} + \frac{\partial v}{\partial y} + \frac{\partial w}{\partial z} = 0$$

yields

$$\frac{\partial^2 \phi}{\partial x^2} + \frac{\partial^2 \phi}{\partial y^2} + \frac{\partial^2 \phi}{\partial z^2} = 0 \qquad (8.3.6)$$

In vector form this is

$$\nabla \cdot \mathbf{v} = -\nabla \cdot \nabla \phi = -\nabla^2 \phi = \nabla^2 \phi = 0 \qquad (8.3.7)$$

and is written $\nabla^2 \phi = 0$. Equation (8.3.6) or (8.3.7) is the *Laplace equation*. Any function ϕ that satisfies the Laplace equation is a possible irrotational fluid-flow case. As there are an infinite number of solutions to the Laplace equation, each of which satisfies certain flow boundaries, the main problem is the selection of the proper function for the particular flow case.

Because ϕ appears to the first power in each term, Eq. (8.3.6) is a linear equation and the sum of two solutions also is a solution. For example, if ϕ_1 and ϕ_2 are solutions of Eq. (8.3.6), then $\phi_1 + \phi_2$ is a solution; thus,

$$\nabla^2 \phi_1 = 0 \qquad \nabla^2 \phi_2 = 0$$

and

$$\nabla^2 (\phi_1 + \phi_2) = \nabla^2 \phi_1 + \nabla^2 \phi_2 = 0$$

The same is true if ϕ_1 is a solution. $C\phi_1$ is a solution if C is a constant.

EXERCISES

8.3.1 Select the value of ϕ that satisfies continuity: (a) $x^2 + y^2$; (b) $\sin x$; (c) $\ln (x + y)$; (d) $x + y$; (e) none of these answers.

8.3.2 In irrotational flow of an ideal fluid (a) a velocity potential exists; (b) all particles must move in straight lines; (c) the motion must be uniform; (d) the flow is always steady; (e) the velocity must be zero at a boundary.

8.3.3 A function ϕ that satisfies the Laplace equation (a) must be linear in x and y; (b) is a possible case of rotational fluid flow; (c) does not necessarily satisfy the continuity equation; (d) is a possible fluid-flow case; (e) is none of these answers.

8.3.4 If both ϕ_1 and ϕ_2 are solutions of the Laplace equation, which of the following is also a solution? (a) $\phi_1 - 2\phi_2$; (b) $\phi_1\phi_2$; (c) ϕ_1/ϕ_2; (d) ϕ_1^2; (e) none of these answers.

8.3.5 Select the relation that must hold if the flow is irrotational

(a) $\dfrac{\partial u}{\partial y} + \dfrac{\partial v}{\partial x} = 0$; (b) $\dfrac{\partial u}{\partial x} = \dfrac{\partial v}{\partial y}$; (c) $\dfrac{\partial^2 u}{\partial x^2} + \dfrac{\partial^2 v}{\partial y^2} = 0$; (d) $\dfrac{\partial u}{\partial y} = \dfrac{\partial v}{\partial x}$;

(e) none of these answers.

8.4   INTEGRATION OF EULER'S EQUATION: BERNOULLI'S EQUATION

Equation (8.2.1) can be rearranged so that every term contains a partial derivative with respect to x. From Eq. (8.3.1)

$$v\frac{\partial u}{\partial y} = v\frac{\partial v}{\partial x} = \frac{\partial}{\partial x}\frac{v^2}{2} \qquad w\frac{\partial u}{\partial z} = w\frac{\partial w}{\partial x} = \frac{\partial}{\partial x}\frac{w^2}{2}$$

and from Eq. (8.3.4)

$$\frac{\partial u}{\partial t} = -\frac{\partial}{\partial x}\frac{\partial \phi}{\partial t}$$

Making these substitutions into Eq. (8.2.1) and rearranging yield

$$\frac{\partial}{\partial x}\left(\frac{p}{\rho} + gh + \frac{u^2}{2} + \frac{v^2}{2} + \frac{w^2}{2} - \frac{\partial \phi}{\partial t}\right) = 0$$

Defining $q^2 = u^2 + v^2 + w^2$† as the square of the speeds yields

$$\frac{\partial}{\partial x}\left(\frac{p}{\rho} + gh + \frac{q^2}{2} - \frac{\partial \phi}{\partial t}\right) = 0 \tag{8.4.1}$$

Similarly, for the y and z directions,

$$\frac{\partial}{\partial y}\left(\frac{p}{\rho} + gh + \frac{q^2}{2} - \frac{\partial \phi}{\partial t}\right) = 0 \tag{8.4.2}$$

$$\frac{\partial}{\partial z}\left(\frac{p}{\rho} + gh + \frac{q^2}{2} - \frac{\partial \phi}{\partial t}\right) = 0 \tag{8.4.3}$$

The quantities within the parentheses are the same in Eqs. (8.4.1) to (8.4.3). Equation (8.4.1) states that the quantity is not a function of x, since the derivative with respect to x is zero. Similarly, the other equations show that the quantity is not a function of y or z. Therefore, it can be a function of t only, say $F(t)$

$$\frac{p}{\rho} + gh + \frac{q^2}{2} - \frac{\partial \phi}{\partial t} = F(t) \tag{8.4.4}$$

In steady flow $\partial \phi/\partial t = 0$ and $F(t)$ becomes a constant E

$$\frac{p}{\rho} + gh + \frac{q^2}{2} = E \tag{8.4.5}$$

The available energy is everywhere constant throughout the fluid. This is Bernoulli's equation for an irrotational fluid.

---

†As noted in Chap. 4, there are several designations for the total velocity. In Chap. 4 total velocity was referred to as V because the velocity at the point in question was derived from a spatially averaged velocity. Here q is used because we refer to total velocity at a point which may vary throughout the flow field.

The pressure term can be separated into two parts, the hydrostatic pressure p_s and the dynamic pressure p_d, so that $p = p_s + p_d$. Inserting in Eq. (8.4.5) gives

$$gh + \frac{p_s}{\rho} + \frac{p_d}{\rho} + \frac{q^2}{2} = E$$

The first two terms can be written

$$gh + \frac{p_s}{\rho} = \frac{1}{\rho}(p_s + \gamma h)$$

with h measured vertically upward. The expression is a constant, since it expresses the hydrostatic law of variation of pressure. These two terms may be included in the constant E. After dropping the subscript on the dynamic pressure, there remains

$$\frac{p}{\rho} + \frac{q^2}{2} = E \tag{8.4.6}$$

This simple equation permits the variation in pressure to be determined if the speed is known or vice versa. Assuming both the speed q_0 and the dynamic pressure p_0 to be known at one point,

$$\frac{p_0}{\rho} + \frac{q_0^2}{2} = \frac{p}{\rho} + \frac{q^2}{2}$$

or

$$p = p_0 + \frac{\rho q_0^2}{2}\left[1 - \left(\frac{q}{q_0}\right)^2\right] \tag{8.4.7}$$

Example 8.2

A submarine moves through water at 30 ft/s. At a point A on the submarine 5 ft above the nose, the velocity of the submarine relative to the water is 50 ft/s. Determine the dynamic-pressure difference between this point and the nose, and determine the difference in total pressure between the two points.

Solution

If the submarine is stationary and the water is moving past it, the velocity at the nose is zero and the velocity at A is 50 ft/s. By selecting the dynamic pressure at infinity as zero, from Eq. (8.4.6)

$$E = 0 + \frac{q_0^2}{2} = \frac{30^2}{2} = 450 \text{ ft·lb/slug}$$

For the nose

$$\frac{p}{\rho} = E = 450 \qquad p = 450(1.935) = 870 \text{ lb/ft}^2$$

For point A

$$\frac{p}{\rho} = E - \frac{q^2}{2} = 450 - \frac{50^2}{2} \quad \text{and} \quad p = 1.935\left(\frac{30^2}{2} - \frac{50^2}{2}\right) = -1548 \text{ lb/ft}^2$$

Therefore, the difference in dynamic pressure is

$$-1548 - 870 = -2418 \text{ lb/ft}^2$$

The difference in total pressure can be obtained by applying Eq. (8.4.5) to point A and to the nose n as

$$gh_A + \frac{p_A}{\rho} + \frac{q_A^2}{2} = gh_n + \frac{p_n}{\rho} + \frac{q_n^2}{2}$$

Hence,

$$p_A - p_n = \rho\left(gh_n - gh_A + \frac{q_n^2 - q_A^2}{2}\right) = 1.935\left(-5g - \frac{50^2}{2}\right) = -2740 \text{ lb/ft}^2$$

It can also be reasoned that the actual pressure difference varies by 5γ from the dynamic pressure difference since A is 5 ft above the nose, or

$$-2418 - 5(62.4) = -2740 \text{ lb/ft}^2$$

EXERCISES

8.4.1   Euler's equations of motion can be integrated when it is assumed that (*a*) the continuity equation is satisfied; (*b*) the fluid is incompressible; (*c*) a velocity potential exists and the density is constant; (*d*) the flow is rotational and incompressible; (*e*) the fluid is nonviscous.

8.4.2   The Bernoulli equation in steady ideal-fluid flow states that (*a*) the velocity is constant along a streamline; (*b*) the energy is constant along a streamline but may vary across streamlines; (*c*) when the speed increases, the pressure increases; (*d*) the energy is constant throughout the fluid; (*e*) the net flow rate into any small region must be zero.

8.4.3   An unsteady-flow case may be transformed into a steady-flow case (*a*) regardless of the nature of the problem; (*b*) when two bodies are moving toward each other in an infinite fluid; (*d*) when an unsymmetrical body is rotating in an infinite fluid; (*e*) when a single body translates in an infinite fluid; (*f*) under no circumstances.

8.5   STREAM FUNCTIONS AND BOUNDARY CONDITIONS

Two stream functions are defined. One is for two-dimensional flow, where all lines of motion are parallel to a fixed plane, say the xy plane, and the flow is identical in each of these planes. The other is for three-dimensional flow with axial symmetry, that is, all flow lines are in planes intersecting the same line or axis, and the flow is identical in each of these planes.

Two-Dimensional Stream Function

If A and P represent two points in one of the flow planes, for example, the xy plane (Fig. 8.2), and if the plane has unit thickness, the rate of flow across any two lines

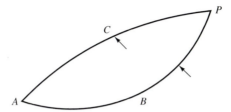

Figure 8.2   Fluid region showing the positive flow direction used in the
definition of a stream function.

ACP and *ABP* must be the same if the density is constant and no fluid is created or destroyed within the region as a consequence of continuity. If *A* is a fixed point and *P* a movable point, the flow rate across any line connecting the two points is a function of the position of *P*. If this function is ψ, and if it is taken as a sign convention that it denotes the flow rate from right to left as the observer views the line from *A* looking toward *P*, then

$$\psi = \psi(x, y)$$

is defined as the stream function.

   If ψ_1 and ψ_2 represent the values of stream function at points P_1 and P_2 (Fig. 8.3), respectively, then $\psi_2 - \psi_1$ is the flow across $P_1 P_2$ and is independent of the location of *A*. Taking another point *O* in the place of *A* changes the values of ψ_1 and ψ_2 by the same amount, namely, the flow across *OA*. Then ψ is indeterminate by an arbitrary constant.

   The velocity components *u*, *v* in the *x*, *y* directions can be obtained from the stream function. In Fig. 8.4*a*, the flow $\delta\psi$ across $\overline{AP} = \delta y$, from right to left, is

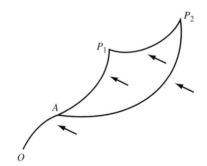

Figure 8.3   Flow between two points
in a fluid region.

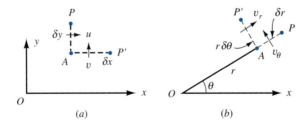

Figure 8.4   Selection of path to show relation of velocity
components to stream function.

$-u\,\delta y$, or

$$u = -\frac{\delta \psi}{\delta y} = -\frac{\partial \psi}{\partial y} \tag{8.5.1}$$

and similarly

$$v = \frac{\delta \psi}{\delta x} = \frac{\partial \psi}{\partial x} \tag{8.5.2}$$

In words, the partial derivative of the stream function with respect to any direction gives the velocity component $+90°$ (counterclockwise) to that direction. In plane polar coordinates

$$v_r = -\frac{1}{r}\frac{\partial \psi}{\partial \theta} \qquad v_\theta = \frac{\partial \psi}{\partial r}$$

from Fig. 8.4b.

When the two points P_1 and P_2 of Fig. 8.3 lie on the same streamline, $\psi_1 - \psi_2 = 0$ as there is no flow across a streamline. Hence, a streamline is given by $\psi = $ constant. Comparing Eqs. (8.3.4) with Eqs. (8.5.1) and (8.5.2) leads to

$$\frac{\partial \phi}{\partial x} = \frac{\partial \psi}{\partial y} \qquad \frac{\partial \phi}{\partial y} = -\frac{\partial \psi}{\partial x} \tag{8.5.3}$$

These are the Cauchy-Riemann equations.

By Eqs. (8.5.3), a stream function can be found for each velocity potential. If the velocity potential satisfies the Laplace equation, the stream function also satisfies it. Hence, the stream function can be considered as the velocity potential for another flow case.

Stokes Stream Function for Axially Symmetric Flow

In any one of the planes through the axis of symmetry select two points A and P such that A is fixed and P is variable. Draw a line connecting AP. The flow through the surface generated by rotating AP about the axis of symmetry is a function of the position of P. Let this function be $2\pi\psi$, and let the axis of symmetry be the x axis of a cartesian system of reference. Then ψ is a function of x and $\hat{\omega}$, where

$$\hat{\omega} = \sqrt{y^2 + z^2}$$

is the distance from P to the x axis. The surfaces $\psi = $ constant are stream surfaces.

To find the relation between ψ and the velocity components u and v' parallel to the x axis and the $\hat{\omega}$ axis (perpendicular to the x axis), respectively, a procedure similar to that for two-dimensional flow is employed. Let PP' be an infinitesimal step parallel first to $\hat{\omega}$ and then to x; that is, $PP' = \delta\hat{\omega}$ and then $PP' = \delta x$. The resulting relations between stream function and velocity are given by

$$-2\pi\hat{\omega}\,\delta\hat{\omega}\,u = 2\pi\,\delta\psi \qquad \text{and} \qquad 2\pi\hat{\omega}\,\delta x\,v' = 2\pi\,\delta\psi$$

Solving for u, v' gives

$$u = -\frac{1}{\hat{\omega}}\frac{\partial \psi}{\partial \hat{\omega}} \qquad v' = \frac{1}{\hat{\omega}}\frac{\partial \psi}{\partial x} \tag{8.5.4}$$

The same sign convention is used as in the two-dimensional case.

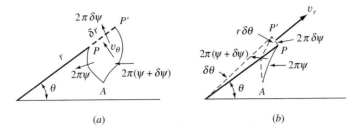

Figure 8.5 Displacement of *P* to show the relation between velocity components and Stokes stream function.

The relations between the stream function and the potential function are

$$\frac{\partial \phi}{\partial x} = \frac{1}{\hat{\omega}} \frac{\partial \psi}{\partial \hat{\omega}} \qquad \frac{\partial \phi}{\partial \hat{\omega}} = -\frac{1}{\hat{\omega}} \frac{\partial \psi}{\partial x} \qquad (8.5.5)$$

In three-dimensional flow with axial symmetry, ψ has the dimensions $L^3 T^{-1}$, or volume per unit time.

The stream function is used for flow about bodies of revolution that are frequently expressed most readily in spherical polar coordinates. Let r be the distance from the origin and θ be the polar angle; the meridian angle is not needed because of axial symmetry. From Figs. 8.5*a* and *b*

$$2\pi r \sin\theta \, \delta r \, v_\theta = 2\pi \, \delta\psi \qquad -2\pi r \sin\theta \, r \, \delta\theta \, v_r = 2\pi \, \delta\psi$$

from which

$$v_\theta = \frac{1}{r \sin\theta} \frac{\partial \psi}{\partial r} \qquad v_r = -\frac{1}{r^2 \sin\theta} \frac{\partial \psi}{\partial \theta} \qquad (8.5.6)$$

and

$$\frac{1}{\sin\theta} \frac{\partial \psi}{\partial \theta} = r^2 \frac{\partial \phi}{\partial r} \qquad \frac{\partial \psi}{\partial r} = -\sin\theta \frac{\partial \phi}{\partial \theta} \qquad (8.5.7)$$

These expressions are useful in dealing with flow about spheres, ellipsoids, and disks and through apertures.

Boundary Conditions

At a fixed boundary the velocity component normal to the boundary must be zero at every point on the boundary (Fig. 8.6)

$$\mathbf{q} \cdot \mathbf{n}_1 = 0 \qquad (8.5.8)$$

Figure 8.6 Notation for the boundary condition at a fixed boundary.

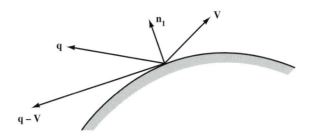

Figure 8.7   Notation for the boundary condition at a moving boundary.

where $\mathbf{n}_1$ is a unit vector normal to the boundary. In scalar notation this is easily expressed in terms of the velocity potential

$$\frac{\partial \phi}{\partial n} = 0 \tag{8.5.9}$$

at all points on the boundary. For a moving boundary (Fig. 8.7), where the boundary point has the velocity $\mathbf{V}$, the fluid velocity component normal to the boundary must equal the velocity of the boundary normal to the boundary; thus,

$$\mathbf{q} \cdot \mathbf{n}_1 = \mathbf{V} \cdot \mathbf{n}_1 \tag{8.5.10}$$

or

$$(\mathbf{q} - \mathbf{V}) \cdot \mathbf{n}_1 = 0 \tag{8.5.11}$$

For two fluids in contact, a dynamical boundary condition is required, that is, the pressure must be continuous across the interface.

A stream surface in steady flow (fixed boundaries) satisfies the condition for a boundary and can be taken as a solid boundary.

EXERCISES

8.5.1   The Stokes stream function applies to (*a*) all three-dimensional ideal-fluid-flow cases; (*b*) ideal (nonviscous) fluids only; (*c*) irrotational flow only; (*d*) cases of axial symmetry; (*e*) none of these cases.

8.5.2   The Stokes stream function has the value $\psi = 1$ at the origin and the value $\psi = 2$ at (1, 1, 1). The discharge through the surface between these points is (*a*) 1; (*b*) π; (*c*) 2π; (*d*) 4; (*e*) none of these answers.

8.5.3   The two-dimensional stream function (*a*) is constant along an equipotential surface; (*b*) is constant along a streamline; (*c*) is defined for irrotational flow only; (*d*) relates velocity and pressure; (*e*) is none of these answers.

8.5.4   In two-dimensional flow $\psi = 4$ ft^2/s at (0, 2) and $\psi = 2$ ft^2/s at (0, 1). The discharge between the two points is (*a*) from left to right; (*b*) 4π cfs/ft; (*c*) 2 cfs/ft; (*d*) $1/\pi$ cfs/ft; (*e*) none of these answers.

8.5.5   The boundary condition for steady flow of an ideal fluid is that the (*a*) velocity is zero at the boundary; (*b*) velocity component normal to the boundary is zero; (*c*) velocity component tangent to the boundary is zero; (*d*) boundary surface must be stationary; (*e*) continuity equation must be satisfied.

8.6   TWO-DIMENSIONAL FLOWS

Flow Net

In general, distributions of ψ and ϕ are obtained by solving Laplace's equation. For irregular geometries numerical methods centered on relaxation methods [1] are employed.

For the flow examples in this section a number of exact solutions for flows with relatively simple geometries and boundary conditions can be obtained. Functions describing the spatial distribution of ψ and ϕ at every point in the flow field are obtained. In order to visualize the resulting stream and potential function distributions it is customary to create a *flow net,* which is composed of a family of lines (or contours) of constant ϕ and lines (or contours) of constant ψ. A line (or contour) of constant ϕ is called an *equipotential* line, and it can be readily shown that the velocity vector is everywhere normal to the equipotential line. A line (or contour) of constant ψ is everywhere tangent to the velocity vector and will always intersect an equipotential line at right angles to it. In other words streamlines and equipotential lines are *orthogonal*. In drawing the flow net it is customary (Fig. 8.8) to let the change in constant between the adjacent equipotential lines or corresponding streamlines be constant. With reference then to Fig 8.8 the velocity u_s can be found exactly by inserting the position coordinates in the functions for $\phi(x, y)$ and $\psi(x, y)$ and differentiating. Alternatively, it can be estimated from the flow net as

$$u_s \approx -\frac{\Delta\phi}{\Delta s} = -\frac{-\Delta c}{\Delta s} = \frac{\Delta c}{\Delta s}$$

Similarly for v_s

$$v_s \approx \frac{\Delta\psi}{\Delta n} = \frac{\Delta c}{\Delta n}$$

In the limit as Δn and Δs approach zero, the functional estimate from the exact solution is obtained. The accompanying dynamic pressure head can be found by use of the Bernoulli equation [Eq. (8.4.6)].

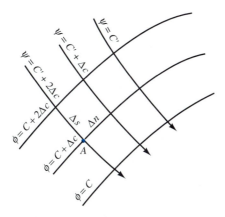

Figure 8.8   Elements of a flow net.

Because of the similarity of the differential equations describing groundwater flow and irrotational flow, the flow net can be used to determine streamlines and lines of constant piezometric head $(h + p/\gamma)$ for percolation through homogeneous porous media. Therefore, the following flow cases can also be interpreted in terms of the highly rotational, slow, viscous flow through a porous medium.

Two simple flow cases that can be interpreted for flow along straight boundaries are first examined. Then the source, vortex, doublet, uniform flow, and flow around a cylinder, with and without circulation, are discussed.

Flow around a Corner

The potential function

$$\phi = A(x^2 - y^2)$$

has as its stream function

$$\psi = 2Axy = A r^2 \sin 2\theta$$

in which r and θ are polar coordinates. It is plotted for equal-increment changes in ϕ and ψ in Fig. 8.9. Conditions at the origin are not defined, as it is a stagnation point. As any of the streamlines may be taken as fixed boundaries, the plus axes may be taken as walls, yielding flow into a 90° corner. The equipotential lines are hyperbolas having axes coincident with the coordinate axes and asymptotes given by $y = \pm x$. The streamlines are rectangular hyperbolas having $y = \pm x$ as axes and the coordinate axes as asymptotes. From the polar form of the stream function it is noted that the two lines $\theta = 0$ and $\theta = \pi/2$ are the streamline $\psi = 0$.

This case can be generalized to yield flow around a corner with angle α. By examining

$$\phi = A r^{\pi/\alpha} \cos \frac{\pi\theta}{\alpha} \qquad \psi = A r^{\pi/\alpha} \sin \frac{\pi\theta}{\alpha}$$

it is noted that the streamline $\psi = 0$ is now given by $\theta = 0$ and $\theta = \alpha$. Two flow nets are shown in Fig. 8.10 for $\alpha = 225°$ and $\alpha = 45°$.

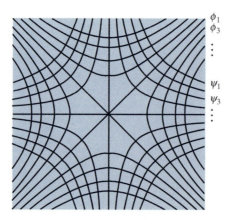

Figure 8.9   Flow net for flow around a 90° bend.

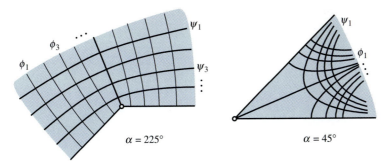

ψ_1

ϕ_3

ϕ_1

ψ_3

ψ_1

ϕ_1

$\alpha = 225°$

$\alpha = 45°$

Figure 8.10   Flow net for flow along two inclined surfaces.

Source and Sink

A line normal to the xy plane, from which fluid is imagined to flow uniformly in all directions *at right angles* to it, is a source. It appears as a point in the customary two-dimensional flow diagram. The total flow per unit time and unit length of line is called the *strength* of the source. As the flow is in radial lines from the source, the velocity a distance r from the source is determined by the strength divided by the flow area of the cylinder, or $2\pi\mu/2\pi r$, in which the strength is $2\pi\mu$. Then, since by Eq. (8.3.4) the velocity in any direction is given by the negative derivative of the velocity potential with respect to the direction,

$$-\frac{\partial\phi}{\partial r} = \frac{\mu}{r} \qquad \frac{\partial\phi}{\partial\theta} = 0$$

and

$$\phi = -\mu \ln r$$

is the velocity potential, in which r is the distance from the source. This value of ϕ satisfies the Laplace equation in two dimensions.

The streamlines are radial lines from the source, that is,

$$\frac{\partial\psi}{\partial r} = 0 \qquad -\frac{1}{r}\frac{\partial\psi}{\partial\theta} = \frac{\mu}{r}$$

From the second equation

$$\psi = -\mu\theta$$

Lines of constant ϕ (equipotential lines) and constant ψ are shown in Fig. 8.11. A *sink* is a negative source, a line into which fluid is flowing.

Vortex

For the flow case given by selecting the stream function for the source as a velocity potential

$$\phi = -\mu\theta \qquad \psi = \mu \ln r$$

which also satisfies the Laplace equation, it is seen that the equipotential lines are radial lines and the streamlines are circles. The velocity is in a tangential direction

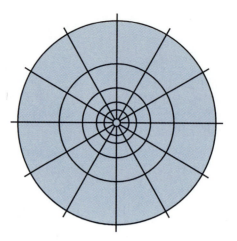

Figure 8.11   Flow net for source or vortex.

only, since $\partial\phi/\partial r = 0$. It is

$$q = -\frac{1}{r}\frac{\partial\phi}{\partial\theta} = \frac{\mu}{r}$$

since $r\,\delta\theta$ is the length element in the tangential direction.

In referring to Fig. 8.12, the *flow along a closed curve* is called the *circulation*. The flow along an element of the curve is defined as the product of the length element δs of the curve and the component of the velocity tangent to the curve, $q\cos\alpha$. Hence, the circulation Γ around a closed path C is

$$\Gamma = \int_C q\cos\alpha\,ds = \int_C \mathbf{q}\cdot ds$$

The velocity distribution given by the equation $\phi = -\mu\theta$ is for the *vortex* and is such that the circulation around any closed path that contains the vortex is constant. The value of the circulation is the strength of the vortex. By selecting any circular path with radius r to determine the circulation, $\alpha = 0°$, $q = \mu/r$, and $ds = r\,d\theta$; hence,

$$\Gamma = \int_C q\cos\alpha\,ds = \int_0^{2\pi}\frac{\mu}{r}r\,d\theta = 2\pi\mu$$

At the point $r = 0$, $q = \mu/r$ goes to infinity; hence, this point is called a singular point. Figure 8.11 shows the equipotential lines and streamlines for the vortex.

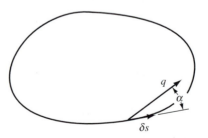

Figure 8.12   Notation for the definition of circulation.

Doublet

The two-dimensional doublet is defined as the limiting case as a source and sink of equal strength approach each other so that the product of their strength and the distance between them remains a constant $2\pi\mu$; μ is called the *strength* of the doublet. The axis of the doublet is from the sink toward the source, that is, the line along which they approach each other.

In Fig. 8.13 a source is located at $(a,0)$ and a sink of equal strength at $(-a,0)$. The velocity potential for both, at some point P, is

$$\phi = -m \ln r_1 + m \ln r_2$$

with r_1 and r_2 measured from source and sink, respectively, to the point P. Thus, $2\pi m$ is the strength of the source and sink. To take the limit as a approaches zero for $2am = \mu$, the form of the expression for ϕ must be altered. The terms r_1 and r_2 can be expressed in terms of the polar coordinates r and θ by the cosine law:

$$r_1^2 = r^2 + a^2 - 2ar\cos\theta = r^2\left[1 + \left(\frac{a}{r}\right)^2 - 2\frac{a}{r}\cos\theta\right]$$

$$r_2^2 = r^2 + a^2 + 2ar\cos\theta = r^2\left[1 + \left(\frac{a}{r}\right)^2 + 2\frac{a}{r}\cos\theta\right]$$

Rewriting the expression for ϕ with these relations gives

$$\phi = -\frac{m}{2}(\ln r_1^2 - \ln r_2^2) = -\frac{m}{2}\left\{\ln r^2 + \ln\left[1 + \left(\frac{a}{r}\right)^2 - 2\frac{a}{r}\cos\theta\right]\right.$$
$$\left. - \ln r^2 - \ln\left[1 + \left(\frac{a}{r}\right)^2 + 2\frac{a}{r}\cos\theta\right]\right\}$$

The series expression

$$\ln(1+x) = x - \frac{x^2}{2} + \frac{x^3}{3} - \frac{x^4}{4} + \cdots$$

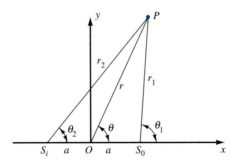

Figure 8.13   Notation for the derivation of a two-dimensional doublet.

leads to

$$
\phi = -\frac{m}{2}\left\{ \left[\left(\frac{a}{r}\right)^2 - 2\frac{a}{r}\cos\theta - \frac{1}{2}\left[\left(\frac{a}{r}\right)^2 - 2\frac{a}{r}\cos\theta\right]^2 \right. \right.
$$

$$
+ \frac{1}{3}\left[\left(\frac{a}{r}\right)^2 - 2\frac{a}{r}\cos\theta\right]^3 - \cdots - \left[\left(\frac{a}{r}\right)^2 + 2\frac{a}{r}\cos\theta\right]
$$

$$
\left. \left. + \frac{1}{2}\left[\left(\frac{a}{r}\right)^2 + 2\frac{a}{r}\cos\theta\right]^2 - \frac{1}{3}\left[\left(\frac{a}{r}\right)^2 + 2\frac{a}{r}\cos\theta\right]^3 + \cdots \right\}\right.
$$

After simplifying

$$
\phi = 2am\left[\frac{\cos\theta}{r} + \left(\frac{a}{r}\right)^2\frac{\cos\theta}{r} - \left(\frac{a}{r}\right)^4\frac{\cos\theta}{r} - \frac{4}{3}\left(\frac{a}{r}\right)^2\frac{\cos^3\theta}{r} + \cdots \right]
$$

If $2am = \mu$, and if the limit is taken as a approaches zero, then

$$
\phi = \frac{\mu\,\cos\theta}{r}
$$

which is the velocity potential for a two-dimensional doublet at the origin, with the axis in the $+x$ direction.

Using the relations

$$
v_r = -\frac{\partial\phi}{\partial r} = -\frac{1}{r}\frac{\partial\psi}{\partial\theta} \qquad v_\theta = -\frac{1}{r}\frac{\partial\phi}{\partial\theta} = \frac{\partial\psi}{\partial r}
$$

gives for the doublet

$$
\frac{\partial\psi}{\partial\theta} = -\frac{\mu\,\cos\theta}{r} \qquad \frac{\partial\psi}{\partial r} = \frac{\mu}{r^2}\sin\theta
$$

After integrating,

$$
\psi = -\frac{\mu\,\sin\theta}{r}
$$

is the stream function for the doublet. The equations in cartesian coordinates are

$$
\phi = \frac{\mu x}{x^2 + y^2} \qquad \psi = -\frac{\mu y}{x^2 + y^2}
$$

Rearranging gives

$$
\left(x - \frac{\mu}{2\phi}\right)^2 + y^2 = \frac{\mu^2}{4\phi^2} \qquad x^2 + \left(y + \frac{\mu}{2\psi}\right)^2 = \frac{\mu^2}{4\psi^2}
$$

The lines of constant ϕ are circles through the origin with centers on the x axis, and the streamlines are circles through the origin with centers on the y axis, as shown in Fig. 8.14. The origin is a singular point where the velocity goes to infinity.

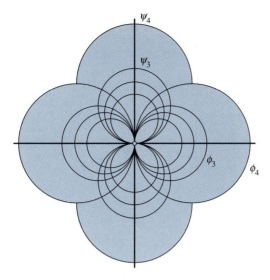

Figure 8.14 Equipotential lines and streamlines for the two-dimensional doublet.

Uniform Flow

Uniform flow in the $-x$ direction, $u = -U$, is expressed by

$$\phi = Ux \qquad \psi = Uy$$

and in polar coordinates as

$$\phi = Ur \cos \theta \qquad \psi = Ur \sin \theta$$

Flow around a Circular Cylinder

The addition of the flow due to a doublet and a uniform flow results in flow around a circular cylinder; thus,

$$\phi = Ur \cos \theta + \frac{\mu \cos \theta}{r} \qquad \psi = Ur \sin \theta - \frac{\mu \sin \theta}{r}$$

As a streamline in steady flow is a possible boundary, the streamline $\psi = 0$ is given by

$$0 = \left(Ur - \frac{\mu}{r}\right) \sin \theta$$

which is satisfied by $\theta = 0$, π, or by the value of r that satisfies

$$Ur - \frac{\mu}{r} = 0$$

If this value is $r = a$, which is a circular cylinder, then

$$\mu = Ua^2$$

and the streamline $\psi = 0$ is the x axis and the circle $r = a$. The potential and stream functions for uniform flow around a circular cylinder of radius a are, by substitution of the value of μ, given by

$$\phi = U\left(r + \frac{a^2}{r}\right)\cos\theta \qquad \psi = U\left(r - \frac{a^2}{r}\right)\sin\theta$$

for the uniform flow in the $-x$ direction. The equipotential lines and streamlines for this case are shown in Fig. 8.15.

The velocity at any point in the flow can be obtained from either the velocity potential or the stream function. On the surface of the cylinder the velocity is necessarily tangential and is expressed by $\partial\psi/\partial r$ for $r = a$; thus,

$$q\bigg|_{r=a} = U\left(1 + \frac{a^2}{r^2}\right)\sin\theta\bigg|_{r=a} = 2U\sin\theta$$

The velocity is zero (stagnation point) at $\theta = 0, \pi$ and has maximum values of $2U$ at $\theta = \pi/2, 3\pi/2$. For the dynamic pressure zero at infinity, with Eq. (8.4.7) for $p_0 = 0, q_0 = U$,

$$p = \frac{\rho}{2}U^2\left[1 - \left(\frac{q}{U}\right)^2\right]$$

which holds for any point in the plane except the origin. For points on the cylinder

$$p = \frac{\rho}{2}U^2(1 - 4\sin^2\theta)$$

The maximum pressure, which occurs at the stagnation points, is $\rho U^2/2$; and the minimum pressure, at $\theta = \pi/2, 3\pi/2$, is $-3\rho U^2/2$. The points of zero dynamic pressure are given by $\sin\theta = \pm\frac{1}{2}$, or $\theta = \pm\pi/6, \pm5\pi/6$. A cylindrical pitot-static tube is made by providing three openings in a cylinder, at 0 and $\pm30°$, as the difference in pressure between 0 and $\pm30°$ is the dynamic pressure $\rho U^2/2$.

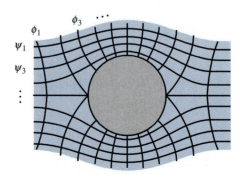

Figure 8.15   Equipotential lines and stream-lines for flow around a circular cylinder.

The drag on the cylinder is shown to be zero by integration of the x component of the pressure force over the cylinder; thus,

$$\text{Drag} = \int_0^{2\pi} pa \cos \theta \, d\theta = \frac{\rho a U^2}{2} \int_0^{2\pi} (1 - 4 \sin^2 \theta) \cos \theta \, d\theta = 0$$

Similarly, the lift force on the cylinder is zero.

Flow around a Circular Cylinder with Circulation

The addition of a vortex to the doublet and the uniform flow results in flow around a circular cylinder with circulation and

$$\phi = U\left(r + \frac{a^2}{r}\right)\cos \theta - \frac{\Gamma}{2\pi}\theta \qquad \psi = U\left(r - \frac{a^2}{r}\right)\sin \theta + \frac{\Gamma}{2\pi}\ln r$$

The streamline $\psi = (\Gamma/2\pi)\ln a$ is the circular cylinder $r = a$. At great distances from the origin, the velocity remains $u = -U$, showing that flow around a circular cylinder is maintained with addition of the vortex. Some of the streamlines are shown in Fig. 8.16.

The velocity at the surface of the cylinder, which is necessarily tangent to the cylinder, is

$$q = \left.\frac{\partial \psi}{\partial r}\right|_{r=a} = 2U \sin \theta + \frac{\Gamma}{2\pi a}$$

Stagnation points occur when $q = 0$, that is,

$$\sin \theta = -\frac{\Gamma}{4\pi U a}$$

When the circulation is $4\pi U a$, the two stagnation points coincide at $r = a$, $\theta = -\pi/2$. For larger circulation, the stagnation point moves away from the cylinder.

The pressure at the surface of the cylinder is

$$p = \frac{\rho U^2}{2}\left[1 - \left(2 \sin \theta + \frac{\Gamma}{2\pi a U}\right)^2\right]$$

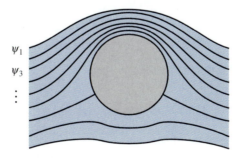

ψ_1
ψ_3
$\vdots$

Figure 8.16   Streamlines for flow around a
circular cylinder with circulation.

The drag again is zero. The lift, however, becomes

$$\text{Lift} = -\int_0^{2\pi} pa \sin\theta \, d\theta$$

$$= -\frac{\rho a U^2}{2} \int_0^{2\pi} \left[1 - \left(2\sin\theta + \frac{\Gamma}{2\pi a U}\right)^2\right] \sin\theta \, d\theta = \rho U \Gamma$$

showing that the lift is directly proportional to the density of fluid, the approach velocity U, and the circulation Γ. This thrust, which acts at right angles to the approach velocity, is referred to as the *Magnus effect*. The Flettner rotor ship was designed to utilize this principle by mounting circular cylinders with axes vertical on the ship and then mechanically rotating the cylinders to provide circulation. Airflow around the rotors produces the thrust at right angles to the relative wind direction. The close spacing of streamlines along the upper side of Fig. 8.16 indicates that the velocity is high there and that the pressure must then be correspondingly low.

The theoretical flow around a circular cylinder with circulation can be transformed [2] into flow around an airfoil with the same circulation and the same lift. The airfoil develops its lift by producing a circulation around it due to its shape. It can be shown [2] that the lift is $\rho U \Gamma$ for any cylinder in two-dimensional flow. The angle of inclination of the airfoil relative to the approach velocity (angle of attack) greatly affects the circulation. For large angles of attack the flow does not follow the wing profile, and the theory breaks down.

It should be mentioned that all two-dimensional ideal-fluid-flow cases can be conveniently handled by complex-variable theory and by a system of *conformal mapping*, which transforms a flow net from one configuration to another by a suitable complex-variable mapping function.

Example 8.3

A source with strength 0.2 m³/s·m and a vortex with strength 1 m²/s are located at the origin. Determine the equations for velocity potential and stream function. What are the velocity components at $x = 1$ m, $y = 0.5$ m?

Solution

The velocity potential for the source is

$$\phi = -\frac{0.2}{2\pi} \ln r \qquad \text{m}^2/\text{s}$$

and the corresponding stream function is

$$\psi = -\frac{0.2}{2\pi} \theta \qquad \text{m}^2/\text{s}$$

The velocity potential for the vortex is

$$\phi = -\frac{1}{2\pi} \theta \qquad \text{m}^2/\text{s}$$

and the corresponding stream function is

$$\psi = \frac{1}{2\pi} \ln r \qquad \text{m}^2/\text{s}$$

Adding the respective functions gives

$$\phi = -\frac{1}{\pi}\left(0.1 \ln r + \frac{\theta}{2}\right) \quad \text{and} \quad \psi = -\frac{1}{\pi}\left(0.1\theta - \frac{1}{2} \ln r\right)$$

The radial and tangential velocity components are

$$v_r = -\frac{\partial\phi}{\partial r} = \frac{1}{10\pi r} \qquad v_\theta = -\frac{1}{r}\frac{\partial\phi}{\partial\theta} = \frac{1}{2\pi r}$$

At $(1, 0.5)$, $r = \sqrt{1^2 + 0.5^2} = 1.117$ m, $v_r = 0.0285$ m/s, and $v_\theta = 0.143$ m/s.

A circular cylinder 2 m in diameter and 20 m long is rotating at 120 rpm in the positive direction (counterclockwise) about its axis. Its center is at the origin of a cartesian coordinate system. Wind at 10 m/s blows over the cylinder in the positive x direction; $t = 20°C$ and $p = 100$-kPa abs. Determine the lift on the cylinder.

Example 8.4

Solution

The stagnation point has $\psi = 0$. By selecting increments of R, θ can be determined from

$$\sin\theta = \frac{-\Gamma}{2\pi U}\frac{\ln R}{R - 1/R}$$

The lift is given by $\rho U \Gamma L$.

EXERCISES

8.6.1   Select the relation that must hold in two-dimensional, irrotational flow
(a) $\dfrac{\partial\theta}{\partial x} = \dfrac{\partial\psi}{\partial y}$;   (b) $\dfrac{\partial\phi}{\partial x} = -\dfrac{\partial\psi}{\partial y}$;   (c) $\dfrac{\partial\phi}{\partial y} = \dfrac{\partial\psi}{\partial x}$;   (d) $\dfrac{\partial\phi}{\partial x} = \dfrac{\partial\phi}{\partial y}$; (e) none of these answers.

8.6.2   A source in two-dimensional flow (a) is a point from which fluid is imagined to flow outward uniformly in all directions; (b) is a line from which fluid is imagined to flow uniformly in all directions at right angles to it; (c) has a strength defined as the speed at unit radius; (d) has streamlines that are concentric circles; (e) has a velocity potential independent of the radius.

8.6.3   The two-dimensional vortex (a) has a strength given by the circulation around a path enclosing the vortex; (b) has radial streamlines; (c) has a zero circulation around it; (d) has a velocity distribution that varies directly as the radial distance from the vortex; (e) creates a velocity distribution that has rotation throughout the fluid.

8.7   WATER WAVES: A MOVING BOUNDARY PROBLEM

As suggested at the close of Sec. 8.5, the two basic features of the exact solutions in Sec. 8.6 may not hold. In fact not only may the flow be unsteady but many practical problems are marked by flow fields where the boundaries themselves are deforming or moving in time. Moving boundaries are most often confined to regions where sharp if not discontinuous density gradients occur in the fluid, and most of these problems involve the analysis and prediction of waves at the interface or moving boundary. Examples include the ubiquitous water waves seen at the beach (an air-water interface), the wave-like deformation of the fluidized sediments comprising the bottom of a river or channel during *bedload* transport, or the mountainous wind-driven dunes in the Sahara Desert.

The earliest approaches to the problem of predicting waves were due to Airy [3] and Stokes [4] who formulated and solved the problem of water waves via a time-varying potential function. The resulting linear wave theory approach to the problem of the free surface is in widespread design use today by coastal engineers.

Figure 8.17 contains a schematic of the two-dimensional vertical plane flow field and terminology. While we are not concerned with how the wave surface deformation arose, it is noted that a *progressive* wave is depicted moving to the right with a wave speed or *celerity, C.* The water depth from the bottom to the undisturbed or still water level (SWL) is d and for this section is assumed constant. The time- and space-varying deformation of the free surface is measured relative to the SWL and is denoted by $\eta(x, y, t)$. The distance between the maximum height (the crest) and the minimum height (the trough) is called the *wave height, H.* An observer fixed at the origin would see the wave form repeat itself over one period, T, which yields the wavelength, L, or distance between successive crests, as

$$L = CT$$

where C is the wave speed. The unit normals, $\mathbf{N}$ and $\mathbf{n}$, correspond to the bottom and surface, respectively.

The first and most fundamental assumption is frictionless or irrotational flow, in which case a potential function solution can be found by integrating the Laplace equation [Eqs. (8.3.6) and (8.3.7)], that is,

$$\nabla^2 \phi = 0$$

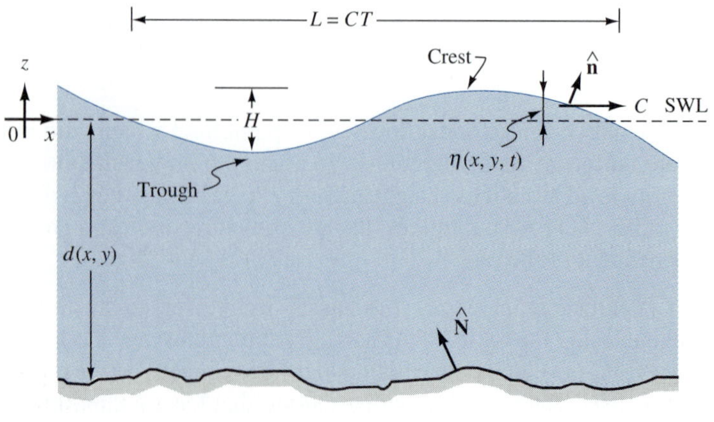

Figure 8.17   Water wave definition sketch.

The following boundary conditions are required. With reference to Eq. (8.5.8) and Fig. 8.6 the fluid particles cannot cross the solid boundary at the bottom. Therefore,

$$\mathbf{q} \cdot \mathbf{N} = \frac{\partial \phi}{\partial N} = 0 \tag{8.7.1}$$

At the *free surface* the boundary is moving and its position can only be found by solution of the problem; therefore, the problem is highly nonlinear. With regard to Eq. (8.5.10) and Fig. 8.7 the *kinematic boundary condition* says that a fluid particle at the surface must stay at the surface $z = \eta$, that is,

$$\mathbf{q} \cdot \mathbf{n} = \mathbf{V} \cdot \mathbf{n} \tag{8.7.2}$$

This boundary condition can also be written as

$$\mathbf{q} \Big|_{z=\eta} = -\nabla \phi \Big|_{z=\eta} = -\frac{\partial \phi}{\partial n} \Big|_{z=\eta} \tag{8.7.3}$$

Note again that this boundary condition applies at the free surface $z = \eta$ whose position is unknown.

The final condition is the *dynamic surface condition*. Here the pressure on the water surface is assumed to be gage pressure zero, and because the velocity vector at the free surface is everywhere tangent to it, the free surface is therefore a streamline and Bernoulli's equation applies [Eq. (8.4.4)], that is,

$$\frac{p}{\rho} + g\eta + \frac{q^2}{2} - \frac{\partial \phi}{\partial t} = F(t)$$

If the datum is taken as the free surface, then $F(t) = 0$, and if the pressure is at gage, then the equation becomes

$$\frac{q^2}{2} + g\eta - \frac{\partial \phi}{\partial t} = 0 \tag{8.7.4}$$

At this point the problem is highly nonlinear, due to the unknown position of the free surface and the presence of the kinetic energy term ($q^2/2$) in the dynamic surface condition.

Two simplifying assumptions are now made to linearize the problem. First, the small amplitude assumption is invoked where η is assumed to be much smaller than the wavelength, L. In so doing the kinematic condition [Eq. (8.7.3)] can now be approximated by application at $z = 0$, the SWL, instead of the free surface, that is,

$$\mathbf{q} \Big|_{z=\eta} \approx \frac{\partial \eta}{\partial t} = \frac{\partial \phi}{\partial z} \Big|_{z=0} \tag{8.7.5}$$

The second assumption linearizes the dynamic condition by neglecting the term $q^2/2$ on the basis of its small size relative to the other terms. Subsequently,

$$\eta = -\frac{1}{g} \frac{\partial \phi}{\partial t} \Big|_{z=0} \tag{8.7.6}$$

Ocean and coastal engineering textbooks review these assumptions in greater detail but indicate that these assumptions are valid for $H/L \leq 1/50$. Almost 50 percent of the full wave energy distribution falls into the linear category.

The two linearized free surface conditions [Eqs. (8.7.5) and (8.7.6)] can now be put together by elimination of η to get

$$\frac{\partial \phi}{\partial z} + \frac{1}{g}\frac{\partial^2 \phi}{\partial t^2} = 0$$

which is approximately valid at $z = 0$. Equation (8.7.6) along with the Laplace equation [Eqs. (8.3.7) and (8.7.1)] are now integrated to give the following potential function for a right running wave

$$\phi(x, z, t) = \left(\frac{\pi H}{kT}\right)\frac{\cosh[k(z + d)]}{\cosh(kd)}\sin(kx - \omega t) \qquad (8.7.7)$$

where $k = 2\pi/L$ is called the *wave number* and $\omega = 2\pi/T$ is called the *circular frequency*. By inserting Eq. (8.7.7) back into Eq. (8.7.6), the free surface position is also found for the right running wave

$$\eta(x, t) = \frac{H}{2}\cos(kx - \omega t) \qquad (8.7.8)$$

In order to determine the speed of the wave, the argument $(kx - \omega t)$ may be written as $k(x - Ct)$ where C is the wave speed. The wave speed C required to keep a position on the wave envelope fixed is found by substituting Eq. (8.7.7) into the boundary condition Eq. (8.7.6) and evaluating it at $z = 0$. The resulting wave speed is given by

$$C^2 = \frac{g}{k}\tanh kd \qquad (8.7.9)$$

Therefore, wave speed is a function of both depth and wavelength. When the relative depth $d/L > 1/2$, then a *deep water* condition results where

$$C^2 = \frac{g}{k} \qquad (8.7.10)$$

When $d/L < 1/20$, then a *shallow water* wave results with

$$C^2 = gd \qquad (8.7.11)$$

Finally the local fluid velocities are found by differentiating the potential function, yielding

$$u(x, z, t) = -\frac{\partial \phi}{\partial x} = \frac{HgT}{2L}\frac{\cosh[k(z + d)]}{\cosh(kd)}\cos(kx - \omega t) \qquad (8.7.12)$$

$$w(x, z, t) = -\frac{\partial \phi}{\partial z} = \frac{HgT}{2L}\frac{\sinh[k(z + d)]}{\cosh(kd)}\sin(kx - \omega t) \qquad (8.7.13)$$

Figure 8.18 contains a schematic of the total velocity vector at two depths in the water column for various positions (fixed values of the argument $kx - \omega t$) during a complete wave. As noted in Example 4.3, the velocity functions are periodic functions with a depth-varying amplitude function, that is,

$$u = A(z)\cos(kx - \omega t)$$
$$w = B(z)\sin(kx - \omega t)$$

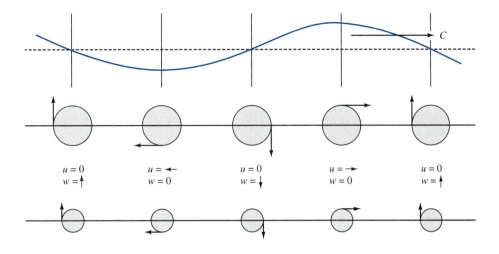

Figure 8.18   Total velocity vectors for two depths versus horizontal position.

Therefore, for each fixed position or argument the velocity vectors have the same direction but diminish in magnitude with increasing depth towards the bottom.

At each fixed depth the total velocity vector changes orientation with position under the wave. When $kx - \omega t = 0, \pi,$ or 2π radians, the maximum and minimum elevations occur and the total velocity vector is horizontal, being positive or right running at the crest and negative or left running at the trough. At $\pi/2$ or $3\pi/2$ when $\eta = 0$, the maximum vertical velocities are obtained, positive for $\pi/2$, negative for $3\pi/2$.

Looking at this figure it is noted that the total velocity vectors are moving in a counterclockwise progression from 0 to 2π. It further appears as if there is no net fluid parcel movement in the direction of wave advance. This is no illusion because a particle path analysis based upon the integration of the velocity field would yield just such a result. As the wave moves through the fluid no net fluid mass transport occurs. The only transport that occurs is the transmission of kinetic and potential energy. More elaborate discussion of these aspects can be found in coastal or ocean engineering textbooks [5, 6].

EXERCISES

8.7.1   At the water surface of a progressive wave (*a*) the water surface is a streamline; (*b*) the velocity is nonzero and specified by the kinematic boundary condition; (*c*) it is assumed for initial purposes that the pressure does not vary; (*d*) the Bernoulli equation applies; (*e*) all of the above.

8.7.2   The linearized, small-amplitude wave assumptions (*a*) are valid for 50 percent of the wave spectrum of engineering significance; (*b*) assume that the free surface height is much smaller than the wavelength; (*c*) assume that the kinetic energy head term of the Bernoulli equation applied at the surface is quite small relative to the other terms; (*d*) limits our wave analysis to waves in water less than 5 meters deep; (*e*) all except (*d*).

8.7.3   With regard to wave velocity and speed (*a*) the horizontal and vertical velocities are 90° out of phase; (*b*) the deepwater wave speed is linearly proportional to wavelength; (*c*) the shallow water wave speed is proportional to depth; (*d*) the shallow water wave speed of a wave in 2 meters of depth is 0.71 times slower than the same shallow water wave in 3 meters of water; (*e*) a tsunami with a wavelength of 4000 km created in water 5 km deep travels as a shallow water wave with a speed of 221 m/s; (*f*) only (*a*) and (*e*).

PROBLEMS

8.1   Compute the gradient of the following two-dimensional scalar functions:
(*a*) $\phi = -2 \ln(x^2 + y^2)$ (*b*) $\phi = Ux + Vy$ (*c*) $\phi = 2xy$.

8.2   Compute the divergence of the gradients of ϕ found in Prob. 8.1.

8.3   Compute the curl of the gradients of ϕ found in Prob. 8.1.

8.4   For $\mathbf{q} = \mathbf{i}(x + y) + \mathbf{j}(y + z) + \mathbf{k}(x^2 + y^2 + z^2)$ find the components of rotation at $(2, 2, 2)$.

8.5   Derive the following equation of continuity for two-dimensional flow in polar coordinates by equating the net efflux from a small polar element to zero (Fig. 8.19).

$$\frac{\partial v_r}{\partial r} + \frac{v_r}{r} + \frac{1}{r}\frac{\partial v_\theta}{\partial \theta} = 0$$

8.6   The *x* component of velocity is $u = x^2 + z^2 + 5$ and the *y* component is $v = y^2 + z^2$. Find the simplest *z* component of velocity that satisfies continuity.

8.7   A velocity potential in two-dimensional flow is $\phi = y + x^2 - y^2$. Find the stream function for this flow.

8.8   The two-dimensional stream function for a flow is $\psi = 9 + 6x - 4y + 7xy$. Find the velocity potential.

8.9   Derive the partial differential equations relating ϕ and ψ for two-dimensional flow in plane polar coordinates.

8.10   From the continuity equation in polar coordinates in Prob. 8.5, derive the Laplace equation in the same coordinate system.

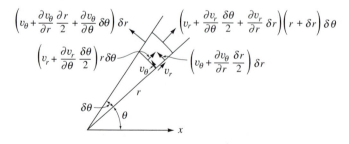

Figure 8.19   Problems 8.5 and 8.10.

8.11   Does the function $\phi = 1/r$ satisfy the Laplace equation in two dimensions? Is it satisfied in three-dimensional flow?

8.12   Use the equations developed in Prob. 8.9 to find the two-dimensional stream function for $\phi = \ln r$.

8.13   Find the Stokes stream function for $\phi = 1/r$.

8.14   For the Stokes stream function $\psi = 9r^2 \sin^2 \theta$, find ϕ in cartesian coordinates.

8.15   In Prob. 8.14 what is the discharge between stream surfaces through the points $r = 1, \theta = 0$ and $r = 1, \theta = \pi/4$?

8.16   Write the boundary conditions for steady flow around a sphere, of radius a, at its surface and at infinity.

8.17   A circular cylinder of radius a has its center at the origin and is translating with velocity V in the y direction. Write the boundary condition in terms of ϕ that is satisfied at its surface and at infinity.

8.18   A circular cylinder 8 ft in diameter rotates at 500 rpm. In an airstream, $\rho = 0.002$ slug/ft^3 and moving at 400 ft/s, what is the lift force per foot of cylinder, assuming 90-percent efficiency in developing circulation from the rotation?

8.19   Show that if two stream functions ψ_1 and ψ_2 satisfy the Laplace equation, then $\nabla^2 \psi = 0$ for $\psi = \psi_1 + \psi_2$.

8.20   Show that if u_1, v_1 and u_2, v_2 are the velocity components of two velocity potentials ϕ_1 and ϕ_2 which satisfy the Laplace equation, then for $\phi = \phi_1 + \phi_2$ the velocity components are $u = u_1 + u_2$ and $v = v_1 + v_2$.

8.21   A two-dimensional source is located at $(1, 0)$ and another one of the same strength is located at $(-1, 0)$. Construct the velocity vector $(0, 0)$, $(0, 1)$, $(0, -1)$, $(0, -2)$, and $(1, 1)$. *Hint:* Use the results of Prob. 8.20 to draw the velocity components by adding the individual velocity components induced at the point in question by each source, without regard to the other, due to its strength and location.

8.22   Determine the velocity potential for a source located at $(1, 0)$. Write the equation for the velocity potential for the source system described in Prob. 8.21.

8.23   Draw a set of streamlines for each of the sources described in Prob. 8.21 and from this diagram construct the streamlines for the combined flow. *Hint:* For each of the sources draw streamlines separated by an angle of $\pi/6$. Finally, combine the intersection points of those rays for which $\psi_1 + \psi_2$ is constant.

8.24   Does the line $x = 0$ form a line in the flow field described in Prob. 8.21 for which there is no velocity component normal to it? Is this line a streamline? Could this line be the trace of a solid plane lamina that was submerged in the flow? Does the velocity potential determined in Prob. 8.22 describe the flow in the region $x > 0$ for a source located at a distance of unity from a plane wall? Justify your answers.

8.25   Determine the equation for the velocity on the line $x = 0$ for the flow described in Prob. 8.21. Find an equation for the pressure on the surface whose trace is $x = 0$. What is the force on one side of this plane due to the source located at distance of unity from it? Water is the fluid.

8.26   In two-dimensional flow what is the nature of the flow given by $\phi = 7x + 2 \ln r$?

8.27   Use a method similar to that suggested in Prob. 8.23 to draw the potential lines for the flow given in Prob. 8.26.

8.28   Use the suggestion in Prob. 8.23 to draw a flow net for a flow consisting of a source and a vortex which are located at the origin. Use the same value for μ in both the source and the vortex.

8.29   A source discharging 20 cfs/ft is located at $(-1, 0)$, and a sink of twice the strength is located at $(2, 0)$. For dynamic pressure at the origin of 100 lb/ft^2, $\rho = 1.8$ slug/ft^3, find the velocity and dynamic pressure at $(0, 1)$ and $(1, 1)$.

8.30   Select the strength of the doublet needed to portray a uniform flow of 20 m/s around a cylinder of 2-m radius.

8.31   Develop the equations for flow around a Rankine cylinder formed by a source, an equal sink, and a uniform flow. If $2a$ is the distance between source and sink, their strength is $2\pi\mu$, and U is the uniform velocity, develop an equation for length of the body.

8.32   Calculate the cartesian coordinates of the points of intersection of streamlines and equipotential lines for Fig. 8.10; $\alpha = 225°$. Let $A = 1$ and $\Delta\phi = \Delta\psi = 1$.

8.33   Determine the cartesian coordinates of the intersections of the flow net of Fig. 8.15, first quadrant only; $U = 2$, $R = 1$, $\psi = 0, 1, \ldots, 4$, and $\phi = 0, 1, \ldots, 6$.

8.34   Show that the velocities given by Eqs. (8.7.12) and (8.7.13) satisfy the continuity equation.

8.35   If the water in a U-tube (Fig. 8.20) is forced to oscillate due to an applied pressure on one of the legs of the tube, derive the free surface kinematic boundary condition in the other leg. Assume a sinusoidal form for the applied pressure force.

8.36   Show in detail how the wave speed in Eq. (8.7.9) is given by Eq. (8.7.11) in shallow water. *Hint:* $e^{\pm kd} \approx 1 \pm kd + O(kd)^2$.

8.37   Show in detail how the wave speed given in Eq. (8.7.9) is transformed to $C^2 = g/k$ for deep water.

8.38   Tsunamis occur as the result of a sudden displacement of the sea floor due to an earthquake. Using linear wave theory estimate the velocity of the wave created if the sea floor drops by 1 ft, which correspondingly results in a surface amplitude of 1 ft. Assume the basin depth is 15,000 ft. *Hint:* A tsunami is essentially a "single" wave with infinite wavelength, so it might help to determine its period.

8.39   Plot the time trace, for one period, of the u and w velocities at a point 2 ft below the free surface and 0.2 ft ahead of the wave crest. Use $t = 0$.

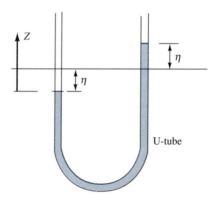

Figure 8.20   Problem 8.35.

8.40 In a wave tank 600 ft long and 60 ft deep filled with fresh water to the 20-ft mark, a 1-ft-high linear wave is generated with a period of 4 sec by a wave paddle. Determine the wave speed (celerity), wave length, wave number, and wave frequency.

8.41 Verify that Eq. (8.7.7) is in fact a solution to the Laplace equation.

8.42 A standing wave results when two progressive waves meet when traveling in opposite directions. This would occur in a basin where a right traveling wave reflects off the right wall and produces a left traveling wave which meets more oncoming right traveling waves. The standing wave is a simple model used in uniformly deep lakes, estuaries, and harbors where the standing waves produced by winds, or possibly earthquakes, are called *seiches*. Use linear wave theory to find the mathematical form of the simplest standing wave.

8.43 Write the potential function $\phi(x, z, t)$ in Eq. (8.7.7) in terms of C, L, and T. Write the velocities of Eqs. (8.7.12) and (8.7.13) in the same form.

REFERENCES

1. J.H. Ferziger and M. Perić, *Computational Methods for Fluid Dynamics,* Springer Verlag, Berlin, 1996.

2. V.L. Streeter, *Fluid Dynamics,* pp. 137–155, McGraw-Hill, New York, 1948.

3. G.B. Airy, "On Tides and Waves," in *Encyclopaedia Metropolitana,* vol. 5, p. 289, 1845.

4. G.G. Stokes, "On the Theory of Oscillatory Waves," in *Transactions of the Cambridge Philosophical Society,* vol. 8, pp. 441–455, 1847.

5. B. Kinsman, *Wind Waves, Their Generation and Propagation on the Ocean Surface,* Prentice-Hall, Inc., New Jersey, 1965.

6. U.S. Army Corps of Engineers, *Shore Protection Manual,* Superintendent of Documents, U.S. Government Printing Office, Washington, D.C., 1984.

chapter

9

Transport by Advection and Diffusion

The flow fields discussed in the previous chapters give rise to or are affected by the transport of heat or mass. In this chapter simple heat and mass transport fields are identified, and the corresponding equations from Chaps. 3 and 4 are solved to give practical distributions of temperature and concentration. As in the previous chapters the focus of this chapter will be on the mass and heat distribution in liquids, with specialized gas transfer principles introduced in connection with phase changes such as for the case of evaporation. Elementary multicomponent and multiphase flows are considered in Chap. 14 as are the detailed effects of interfacial exchange at boundaries.

The simplest transport mechanisms are considered in this chapter: molecular diffusion and advection or convection. From Eqs. (4.7.3) and (4.8.8) molecular diffusion of heat and mass are parameterized by the thermal diffusivity ($\alpha = k/\rho c_p$) and the diffusion coefficient ($\mathscr{D}$), both with dimensions of [L^2/t]. From these dimensional structures it is noted that pure diffusion is a very time-consuming transport process, because it can be inferred that a length scale for diffusion over a given time, T_o, is proportional to $(\mathscr{D}T_o)^{1/2}$ while a corresponding diffusion time scale over a distance L_o is proportional to $L_o^2/\mathscr{D}$. As an example, the typical diffusion coefficient for chlorine or oxygen in water is on the order of 10^{-5} cm^2/s. Therefore the diffusion time scale is on the order of 10^9 seconds or approximately 32 years. Clearly molecular diffusion is a very slow but persistent transport agent.

In contrast advection, the simple carrying of mass or heat by the fluid velocity, is a much more powerful transport agent. If a channel velocity is 14 cm/s, typical of many river currents, the *travel time* of a nonsettling or neutrally buoyant particle would be approximately 7 seconds to travel the same one meter distance. Therefore advection, convection, and the turbulent fluxes derived from them will be quite important transport agents. Molecular diffusion will be quite important in cases where the velocity is at or near zero, which includes all solids and fluid flows very near solid walls.

9.1 STEADY MOLECULAR DIFFUSION AND CONDUCTION

From Eqs. (4.7.4) and (4.8.10) the diffusion equations for heat and mass transfer are found by eliminating the advection-convection terms and (for the time being) setting the source and sink terms to zero. The equations become

$$\frac{\partial T}{\partial t} = \alpha \, \nabla^2 T = \alpha \left(\frac{\partial^2 T}{\partial x^2} + \frac{\partial^2 T}{\partial y^2} + \frac{\partial^2 T}{\partial z^2} \right) \qquad (9.1.1)$$

and

$$\frac{\partial C}{\partial t} = \mathscr{D} \, \nabla^2 C = \mathscr{D} \left(\frac{\partial^2 C}{\partial x^2} + \frac{\partial^2 C}{\partial y^2} + \frac{\partial^2 C}{\partial z^2} \right) \qquad (9.1.2)$$

If the heat and mass transport conditions are uniform in a yz plane perpendicular to the x direction then a one-dimensional form is identified as

$$\frac{\partial T}{\partial t} = \alpha \frac{\partial^2 T}{\partial x^2} \qquad (9.1.3a)$$

$$\frac{\partial C}{\partial t} = \mathscr{D} \frac{\partial^2 C}{\partial x^2} \qquad (9.1.3b)$$

The simplest possible circumstance is one-dimensional *steady state diffusion* whenever all time derivatives are zero.

Steady Heat Conduction

Consider a thin slab of metal separating two nonmoving liquid baths as in Fig. 9.1. The liquid A temperature steadily imposes a temperature T_A on the left face of the slab, while the liquid B maintains a constant temperature T_B on the right face. The variation of temperature through the slab is steady over a long time period as the boundary conditions are steady. Eq. (9.1.3a) is reduced to a one-dimensional steady state form as

$$\alpha \frac{\partial^2 T}{\partial x^2} = \alpha \frac{d^2 T}{dx^2} = 0 \qquad \longrightarrow \qquad \frac{d^2 T}{dx^2} = 0 \qquad (9.1.4)$$

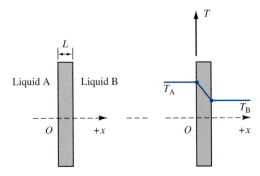

Figure 9.1   Thin slab of metal separating two fluids of differing temperature.

Since the temperature variation in the yz plane is assumed uniform and since the only dependent variable is x, this equation becomes a second-order ordinary differential equation. The solution is

$$T(x) = C_1 x + C_2$$

Since there are two integration coefficients, two boundary conditions must be applied: $T(x = 0) = T_A$ and $T(x = L) = T_B$. Therefore, $C_2 = T_A$ and $C_1 = (T_B - T_A)/L$, and

$$T(x) = \left[\frac{(T_B - T_A)}{L} \right] x + T_A \tag{9.1.5}$$

The heat flux through the slab per unit area is found from Eq. (3.9.2) as

$$N_{Tx} = \frac{q_T}{A} = -k \frac{dT}{dx}$$

and from Eq. (9.1.5) above

$$N_{Tx} = -k \frac{d}{dx} \left[\left(\frac{T_B - T_A}{L} \right) x \right] = -k \frac{(T_B - T_A)}{L} \tag{9.1.6}$$

Therefore, the flux is constant. If the surface area is finite, such as at the wall of a house, then both sides may be multiplied by the area to determine the total heat flux as

$$q_T = kA \frac{(T_A - T_B)}{L} = \frac{T_A - T_B}{R}$$

and the ratio L/kA is called the *thermal resistance, R*. Finally, if the slab consists of m layers each with its own thickness L_i and conductivity k_i but still subjected to the same overall temperature drop $T_A - T_B$, then the constant heat flux is maintained through each material and

$$q_T = \frac{T_A - T_B}{\sum\limits_{i=1}^{m} R_i} \tag{9.1.7}$$

where

$$R_i = \frac{L_i}{k_i A}$$

Similar approaches for tubes and hollow spheres are possible. In a pipe with inside and outside wall temperatures of T_A and T_B, respectively, the heat flux is along the radial direction, r, and is given by

$$\frac{q_T}{A} = -k \frac{dT}{dr}$$

If the cross sectional area normal to the heat flux is given by

$$A = 2\pi r L$$

then

$$q_T = \frac{T_A - T_B}{R}$$

where

$$R = \frac{\ln(r_B/r_A)}{2\pi kL}$$

The temperature distribution is given by

$$T(r) = T_A - \left[\frac{T_A - T_B}{\ln(r_B/r_A)}\right] \ln\left(\frac{r}{r_A}\right) \qquad (9.1.8)$$

Similar formulas for temperature distributions in hollow spheres of radii r_A and r_B are also possible [1]†

$$T(r) = T_A - \left[\frac{T_A - T_B}{\frac{1}{r_A} - \frac{1}{r_B}}\right]\left(\frac{1}{r_A} - \frac{1}{r}\right) \qquad (9.1.9)$$

and

$$q_T = \frac{4\pi k(T_A - T_B)}{\left(\frac{1}{r_A} - \frac{1}{r}\right)}$$

Example 9.1

A house basement wall is comprised of a layer of concrete block 19.7 cm thick and oak paneling 1.27 cm thick separated by a layer of fiberglass insulation. The temperature of the earth around the foundation is assumed steady at 12.7°C (285.7K), while the temperature of the stagnant (nonmoving) air in the basement is 22.7°C (295K). What insulation thickness is required such that there is no heat loss from the basement? The conductivities (k) for the materials are 0.208 W/m·K for the oak, 0.762 for the concrete, and 0.0310 for the insulation. Calculate the solution for a 1-m² area. What is the temperature at the interface between the insulation and the concrete block?

Solution

From Eq. (9.1.7) the resistances are calculated as follows:

$$R_{block} = R_b = \frac{L_b}{k_b A} = \frac{0.197}{0.762(1)} = 0.26 \text{ K/W}$$

$$R_{insulation} = R_i = \frac{L_i}{k_i A} = \frac{L_i}{0.031(1)} = 32.26 L_i \text{ K/W}$$

$$R_{panel} = R_p = \frac{L_p}{k_p A} = \frac{0.0127}{0.208(1)} = 0.06 \text{ K/W}$$

and are then inserted into the corresponding heat flux equation

$$q_T = \frac{T_{basement} - T_{earth}}{R_b + R_i + R_p} = \frac{(295 - 285.7) \text{ K}}{(0.26 + 32.26 L_i + 0.06) \text{ K/W}}$$

For no heat loss or gain the flux must be 0, and the above equation is solved for the insulation thickness, L_i. However it is clearly noted that the *only* solution permitted where there is no heat flux is the one where the inside and outside temperatures are

| †Numbered references will be found at the end of this chapter.

equal. All materials conduct heat, the differences being how much or how little each conducts. Therefore, we will assume a very small heat loss on the order of 0.5 W or J/s; then

$$0.5 = \frac{9.3}{(0.32 + 32.26L_i)}$$

or

$$L_i = 0.566 \text{ m} = 56.6 \text{ cm}$$

To find the temperature on the inside of the cinder block T_{ib}, the heat flux, 0.5 W, is constant in all three material types. Therefore,

$$q_T = 0.5 \text{ W} = \frac{k_b A}{L_b}(T_{ib} - 285.7 \text{ K})$$

$$= \frac{0.762(1)}{0.197}(T_{ib} - 285.7 \text{ K})$$

or

$$T_{ib} = \frac{0.5(0.197)}{0.762} + 285.7 \text{ K} = 285.8 \text{ K}$$

Clearly the block provides very little insulation. The temperature at the inside of the insulation is 294.9 K. Therefore, the bulk of the temperature difference is sustained by the insulation (9.1 K) while the block and panel combined account for 0.2 K of the temperature difference.

Mass Diffusion: Low Concentration

Before proceeding to the steady-state one-dimensional diffusion solution, it is necessary to remember that a mixture of mass species involves at a minimum two components and is termed a *binary* mixture. In liquids, but not gases, the molecules are quite closely spaced. Therefore, Fickian diffusion of a species through a liquid will be much slower than for a gas. Diffusivities for liquids therefore are quite often a function of the concentration of the diffusing species and are typically 10^4 to 10^5 times lower than for gases. However, due to the higher species concentration in liquids the fluxes of liquids and gases are closer in magnitude.

The fluxes for a binary mixture of two components A and B are found from Eqs. (4.8.5) and (4.8.7) and are, for the z direction,

$$N_{A_z} = wC_A - \mathcal{D}_{AB}\frac{\partial C_A}{\partial z} \qquad (9.1.10a)$$

$$N_{B_z} = wC_B - \mathcal{D}_{BA}\frac{\partial C_B}{\partial z} \qquad (9.1.10b)$$

For a binary mixture the total velocity w is

$$w = \frac{1}{C}(C_A w_A + C_B w_B) = \frac{1}{\rho}(C_A w_A + C_B w_B)$$

but since $N_{A_z} = C_A w_A$ and $N_{B_z} = C_B w_B$, then Eqs. (9.1.10a and b) become

$$N_{A_z} = -\mathcal{D}_{AB} \frac{\partial C_A}{\partial z} + \frac{C_A}{\rho}(N_{A_z} + N_{B_z}) \tag{9.1.11a}$$

$$N_{B_z} = -\mathcal{D}_{BA} \frac{\partial C_B}{\partial z} + \frac{C_B}{\rho}(N_{A_z} + N_{B_z}) \tag{9.1.11b}$$

From Eq. (1.5.5) $\omega_A = C_A/\rho$ and $\omega_B = C_B/\rho$.

The first terms in Eq. (9.1.11) are the diffusion terms, while the last terms are the fluxes resulting from advection or bulk flow. To begin the description of steady state mass diffusion then will require plausible assumptions about the importance of advection (N_{A_z} and N_{B_z}) as transport agents. The bulk transport term for the flux N_{A_z} can become small under a variety of circumstances. The first is where each species has a finite advection velocity w_A and w_B but the fluxes are equal and opposite, that is, $N_{A_z} = -N_{B_z}$. The resulting net advective flux is zero, and this is the case of *counter diffusion*. The second is where one species, say B, is assumed to be a stagnant substance such that $N_{B_z} = 0$. With N_{A_z} remaining in the bulk transport term, it is assumed that $C_A/\rho = \omega_A$ is very small and the term $\omega_A N_{A_z}$ (Eq. 9.1.11a) becomes unimportant compared to the diffusive or gradient flux term. This *low concentration* assumption coupled with the stagnant nature of the species B results in a simplified steady flux equation. From Eqs. (9.1.3b) and (9.1.11a) written in the z direction

$$N_{A_z} = -\mathcal{D}_{AB} \frac{\partial C_A}{\partial z}$$

$$\frac{\partial}{\partial z} N_{A_z} = \frac{d}{dz} N_{A_z} = -\frac{d}{dz}\left(\mathcal{D}_{AB} \frac{dC_A}{dz}\right) = \mathcal{D}_{AB} \frac{d^2 C_A}{dz^2} \quad \rightarrow \quad \frac{d^2 C_A}{dz^2} = 0$$

The boundary conditions for concentration are fixed at either end of the z domain, that is, $C_A(z = z_1) = C_{A_1}$ and $C_A(z = z_2) = C_{A_2}$. Then the solution becomes

$$C_A(z) = C_1 z + C_2 \tag{9.1.12a}$$

$$C_A(z) = \left[\frac{C_{A_2} - C_{A_1}}{z_2 - z_1}\right] z + \frac{C_{A_1} z_2 - C_{A_2} z_1}{z_2 - z_1} \tag{9.1.12b}$$

If $z_1 = 0$ and $z_2 = L$, then the form of Eq. (9.1.5) is recovered. The flux is given by

$$N_{A_z} = -\mathcal{D}_{AB} \frac{\partial C_A}{\partial z} = -\mathcal{D}_{AB} \left[\frac{C_{A_2} - C_{A_1}}{z_2 - z_1}\right] \tag{9.1.13}$$

Consequently the low concentration diffusion of a species through a stagnant medium results in linear concentration profiles with a constant flux and becomes the transport analog to the heat conduction transport cases discussed in the previous section. Solutions for the various geometries discussed in the previous subsection apply here as well.

High Concentration Diffusion Through a Stagnant Medium

By undoing the low concentration assumption in the previous subsection, the first impact of convection-advection can be examined. The steady state one-dimensional equation still applies, that is,

$$\frac{d}{dz} N_{A_z} = 0$$

However, from Eq. (9.1.11a)

$$N_{A_z} = -\mathcal{D}_{AB} \frac{dC_A}{dz} + \omega_A N_{A_z}$$

or

$$N_{A_z} = -\rho\mathcal{D}_{AB} \frac{d\omega_A}{dz} + \omega_A N_{A_z}$$

The solution for N_{A_z} is formed by first grouping terms

$$N_{A_z} = -\frac{\rho\mathcal{D}_{AB}}{(1-\omega_A)} \frac{d\omega_A}{dz}$$

and then integrating by use of the previous boundary condition

$$\int_{z_1}^{z_2} N_{A_z}\, dz = -\rho\mathcal{D}_{AB} \int_{\omega_{A_1}}^{\omega_{A_2}} \frac{d\omega_A}{(1-\omega_A)}$$

This gives

$$N_{A_z} = \frac{\rho\mathcal{D}_{AB}}{(z_2 - z_1)} \ln\left[\frac{1 - \omega_{A_2}}{1 - \omega_{A_1}}\right] \tag{9.1.14}$$

The concentration profiles can be found by substituting back into the transport equation and integrating

$$\frac{d}{dz} N_{A_z} = \frac{d}{dz}\left[-\frac{\rho\mathcal{D}_{AB}}{(1-\omega_A)} \frac{d\omega_A}{dz}\right] = 0$$

Integrating twice gives

$$-\ln(1-\omega_A) = C_1 z + C_2 \tag{9.1.15}$$

which after applying the boundary condition gives

$$\left[\frac{1-\omega_A}{1-\omega_{A_1}}\right] = \left[\frac{1-\omega_{A_2}}{1-\omega_{A_1}}\right]^{(z-z_1)/(z_2-z_1)} \tag{9.1.16}$$

Since $\omega_A + \omega_B = 1$, a similar solution for ω_B profiles is readily found from Eq. (9.1.16).

Contrasting the forms of the high and low concentration solutions [Eqs. (9.1.12a) and (9.1.15)] reveals that a logarithmic profile (Fig. 9.2) emerges as the effects of bulk transport begin to influence the physics. By extending Eq. (9.1.15) in a power series as

$$-\ln(1-\omega_A) = \omega_A + \frac{\omega_A^2}{2} + \frac{\omega_A^3}{3} + \cdots$$

it is seen that for very low concentrations, the second order and higher terms are negligible and Eq. (9.1.15) reduces to Eq. (9.1.12a).

Example 9.2

A can of ethanol is left open in a basement. Since the windows are closed the air is stagnant. Assume that the air and the solvent are at 25°C and that the air (Fig. 9.3) has no initial solvent concentration. The air in the basement is at standard atmospheric pressure (101,325 Pa = 101.3 kPa = 1 atm). Over a period of one day how much will the ethanol level drop by evaporation into the air?

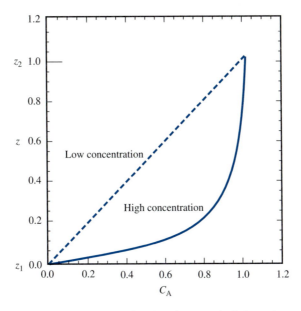

Figure 9.2 Transition from low (linear) to high (logarithmic) concentration solutions.

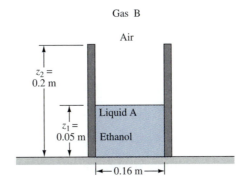

Figure 9.3 Example 9.2.

Solution

From Appendix C the diffusivity for ethanol in air is 0.132 cm²/s at 25°C and one atmosphere of pressure. The solution concept is to integrate Eq. (9.1.14) over the surface area of the tank exposed to gas B and over the time period. The total mass lost over time T_p is

$$M = \int_t^{t+T_p} \int N_{A_z}\, dA\, dt \approx N_{A_z} A T_p = \frac{\rho A T_p \mathscr{D}_{AB}}{(z_2 - z_1)} \ln\left[\frac{1 - \omega_{A_2}}{1 - \omega_{A_1}}\right]$$

Here it is assumed that N_{A_z} is uniform and constant over the surface area and time. Let $T_p = 1\ \text{hr} = 3600\ \text{s}$. In accounting for all the variables, $A = \pi(8)^2 = 201\ \text{cm}^2$, $z_2 - z_1 = 0.2\ \text{m} - 0.05\ \text{m} = 0.15\ \text{m} = 15\ \text{cm}$, and $\omega_{A_z} = 0$. The variables yet to be calculated are ω_{A_1} and ρ.

ω_{A_1} is found via use of Dalton's law. The vapor pressure of ethanol is 7.9 kPa at these temperature and pressure conditions [2]. Therefore, the molar concentration of ethanol gas at the interface is

$$y_{A_1} = \frac{p_{v_{A_1}}}{p} = \frac{7.9 \text{ kPa}}{103.3 \text{ kPa}} = 0.078$$

The molar concentration of air at the interface is therefore $y_{B_1} = 1 - 0.078 = 0.922$. Clearly very little ethanol is available at the interface for diffusion through the stagnant air column.

The mixture density, ρ, must be calculated. With the molecular weights of ethanol (M_A) and air (M_B) being 46 and 29, respectively, the amount of ethanol present in one mole at the interface is

$$(1 \text{ mol})(y_{A_1}) = 0.078 \text{ mol}$$

Since $M_A = 46$, then the mass of ethanol in one mol is

$$0.078 \text{ mol} \frac{(0.046 \text{ kg})}{\text{mol}} = 0.0036 \text{ kg}$$

A similar calculation is used for the air mass at the interface. In one mole the mass of air is

$$(1 \text{ mol})(0.922) \frac{(0.029 \text{ kg})}{\text{mol}} = 0.0267 \text{ kg}$$

Therefore, the total mass present in one mole is $0.0036 \text{ kg} + 0.0267 \text{ kg} = 0.0303 \text{ kg}$. Since one mole has a mass of 0.0303 kg, then the molecular weight must be 30.3 (or 30 if rounded off).

With this knowledge the ideal gas law [Eqs. (1.7.4) and (1.7.7)] is used to calculate the total density, ρ, as

$$\rho = \frac{p}{RT} = \frac{(101{,}325 \text{ N/m}^2)(30)}{\left(8312 \frac{\text{m·N}}{\text{mole kg·K}}\right)(298 \text{ K})} = 1.23 \text{ kg/m}^3$$

The equation for the total mass evaporated during time t_p becomes

$$M = t_p \frac{(1.23 \text{ kg/m}^3)(0.0201 \text{ m}^2)(0.132 \text{ cm}^2/\text{s})\left(10^{-4} \frac{\text{m}^2}{\text{cm}^2}\right)}{(0.15 \text{ m})} \ln\left[\frac{1-0}{1-\omega_{A_1}}\right]$$

ω_{A_1} can be found from y_{A_1}, M_A, y_{B_1}, and M_B as

$$\omega_{A_1} = \frac{y_{A_1} M_A}{y_{A_1} M_A + y_{B_1} M_B} = 0.118$$

Therefore, since $\omega_{A_1} + \omega_{B_1} = 1$, then $\omega_{B_1} = 0.882$.

Completing the calculation of M then for a one-hour period ($t_p = 3600$ s), the total mass that diffuses away from the liquid surface is $2.73(10^{-7})$ kg, which is a loss in fluid height in the container of

$$h = \frac{M}{\rho A} = \frac{(2.73(10^{-7}) \text{ kg})}{(1.23 \text{ kg/m}^3)(0.0201 \text{ m}^2)} = 1.10(10^{-7}) \text{ m}$$

This very small loss would amount to only 0.265 mm per day.

Fluids with higher vapor pressures at these conditions will have a much larger vertical gradient and a correspondingly higher flux and evaporation rate. The reader should try this calculation for water in the same tank. If the evaporation rates are sufficiently high enough that the time rate of loss of liquid causes $z_2 - z_1$ to change over time, then a transient calculation must be performed. This case is treated in more advanced transport phenomena textbooks.

Finally the above example forms the basis for an *Arnold cell,* that is, a device for determining diffusivities by measuring time rate of liquid loss and using the flux equation in an inverse form to accurately determine the molecular diffusion coefficient as

$$\mathscr{D}_{AB} = \frac{h(T_p)(z_2 - z_1)}{T_p \ln\left[\frac{1-\omega_{A_1}}{1-\omega_{A_2}}\right]} \tag{9.1.17}$$

Diffusion with a Chemical Reaction

Within the control volume containing the mixture various chemical reactions may be taking place while diffusion is proceeding. A *homogeneous* reaction is a reaction that occurs uniformly throughout the control volume while a *heterogeneous* reaction takes place typically at an interface such as the solid boundary of the control volume. Heterogeneous reactions are treated in more advanced transport phenomena references. Here the simple homogeneous reaction will be presented.

From Eq. (4.8.4) the steady state, one-dimensional (z) diffusion equation, including a source-sink or reaction term, is

$$\frac{d}{dz}\left(\mathscr{D}_{AB}\frac{dC_A}{dz}\right) + S_i = 0$$

For a sink (destruction or removal) with a first order reaction rate (see Ex. 3.24), $S_i = -k_A C_A$ such that

$$\mathscr{D}_{AB}\frac{d^2 C_A}{dz^2} - k_A C_A = 0 \qquad \rightarrow \qquad \frac{d^2 C_A}{dz^2} - m^2 C_A = 0$$

Here, m is defined as $\sqrt{k_A/\mathscr{D}_{AB}}$. The general form of this solution is

$$C_A(z) = c_1 \exp(mz) + c_2 \exp(-mz)$$

If boundary conditions of the form $C_A(z = z_1 = 0) = C_{A_1}$ and $C_A(z = z_2 = d) = C_{A_2}$ are imposed, the concentration profile becomes

$$C_A(z) = C_{A_2}\operatorname{csch} md \, \sinh mz + C_{A_1}\operatorname{csch} md \, \sinh m(z - d) \tag{9.1.18}$$

The flux N_{A_z} becomes

$$N_{A_z}(z) = m\mathscr{D}_{AB}C_{A_2}\operatorname{csch} md \, \cosh mz - m\mathscr{D}_{AB}C_{A_1}\operatorname{csch} md \, \cosh m(z - d) \tag{9.1.19}$$

What is important to note here is that the flux is no longer constant as it now varies with z. Figure 9.4 contains a series of plots for the following conditions representing the diffusion of dissolved oxygen (DO) through a water surface layer as it is consumed by zooplankton during respiration. The DO concentration at the surface ($z = 0$) is 12 mg/L, while at a depth of 40 cm the DO is 4 mg/L. The various values of the source term k_{DO} are indicated on the drawing. A value of $\mathscr{D}_{AB} = 100 \text{ cm}^2/\text{s}$ is used. It is noticed that a stronger respiration rate increases the curvature of the profile and localizes the zone of maximum gradient and curvature near the surface $z_1 = 0$.

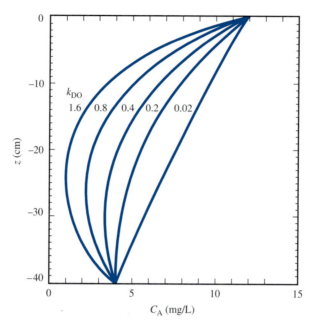

Figure 9.4   Concentration of DO due to diffusion and uptake by zooplankton.

Other Boundary Condition Combinations

The one-dimensional steady state diffusion equation applied in this section is a second-order equation and must be integrated twice with respect to the independent variable z (or x). Therefore, the resulting two integration coefficients require two *boundary conditions* to complete the integration. This section has employed concentration values at the end of the domain (L) or case a below. Other possible combinations are listed as

Case a	$C(z = z_1) = C_{Left} = C_L$	$C(z = z_2 = z_1 + L) = C_{Right} = C_R$
Case b	$C(z = z_1) = C_L$	$\mathscr{D}\partial C/\partial z(z = z_2) = N_R$
Case c	$\mathscr{D}\partial C/\partial z(z = z_1) = N_L$	$C(z = z_2) = C_R$
Case d	$\mathscr{D}\partial C/\partial z(z = z_1) = N_L$	$\mathscr{D}\partial C/\partial z(z = z_2) = N_R$
Case e	$C(z = z_1) = C_L$	$\mathscr{D}\partial C/\partial z(z = z_1) = N_L$

Therefore, it is possible to specify either the flux or the concentration at either end of the domain for this second-order diffusion equation. In certain rare cases it is possible to specify both the concentration and the flux at the upstream boundary. In all cases the number of boundary conditions required to be specified in each coordinate direction equals the order of the highest derivative in that coordinate direction and the highest order differential term allowed in the boundary condition is one less than the highest order differential in the governing equation.

EXERCISES

9.1.1   Molecular mass diffusion or heat conduction (*a*) is a result of fluid velocity carrying the heat or mass; (*b*) is always from high temperature to low temperature; (*c*) flux is always uniform in all three coordinate directions; (*d*) always results in a linear concentration variation between high and low concentration at steady state; (*e*) occurs at rates the same order of magnitude as advective transport.

9.1.2 Thermal resistance (*a*) is a property of the material; (*b*) varies inversely as a function of the length of the object through which heat is conducted; (*c*) is infinite for house insulation; (*d*) varies linearly in proportion to the surface area of the object; (*e*) is a function of the material, area, and length of the object.

9.1.3 Steady state mass diffusion can occur when (*a*) the concentrations of the two components of a binary mixture are high; (*b*) counter diffusion exists; (*c*) one medium in the mixture is stagnant; (*d*) one diffusing substance is of low concentration; (*e*) *b* and *c*; (*f*) *b*, *c*, or *d*.

9.2   ADVECTION AND CONVECTION: BULK APPROACHES

The fluid in contact with the wall in Fig. 9.1 was not moving and the temperature could be independently specified as T_A and T_B. Suppose now that the fluid on either side was moving and a thermistor probe measured the temperature variation in the fluid to slab cross section. The data would show (schematically) that in a thin fluid region near the wall (Fig. 9.5) the temperature of the fluid near the wall

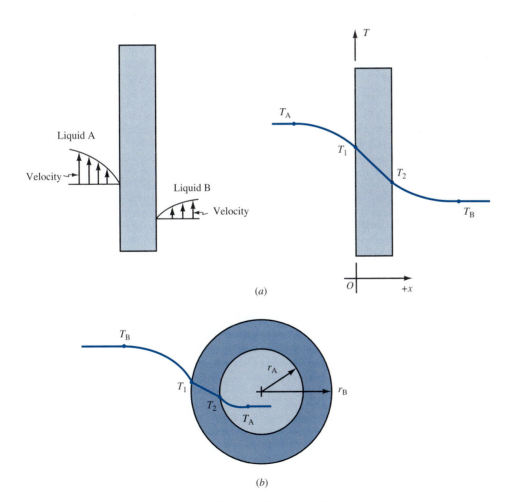

Figure 9.5   Temperature variation in a fluid very close to a wall.

varies as the wall is approached on either side. This near-wall change in temperature results from convection or advection introduced by the motion of the fluid over the surface. Two approaches to parameterizing the velocity effects are possible: the *bulk approach,* or the mechanics-based *boundary layer* approach. In the former case the overall heat or mass transfer rates are calculated from average values of the flow field and its geometry. The latter approach computes the heat or mass transfer rates for an elemental area of the geometry and then integrates to find the overall, or bulk, transfer rate for the entire flow. In the former case the correlation between the average variables is expressed by a convective heat or mass transfer coefficient, typically found by experiment. In the latter case the coefficient can be found (hopefully) by direct mathematical analysis followed by laboratory or field experiment validation. Bulk approaches are reviewed in this section.

Heat Transfer

For steady advection or convection conditions the bulk heat transport is given by

$$q_T = hA(T_s - T_f) \tag{9.2.1}$$

where q_T is the flux; A is the cross-sectional area through which the heat is moving (perpendicular to the flux direction); T_s is the temperature of the surface of the medium in contact with the moving fluid; and T_f represents some as yet unspecified bulk or average temperature of the fluid some distance from the surface. The position of T_f is a function of the overall flow geometry. If $T_s > T_f$, then heat escapes the medium, while if $T_f > T_s$, then heat is transferred from the fluid to the medium. The variable h is called the *convective heat transfer coefficient* and attains a single overall value when using the bulk analysis approach. Intuition would suggest that since there exists an almost infinite variety of flow fields and geometries, there would be a like number of bulk transfer coefficients.

The problem depicted in Fig. 9.5a is a combined conduction-convection problem which is analyzed as follows. In the steady flow case the key principle is that the flux perpendicular to the surface is constant through the system. Therefore,

$$q_T = -h_A A(T_1 - T_A) = \frac{kA}{L}(T_1 - T_2) = h_B A(T_2 - T_B) \tag{9.2.2}$$

The minus sign in the convective heat transfer occurs due to the sign convention for flux, that is, it is defined as positive relative to and away from the surface. In this case the flux on the left-hand side of the slab is in the $+x$ direction which is the opposite of a positive flux which woud be directed away from the surface in the $-x$ direction.

Assembling Equation (9.2.2) into the overall heat transfer form, as in Sec. 9.1, gives

$$q_T = \frac{T_A - T_B}{\left(\frac{1}{h_A A} + \frac{L}{kA} + \frac{1}{h_B A}\right)} = \frac{T_A - T_B}{\sum R} = \frac{\Delta T}{\sum R}$$

The *overall heat transfer coefficient,* H_T, is defined for the combined system by assuming the following form

$$q_T = H_T A \Delta T \tag{9.2.3}$$

where

$$H_T = \left[\cfrac{1}{\frac{1}{h_A} + \frac{L}{k} + \frac{1}{h_B}} \right]$$

(9.2.4a)

and

$$\Delta T = T_A - T_B$$

(9.2.4b)

The same principle can be applied to a conduit as labeled in Fig. 9.5b. Again the overall flux q_T is found from the series solution with combined conduction and convection as

$$q_T = \frac{T_A - T_B}{\sum R} = \frac{T_A - T_B}{\frac{1}{h_B A_B} + \frac{r_A - r_B}{k A_m} + \frac{1}{h_A A_A}}$$

(9.2.5)

In Eq. (9.2.5) $A_B = 2\pi r_B L$, where L is the conduit length; $A_A = 2\pi r_A L$; and A_m equals the log-mean area, that is,

$$A_m = \frac{A_A - A_B}{\ln(A_A/A_B)}$$

The overall heat transfer coefficient is given by

$$H_T = \frac{1}{\frac{1}{h_B} + \frac{(r_A - r_B)A_B}{k A_m} + \frac{A_B}{h_A A_A}}$$

if it is based upon the inside area of the tube. A similar expression can be found for H_T based upon the outside area.

Dimensionless Heat Transfer Groups and Parameter Specification

At first glance it might seem that the heat transfer coefficient, h, would be as easy to select for a problem as were the diffusion coefficients, that is, one simply looks up the diffusivities in a standard reference textbook or handbook (e.g., Ref. [2]), making appropriate adjustments for temperature or other factors. The heat transfer coefficient, however, is not a property of the fluid and therefore is not often a simple constant. It varies from point to point in the flow field and is quite dependent upon the geometry of the problem, as well as the origin of the fluid motion and intensity of the motion relative to viscosity.

In one sense the use of the bulk or overall average value of h seems to reduce some of the above complexities, but other sources of empiricism arise, particularly in choosing the correct location for obtaining the value of T_A (or T_B). In the two advection-diffusion example problems the locations are quite different for each geometry, and the form of the temperature used in the formula is often a function of the geometry as well. For example, in flat plate analysis, T_f in Eq. (9.2.1) is taken as the temperature of the fluid at some point quite far away ($\rightarrow \infty$) from the surface, s. Therefore, $q_T = hA(T_s - T_\infty)$ for the plate. For a pipe, the convection at the outside wall may be treated as a plate with $T_f \rightarrow T_\infty$. However, inside the pipe the flow geometry restricts the growth of the internal boundary layer, and T_f is taken as the cross-sectional average or bulk temperature of the conduit fluid.

Therefore, the majority of the heat transfer coefficient data have been derived from carefully performed laboratory experiments where geometries are restricted to simple, easily configured shapes or those of industrial or practical value. The vast quantities of data for h then are limited to the geometry of the flow field and are presented in their most compressed form via the use of nondimensional groups (see Chap. 5).

The Reynolds number (among others) is an important nondimensional number that relates the strength of inertia to friction in the momentum equation. Might there be an analogous nondimensional grouping for heat transfer? If the fluid moving near the wall is again examined, the strength of the convection-advection can be compared to conduction by noting the equivalence of the heat flux through the system. Therefore, at the wall surface ($x = 0$)

$$q_T = -hA(T_1 - T_A) = -kA\frac{\partial}{\partial y}(T - T_1)\Big|_{x=0}$$

The ratio of heat transfer to diffusion coefficients is found to be

$$\frac{h}{k} = \frac{\frac{\partial}{\partial y}(T - T_1)\Big|_{x=0}}{(T - T_1)}$$

and can be made dimensionless with multiplication by a length scale L, that is,

$$\frac{hL}{k} = \frac{\frac{\partial}{\partial y}(T - T_1)\Big|_{x=0}}{(T - T_1)L} = \mathbf{N}_u \tag{9.2.6}$$

This dimensionless ratio is called the *Nusselt number* (see Sec. 5.4) and reflects the relative strength of advection to diffusion.

Since viscosity, ν, or momentum diffusion by friction and the heat diffusivity have different magnitudes, the *Prandtl number* is used to reflect the ratio, that is, $\mathbf{P}_r = \nu/\alpha$.

In *forced convection* or advection problems the *Stanton number*, $\mathbf{S}_t$, is defined as

$$\mathbf{S}_t = \frac{h}{\rho v c_p} \tag{9.2.7}$$

where v is the representative velocity of the flow field used to parameterize the Reynolds number. This number is also formed from $\mathbf{S}_t = \mathbf{N}_u/\mathbf{R}\,\mathbf{P}_r$. In *natural convection* problems the flow velocity, v, is induced by the density instabilities in the flow. If density is therefore linearly related to the temperature difference ΔT, then $\rho = \rho_o(1 - \beta\Delta T)$ (see Chap. 1, Sec. 6) and the *Grashof number*, $\mathbf{G}_r$ (see Eq. 5.4.4) arises as a controlling dimensionless group in the correlation for h. Therefore, from the laboratory perspective the Nusselt number, $\mathbf{N}_u$, is the dimensionless group that must be related to ($\mathbf{R}, \mathbf{P}_r$) for forced convection or ($\mathbf{G}_r, \mathbf{P}_r$) for natural convection. Additionally the Stanton number could also be correlated with ($\mathbf{R}, \mathbf{P}_r$) for the forced convection case. Data for various values of h are therefore expressed in terms of the $\mathbf{R}, \mathbf{P}_r, \mathbf{N}_u$, and $\mathbf{G}_r$ dimensionless groups.

As an example of the use of the nondimensional parameterizations, consider the specification of the temperature of the fluid moving inside a horizontal, constant-diameter (D) pipe as in Fig. 9.6. An elemental control volume (Fig. 9.6a) analysis of a differential length of pipe dx reveals that the net heat flux (q_T) through the walls of the pipe is balanced by the net advection of thermal energy out of the pipe by the fluid motion, that is,

$$\rho V\frac{\pi D^2}{4}c_p[T(x + dx) - T(x)] = \rho V\frac{\pi D^2}{4}c_p dT = \pi Dq dx \tag{9.2.8}$$

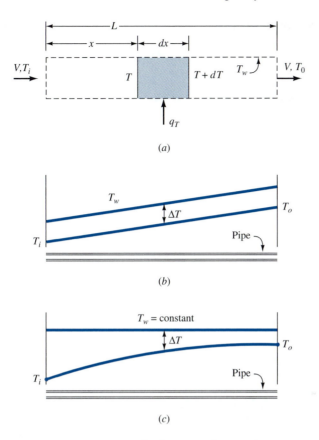

Figure 9.6 Temperature distribution in a horizontal, constant-diameter pipe.

The heat flux q_T is defined by Eq. (9.2.1) as

$$q_T = h(T_w - T(x)) \qquad \text{(9.2.9)}$$

where h is the convective heat transfer coefficient, T_w is the inside wall temperature of the pipe, and $T(x)$ is the cross-sectional average temperature of the fluid which will vary according to position, x, along the pipe. After some algebra and taking the limit as $\Delta x \to 0$

$$\left[\frac{1}{T_w - T(x)} \right] \frac{dT}{dx} = \frac{4h}{\rho c_p V D} \qquad \text{(9.2.10)}$$

This equation may be integrated for two conditions, constant heat flux or constant wall temperature.

The constant heat flux case implies that the terms $T_w - T(x)$, h, and q_T in Eq. (9.2.9) are constants; therefore, Eq. (9.2.10) integrates from the inlet $x = 0$ ($T = T_i$) to the outlet $x = L$ ($T = T_o$) as

$$\frac{T_o - T_i}{T_w - T} = \frac{T_o - T_i}{\Delta T} = 4 \frac{L}{D} \left(\frac{h}{\rho c_p V_m} \right) = 4 \frac{L}{D} \mathbf{S}_t \qquad \text{(9.2.11)}$$

where $\mathbf{S}_t$ is the Stanton number. The resulting temperature variation is schematized as in Fig. 9.6b.

For the constant temperature case, T_w is constant, and Eq. (9.2.10) becomes

$$\frac{1}{T_w - T}\frac{dT}{dx} = 4\frac{\mathbf{S}_t}{D}$$

By integration over the entire length of pipe

$$\ln\left[\frac{T_w - T_i}{T_w - T_o}\right] = 4\frac{L}{D}\mathbf{S}_t \tag{9.2.12}$$

Therefore, a logarithmic temperature variation occurs over L (Fig. 9.6c). In a manner similar to the pipe cross-section analysis presented earlier, the *log-mean temperature difference* is defined as

$$\Delta T_{lm} = \frac{T_o - T_i}{\ln\left[\dfrac{T_w - T_i}{T_w - T_o}\right]}$$

and Eq. (9.2.12) becomes

$$\frac{T_o - T_i}{\Delta T_{lm}} = 4\frac{L}{D}\mathbf{S}_t \tag{9.2.13}$$

More explicit examples of design parameterization will follow.

Convective and Advective Mass Transfer and Dimensionless Groups

In direct analogy to the heat transfer case and with all the attendant difficulties, it is possible to describe the convective or advective mass flux for species A by the following formula

$$q_{mA} = h_m A(C_{A,s} - C_{A,f}) \tag{9.2.1}$$

Here h_m is the mass transfer coefficient and $C_{A,s}$ and $C_{A,f}$ are the concentrations of species A at the surface of the interface, s, and far-field fluid, f, positions, respectively. Again the location specification of $C_{A,f}$ becomes dependent upon the flow geometry, dynamics, and properties of the flow. For example, for the previous boundary layer example the far-field concentration is taken outside the boundary layer, that is, $C_{A,f} \to C_{\infty,f}$, where again the interior conduit value is the bulk or cross-section average concentration. $C_{A,s}$ is the concentration value at the solid-fluid interface which is in equilibrium for the given pressure and temperature of the system.

As per the previous section, the relevant dimensionless groups are derived by considering the relative strength of advection to diffusion for a simple system with constant flux. At the solid-fluid interface the values are equal and

$$q_m = -h_m A(C_{A,s} - C_{A,f}) = -\mathscr{D}_{AB}\, A\, \frac{d}{dx}(C_A - C_{A,s})\Big|_{x=0}$$

Therefore,

$$\frac{h_m L}{\mathscr{D}_{AB}} = \frac{-\dfrac{d}{dx}(C_A - C_{A,s})\Big|_{x=0}}{(C_{A,s} - C_{A,\infty})/L} \tag{9.2.14}$$

This dimensionless grouping is called the *Sherwood number, S_h,* or the *mass transfer Nusselt number, $N_{u_{AB}}$.* The ratio of mass diffusivity to momentum diffusivity, $\nu/\mathcal{D}_{AB}$, is defined as the *Schmidt number, S_c.* Therefore, the laboratory correlation to be quantified is $N_{u_{AB}} = S_h = f(R_e, S_c)$.

A Summary of Common Bulk Expressions

While it is not exhaustive, this section summarizes several of the bulk representations of the heat and mass transfer coefficients required to parameterize the effects of convection-advection. The simplest geometries are flat plates, conduits, and spheres, and in this section the overall coefficients are parameterized as opposed to the point by point variations.

1. *Flat plate.* Laminar flow over a smooth plate of length L occurs for a Reynolds number less than 10^5. The Nusselt number function [3] is given by

$$N_u = 0.664\, R_L^{0.5}\, P_r^{0.333} \tag{9.2.15}$$

where the Reynolds number is based upon the length of the plate.

For a Reynolds number greater than 10^5 or for a rough surface with a Reynolds number greater than 10^4, a turbulent flow relationship is established and

$$N_u = 0.0366\, R_L^{0.8}\, P_r^{0.333} \tag{9.2.16}$$

For laminar mass transfer of a single species over a flat plate [4]

$$S_h = 0.664\, R_L^{0.5}\, S_c^{0.333} \tag{9.2.17}$$

For turbulent flow it is possible to invoke the *transfer analogy* between mass, momentum, and energy and state that

$$S_h = 0.0366\, R_L^{0.8}\, S_c^{0.333} \tag{9.2.18}$$

The concept of transfer analogies is discussed in many references, for example, Ref. [5], and is useful in establishing parameterizations for new problems from existing solutions for the same geometry for a different transfer property. Such adaptations are permitted if the properties of the system are constant and no energy or mass is created in the flow field. This implies no sources or sinks due to radiation, biological and chemical reaction, or viscous dissipation.

2. *Conduit flow.* Laminar flow inside a conduit with constant wall temperature T_w is described for a Reynolds number (based on the diameter D) less than 2100 [1]

$$N_u = 1.86 \left(R P_r \frac{D}{L} \right)^{0.333} \left(\frac{\mu_b}{\mu_w} \right)^{0.14} \tag{9.2.19}$$

where L is the pipe length, μ_w is the viscosity at the wall temperature, and μ_b is the viscosity at the cross-sectional average temperature. The equation is valid to a value of $R P_r D/L > 10$. From the above reference the transfer coefficient for turbulent pipe flow is given by

$$N_u = 0.027 R^{0.8} P_r^{0.333} \left(\frac{\mu_b}{\mu_w} \right)^{0.14} \tag{9.2.20}$$

which is valid for $L/D > 60$ and larger differences between the wall and bulk temperature (ΔT). For smaller values of ΔT Sleicher and Rouse [6] recommend

$$\mathbf{N}_u = 0.023\mathbf{R}^{0.8}\mathbf{P}_r^n \tag{9.2.21}$$

where $n = 0.4$ for heating ($T_w > T$) and 0.3 for $T_w < T$.

For mass transfer in a conduit the formula due to Harriott and Hamilton [7] is valid

$$\mathbf{S}_h = 0.0096\mathbf{R}^{0.913}\mathbf{S}_c^n 0.346 \tag{9.2.22}$$

3. *Spheres.* Heat and mass transfer resulting from flow past a single sphere is reviewed in Refs. [4, 8]. For a single sphere [9]

$$\mathbf{N}_u = 2 + [0.4\mathbf{R}^{0.5} + 0.06\mathbf{R}^{0.667}]\mathbf{P}_r^{0.4}\left(\frac{\mu}{\mu_w}\right)^{0.14} \tag{9.2.23}$$

where the Reynolds number is based upon the particle diameter and the free stream or settling velocity. The wall viscosity and free stream viscosity, μ_w and μ, respectively, are important if the difference between the wall and free stream temperatures is large. This equation is valid for a particle Reynolds number of less than $7.6(10^4)$ and $1.0 < \mu/\mu_w \leq 3.2$.

The mass transfer correlation is found in Ref. [10] as

$$\mathbf{S}_h = 2 + 0.6\mathbf{S}_c^{0.333}\mathbf{R}^{0.5} \tag{9.2.24}$$

In both cases it is noted that if the fluid is inert, then $\mathbf{R} = 0$ and $\mathbf{N}_u = \mathbf{S}_h = 2$.

Example 9.3

In Ex. 3.23 calculate the wall temperature at the outlet (point 2) and heat transfer coefficient for the heat exchanger in question. $T_1 = 90°C$, $q = 5$ kJ/m^2·s.

Solution

In the 10-cm-diameter pipe, $\dot{m}_1 = \dot{m}_2 = 1.0$ kg/s. The average velocity is

$$V = \frac{\dot{m}}{\rho A} = \frac{1.0 \text{ kg/s}}{(965.3 \text{ kg/m}^3)\,\pi\,(0.05 \text{ m})^2} = 0.132 \text{ m/s}$$

The Reynolds number is

$$\mathbf{R} = \frac{VD}{\nu} = \frac{(0.132)(0.1)}{0.328(10^{-6})} = 4(10^4)$$

and, therefore, the flow is turbulent. From Eq. (9.2.21)

$$\mathbf{N}_u = \frac{hD}{k} = 0.023\mathbf{R}^{0.8}\mathbf{P}_r^{0.3}$$

and so

$$h = k\frac{0.023}{0.1}[4(10^4)]^{0.8}(2)^{0.3}$$

Here we select the Prandtl number to be 2.0 based on Appendix C and k is 0.68 W/m·K. Therefore, h is estimated to be 925 J/m^2·s·K.

At the outlet the bulk or cross-sectional average temperature was estimated to be 82.5°C. Therefore, the wall temperature is estimated from the convective heat transfer formula for the last meter of pipe length

$$q(x = L = 20 \text{ m}) = -h(T_w - T_2(x = L))$$

Solving for $T_w(x = L)$ gives

$$T_w(x = L = 20 \text{ m}) = -\frac{q}{h} + T_2(x = L)$$

$$= -\frac{5000 \text{ J/m}^2 \cdot \text{s}}{925 \text{ J/m}^2 \cdot \text{s} \cdot \text{K}} + 82.5°C$$

Therefore, $T_w = 77.1°C$.

Dry air flows over the surface of a small pond which is 200 m long and 35 m wide. Assume that the air is at standard atmospheric pressure (760-mm Hg = 10.34-m H_2O). The air temperature and water temperature are in equilibrium at 25°C, and the average wind velocity is 8 m/s. By using the flat plate turbulent transfer analogy, estimate the evaporation rate or flux of water vapor from the surface.

Example 9.4

Solution

For dry air at standard pressure (101.3 kPa) the viscosity is $1.46(10^{-5})$ m²/s and the mass diffusivity of water vapor in air is 0.242 cm²/s or $2.42(10^{-5})$ m²/s. The vapor pressure of water at 25°C is 3227 N/m².

The Reynolds number for the 200-m fetch length is

$$\mathbf{R} = \frac{uL}{\nu} = \frac{(8 \text{ m/s})(200 \text{ m})}{1.46(10^{-5}) \text{m}^2/\text{s}} = 1.1(10^8)$$

which is typical for geophysical or natural boundary layer flow. The Schmidt number is $\nu/\mathcal{D}_{AB}$ or 0.6, and thus the Sherwood number can be found from Eq. (9.2.18) as

$$\mathbf{S}_h = \frac{h_m L}{\mathcal{D}_{AB}} = 0.0366\mathbf{R}_L^{0.8}\mathbf{S}_c^{0.333}$$

Therefore,

$$h_m = 0.0366\frac{\mathcal{D}_{AB}}{L}\mathbf{R}_L^{0.8}\mathbf{S}_c^{0.333}$$

$$= 0.0366\frac{2.42(10^{-5}) \text{ m}^2/\text{s}}{200 \text{ m}}[1.1(10^8)](0.6)^{0.333}$$

$$= 0.1 \text{ m/s}$$

The evaporation flux is found from Eq. (9.2.8) by assuming that $C_{A,f}$ is the water vapor concentration far from the surface, that is, $C_{A,\infty}$. Therefore,

$$q_m = h_m A(C_{A,s} - C_{A,\infty})$$

For the sake of discussion $C_{A,\infty}$ is assumed here to equal zero. The water vapor concentration at the surface is found from

$$C_{A,s} = \frac{p_A}{TR} = \frac{3227 \text{ N/m}^2}{(302 \text{ K})\left(287 \frac{\text{N} \cdot \text{m}}{\text{kg} \cdot \text{K}}\right)} = 0.037 \text{ kg/m}^3$$

Therefore, the flux away from the entire surface of the pond is

$$q_m = (0.1 \text{ m/s})(200 \text{ m})(35 \text{ m})(0.037 - 0) \text{ kg/m}^3$$

$$= 25.9 \text{ kg/s of water vapor}$$

This is the equivalent to 0.026 m^3 of liquid water lost over the entire surface per second or $3.7(10^{-6})$ m of water lost from the pond per second. This value is a bit of an overestimate primarily as a result of the fact that $C_{A,\infty}$ is rarely if ever 0.

EXERCISES

9.2.1   The advective heat flux (*a*) is uniform in all coordinate directions; (*b*) is proportional to the temperature difference between two points; (*c*) is a vector quantity; (*d*) is a property of the medium; (*e*) *b* and *c*.

9.2.2   The overall heat or mass transfer coefficient (*a*) includes heat transfer due to a moving medium as well as conduction or diffusion; (*b*) is inversely proportional to the convective heat transfer coefficient; (*c*) varies linearly with cross sectional area perpendicular to the flux; (*d*) is proportional to the thermal or diffusive resistance, *R*; (*e*) *a* and *b*.

9.2.3   The Nusselt and Sherwood numbers (*a*) relate the strength of advection to density gradients; (*b*) are similar or analogous; (*c*) relate the strength of diffusive transport to advective transport; (*d*) are proportional to the Reynolds number; (*e*) *b* and *c*.

9.3   LAMINAR BOUNDARY LAYER TRANSPORT

In Sec. 7.2 the momentum transport boundary layer was investigated with regard to determining the exact value of the shear stress, τ_o, at the interface between the flat plate and the laminar flow field [Eqs. (7.2.6)–(7.2.8)]. These equations were subsequently integrated to determine the total force or drag on the plate [Eqs. (7.2.9) and (7.2.10)]. Exact solutions are similarly available for laminar heat and mass species transport over flat plates, which can also be analyzed to give exact solutions for the convective heat and mass transport coefficients and corresponding fluxes. The procedure discussed in Sec. 7.2 is extended to meet this objective and is referred to as the von Kármán integral analysis technique.

Heat Transfer

A control volume is defined as in Fig. 9.7 and the following assumptions are made: the flow is steady and laminar and contains no vertical acceleration, and the temperature of the plate, T_s, is lower than the free stream temperature, T_∞, beyond the thermal boundary layer, δ_T. To have $T_s \rightarrow T_\infty$ would introduce free convection by vertical acceleration of unstable fluid parcels. The following boundary conditions apply:

$$T = T_s \ @ \ y = 0; \qquad \frac{\partial T}{\partial y} = 0 \ @ \ y = \delta_T$$

$$T = T_\infty \ @ \ y = \delta_T; \qquad \frac{\partial^2 T}{\partial y^2} = 0 \ @ \ y = 0$$

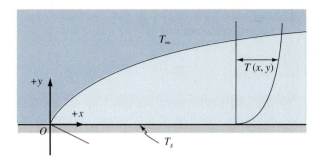

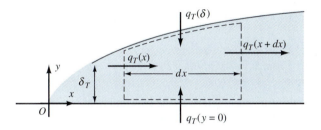

Figure 9.7 Control volume definition for boundary layer heat transfer.

Again, as in the momentum boundary layer, a power series expansion for the temperature profile $T(y)$ is sought as

$$T(y) - T_s = c_1 + c_2 y + c_3 y^2 + c_4 y^3$$

which after application of the boundary conditions becomes

$$\frac{T(y) - T_s}{T_\infty - T_s} = \frac{3}{2}\frac{y}{\delta_T} - \frac{1}{2}\left(\frac{y}{\delta_T}\right)^3 \tag{9.3.1}$$

In contrast, from Sec. 7.2 the velocity profile is given by

$$\frac{u(y)}{u} = \frac{3}{2}\frac{y}{\delta} - \frac{1}{2}\left(\frac{y}{\delta}\right)^3 \tag{9.3.2}$$

The thermal (δ_T) and momentum boundary layer heights are related by the Prandtl number

$$\frac{\delta}{\delta_T} = \mathbf{P}_r^{1/3} \tag{9.3.3}$$

Equation (7.2.7) gives the growth of δ as a function of the distance x from the leading edge and the Reynolds number at the distance x, $\mathbf{R} = ux/\nu$.

The control volume analysis allows the heat transfer coefficient to be evaluated. The four boundary heat fluxes must balance; therefore, $q_T(y = 0) = q_T(x + dx) - q_T(x) - q_T(\delta)$ and

$$-k\frac{\partial T}{\partial y}\bigg|_{y=0} dx(1) = \int_0^{\delta_T} \rho c_p\, u(x + dx, y)\, T(x + dx, y)\, dy$$

$$- \int_0^{\delta_T} \rho c_p\, u(x, y)\, T(x, y)\, dy - dx \int_0^{\delta_T} \rho c_p\, u(x, y)\, T_\infty\, dy$$

By assuming ρc_p to be constant, dividing by the plate surface area $dx(1)$, and taking the limit as $dx \rightarrow 0$, the following results

$$-\frac{k}{\rho c_p}\frac{\partial T}{\partial y}\bigg|_{y=0} = \frac{d}{dx}\int_0^{\delta_T} u(y)(T_\infty - T(y))\,dy \qquad (9.3.4)$$

Insertion of the velocity and temperature profiles [Eqs. (9.3.1) and (9.3.2)] into Eq. (9.3.4) gives the following equation for the Nusselt number at any point x along the plate

$$\mathbf{N}_{ux} = \frac{h(x)\,x}{k} = 0.36\mathbf{R}_x^{0.5}\mathbf{P}_r^{0.333} \qquad (9.3.5)$$

In going from Eq. (9.3.4) to Eq. (9.3.5), the Newton rate equation has been used, that is,

$$q(x, y = 0) = h(x)(T_s - T_\infty) = -k\frac{\partial T}{\partial y}\bigg|_{y=0}$$

As seen this solution parameterizes the flux and heat transfer coefficient at each point along the plate. The design form however requires a bulk or average representation which can be found as

$$q = hA(T_s - T_\infty) = \int h(x)(T_s - T_\infty)\,dA$$

Substitution of Eq. (9.3.5) into this equation and integrating over the width (w) and length (L) of the plate yield the bulk form of the Nusselt number

$$\mathbf{N}_{uL} = \frac{hL}{k} = 0.72\mathbf{R}_L^{0.5}\mathbf{P}_r^{0.333} \qquad (9.3.6)$$

It is noted that the exact solution approach based upon the Blasius flat plate solution [11] gives

$$\mathbf{N}_{uL} = 0.664\mathbf{R}_L^{0.5}\mathbf{P}_r^{0.333} \qquad (9.2.15)$$

These two solutions differ by 9 percent. The approximate integral approach is a quite useful technique when profile solutions are not known *a priori*.

Mass Transfer

Using the same set of restrictions imposed for the laminar thermal boundary layer, the von Kármán integral analysis approximation can be applied to the analysis of the concentration boundary layer (δ_c) and mass transfer coefficient (h_m). An additional caveat imposed for this analysis is that there are no sources or sinks within the control volume.

The profile is assumed to be of the form

$$C - C_s = d_1 + d_2 y + d_3 y^2 + d_4 y^3$$

and subject to the following boundary conditions

$$C = C_s \ @ \ y = 0; \qquad \frac{\partial C}{\partial y} = 0 \ @ \ y = \delta_c$$

$$C = C_\infty \ @ \ y = \delta_c; \qquad \frac{\partial^2 C}{\partial y^2} = 0 \ @ \ y = 0$$

The profile becomes

$$\frac{C(y) - C_s}{C_\infty - C_s} = \frac{3}{2}\left(\frac{y}{\delta_c}\right) - \frac{1}{2}\left(\frac{y}{\delta_c}\right)^3 \tag{9.3.7}$$

The concentration boundary layer height (δ_c) is related to the momentum boundary layer height by

$$\delta/\delta_c = \mathbf{S}_c^{1/3} \tag{9.3.8}$$

A control volume analysis identical to the one for the thermal layer reveals that

$$h_{mx}(C_s - C_\infty) = \frac{d}{dx}\int_0^{\delta_c} [C(x, y) - C_\infty]u(x, y)\,dy \tag{9.3.9}$$

Substitution of the appropriate profile into Eq. (9.3.9) and integrating yield a nondimensional relationship for the mass transfer coefficient at each location x along the plate

$$\mathbf{S}_{h_x} = \frac{x\,h_{mx}}{\mathcal{D}} = 0.36\mathbf{R}_x^{0.5}\mathbf{S}_c^{0.333} \tag{9.3.10}$$

The overall or bulk Sherwood number is found by integrating over the length and width of the plate and is equal to

$$\mathbf{S}_h = \frac{L\,h_m}{\mathcal{D}} = 0.72\mathbf{R}_L^{0.5}\mathbf{S}_c^{0.333} \tag{9.3.11}$$

We conclude this section by noting that for most geophysical flows laminar boundary layers are quite thin and occur over small spatial extents. Many engineered processes, however, design laminar boundary layer mechanics into the process. The next section revisits the problem of turbulence and how to parameterize its effects on heat and mass transport.

Example 9.5

Returning to Ex. 9.4, determine the spatial extent of the laminar water vapor boundary layer and the thickness just as it makes the transition from laminar to turbulent flow.

Solution

The momentum boundary layer will undergo a transition to a turbulent layer at the distance where $\mathbf{R}_x = Ux/\nu \geq 2.5(10^5)$. Much lower critical Reynolds numbers exist as the surface roughness increases. From the previous example then

$$x = \frac{\mathbf{R}_x\nu}{U} = \frac{2.5(10^5)\,1.46(10^{-5})\ \text{m}^2/\text{s}}{8\ \text{m/s}} = 0.45\ \text{m}$$

The boundary layer thicknesses are related by

$$\delta/\delta_c = \left(\frac{\nu}{\mathcal{D}}\right)^{1/3} = (0.6)^{1/3} = 0.84$$

Therefore, $\delta_c = 1.18\delta$.

From Equation (7.2.7)

$$\delta = \frac{4.65\,x}{\sqrt{\mathbf{R}_x}} = \frac{4.65(0.45)}{\sqrt{\frac{8(0.45)}{1.46(10^{-5})}}} = 0.42(10^{-2})\ \text{m} = 0.42\ \text{cm}$$

Therefore, δ_c at the transition point is

$$1.18(0.42\ \text{cm}) = 0.5\ \text{cm}$$

EXERCISE

9.3.1   For a laminar boundary layer analysis of heat or mass diffusion (*a*) the normal flux at the wall, $y = 0$, is zero; (*b*) the boundary layer height grows linearly with distance downstream on a flat plate; (*c*) the Nusselt or Sherwood numbers are proportional to the square root of the Reynolds number; (*d*) the concentration or temperature profile is a quadratic equation with the variable being the distance away from the wall; (*e*) the boundary layer heights are large.

9.4   TURBULENT TRANSPORT RELATIONS

As noted in Chap. 3 turbulent flow is described as an unsteady swirling motion of the fluid which has a strong effect on the distribution of momentum, heat, and mass species. As noted in Sec. 6.4 analysis procedures based upon averaging out the effect of the fine scale turbulence lead to a complex set of Reynolds equations governing the distribution of the averaged variables. Chap. 6 concentrated on the turbulent velocity fluctuations for pipe and channel flow. In this section we extend these concepts to heat and mass transport.

Averaging of Governing Equations

The equations for heat and mass species, Eqs. (4.7.4*b*) and (4.8.10*a*), respectively, are analyzed by Reynolds averaging. The turbulent temperature (T') and concentration (C') are defined relative to the mean quantity denoted by the overbar;

$$T(\mathbf{x}, t) = \overline{T}(\mathbf{x}, t) + T'(\mathbf{x}, t) \tag{9.4.1a}$$

$$C(\mathbf{x}, t) = \overline{C}(\mathbf{x}, t) + C'(\mathbf{x}, t) \tag{9.4.1b}$$

As before the average is defined as

$$\overline{T}(\mathbf{x}, t) = \frac{1}{T_o} \int_t^{t+T_o} T(\mathbf{x}, t)\, dt$$

It is noticed that even though the variables have been averaged it will be the case that even the mean flow variables can vary both in time and space. Time variation in the mean flow variables is an ever-present phenomenon in geophysical flows.

The heat and mass transfer equations are averaged retaining the continuity equation. The heat equation, therefore, is

$$\rho c_p \left[\frac{\partial \overline{T}}{\partial t} + \nabla \cdot \overline{\mathbf{v}T} \right] = k\nabla^2 \overline{T} + \overline{S}_{\mathrm{T}} \tag{9.4.2a}$$

and the mass equation is

$$\frac{\partial \overline{C}}{\partial t} + \nabla \cdot \overline{\mathbf{v}C} = \mathscr{D}\nabla^2 \overline{C} + \overline{S}_{\mathrm{C}} \tag{9.4.2b}$$

Reynolds rules of averaging, used in Sec. 6.4, are employed to decompose the analogous advection terms in Eq. (9.4.2). For two functions *a* and *b*

$$\overline{ab} = \overline{(\overline{a} + a')(\overline{b} + b')} = \overline{\overline{a}\,\overline{b}} + \overline{a'b'} \approx \overline{a}\,\overline{b} + \overline{a'b'}$$

Therefore, the turbulent heat and mass transport equations become, respectively,

$$\rho c_p \left[\frac{\partial \overline{T}}{\partial t} + \frac{\partial}{\partial x}(\overline{u}\,\overline{T} + \overline{u'T'}) + \frac{\partial}{\partial y}(\overline{v}\,\overline{T} + \overline{v'T'}) + \frac{\partial}{\partial z}(\overline{w}\,\overline{T} + \overline{w'T'}) \right] = k\nabla^2 \overline{T} + \overline{S}_T^*$$

(9.4.3a)

$$\frac{\partial \overline{C}}{\partial t} + \frac{\partial}{\partial x}(\overline{u}\,\overline{C} + \overline{u'C'}) + \frac{\partial}{\partial y}(\overline{v}\,\overline{C} + \overline{v'C'}) + \frac{\partial}{\partial z}(\overline{w}\,\overline{C} + \overline{w'C'}) = \mathscr{D}\nabla^2 \overline{C} + \overline{S}_C$$

(9.4.3b)

In order to preserve the *similarity* of these equations to their unaveraged counterparts, the correlation terms are subtracted from both sides of the equation, resulting in a heat equation

$$\rho c_p \frac{D\overline{T}}{Dt} =$$

$$\frac{\partial}{\partial x}\left[k\frac{\partial \overline{T}}{\partial x} - \rho c_p \overline{u'T'} \right] + \frac{\partial}{\partial y}\left[k\frac{\partial \overline{T}}{\partial y} - \rho c_p \overline{v'T'} \right] + \frac{\partial}{\partial z}\left[k\frac{\partial \overline{T}}{\partial z} - \rho c_p \overline{w'T'} \right] + \overline{S}_T^*$$

(9.4.4a)

and mass equation

$$\frac{D\overline{C}}{Dt} = \frac{\partial}{\partial x}\left[\mathscr{D}\frac{\partial \overline{C}}{\partial x} - \overline{u'C'} \right] + \frac{\partial}{\partial y}\left[\mathscr{D}\frac{\partial \overline{C}}{\partial y} - \overline{v'C'} \right] + \frac{\partial}{\partial z}\left[\mathscr{D}\frac{\partial \overline{C}}{\partial z} - \overline{w'C'} \right] + \overline{S}_C$$

(9.4.4b)

The heat and mass equations can also be rewritten in turbulent flux form, respectively, as

$$\frac{d\overline{T}}{dt} + \nabla \cdot \overline{\mathbf{N}}_T = \frac{\overline{S}_T^*}{\rho C_p} = S_T$$

(9.4.5a)

$$\frac{d\overline{C}}{dt} + \nabla \cdot \overline{\mathbf{N}}_C = S_C$$

(9.4.5b)

Here $\overline{\mathbf{N}}_T$ and $\overline{\mathbf{N}}_C$ are the total transport fluxes for heat and mass, respectively

$$\overline{\mathbf{N}}_T = N_{T_x}\mathbf{i} + N_{T_y}\mathbf{j} + N_{T_z}\mathbf{k}$$

(9.4.6a)

$$= \left(\overline{u}\,\overline{T} + \overline{u'T'} - \alpha\frac{\partial \overline{T}}{\partial x} \right)\mathbf{i} + \left(\overline{v}\,\overline{T} + \overline{v'T'} - \alpha\frac{\partial \overline{T}}{\partial y} \right)\mathbf{j} + \left(\overline{w}\,\overline{T} + \overline{w'T'} - \alpha\frac{\partial \overline{T}}{\partial z} \right)\mathbf{k}$$

$$\overline{\mathbf{N}}_C = N_{C_x}\mathbf{i} + N_{C_y}\mathbf{j} + N_{C_z}\mathbf{k}$$

(9.4.6b)

$$= \left(\overline{u}\,\overline{C} + \overline{u'C'} - \mathscr{D}\frac{\partial \overline{C}}{\partial x} \right)\mathbf{i} + \left(\overline{v}\,\overline{C} + \overline{v'C'} - \mathscr{D}\frac{\partial \overline{C}}{\partial y} \right)\mathbf{j} + \left(\overline{w}\,\overline{C} + \overline{w'C'} - \mathscr{D}\frac{\partial \overline{C}}{\partial z} \right)\mathbf{k}$$

In direct analogy to the Reynolds stresses defined in Chap. 6.4, the correlation terms (e.g., $\overline{u'T'}$) or turbulent fluxes are called *Reynolds fluxes*. Terms of the form $\overline{u}\,\overline{C}$, $\overline{u}\,\overline{T}$, etc., are the advective fluxes while the gradient terms involving the thermal or mass diffusivity are the molecular diffusion fluxes.

Eddy Diffusivities

The Reynolds fluxes as well as the stresses must be *closed.* This ubiquitous piece of jargon refers to the necessity of relating the Reynolds fluxes to terms involving the mean flow variables ($\overline{\mathbf{v}}$, $\overline{C}$, $\overline{T}$, etc.). The earliest and most enduring closure was presented by G. Taylor [12] who argued that in a turbulence field far from boundaries it was possible to relate the turbulent fluxes to the spatial gradient of the mean flow variable. For example,

$$-\overline{u'T'} = E_{T_x}\frac{\partial \overline{T}}{\partial x}$$

$$-\overline{w'C'} = E_{C_z}\frac{\partial \overline{C}}{\partial z}$$

E_{T_x} is defined as the *eddy diffusivity for heat* and E_{C_z} is the *eddy diffusivity for the mass species.* These coefficients are not properties of the fluid as they depend upon the geometry of the flow field, the intensity of the turbulence in the flow field, and the possible effect of stratification. Therefore, E_T and E_C are field variables which can vary over time and space.

From Eqs. (9.4.4*a*) and (9.4.4*b*) and with the eddy diffusivity gradient closure, the heat and mass equations, respectively, are

$$\frac{\partial \overline{T}}{\partial t} + \nabla \cdot \left(\overline{\mathbf{v}}\overline{T}\right) = \frac{\partial}{\partial x}\left[(\alpha + E_{T_x})\frac{\partial \overline{T}}{\partial x}\right] + \frac{\partial}{\partial y}\left[(\alpha + E_{T_y})\frac{\partial \overline{T}}{\partial y}\right] \qquad \textbf{(9.4.7a)}$$

$$+ \frac{\partial}{\partial z}\left[(\alpha + E_{T_z})\frac{\partial \overline{T}}{\partial z}\right] + \overline{S}_T$$

$$\frac{\partial \overline{C}}{\partial t} + \nabla \cdot \left(\overline{\mathbf{v}}\overline{C}\right) = \frac{\partial}{\partial x}\left[(\mathscr{D} + E_{C_x})\frac{\partial \overline{C}}{\partial x}\right] + \frac{\partial}{\partial y}\left[(\mathscr{D} + E_{C_y})\frac{\partial \overline{C}}{\partial y}\right] \qquad \textbf{(9.4.7b)}$$

$$+ \frac{\partial}{\partial z}\left[(\mathscr{D} + E_{C_z})\frac{\partial \overline{C}}{\partial z}\right] + \overline{S}_C$$

As with the averaged turbulent momentum equations, the overbar is traditionally dropped from further use and the average variables are understood to exist in the equation.

Comparing the turbulent eddy diffusivities to the eddy viscosities yields a *turbulent Prandtl number* (ϵ/E_T) or a *turbulent Schmidt number* (ϵ/E_C) but in practice these are useful concepts only in engineered flows where the eddy diffusivities and viscosities are less transient and often more spatially uniform than their geophysical counterparts.

In the subsequent sections of this chapter turbulent transport will be analyzed. The eddy diffusivity functions must be specified according to the type of turbulence field being addressed. Therefore the next section starts with the simplest turbulent transport field description, that is, constant diffusivities. The resulting *turbulent diffusion* descriptions are useful building blocks for more complex flows.

EXERCISES

9.4.1   The eddy diffusivities for turbulent heat or mass transport (*a*) are properties of the fluid; (*b*) are constant; (*c*) result from a spatial averaging procedure; (*d*) are relatively the same order of magnitude as the viscosities; (*e*) none of the above.

9.4.2 Reynolds fluxes (*a*) are always turbulent; (*b*) consist of both laminar and turbulent contributions; (*c*) are uniformly distributed in space; (*d*) are steady; (*e*) were defined by G. Taylor in 1921.

9.5 TURBULENT DIFFUSION

As introduced in the previous section, the gradient or diffusion approximation is used extensively to describe the turbulent transport fluxes in terms of mean flow variables. As noted by many [13, 14] other types of closures are available but in contrast to the gradient diffusion approximation are quite complex and useable primarily in complex computer models of the flow and transport. The more advanced closure models are necessary when considering flows that have highly irregular geometries, are transient in both the mean and turbulent quantities, and are fully coupled in that heat or mass can introduce strong stratification or density gradients in the flows thereby altering the flow and circulation pattern. This collection of requirements for use of advanced turbulent closure models contains descriptors for geophysical flows, and such complex flows and turbulence models are dealt with in courses in atmospheric sciences, oceanography, and even astrophysics.

Diffusion

Our attention will focus on engineered flows which are reasonably steady and have uniform geometry. For these cases the gradient diffusion approximation is an excellent closure. The simplest closure, the constant eddy diffusivity, is considered first. Before proceeding to turbulent diffusion, however, it is instructive to reconsider the case of molecular diffusion. For the one-dimensional case consider a mass, M, of particles which are marked such that their paths can be traced over time. With regard to Fig. 9.8, at time $t = 0$ the mass is deposited at the origin and diffusion will cause material to move away from the origin in either direction. The equation governing this motion is

$$\frac{\partial C}{\partial t} = \mathcal{D} \frac{\partial^2 C}{\partial x^2} \qquad (9.5.1)$$

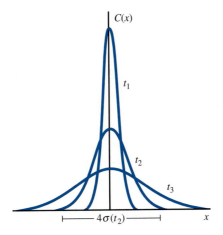

Figure 9.8 Schematic of turbulent diffusion in time.

and has a solution of the form

$$C(x, t) = \frac{B}{\sqrt{t}} e^{-x^2/4\mathcal{D}t}$$

(9.5.2)

At every time point the mass per unit width in the system is conserved. Therefore, regardless of the spatial distribution at any time, t, we have

$$M = \int_{-\infty}^{\infty} C(x, t)\, dx$$

(9.5.3)

Substituting Eq. (9.5.2) into Eq. (9.5.3), integrating, and using the fact that

$$\int_{-\infty}^{\infty} e^{-y^2}\, dy = \sqrt{\pi}$$

the coefficient of integration, B, is obtained. Therefore,

$$C(x, t) = \frac{M}{\sqrt{4\pi\mathcal{D}t}} e^{-x^2/4\mathcal{D}t}$$

(9.5.4)

It is useful to characterize geometrical details of the spreading process, the simplest measures being the width of the curve, the time it takes the concentration curve to reach a given position, or the speed of the curve spreading. In a simple-minded way the width or spread of a curve can be formed by computing the variance, σ^2, of the distribution about the mean, that is,

$$\sigma^2(t)M = \int_{-\infty}^{\infty} (x - \mu)^2 C(x, t)\, dx$$

(9.5.5)

It is noted that the variance will be a function of time. For the distribution here, $\mu = 0$, as the function is symmetric. Substituting Eq. (9.5.5) into Eq. (9.5.4) and dividing both sides by M give

$$\sigma^2(t) = \frac{1}{2\sqrt{\pi\mathcal{D}t}} \int_{-\infty}^{\infty} x^2 e^{-x^2/4\mathcal{D}t}\, dx = 2\mathcal{D}t$$

(9.5.6)

Therefore, the width, as measured by the variance, grows linearly with time for diffusion processes marked by a constant diffusion coefficient. It is also noticed that the time rate of change or *spreading rate* or *growth rate* is constant

$$\frac{d}{dt}\sigma^2(t) = 2\mathcal{D}$$

(9.5.7)

Indeed as the derivation in Fischer et al. [15] points out Eq. (9.5.7) can be derived independently of the shape of the concentration curve. To show this, multiply both sides of Eq. (9.5.1) by x^2 and integrate both sides of the equation with respect to dx over the interval $-\infty < x < \infty$.

One further consequence comes from the integration of Eq. (9.5.7) between two time periods

$$\int_1^2 d\sigma^2 = 2\mathcal{D} \int_1^2 dt \quad \Longrightarrow \quad \sigma_2^2 - \sigma_1^2 = 2\mathcal{D}(t_2 - t_1)$$

(9.5.8)

Therefore, if the variance of a concentration patch can be sampled at two different times, then the diffusion coefficient can be directly ascertained as

$$\mathcal{D} = \frac{(\sigma_2^2 - \sigma_1^2)}{2(t_2 - t_1)} \qquad (9.5.9)$$

Analogy to Gaussian Probability Distribution

It is noteworthy and not coincidental that the form of Eq. (9.5.4) is similar to the normal probability distribution. This has been noted since Albert Einstein's derivation of Eq. (9.5.4) in 1905 (see the discussion in Ref. [4]). The normal probability (P) distribution for a variable n is given by

$$P(n) = \frac{1}{\sqrt{2\pi}\sigma} e^{-n^2/2\sigma^2}$$

Therefore, if $n \rightarrow x$, $\sigma^2 = 2\mathcal{D}t$ in Eq (9.5.5), $M = 1$, and the mean of the distribution is centered about $x = 0$, then Eq. (9.5.4) is recovered. Knowing this information is helpful in field sampling. As hinted in Eq. (9.5.9) the size of the concentration cloud is often used to help in determining the diffusion coefficient or later on the eddy diffusivity. A convenient measure of size can be adapted from the normal distribution as it is well known that 95 percent of the total area under the curve is contained within a width of $\pm 2\sigma$ (4σ total) measured from the mean of the curve (see Fig. 9.8). From another perspective if we know the diffusion coefficient or eddy diffusivity, the size (width) of the cloud, $w_c(t)$, at any time t is

$$w_c(t) = 4\sigma = 4\sqrt{2\mathcal{D}t} \qquad (9.5.10)$$

The analogy to probability distributions may be extended beyond just the normal distribution but for now only the concept of generalized moments is useful. These are defined as follows:

$$\text{zeroth moment} = M_0 = M = \int_{-\infty}^{\infty} C(x, t)\, dx$$

$$\text{first moment} = M_1 = \int_{-\infty}^{\infty} x C(x, t)\, dx$$

$$\text{second moment} = M_2 = \int_{-\infty}^{\infty} x^2 C(x, t)\, dx$$

Any moment beyond the second can readily be made by incrementing the power of the exponent on the independent variable in the integrand. For the normal distribution and its concentration distribution analogy in Eq. (9.5.4), the mean, μ, and its variance, σ^2, are found from

$$\mu = \frac{M_1}{M_0}$$

while

$$\sigma^2 = \frac{\int_{-\infty}^{\infty}(x - \mu)^2\, C(x, t)\, dx}{\int_{-\infty}^{\infty} C(x, t)\, dx} = \frac{M_2}{M_0} - \left(\frac{M_1}{M_0}\right)^2 = \frac{M_2}{M_0} - \mu^2 \qquad (9.5.11)$$

Turbulent Diffusion Basics

As has been stressed in a number of places the various turbulent transport fluxes (e.g., $\overline{u'T'}$ or $\overline{w'C'}$) possess several attributes. First, they range in size from fluctuations the size of the flow geometry down to the smallest turbulent scale termed the *Kolmogorov scale*. Second, they are quite transient and random in the sense that for the same mean flow, the details of an individual turbulent fluctuation most likely will not be repeatable, but the statistical characteristics of the random nature will be. Third, the random swirling nature of the flow results in intermittent but large spatial gradients in variables as packets of fluid with quite different characteristics are brought in contact by the turbulence. Fourth, the continuous introduction of packets of high and low concentration by the turbulence will result in an overall dilution of zones of high concentration via a spreading or mixing behavior.

The energy to create turbulence is typically put into the system at scales the size of the entire flow geometry (e.g., a wind storm over a lake or pond). Viscosity is the agent acting to dissipate the turbulent energy in the form of heat, and it operates on the smallest possible scales of the flow field. In between the creation and dissipation turbulent scales there is an orderly *cascade* of turbulent energy by which energy is nonlinearly transmitted from the large to small turbulent fluctuations or scales by a continuously varying (i.e., decreasing) sequence of turbulent eddies. For steady mean flow the turbulent energy creation is in *equilibrium* with the rate of destruction, and therefore, the rate at which energy is transmitted to the smallest turbulent eddies is constant.

As noted above, the smallest turbulent scales are termed the *Kolmogorov* or *dissipation scales* for length, time, and velocity, whose values are on the order of 0.1 cm, 0.1–1.0 s, and 0.1 cm/s, respectively. While useful characterizations of scale, they do not provide much useful information about the gradient analogy for turbulent mixing and ultimately the applicability of turbulent diffusion. To do so requires the use of other measures for characterizing the impact and size of the eddies.

Suppose, as in Fig. 9.9, a mass of particles is released at a point in a turbulence field. The turbulence field will be very far away from boundaries and shall

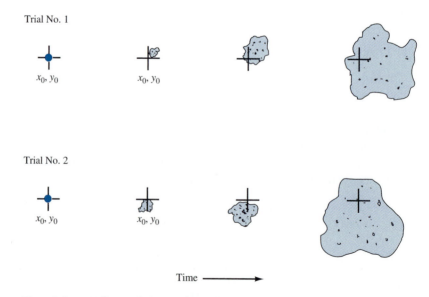

Figure 9.9   Diffusion of a mass of particles in time.

be further defined such that all the statistical measures about the spatial and temporal structure of the fluctuations shall be constant. Therefore the turbulence will be considered *homogeneous,* that is, the statistics of the turbulent fluctuations are constant in one particular direction and *isotropic,* that is, spatial statistical structure in all three directions is constant. Further, the velocity turbulence fluctuations will be *stationary,* meaning the statistics of the turbulent fluctuations do not change over time. In Fig. 9.9 a clump of particles of mass M is placed at the origin of the turbulence field and the larger scale turbulent fluctuations will begin to distort, shear, and spread the particles as time proceeds in trial 1. If the experiment were to be repeated, as in trial 2, the shape of the mass M at each time step would be quite different. However, in recognizing the random nature of the turbulence field the statistical or ensemble averages of the evolving or spreading masses over a repeated number of trials will permit mathematical elaboration. The reader is referred to the discussion in Refs. [4, 13–16] for more elaborate discussions of the ensemble averaging process.

During the initial stages of the particle motion the individual particles are tightly bunched, that is, their motion is constrained by neighboring particles; consequently the particles have not adjusted to the turbulence field or reached equilibrium with it. Over longer time periods the mass has pulled apart or spread through the action of turbulent shear stress, and the particles have adjusted to, or are in equilibrium with, the turbulent field. These two periods can be determined using the concept of the correlation coefficient and their associated time and length scales.

The Lagrangian correlation coefficient R for the velocity fluctuation u' is found from the following general formulas (see Ref. [4])

$$R_{L_x}(\tau) = \frac{\overline{u'(t)u'(t + \tau)}}{\sqrt{\overline{u'^2(t)}}\sqrt{\overline{u'^2(t + \tau)}}} = \frac{\left\langle \overline{u'(t)u'(t + \tau)} \right\rangle}{\langle u'^2 \rangle} \tag{9.5.12}$$

The brackets denote the ensemble average over all trials, the overbar signifies the time average, the L denotes the Lagrangian formulation, and the x denotes the coordinate. Since the velocities are correlated with themselves, R_{L_x} is called the *autocorrelation function.* Figure 9.10 (modified slightly from Ref. [4]) depicts the transition of the correlation coefficient over time from $\tau \sim 0$, where the correlation is 1.0, to $\tau \rightarrow \infty$, where no correlation exists at all. The bottom portion of the figure shows a typical autocorrelation function corresponding to this behavior and defines the associated time scale as

$$T_{L_x} = \int_0^\infty R_{L_x}(\tau)\, d\tau \tag{9.5.13}$$

As shown in Ref. [15]† the G. Taylor analysis [12] of the cloud growth shows that the rate of ensemble-averaged variance growth of the cloud is

$$\sigma_x^2(t) = 2\langle \overline{u'^2} \rangle \int_0^t (t - \tau) R_{L_x}(\tau)\, d\tau \tag{9.5.14}$$

Similar expressions can be found for the y and z direction. Based upon the treatment in the above references or Csanady [17], two distinct behaviors can be identified for the behavior of $\sigma_x^2(t)$, that is, *small time* where $t \rightarrow 0$ and $t < T_{L_x}$ and *long time* where $t \rightarrow \infty$ and $t > T_{L_x}$. Table 9.1 summarizes the results. Therefore, for short times the variance grows quickly in proportion to t^2, while for long times

---

| †There are numerous references in addition to this one.

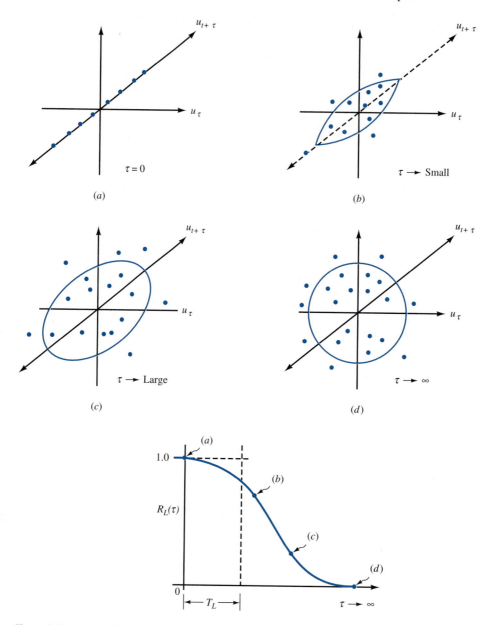

Figure 9.10   Correlation coefficient evolution in time. (*a*) Perfect correlation. (*b*) High correlation.
(*c*) Low correlation. (*d*) No correlation.

TABLE 9.1   Diffusion variance regimes

Region	Criteria (time)	Criteria (length)	Variance growth
Small time	$t \to 0,\, t < T_{L_x}$	$L^2 < 2l_L^2$	$\sigma_x^2(t) = \langle \overline{u'^2} \rangle t^2$
Long time	$t \to \infty,\, t > T_{L_x}$	$L^2 > 2l_L^2$	$\sigma_x^2(t) = 2\langle \overline{u'^2} \rangle t T_{L_x}$

the variance grows linearly with time, which is identical to the molecular diffusion case analyzed previously.

Given a full three-dimensional turbulence field $T_L = (T_{L_x} + T_{L_y} + T_{L_z})/3$ and this average time scale can be used to identify the corresponding Lagrangian length scale as

$$l_L^2 = \left\langle \overline{u'^2} \right\rangle T_L^2 \tag{9.5.15}$$

Here l_L is the estimated distance a fluid particle will travel before losing its memory of the initial velocity. By use of the long-time relationship (Table 9.1) for ensemble-averaged variance growth the ensemble average size of the cloud, $L(t)$, can be defined as

$$L^2(t) \approx 2\left\langle \overline{u'^2} \right\rangle T_L t \tag{9.5.16}$$

which by insertion into Eq. (9.5.14) allows the size of the cloud required for the diffusion approximation to hold (i.e., long time) as

$$L^2 > 2 l_L^2 \tag{9.5.17}$$

In summary then, the gradient diffusion approximation for turbulence and the corresponding closure is applicable for material placed in the flow field for a time longer than the Lagrangian time scale and has a size greater than the Lagrangian length scale.

Selected Solutions

Exact solutions to the linear diffusion or advection-diffusion equation with constant velocity are ubiquitous with the books by Carlslaw and Jaeger [18] and Crank [19] forming virtual encyclopedias of such solutions. The solutions presented here are selected due to their widespread applicability to environmental problems. The procedures for creating the solution in the main are found in the original references.

1. *Time-varying initial condition.* The applicable governing equation is the one-dimensional form of Eq. (9.4.7b)

$$\frac{\partial C}{\partial t} = (\mathcal{D} + E_{C_x})\frac{\partial^2 C}{\partial x^2} = D_x \frac{\partial^2 C}{\partial x^2}$$

subject to the initial condition that the concentration is zero throughout the domain $0 \le x < \infty$ at time $t = 0$. At time $t = 0$, the boundary condition at $x = 0$ is such that the concentration at $x = 0$ is raised to C_0, that is, $C(x = 0, t \ge 0) = C_0$. The second boundary condition applies at $x \to \infty$ and is essentially a statement that $C(x \to \infty, t) = 0$. The solution is found by a similarity transformation method analogous to the boundary layer transformation and the result is

$$C(x, t) = C_0\left(1 - \text{erf}\left[\frac{x}{\sqrt{4D_x t}}\right]\right) \tag{9.5.18}$$

In this equation $D_x = \mathcal{D} + E_{C_x}$ will be called the *turbulent diffusion coefficient.* In practice its value will be close if not equal to the value of the eddy diffusivity.

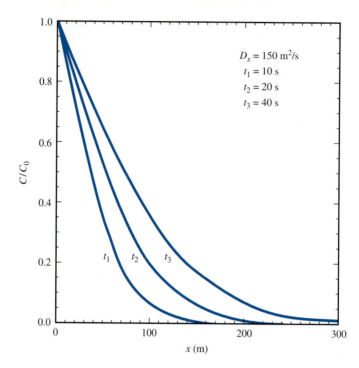

Figure 9.11   Solution behavior of Eq. (9.5.18) for increasing values of time.

The *error function* for an argument α is defined as

$$\text{erf}(\alpha) = \frac{2}{\sqrt{\pi}} \int_0^\alpha e^{-\beta^2}\, d\beta \qquad (9.5.19)$$

The complementary error function is defined such that $\text{erfc}(x) = 1 - \text{erf}(x)$.

Figure 9.11 contains a schematic of the solution behavior for various values of increasing time t. It is noted that the mass in the domain is increasing over time and is due to the fact that diffusive flux is finite. The total mass in the system at any time t_x should be equal to the integral of the flux from $0 \rightarrow t_x$, that is,

$$\int_0^{t_x} N_{C_x}\, dt = M(t_x) = \int_0^\infty C(x, t_x)\, dx$$

2. *Two- and three-dimensional diffusion.* The basic building block for analyzing jets and plumes from wastewater effluents or smokestacks is the two-dimensional diffusion equation which becomes

$$\frac{\partial C}{\partial t} = D_x \frac{\partial^2 C}{\partial x^2} + D_y \frac{\partial^2 C}{\partial y^2} \qquad (9.5.20)$$

The presence of two turbulent diffusion coefficients suggests that the turbulence field is homogeneous in x and y, respectively, but not isotropic. This is not a problem when using the gradient diffusion approximation except insofar as a different diffusion coefficient is necessary. The solution condition is satisfied by placing a mass, M, of material at the origin of the xy system. As noted in Fischer et al. [15] the solution can be found by a separation concept wherein $C(x, y, t) = C_1(x, t)\, C_2(y, t)$ is substituted into the governing equation and a pair

413

of one-dimensional diffusion equations is achieved. The solution is

$$C(x, y, t) = \frac{M}{4\pi t \sqrt{D_x D_y}} \exp\left\{-\frac{x^2}{4D_x t} - \frac{y^2}{4D_y t}\right\}$$ (9.5.21)

The three-dimensional analogy is

$$C(x, y, z, t) = \frac{M}{4\pi t \sqrt{D_x D_y D_z}} \exp\left\{-\frac{x^2}{4D_x t} - \frac{y^2}{4D_y t} - \frac{z^2}{4D_z t}\right\}$$ (9.5.22)

Figure 9.12 contains a schematic of the growth in time of the line of constant concentration (isopleth) coincident to the $\pm 2\sigma$ (4σ) width of the diffusing cloud. It is noticed that unequal diffusion coefficients result in a nonsymmetric distribution about the origin.

3. *Advection and diffusion.* In the x direction a constant velocity, u, can advect or carry the diffusing cloud downstream. The basic presumption required to extend the diffusion analysis is that the velocity does not distort the turbulence field statistics. In so doing a new moving coordinate system $x_* = x - ut$ can be defined which reduces the advection diffusion equation to a pure diffusion equation and for which all existing diffusion solutions may be adapted. Consider the following lateral diffusion problem (Fig. 9.13) first detailed in Fischer et al. [15]. Two fluid streams with velocity u are blended, at the origin $x = 0$; one stream has no concentration while the other stream has a concentration C_0. The steady advection diffusion equation becomes

$$u\frac{\partial C}{\partial x} = D_y \frac{\partial^2 C}{\partial y^2}$$

The horizontal diffusion term has not been included because the gradients in the horizontal are quite small in contrast to those in the y direction. The boundary

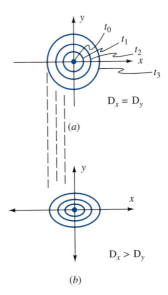

Figure 9.12 Schematic of the growth in time of a concentration line source for (a) equal diffusivities and (b) differing diffusivities.

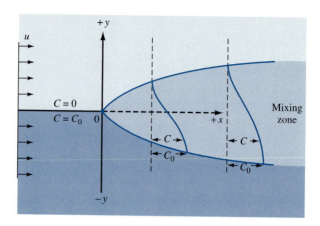

Figure 9.13   Lateral diffusion at the interface of two streams
with different concentrations.

conditions are (Fig. 9.13)

$$C(x = 0, y) = \begin{cases} 0 & y > 0 \\ C_0 & y < 0 \end{cases}$$

$$C(x, y \to \infty) = 0$$

$$C(x, y \to -\infty) = C_0$$

The solution is

$$C(x, y) = \frac{C_0}{2}\left\{1 - \mathrm{erf}\left(\frac{y}{\sqrt{4D_y x/u}}\right)\right\} \tag{9.5.23}$$

Figure 9.13 contains a schematic of the resulting *mixing zone* for the fluids.

Instead of blending two continuous streams, suppose as in French [20] that mass, M, is continuously injected into the two-dimensional (horizontal) flow field dominated by an advection velocity u in the x direction at the time rate denoted by $\dot{M}$. In this case Eq. (9.5.21) still applies and the solution becomes

$$C(x, y) = \frac{\dot{M}}{u\sqrt{4\pi x D_y/u}}\exp\left(-\frac{y^2 u}{4x D_y}\right) \tag{9.5.24}$$

4. *Transient advection plus diffusion.* We return now to the case in the first exact solution [Eq. (9.5.18)] and ask what the impact of advection, u, would be on the solution. The same initial and boundary conditions apply as in the pure diffusion case and the solution [20] is

$$C(x, y) = \frac{C_0}{2}\left[\mathrm{erfc}\left(\frac{x - ut}{\sqrt{4D_x t}}\right) + \mathrm{erfc}\left(\frac{x + ut}{\sqrt{4D_x t}}\right)\exp\left(\frac{ux}{D_x}\right)\right] \tag{9.5.25}$$

5. *Transient diffusion with first-order reaction.* Finally, consider the same problem as in the first solution example, but in addition to diffusion there is a first-order reaction rate (decay in this case) parameterized as in Chap. 3 as $-k_1 C$. The governing equation becomes

$$\frac{\partial C}{\partial t} = D_x \frac{\partial^2 C}{\partial x^2} - k_1 C \tag{9.5.26}$$

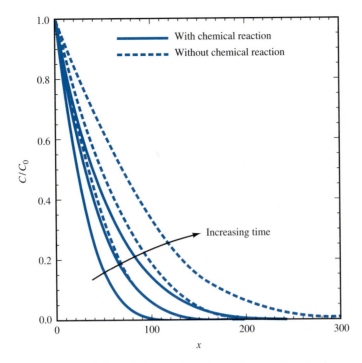

Figure 9.14 Solution behavior of transient diffusion with and without a first-order reaction.

The boundary and initial conditions are as in solutions 1 and 4, $C(x = 0, t) = C_0$, $C(x \to \infty, t) = C(x, t \le 0) = 0$, and the solution is

$$C(x, t) = \frac{C_0}{2} \exp(-x\sqrt{k_1/D_x}) \text{erfc}\left(\frac{x}{\sqrt{2D_x t}} - \sqrt{k_1 t}\right) \tag{9.5.27}$$

$$+ \frac{C_0}{2} \exp(x\sqrt{k_1/D_x}) \text{erfc}\left(\frac{x}{\sqrt{2D_x t}} + \sqrt{k_1 t}\right)$$

The total amount of mass of concentration C put into the domain per unit area by diffusive flux at $x = 0$ up to time t_x (see Prob. 1) is

$$M(t_x) = C_0\sqrt{D_x/k_1} \left\{ (k_1 t_x + \tfrac{1}{2}) \text{erf}(\sqrt{k_1 t_x}) + \frac{\exp(-k_1 t_x)}{\sqrt{\pi k_1 t_x}} \right\} \tag{9.5.28}$$

Figure 9.14 contains a schematic of how the solution behaves. This is essentially Eq. (9.5.18) as depicted in Fig. 9.11 with the addition of a set of solutions for the same C_0 and D_x but an additional chemical decay. As noted the effect of the decay rate is to confine the steep concentration gradients to a region ever closer to the surface or origin $x = 0$. This region becomes quite thin as the rate increases.

This model has considerable historical significance as it serves as the basis for the *penetration theory* of interphase mass transfer at boundaries.

A series of example problems is compiled in Sec. 9.7, but before proceeding it is necessary to complete the discussion of turbulent diffusion by discussing the nature of turbulent diffusion as it is affected by walls and boundaries. Therefore, the next section concentrates on boundary layer and channel transport.

EXERCISES

9.5.1   The spreading of a point source of material (or heat) by molecular diffusion (*a*) gives concentration distributions that vary linearly with time; (*b*) has a constant spreading rate; (*c*) has a "width" which varies exponentially with time; (*d*) has a shape which is identical to a Gaussian probability distribution; (*e*) *b* and *d*.

9.5.2   Turbulent diffusion of a cloud of particles can be determined with a constant diffusivity gradient approximation (*a*) when the length of time of the diffusion process is longer than the autocorrelation function of the process; (*b*) directly related to cloud distribution variance; (*c*) when the length of time of the diffusion process is longer than the Lagrangian diffusion time, T_{L_x}; (*d*) when the variance growth is proportional to time and *t* is greater than T_x; (*e*) never.

9.5.3   The exact diffusion solutions catalogued in Sec. 9.5 are valid (*a*) for very simple geometries; (*b*) for constant turbulent diffusivities; (*c*) for either molecular diffusion or turbulent diffusion with constant eddy diffusivities; (*d*) for both steady and transient conditions; (*e*) all of the above.

9.6   CHANNEL DIFFUSION AND DISPERSION

The solutions in Sec. 9.5 and selected examples in Sec. 9.7 are based upon the diffusion process occurring in a specialized turbulence field (isotropic and homogeneous) well away from any boundaries. With the exception of free jets such as smoke or exhaust stacks in the atmosphere, most jet flows encounter and are affected by boundaries. Such is the case for wastewater or industrial process effluent discharges in rivers, estuaries, or lakes. As stressed in another portion of this text, the presence of boundaries gives rise to boundary layers which in turn cause the flow field and its turbulence characteristics to be different in all three possible coordinate directions. Therefore, one possible effect of walls is to cause the turbulence to be homogeneous, but not isotropic, in each of the three directions. It is anticipated that the eddy diffusivities will then be quite different in each of the channel directions. Therefore, the first objective of this section is to learn how to predict the eddy diffusivities in turbulent channel flow and the impact of their differences in magnitude on the turbulent mixing in the channel.

A second possible impact of walls is that the plume will not be free to spread in all three directions, that is, vertically, laterally, and downstream. Figure 9.15 is a schematic of the spreading process of river effluent discharged in the center of a stream. The plume boundaries, defined by our previous variance description (4σ), will eventually intersect not only the bottom (region 1) but the side walls of the channel (region 2). In region 3 and beyond the transport can still be described by simple one-dimensional models, but a different mechanism called *dispersion* will be defined to do so. In many ways its constructs are similar to eddy diffusivities except that the definition will be based upon the need to "close" terms resulting from *spatial averages* as opposed to the temporally averaged turbulent fluxes of the previous chapters.

Vertical Eddy Diffusivity

From the schematic in Fig. 9.15 and the knowledge that $d \ll W$, it is anticipated that gradients of momentum and transport will be sharper in the vertical direction

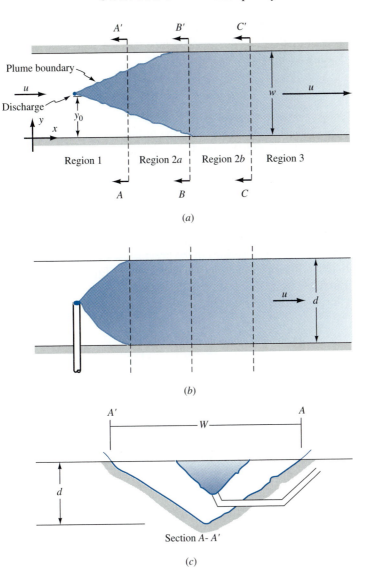

Figure 9.15 Turbulent diffusion and dispersion in a stream from a center discharged effluent. (a) Plane view. (b) Side view. (c) Cross-section, down-channel view.

than the transverse direction. Therefore, it is reasonable to expect that the vertical turbulent diffusion parameterized by the vertical eddy diffusivity dominates region 1 and that the mixing or spreading throughout the vertical plane will occur more quickly than in the lateral dimension.

Elder [21] was the first to find an expression for the vertical eddy diffusivity, E_z, based upon exact boundary layer solutions for turbulent channel flow. Here, a nondimensional vertical plane turbulent boundary layer solution of the form

$$\frac{u(z)}{u_*} = \frac{1}{k}\left(1 + \ln\frac{z}{d}\right) \tag{9.6.1}$$

was assumed, where z is measured vertically up from the bottom extending to height d. Implied in this form is that the flow field is infinitely wide. As noted in Chap. 6,

k is the von Kármán coefficient and u_* is the friction velocity such that at $z = 0$, $u = u_*$. By adapting the procedure used in pipe flow, a shear stress distribution over the entire depth was found [21] to be

$$\tau = \rho \eta \frac{\partial u}{\partial z} = \tau_0 \left(1 - \frac{z}{d}\right)$$

(9.6.2)

where η is the vertical eddy viscosity for the turbulent boundary layer [Eq. (6.4.9)] and τ_0 is the shear stress at the bottom ($z = 0$). The eddy viscosity can be solved for

$$\eta(z) = k \frac{z}{d} \left(1 - \frac{z}{d}\right) u_* d$$

(9.6.3)

To find the eddy diffusivity, Elder assumed similarity between the turbulent transport of momentum and mass and set $\eta(z) = E_z(z)$. In both cases it is noted that unlike the case for the constant diffusivities in the previous section, this diffusivity behaves analogously to the Prandtl mixing length formulation [Eq. (6.4.10)] and varies with distance from the bottom, that is, it is not constant.

To put this in a form more useable for design guidance, Elder provided a vertical average form as

$$\overline{E}_z^\bullet = \frac{1}{d} \int_0^d E_z(z)\, dz = k u_* d \int_0^d \frac{z}{d}\left(1 - \frac{z}{d}\right) dz = 0.067\, u_* d$$

(9.6.4)

Here the — • symbol is used for the spatial average to distinguish it from the temporal average overbar. It is seen that the average or bulk vertical eddy diffusivity is only a function of depth and friction velocity, not the average channel flow velocity. For a 5-m-deep channel with a 2-cm/s friction velocity, $\overline{E}_z^\bullet = 134$ cm²/s or 0.0134 m²/s. Contrast this value with a typical molecular diffusion coefficient of the order $1.0(10^{-5})$ cm²/s for various solutes in water [2].

Transverse Eddy Viscosity

Lateral spreading is parametrized by the transverse eddy diffusivity, E_y and is the dominant mechanism in region 2. The extension of the concept in Eq. (9.5.14) to the y or lateral direction suggests in part that the transverse velocity will play a role in the lateral spreading rate. However, no exact solution for this velocity distribution is available [15, 20]. Therefore, this information has been compiled in an empirical form. The average lateral eddy diffusivity is of the form

$$\overline{E}_y^\bullet(y) = c_e u_* d$$

(9.6.5)

As the above two references point out, substantial field data on lateral spreading must be (and has been) analyzed and from the information in these references the correlation coefficient c_e is

$$\overline{E}_y^\bullet = (0.15 \pm 0.075) u_* d$$

(9.6.6)

for straight rectangular channels, and

$$\overline{E}_y^\bullet = (0.60 \pm 0.30) u_* d$$

(9.6.7)

for more natural channels containing meanders and boundary roughness or irregularities. If engineered irregularities such as bank protection, groins, or sea walls exist on the boundary, then $\overline{E}_y^{\bullet}$ will be greater still. It is intuitive that geometrical irregularity and roughness will increase $\overline{E}_y^{\bullet}$ as they create more energetic mixing through the creation of large scale turbulent eddies.

Time to Complete Mixing

By comparing Eqs. (9.6.6) and (9.6.7) to Eq. (9.6.4), it is seen that for smooth straight channels $\overline{E}_y^{\bullet}$ is nearly 2.25 times greater than $\overline{E}_z^{\bullet}$ while for natural channels $\overline{E}_y^{\bullet}$ is almost 10 times greater than $\overline{E}_z^{\bullet}$. Although the transverse diffusivity is a more powerful mixing agent, the small vertical depth allows mixing to occur over the full depth much sooner than complete lateral spreading. An estimate of this mixing time scale can be obtained on dimensional grounds

$$\overline{E}_z^{\bullet} \rightarrow \frac{d^2}{t_z} \qquad \overline{E}_y^{\bullet} \rightarrow \frac{W^2}{t_y}$$

Here d and W are the depth and width, respectively, and t_z and t_y are the estimated time scales required for vertical and lateral spreading, respectively, to result in the material occupying the full depth (end of region 1, Fig. 9.15) and full width (end of region 2, Fig. 9.15). Using the expression for the diffusivities, then

$$t_z \sim \frac{d}{0.067 u_*} \sim 15 d / u_* \qquad\qquad \text{(9.6.8)}$$

$$t_y \sim 1.667 \frac{W^2}{d u_*}$$

If W is considered to fall in the range of 10 to 100 times larger than d, a range of mixing times occurs. For $10d \sim W \rightarrow$ narrow channel,

$$t_y = \frac{1.667(10^2)d}{u_*} \qquad\qquad \text{(9.6.9)}$$

and for $100d \sim W \rightarrow$ wide channel,

$$t_y = \frac{1.667(10^4)d}{u_*} \qquad\qquad \text{(9.6.10)}$$

In the narrow case then, t_y is approximately 11 times longer than t_z, and for the wide channel t_y is approximately 1100 times longer than t_z. Therefore, full depth vertical mixing takes a very short time to complete relative to lateral mixing even though the lateral diffusivities are over tenfold greater.

Dispersion

Essentially, dispersion is an artifact of areal or spatial averaging and was first described by Taylor [22] and Aris [23]. Reviews of dispersion are found in Refs. [15, 24–25]. The one-dimensional area average formulation of the contaminant transport or momentum and heat equation is derived mathematically from the recognition that the vertical and cross channel variations over a long and complex river channel (or pipe) are important transport agents. The basic operation involved in the derivation is to create a governing partial differential equation or control volume equation which is derived in terms of area averaged variables. This average is formally

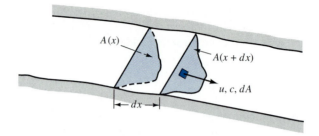

Figure 9.16   Control volume definition in a channel used for areal averaging.

defined for a variable $\alpha(\mathbf{x}, t)$ as

$$\overline{\alpha}_z^{\bullet}(x, t)A(x) = \int \alpha(\mathbf{x}, t)\, dA$$

This equation is very familiar as it was a basic feature of the control volume derivation for mass, momentum, and energy. While more elegant derivations are found in the literature, we shall introduce dispersion by returning to the control volume.

Consider the river channel control volume in Fig. 9.16. The control volume boundaries are perpendicular to the shore and the flow is perpendicular to the area. From Eq. (3.9.8) and with the assumptions in Sec. 3.3, the control volume equation is

$$\frac{\partial}{\partial t}\int C\, d\mathcal{V} + \int_{x+dx} C(x + dx, y, z, t)\, u(x + dx, y, z, t)\, dA$$

$$- \int_{x} C(x, y, z, t)\, u(x, y, z, t)\, dA = 0$$

The quantities at $x + dx$ can be related to those at x by using a Taylor's series expansion. Therefore,

$$\frac{\partial}{\partial t}\int C\, dA\, dx + \left[\frac{\partial}{\partial x}\int Cu\, dA\right]dx = 0$$

where $d\mathcal{V} = dA\, dx$. Remembering the definition of the area average, the above equation is written as

$$\frac{\partial}{\partial t}(\overline{C}^{\bullet}A)\, dx + \left[\frac{\partial}{\partial x}\overline{CuA}^{\bullet}\right]dx = 0 \tag{9.6.11}$$

After division by dx, there remains the problem identical to the problem from which the kinetic energy [Eq. (3.4.14)] and momentum [Eq. (3.6.8)] correction factors are derived, that is,

$$\overline{CuA}^{\bullet} \neq \overline{C}^{\bullet}\overline{u}^{\bullet}A$$

Again, the overbar notation — ● specifically refers to the area or space average value as distinct from the temporal average denoted by the overbar. Instead of using the correction factor concept, however, current practice is to use the decomposition for the nonlinear terms as described for the Reynolds equation in Sec. 6.4. Therefore, the departures from the space average are defined and denoted by the double prime

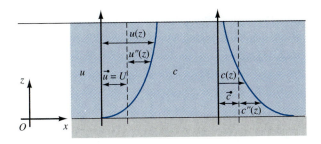

Figure 9.17 Definition of spatial averaging decomposition variables.

and applied to the average of the resulting decomposed product (Fig. 9.17), that is,

$$\alpha(x, y, z, t) = \overline{\alpha}^{\bullet}(x, t) + \alpha''(x, y, z, t)$$

and

$$\overline{CuA}^{\bullet} \approx \overline{C}^{\bullet}\overline{u}^{\bullet}A + \overline{C''u''}^{\bullet}A \qquad (9.6.12)$$

Equation (9.6.12) is substituted into Eq. (9.6.11) and becomes

$$\frac{\partial}{\partial t}(\overline{C}^{\bullet}A) + \frac{\partial}{\partial x}(\overline{C}^{\bullet}\overline{u}^{\bullet}A) + \frac{\partial}{\partial x}(\overline{C''u''}^{\bullet}A) = 0 \qquad (9.6.13)$$

It should be remembered that $\overline{u}^{\bullet}$ is the cross-section average velocity which was referred to as U or V in Chap. 3. If the continuity equation is subtracted, the final equation becomes

$$\frac{\partial \overline{C}^{\bullet}}{\partial t} + \overline{u}^{\bullet}\frac{\partial \overline{C}^{\bullet}}{\partial x} + \frac{\partial \overline{C''u''}^{\bullet}}{\partial x} = 0 \qquad (9.6.14)$$

The spatial correlation term $\overline{C''u''}^{\bullet}$ also requires closure or specification in terms of mean flow variables. Taylor [22] and Aris [23] were able to show that once again the gradient diffusion approximation was valid; therefore,

$$\overline{C''u''}^{\bullet} = -K\frac{\partial \overline{C}^{\bullet}}{\partial x}$$

where K is defined as the *dispersion coefficient*. Noting that $\overline{u}^{\bullet} = U$ from Chap. 3, the final equation becomes

$$\frac{\partial \overline{C}^{\bullet}}{\partial t} + U\frac{\partial \overline{C}^{\bullet}}{\partial x} = K\frac{\partial^2 \overline{C}^{\bullet}}{\partial x^2} = 0 \qquad (9.6.15)$$

If the effects of turbulent diffusion and molecular diffusion are included, the derivation becomes quite laborious. However, the result is all lumped into the dispersion term as

$$K\frac{\partial \overline{C}^{\bullet}}{\partial x} = -\overline{C''u''}^{\bullet} + (\overline{E}_x^{\bullet} + \mathcal{D})\frac{\partial \overline{C}^{\bullet}}{\partial x}$$

However, as will be seen later K is a good deal larger than $\overline{E}_x^{\bullet}$ and $\mathcal{D}$, and these additive effects are typically ignored.

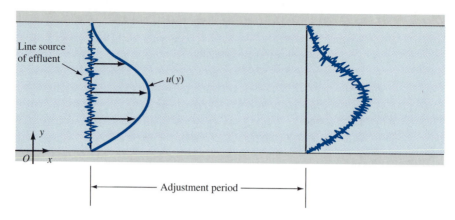

Figure 9.18   Distribution of flow variables before and after an adjustment period from the top view of the channel.

The two remaining questions become the same as in the eddy diffusivity analysis, that is, when is the gradient approximation valid for the dispersion relationship and how is K estimated?

As to the first question it has been seen that the gradient diffusion approximation applies after an initial period of adjustment which results in a distribution that is essentially normally distributed about the mean flow variables. This adjustment period was parameterized by the Lagrangian length l_L and time T_L scales. The dispersion references in this section clearly demonstrate that the same period of adjustment is necessary and that its completion is marked by a condition where the fluctuations in the dependent variables about the *spatial mean* are normally distributed. Figure 9.18 depicts a schematic of the pre- and postadjustment periods. With the adjustment period completed and the gradient diffusion applicable, the same relation between the variance and the dispersion relation holds, that is, $d\sigma^2/dt = 2K$. The other relations derived for the turbulent diffusion case will also apply for the dispersion zone with K substituted for the diffusion coefficient $\mathcal{D}$.

For laminar flow an estimate of the time required for a normal distribution to obtain in dispersion transport over spatial domain d is approximately $t > 0.4d^2/\mathcal{D}$. The time and length required to achieve this state in turbulent conditions (region 3, Fig. 9.15) are not nearly as straightforward as the laminar case and are estimated by the method in Fischer et al. [15]. It is made more difficult by the trade-off between downstream advection and the intensity of transverse mixing (E_y). If a rectangular channel is assumed and a continuous mass input rate occurs ($\dot{M}$) at a vertical line source in the channel centerline, then the steady state two-dimensional solution for concentration is

$$\frac{C}{C_0} = \frac{1}{(4\pi x^*)^{1/2}} \sum_{n=-\infty}^{\infty} \{\exp[-(y^* - 2n - y_0^*)^2/4x^*] + \exp[-(y^* - 2n + y_0^*)^2/4x^*]\}$$

$$(9.6.16)$$

where $C_0 = \dot{M}/UWd$; $x^* = xE_y/UW^2$ and $y^* = y/W$. The source is located at $y = y_0$. By analyzing the concentration down the centerline of the channel, complete mixing is defined as that distance downstream, L_m, where the concentration is within 5 percent of its cross-section mean everywhere on the cross section. This distance is

$$L_m \geq \frac{0.1UW^2}{E_y}$$

$$(9.6.17)$$

TABLE 9.2   Mixing coefficient comparison

Narrow channel	Equation number	Wide channel
$E_y = 0.6\,u_*d$ $\quad = 0.06\ \mathrm{m^2/s}$	9.6.7	$E_y = 0.06\ \mathrm{m^2/s}$
$K = 0.011 U^2 W^2 / u_* d$ $\quad = 2.75\ \mathrm{m^2/s}$	9.6.20	$K = 275\ \mathrm{m^2/s}$
$t_y \approx 0.1 W^2 / E_y = L_m / U$ $\quad = 1.16\ \mathrm{hr}$		$t_y = 116\ \mathrm{hr}$
$L_m \approx 0.1 U W^2 / E_y$ $\quad = 0.42\ \mathrm{km}$	9.6.17	$L_m = 42\ \mathrm{km}$

If the discharge is placed on the side of the bank, then W is replaced by $2W$, and the corresponding channel length to complete mixing is

$$L_m \geq \frac{0.4 U W^2}{E_y} \tag{9.6.18}$$

The procedures for estimating the dispersion coefficient are quite numerous; each is limited to a particular series of assumptions about the flow physics and geometry. Elder [21] once again provided an early estimate for channels which was based upon the assumption that the fluctuations about the vertical average were solely responsible for dispersion. Extending the analysis for E_z presented earlier, he demonstrated that

$$K = 5.93\,u_*\,d \tag{9.6.19}$$

where d is again the depth.

Fischer [27] presented a much more in-depth application of Taylor's work to channels and actually derived a procedure for exactly calculating K from a channel whose flow is very accurately known. French [20] presents the method in detail. Recognizing the need for quicker estimates for streams not well instrumented Fischer et al. [15] presented the following empirical formula based upon a number of field experiments

$$K = \frac{0.011 U^2 W^2}{u_* d} \tag{9.6.20}$$

They noted that the answers are correct to within a factor of four which, when considering that field measurements themselves are only valid to within a factor of two, is acceptable.

If the conditions used in the previous example comparing $\mathcal{D}$, E_z, and E_y are applied to Eq. (9.6.18) for an average velocity of 0.1 m/s, the dispersion coefficient and related scales are as listed in Table 9.2.

It is noted that the dispersion coefficient even for this narrow channel is more than one order of magnitude larger than E_y, which as we have already seen is an order of magnitude larger than E_z and 10^5 times larger than $\mathcal{D}$.

EXERCISES

9.6.1   Vertical mixing is marked by (a) very high eddy viscosities; (b) E_z being constant throughout the channel depth; (c) an E_z that is smaller than the lateral eddy diffusivity, E_y, but with much shorter vertical mixing times than in the case of lateral mixing; (*d*) depth-averaged eddy diffusivities proportional to time variation and friction velocity, u_*; (*e*) none of the above.

9.6.2   All else being equal a channel ten times wider than another will have (*a*) a 100-fold greater time to lateral mixing; (*b*) a tenfold greater mixing coefficient; (*c*) a tenfold greater dispersion coefficient; (*d*) equal flow rates; (*e*) *a* and *d*.

9.7   APPLICATIONS OF DIFFUSION AND DISPERSION TECHNIQUES

Estimation of Friction Velocity

As can be readily seen in Sec. 9.6 the key variable in the estimates of the eddy diffusivities and dispersion coefficients is the friction velocity, u_*. For steady flow down a mildly sloped channel an estimate of u_* ($= \sqrt{\tau_0/\rho}$) can be obtained.

Consider a control volume of a channel as indicated in Fig. 9.19. The bottom slope is extremely small such that $S_o \sim \tan\theta \sim \theta$. Gravity is the only body force acting on the control volume denoted in the figure. A local coordinate system is embedded in the bottom, and therefore, the angle between the acceleration due to gravity and *y* is also θ. With steady conditions, the channel flow or average velocity is the same everywhere. Therefore, the momentum exchange vectors (M_{1x} and M_{2x}) in the channel coordinate direction, *x* direction, are equal and cancel each other. The pressure forces on either end ($F_{p_{1x}}$ and $F_{p_{2x}}$) are equal and also cancel each other. Therefore, in the *x* direction the force balance becomes the component of weight of the fluid in the control volume as balanced by the bottom friction force, F_{τ_x},

$$W_x - F_{\tau_x} = 0$$

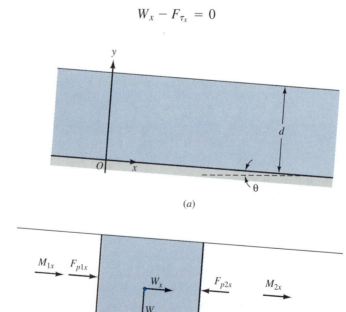

Figure 9.19   Control volume definition for a mildly sloped channel.

In terms of the given variables and assuming a unit width, we have

$$\rho g \Delta x\, d\,(1) \sin\theta - \tau_o \Delta x\,(1)$$

In the above $\Delta x\, d\,(1)$ is the volume and $\Delta x\,(1)$ is the surface area of the control volume in contact with the bottom. The bottom stress in terms of the slope and depth becomes

$$\tau_o = \gamma d S_o \qquad\qquad (9.7.1)$$

or

$$u_* = \sqrt{g d S_o} \qquad\qquad (9.7.2)$$

Example 9.6

For the 5-m-deep channel in the previous section the friction velocity was stated to be 2 cm/s. To what channel slope would this condition correspond?

Solution

From Eq. (9.7.2)

$$S_o = \frac{u_*^2}{gd}$$

Thus,

$$S_o = \frac{(0.02 \text{ m/s})^2}{(9.806 \text{ m/s}^2)(5 \text{ m})} = 8.2(10^{-6})$$

The condition in Ex. 9.6 corresponds to almost flat bottom conditions, such as in *estuaries,* or the confluence of a river and ocean. More typically mild slope conditions in rivers apply to slopes as large as 0.003 to 0.004. Values much greater than this often result in nonhydrostatic pressure fields which require *steep slope* hydraulics, which is a subject beyond the scope of this text.

Mixing Coefficients

Example 9.7

During a storm a portion of the Cuyahoga River entering Lake Erie of the Laurentian Great Lakes has a slope of 0.002, a depth of 4 m, and during test conditions a steady average velocity of 0.8 m/s. What are the lateral and vertical eddy diffusivities, the dispersion coefficient, and the downchannel length required for the end of region 2 for both a centerline and side bank discharge? Assume the channel width of 110 m.

Solution

The functions listed in Table 9.2 all require an estimate of the friction velocity. Therefore,

$$u_* = \sqrt{g d S_o} = 0.28 \text{ m/s}$$

The values for the various mixing coefficients are

$$E_y = 0.6du_* = 0.67 \text{ m}^2/\text{s}$$

$$E_z = 0.067du_* = 0.075 \text{ m}^2/\text{s}$$

$$K = \frac{0.011U^2W^2}{du_*} = 76.0 \text{ m}^2/\text{s}$$

The channel lengths to complete mixing (end of region 2) for centerline (c) and side (s) discharges are

$$L_c = \frac{0.1UW^2}{E_y} = \frac{0.1(0.8 \text{ m/s})(110 \text{ m})^2}{0.67 \text{ m}^2/\text{s}} = 1447 \text{ m} = 1.447 \text{ km}$$

$$L_s = 4L_c = 5790 \text{ m} = 5.79 \text{ km}$$

Finally an estimate for the time required to complete mixing from the centerline discharge is found from

$$\frac{L_c}{U} = t_c = \frac{0.1W^2}{E_y} = \frac{0.1(110 \text{ m})^2}{0.67 \text{ m}^2/\text{s}} = 1805 \text{ s} = 0.5 \text{ h}$$

Conservative Tracer Plume and Geometry

Example 9.8

Assume the same condition as in the previous problem and further assume a discharge in the center of the channel. Material is discharged at the rate of 1.0 m³/s and the concentration of pollutant in the discharge is 450 mg/L. Determine the pollutant concentration and the plume width at 125-, 250- and 500-m downstream.

Solution

From the definition of width given in Eq. (9.5.10), the plume width at a point in the downstream direction is estimated by

$$w_P(x) = 4\sigma = 4\sqrt{2E_y t} = 4\sqrt{\frac{2E_y x}{U}}$$

where time and space are related by $x/t = U$. Therefore, the downstream centerline discharge plume widths become

$$w_c(125) = 4\sqrt{\frac{2(0.67 \text{ m}^2/\text{s})(125 \text{ m})}{0.8 \text{ m/s}}} = 57.9 \text{ m}$$

$$w_c(250) = 4\sqrt{\frac{2(0.67 \text{ m}^2/\text{s})(250 \text{ m})}{0.8 \text{ m/s}}} = 87.9 \text{ m}$$

$$w_c(500) = 115.8 \text{ m}$$

It is therefore seen that at $x = 500$-m downstream, the plume has just reached the channel sides. The estimated downstream distance, x_c, for the plume to just reach the walls is

$$\sqrt{x_c} = \frac{W}{4\sqrt{\frac{2E_y}{U}}} = \frac{110 \text{ m}}{4\sqrt{\frac{2(0.67 \text{ m}^2/\text{s})}{0.8 \text{ m/s}}}} \quad \Rightarrow \quad x_c = 451 \text{ m}$$

It is interesting to contrast the distance downstream to have the initial edge of the plume touch the shore to the distance $L_c = 1447$ m which was the distance downstream required for complete cross-section mixing. It is noticed that an additional 996 m of downchannel mixing is required to have the cross-channel concentration to within 5 percent of the cross-channel mean concentration. Therefore, region 2 consists of two subregions: zone 2*a* which is a plume, already vertically mixed and spreading due to transverse eddy diffusivities, and zone 2*b* which is marked by the plume having intersected the channel sides but not yet achieving the condition required for the dispersion concept to apply.

From Eq. (9.5.24) the maximum concentration occurs along the centerline ($y = 0$), that is,

$$C(x, y = 0) = C_{max}(x) = \frac{\dot{M}}{U\sqrt{4\pi E_y/U}} \qquad (9.7.3)$$

The mass flow rate $\dot{M} = Q_d C_d$, where Q_d is the flow rate of the discharge pipe and C_d is the discharge concentration. For the given condition

$$\dot{M} = (1 \text{ m}^3/\text{s})(450 \text{ mg/L})\left(10^{-3} \frac{\text{kg/m}^3}{\text{mg/L}}\right) = 0.45 \text{ kg/s}$$

Therefore, from Eq. (9.7.3)

$$C_{max}(x = 125 \text{ m}) = \frac{0.45 \text{ kg/s}}{0.8 \text{ m/s}\sqrt{\frac{4\pi(125 \text{ m})(0.67 \text{ m}^2/\text{s})}{0.8 \text{ m/s}}}} = 0.0155 \text{ kg/m}^3 = 15.5 \text{ mg/L}$$

and

$$C_{max}(x = 250 \text{ m}) = 0.011 \text{ kg/m}^3 = 11 \text{ mg/L}$$

We can also ask what the concentration is as the plume boundary just touches the shore at $x_c = 451$ m. This can be estimated by assuming $y = 2\sigma = w_p/2$ at $x_c = 451$ m and calculating from Eq. (9.5.24) as

$$C(x_c = 451, y = 2\sigma = 55 \text{ m}) = \left[\frac{\dot{M}}{U\sqrt{\frac{4\pi(451 \text{ m})(0.67 \text{ m}^2/\text{s})}{0.8 \text{ m/s}}}}\right]\exp\left\{\frac{-y^2 U}{4(451 \text{ m})E_y}\right\}$$

$$= [C_{max}(x_c = 451 \text{ m})]\exp\left\{\frac{(-0.55 \text{ m})^2(0.8 \text{ m})}{4(451 \text{ m})(0.67 \text{ m}^2/\text{s})}\right\}$$

$$= (0.0082 \text{ kg/m}^3)(0.135)$$

$$= 0.0011 \text{ kg/m}^3 = 1.1 \text{ mg/L}$$

The fact that there is some pollutant concentration at the width corresponding to the stream bank is a consequence of the definition of the plume width as being 4σ. This only accounts for 95 percent of the mass, implying that some pollutant mass will exist beyond the $\pm 2\sigma$ width.

The Initial Adjustment Period

The previous example concentrated on region 2 which was marked by lateral mixing and downstream advection, as noted in Fig. 9.15. The spreading appears to be a two-dimensional process and the predictions in Ex. 9.8 were based upon such an equation. The dispersion concept discussed in Sec. 9.6 stated that a one-dimensional advection diffusion equation can also be used for the same region based upon *area*-averaged variables, varying only in time (t) and channel distance (x). Therefore, two- or three-dimensional spreading of a passive scalar can be modeled as a one-dimensional process. As discussed in Sec. 9.6 this gradient diffusion approximation with a dispersion coefficient can only apply after an initial period of adjustment, after which fluctuations about the cross-sectional average of the variables in a plane moving with the mean velocity U are normally distributed. The question becomes how far downstream is the initial adjustment length. From Chatwin [28] it is empirically argued that for dispersion resulting from laminar spreading in a vertical plane (depth d) marked by longitudinal cross-sectional average velocity U and molecular diffusion coefficient $\mathcal{D}$, the initial adjustment period t_p is given by

$$t_p > \frac{0.4d^2}{\mathcal{D}} \tag{9.7.4}$$

From Fischer et al. [15] the concept was extended to turbulent lateral spreading (E_y) over a width W, that is,

$$t_p > \frac{0.4W^2}{E_y} \tag{9.7.5}$$

A formula in terms of distance is found by $x/t = U$ and

$$x_p > \frac{0.4W^2U}{E_y} \tag{9.7.6}$$

This value is often expressed in nondimensional form [Eq. (9.6.16)] as x_p^*

$$x_p^* = \frac{xE_y}{UW^2} > 0.4 \tag{9.7.7}$$

Example 9.9

What are the time and distance conditions for the initial adjustment period in Ex. 9.8?

Solution

For the adjustment period

$$t_p > \frac{0.4W^2}{E_y} = \frac{0.4(110 \text{ m})^2}{0.67 \text{ m}^2/\text{s}} = 7223.8 \text{ s} = 2.0 \text{ h}$$

For the adjustment channel length

$$x_p > \frac{0.4W^2U}{E_y} = \frac{0.4(110 \text{ m})^2(0.8 \text{ m/s})}{0.67 \text{ m}^2/\text{s}} = 5780 \text{ m} = 5.78 \text{ km}$$

In comparison to x_c from the previous example and L_c from Ex. 9.7, it is seen that the distance over which the plume width just grazes the shoreline (x_c) is quite modest, being just 451 m. The length to complete mixing ($L_c = 1447$ m) is a good deal less than the channel length required for the adjustment to normality ($x_p = 5780$ m). It must be remembered that the length to complete mixing is a definition based upon the intensity of the average concentration at the wall being within 5 percent of the cross-section average. The adjustment to normality condition refers to the statistics of the fluctuations about the cross-section average being distributed in a special fashion (i.e., normally distributed). These are two different criteria, although in practice they are often confused.

PROBLEMS

9.1 A pane of window glass with an area of 0.5 m² has thermal conductivity $k =$ 0.87 W/m·K. The outer-surface temperature is 28.5°C and the inner-surface temperature is 20°C. If the window is 6 mm thick, calculate the rate of heat transfer through the window. What is its thermal resistance?

9.2 A house has a roof with surface temperature 75°F and total area of 4200 ft². The temperature of the ambient air is 20°F. If the unit average convective heat transfer coefficient is 1.8 Btu/h·ft²·°F, determine the rate of heat transfer between the roof and the air. In what direction does the heat flow?

9.3 The wall of a furnace consists of two layers of different materials. The inner layer is 10 mm thick with $k_1 = 35$ W/m·K and the outer layer is 12 cm thick with $k_2 = 3.15$ W/m·K. The inner surface temperature is maintained at 950 K while the outer surface temperature is at 380 K. Find the heat flux through the furnace wall and the temperature at the interface between the two layers.

9.4 The outer surface of a 20-cm-thick concrete wall with thermal conductivity $k = 0.65$ W/m·K is exposed to a cold wind at -4.5°C and the convection heat transfer coefficient is 35 W/m²·K. At the calm side, the air temperature is 12°C and the convection heat transfer coefficient is 14 W/m²·K. Find the heat flux through the wall.

9.5 Bricks having dimensions $20 \times 10 \times 8$ cm, are used in the construction of a furnace wall. Two kinds of material are available. The first one has a maximum temperature limit of 850°C and a thermal conductivity of 1.25 Btu/h·ft·°F, and the other has a maximum temperature limit of 580°C and a thermal conductivity of 0.8 Btu/h·ft·°F. If the permissible heat flux through the furnace wall is 350 Btu/h·ft², determine the most economical design for the furnace wall assuming that the bricks cost the same and that they may be laid in any manner. The inner temperature of the furnace wall is 850°C, while the outer surface is maintained at 200°C.

9.6 Air at 95°C is flowing over a 20×80 cm flat plate. The temperature of the plate is 22°C and the heat flux is 150 W. Find the average heat transfer coefficient between the air and the plate.

9.7 For the plate in Prob. 9.6 the heat transfer coefficient is given by $h_c(x) = 16.78\, x^{-2/3}$ W/m²·K where x is the distance from the leading edge of the plate. (*a*) Find the average heat transfer coefficient ($\bar{h}_c$). (*b*) Determine the heat flux between the plate and the air.

9.8 Heat is transferred from the inside of a room to the outside air at -4°C. The inside surface of the room walls have a unit surface conductance of 16.7 W/m²·K and an outside surface conductance of 32.5 W/m²·K. The walls have unit thermal conductance of 2.35 W/m²·K. (*a*) Find the temperature at the outside surface of the walls. (*b*) Find the heat flux through each wall.

9.9 Saturated steam flows inside a pipe at 120 psi. The unit surface conductances for the inside and outside surfaces are 428.5 Btu/h·ft²·°F and 6.54 Btu/h·ft²·°F, respectively. The pipe itself has unit surface conductance of 855 Btu/h·ft²·°F. Find the temperature at the outside surface of the pipe if the pipe is in a room with temperature 82°F.

9.10 The inside temperature of a submarine, 35 ft in diameter and 250 ft long, is to be maintained at 68°F. The inside unit surface conductance is 3.15 Btu/h·ft²·°F. When the submarine is not moving, the outside unit surface conductance is

16.5 Btu/h·ft²·°F, while when moving at maximum speed the outside unit surface conductance is 123 Btu/h·ft²·°F. During operation, sea water temperatures vary from 28 to 60°F (at maximum speed). The submarine wall is constructed of 0.75 in. of stainless steel on the outside, a 1.3-in. layer of fiberglass insulation, and a 0.25-in. thickness of aluminum at the inside. Determine the minimum size, in kW, of the heating unit required to keep the inside temperature at 68°F.

9.11   A long heating rod, with a cross-sectional area of 10 cm², is immersed in oil at 92°C. An electric current through the rod uniformly generates heat at a rate of 900 kW/m³. What is the unit surface conductance if the temperature of the heater is to be below 180°C? The heating rod is made from material with thermal conductivity of 72 W/m·K at 180°C.

9.12   Saturated steam at 105°C flows inside a pipe, made of copper, with an inside diameter of 0.15 m and an outside diameter of 0.17 m. The pipe is set in an environment of 34°C and the heat transfer coefficient of the outside surface conductance is 19.3 W/m²·K. (*a*) Find the heat loss through the copper pipe. (*b*) Insulation 4 cm thick, with $k = 0.30$ W/m·K, is used in order to reduce the heat losses through the copper pipe. What is the heat loss through the pipe in this case? (*c*) What is the percentage of the heat loss reduction?

9.13   A long hollow cylinder, of inside radius R_1 and outside radius R_2, is made of material with a thermal conductivity that varies linearly with temperature. The temperature at the inner surface is T_1, while the temperature at the outer surface is T_2. Show that the rate that heat is transferred by conduction through the cylinder is given by

$$q = \bar{k} L \bar{A} \frac{T_1 - T_2}{R_1 - R_2}$$

where $\bar{A} = 2\pi(R_1 - R_2)/\ln(R_2/R_1)$, $\bar{k} = k_0[1 + \alpha(T_1 + T_2)/2]$, L is the length of the cylinder, and α is a constant.

9.14   A hollow sphere with an inside radius R_1 and an outside radius R_2, is covered with a layer of insulation of outside radius R_3. Find the rate of heat transfer through the sphere in terms of R_1, R_2, R_3, the temperatures, the thermal conductivities, and the heat transfer coefficients.

9.15   A thin-walled cylindrical container, 4.2 ft in diameter, is filled to a depth of 6 ft with water at 78°F, as shown in Fig. 9.20. Determine the time required to heat the water from 78°F to 145°F if the container is suddenly immersed in oil at 250°F. The overall heat transfer coefficient between the water and the oil is 75 Btu/h·ft²·°F.

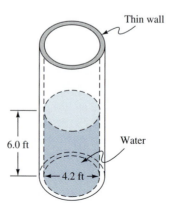

Figure 9.20   Problem 9.15.

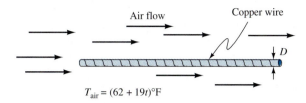

$T_{air} = (62 + 19t)°F$

Figure 9.21   Problem 9.16.

9.16   A 0.06-in.-diameter and 1-ft-long copper wire, shown in Fig. 9.21, is placed in a stream of air whose temperature is given by $T_{air} = (62 + 19t)$, in °F, where t is the time in seconds. If the initial temperature of the wire is 42°F, plot its temperature versus time, from $t_0 = 0$ s to $t_f = 200$ s. The unit surface conductance between the air and the wire can be taken as 9.2 Btu/h·ft²·°F.

9.17   A wire of radius R (Fig. 9.22) emerges from a dye with a velocity U_0 at a temperature T_0 above ambient. Find the steady state temperature distribution along the wire if the exposed length downstream from the dye is quite long.

9.18   Air at 29°C containing water vapor with a partial pressure of 0.3 atm is flowing over a swimming pool. The pool has a temperature of 21°C. If the pool has an area of 10×10 m², find the rate at which mass is transferred from the pool. The average mass transfer coefficient can be taken as 0.0034 m/s.

9.19   Find the Reynolds number for flow over a tube from the following data: $d = 20$ cm, $U_0 = 2.5$ ft/s, $\rho = 450$ kg/m³, and $\mu = 75$ lb$_m$/ft·hr.

9.20   Find the Prandtl number from the following data: $\mu = 15.2(10)^{-4}$ N·s/m², $c_p = 2000$ J/kg·K, and $k = 2.38$ Btu/ft·hr·°F.

9.21   Find the Nusselt number for flow over a tube from the following data: $d = 20$ cm, $k = 0.42$ W/m·K, and $\bar{h} = 38$ Btu/hr·ft²·°F.

9.22   Find the Stanton number for flow over a tube from the following data: $d = 20$ cm, $U_0 = 4.13$ m/s, $\rho = 1530$ lb$_m$/ft³, $\mu = 12.8(10)^{-4}$ N·s/m², $c_p = 4000$ J/kg·K, and $\bar{h} = 4.32$ Btu/hr·ft²·°F.

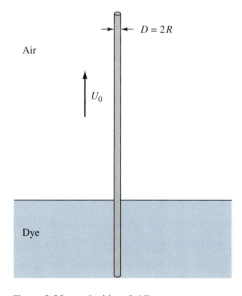

Figure 9.22   Problem 9.17.

9.23   Air is flowing over a 3.2-m-long flat plate, and the average Reynolds number is estimated to be $4.32(10^6)$. The average Nusselt number was found to be equal to 4500. What is the average heat transfer coefficient if oil at 38°C is flowing over the same plate with the same **R**?

9.24   Air at 22°C and at a pressure of 120 kPa is flowing over a flat plate with a velocity of 4.5 m/s. If the plate is 50 cm wide and the surface in contact with the air is at 82°C, calculate the following quantities at $x = 42$ cm: (*a*) the local friction coefficient; (*b*) the average friction coefficient; (*c*) the drag force; (*d*) the local convective heat transfer coefficient; (*e*) the average convective heat transfer coefficient; (*f*) the rate of heat transfer.

9.25   For a flow over a flat plate derive the relationship between the thermal and hydrodynamic boundary layer thickness and the Prandtl number. Assume linear velocity and temperature distributions in the boundary.

9.26   Water is flowing between two parallel flat plates (Fig. 9.23), spaced 6 cm apart, with an average velocity of 26 m/s. Find the distance, l, from the entrance where the two boundary layers meet.

9.27   As air flows over a sheet of ice, the ice melts and dissolves into the air. If the average convection heat transfer coefficient, $\bar{h}_L$, is 72 W/m²·K, find the average convection mass transfer.

9.28   Two surfaces of a plane wall (Fig. 9.24) are maintained at temperatures T_o and T_L, respectively. If the wall thickness is L and its thermal conductivity is given by $k = k_0(1 + a_1T + a_2T^2)$, find the heat flux through the wall.

9.29   Repeat Prob. 9.28 if, in addition to the conditions described, the cross sectional area decreases linearly from A_0 at $x = 0$ to A_L at $x = L$.

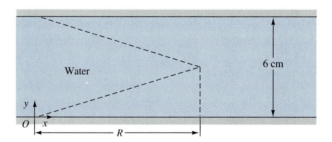

Figure 9.23   Problem 9.26.

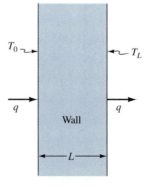

Figure 9.24   Problem 9.28.

9.30 A steel pipe of 1.90-in.-outside diameter and 1.61-in.-inside diameter is to carry water at 52°F. Two layers of insulation are to be used, a 1.2-in.-thick layer of 85-percent magnesia, $k = 0.034$ Btu/h·ft·°F, and a 1.0-in.-thick layer of glass fiber, $k = 0.02$ Btu/h·ft·°F. The ambient air is at a temperature of 110°F. The inside and outside surfaces have convective heat transfer coefficients of 135 Btu/h·ft²·°F and 8 Btu/h·ft²·°F, respectively. (*a*) If minimal heat loss is desired, which material should be placed next to the surface of the pipe? (*b*) What is the heat flux through the pipe's surface area?

9.31 A steel pipe with a 1-in.-nominal diameter is immersed in water at 32°C. The outside surface of the pipe has a 180°C temperature. The convective heat transfer coefficient between the surface of the pipe and the water is 2.32 Btu/h·ft²·°F. If the heat loss is to be reduced by half with the addition of insulation with thermal conductivity of 0.092 Btu/h·ft·°F, find the thickness of the insulation.

9.32 For the conditions of Prob. 9.31, the heat transfer coefficient varies according to $h_{ins} = 0.7 d_{ins}^{-2/3}$, in units of Btu/h·ft²·°F, where d_{ins} is the outside diameter of the insulation in feet. Find the thickness of the insulation that will reduce the heat loss to one-half of that for the bare pipe.

9.33 For steady state heat conduction through a hollow cylinder as derived in Prob. 9.13, show that for a hollow cylindrical element, $\bar{A}$ satisfies the equations for steady state radial heat transfer.

9.34 For steady state heat conduction through a hollow cylinder as derived in Prob. 9.13, find the resulting percent error if the arithmetic mean area, $\pi(R_1 + R_2)$, is used instead of the logarithmic mean, for values of R_2/R_1 of 2, 5, and 7.

9.35 Evaluate the following parameters at $T = 62°C$, for air, water, and glycerin: $Lu_0\rho/\mu$, $\mu c_p/k$, hL/k, and $h/\rho c_p u_0$. The values of L, u_0, and h can be taken as 1 m, 22 m/s, and 52 W/m²·K, respectively.

9.36 A fluid is flowing parallel to a flat plate, as shown in Figure 9.25. At a distance L from the leading edge of the plate, the fluid and the plate have the same temperature, while for $x > L$ the plate has constant temperature T_s, where $T_s > T_\infty$. If the velocity and temperature distributions for both the hydrodynamic and the thermal boundary layers are described by cubic profiles, show that the ratio $\xi = \delta_T/\delta$ can be expressed as

$$\xi = \frac{\delta_T}{\delta} \approx P_r^{-1/3} \cdot \left[1 - \left(\frac{L}{x} \right)^{3/4} \right]^{1/3}$$

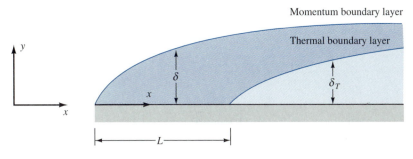

Figure 9.25   Problem 9.36.

9.37   For the conditions described in Prob. 9.36 show that

$$N_{u_x} \approx 0.33 \frac{P_r^{1/3} \cdot R_x^{1/2}}{\left[1 - \left(\frac{L}{x}\right)^{3/4}\right]^{1/3}}$$

9.38   Consider flow over a flat plate with constant free stream velocity. If the velocity and temperature distributions for both the hydrodynamic and the thermal boundary layers are described by linear profiles, find the expressions for the local Nusselt number, N_{u_x} in terms of R_x and P_r.

9.39   Repeat Prob. 9.38 with velocity and temperature distributions given as

$$U = \alpha_0 + \alpha_1 y + \alpha_2 y^2 \qquad T - T_s = \beta_0 + \beta_1 y + \beta_2 y^2$$

9.40   Repeat Prob. 9.38 with velocity and temperature distributions given as

$$U = \alpha_1 \sin(\beta_1 y) \qquad T - T_s = \alpha_2 \sin(\beta_2 y)$$

9.41   Repeat Prob. 9.38 with velocity and temperature distributions given as

$$u/u_\infty = (y/\delta)^{1/7} \qquad \frac{T - T_s}{T_\infty - T_s} = (y/\delta_T)^{1/7}$$

9.42   A gas mixture has the following composition: CH_4, 0.77; N_2, 0.14; C_2H_6, 0.05; and CO_2, 0.04. For this gas mixture find (a) the molar fraction of N_2 and C_2H_6; (b) the weight fraction of CO_2; (c) the partial pressures of the gas mixture components if the total pressure of the gas mixture is 1235 kPa.

9.43   A certain gas mixture is contained in a 22-m^3 container. The gas mixture has the following composition: H_2, 12.5 percent; N_2, 53.4 percent; and CO_2, 34.1 percent. If the pressure in the container is 1450 kPa, at 32°C, find (a) the molar fraction of CO_2; (b) the volume fraction of H_2; (c) the weight of the mixture; (d) the mass density of N_2; (e) the partial pressures of the gas mixture components.

9.44   For a binary gas mixture of species A and B, show that (a) $\mathcal{D}_{AB} = \mathcal{D}_{BA}$; (b) $\mathbf{J}_A + \mathbf{J}_B = 0$; (c) $\mathbf{n}_A + \mathbf{n}_B = \rho v$; (d) $\mathbf{N}_A + \mathbf{N}_B = c\mathbf{V}$.

9.45   For a binary mixture of species A and B, show that:

$$\omega_A = \frac{x_A M_A}{x_A M_A + x_B M_B}$$

and

$$d\omega_A = \frac{M_A M_B \, dx_A}{(x_A M_A + x_B M_B)^2}$$

9.46   For a binary mixture of species A and B, show that:

$$x_A = \frac{\omega_A/M_A}{\omega_A/M_A + \omega_B/M_B}$$

and

$$dx_A = \frac{d\omega_A}{M_A M_B(\omega_A/M_A + \omega_B/M_B)^2}$$

9.47   125 m^3/min of dry air (O_2, 20.95 percent; N_2, 78.08 percent; CO_2, 0.03 percent; Ar, 0.93 percent; and other gases 0.01 percent) at 29°C is mixed with a stream of a certain gas mixture with composition N_2, 32 percent; O_2, 40 percent; and H_2,

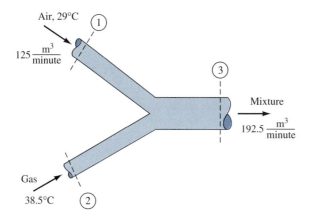

Figure 9.26 Problem 9.47.

28 percent, at 38.5°C as shown in Figure 9.26. The total pressures at inlets (1) and (2) are 1234 and 980 kPa, respectively (see Fig. 9.26). Mixing occurs adiabatically and at steady state. If the outlet flow discharge is 192.5 m³/min determine (*a*) the mass flow rates for the dry air and the gas mixture; (*b*) the pressure at the outlet; (*c*) the composition of the resulting gas mixture at the outlet; (*d*) the temperature at the outlet.

9.48 A gas mixture of O_2 and CO_2 at 28° and a total pressure of 1450 kPa is flowing in a pipe of 0.3-m diameter. If $x_{O_2} = 0.48$, $u_{O_2} = 0.23$ m/s, and $u_{CO_2} = 0.14$ m/s, calculate (*a*) x_{CO_2}; (*b*) $\rho_{mixture}$, ρ_{O_2}, and ρ_{CO_2}; (*c*) $c_{mixture}$, c_{O_2}, and c_{CO_2}; (*d*) the average flow velocity in the pipe and the gas flow rate; (*e*) J_{O_2} and J_{CO_2}.

9.49 The most commonly used disinfectant in wastewater treatment plants is chlorine gas (Cl_2). When chlorine gas is added to water, the following reaction takes place

$$Cl_2 + H_2O \longleftrightarrow HOCl + H^+ + Cl^-$$

What is the time required for 1 mole of Cl_2 to diffuse through a 0.6-cm-thick stagnant water film at 19°C when the Cl_2 concentration levels are 0.032 mol/m³ on one edge of the film and zero on the other edge?

9.50 A 2-cm-diameter spherical mothball (naphthalene) is suspended in stagnant air. The naphthalene has a molecular weight of 128 g/mol and a vapor pressure of 0.13 kPa at 74°C. The pressure and the temperature of the system are 100 kPa and 74°C, respectively. What is the amount of naphthalene which enters the gas phase in one day?

9.51 Two rigid, insulated large tanks are interconnected by a 1-m-long circular duct, 12 cm in diameter. The temperature and pressure in both tanks are 19°C and 100 kPa, respectively. The first tank contains a uniform gas mixture of 52-percent CO_2 and 48-percent NH_3 and the other tank contains a uniform gas mixture of 23-percent CO_2 and 77-percent NH_3. What is the rate of NH_3 transfer between the two tanks? Assume steady state mass transfer.

9.52 Repeat Prob. 9.51 for a conical duct, with diameters of 6 and 14 cm, respectively, at the duct's two ends. Assume in this case that the NH_3 diffuses in the direction of the decreasing diameter.

9.53 A pan of length 4 m, which is quite wide, contains water at a uniform depth of 2 cm. The water has a temperature of 18°C, and the total pressure on the system is

1 atm. Air is flowing over the pan at a speed of 7 m/s. The transition from laminar to turbulent flow occurs near $\mathbf{R}_x = 3(10^5)$. Assuming a mass diffusivity of $0.3(10^{-4})$ m^2/s, find the time required to evaporate all of the water in the pan.

9.54   Assuming linear velocity and concentration profiles in the laminar boundary layer over a flat plate, derive the relationship between the momentum boundary layer thickness, δ; the concentration boundary layer thickness, δ_c; and the Schmidt number, $\mathbf{S}_c$.

9.55   A round jet of discolored water enters a water tank of the same density at a velocity of 10 ft/s and jet diameter of 4 in. If, in addition to the variation with axial distance, the proportionality factor for the maximum time-average velocity is $6.2D$, what is the maximum velocity at $x = 5$ ft?

9.56   If a vertical water jet at the free surface was used to scour sediment from a stream bottom, what maximum diameter water jet would be allowed? The required scour velocity is 2 ft/s, the water depth is 6 ft, and the jet discharge is 0.5 ft^3/s. Use the proportionality factor in Prob. 9.55.

9.57   A pollutant is released in quiescent fluid in a long uniform tank at $x = 0$ and $t = 0$. The tank cross-sectional area is 25 m^2, $\mathcal{D} = 10^{-9}$ m^2/s, and $M = 1000$ kg. What is the concentration at $x = 0$ at $t = 4$ days and at $t = 30$ days? What is the concentration at $x = 1$ m and $t = 365$ days?

9.58   At an instant, 20 kg of salt is released into a stream. What will the distribution of the salt be 30 min later? The stream's velocity is 0.4 m/s; its cross-sectional area is 10 m^2; and the dispersion coefficient is 40 m^2/s.

9.59   Determine the longitudinal dispersion coefficient, K, for a 3-in.-diameter pipeline carrying water at 6 ft/s. $T = 60°F$.

9.60   A fluid is agitated so that the kinematic eddy viscosity increases linearly from zero ($y = 0$) at the bottom of the tank to 0.2 m^2/s at $y = 600$ mm. For uniform particles with fall velocities of 300 mm/s in still fluid, find the concentration at $y = 350$ mm if its discharge is 10 per liter at $y = 600$ mm.

9.61   In Prob. 9.57 determine the concentration at $x = 0, 1, 2, 3$, and 4 cm for intervals of 1 day up to 10 days.

9.62   In Prob. 9.58, after 10 min determine the concentration at 40-m intervals around the section of maximum concentration.

9.63   Repeat Prob. 9.54 with velocity and concentration distributions given as

$$u = \alpha_0 + \alpha_1 y + \alpha_2 y^2 \qquad C - C_s = \beta_0 + \beta_1 y + \beta_2 y^2$$

9.64   Repeat Prob. 9.54 with velocity and concentration distributions given as

$$u = \alpha_0 + \alpha_1 y + \alpha_2 y^2 + \alpha_3 y^3 \qquad C - C_s = \beta_0 + \beta_1 y + \beta_2 y^2 + \beta_3 y^3$$

9.65   Derive Eq. (9.3.6) using a cubic concentration profile in the von Kármán integral analysis approximation.

9.66   The following concentration profile

$$C - C_s = \alpha_1 y e^{\alpha_2 y}$$

is suggested for use in the von Kármán integral analysis approximation. Is this suggested profile a reasonable selection?

9.67   Repeat Prob. 9.66 given the following concentration profile

$$C - C_s = \alpha_1 y + \alpha_2 \sin(\alpha_3 y)$$

9.68   In turbulent flows the heat transfer equation is averaged over time using Reynolds averaging. What additional terms are introduced?

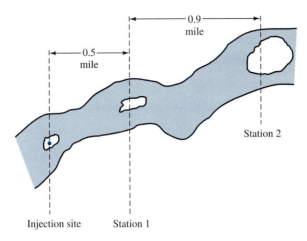

Figure 9.27   Problems 9.69 through 9.73.

9.69 A tracer is injected into the Scioto River, and the resulting tracer cloud (see Fig. 9.27) is sampled as it passes two downstream stations, X_1 and X_2. The two stations are 0.9 mile apart. The results are summarized in the following table.

Station X_1		Station X_2	
t (min)	$C(x_1, t)$ (mg/L)	t (min)	$C(x_2, t)$ (mg/L)
0	0	50	0
2	0.32	53	0.11
4	0.71	56	0.20
6	0.99	59	0.39
8	1.20	62	0.48
10	1.18	65	0.54
12	1.09	68	0.54
14	0.82	71	0.47
16	0.71	74	0.40
18	0.61	77	0.33
20	0.50	80	0.28
22	0.42	83	0.20
24	0.38	86	0.15
26	0.29	89	0.10
28	0.20	92	0.07
30	0.18	95	0.07
32	0.12	98	0.04
34	0.10	101	0.03
36	0.08	104	0.02
38	0.05	107	0.01
40	0.03	110	0.01
42	0	113	0

For the two stations X_1 and X_2 plot the concentration versus time and calculate the time-averaged concentrations by graphical integration.

9.70 For the concentration data in stations X_1 and X_2 presented in Prob. 9.69, calculate the concentration fluctuations $C'(x, t)$. Plot $C'(x, t)$ versus time and verify that $\overline{C'}(x, t) = 0$.

9.71   In analogy with Eq. (9.5.9), the dispersion coefficient, K, can be estimated using field data by the equation

$$K = 0.5\,\bar{u}^2 \frac{d\sigma^2}{dt}$$

Using the field data for the Scioto River presented in Prob. 9.69, estimate a value for the dispersion coefficient K.

9.72   Another way of estimating the dispersion coefficient from field data is based on the solution of Eq. (9.4.15) for a pulse input tracer. The equation has a solution of the form

$$C(x, t) = \frac{M}{A\sqrt{4\pi K t}} \exp\left[\frac{-(x - ut)^2}{4Kt}\right]$$

Rearranging the above equation and after taking logarithms of both sides we obtain

$$\log(C\sqrt{t}) = \alpha_0 + \alpha_1 \frac{(x - \bar{u}t)^2}{t}$$

where M is the weight of the tracer, A is the flow area, $\alpha_0 = \log(M/A\sqrt{4\pi K})$, and $\alpha_1 = -\log(e)/4K$. If the station X_1 is 0.5 mile downstream from the injection location and the time of injection is 10 AM, estimate a value for the dispersion coefficient K, from the data given in Prob. 9.69. Assume an average flow velocity for the Scioto River of 0.35 m/s.

9.73   Repeat Prob. 9.72 for station X_2 and compare your results with those in Prob. 9.72.

9.74   In turbulent flows the mass transfer equation is averaged over time using Reynolds averaging. What additional terms are introduced?

9.75   Using the transformation $\eta = x/\sqrt{D_x t}$ and the specified initial and boundary conditions, derive Eq. (9.5.18). *Hint:* transform the one-dimensional mass transfer equation to an ordinary differential equation with η as the independent variable.

9.76   Given the boundary and initial conditions

$$C = C_0 \quad \text{at} \quad t = 0 \quad \text{for all } z$$
$$C = C_s \quad \text{at} \quad z = 0 \quad \text{for all } t$$
$$C = C_0 \quad \text{as} \quad z \to \infty \quad \text{for all } t$$

find the solution of the one-dimensional diffusion equation

$$\frac{\partial C}{\partial t} = D\frac{\partial^2 C}{\partial z^2}$$

(*Hint:* the above equation describes the transient mass diffusion in a semifinite medium).

9.77   Derive the solution for the one-dimensional heat transfer equation that describes the transient heat transfer in a semifinite medium (by analogy to the solution obtained in Prob. 9.76).

9.78   Find the depth in saturated soil at which the annual temperature variation will be 18.5 percent of that at the surface.

9.79   A straight, smooth, rectangular channel is 120 m wide and its bottom slope is 0.002. Water is flowing at a depth of 2.5 m with an average velocity of 0.56 m/s.

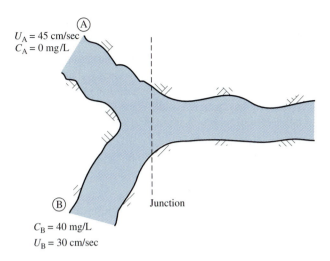

$U_A = 45$ cm/sec
$C_A = 0$ mg/L

Junction

$C_B = 40$ mg/L
$U_B = 30$ cm/sec

Figure 9.28   Problem 9.82.

If the water temperature is 22°C, estimate the bottom shear stress and determine the type of the flow.

9.80   For the conditions described in Prob. 9.79 and using the appropriate equations, estimate the vertical and lateral eddy diffusivities in the channel. If a pollutant point source is located at the bank of the channel, determine the time and channel length required to complete mixing.

9.81   A chemical industry site discharges 0.15 m³/s of effluent which contains 150 g/m³ of a chemical. The effluent is discharged near the centerline of a wide, meandering natural channel. The average water depth in the channel is 9 m, the bed slope is 0.002, and the average flow velocity is 50 cm/s. Determine the width of the plume and the maximum concentration at a distance 1.4 km downstream from the point of discharge.

9.82   Two streams, A and B, which flow together at a smooth junction (Fig. 9.28) have quite different chemical contents. For a distance downstream of the junction the two streams converge to a 350-m rectangular channel with a water depth of 10 m. Both streams have rectangular cross sections, with stream A having a width of 250 m and water depth of 5 m, and stream B having a width of 400 m and water depth of 9 m. The average flow velocities for streams A and B are 45 cm/s and 30 cm/s, respectively. Stream A has no chemical contents while stream B contains 40 mg/L of a chemical. The average temperatures in the two streams A and B are 28°C and 20°C, respectively. For the conditions described and assuming that $S_o = 0.0018$, determine: (*a*) the maximum concentration of the chemical in the channel downstream from the junction, assuming complete mixing; (*b*) the length of the channel required to provide complete mixing between the two streams (assume that the channel has no meanders); (*c*) the temperature and flow velocity far downstream from the junction.

9.83   Ten kilograms of rhodamine dye (SG $= 1.0$) are dropped on the water surface and at the centerline of a rectangular channel which is 150 ft wide and the water depth is 10 ft. If the 0.0014-sloped channel conveys water at 1.5 ft/s, estimate the value of the longitudinal dispersion coefficient using Eq. (9.6.20). Determine the maximum concentration and the width of the plume 5 miles downstream from the injection point. Has complete mixing been reached at the downstream location specified? If not, determine the distance from the injection point required for complete mixing.

9.84   A shoe factory discharges neutrally buoyant waste at the side of a slowly meandering natural channel, which carries water at a depth of 7 ft and an average flow velocity of 3.2 ft/s. At a point 20 miles downstream the injection point observations show that complete mixing has been reached. What is the average width of the channel's cross-sectional area and its flow rate in ft^3/s? Assume $S_o = 0.0013$.

9.85   A rectangular channel of slope 0.00085 carries water at an average velocity of 35 cm/s. The bottom friction velocity is found to be 20 cm/s. If the times to complete vertical and transverse mixing are equal, find the flow rate in the channel. What are the vertical and transverse eddy viscosities?

9.86   Find the time and the distance required for the initial adjustment period for the conditions described in Prob. 9.81.

9.87   Repeat Prob. 9.86 for the conditions described in Prob. 9.84.

9.88   Given the length to complete mixing, L_m; the transverse mixing coefficient, E_y; and the longitudinal dispersion coefficient, K, find an expression for the width, W, for a rectangular channel.

REFERENCES

1.   C. Geankopolis, *Transport Processes and Unit Operation,* 3rd ed., Prentice Hall, NJ, 1993.

2.   D. Lide, ed., *Handbook of Chemistry and Physics,* 14th ed., CRC Press, Florida, 1993.

3.   F. Kreith and W. Black, *Basic Heat Transfer,* Harper and Row, New York, 1990.

4.   T. Sherwood, R. Pickford, and C. White, *Mass Transfer,* McGraw Hill, New York, 1975.

5.   J. Welty, C. Wicks, and R. Wilson, *Fundamentals of Momentum, Heat and Mass Transfer,* 3rd ed., John Wiley and Sons, New York, 1984.

6.   C. Sleicher and M. Rouse, "A Convenient Correlation for Heat Transfer to Constant and Variable Property Fluids in Turbulent Pipe Flow," *Int. J. Heat and Mass Transfer,* 18, 677–683, 1975.

7.   P. Harriot and R. Hamilton, "Solid-Liquid Mass Transfer in Turbulent Pipe Flow," *Chem. Eng. Sci.,* 20, pp. 1073–1078, 1965.

8.   R. Brodkey and D. Hershey, *Transport Phenomena, A Unified Approach,* McGraw Hill, New York, 1988.

9.   S. Whitaker, "Forced Convection Heat Transfer Correlations for Flow in Pipes, Past Flat Plates, Single Cylinders, Single Spheres, and for Flow in Packed Beds and Tube Bundles," *A. I. Ch. E.,* 18, pp. 361–371, 1972.

10.   D. Kunii and O. Levenspiel, *Fluidization Engineering,* J. Wiley, New York, 1969.

11.   H. Schlicting, *Boundary Layer Theory,* 7th ed., McGraw Hill, New York, 1979.

12.   G. Taylor, "Diffusion by Continuous Movements," *Proc. London Math. Soc.,* Ser. A, 20, pp. 196–211, 1921.

13.   W. McComb, *The Physics of Fluid Turbulence,* Oxford Science Publ., Clarendon Press, Oxford, 1990.

14. M. Lesieur, *Turbulence in Fluids,* 2nd ed., M. Nijhoff, Dordrecht, 1991.

15. H. Fischer, E. List, R. Koh, J. Imberger, and N. Brooks, *Mixing in Inland and Coastal Waters,* Academic Press, New York, 1979.

16. H. Tennekes and J. Lumley, *A First Course in Turbulence,* MIT Press, Cambridge, MA, 1972.

17. G. Csanady, *Turbulent Diffusion in the Environment,* D. Reidel, Dordrecht, 1973.

18. H. Carlslaw and J. Jaeger, *Conduction of Heat in Solids,* Dover, NY, 1973.

19. J. Crank, *Mathematics of Diffusion,* Clarendon Press, Oxford, 1984.

20. R. French, *Open Channel Hydraulics,* McGraw Hill, New York, 1985.

21. J. W. Elder, "The Dispersion of Marked Fluid in Turbulent Shear Flow," *J. Fluid Mech.,* 5, pp. 544–560, 1959.

22. G. Taylor, "The Dispersion of Soluble Matter in a Solvent Flowing Slowly Through a Tube," *Proc. R. Soc. London,* Ser. A, 219, pp. 186–203, 1953.

23. R. Aris, "On the Dispersion of a Solute in a Fluid Flowing Through a Tube," *Proc. R. Soc. London,* Ser. A, 235, pp. 67–77, 1956.

24. E. Holley, D. Harleman, and H. Fischer, "Dispersion in Homogeneous Estuary Flow," *J. Hydraulics Div.,* ASCE, 96, pp. 1691–1709, 1970.

25. E. Holley, J. Siemons, and G. Abraham, "Some Aspects of Analyzing Transverse Diffusion in Rivers," *J. Hydraulic Res.,* 10, pp. 27–57, 1972.

26. P. Chatwin and C. Allen, "Mathematical Models of Dispersion in Rivers and Estuaries," *Ann. Rev. Fluid Mech.,* 17, pp. 119–150, 1985.

27. H. Fischer, "The Mechanics of Dispersion in Natural Streams," *J. Hydraulics Div.,* ASCE, 93, p. 187, 1967.

28. P. Chatwin, "The Approach to Normality of the Concentration Distribution of a Solute in a Solvent Flowing along a Straight Pipe," *J. Fluid Mech.,* 43, pp. 321–352, 1970.

ADDITIONAL READING

Bennet, C., and J. Myers: *Momentum, Heat and Mass Transfer,* 3rd ed., McGraw Hill, New York, 1974.

Eckart, E., and R. Drake: *Analysis of Heat and Mass Transfer,* McGraw Hill, New York, 1972.

Hemond, H., and E. Fechner: *Chemical Fate and Transport in the Environment,* Academic Press, New York, 1994.

Kay, J., and R. Nedderman: *Fluid Mechanics and Transfer Processes,* Cambridge U. Press, New York, 1985.

2

APPLICATIONS OF FLUID MECHANICS AND TRANSPORT

In Part 1, the fundamental concepts and equations have been developed and illustrated by many examples and simple applications. In Part 2, after the initial chapter on measurement and elementary data analysis, the remaining chapters present important applications on turbomachinery, steady and unsteady closed conduit flows, open channel flows, and transport phenomena.

chapter
10

Measurements

Fluid measurements include determination of elevation, pressure, velocity, temperature, and concentration. In addition the rates of heat, mass, and momentum transport are also required and typically inferred from the previously mentioned measured data. Modern instrumentation systems permit quite large data sets to be measured at high sampling rates, and therefore, this chapter begins with a brief discussion of the elements, functions, and terminology of the measurement system. The emphasis of the remainder of the chapter is on the various procedures for direct measurement of the variables listed above followed by direct measurement techniques for flow rates in conduits and open channels and indirect methods for inferring flow rates and mass fluxes.

10.1 SYSTEM ATTRIBUTES AND FUNCTIONS

The various dependent and independent variables defined in the previous nine chapters will at some point need to be measured. Whether in laboratory representations of the prototype or the actual prototype itself, measurements form the basis for evaluating the correlations between fluid variables in the theories or for establishing data bases for numerical experiments.

A measurement system is typically comprised of four functions (Fig. 10.1a–d) whose integrated effect or performance must contain several attributes. First, the system must meet the *precision* goals of the project and be able to do so during the entire deployment of the system. The definition of precision includes both the *resolution* of the instrument and its *accuracy*. The resolution is the smallest quantity

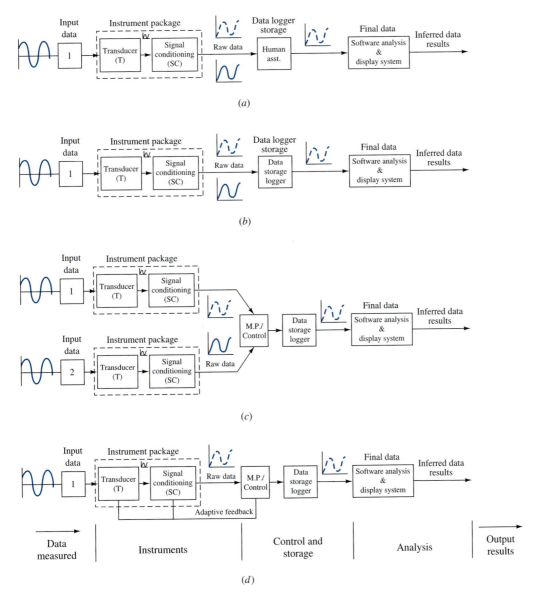

Figure 10.1   Attributes of a measurement system

that can be measured by the device while the accuracy refers to the fractional error in the instrument. These precision characteristics are predetermined by the analysts as necessary to meet the reporting goals of the project to a quantifiable confidence level. Once the instrument is deployed these precision attributes must remain *stable,* that is, they must not deviate beyond these limits for the range of implementation conditions (temperature, etc.) expected during the instrument's deployment. Instruments that are not stable will require considerable checking during data collection, which is costly and interrupts data collection.

A second attribute, an extension of the first one, is that system *calibration* should be readily obtained and, once again, stable. Calibration, the act of determining the precision attributes, consists of a number of steps, and it is desirable that the calibration steps be required only at the beginning and the end of the experiment and not during the experiment. The calibration should straightforwardly elicit the error and uncertainty estimates in a quantitative form which can be used or carried through all subsequent analyses with the data.

Third, the system should be *reliable* in that the *transducers* and associated signal analyser should collect and store the data, without malfunction or data dropouts, at the predetermined levels of precision, error, and uncertainty for the length of the instrument's deployment. Many, if not most, modern analysis techniques require that a *time series* of data be collected at regular intervals over the deployment. Any data dropouts intrude on the ability to perform the desired analyses. Long-term unattended measurements made with mechanical or electrochemical transducers are subject to failure of the moving parts, are sometimes unreliable, and therefore often require human intervention and checking. Repeated data collection by direct human activity is quite reliable over the short term but not so over the long term without considerable scheduling of teams of often costly personnel.

Fourth, and finally, the data collected by the system should come in a form that is amenable to *digital* storage, analysis, and reuse. The availability of low-cost microprocessors and computer technology, coupled with the vast amount of data required for accurate analyses, dictates the use of modern digital methods and hardware. The days of the research assistant jotting down individual measured numbers in a field or lab notebook are, at least for fluid mechanics, undesirable data collection and transmission procedures.

In Fig. 10.1*a–d* the block system diagrams for increasingly sophisticated measurement systems are displayed. As noted there are as many as five functions a measurement system must address. Common to all are the *data input,* the *instrument package,* and a *data logger* or *storage system.* Two other components a system can have include a *microcontroller* or *microprocessor* for multiinstrument control or adaptive sampling, and a postcollection *software analysis and display system* for data analysis and creation of inferred data.

Data Input

With the exception of certain engineered laminar flows the flow field variables to be measured vary temporally and spatially over the entire flow domain. The length and time scales of the smallest nonmolecular fluctuations to be measured are called the *Taylor microscales,* and for turbulent flows they are approximately 0.01 second and 0.01 cm. The largest length and time scales of the signal variability to be measured are dictated by the size of the flow field geometry.

The fluid mechanics and transport variables to be measured can be classified into three broad categories:

1. *Directly measured* variables such as temperature, velocity, elevation, or concentration.

2. *Integrated measures* such as total heat content, total mass, or total energy in a sampling volume.

3. *Inferred data* which are data compiled from combinations of other measured data.

This last category is often used when no instruments are available to make the required measurements directly, an example being the measurement of advective or turbulent flux of sediment or pollutants.

The experimenter should be able to characterize the desired degree of variability to be resolved in the flow field, as it directly dictates the precision that the instrument must have and the sampling frequency or rate at which the instrument must collect data. Furthermore, the magnitude of the maximum values to be measured should also be known *a priori* as the full-scale capabilities of the instrument must "comprehend" the anticipated values. As a conceptual example an experimenter is asked to measure the wave height profile at a point near a beach. The minimum wave period, T, is anticipated to be 2 seconds and wave height data must be measured with a precision of ± 0.1 cm. It is necessary to take three equispaced independent measurements every one-quarter wavelength ($T/4$). Therefore, the sampling frequency of the instrument must be at least six hertz or once every 0.167 second. It is expected that the maximum expected free surface height above the still water level will be 2 m; therefore, the full-scale instrument capability is ± 2 m.

Instrument Packages

The device to make the measurement consists of two components: a *transducer* or *sensor* and typically a *signal conditioning* package for preparing the signal for further use by a data logger or direct readout by the experimenter.

The *transducer* [1]† is the primary sensing element of the variables in the flow field. It receives energy from the flow field and produces an output which is directly dependent on the measured quantity. Transducers may be either *active* or *passive,* the distinction being drawn by whether the output energy for the sensor is derived almost entirely from the flow field (passive) or from the electronics package (active). An example of a passive transducer would be the speedometer in a car which derives its output from the rotation of the car wheels. An example of an active transducer is a sonar (acoustic) device used by submarines to detect other submarines or submerged land forms.

With the exception of elevation or geometry measurements, which are made directly, most traditional sensors are electromechanical in nature and their basis depends upon *energy conversion* from one form to a form more readily measurable. All electromechanical devices are passive. Recent instruments whose operating basis depends upon *energy propagation* instead of energy conversion are now available for making quite accurate velocity and particulate concentration measurements. These instruments are active transducers based upon acoustic, laser, or electromagnetic radiation transmissions.

The signal conditioning component essentially translates the energy conversion or the energy propagation into an electronic signal. The intensity of the conversion or

---

| †Numbered references will be found at the end of this chapter.

propagation is then related to the system output voltage. *Linearity* of the relationship between the output voltage and the energy conversion or propagation is a highly desired goal as it permits stable calibrations. Systems where the output voltage levels and conversion rates are continuously related are called *analog* systems.

A second purpose of the signal conditioning component is to further process the signal so that it meets the accuracy, precision, and range characteristics published by the manufacturer. These operations consist of filtering, averaging, clipping, and other electronic processes for sharpening or clarifying the raw data. This stage was traditionally done in an analog fashion with a final continuous readout on a gage which an analyst would note in a research or lab notebook. This elementary system is depicted in Fig. 10.1*a*.

Most of the conditioning operations are now done *digitally* where an *analog to digital* (A to D) conversion is performed on the continuous voltage signal by an integrated circuit. In this fashion data are stored at discreet intervals whose frequencies equal or exceed the sampling frequencies required by the experiment. All filtering and averaging operations are then done digitally. Figure 10.1*a* also depicts this output option.

Storage and Control

The output of the instrument package is the raw data in either an analog or digital form. Both the design of experiments and the advances of instruments themselves result in remarkably large data sets. For example the field measurements required to verify boundary layer predictions of entrainment and scour rates from the bottom of the water column required over 50 Mbytes of data be obtained and stored in just two days of field experiments at a 4-Hz sampling rate. A *data logger* achieves this purpose. Modern data loggers can accept data in either digital or analog form; therefore, they contain A to D conversion capabilities. Storage can be on floppy disks (2 Mbytes), floptical (20 Mbytes), optical disks (1 Gbyte), or in various tape formats. Figure 10.1*b* depicts this block diagram. Multiple measurements are common experimental practice and data loggers can accept multiple instruments as in Fig. 10.1*c*. As an example, the boundary layer experiment mentioned above required simultaneous measurement and logging of 4 velocity measurements, 1 temperature and pressure measurement, and 100 sediment concentration measurements at the 4-Hz sampling frequency.

Modern data loggers may be powered with the standard AC voltage found in laboratories or with low-power DC voltage from batteries which allows unattended field operation. Field units can also be equipped to communicate to a central workstation via satellite or cellular telephone communication. Satellite or telephone communication requires a higher voltage power supply which batteries can provide; however solar panels are more effective.

The essential attribute of the control system is software-programmable implementation of sampling. It is possible to create *control* systems from *microcontroller* integrated circuits which limit the use of simultaneous arithmetic operations, or from *microprocessors* which allow much more complex sampling regimes to be configured and implemented. A microprocessor such as the familiar INTEL X86 or Pentium circuits essentially permits a computer-controlled instrument environment to exist where sampling frequencies are configured in the software and data are transmitted through standardized interfaces and stored on common storage devices. Arithmetic operations also can be programmed in the control software such that the microprocessor can perform calculations with the data or configure *conditional* or *adaptive sampling*. Conditional sampling occurs when groups of measured variables have achieved critical values and then the instruments are turned on to sample. Alternatively, sampling may be at a preset frequency which becomes more frequent when groups of variables have reached trigger levels. Arithmetic operations also allow on-board processing of raw data to proceed simultaneously with sampling.

Microcontrollers, such as those associated with the HPC16000 series, are also software-programmable systems for instrument control that differ from microprocessors in that arithmetic operations cannot be readily performed. Therefore, data cannot be analyzed *on-board* during the course of an experiment. All other software control is permitted.

Analysis and Results

Typically analysis is the least automated portion of the system, and the final function is to analyze the measured data or create new data from the measured data. A number of general analysis software packages exist such as Mathcad, Matlab, SAS, or SYSTAT for routine analysis and display. However, more specialized programs are usually required and need to be composed by the analyst.

Error Estimation

The act of measuring data or using data incurs error, which is defined as the difference between the true value and the measured or derived quantity. It is common to report *measurement error* as a fractional quantity relative to the true value. For instance, if the true river velocity, u, is 15 cm/sec and the measured quantity, u_m, is 14.4 cm/sec, then the *absolute error, e*, is 0.6 cm/sec and the *relative error, ε*, is

$$\varepsilon = e/u = -0.6/15 = -0.04 \quad \text{or} \quad \varepsilon = 4 \text{ percent}$$

Errors may be from either over- or underestimates of the true value and are either *systematic* or *random* in origin. A systematic error is one that is predictably and repeatedly observed and therefore can be parameterized and accounted for in the calibration. An example of a systematic error would be a current meter that always reads 0.1 cm/sec when there is no river velocity, that is, a *zero offset* error. A random error is one where the results from a number of repeated observations under the same conditions give data that are all slightly different but cluster about a predictable average. The goal of the instrument engineer is to properly design the system and calibrate it such that only random errors are present.

Error estimates are treated in a number of textbooks (see, e.g., Refs. [2, 3]). There are two procedures available for determining error estimates. First, the overall instrument performance may be derived by application to standard or accepted data sets. Second, instrument error can be determined by assessing the error at each stage of the measurement process followed by compilation into a single value for the entire process. The second approach is used when standardized data are lacking. Ideally both estimates should be consistent.

While the mathematical details can be found in the above references, the following is accepted practice in estimating error. Suppose the instrument sampling the river velocity must perform three operations in order to achieve the measured result and relative error occurs in all three stages. The overall worst case error would occur if the relative errors simply added, that is,

$$\varepsilon = \varepsilon_a + \varepsilon_b + \varepsilon_c \tag{10.1.1}$$

Based upon a mean squared error analysis and assuming the relative errors are un-correlated, it is more typical to estimate the *expected error* by

$$\varepsilon = (\varepsilon_a^2 + \varepsilon_b^2 + \varepsilon_c^2)^{1/2} \tag{10.1.2}$$

If ε_b and ε_c are correlated, then

$$\varepsilon = [\varepsilon_a^2 + (\varepsilon_b + \varepsilon_c)^2]^{1/2} \tag{10.1.3}$$

Unlike laboratory experiments where repeated observations or trials can be obtained under identical circumstances, field data rarely if ever are repeatable. Therefore, error estimates for field data must be estimated from Eqs. (10.1.2) or (10.1.3). If repeated observations of each data point are possible, then random errors can be minimized through the estimate of the *most probable error.* Continuing the river flow example, a laboratory model of the river is used and the river velocity, u, at a project milepost is sampled N times for the same inflow and elevation condition. The most probable error for u is based upon the assumption of a normal probability distribution for the errors about the average value $\bar{u}$.

The average of the measured values of river velocity, u_{mi}, is

$$\bar{u}_m = \frac{\sum_{i=1}^{N} u_{mi}}{N} \tag{10.1.4}$$

where N is the number of observations. The error is estimated as

$$e_{\bar{u}} = \sum \left(\frac{1}{N^2}\right)\sigma^2 = \frac{\sigma}{N^{1/2}} \tag{10.1.5}$$

where σ is the standard deviation

$$\sigma = \left[\frac{\sum (u_{mi} - \bar{u}_m)}{N}\right]^{1/2} \tag{10.1.6}$$

In many cases the long-term average value is zero, and errors may be either positive or negative. In order to estimate deviation under these conditions, the *root-mean-square* (rms) value can be calculated

$$e_{rms} = \left[\frac{1}{N}\sum_{i=1}^{N} (u_{mi} - \bar{u}_m)^2\right]^{1/2} \tag{10.1.7}$$

To account for the bias introduced in Eq. (10.1.5) by the fact that the true river velocity is not known, the most probable error is adjusted upward slightly as

$$e_{\bar{u}} = \frac{\sigma}{(N-1)^{1/2}} \tag{10.1.8}$$

Clearly if repeated laboratory observations are possible, then the more observations, N, that can be obtained the smaller the most probable error becomes. However, each time the error is reduced by a factor of 10, the number of readings must increase by a factor of 100.

What follows in the rest of this chapter is a summary of the various transducers for measuring velocity, pressure, flow rates, temperatures, and concentrations. For a more complete discussion of measurement systems the reader is referred to other textbooks (e.g., Refs. [1, 4]). The review here will include more detailed treatments of enduring transducers supplemented by briefer descriptions of new or research-level instrumentation developed for more rapid and precise measurements than required for operational conditions.

EXERCISES

10.1.1 If a zero offset of 0.2°C is noted in a temperature sensor, then (*a*) the random error is twice the magnitude of the measurement error; (*b*) the systematic error is 0.2 degree; (*c*) the relative error is 2 percent; (*d*) none of the above.

10.1.2 Fine sediment concentration measurements are made under repeated conditions at the same point in the flow field. They are 10.1, 10.35, 9.96, 9.97, and 10.01 mg/L. The estimated error (*a*) is independent of the number of observations; (*b*) is inversely proportional to the number of observations; (*c*) equals 10.07 mg/L; (*d*) equals 0.07 mg/L; (*e*) the biased estimate percent error is lower than the standard error estimate.

10.2 PRESSURE MEASUREMENT

Direct pressure measurement is often required for conduit or piping systems. However, for most civil, environmental, or agricultural engineering applications, pressure measurements are often used to measure elevations of liquid levels or in many devices that determine the velocity or flow rate of a fluid stream. The use of pressure as a surrogate for elevation or velocity measurements is allowed by the fundamental relationship between these variables in the energy equation.

Piezometers and Static Tubes

The static pressure of a fluid in motion is its pressure when the velocity is undisturbed by the measurement. Figure 10.2*a* indicates one method of measuring static pressure, the *piezometer* opening. When the flow is parallel, as indicated, the pressure variation is hydrostatic normal to the streamlines; hence, by measuring the pressure at the wall the pressure at any other point in the cross section can be determined. The piezometer opening should be small, with the length of the opening at least twice the

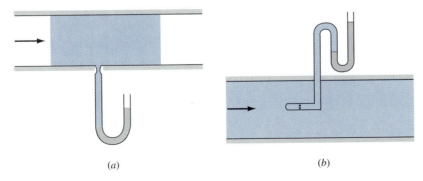

(*a*)   (*b*)

Figure 10.2   Static-pressure-measuring devices. (*a*) Piezometer opening. (*b*) Static tube.

diameter, and should be normal to the surface, with no burrs at its edges, because small eddies would form and distort the measurement. A small amount of rounding of the opening is permissible. Since any slight misalignment or roughness at the opening may cause errors in measurement, it is advisable to use several piezometer openings connected together into a *piezometer ring*. When the surface is rough in the vicinity of the opening, the reading is unreliable. For small irregularities it may be possible to smooth the surface around the opening. The piezometric opening is connected to a manometer, a micromanometer, or any of several types of electronic transducers.

For rough surfaces, the *static tube* (Fig. 10.2b) can be used. It consists of a tube that is directed upstream with the end closed. It has radial holes in the cylindrical portion downstream from the nose. The flow is presumed to be moving by the openings as if it were undisturbed. There are disturbances, however, due to both the nose and the right-angled leg that is normal to the flow. The static tube should be calibrated, as it may read too high or too low. If it does not read true static pressure, the discrepancy Δh normally varies as the square of the velocity of flow by the tube, that is

$$\Delta h = C \frac{v^2}{2g}$$

in which C is determined by towing the tube in still fluid where pressure and towing velocity are known or by inserting it into a smooth pipe that contains a piezometer ring.

Such tubes are relatively insensitive to the Reynolds number and to Mach numbers below unity. Their alignment with the flow is not critical, so that an error of but a few percent is to be expected for a yaw misalignment of 15 degrees.

Elastic and Piezoelectric Transducers

A number of commonly used elastic pressure transducers depend on the distortion of a flexible metallic element by the pressure to be measured. The distortion can be very accurately and quickly measured. Typical designs include the diaphragm type, the bourdon tube or gage, or the bellows design. Figure 10.3 contains a schematic of

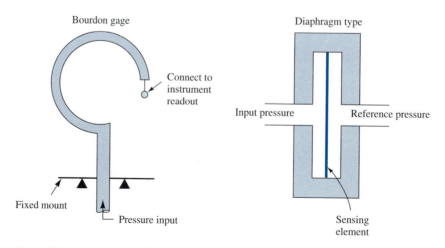

Figure 10.3   Schematic of bourdon tube and diaphragm gage.

the bourdon and diaphragm gages. In both cases the distortion results in the motion of the metallic unit: the free end of the bourdon tube attempting to straighten with increasing pressure, or the diaphragm's clamped sensing element deforming in response to the imposed pressure difference. While the bourdon tube often connects the free end to a mechanical readout system, the diaphragm device may use a strain gage readout system attached to the diaphragm. The distortion of the strain gage will change the resistance of the unit to a known imposed current which in turn can be measured by a Wheatstone bridge type of circuit. More recently, light-emitting or photodiodes have been used to optically sense the degree of diaphragm distortion.

For extremely rapid response, accurate measurements necessary for sampling at very high rates with great endurance and durability, *piezoelectric* transducers are used. A solid material that generates an electric signal when deformed is termed piezoelectric and either naturally occurring crystals (e.g., quartz), ferroceramics (barium), or synthetic materials are used in transducers. A typical piezoelectric transducer has a small size (footprint) compared to the diaphragm device and creates minimal flow field disruption with less bias in the resulting data.

EXERCISES

10.2.1 A piezometer opening is used to measure (*a*) the pressure in a static fluid; (*b*) the velocity in a flowing stream; (*c*) the total pressure; (*d*) the dynamic pressure; (*e*) the undisturbed fluid pressure.

10.2.2 A static tube is used to measure (*a*) the pressure in a static fluid; (*b*) the velocity in a flowing stream; (*c*) the total pressure; (*d*) the dynamic pressure; (*e*) the undisturbed fluid pressure.

10.2.3 The piezolectric properties of quartz are used to measure (*a*) temperature; (*b*) density; (*c*) velocity; (*d*) pressure; (*e*) none of these.

10.2.4 Water for a pipeline was diverted into a weigh tank for exactly 10 min. The increased weight in the tank was 4765 lb. The average flow rate, in gallons per minute, was (*a*) 66.1; (*b*) 57.1; (*c*) 7.95; (*d*) 0.13; (*e*) none of these.

10.2.5 A rectangular tank with cross-sectional area of 8 m^2 was filled to a depth of 1.3 m by a steady flow of liquid for 12 min. The rate of flow, in liters per second, was (*a*) 14.44; (*b*) 867; (*c*) 901; (*d*) 6471; (*e*) none of these.

10.3 ELEVATION MEASUREMENT

A seemingly simple but very important measurement is the measurement of the water (or liquid) surface elevation above a reference datum. While many applications involving the energy equation require only knowledge of the elevation difference between two reservoirs, other measurements such as wave heights, tidal elevation, and storm surge heights require absolute heights to be obtained.

Direct Measures

For occasional direct measurements where accuracy is not of paramount importance simple elevation markings painted on the side of a ship hull or elevation marks at a reservoir outlet suffice. Quite accurate direct elevation measurements required

Hook gage

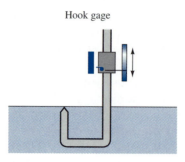

Figure 10.4   Hook gage.

for highly precise laboratory experiments require a hook gage (Fig. 10.4) which is connected to a vernier-caliper or electronic readout system and provides readouts accurate to 0.1 mm. Hook gages are used in order to minimize the biasing effects of surface tension, which would occur if the point gage simply dropped down to the surface from above. Most laboratory experiments are performed for steady mean flow conditions where the water levels are steady. Hook gages are perfectly suited for such steady or very slowly varying laboratory conditions.

The continuous measurement of slowly varying water levels in the field is found in three important activities: the measurement of river stage required for estimating flow rates, of tidal elevations along ocean coastlines, and of storm surge and seiche elevations. Figure 10.5 contains a schematic of a typical water level gage used in a United States Geological Survey stream gaging installation. The device is based upon measuring elevations via pressure measurements or direct elevation measurements.

The system normally utilizes a stilling basin to eliminate or physically filter out high frequency water level changes due to wind waves. While readouts are continu-

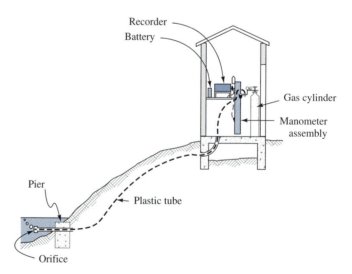

Figure 10.5   Typical U.S. Geological Survey stream gaging installation.

ously obtained, they are typically averaged over an hour and the hourly average value is stored for the permanent record. Many of these gages are configured to report automatically to a centralized computer facility. In this fashion data can be collected on open coastline tidal elevations or storm surges as used for example by the United States National Weather Service to forecast river flooding conditions.

Direct measurements of high frequency elevation changes, for example, wind wave data, require more specialized devices called *electronic staff gages*. These devices are electronic elements that resemble thick (approximately 2-mm) wires and are constructed such that their electronic properties change as a function of the height of the water on the element. The staff is placed vertically in the water body and each end is connected to a circuit package. Staffs may be either resistance-based or capacitance-based. The water level changes cause an almost instantaneous change in either the resistance or capacitance of the system, which in turn will cause a measureable voltage or current change in the system. Of course these devices can be used to measure not only high frequency fluctuations but slowly varying steady deviations as well. Therefore, not only are they relatively inexpensive but they are quite broadly applicable.

Pressure-Based Measurement

Any of the pressure transducers mentioned in Sec. 10.2 can be used to measure the average or steady water surface level via the use of the hydrostatic pressure equation. More typically pressure transducers are used to measure higher frequency fluctuations. For instance the U.S. Army Corps of Engineers maintain a series of permanent wave height and direction meters around the continental shelf region of the United States and the Gulf of Mexico which are based upon simultaneous high frequency measurements of pressure and horizontal velocity. The velocity data allow determination of the wave propagation direction while a pressure transducer (piezoelectric-quartz) measures the wave heights. Linear wave theory (Sec. 8.7) coupled with Bernoulli's equation is used to relate water surface elevation, η, to the subsurface gage pressure as

$$p(z^*, t) = \rho g[\eta(t)K_{z^*} - z^*]$$ (10.3.1)

where z^* is the depth below the still water level where p is measured and K_{z^*} is the pressure response factor

$$K_{z^*} = \frac{\cosh[2\pi(z^* + d)/L]}{\cosh[2\pi d/L]}$$

Equation (10.3.1) is solved for η as

$$\eta(t) = \frac{C_o}{K_{z^*}}\left[\frac{p(z^*, t)}{\rho g} + z^*\right]$$ (10.3.2)

If waves are nonlinear, then the linear wave theory basis of the measurements must be extended. For weakly nonlinear conditions a simple laboratory calibrated coefficient, C_o, is used by multiplication of the right-hand side of Eq. (10.3.2); its value is 1.0 for linear waves and greater than one for nonlinear conditions.

EXERCISES

10.3.1   If a quartz pressure transducer and staff gage are used to simultaneously measure the same elevation, then (*a*) the four-stage pressure transducer system will have a higher expected error than the four-stage staff gage; (*b*) the pressure gage will have a quicker response time than the staff gage; (*c*) the pair of gages should yield the same result if compared upon the same basis; (*d*) the four-stage pressure gage system has uncorrelated errors; (*e*) none of the above.

10.3.2   Water level measurements made with a pressure transducer are (*a*) linearly related to measured pressure at depth; (*b*) require small adjustments for signal attenuation with depth; (*c*) are related using Bernoulli's equation; (*d*) all of the above.

10.4   TEMPERATURE MEASUREMENT

In addition to a standard bulb–based thermometer, temperature measurements are based upon bimetallic, thermocouple, or thermistor sensors. *Bimetallic sensors,* commonly used in heating control systems, are formed of two different materials with quite different thermal expansion coefficients. *Thermocouple* sensors again are based upon two dissimilar materials in contact with each other but in this case a small electric potential difference develops across the sensor which is a function of the temperature. The resulting differential voltage is easily measured. Thermocouples are quite small and relatively low cost.

Thermistors are based upon the relationship between the resistance to an electrical current and the resulting heat generated by the sensor. Early forms of semiconductor resistor–type thermistors have a highly nonlinear relationship between resistance and temperature and careful laboratory calibration is required. The recent use of diodes or transistor-based thermistor sensors has resulted in transducers with more favorable linear response. The application temperature range, precision, and sampling frequency capabilities of thermistors are quite high and the cost is relatively low. Their small size considerably reduces sampling bias from obstructed or disturbed flow patterns.

10.5   VELOCITY MEASUREMENT

Velocity magnitude and direction measurements are of critical importance, both as stand-alone data about conditions at the sampling point and as data to be integrated across a plane such as a river cross section in order to determine the volume or mass flow rate at the section. Rate measuring devices will be discussed in the ensuing two sections. As discussed earlier the wide range of scales of velocity fluctuations encountered in the flow field, especially geophysical scale flows, presents a challenging problem in selecting the type of device to use. Widely available and inexpensive devices are available that measure the velocity at a point and can readily obtain data about slowly varying and nonturbulent or nongravity wave-dominated flows. Much more precision is required to sample all the high-speed turbulent- and wave-induced fluctuations. Most applications of velocity measurements required for rate device calibration require only the first level of velocity measurement. However, the increasing demand for more precise parameterization of pollutant and sediment or particulate transport has resulted in more widespread measurement of turbulence and/or wave fluctuations.

The simplest measurements are point measurements made with electromechanical devices placed directly in the flow field. These are termed *invasive* sensors or sampling devices (as opposed to *remote* sensors). Invasive devices often do not measure velocity directly but another quantity which is more readily measureable and easily related to velocity. Typically these devices record the total velocity magnitude

$$V = (u^2 + v^2 + w^2)^{1/2}$$

at a point and the direction of the total velocity vector. The cost of the instrument climbs quickly as the resolution and precision requirements increase, the degree of disruption drops, and the demand for individual velocity data for u, v, and w increases. A number of invasive devices are reviewed next.

The Pitot Tube

The pitot tube is one of the most accurate and enduring methods of measuring velocity. In Fig. 10.6 a glass tube or hypodermic needle with a right-angled bend is used to measure the velocity v in an open channel. The tube opening is directed upstream so that the fluid flows into the opening until the pressure builds up in the tube sufficiently to withstand the impact of velocity against it. Directly in front of the opening the fluid is at rest. The streamline through 1 leads to the point 2, called the *stagnation point,* where the fluid is at rest, and there divides and passes around the tube. The pressure at 2 is known from the liquid column in the tube. Bernoulli's equation, applied between points 1 and 2, produces

$$\frac{v_1^2}{2g} + \frac{p_1}{\gamma} = \frac{p_2}{\gamma} = h_0 + \Delta h$$

since both points are at the same elevation. As $p_1/\gamma = h_0$, the equation reduces to

$$\frac{v_1^2}{2g} = \frac{v^2}{2g} = \Delta h \qquad (10.5.1)$$

or

$$v = \sqrt{2g\Delta h} \qquad (10.5.2)$$

Practically, it is very difficult to read the height Δh from a free surface.

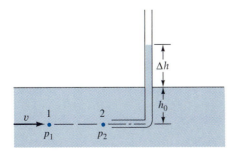

Figure 10.6   Simple pitot tube.

The pitot tube measures the stagnation pressure, which is also referred to as the *total pressure*. The total pressure is composed of two parts, the static pressure h_0 and the dynamic pressure Δh, expressed in length of a column of the flowing fluid (Fig. 10.6). The dynamic pressure is related to velocity head by Eq. (10.5.1).

By combining the static-pressure measurement and the total-pressure measurement, that is, measuring each and connecting to opposite ends of a differential manometer, the dynamic pressure head is obtained. Figure 10.7a illustrates one arrangement. Bernoulli's equation applied from 1 to 2 is

$$\frac{v_1^2}{2g} + \frac{p_1}{\gamma} = \frac{p_2}{\gamma} \tag{10.5.3}$$

The equation for the manometer, in units of length of water, is

$$\frac{p_1}{\gamma}S + kS + R'S_0 - (k + R')S = \frac{p_2}{\gamma}S$$

Simplifying gives

$$\frac{p_2 - p_1}{\gamma} = R'\left(\frac{S_0}{S} - 1\right) \tag{10.5.4}$$

Substituting for $(p_2 - p_1)/\gamma$ in Eq. (10.5.3) and solving for v results in

$$v_1 = v = \sqrt{2gR'\left(\frac{S_0}{S} - 1\right)} \tag{10.5.5}$$

The pitot tube is also insensitive to flow alignment, and an error of only a few percent occurs if the tube has a yaw misalignment of less than 15 degrees.

The static tube and pitot tube can be combined into one instrument, called a *pitot-static tube* (Fig. 10.7b). Analyzing this system in a manner similar to that in Fig. 10.7a shows that the same relations hold; Eq. (10.5.4) expresses the velocity, but the uncertainty in the measurement of static pressure requires a corrective coefficient C to be applied

$$v = C\sqrt{2gR'\left(\frac{S_0}{S} - 1\right)} \tag{10.5.6}$$

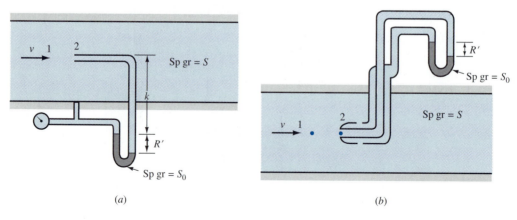

(a)   (b)

Figure 10.7   Velocity measurement. (a) Pitot tube and piezometer opening. (b) Pitot-static tube.

A particular form of pitot-static tube with a blunt nose, the *Prandtl tube,* has been so designed that the disturbances due to nose and leg cancel, leaving $C = 1$ in the equation. For other pitot-static tubes the constant C must be determined by calibration.

The pitot-static tube can be used for velocity determinations in compressible flow. In Fig. 10.7b the velocity reduction from free-stream velocity at 1 to zero at 2 takes place very rapidly without significant heat transfer. Friction plays a very small part, so that the compression can be assumed to be isentropic. Applying the Bernoulli equation and the ideal gas law to points 1 and 2 of Fig. 10.7b with $V_2 = 0$ gives

$$\frac{V_1^2}{2} = c_p T_1 \left[\left(\frac{p_2}{p_1} \right)^{\frac{k-1}{k}} - 1 \right] = c_p T_2 \left[1 - \left(\frac{p_1}{p_2} \right)^{\frac{k-1}{k}} \right] \tag{10.5.7}$$

in which c_p is the specific heat of the gas at constant pressure. The static pressure p_1 can be obtained from the side openings of the pitot tube and the stagnation pressure from the impact opening leading to a simple manometer. Alternatively, $p_2 - p_1$ can be found from the differential manometer. If the tube is not designed so that the true static pressure is measured, it must be calibrated and the true static pressure computed.

Electromechanical Current Meters

This type of sensor is most often used to measure nonturbulent variations in velocity. Figures 10.8a and 10.8b are examples of the popular rotating cup type of meters. Figure 10.8a is a current meter designed for surface water measurements. The cups are so shaped that the drag varies with orientation, causing a relatively slow rotation.

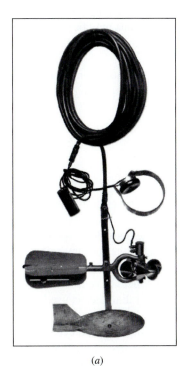

(a)

(b)

Figure 10.8 Velocity-measuring devices. (a) Price current meter for liquids (*W. and L. E. Gurley*). (b) Air anemometer (*Taylor Instrument Co.*).

With an electric circuit a signal is detected and velocity is related to the number of revolutions. The number of signals in a given time is a function of the velocity. The meters are calibrated by towing them through liquid at known speeds. For measuring high-velocity flow a current meter with a propeller as rotating element is used, as it offers less resistance to the flow.

Air velocities are measured with cup type or vane (propeller) anemometers (Fig. 10.8b) which drive generators indicating air velocity directly or drive counters indicating the number of revolutions. By designing the vanes to have very low inertia, employing precision bearings and optical tachometers which effectively take no power to drive them, anemometers can be made to read very low air velocities. They can be sensitive enough to measure the convection air currents which the human body causes by its heat emission to the atmosphere.

For measuring conduit flows, a derivative of the rotating cup device, called a *positive displacement meter,* is used. It has pistons or partitions which are displaced by the flowing fluid and a counting mechanism that records the number of displacements in any convenient unit, such as liters or cubic feet.

A common meter is the *disk meter* or *wobble meter* used on many domestic water-distribution systems. The disk oscillates in a passageway so that a known volume of fluid moves through the meter for each oscillation. A stem normal to the disk operates a gear train, which in turn operates a counter. In good condition, these meters are accurate to within 1 percent. When they are worn, the error may be very large for small flows, such as those caused by a leaky faucet.

The flow of natural gas at low pressure is usually measured by a volumetric meter with a traveling partition. The partition is displaced by gas inflow to one end of the chamber in which it operates, and then, by a change in valving, it is displaced to the opposite end of the chamber. The oscillations operate a counting mechanism.

Oil flow or high-pressure gas flow in a pipeline is frequently measured by a rotary meter, in which cups or vanes move about an annular opening and displace a fixed volume of fluid for each revolution. Radial or axial pistons may be so arranged that the volume of continuous flow through them is determined by rotations of a shaft.

Positive-displacement meters normally have no timing equipment to measure the rate of flow. The rate of steady flow can be determined by recording the time for displacement of a given volume of fluid.

Electromagnetic Current Meters

Electromagnetic current meter (ECM) devices depend on being able to relate the distortion of a magnetic field to the turbulent velocity flowing over a sensing head possessing a regular geometry. Figure 10.9 contains a picture of a spherical element meter by Marsh-McBirney. In marine applications, the devices developed in England at the Institute of Oceanographic Sciences (the Colebrooke sensor), and based on the work of Thorpe et al. [5], continue to be frequently used, e.g., Heathershaw [6] and Soulsby [7]. The Marsh-McBirney device has a spherical sensor with a diameter of 4.0 cm, while a disk-shaped sensor with a 10-cm diameter is used in the Colebrooke sensor.

The exhaustive series of tests by Aubrey and Trowbridge [8] indicate that for pure steady flow, pre- and postdeployment calibration were necessary due to zero drift and offset problems from biological fouling. Further, a two-linear segment relation between voltage and velocity was necesary in the calibration. With these in

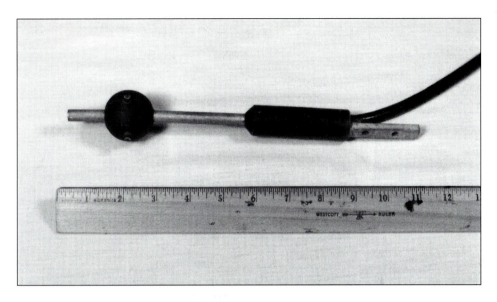

Figure 10.9 Electromagnetic current meter; scale in inches.

mind, the following results are noted:

1. Sampling in current-dominated flows is quite satisfactory with rms errors on the order of 1 to 5 cm/sec.

2. The response to wave-like or oscillatory conditions was also determined to be satisfactory with rms errors of 1 to 2 cm/sec.

3. Combined wave and current flow results were much more sensitive to flow regime, the cosine direction response developed a bump at lower Reynolds numbers, and rms errors had a range of 1 to 5 cm/sec.

In attempting to use these devices for estimates of turbulent boundary layer parameters such as u_*, adequate results may be achieved when free stream turbulence is not large relative to the mean velocity magnitude. The sensor head (approximately 13 cm) size precludes sensing any closer than three diameters away from a boundary which suggests that boundary layer applications must be done with caution.

Finally, the existing ECMs only measure two orthogonal components; therefore, two current meters must be placed orthogonal to each other and in close proximity to each other if full three-dimensional velocity data are required. In such a case the sensors' magnetic fields might interact, suggesting that larger instrument separations are necessary. However, increasing the separation between heads effectively precludes the measurements from being considered to have been obtained at the same point. Full three-dimensional data at a point are quite difficult to resolve with these devices.

Hot-Wire and Hot-Film Anemometers

Long used in laboratory examinations for turbulent flows (i.e., since Ludwig [9]), the heat transfer characteristics of the small thin-wire or film-coated probe (Fig. 10.10)

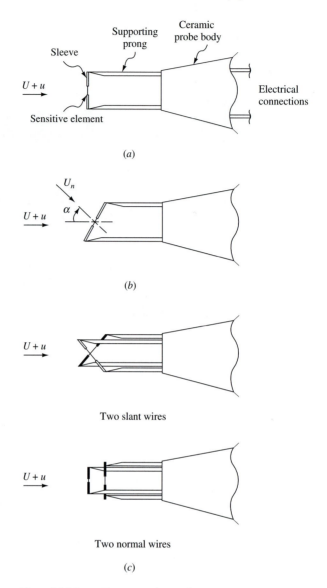

Figure 10.10   Hot-wire probe configurations.

can be precisely related to the local turbulent flow over the probe. The deployment of hot-wire anemometers is primarily confined to wind tunnels due to the fragile nature of the heated wire. The marine use of hot-film anemometers was first implemented by Gust [10] with sensors of quartz-coated platinum film developed to provide protection against damage from sediment particle impact. Collection frequencies of 30 Hz are routinely possible but 3–4 Hz frequencies are all that are necessary to resolve 90 percent of the kinetic energy. Data are reported with 0.01 cm/sec accuracy. Gust and Weatherly [11] suggest that the sensors cannot be used closer than 5 mm to a boundary, but this places it closer than most of the other reviewed probes and effectively provides the possibility of data collection in complex boundary layers. The extremely small dimensions of the wire and the thin nature of the operating film often give rise to concerns about ruggedness in heavily sediment-laden flows. At this time such concerns have not been completely resolved.

Laser Doppler Velocimetry

The laser doppler velocimeter (LDV) technology has been well used in laboratory turbulence studies since the work of Yeh and Cumins [12] with reviews of laboratory applications found in Buchave et al. [13] or the monograph by Drain [14]. The towed system of Veth [15] represents one of the first implementations of this technology for marine sampling. A considerable body of work on the development of a laser ocean-deployment system is due to Agrawal and his coworkers [16–19]. It is believed that this is the first autonomous, submersible, free-standing laser doppler velocimeter available.

The operating principle (Fig. 10.11) depends upon the presence of natural particles or other seeded scatterers traveling with the fluid velocity in the flow field. When the scatterers pass through the intersection of two orthogonal coherent beams of laser light, the resulting backscattered light is frequency modulated in proportion to the velocity. The device may be constructed in either backscatter or forwardscatter fashion, but present systems favor the backscatter form, particularly due to its lower physical profile and reduced flow field disruption. Agrawal and Belting [18] report the capability of undisturbed measurements as close as 3 cm to a boundary. The device is noninvasive, possesses favorable linearity over a very wide response range, and is quite stable over a variety of environmental conditions. The devices do not work well when low concentrations of scatterers exist.

Acoustic Devices

With the introduction of inexpensive acoustic transducers and the very precise knowledge of the speed of sound, the last decade has spawned a new generation of inexpensive, durable, and accurate acoustic flow meter devices. In a still fluid the

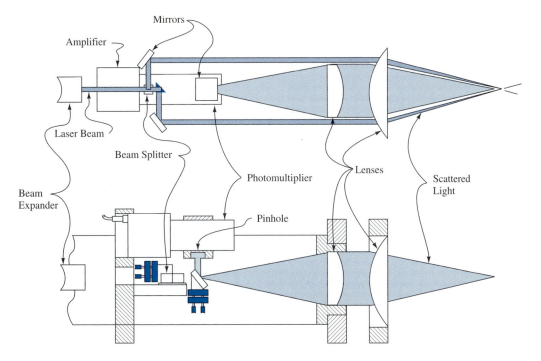

Figure 10.11   Laser doppler velocimeter.

speed of sound is known precisely as a function of temperature and salinity. The time of travel of sound between two points separated by a distance L in freshwater is $t_o = L/C_s$, where C_s is the speed of sound. For freshwater, $C_s = 1500$ m/sec. If the fluid velocity is moving with magnitude V, then the time of flight is given by (using a Taylor's series expansion)

$$t = \frac{L}{C_s + V} \approx \frac{L}{C_s}\left(1 - \frac{V}{C_s}\right) \tag{10.5.8}$$

The velocity can be found from

$$\Delta t = t_0 - t = \frac{LV}{C_s^2} \tag{10.5.9}$$

Therefore, by measuring the time of flight of an acoustic pulse across a flow field sampling volume and comparing that time to the transit time across the same volume in a still fluid, the velocity can be determined. With the sampling volumes being small and C_s^2 being large, the resulting estimates for V are quite sensitive to small errors. To address this a forward-backward system (Fig. 10.12) is employed which senses a frequency difference, Δf, that can be directly related to velocity. This system cancels out the dependence on C_s and reduces error, that is,

$$\Delta f = \frac{2V \cos \theta}{L}$$

where θ is the angle between the direction of acoustic beam propagation and the direction of flow velocity, V. θ is well known *a priori* for conduit flows, and a variety of embedded or clamp-on ultrasonic flowmeters have been available for conduits for some time.

Systems based upon this acoustic sampling concept are now being used to measure average river channel velocities in control sections of the river with well-behaved geometries.

For field installations, the direction of flow is unknown as it continuously varies. Therefore, the capability of measuring full three-dimensional velocities is important. Figure 10.13 is a schematic of an acoustic current meter (Williams et al. [20]) which measures the three-dimensional turbulent velocity field by sensing the instantaneous frequency shifts across a series of four orthogonal ray paths in the sampling volume. The transducers are 5-MHz acoustic devices that allow sampling frequencies as high as 10 Hz with a precision of ± 0.01 cm/sec.

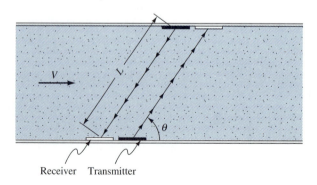

Figure 10.12   Schematic of conduit ultrasonic flowmeter.

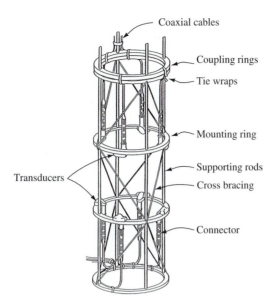

Figure 10.13   Acoustic current meter.

Another field device used in both atmospheric and marine installations is the acoustic doppler current profiler. Instead of a measurement at one point a whole profile of three-dimensional currents is sampled from one central transducer and processor placed at the end of the profile. Due to an optimization of resolution, acoustic frequency and resolution requirements these devices can sample only currents, and not the turbulent characteristics. These devices are quite cost-effective on a cost per data returned basis, are noninvasive, and are quite reliable because of the acoustic (nonmechanical) basis.

EXERCISES

10.5.1   The simple pitot tube measures the (*a*) static pressure; (*b*) dynamic pressure; (*c*) total pressure; (*d*) velocity at the stagnation point; (*e*) difference in total and dynamic pressure.

10.5.2   A pitot-static tube ($C = 1$) is used to measure air speeds. With water in the differential manometer and a gage difference of 3 in., the air speed for $\gamma = 0.0624$ lb/ft^3, in feet per second, is (*a*) 4.01; (*b*) 15.8; (*c*) 24.06; (*d*) 127; (*e*) none of these.

10.5.3   The pitot-static tube measures the (*a*) static pressure; (*b*) dynamic pressure; (*c*) total pressure; (*d*) difference in static and dynamic pressure; (*e*) difference in total and dynamic pressure.

10.5.4   The velocity of a known flowing gas may be determined from measurement of (*a*) static and stagnation pressure only; (*b*) static pressure and temperature only; (*c*) static and stagnation temperature only; (*d*) stagnation temperature and stagnation pressure only; (*e*) none of these answers.

10.5.5   The hot-wire anemometer is used to measure (*a*) pressure in gases; (*b*) pressure in liquids; (*c*) wind velocities at airports; (*d*) gas velocities; (*e*) liquid discharges.

10.5.6   A piston-type displacement meter has a volume displacement of 35 cm^3 per revolution of its shaft. The discharge, in liters per minute, for 1000 rpm is (*a*) 1.87; (*b*) 4.6; (*c*) 35; (*d*) 40.34; (*e*) none of these.

10.5.7   Velocity in a pipe or conduit system can be measured acoustically by knowing the pipe geometry and (*a*) the water temperature and speed of sound; (*b*) the direction of acoustic beam propagation; (*c*) the separation distance between the transducers; (*d*) the use of two transducers; (*e*) all of the above.

10.6   RATE DEVICES: ORIFICES

A *rate meter* is a device that determines, generally by a single measurement, the quantity (weight or volume) per unit time that passes a given cross section. Included among rate meters are the orifice, nozzle, venturi meter, rotometer, and weir. The orifice is discussed in this section. The venturi meter, nozzle, and some other closed-conduit devices are discussed in Sec. 10.7, and the free surface counterpart, the weir, is discussed in Sec. 10.8.

Orifice in a Reservoir

An orifice can be used for measuring the rate of flow out of a reservoir or through a pipe. An orifice in a reservoir or tank may be in the wall or in the bottom. It is an opening, usually round, through which the fluid flows, as in Fig. 10.14. It may be square-edged, as shown, or rounded, as in Fig. 3.13. The area of the orifice is the area of the opening. With the square-edged orifice, the fluid jet contracts during a short distance of about one-half diameter downstream from the opening. The portion of the flow that approaches along the wall cannot make a right-angled turn at the opening and therefore maintains a radial velocity component that reduces the jet area. The cross section where the contraction is greatest is called *vena contracta*. The streamlines are parallel throughout the jet at this section, and the pressure is atmospheric.

The head *H* on the orifice is measured from the center of the orifice to the free surface. The head is assumed to be held constant. Bernoulli's equation applied from

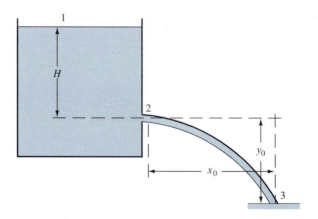

Figure 10.14   Orifice in a reservoir.

a point 1 on the free surface to the center of the vena contracta, point 2, with local atmospheric pressure as datum and point 2 as elevation datum, neglecting losses, is written as

$$\frac{V_1^2}{2g} + \frac{p_1}{\gamma} + z_1 = \frac{V_2^2}{2g} + \frac{p_2}{\gamma} + z_2$$

Inserting the values gives

$$0 + 0 + H = \frac{V_2^2}{2g} + 0 + 0$$

or

$$V_2 = \sqrt{2gH} \qquad (10.6.1)$$

This is only the *theoretical* velocity, because the losses between the two points were neglected. The ratio of the *actual* velocity V_a to the theoretical velocity V_t is called the *velocity coefficient* C_v, that is,

$$C_v = \frac{V_a}{V_t} \qquad (10.6.2)$$

and hence

$$V_{2a} = C_v\sqrt{2gH} \qquad (10.6.3)$$

The actual discharge Q_a from the orifice is the product of the actual velocity at the vena contracta and the area of the jet. The ratio of the jet area A_2 at the vena contracta to the area of orifice A_0 is symbolized by another coefficient, called the *coefficient of contraction* C_c, that is,

$$C_c = \frac{A_2}{A_0} \qquad (10.6.4)$$

The area at the vena contracta is $C_c A_0$. The actual discharge is thus

$$Q_a = C_v C_c A_0 \sqrt{2gH} \qquad (10.6.5)$$

It is customary to combine the two coefficients into a *discharge coefficient* C_d as

$$C_d = C_v C_c \qquad (10.6.6)$$

from which

$$Q_a = C_d A_0 \sqrt{2gH} \qquad (10.6.7)$$

There is no way to compute the losses between points 1 and 2; hence, C_v must be determined experimentally. It varies from 0.95 to 0.99 for the square-edged or rounded orifice. For most orifices, such as the square-edged one, the amount of contraction cannot be computed and test results must be used. There are several methods for obtaining one or more of the coefficients. By measuring area A_0, the head H, and the discharge Q_a (by gravimetric or volumetric means), C_d is obtained from Eq. (10.6.7). Determination of either C_v or C_c then permits determination of the other by Eq. (10.6.6).

Trajectory Method By measuring the position of a point on the trajectory of the free jet downstream from the vena contracta (Fig. 10.14), the actual velocity V_a can be determined if air resistance is neglected. The x component of velocity does not change; therefore, $V_a t = x_0$, in which t is the time for a fluid particle to travel from the vena contracta to point 3. The time for a particle to drop a distance y_0 under the action of gravity when it has no initial velocity in that direction is expressed by $y_0 = gt^2/2$. After t is eliminated in the two relations,

$$V_a = \frac{x_0}{\sqrt{2y_0/g}}$$

With V_{2t} determined by Eq. (10.6.1), the ratio $V_a/V_t = C_v$ is known.

Direct Measuring of V_a   With a pitot tube placed at the vena contracta, the actual velocity V_a is determined.

Direct Measuring of Jet Diameter   With outside calipers, the diameter of the jet at the vena contracta can be approximated. This is not a precise measurement and in general is less satisfactory than the other methods.

Use of the Momentum Equation   When the reservoir is small enough to be suspended on knife-edges, as in Fig. 10.15, it is possible to determine the force F that creates the momentum in the jet. With the orifice opening closed, the tank is leveled by adding or subtracting weights. With the orifice discharging, a force creates the momentum in the jet and an equal and opposite force F' acts against the tank. By addition of sufficient weights W, the tank is again leveled. From the figure, $F' = Wx_0/y_0$. With the momentum equation,

$$\Sigma F_x = \frac{Q\gamma}{g}(V_{x_{\text{out}}} - V_{x_{\text{in}}}) \qquad \text{or} \qquad \frac{Wx_0}{y_0} = \frac{Q_a\gamma V_a}{g}$$

since $V_{x_{\text{in}}}$ is zero and V_a is the final velocity. Since the actual discharge is measured, V_a is the only unknown in the equation.

Losses in Orifice Flow

The head loss in flow through an orifice is determined by applying the energy equation with a loss term for the distance between points 1 and 2 (Fig. 10.14),

$$\frac{V_{1a}^2}{2g} + \frac{p_1}{\gamma} + z_1 = \frac{V_{2a}^2}{2g} + \frac{p_2}{\gamma} + z_2 + \text{losses}$$

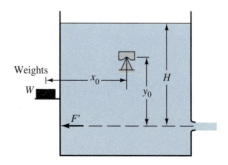

Figure 10.15   Momentum method for determination of C_v and C_c.

Substituting the values for this case gives

$$\text{Losses} = H - \frac{V_{2a}^2}{2g} = H(1 - C_v^2) = \frac{V_{2a}^2}{2g}\left(\frac{1}{C_v^2} - 1\right) \qquad \text{(10.6.8)}$$

in which Eq. (10.6.3) has been used to obtain the losses in terms of H and C_v or in terms of V_{2a} and C_v.

Example 10.1

A 75-mm-diameter orifice under a head of 4.88 m discharges 8900-N water in 32.6 s. The trajectory was determined by measuring $x_0 = 4.76$ m for a drop of 1.22 m. Determine C_v, C_c, C_d, the head loss per unit weight, and the power loss.

Solution

The theoretical velocity V_{2t} is

$$V_{2t} = \sqrt{2gH} = \sqrt{2(9.806)(4.88)} = 9.783 \text{ m/s}$$

The actual velocity is determined from the trajectory. The time to drop 1.22 m is

$$t = \sqrt{\frac{2y_0}{g}} = \sqrt{\frac{2(1.22)}{9.806}} = 0.499 \text{ s}$$

and the velocity is expressed by

$$x_0 = V_{2a}t \qquad V_{2a} = \frac{4.76}{0.499} = 9.539 \text{ m/s}$$

Then

$$C_v = \frac{V_{2a}}{V_{2t}} = \frac{9.539}{9.783} = 0.975$$

The actual discharge Q_a is

$$Q_a = \frac{8900}{9806(32.6)} = 0.0278 \text{ m}^3/\text{s}$$

With Eq. (10.6.7)

$$C_d = \frac{Q_a}{A_0\sqrt{2gH}} = \frac{0.0278}{\pi(0.0375^2)\sqrt{2(9.806)(4.88)}} = 0.643$$

Hence, from Eq. (10.6.6),

$$C_c = \frac{C_d}{C_v} = \frac{0.643}{0.975} = 0.659$$

The head loss, from Eq. (10.6.8), is

$$\text{Loss} = H(1 - C_v^2) = 4.88(1 - 0.975^2) = 0.241 \text{ m·N/N}$$

The power loss is

$$Q\gamma(\text{loss}) = 0.0278(9806)(0.241) = 65.7 \text{ W}$$

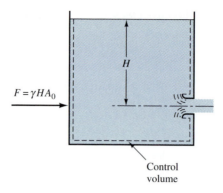

Figure 10.16   The Borda mouthpiece.

The use of the Borda mouthpiece (Fig. 10.16), a short, thin-walled tube about one diameter long that projects into the reservoir (reentrant), permits the application of the momentum equation to derive a relation between C_v and C_d. The velocity along the wall of the tank is almost zero at all points; hence, the pressure distribution is hydrostatic. With the component of force exerted on the liquid by the tank parallel to the axis of the tube, there is an unbalanced force due to the opening, which is $\gamma H A_0$. The final velocity is V_{2a}, the initial velocity is zero, and Q_a is the actual discharge. Then

$$\gamma H A_0 = Q_a \frac{\gamma}{g} V_{2a}$$

and

$$Q_a = C_d A_0 \sqrt{2gH} \qquad V_{2a} = C_v \sqrt{2gH}$$

Substituting for Q_a and V_{2a} and simplifying leads to

$$1 = 2 C_d C_v = 2 C_v^2 C_c$$

Orifice in a Pipe

The square-edged orifice in a pipe (Fig. 10.17) causes a contraction of the jet downstream from the orifice opening. For incompressible flow, Bernoulli's equation

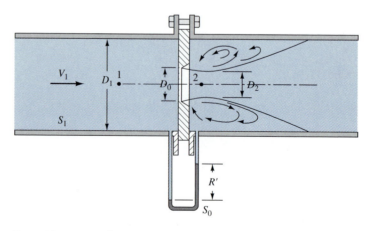

Figure 10.17   Orifice in a pipe.

applied from section 1 to the jet at its vena contracta, section 2, is

$$\frac{V_{1t}^2}{2g} + \frac{p_1}{\gamma} = \frac{V_{2t}^2}{2g} + \frac{p_2}{\gamma}$$

The continuity equation relates V_{1t} and V_{2t} with the contraction coefficient $C_c = A_2/A_0$ as

$$V_{1t}\frac{\pi D_1^2}{4} = V_{2t}C_c\frac{\pi D_0^2}{4} \tag{10.6.9}$$

After eliminating V_{1t},

$$\frac{V_{2t}^2}{2g}\left[1 - C_c^2\left(\frac{D_0}{D_1}\right)^4\right] = \frac{p_1 - p_2}{\gamma}$$

and solving for V_{2t}, the result is

$$V_{2t} = \sqrt{\frac{2g(p_1 - p_2)/\gamma}{1 - C_c^2(D_0/D_1)^4}}$$

Multiplying by C_v to obtain the actual velocity at the vena contracta gives

$$V_{2a} = C_v\sqrt{\frac{2(p_1 - p_2)/\rho}{1 - C_c^2(D_0/D_1)^4}}$$

and, finally, multiplying by the area of the jet, C_cA_0, produces the actual discharge Q as

$$Q = C_dA_0\sqrt{\frac{2(p_1 - p_2)/\rho}{1 - C_c^2(D_0/D_1)^4}} \tag{10.6.10}$$

in which $C_d = C_vC_c$. In terms of the gage difference R', Eq. (10.6.10) becomes

$$Q = C_dA_0\sqrt{\frac{2gR'(S_0/S_1 - 1)}{1 - C_c^2(D_0/D_1)^4}} \tag{10.6.11}$$

Because of the difficulty in determining the two coefficients separately, a simplified formula is generally used,

$$Q = CA_0\sqrt{\frac{2\Delta p}{\rho}} \tag{10.6.12}$$

or its equivalent,

$$Q = CA_0\sqrt{2gR'\left(\frac{S_0}{S_1} - 1\right)} \tag{10.6.13}$$

Values of C are given in Fig. 10.18 for the orifice.

Unsteady Orifice Flow from Reservoirs

In the orifice situations considered, the liquid surface in the reservoir has been assumed to be held constant. An unsteady-flow case of some practical interest exists

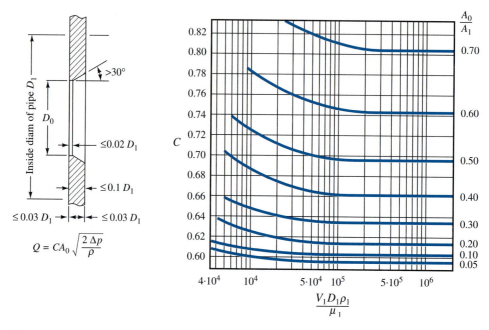

Figure 10.18   VDI orifice and discharge coefficients (*Ref. 11 in NACA Tech. Mem. 952*).

when determining the time to lower the reservoir surface a given distance. Theoretically, Bernoulli's equation applies only to steady flow, but if the reservoir surface drops slowly enough, the error from using Bernoulli's equation is negligible. The volume discharged from the orifice in time δt is $Q\,\delta t$, which must just equal the reduction in volume in the reservoir in the same time increment (Fig. 10.19), $A_R(-\delta y)$, in which A_R is the area of the liquid surface at height y above the orifice. Equating the two expressions gives

$$Q\,\delta t = -A_R\,\delta y$$

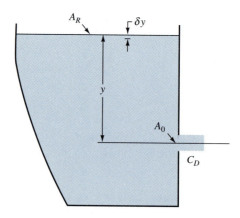

Figure 10.19   Notation for falling head.

Solving for δt and integrating between the limits $y = y_1, t = 0$ and $y = y_2, t = t$ yields

$$t = \int_0^t dt = -\int_{y_1}^{y_2} \frac{A_R\, dy}{Q}$$

The orifice discharge Q is $C_d A_0 \sqrt{2gy}$. After substitution for Q,

$$t = -\frac{1}{C_d A_0 \sqrt{2g}} \int_{y_1}^{y_2} A_R y^{-1/2} dy$$

When A_R is known as a function of y, the integral can be evaluated. Consistent with other SI or USC units, t is in seconds. For the special case of a tank with constant cross section,

$$t = -\frac{A_R}{C_d A_0 \sqrt{2g}} \int_{y_1}^{y_2} y^{-1/2} dy = \frac{2A_R}{C_d A_0 \sqrt{2g}}(\sqrt{y_1} - \sqrt{y_2})$$

Example 10.2

A tank has a horizontal cross-sectional area of 2 m² at the elevation of the orifice, and the area varies linearly with elevation so that it is 1 m² at a horizontal cross section 3 m above the orifice. For a = 100-mm-diameter orifice, $C_d = 0.65$, compute the time, in seconds, to lower the surface from 2.5 to 1 m above the orifice.

Solution

$$A_R = 2 - \frac{y}{3} \text{ m}^2$$

and

$$t = -\frac{1}{0.65\pi(0.05^2)\sqrt{2(9.806)}} \int_{2.5}^{1} \left(2 - \frac{y}{3}\right) y^{-1/2} dy = 73.8 \text{ s}$$

Example 10.3

A reservoir of variable area is drained by a 150-mm-diameter short pipe with a valve attached. The valve is being adjusted so that the loss (in velocity heads) for the piping system is

$$K = 1.5 + 0.04t + 0.0001t^2$$

with t in seconds. The reservoir area is given by

$$A = 4 + 0.1y + 0.01y^2 \text{ m}^2$$

where y is the elevation of the reservoir surface above the centerline of the valve. If $y = 20$ m at $t = 0$, determine y, A, K, and the discharge Q for 300 s.

Solution

$$Q\, dt = -A\, dy \qquad\qquad y + 0 + 0 = \frac{KV^2}{2g} = \frac{KQ^2}{2gA_0^2}$$

$$dy = -\frac{Q}{A} dt \qquad\qquad Q = A_0 \sqrt{\frac{2gy}{K}}$$

$$dy = -\frac{A_0}{A}\sqrt{\frac{2gy}{K}} dt$$

A spreadsheet may be used to solve the differential equation. If second-order Runge Kutta (see Appendix B and the Web page) is used with $dt = 7.5$ s, the following results are obtained.

t	y	Q	A	K
0	20.000	0.286	10.000	1.500
105	17.982	0.127	9.032	6.803
210	16.746	0.085	8.479	14.310
300	15.939	0.066	8.134	22.500

EXERCISES

10.6.1 The actual velocity at the vena contracta for flow through an orifice from a reservoir is expressed by (a) $C_v\sqrt{2gH}$; (b) $C_c\sqrt{2gH}$; (c) $C_d\sqrt{2gH}$; (d) $\sqrt{2gH}$; (e) $C_v V_a$.

10.6.2 A fluid jet discharging from a 20-mm-diameter orifice has a diameter 17.5 mm at its vena contracta. The coefficient of contraction is (a) 1.31; (b) 1.14; (c) 0.875; (d) 0.766; (e) none of these answers.

10.6.3 The ratio of actual discharge to theoretical discharge through an orifice is

(a) $C_c C_v$ (b) $C_c C_d$ (c) $C_v C_d$ (d) $\dfrac{C_d}{C_v}$ (e) $\dfrac{C_d}{C_c}$

10.6.4 The losses in orifice flow are (a) $\dfrac{1}{C_c^2}\left(\dfrac{V_{2a}^2}{2g} - 1\right)$ (b) $\dfrac{V_{2t}^2}{2g} - \dfrac{V_{2a}^2}{2g}$

(c) $H(C_v^2 - 1)$ (d) $H - \dfrac{V_{2t}^2}{2g}$ (e) none of these answers

10.6.5 For a liquid surface to lower at a constant rate, the area of reservoir A_R must vary with head y on the orifice, as (a) $\sqrt{y}$; (b) y; (c) $1/\sqrt{y}$; (d) $1/y$; (e) none of these answers.

10.6.6 A 50-mm-diameter Borda mouthpiece discharges 7.68 L/s under a head of 3 m. The velocity coefficient is (a) 0.96; (b) 0.97; (c) 0.98; (d) 0.99; (e) none of these answers.

10.7   VENTURI METER, NOZZLE, AND OTHER CONDUIT RATE MEASURING DEVICES

Venturi Meter

The venturi meter is used to measure the rate of flow in a pipe. It is generally a casting (Fig. 10.20) consisting of (1) an upstream portion, which is the same size as the pipe, has a bronze liner, and contains a piezometer ring for measuring static pressure; (2) a converging conical region; (3) a cylindrical throat with a bronze liner containing a piezometer ring; and (4) a gradually diverging conical region leading to a cylindrical section the size of the pipe. A differential manometer is attached to the two piezometer rings. The size of a venturi meter is specified by the pipe

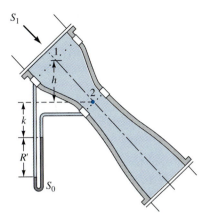

Figure 10.20 Venturi meter.

and throat diameter, for example, a 6- by 4-in. venturi meter fits a 6-in.-diameter pipe and has a 4-in.-diameter throat. For accurate results the venturi meter should be preceded by at least 10 diameters of straight pipe. In the flow from the pipe to the throat, the velocity is greatly increased and the pressure correspondingly decreased. The amount of discharge in incompressible flow is shown to be a function of the manometer reading.

The pressures at the upstream section and throat are *actual pressures,* and the velocities from Bernoulli's equation are *theoretical velocities.* When losses are considered in the energy equation, the velocities are *actual velocities.* First, with the Bernoulli equation (i.e., without a head-loss term) the theoretical velocity at the throat is obtained. Multiplying this by the velocity coefficient, C_v, gives the actual velocity. Then the actual velocity times the actual area of the throat determines the actual discharge. From Fig. 10.20

$$\frac{V_{1t}^2}{2g} + \frac{p_1}{\gamma} + h = \frac{V_{2t}^2}{2g} + \frac{p_2}{\gamma} \tag{10.7.1}$$

in which elevation datum is taken through point 2. V_1 and V_2 are average velocities at sections 1 and 2, respectively; hence, α_1 and α_2 are assumed to be unity. With the continuity equation $V_1 D_1^2 = V_2 D_2^2$,

$$\frac{V_1^2}{2g} = \frac{V_2^2}{2g}\left(\frac{D_2}{D_1}\right)^4 \tag{10.7.2}$$

which holds for either the actual velocities or the theoretical velocities. Equation (10.7.1) can be solved for V_{2t} as

$$\frac{V_{2t}^2}{2g}\left[1 - \left(\frac{D_2}{D_1}\right)^4\right] = \frac{p_1 - p_2}{\gamma} + h$$

and

$$V_{2t} = \sqrt{\frac{2g[h + (p_1 - p_2)/\gamma]}{1 - (D_2/D_1)^4}} \tag{10.7.3}$$

Introducing the velocity coefficient $V_{2a} = C_v V_{2t}$ gives

$$V_{2a} = C_v \sqrt{\frac{2g[h + (p_1 - p_2)/\gamma]}{1 - (D_2/D_1)^4}}$$

(10.7.4)

After multiplying by A_2, the actual discharge Q is determined to be

$$Q = C_v A_2 \sqrt{\frac{2g[h + (p_1 - p_2)/\gamma]}{1 - (D_2/D_1)^4}}$$

(10.7.5)

The gage difference R' can now be related to the pressure difference by writing the equation for the manometer. In units of length of water where S_1 is the specific gravity of flowing fluid and S_0 the specific gravity of manometer liquid

$$\frac{p_1}{\gamma}S_1 + (h + k + R')S_1 - R'S_0 - kS_1 = \frac{p_2}{\gamma}S_1$$

Simplifying gives

$$h + \frac{p_1 - p_2}{\gamma} = R'\left(\frac{S_0}{S_1} - 1\right)$$

(10.7.6)

By substituting into Eq. (10.7.5),

$$Q = C_v A_2 \sqrt{\frac{2gR'(S_0/S_1 - 1)}{1 - (D_2/D_1)^4}}$$

(10.7.7)

which is the venturi meter equation for incompressible flow. The contraction coefficient is unity; hence, $C_v = C_d$. It should be noted that h has dropped out of the equation. The discharge depends upon the gage difference R' regardless of the orientation of the venturi meter; whether it is horizontal, vertical, or inclined, exactly the same equation holds.

C_v is determined by calibration, that is, by measuring the discharge and the gage difference and solving for C_v, which is usually plotted against the Reynolds number. Experimental results for venturi meters are given in Fig. 10.21. They are

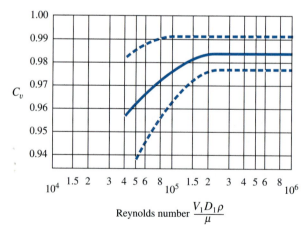

Figure 10.21   Coefficient C_v for venturi meters. *(Fluid Meters: Their Theory and Application, 6th ed., American Society of Mechanical Engineers, 1971).*

applicable to diameter ratios D_2/D_1 from 0.25 to 0.75 within the tolerances shown by the dotted lines. Where feasible, a venturi meter should be so selected that its coefficient is constant over the range of Reynolds numbers for which it is to be used.

The coefficient may be slightly greater than unity for venturi meters that are unusually smooth inside. This does not mean that there are no losses; it results from neglecting the kinetic-energy correction factors α_1 and α_2 in the Bernoulli equation. Generally, α_1 is greater than α_2, since the reducing region acts to make the velocity distribution uniform over section 2.

The venturi meter has a low overall loss owing to the gradually expanding conical region, which aids in reconverting the high kinetic energy at the throat into pressure energy. The loss is about 10 to 15 percent of the head change between sections 1 and 2.

Venturi Meter for Compressible Flow

The theoretical flow of a compressible fluid through a venturi meter is substantially isentropic and is obtained from Eqs. (4.3.4) and (4.5.9), and the expression $\rho v^k =$ constant. When multiplied by C_v, the velocity coefficient, the mass flow rate is

$$\dot{m} = C_v A_2 \sqrt{\frac{[2k/(k-1)]\, p_1 \rho_1\, (p_2/p_1)^{2/k}\, [1 - (p_2/p_1)^{(k-1)/k}]}{1 - (p_2/p_1)^{2/k}\, (A_2/A_1)^2}} \qquad \textbf{(10.7.8)}$$

The velocity coefficient is the same as for liquid flow.

Flow Nozzle

The ISA (Instrument Society of America) flow nozzle (originally the VDI flow nozzle) is shown in Fig. 10.22. It has no contraction of the jet other than that of the nozzle opening; therefore, the contraction coefficient is unity.

Equations (10.7.5) and (10.7.7) hold equally well for the flow nozzle. For a horizontal pipe ($h = 0$), Eq. (10.7.5) can be written

$$Q = C A_2 \sqrt{\frac{2\Delta p}{\rho}} \qquad \textbf{(10.7.9)}$$

in which

$$C = \frac{C_v}{\sqrt{1 - (D_2/D_1)^4}} \qquad \textbf{(10.7.10)}$$

and $\Delta p = p_1 - p_2$. The value of coefficient C in Fig. 10.22 is used in Eq. (10.7.9). When the coefficient given in the figure is to be used, it is important that the dimensions shown be closely adhered to, particularly in the location of the piezometer openings (two methods are shown) for measuring pressure drop. At least 10 diameters of straight pipe should precede the nozzle.

The flow nozzle costs less than the venturi meter. It has the disadvantage that the overall losses are much higher because of the lack of guidance of the jet downstream from the nozzle opening.

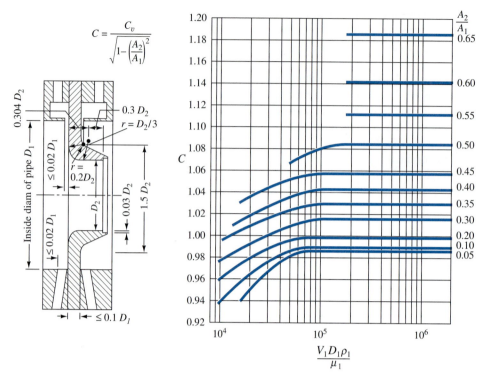

Figure 10.22 ISA (VDI) flow nozzle and discharge coefficient. *(Ref. 11 in NACA Tech. Mem. 952).*

Example 10.4. Determine the flow through a 6-in.-diameter waterline that contains a 4-in.-diameter flow nozzle. The mercury-water differential manometer has a gage difference of 10 in. The water temperature is 60°F.

Solution

From the data given, $S_0 = 13.6$, $S_1 = 1.0$, $R' = 10/12 = 0.833$ ft, $A_2 = \pi/36 = 0.0873$ ft^2, $\rho = 1.938$ slugs/ft^3, and $\mu = 2.359 \times 10^{-5}$ lb·s/ft^2. Substituting Eq. (10.7.10) into Eq. (10.7.7) gives

$$Q = CA_2\sqrt{2gR'\left(\frac{S_0}{S_1} - 1\right)}$$

From Fig. 10.22 and for $A_2/A_1 = (4/6)^2 = 0.444$, assume that the horizontal region of the curves applies. Hence, $C = 1.056$. Then compute the flow and the Reynolds number as

$$Q = 1.056(0.0873)\sqrt{64.4(0.833)\left(\frac{13.6}{1.0} - 1.0\right)} = 2.40 \text{ cfs}$$

Then

$$V_1 = \frac{Q}{A_1} = \frac{2.40}{\pi/16} = 12.21 \text{ ft/s}$$

and

$$\mathbf{R} = \frac{V_1 D_1 \rho}{\mu} = \frac{12.21(1.938)}{2(2.359 \times 10^{-5})} = 502,000$$

The chart shows the value of C to be correct; therefore, the discharge is 2.40 cfs.

Elbow Meter

The elbow meter for incompressible flow is one of the simplest flow-rate measuring devices. Piezometer openings on the inside and on the outside of the elbow are connected to a differential manometer. Because of centrifugal force at the bend, the difference in pressures is related to the discharge. A straight calming length should precede the elbow, and for accurate results the meter should be calibrated in place [21]. As most pipelines have an elbow, it can be used as the meter. After calibration the results are as reliable as with a venturi meter or a flow nozzle.

Rotameter

The rotameter is a variable-area meter that consists of an enlarging transparent tube and a metering "float" (actually heavier than the liquid) that is displaced upward by the upward flow of fluid through the tube. The tube is graduated to read the flow directly. Notches in the float cause it to rotate and thus maintain a central position in the tube. The greater the flow, the higher the position the float assumes.

EXERCISES

10.7.1 Which of the following measuring instruments is a rate meter? (*a*) Current meter; (*b*) disk meter; (*c*) hot-wire anemometer; (*d*) pitot tube; (*e*) venturi meter.

10.7.2 The discharge coefficient for a 40- by 20-mm venturi meter at a Reynolds number of 200,000 is (*a*) 0.95; (*b*) 0.96; (*c*) 0.973; (*d*) 0.983; (*e*) 0.992.

10.7.3 Select the correct statement: (*a*) The discharge through a venturi meter depends upon Δp only and is independent of orientation of the meter. (*b*) A venturi meter with a given gage difference R' discharges at a greater rate when the flow is vertically downward through it than when the flow is vertically upward. (*c*) For a given pressure difference the equations show that the discharge of gas is greater through a venturi meter when compressibility is taken into account than when it is neglected. (*d*) The coefficient of contraction of a venturi meter is unity. (*e*) The overall loss is the same in a given pipeline whether a venturi meter or a nozzle with the same D_2 is used.

10.8   OPEN-CHANNEL RATE DEVICES

Weirs

A number of conduit obstruction-based rate meters have counterparts in free surface channel flow. For instance the Parshall flume is analogous to the venturi meter while

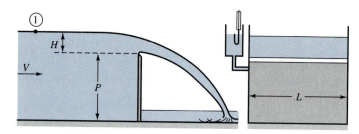

Figure 10.23   Sharp-crested rectangular weir.

weirs are analogous to orifices. A weir is an obstruction in the channel that causes the liquid to back up behind it and flow over it. By measuring the height of the upstream liquid surface, the rate of flow is determined. Weirs constructed from a sheet of metal or other material so that the jet, or *nappe,* springs free as it leaves the upstream face are called *sharp-crested weirs.* Other weirs, such as the *broad-crested weir,* support the flow in a longitudinal direction.

The sharp-crested rectangular weir (Fig. 10.23) has a horizontal crest. The nappe is contracted at top and bottom as shown. An equation for discharge can be derived if the contractions are neglected. Without contractions the flow appears as in Fig. 10.24. The nappe has parallel streamlines with atmospheric pressure throughout.

Bernoulli's equation applied between 1 and 2 (Fig 10.24) is

$$H + 0 + 0 = \frac{v^2}{2g} + H - y + 0$$

in which the velocity head at section 1 is neglected. Solving for v gives

$$v = \sqrt{2gy}$$

The theoretical discharge Q_t is

$$Q_t = \int v \, dA = \int_0^H vL \, dy = \sqrt{2g} \, L \int_0^H y^{1/2} \, dy = \frac{2}{3} \sqrt{2g} \, LH^{3/2}$$

in which L is the width of the weir. Experiment shows that the exponent of H is correct but the coefficient is too great. The contractions and losses reduce the actual discharge to about 62 percent of the theoretical, or

$$Q = \begin{cases} 3.33LH^{3/2} & \text{USC units} \\ 1.84LH^{3/2} & \text{SI units} \end{cases} \qquad (10.8.1)$$

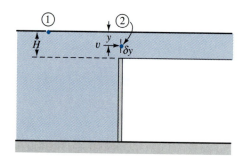

Figure 10.24   Weir nappe without contractions.

When the weir does not extend completely across the width of the channel, it has *end contractions,* which are illustrated in Fig. 10.25a. An empirical correction for the reduction of flow is accomplished by subtracting $0.1H$ from L for each end contraction. The weir in Fig 10.23 is said to have its end contractions *suppressed.*

The head H is measured upstream from the weir a sufficient distance to avoid the surface contraction. A hook gage mounted in a stilling pot connected to a piezometer opening determines the water-surface elevation from which the head is determined.

When the height P of a weir (Fig. 10.23) is small, the velocity head at 1 cannot be neglected. A correction may be added to the head,

$$Q = CL\left(H + \alpha\frac{V^2}{2g}\right)^{3/2} \tag{10.8.2}$$

in which V is velocity and α is greater than unity, usually taken as about 1.4, which accounts for the nonuniform velocity distribution. Equation (10.8.2) must be solved for Q by trial, since both Q and V are unknown. As a first trial, the term $\alpha V^2/2g$ can be neglected to approximate Q. With this trial discharge, a value of V is computed since

$$V = \frac{Q}{L(P + H)}$$

For small discharges the V-notch weir is particularly convenient. The contraction of the nappe is neglected, and the theoretical discharge is computed (Fig. 10.25b) as follows:

1. The velocity at depth y is $v = \sqrt{2gy}$, and the theoretical discharge is

$$Q_t = \int v\, dA = \int_0^H vx\, dy$$

2. By similar triangles x can be related to y

$$\frac{x}{H - y} = \frac{L}{H}$$

3. After substituting for v and x,

$$Q_t = \sqrt{2g}\frac{L}{H}\int_0^H y^{1/2}(H - y)\, dy = \frac{4}{15}\sqrt{2g}\frac{L}{H}H^{5/2}$$

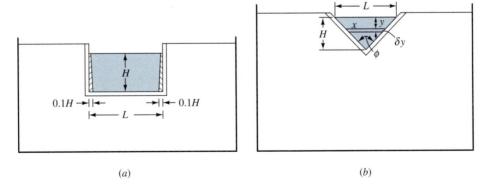

(a)	(b)

Figure 10.25 Weirs: (a) horizontal with end contractions and (b) V-notch weir.

4. Expressing L/H in terms of the angle ϕ of the V notch gives

$$\frac{L}{2H} = \tan\frac{\phi}{2}$$

Hence,

$$Q_t = \frac{8}{15}\sqrt{2g}\,\tan\frac{\phi}{2}H^{5/2}$$

5. The exponent in the equation is approximately correct, but the coefficient must be reduced by about 42 percent because of the neglected contractions. An approximate equation for a 90°-V-notch weir is

$$Q = \begin{cases} 2.50H^{2.50} & \text{USC units} \\ 1.38H^{2.50} & \text{SI units} \end{cases} \qquad \text{(10.8.3)}$$

Experiments show that the coefficient is increased by roughening the upstream side of the weir plate, which causes the boundary layer to grow thicker. The greater amount of slow-moving liquid near the wall is more easily turned, and hence there is less contraction of the nappe.

The broad-crested weir (Fig. 10.26a) supports the nappe so that the pressure variation is hydrostatic at section 2. Bernoulli's equation applied between points 1 and 2 can be used to find the velocity v_2 at height z, neglecting the velocity of approach, as

$$H + 0 + 0 = \frac{v_2^2}{2g} + z + (y - z)$$

In solving for v_2,

$$v_2 = \sqrt{2g(H - y)}$$

and z drops out; hence, v_2 is constant at section 2. For a weir of width L normal to the plane of the figure, the theoretical discharge is

$$Q = v_2 L y = Ly\sqrt{2g(H - y)} \qquad \text{(10.8.4)}$$

A plot of Q as abscissa against the depth y as ordinate, for constant H, is given in Fig. 10.26b. The depth is shown to be that which yields the maximum discharge by the following reasoning.

A gate or other obstruction placed at section 3 of Fig.10.26a can completely stop the flow by making $y = H$. If a small flow is permitted to pass section 3 (holding

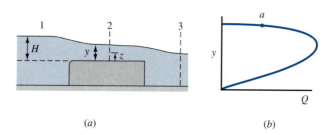

(a)   (b)

Figure 10.26   Broad-crested weir.

H constant), the depth y becomes a little less than H and the discharge is, for example, as shown by point a on the depth-discharge curve. By further lifting of the gate or obstruction at section 3, the discharge-depth relation follows the upper portion of the curve until the maximum discharge is reached. Any additional removal of downstream obstructions, however, has no effect upon the discharge, because the velocity of flow at section 2 is $\sqrt{gy}$, which is exactly the speed an elementary wave can travel in still liquid of depth y. Hence, the effect of any additional lowering of the downstream surface elevation cannot travel upstream to affect further the value of y, and the discharge occurs at the maximum value. This depth y is called the *critical depth* and is discussed in Sec. 13.5. The speed of an elementary wave is derived in Sec. 13.12.

By taking dQ/dy and with the result set equal to zero, for constant H,

$$\frac{dQ}{dy} = 0 = L\sqrt{2g(H-y)} + Ly\frac{1}{2}\frac{-2g}{\sqrt{2g(H-y)}}$$

Solving for y gives

$$y = \frac{2}{3}H$$

Inserting the value of H, that is, $3y/2$, into the equation for velocity v_2 gives

$$v_2 = \sqrt{gy}$$

and substituting the value of y into Eq. (10.8.4) leads to

$$Q_t = \begin{cases} 3.09LH^{3/2} & \text{USC units} \\ 1.705LH^{3/2} & \text{SI units} \end{cases} \qquad \text{(10.8.5)}$$

Experiments show that, for a well-rounded upstream edge, the discharge is

$$Q = \begin{cases} 3.03LH^{3/2} & \text{USC units} \\ 1.67LH^{3/2} & \text{SI units} \end{cases} \qquad \text{(10.8.6)}$$

which is within 2 percent of the theoretical value. The flow, therefore, adjusts itself to discharge at the maximum rate.

Since viscosity and surface tension have a minor effect on the discharge coefficients of weirs, a weir should be calibrated with the liquid that it will measure.

Tests on a 60°-V-notch weir yield the following values of head H on the weir and discharge Q:

Example 10.5

H, ft	0.345	0.356	0.456	0.537	0.568	0.594	0.619	0.635	0.654	0.665
Q, cfs	0.107	0.110	0.205	0.303	0.350	0.400	0.435	0.460	0.490	0.520

Use the least squares algorithm to determine the constants in $Q = CH^m$ for this weir.

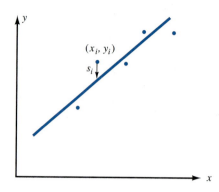

Figure 10.27   Log-log plot of Q versus H for V-notch weir.

Solution

Taking the logarithm of each side of the equation

$$\ln Q = \ln C + m \ln H \qquad \text{or} \qquad y = B + mx$$

it is noted that the best value of B and m are needed for a straight line through the data when plotted on log-log paper.

By the theory of least squares, the best straight line through the data points is the one yielding a minimum value of the sums of the squares of vertical displacements of each point from the line, or from Fig. 10.27,

$$F = \sum_{i=1}^{n} s_i^2 = \sum_{i=1}^{n} [y_i - (B + mx_i)]^2$$

where n is the number of experimental points. To minimize F, $\partial F/\partial B$ and $\partial F/\partial m$ are taken and set equal to zero, yielding two equations in the two unknowns B and m as

$$\frac{\partial F}{\partial B} = 0 = 2\Sigma[y_i - (B + mx_i)](-1)$$

from which

$$\Sigma y_i - nB - m\Sigma x_i = 0 \tag{1}$$

and

$$\frac{\partial F}{\partial m} = 0 = 2\Sigma[y_i - (B + mx_i)](-x_i)$$

or

$$\Sigma x_i y_i - B\Sigma x_i - m\Sigma x_i^2 = 0 \tag{2}$$

Solving Eqs. (1) and (2) for m gives

$$m = \frac{\Sigma x_i y_i/\Sigma x_i - \Sigma y_i/n}{\Sigma x_i^2/\Sigma x_i - \Sigma x_i/n} \qquad B = \frac{\Sigma y_i - m\Sigma x_i}{n}$$

These equations are readily solved by a calculator or a simple spreadsheet program. The answer for the data of this problem is $m = 2.437$ and $C = 1.395$.

Measurement of River Flow

Daily or hourly records of river discharge over long periods of time are essential for water resources planning or protection against floods. Repeated measurements of discharge by determining the velocity distribution over a cross section of the river is costly. To avoid the cost and still obtain daily records, *control sections* are established where the river channel is stable, that is, with little change in bottom or sides of the stream bed. The control section is frequently at a break in slope of the river bottom where it becomes steeper downstream.

The purpose of the control section is to establish a precise relationship between flow and elevation, a much more easily measured variable. The volume flow rate at a fixed cross section of the channel is found from

$$Q(t) = \int \mathbf{v} \cdot d\mathbf{A} = \int u(y, z, t)\, dA \qquad \textbf{(10.8.7)}$$

Here u is the velocity component in the x direction which is perpendicular to the cross section and directed downstream (Figure 10.28a). Therefore, the most direct data on $Q(t)$ would result from breaking the cross section up into a number of incremental

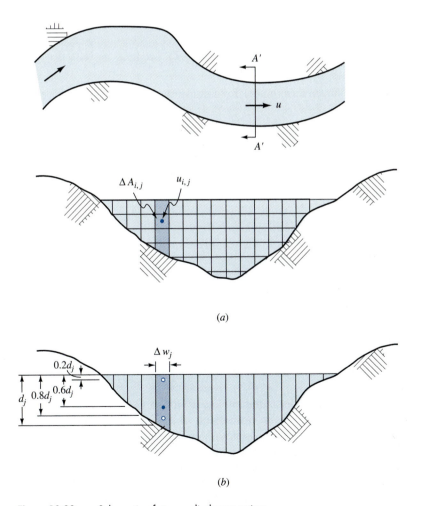

(a)

(b)

Figure 10.28   Schematic of stream discharge rating.

areas of size $\Delta A_{i,j}$, measuring the downchannel velocity, $u_{i,j}$, in each incremental area and summing. For example,

$$Q = \sum_{i,j} u_{i,j}\,\Delta A_{i,j}$$

Implicit in this prescription is that the $u_{i,j}$ in each area represents the cross-sectional average for each $\Delta A_{i,j}$. Further it is assumed that the time it takes the data collection team to collect the data over all of the cross sections is smaller than the time it takes for any appreciable changes in the flow rate to occur.

One can simplify the velocity data collection requirements considerably by taking advantage of the well-known open channel velocity profile and its relationship to the average velocity as detailed in Ex. 6.4. In this example it was learned that a velocity measurement at $z/d = 0.4$ ($z/d = 0.6$ if z is measured from the surface to the bottom) or the average of two velocity measurements at $z/d = 0.2$ and 0.8 would give the average velocity, $\bar{u}$. Therefore, the cross section (Figure 10.28b) is broken into equal-width water-column increments (Δw_j), the velocity is sampled at the two normalized depths within each increment, and the products summed as

$$Q = \sum_{j=1}^{M} dQ_j = \sum_{j=1}^{M} \bar{u}_j\,\Delta w_j\,d_j \tag{10.8.8}$$

$$\approx \sum_{j=1}^{M} \frac{1}{2}[u_j\,(z/d_j = 0.2) + u_j(z/d_j = 0.8)]\,\Delta w_j\,d_j$$

Using the one-point measurement for the average velocity gives the estimate for Q as

$$Q = \sum_{j=1}^{M} \bar{u}_j\,\Delta w_j\,d_j \approx \sum_{j=1}^{M} \bar{u}_j\left(\frac{z}{d_j} = 0.6\right)\Delta w_j\,d_j \tag{10.8.9}$$

Typically each incremental width is no more than 10 percent of the total cross-sectional width.

At the same time the flow is being measured, a staff gage or pressure transducer, mounted at the control section, measures the water column height or stage. Therefore, the flow measured by the sampling team is related to the height of the water during the sampling. When a stream has never been rated, pairs of flow and height data must be collected which span the entire stream discharge response from low flow to flood flow. These data are used to establish a stage-discharge curve. The stage discharge curve then allows rapid measurements of flow by simple recording of water level elevation, a very simple and economical measurement to make. Subsequent spot checks must be performed to determine whether the original calibration continues to hold. If a stable control section has been used, very little drift in the calibration will be observed. For unstable cross sections the calibration could change over a matter of days.

Errors in this inferred volume rate occur in two ways. The first is the standard instrument errors in the three measurements of u, Δw, and d. The second error is in the implied use of the numerical approximation of the integral in Eq. (10.8.7) by the trapezoidal integration method used in Eq. (10.8.8). The methods in Eqs. (10.1.2) and (10.1.3) are used to estimate the instrument error. Textbooks on numerical methods (e.g., [22]) indicate that the numerical approximation error is of the order $O(\Delta w^2)$ for equal-spaced width intervals. Therefore, more incremental water column slices in the cross section reduce the approximation error.

EXERCISES

10.8.1 The discharge through a V-notch weir varies as (*a*) $H^{-1/2}$; (*b*) $H^{1/2}$ (*c*) $H^{3/2}$; (*d*) $H^{5/2}$; (*e*) none of the answers.

10.8.2 The discharge of a rectangular sharp-crested weir with end contractions is less than for the same weir with end contractions suppressed by (*a*) 5 percent; (*b*) 10 percent; (*c*) 15 percent; (*d*) no fixed percent; (*e*) none of these answers.

10.8.3 In the segmentation of a river for a stage calibration, the initial width of each segment is 10 m long. Using a width of 5 m, a doubling of the number of cells (*a*) will lower the expected error; (*b*) will decrease the numerical integration error fourfold; (*c*) will increase the estimated error due to the increased number of measurements; (*d*) none of the above.

10.9   PARTICLE CONCENTRATION MEASUREMENTS

The possible combination of particulate or dissolved mass species being carried along by the moving fluid are endless, ranging from esoteric compounds occurring in a chemical engineering unit process to the simple transport of a raindrop or sediment particle. An exhaustive review of all the procedures for measuring concentrations of these species is prohibitive. Attention is directed here to a specialized subset of concentrations deriving from particles transported in the moving fluid. Even with this focus a complete review is not possible. A *particle* is defined here as a small but coherent mass comprised of another material or a mass of another phase of the fluid. Examples of the former include sand carried in rivers and streams or coal dust carried in the atmosphere. Examples of the latter include raindrops or clouds in the saturated atmosphere. In all cases it is likely that the individual particles comprising a sampled volume will have a random size distribution which, fortunately, tends towards a predictable distribution of probable sizes. Furthermore, it is also the case that each particle's mass will be greater than the mass of fluid it displaces and that the particles will settle or move relative to the bulk fluid motion. Therefore, the variables of most measurement interest are the concentration or mass per unit volume of the particles and the size distribution of the particles comprising the sample.

Particle Size

Potentially available equipment for making simultaneous measurements of both size and concentration variables using the same apparatus does not exist at this time. Rather, samples of the mixture whose concentration is being measured must be collected at the measurement point and subjected to size-class analysis by subsequent laboratory analysis.

The particle size distributions encountered in everyday activities span from 0.001 μm (one μm is one micron which is 10^{-6} m) to 10^4 μm or 1 cm. The smallest particles are typically coal and metallurgical dusts and viruses while the largest particles are gravel found in stream beds or quarries. Figure 10.29 [23] contains a map of the various types of particles, their size ranges, and their classification labels. The size chart is displayed in microns. A phi unit system is also used, which is a power-of-two-based system defined such that the diameter in phi units

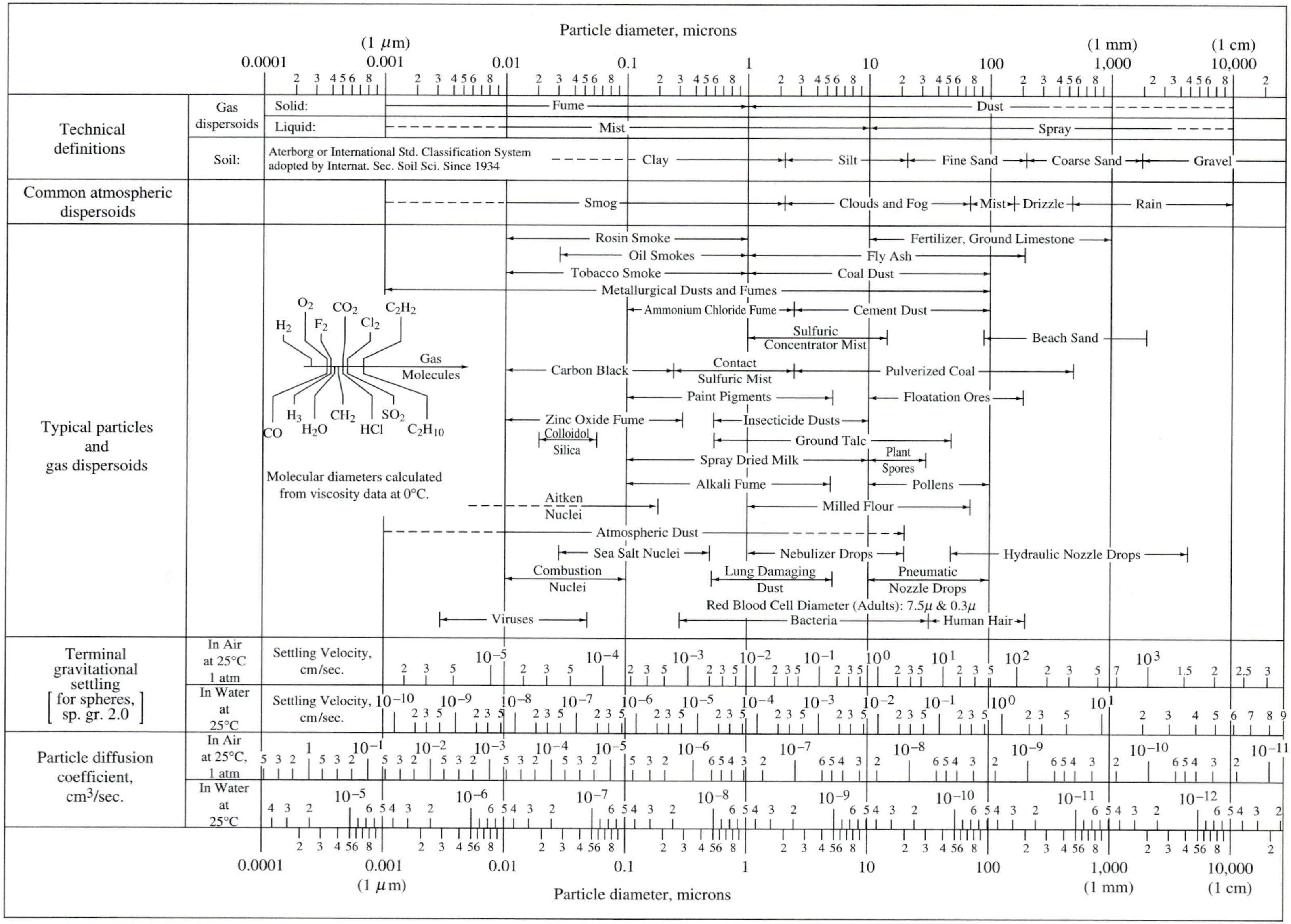

Figure 10.29 Map of various types of particles, their sizes, and classification [redrawn from Ref. 23].

(ϕ) is related to the diameter (d) in millimeters by

$$\phi = -\log_2 d \qquad (10.9.1)$$

It is also noted that most particles are very irregularly shaped and that the diameter measure above and in the figure refers to the diameter of a sphere of volume equivalent to the irregularly shaped particle. Figure 10.30 contains a map of various types of instrumentation available to make the particle size distribution measurements. As can readily be seen, there is considerable overlap in the size ranges that each can measure, and various handbooks of measurement methods (e.g., *The National Handbook of Recommended Methods for Water-Data Acquisition,* U.S. Geological Survey, 1977) suggest that consistency between size measurements made with multiple instrument types is highly recommended for high-quality data results. Time and expense limitations often prohibit such redundant measurements.

The operating principles upon which these devices are based is related to some particularly dominant physical manifestation of the particle motion or, as seen in other instruments, the description of an energy transmission or propagation. At the very smallest particle sizes the molecular diffusion of a particle is highly size dependent. Therefore, a number of particle size measurement systems are based upon methods for distinguishing the diffusion characteristics. At the largest scales the differential settling rate or submerged mass of the particle is used to distinguish the size class either through the use of visual accumulation tubes or simple sieving. At the intermediate size ranges a variety of electromagnetic, acoustic, or laser- and light-based devices are in use. Here the degree of energy propagation or reflection is measured on a particle by particle basis and related to particle size by mechanistic theories of scattering.

No matter which operating principle is used, each instrument essentially provides a histogram of data. The histogram is comprised of equal-size particle intervals

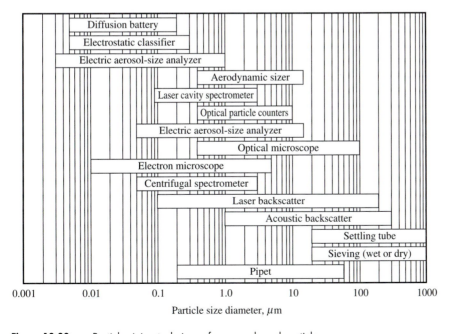

Figure 10.30   Particle sizing techniques for aerosols and particles.

which span the anticipated range of sizes in the sample. Then the particle sizer measures either the number of particles in each size range or the mass of the particles in each size range. Cumulative histograms can also be constructed from the particle size histograms. With the histogram information available, various descriptors of the distribution are possible. First, if the histogram is tightly clustered about one dominant size, then the sample is referred to as *well sorted* while a wide range in the sample histogram is referred to as *well graded*. The *median diameter* (M_d) and *mean diameter* (M) are common size descriptors. The median diameter is the particle size for which there is an equal number by weight of smaller and larger particles. The equivalent spherical median diameter is referred to then as the d_{50} particle size in millimeters or ϕ_{50} in phi units. Many histograms can be analyzed for their distribution characteristics using statistical tests. A wide variety of particle size histograms are classified as originating from *log-normal* size distributions which are fully characterized by the mean and standard deviation of the distribution. If a size is reasonably well described by the log-normal distribution, then the following statistics (based upon diameter in phi units) are known. The mean diameter is found as

$$M_\phi = \frac{\phi_{16} + \phi_{50} + \phi_{84}}{3} \qquad \text{(10.9.2)}$$

The standard deviation is

$$\sigma_\phi = \frac{\phi_{84} - \phi_{16}}{2} \qquad \text{(10.9.3)}$$

and the skewness, a measure of distribution symmetry, is given by

$$S_\phi = \frac{M_\phi - M_{d\phi}}{\sigma_\phi} \qquad \text{(10.9.4)}$$

In these definitions ϕ_{84} refers to the diameter in phi units by which 84 percent of the sample is heavier (dry weight). Therefore, ϕ_{84} is a very small particle and ϕ_{16} is a larger particle.

Unlike velocity and temperature measurements, which can be made quite rapidly by sensors placed directly and unobtrusively in the flow field, particle size analyses are laboratory-based results which require slow, careful analysis on samples brought in from the field or sampling site.

Concentration Data

The earliest and most routine concentration measurements were derived from particle sizing activities. In these laboratory-based observations, a mixture of known volume is analyzed by evaporating or otherwise eliminating the fluid, burning off the organic material that may be part of the sample but not part of the particle-based material, and then weighing the remaining sample. In this fashion the weight or mass per unit volume of mixture is determined. This technique is quite time consuming and does not allow concentration measurements to be made in rapidly varying turbulent conditions which are of environmental concerns.

Direct, time series, concentration measurements are obtained in the field via the use of light- or acoustics-based devices and once again the degree of disruption of their propagation can be related to the particle concentration and forms the basis for the measurement. The earliest devices for rapid in-situ waterborne concentration measurements were *nephelometers, transmissometers,* or *turbidimeters,* which have a light source of known power and spectral content that is sent across a sampling volume to a photodiode receiver [24]. These forwardscatter devices then relate the energy reduction to concentration. The light source can be the full visible band spectrum, or more recent versions have used light-emitting diodes (LEDs) in the red, green, and blue bands to do a more focused measurement on selected particle sizes. The sampling volume size (a 25-cm path length is typical) must be related to the anticipated concentration because if the receiver is too far away and the concentrations are too large, then all the light will be attenuated before it is sensed at the receiver.

The forwardscatter concept requires rather bulky instrumentation and a larger sampling volume. Optical backscatter devices [25, 26] have been developed to remedy both problems and thereby decrease flow disruption, especially near walls or boundaries, and increase sampling frequency. Figure 10.31 is a schematic of such a device. An infrared LED is used to provide optimal grain size resolution, and because other short path length devices can be placed nearby, quite fine spatial resolution is possible.

Acoustic backscatter devices (Fig. 10.32) have recently been introduced for in-situ measurements and behave in much the same way. A highly collimated burst of sound (10^{-5} second in duration) is sent over the sampling volume and over the ensuing one second the time trace of the sound level reflected back from the sampling volume is measured. Because the speed of sound is extremely well known and because the sound scattering is primarily unaffected by neighboring particles, the received acoustic energy from each location in the profile can be determined and related to concentration [27]. Therefore, a spatial *profile* of concentration can be measured with quite fine resolution. For reasonable sediment sizes, 3-Mhz sound has been used with 1.5-m^3 sampling volume and 1.14-cm sampling intervals, and therefore, over 100 concentration measurements are obtained at each sampling. Spatial distributions of data that have been instantaneously collected are quite rare.

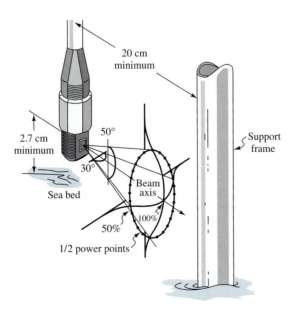

Figure 10.31   Optical backscatter device.

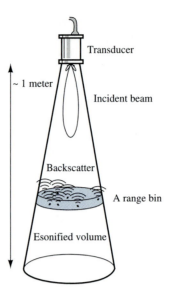

Figure 10.32   Acoustic backscatter device.

Flux or Transport Rates

In direct analogy to the earlier sections on conduit and free surface rate devices it is possible to attempt such data measurement for mass species flux or transport rates. Here, however, the results are not nearly as reliable as they were in the case of the orifice or nozzle. Two types of transport rate data exist, the *extensive* form or *integrated rate,* and the *intensive* form or the *flux rate.* The intensive quantity or flux rate for a single species, **N**, is defined from Eq. (4.8.7) in terms of the mixture velocity vector, **v**, as

$$\mathbf{N} = -\mathscr{D}\,\nabla C + \mathbf{v}C$$

It is a vector quantity having the dimensions of mass per unit time moving through a unit area perpendicular to the flow ($M/L^2/t$). If the total velocity is moving in the streamline direction, s, then the flux in the streamline direction is

$$N_s = -\mathscr{D}\,\frac{\partial C}{\partial s} + v_s C$$

The extensive quantity or the integrated rate is essentially the total mass per unit time moving through a defined area such as a channel cross section, that is,

$$\dot{m} = \int \mathbf{N}\cdot\hat{n}\,dA \tag{10.9.5}$$

where the dimensions are (M/t) and $\hat{n}$ is the unit normal vector of the surface area. For example, if a channel cross section was defined and the velocity perpendicular to it at the cross section was given by $u(y, z)$, then the *total loading rate* of sediment, C, passing through the cross section is given by

$$\dot{m} = \int \left(-\mathscr{D}\,\frac{\partial C}{\partial x} + uC\right)dA$$

Typically the molecular or Fickian diffusion flux is quite small compared to the advective component in turbulent flows. Therefore,

$$\dot{m} = \int uC \, dA \tag{10.9.6}$$

By use of the Reynolds averaging procedures presented in Eqs. (6.4.1) to (6.4.3) the *turbulent loading rate* is calculated as

$$\overline{m}(t) = \int \overline{uC} \, dA = \int (\overline{u}\,\overline{C} + \overline{u'C'}) \, dA \tag{10.9.7}$$

and consists of advective and turbulent flux contributions.

There are no unified, self-contained, commercially available instruments for making either the flux or integrated transport measurements. These data must be inferred from independent measurements of velocity and concentration which are then numerically combined during the software analysis and display portion of the experiment. Compounding these difficulties, unlike the momentum boundary layer theories which provided the theoretical guidance for ready measurement of the average velocity in the water column, there is no available unified concentration boundary layer theory for determining the equivalent average concentration with a one- or two-point measurement. Furthermore, in turbulent flows the measurement technology will most often allow only the advection component to be measured but not the integrated Reynolds fluxes ($\int u'C' \, dA$).

Mass transport data, therefore, are very difficult to measure. Yet they are among the most important data required for industrial process and environmental management today.

EXERCISES

10.9.1 Clays, tobacco smoke, combustion particles, and red blood cells (*a*) all have settling velocities in the range of $1(10^{-2})$ to $5(10^0)$ cm/s; (*b*) vary from 0.1 to 1.0 micron in equivalent spherical particle diameter; (*c*) all range in particle size from 9 to 14 ϕ-units; (*d*) have equivalent characteristics to atmospheric cloud and fog aerosols; (*e*) all of the above.

10.9.2 For the particles in the above exercise, the particle sizes are measured by (*a*) sieving; (*b*) laser backscatter; (*c*) electrostatic classifier; (*d*) electron microscope; (*e*) *b* and *d*.

10.9.3 The total flux of dissolved oxygen at a cross section in a river is (*a*) dominated by the turbulent and diffusive flux terms; (*b*) linearly related to the average velocity; (*c*) dominated by the turbulent and advective components during flood conditions; (*d*) sensitive to measurement errors in the turbulent flux terms; (*e*) *c* and *d*.

10.10 MEASUREMENT OF VISCOSITY

The treatment of fluid measurement is concluded with a discussion of methods for determining viscosity. Viscosity can be measured in a number of ways: (1) by Newton's law of viscosity, (2) by the Hagen-Poiseuille equation, and (3) by methods that require calibration with fluids of known viscosity.

By measuring the velocity gradient du/dy and the shear stress τ, in Newton's law of viscosity [Eq. (1.2.1)],

$$\tau = \mu \frac{du}{dy} \tag{10.10.1}$$

the dynamic or absolute viscosity can be computed. This is the most basic method, because it determines all the other quantities in the defining equation for viscosity. By means of a cylinder that rotates at a known speed with respect to an inner concentric stationary cylinder, du/dy is determined. By measuring the torque on the stationary cylinder, the shear stress can be computed. The ratio of shear stress to rate of change of velocity expresses the viscosity.

A schematic view of a concentric-cylinder viscometer is shown in Fig. 10.33a. When the speed of rotation is N rpm and the radius is r_2, the fluid velocity at the surface of the outer cylinder is $2\pi r_2 N/60$. With clearance b

$$\frac{du}{dy} = \frac{2\pi r_2 N}{60b}$$

The equation is based on $b \ll r_2$. The torque T_c on the inner cylinder is measured by a torsion wire from which it is suspended. By attaching a disk to the wire, its rotation can be determined by a fixed pointer. If the torque due to fluid below the bottom of the inner cylinder is neglected, the shear stress is

$$\tau = \frac{T_c}{2\pi r_1^2 h}$$

Substituting into Eq. (10.10.1) and solving for the viscosity yields

$$\mu = \frac{15 T_c b}{\pi^2 r_1^2 r_2 h N} \tag{10.10.2}$$

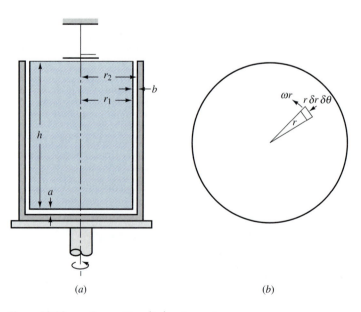

(a)   *(b)*

Figure 10.33   Concentric-cylinder viscometer.

When the clearance a is so small that the torque contribution from the bottom is appreciable, the bottom contribution may be calculated in terms of the viscosity in the following manner.

Referring to Fig. 10.33b,

$$\delta T = r\tau\,\delta A = r\mu\frac{\omega r}{a}r\,\delta r\,\delta\theta$$

in which the velocity change is ωr in the distance a. Integrating over the circular area of the disk and letting $\omega = 2\pi N/60$ leads to

$$T_d = \frac{\mu}{a}\frac{\pi}{30}N\int_0^{r_1}\int_0^{2\pi}r^3\,dr\,d\theta = \frac{\mu\pi^2}{a60}Nr_1^4 \tag{10.10.3}$$

The torque due to the disk and cylinder must equal the torque T in the torsion wire, so that

$$T = \frac{\mu\pi^2Nr_1^4}{a60} + \frac{\mu\pi^2r_1^2r_2hN}{15b} = \frac{\mu\pi^2Nr_1^2}{15}\left(\frac{r_1^2}{4a} + \frac{r_2h}{b}\right) \tag{10.10.4}$$

in which all quantities except μ are known. The flow between the surfaces must be laminar for Eqs. (10.10.2) to (10.10.4) to be valid.

Often the geometry of the inner cylinder is altered to eliminate the torque which acts on the lower surface. If the bottom surface of the inner cylinder is made concave, a pocket of air will be trapped between the bottom surface of the inner cylinder and the fluid in the rotating outer cup. A well-designed cup and a careful filling procedure will ensure that the measured torque will consist of that produced in the annulus between the two cylinders and a minute amount resulting from the action of the air on the bottom surface. Naturally, the viscometer must be provided with a temperature-controlled bath and a variable-speed drive which can be carefully regulated. Such design refinements are needed in order to obtain the rheological diagrams (cf. Fig. 1.2) for the fluid under test.

The measurement of all quantities except μ in the Hagen-Poiseuille equation by a suitable experimental arrangement is another basic method for determining the viscosity. A setup like that in Fig. 10.34 can be used. Some distance is required for the fluid to develop its characteristic velocity distribution after it enters the tube; therefore, the head or pressure must be measured by some means at a point along the tube. The volume V of flow can be measured over a time t where the reservoir surface is held at a constant level. This yields Q, and by determining γ, Δp can be

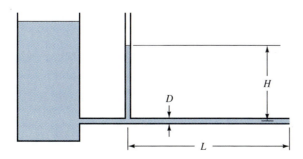

Figure 10.34 Determination of viscosity by flow through a capillary tube.

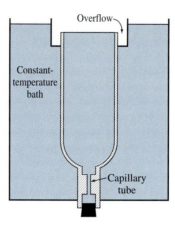

Figure 10.35   Schematic view of Saybolt viscometer.

computed. Then with L and D known, from Eq. (6.3.10a),

$$\mu = \frac{\Delta p \pi D^4}{128 Q L}$$

An adaptation of the capillary tube for industrial purposes is the *Saybolt viscometer* (Fig. 10.35). A short capillary tube is used, and the time is measured for 60 cm^3 of fluid to flow through the tube under a falling head. The time, in seconds, is the Saybolt reading. This device measures kinematic viscosity, evident from a rearrangement of Eq. (6.3.10a). When $\Delta p = \rho g h$, then $Q = \Psi t$. When the terms that are the same regardless of the fluid are separated,

$$\frac{\mu}{\rho t} = \frac{g h \pi D^4}{128 \Psi L} = C_1$$

Although the head h varies during the test, it varies over the same range for all liquids, and the terms on the right-hand side may be considered as a constant of the particular instrument. Since $\mu/\rho = \nu$, the kinematic viscosity is

$$\nu = C_1 t$$

which shows that the kinematic viscosity varies directly as the time t. The capillary tube is quite short, and so the velocity distribution is not established. The flow tends to enter uniformly, and then, owing to viscous drag at the walls, to slow down there and speed up in the center region. A correction in the above equation is needed, which is of the form C/t; hence,

$$\nu = C_1 t + \frac{C_2}{t}$$

The approximate relation between viscosity and Saybolt seconds is expressed by

$$\nu = 0.0022 t - \frac{1.80}{t}$$

in which ν is in Stokes and t in seconds.

For measuring viscosity there are many other industrial methods that generally have to be calibrated for each special case to convert to the absolute units. One consists of several tubes containing "standard" liquids of known graduated viscosities with a steel ball in each of the tubes. The time for the ball to fall the length of the tube depends upon the viscosity of the liquid. By placing the test sample in a similar tube, its viscosity can be approximated by comparison with the other tubes.

The flow of a fluid in a capillary tube is the basis for viscometers of the Oswald-Cannon-Fenske, or Ubbelohde, type. In essence, the viscometer is a U tube one leg of which is a fine capillary tube connected to a reservoir above. The tube is held vertically, and a known quantity of fluid is placed in the reservoir and allowed to flow by gravity through the capillary. The time is recorded for the free surface in the reservoir to fall between two scribed marks. A calibration constant for each instrument takes into account the variation of the capillary's bore from the standard, the bore's uniformity, entrance conditions, and the slight unsteadiness due to the falling head during the 1- to 2-min test. Various bore sizes can be obtained to cover a wide range of viscosities. Exact procedures for carrying out the tests are contained in the standards of the American Society for Testing and Materials.

EXERCISE

10.10.1 A homemade viscometer of the Saybolt type is calibrated by two measurements with liquids of known kinematic viscosity. For $\nu = 0.461$ St, $t = 97$ s, and for $\nu = 0.18$ St, $t = 46$ s, the coefficients C_1 and C_2 in $\nu = C_1 t + C_2/t$ are
(a) $C_1 = 0.005$ (b) $C_1 = 0.0044$ (c) $C_1 = 0.0046$ (d) $C_1 = 0.00317$
    $C_2 = -2.3$ $C_2 = 3.6$ $C_2 = 1.55$ $C_2 = 14.95$
(e) none of these answers.

PROBLEMS

10.1 In a laboratory channel the velocity of the flow at a point is known very accurately to be 11.0 cm/sec. A measuring device is used to take the following data at the point.

Number	1	2	3	4	5	6	7	8	9	10
u (cm/s)	11.72	11.50	11.12	10.91	10.73	10.71	10.82	10.94	10.99	11.12

Compute the absolute, relative, and percent errors.

10.2 For the data given in Prob. 10.1, calculate the mean velocity, the error estimate (most probable), the standard deviation, and the rms error.

10.3 A static tube (Fig. 10.2b) indicates a static pressure that is 1 kPa too low when liquid is flowing at 2 m/s. Calculate the correction to be applied to the indicated pressure for the liquid flowing at 5 m/s.

10.4 Four piezometer openings in the same cross section of a cast-iron pipe indicate the following pressures for simultaneous readings: 43-, 42.6-, 42.4-, and 37-mm Hg. What value should be taken for the pressure?

10.5 To better understand the behavior of the pressure response factor, K_{z^*}, pick a sinusoidal function to represent an idealized pressure signal. Use a wavelength, $L = 2\pi$, and a depth, $d = 10z^*$. Plot the pressure time trace, $p(z^*, t)$, as well as the water level height, $\eta(t)$, on the same graph and compare. For this type of wave, what is the best value of C_0?

10.6 A simple pitot tube (Fig. 10.6) is inserted into a small stream of flowing oil, with $\gamma = 8.6$ kN/m^3, $\mu = 0.065$ Pa s, $\Delta h = 38$ mm, and $h_0 = 125$ mm. What is the velocity at point 1?

10.7 A stationary body immersed in a river has a maximum pressure of 69 kPa exerted on it at a distance of 5.4 m below the free surface. Calculate the river velocity at this depth.

10.8 From Fig 10.7 derive the equation for velocity at point 1.

10.9 In Fig. 10.7 air is flowing ($p = 110$-kPa abs and $t = 5°C$) and water is in the manometer. For $R' = 30$ mm, calculate the velocity of air.

10.10 A pitot-static tube directed into a 4-m/s water stream has a gage difference of 37 mm on a water-mercury differential manometer. Determine the coefficient for the tube.

10.11 A pitot-static tube, $C = 1.12$, has a gage difference of 10 mm on a water-mercury manometer when directed into a water stream. Calculate the velocity.

10.12 A pitot-static tube of the Prandtl type has the following value of gage difference R' for the radial distance from the center of a 3-ft-diameter pipe:

r, ft	0.0	0.3	0.6	0.9	1.2	1.48
R', in.	4.00	3.91	3.76	3.46	3.02	2.40

Water is flowing, and the manometer fluid has a specific gravity of 2.93. Calculate the discharge.

10.13 What would the gage difference be on a water-nitrogen manometer for flow of nitrogen at 200 m/s, using a pitot-static tube? The static pressure is 175-kPa abs, and the corresponding temperature is 25°C. True static pressure is measured by the tube.

10.14 A disk meter has a volumetric displacement of 27 cm^3 for one complete oscillation. Calculate the flow, in liters per minute, for 86.5 oscillations per minute.

10.15 A disk water meter with volumetric displacement of 40 cm^3 per oscillation requires 470 oscillations per minute to pass 0.32 L/s and 3840 oscillations per minute to pass 2.57 L/s. Calculate the percent error, or slip, in the meter.

10.16 A volumetric tank 4 ft in diameter and 5 ft high was filled with oil in 16 min 32.4 s. What is the average discharge in gallons per minute?

10.17 A weight tank receives 75-N liquid, sp gr 0.86, in 14.9 s. What is the flow rate, in liters per minute?

10.18 Determine the equation for trajectory of a jet discharging horizontally from a small orifice with head of 6 m and velocity coefficient of 0.96. Neglect air resistance.

10.19 An orifice of area 30 cm^2 in a vertical plate has a head of 1.1 m of oil, sp gr 0.91. It discharges 6790 N of oil in 79.3 s. Trajectory measurements yield $x_0 = 2.25$ m, $y_0 = 1.23$ m. Determine C_v, C_c, and C_d.

10.20 Calculate Y, the maximum rise of a jet from an inclined plate (Fig. 10.36), in terms of H and α. Neglect losses.

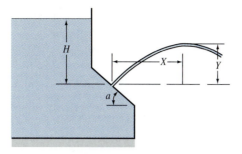

Figure 10.36 Problems 10.20, 10.21, and 10.22.

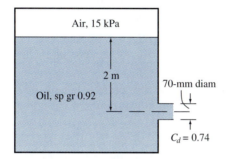

Figure 10.37 Problems 10.25 and 10.26.

10.21 In Fig. 10.36, for $\alpha = 45°$, $Y = 0.48H$. Neglecting air resistance of the jet, find C_v for the orifice.

10.22 Show that the locus of maximum points of the jet of Fig. 10.36 is given by $X^2 = 4Y(H - Y)$ when losses are neglected.

10.23 A 75-mm-diameter orifice discharges 1.80-m³ liquid, $S = 1.07$, in 82.2 s under a 2.75-m head. The velocity at the vena contracta is determined by a pitot-static tube with coefficient 1.0. The manometer liquid is acetylene tetrabromide, $S = 2.96$, and the gage difference is $R' = 1.02$ m. Determine C_v, C_c, and C_d.

10.24 A 100-mm-diameter orifice discharges 44.6-L/s water under a head of 2.75 m. A flat plate held normal to the jet just downstream from the vena contracta requires a force of 320 N to resist impact of the jet. Find C_d, C_v, and C_c.

10.25 Compute the discharge from the tank shown in Fig. 10.37.

10.26 For $C_v = 0.96$ in Fig. 10.37, calculate the losses in meter-newtons per newton and in meter-newtons per second.

10.27 Calculate the discharge through the orifice of Fig. 10.38.

10.28 For $C_v = 0.93$ in Fig. 10.38, determine the losses in joules per newton and in watts.

10.29 In Fig. 10.38 the air pressures are absolute, and isothermal conditions hold as flow occurs. On the left side the air volume is $V_1 = 1$ m³ and $A_1 = 1$ m². On the right side $A_2 = 1.5$ m². Calculate the conditions for 12 s.

10.30 A 4-in.-diameter orifice discharges 1.60-cfs liquid under a head of 11.8 ft. The diameter of the jet at the vena contracta is found by calipering to be 3.47 in. Calculate C_v, C_d, and C_c.

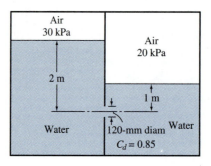

Figure 10.38   Problems 10.27, 10.28, and 10.29.

10.31   A Borda mouthpiece 50 mm in diameter has a discharge coefficient of 0.51. What is the diameter of the issuing jet?

10.32   A 75-mm-diameter orifice, $C_d = 0.82$, is placed in the bottom of a vertical tank that has a diameter of 1.5 m. How long does it take to draw the surface down from 3 to 2.5 m?

10.33   Select the size of orifice that permits a tank of horizontal cross section 1.5 m² to have the liquid surface drawn down at the rate of 160 mm/s for 3.35-m head on the orifice. $C_d = 0.63$.

10.34   A 100-mm-diameter orifice in the side of a 1.83-m-diameter tank draws the surface down from 2.44 to 1.22 m above the orifice in 83.7 s. Calculate the discharge coefficient.

10.35   Select a reservoir of such size and shape that the liquid surface drops 1 m/min over a 3-m distance for flow through a 100-mm-diameter orifice. $C_d = 0.74$.

10.36   In Fig. 10.39 the truncated cone has an angle $\theta = 60°$. How long does it take to draw the liquid surface down from $y = 4$ m to $y = 1$ m?

10.37   Calculate the dimensions of a tank such that the surface velocity varies inversely as the distance from the centerline of an orifice draining the tank. When the head is 300 mm, the velocity of fall of the surface is 30 mm/s, the orifice diameter is 12.5 mm, and $C_d = 0.66$.

10.38   Determine the time required to raise the right-hand surface of Fig. 10.40 by 2 ft.

10.39   How long does it take to raise the water surface of Fig. 10.41 by 2 m? The left-hand surface is a large reservoir of constant water-surface elevation.

10.40   Show that for incompressible flow the losses per unit weight of fluid between the upstream section and throat of a venturi meter are $KV_2^2/2g$ if

$$K = \left[\left(\frac{1}{C_v}\right)^2 - 1\right]\left[\left(1 - \frac{D_2}{D_1}\right)^4\right]$$

10.41   A 4- by 2-m venturi meter carries water at 25°C. A water-air differential manometer has a gage difference of 60 mm. What is the discharge?

10.42   What is the pressure difference between the upstream section and throat of a 150- by 75-mm horizontal venturi meter carrying 50-L/s water at 48°C?

10.43   A 12- by 6-in. venturi meter is mounted in a vertical pipe with the flow upward; 2500-gpm oil, with $S = 0.80$ and $\mu = 0.1$ P, flows through the pipe. The throat section is 6 in. above the upstream section. What is $p_1 - p_2$?

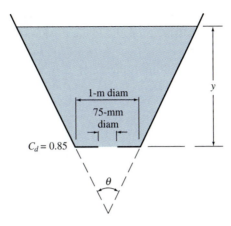

Figure 10.39   Problem 10.36.

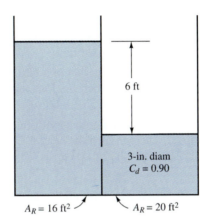

Figure 10.40   Problem 10.38.

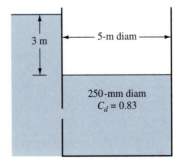

Figure 10.41   Problem 10.39.

10.44 Air flows through a venturi meter in a 55-mm-diameter pipe having a throat diameter of 30 mm, $C_v = 0.97$. For $p_1 = 830$-kPa abs, $t_1 = 15°C$, and $p_2 = 550$-kPa abs, calculate the mass per second flowing.

10.45   Oxygen, with $p_1 = 280$-kPa abs and $t_1 = 50°C$, flows through a 2- by 1-cm venturi meter with a pressure drop of 42 kPa. Find the mass per second flowing and the throat velocity.

10.46   Air flows through an 80-mm-diameter ISA flow nozzle in a 120-mm-diameter pipe; $p_1 = 150$-kPa abs, $t_1 = 5°C$, and a differential manometer with liquid, sp gr 2.93, has a gage difference of 0.8 m when connected between the pressure taps. Calculate the mass rate of flow.

10.47   A 2.5-in.-diameter ISA nozzle is used to measure flow of water at 40°F in a 6-in.-diameter pipe. What gage difference on a water-mercury manometer is required for 300 gpm?

10.48   Determine the discharge in a 300-mm-diameter line with a 160-mm-diameter VDI orifice for water at 20°C when the gage difference is 300 mm on an acetylene tetrabromide (sp gr 2.94)–water differential manometer.

10.49   A 10-mm-diameter VDI orifice is installed in a 25-mm-diameter pipe carrying nitrogen at $p_1 = 8$ atm and $t_1 = 50°C$. For a pressure drop of 140 kPa across the orifice, calculate the mass flow rate.

10.50   Air at 1 atm, $t = 21°C$, flows through a 1-m-square duct that contains a 500-mm-diameter square-edged orifice. With a head loss of 60-mm H_2O across the orifice, compute the flow, in cubic meters per minute.

10.51   A 6-in.-diameter VDI orifice is installed in a 12-in.-diameter oil line, with $\mu = 6$ cP and $\gamma = 52$ lb/ft^3. An oil-air differential manometer is used. For a gage difference of 20 in., determine the flow rate, in gallons per minute.

10.52   A rectangular sharp-crested weir 4 m long with end contractions suppressed is 1.3 m high. Determine the discharge when the head is 200 mm.

10.53   In Fig. 10.23 $L = 2.4$ m, $P = 0.54$ m, and $H = 0.24$ m. Estimate the discharge over the weir. $C = 3.33$.

10.54   A rectangular sharp-crested weir with end contractions is 1.5 m long. How high should it be placed in a channel to maintain an upstream depth of 2.25 m for 0.45-m^3/s flow?

10.55   Determine the head on a 60°-V-notch weir for discharge of 170 L/s.

10.56   Tests on a 90°-V-notch weir gave the following results: $H = 180$ mm, $Q = 19.4$ L/s, $H = 410$ mm, and $Q = 150$ L/s. Determine the formula for the weir.

10.57   A sharp-crested rectangular weir 3 ft long with end contractions suppressed and a 90°-V-notch weir are placed in the same weir box, with the vertex of the 90°-V-notch weir 6 in. below the rectangular weir crest. Determine the head on the V-notch weir (a) when the discharges are equal and (b) when the rectangular weir discharges its greatest amount above the discharge of the V-notch weir.

10.58   A broad-crested weir 1.6 m high and 3 m long has a well-rounded upstream corner. What head is required for a flow of 2.85 m^3/s?

10.59   A sieve analysis of a sand sample obtains the following pairs of data, where the opening size is in mm and the mass retained is in grams. Plot a cumulative frequency distribution on log-normal paper and determine the phi-median diameter, phi-mean diameter, phi-deviation, and phi-skewness.

Size (mm)	2.000	1.414	1.000	0.707	0.500	0.353	0.250	0.177	0.125	0.088	0.062
Mass (gr)	0	0	0.3	1.7	6.2	27.8	24.1	17.7	15.3	5.0	1.9

10.60 Two samples of sand are taken from a site on the U.S. east coast where a large beach fill emplacement has been made. The size range of the sand is determined by sieving using a sonic sifter or some kind of shaker. The results of the grain size analysis are given in the table

Size (ϕ)	−2.25	−1.0	−0.25	0.25	0.75	1.75	2.75	4.0
S1 (% of total)	0.6	0.8	1.4	2.9	30.6	59.2	4.4	0
S2 (% of total)	0.3	0.6	1.9	3.8	36.7	52.8	3.9	0

Determine the median and mean diameters, the standard deviation, and the skewness in phi-units and millimeters.

10.61 A circular disk 180 mm in diameter has a clearance of 0.3 mm from a flat plate. What torque is required to rotate the disk 800 rpm when the clearance contains oil, $\mu = 0.8$ P?

10.62 The concentric-cylinder viscometer (Fig. 10.33a) has the following dimensions: $a = 0.012$ in., $b = 0.05$ in., $r_1 = 2.8$ in., and $h = 6.0$ in. The torque is 20 lb·in. when the speed is 160 rpm. What is the viscosity?

10.63 With the apparatus of Fig. 10.34, $D = 0.5$ mm, $L = 1$ m, $H = 0.75$ m, and 60 cm^3 was discharged in 1 h 30 min. What is the viscosity in poises? $S = 0.83$.

REFERENCES

1. Doeblin, E.O., *Measurement Systems,* McGraw-Hill, New York, 1992.

2. E. Rabinowicz, *An Introduction to Experimentation,* Addison-Wesley, Reading, MA, 1974.

3. P. Bevington and D. Robinson, *Data Reduction and Error Analysis for the Physical Sciences,* McGraw-Hill, New York, 1992.

4. P. Horowitz and W. Hill, *The Art of Electronics,* Cambridge University Press, New York, 1993.

5. S. Thorpe, E. Collins, and D. Gaunt, "An Electromagnetic Current Meter to Measure Turbulent Fluctuations Near the Ocean Floor," *Deep Sea Research,* vol. 20, pp. 933–938, 1973.

6. A. Heathershaw, "Measurements of Turbulence in the Irish Sea Benthic Boundary Layer," in *The Benthic Boundary Layer,* ed. I. N. McCave, Plenum Publ. Co., New York, pp. 11–31, 1976.

7. R. Soulsby, "Measurements of the Reynolds Stress Components Close to a Marine Sand Bank," *Marine Geology,* vol. 98, pp. 7–16, 1981.

8. D. Aubrey and J. Trowbridge, "Kinematic and Dynamic Estimates from Electromagnetic Current Meter Data," *J. Geophys. Res.,* vol. 90, pp. 9137–9146, 1985.

9. Y. Ludwig, "Instrument for Measuring the Wall Shearing Stress of Turbulent Boundary Layers," *Tecn. Memo 1284,* National Advisory Commision for Aeronautics, 22 pp., 1950.

10. G. Gust, "Tools for Oceanic Small-Scale, High Frequency Flows: Metal Clad Hot Wires," *J. Geophys. Res.,* vol. 87, pp. 447–455, 1982.

11.   G. Gust and G. Weatherly, "Velocities, Turbulence and Skin Friction in a Deep-Sea Logarithmic Layer," *J. Geophys. Res.,* vol. 90, pp.4779–4792, 1985.

12.   C. Yeh and H. Cummins, "Localized Fluid Flow Measurements with a Helium Neon Laser Spectrometer," *Applied Physics Letters,* vol. 4, pp 176–178, 1964.

13.   P. Buchave, W. George, and J. Lumley, "The Measurement of Turbulence with the Laser Doppler Velocimeter," *Ann. Rev. Fluid Mech.,* vol. 11, pp. 443–505, 1979.

14.   L. Drain, *The Laser Doppler Techniques,* John Wiley and Sons, Chichester, UK, 1980.

15.   C. Veth, "A Small Scale Velocimeter for Turbulence Studies in the Sea," in *Marine Turbulence,* ed. J. C. Nihoul, Elsevier Press, New York, pp. 303–319, 1980.

16.   Y. Agrawal, "Quadrature Demodulation in Laser Doppler Velocimetry," *Applied Optics,* vol. 23, pp. 1685–1686, 1984.

17.   Y. Agrawal, "A CCD Chirp-Z FFT Doppler Signal Processor for Laser Velocimetry," *Journal of Physics (Engineering),* vol. 17, pp. 458–461, 1984.

18.   Y. Agrawal and C. Belting, "Optical Aspects of Laser Doppler Velocimetry on the Sea Floor and in Surface Layers," *Ocean Optics VIII.* Society of Photo-Optical Instrumentation Engineering, pp. 272–76, 1986.

19.   Y. Agrawal and J. Riley, "Directional Laser Velocimetry Without Frequency Biasing: Part II," *Applied Optics,* vol. 23, pp. 57–60, 1984.

20.   A. Williams III, J. Tochko, R. Koehler, et al. "Measurement of Turbulence in the Oceanic Bottom Boundary Layer with an Acoustic Current Meter Array," *J. Atmos. Oceanic Tech.,* vol. 4, pp. 312–327, 1987.

21.   W. M. Lansford, "The Use of an Elbow in a Pipe Line for Determining the Rate of Flow in a Pipe," *Univ. Ill. Eng. Exp. Stn. Bull.* 289, December 1936.

22.   J. Hoffman, *Numerical Methods for Engineers and Scientists,* McGraw Hill, New York, 1992.

23.   C. E. Lapple, "The Little Things in Life," *Stanford Research Inst. Journal,* vol. 5, pp. 95–102, 1962.

24.   R. Bartz, R. Zanewald, and H. Park, "A Transmissometer for Profiling and Mooring Observations in Water," *Society of Photo-Optical Instrument Engineers,* vol. 160, pp. 102–108, 1978.

25.   J. Downing, R. Sternberg, and C. Lister, "New Instrumentation for the Investigation of Sediment Suspension Processes in the Shallow Marine Environment," *Marine Geology,* vol. 42, pp. 19–34, 1981.

26.   D&A Istruments, "Optical Backscatterance Turbidity Monitor," *Instruction Manual, D&A Instruments and Engineering,* 2428 39th St. N.W., Washington, D.C. 20007, 1988.

27.   C. Libicki, K. Bedford, and J. Lynch, "The Interpretation and Evaluation of a 3 MHz Acoustic Device for Measuring Benthic Boundary Layer Sediment Dynamics," *J. Acoust. Soc. Amer.,* vol. 85, pp. 1501–1511, 1989.

Turbomachinery

To turn a fluid stream or change the magnitude of its velocity requires the application of forces. When a moving vane deflects a fluid jet and changes its momentum, forces are exerted between the vane and jet and work is done by displacement of the vane. Turbomachines make use of this principle: (1) the axial and centrifugal pumps, blowers, and compressors, by continuously doing work on the fluid, add to its energy; (2) the impulse, Francis, and propeller turbines, and the steam and gas turbines continuously extract energy from the fluid and convert it into torque on a rotating shaft; and (3) the fluid coupling and the torque converter, each consisting of a pump and a turbine built together, make use of the fluid to transmit power smoothly. Designing efficient turbomachines utilizes both theory and experimentation. A good design of given size and speed can readily be adapted to other speeds and other geometrically similar sizes by application of the theory of scaled models, as outlined in Sec. 5.6.

Similarity relations are first discussed in this chapter by consideration of homologous units and specific speed. Elementary cascade theory is next taken up before the theory of turbomachines is considered. Water reaction turbines are dealt with before pumps and blowers, followed by the impulse turbine. The chapter closes with a discussion of cavitation.

11.1   HOMOLOGOUS UNITS: SPECIFIC SPEED

Governing Dimensionless Groups

To use scaled models in designing turbomachines, geometric similitude is required as well as geometrically similar velocity vector diagrams at the entrance to, or exit from, the impellers. Viscous effects must unfortunately be neglected, as it is generally impossible to satisfy the above two conditions and have equal Reynolds numbers in the model and prototype. Two geometrically similar units having similar velocity vector diagrams are *homologous*. They will also have geometrically similar streamlines.

The velocity-vector diagram in Fig. 11.1 at the exit from a pump impeller can be used to formulate the condition for similar streamline patterns; β is the blade angle and u is the peripheral speed of the impeller at the end of the vane or blade; v is the velocity of fluid *relative* to the vane and V is the absolute velocity leaving the impeller, the vector sum of u and v; V_r is the radial component of V and is proportional to the discharge; and α is the angle which the absolute velocity makes with u, the tangential direction. According to geometric similitude, β must be the same for two units, and for similar streamlines α also must be the same in each case.

It is convenient to express the fact that α is to be the same in any of the series of turbomachines, called *homologous units*, by relating the speed of rotation, N; the impeller diameter (or other characteristic dimension), D; and the flow rate, Q. For constant α, V_r is proportional to V ($V_r = V \sin\alpha$) and u is proportional to V_r. Hence, the conditions for constant α in a homologous series of units can be expressed as

$$\frac{V_r}{u} = \text{constant}$$

The discharge Q is proportional to $V_r D^2$, since any cross-sectional flow area is proportional to D^2. The speed of rotation N is proportional to u/D. When these values are inserted,

$$\frac{Q}{ND^3} = \text{constant} \tag{11.1.1}$$

expresses the condition in which geometrically similar units are homologous.

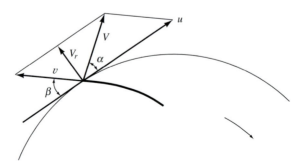

Figure 11.1   Velocity-vector diagram for an exit from a pump impeller.

The discharge Q through homologous units can be related to head H and a representative cross-sectional area A by the orifice formula as

$$Q = C_d A \sqrt{2gH}$$

in which C_d, the discharge coefficient, varies slightly with the Reynolds number and so causes a small change in efficiency with size in a homologous series. The change in discharge with the Reynolds number is referred to as a *scale effect*. Smaller machines, having smaller hydraulic radii of passages, have lower Reynolds numbers and correspondingly higher friction factors; hence, they are less efficient. The change in efficiency from model to prototype may be from 1 to 4 percent. However, in the homologous theory, the scale effect must be neglected, and so an empirical correction for change in efficiency with size is used [see Eq. (11.4.2)]. As $A \sim D^2$, the discharge equation may be

$$\frac{Q}{D^2 \sqrt{gH}} = \text{constant} \qquad \text{(11.1.2)}$$

Eliminating Q between Eqs. (11.1.1) and (11.1.2) gives

$$\frac{gH}{N^2 D^2} = \text{constant} \qquad \text{(11.1.3)}$$

Equations (11.1.1) and (11.1.3) are most useful in determining performance characteristics for one unit from those of a homologous unit of different size and speed.

The same variable groupings may be generated by dimensional analysis. If the pressure rise across the pump, Δp, is viewed as the dependent variable, related to the incompressible flow rate, the pump geometry, and fluid properties, then

$$\Delta p = f_1(\rho, \mu, Q, \omega, D) \qquad \text{(11.1.4)}$$

in which ω is the rotational speed in radians per second. By selecting ρ, ω, and D as repeating variables,

$$\frac{\Delta p}{\rho \omega^2 D^2} = \phi_1\left(\frac{Q}{\omega D^3}, \frac{\rho \omega D^2}{\mu}\right) \qquad \text{(11.1.5)}$$

If Δp is expressed in terms of the head produced by the pump, the *pressure coefficient* on the left-hand side becomes the *head rise coefficient*, $gH/\omega^2 D^2$. If shaft power, $P = T\omega$, is taken as the dependent variable and dimensional analysis is repeated,

$$\frac{P}{\rho \omega^3 D^5} = \phi_2\left(\frac{Q}{\omega D^3}, \frac{\rho \omega D^2}{\mu}\right) \qquad \text{(11.1.6)}$$

Efficiency, $\eta = \Delta p Q/P = \gamma H Q/P$, which is already dimensionless, may also be taken as a dependent variable, yielding

$$\eta = \phi_3\left(\frac{Q}{\omega D^3}, \frac{\rho \omega D^2}{\mu}\right) \qquad \text{(11.1.7)}$$

The last dimensionless number in each of these equations is a rotational Reynolds number that shows the scale effect discussed above. When this is neglected, each parameter is related to the *flow coefficient*. For head rise versus flow

$$\frac{gH}{\omega^2 D^2} = \phi_1\left(\frac{Q}{\omega D^3}\right)$$

(11.1.8)

For shaft power versus flow

$$\frac{P}{\rho\omega^3 D^5} = \phi_2\left(\frac{Q}{\omega D^3}\right)$$

(11.1.9)

For efficiency versus flow

$$\eta = \phi_3\left(\frac{Q}{\omega D^3}\right)$$

(11.1.10)

By convention the rotational speed is generally written as N in rpm and density and gravity are dropped, resulting in the customary parameters, which are not dimensionally pure.

One characteristic curve for a pump in dimensionless form is the plot of $Q/\omega D^3$ as abscissa against $gH/(\omega^2 D^2)$ as ordinate. This curve, obtained from tests on one unit of the series, then applies to all homologous units and can be converted to the usual characteristic curve by selecting the desired values of N and D. If two pumps from the series are operated at the same constant value of the flow coefficient, conditions of dynamic similitude require that

$$\left(\frac{gH}{\omega^2 D^2}\right)_1 = \left(\frac{gH}{\omega^2 D^2}\right)_2$$

(11.1.11)

$$\left(\frac{P}{\rho\omega^3 D^5}\right)_1 = \left(\frac{P}{\rho\omega^3 D^5}\right)_2$$

(11.1.12)

$$\eta_1 = \eta_2$$

(11.1.13)

Example graphs are presented later in this chapter.

Another use of the similitude conditions is to predict the performance of the *same* pump when operated at different rotational speeds. As discussed above our inability to handle the scale effects associated with the rotational Reynolds number forces an empirical correction in efficiency.

Example 11.1

A prototype test of a mixed-flow pump with a 2-m-diameter discharge opening, operating at 225 rpm, resulted in the following characteristics:

H, m	Q, m³/s	η, %	H, m	Q, m³/s	η, %	H, m	Q, m³/s	η, %
18.3	5.663	69	14.48	9.345	87.3	10.67	11.638	82
17.53	6.456	75	13.72	9.769	88	9.91	12.035	79
16.76	7.249	80	12.95	10.251	87.4	9.14	12.403	75
16.0	7.930	83.7	12.19	10.817	86.3	8.38	12.714	71
15.24	8.580	86	11.43	11.213	84.4	7.62	12.997	66.5

What size and synchronous speed (60 Hz) of homologous pump should be used to produce 5.66 m³/s at 18.3-m head at the point of best efficiency? Find the characteristic curves for this case.

Solution

Subscript 1 refers to the 2-m pump. For best efficiency $H_1 = 13.72$, $Q_1 = 9.769$ and $\eta = 88$ percent. With Eqs. (11.1.1) and (11.1.3),

$$\frac{H}{N^2 D^2} = \frac{H_1}{N_1^2 D_1^2} \qquad \frac{Q}{ND^3} = \frac{Q_1}{N_1 D_1^3}$$

or

$$\frac{18.3}{N^2 D^2} = \frac{13.72}{225^2(2^2)} \qquad \frac{5.66}{ND^3} = \frac{9.769}{225(2^3)}$$

After solving for N and D,

$$N = 367 \text{ rpm} \qquad D = 1.417 \text{ m}$$

The nearest synchronous speed (3600 divided by the number of pairs of poles) is 360 rpm. To maintain the desired head of 18.3 m, a new D is necessary. When its size is computed

$$D = \sqrt{\frac{18.3}{13.72} \frac{225}{360}} (2) = 1.444 \text{ m}$$

the discharge at best efficiency is

$$Q = \frac{Q_1 ND^3}{N_1 D_1^3} = 9.767 \frac{360}{225}\left(\frac{1.444}{2}\right)^3 = 5.883 \text{ m}^3/\text{s}$$

which is slightly more capacity than required. With $N = 360$ and $D = 1.444$, equations for transforming the corresponding values of H and Q for any efficiency can be obtained:

$$H = H_1\left(\frac{ND}{N_1 D_1}\right)^2 = H_1\left(\frac{360}{225}\frac{1.444}{2}\right)^2 = 1.335 H_1$$

and

$$Q = Q_1\frac{ND^3}{N_1 D_1^3} = Q_1\left(\frac{360}{225}\right)\left(\frac{1.444}{2}\right)^3 = 0.603 Q_1$$

The characteristics of the new pump are

H, m	Q, m³/s	η, %	H, m	Q, m³/s	η, %	H, m	Q, m³/s	η, %
24.4	3.410	69	19.32	5.628	87.3	14.23	7.008	82
23.4	3.884	75	18.3	5.883	88	13.72	7.247	79
22.34	4.365	80	17.28	6.173	87.4	12.19	7.469	75
21.34	4.775	83.7	16.26	6.514	86.3	11.18	7.656	71
20.33	5.167	86	15.25	6.752	84.4	10.17	7.827	66

The efficiency of the 1.444-m pump might be a fraction of a percent less than that of the 2-m pump, as the hydraulic radii of flow passages are smaller, so that the Reynolds number would be less.

Example 11.2

In Ex. 11.1, is the diameter of 1.444 m the minimum size that would produce at least 5.66 m³/s and at least 18.3-m head at the point of best efficiency?

Solution

Since $H = 18.3$ m was used to find D at 360 rpm and the discharge exceeds the required flow, the required discharge at 400 rpm can be used to determine a smaller diameter (Fig. 11.2).

$$D = D_1\left(\frac{Q}{Q_1}\frac{N_1}{N}\right)^{1/3} = 2\left(\frac{5.66}{9.769}\frac{225}{400}\right)^{1/3} = 1.376 \text{ m}$$

The head produced at best efficiency is

$$H = H_1\left(\frac{N}{N_1}\frac{D}{D_1}\right)^2 = 13.72\left(\frac{400}{225}\frac{1.376}{2}\right)^2 = 20.53 \text{ m}$$

which exceeds the required pump head. It may be noted that this homologous unit requires more power to operate than the size selected in Ex. 11.1 (see Fig. 11.2).

Specific Speed

The specific speed of a homologous unit is a constant widely used in selecting the type of unit and preliminary design. It is usually defined differently for a pump and a turbine.

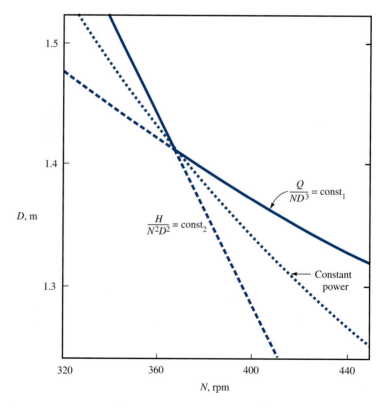

Figure 11.2   Examples 11.1 and 11.2.

The specific speed, N_s, of a homologous series of *pumps* is defined as the speed of a unit of a series of such size that it delivers unit discharge at unit head. It is obtained as follows. Eliminating D in Eqs. (11.1.1) and (11.1.3) and rearranging give

$$\frac{N\sqrt{Q}}{g^{3/4}H^{3/4}} = \text{constant} \qquad (11.1.14)$$

By the definition of specific speed of the pump, the constant N_{sp}, the speed of a unit for $Q = 1$ and $H = 1$, is

$$N_{sp} = \frac{N\sqrt{Q}}{H^{3/4}} \qquad (11.1.15)$$

The specific speed of a series is usually defined for the point of best efficiency, that is, for the speed, discharge, and head that is most efficient.

A dimensionless specific speed ω_{sp} can be defined from Eq. (11.1.14)

$$\omega_{sp} = \frac{\omega\sqrt{Q}}{g^{3/4}H^{3/4}} \qquad (11.1.16)$$

where ω is in radians per second. Its value would remain unchanged for a particular pumping situation whether values are expressed in SI or USC units.

The specific speed of a homologous series of *turbines* is defined as the speed of a unit in the series of such size that it produces unit power with unit head. Since power P is proportional to QH,

$$\frac{P}{\gamma QH} = \text{constant} \qquad (11.1.17)$$

The terms D and Q can be eliminated from Eqs. (11.1.1), (11.1.3), and (11.1.17) to produce

$$\frac{N\sqrt{P/\rho}}{g^{5/4}H^{5/4}} = \text{constant} \qquad (11.1.18)$$

For unit power and head, when the ρ and g terms are dropped, the constant of Eq. (11.1.18) becomes the speed, or the turbine specific speed N_{st} of the series, so that

$$N_{st} = \frac{N\sqrt{P}}{H^{5/4}} \qquad (11.1.19)$$

A dimensionless specific speed ω_{st} for turbines, from Eq. (11.1.18), is

$$\omega_{st} = \frac{\omega\sqrt{P/\rho}}{g^{5/4}H^{5/4}} \qquad (11.1.20)$$

The specific speed of a unit required for a given discharge and head can be estimated from Eqs. (11.1.15) and (11.1.19). For pumps handling large discharges at low heads a high specific speed is indicated; for a high-head turbine producing relatively low power (small discharge) the specific speed is low. Experience has shown that for best efficiency one particular type of pump or turbine is usually indicated for a given specific speed.

Because Eqs. (11.1.15) and (11.1.19) are not dimensionally correct (γ and g have been included in the constant term), the value of specific speed depends on the units involved. For example, in the United States Q is commonly expressed in gallons per minute, millions of gallons per day, or cubic feet per second when referring to specific speeds of pumps.

Centrifugal pumps have low specific speeds, $\omega_{sp} < 1$; mixed-flow pumps have medium specific speeds, $1 < \omega_{sp} < 4$; and axial-flow pumps have high specific speeds, $\omega_{sp} > 4$. Impulse turbines have low specific speeds, $\omega_{st} < 1$; Francis and mixed-flow turbines have medium specific speeds, $1 < \omega_{st} < 7$; and propeller turbines have high specific speeds, $\omega_{st} > 7$.

EXERCISES

11.1.1 Two units are homologous when they are geometrically similar and have (*a*) similar streamlines; (*b*) the same Reynolds number; (*c*) the same efficiency; (*d*) the same Froude number; (*e*) none of these answers.

11.1.2 The following two relations are necessary for homologous units: (*a*) H/ND^3 = constant and Q/N^2D^2 = constant; (*b*) $Q/D^2\sqrt{H}$ = constant and H/N^3D = constant; (*c*) P/QH = constant and H/N^2D^2 = constant; (*d*) $N\sqrt{Q}/H^{3/2}$ = constant and $N\sqrt{P}/H^{3/4}$ = constant; (*e*) none of these answers.

11.1.3 The specific speed of a pump is defined as the speed of a unit (*a*) of unit size with unit discharge at unit head; (*b*) of such size that it requires unit power for unit head; (*c*) of such size that it delivers unit discharge at unit head; (*d*) of such size that it delivers unit discharge at unit power; (*e*) none of these answers.

11.2   ELEMENTARY CASCADE THEORY

Turbomachines either do work on a fluid or extract work from it continuously by having it flow through a series of moving (and possibly fixed) vanes. By an examination of the flow through a series of similar blades or vanes, called a *cascade*, some of the requirements of an efficient system can be developed. Consider, first, flow through the simple fixed cascade system of Fig. 11.3. The velocity vector representing the fluid has been turned through an angle by the presence of the cascade system. A force has been exerted on the fluid, but (neglecting friction effects and turbulence) no work is done on the fluid. Section 3.7 deals with forces on a single vane.

Since turbomachines are rotational devices, the cascade system can be arranged symmetrically around the periphery of a circle, as in Fig. 11.4*a*. If the fluid approaches the fixed cascade in a radial direction, it has its moment of momentum changed from zero to a value dependent upon the mass per unit time flowing, the tangential component of velocity V_t developed, and the radius. From Eq. (3.8.5),

$$T = \rho Q r V_t \tag{11.2.1}$$

Again, no work is done by the fixed-vane system.

Figure 11.3   Simple cascade system.

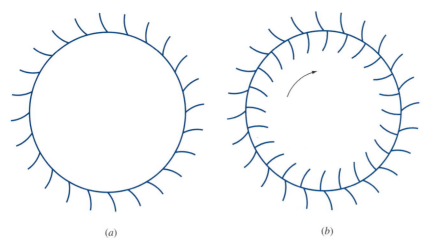

Figure 11.4   Cascades of vanes on periphery of a circular cylinder: (a) stationary vanes and (b) rotating cascade within a fixed cascade.

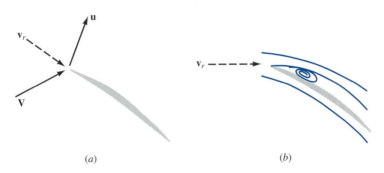

Figure 11.5   Flow onto blades: (a) flow tangent to the blade and (b) flow separation, or shock, with relative velocity not tangent to the leading edge.

Consider now another series of vanes (Fig. 11.4b) rotating within the fixed-vane system at a speed ω. For efficient operation of the system it is important that the fluid flow onto the moving vanes with the least disturbance, that is, in a tangential manner, as illustrated in Fig. 11.5a. When the relative velocity is not tangent to the blade at its entrance, separation may occur, as shown in Fig. 11.5b. The losses tend to increase rapidly (about as the square of the relative velocity) with angle from the tangential and radically impair the efficiency of the machine. Separation also frequently occurs when the approaching relative velocity is tangential to the vane, owing to curvature of the vanes or to expansion of the flow passages, which causes the boundary layer to thicken and come to rest. These losses are called *shock* or *turbulence losses*. When the fluid exits from the moving cascade, it will generally have its velocity altered in both magnitude and direction, thereby changing its moment of momentum and either doing work on the cascade or having work done on it by the moving cascade. In the case of a turbine it is desired to have the fluid leave with no moment of momentum. An old saying in turbine design is "have the fluid enter without shock and leave without velocity."

11.3   THEORY OF TURBOMACHINES

Moment-of-Momentum Equation

Turbines extract useful work from fluid energy; and pumps, blowers, and turbo-compressors add energy to fluids by means of a runner consisting of vanes rigidly attached to a shaft. Since the only displacement of the vanes is in the tangential direction, work is done by the displacement of the tangential components of force on the runner. The radial components of force on the runner have no displacement in a radial direction and hence can do no work.

In turbomachine theory, friction is neglected, and the fluid is assumed to have perfect guidance through the machine, that is, an infinite number of thin vanes, and so the relative velocity of the fluid is always tangent to the vane. This yields circular symmetry and permits the moment-of-momentum equation (Sec. 3.8) to take the simple form of Eq. (3.8.5), for steady flow,

$$T = \rho Q[(rV_t)_{\text{out}} - (rV_t)_{\text{in}}] \tag{11.3.1}$$

in which T is the torque acting on the fluid within the control volume (Fig. 11.6) and $\rho Q(rV_t)_{\text{out}}$ and $\rho Q(rV_t)_{\text{in}}$ represent the moment of momentum leaving and entering the control volume, respectively.

The polar vector diagram is generally used in studying vane relations (Fig. 11.7), with subscript 1 for entering fluid and subscript 2 for exiting fluid. V is the absolute fluid velocity, u is the peripheral velocity of the runner, and v is the fluid velocity relative to the runner. The absolute velocities V and u are laid off from O, and the relative velocity connects them as shown. V_u is designated as the component of

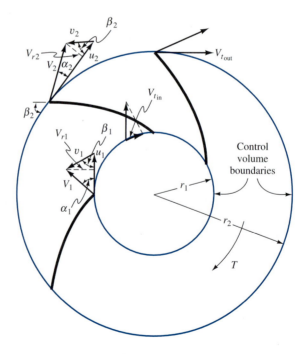

Figure 11.6   Steady flow through a control volume with circular symmetry.

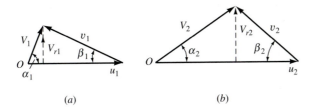

Figure 11.7 Polar vector diagrams. (a) Entrance. (b) Exit.

absolute velocity in the tangential direction, α is the angle the absolute velocity V makes with the peripheral velocity u, and β is the angle the relative velocity makes with $-u$, or it is the *blade angle*, as perfect guidance is assumed. V_r is the absolute velocity component normal to the periphery. In this notation Eq. (11.3.1) becomes

$$T = \rho Q(r_2 V_2 \cos\alpha_2 - r_1 V_1 \cos\alpha_1) \qquad (11.3.2)$$
$$= \rho Q(r_2 V_{u2} - r_1 V_{u1}) = \dot{m}(r_2 V_{u2} - r_1 V_{u1})$$

The mass per unit time flowing is $\dot{m} = \rho Q = (\rho Q)_{\text{out}} = (\rho Q)_{\text{in}}$. In the form above, when T is positive, the fluid moment of momentum increases through the control volume, as expected for a pump. For T negative, the moment of momentum of the fluid decreases, as expected for a turbine runner. When $T = 0$, as in passages where there are no vanes,

$$rV_u = \text{constant}$$

This is *free-vortex* motion, with the tangential component of velocity varying inversely with radius. It is discussed in Sec. 8.6 and compared with the forced vortex in Sec. 2.9.

The wicket gates of Fig. 11.8 are turned so that the flow makes an angle of 45° with a radial line at section 1, where the speed is 4.005 m/s. Determine the magnitude of tangential velocity component V_u over section 2.

Example 11.3

Solution

Since no torque is exerted on the flow between sections 1 and 2, the moment of momentum is constant and the motion follows the free-vortex law

$$V_u r = \text{constant}$$

At section 1

$$V_{u1} = 4.005 \cos 45° = 2.832 \text{ m/s}$$

Then

$$V_{u1} r_1 = (2.832 \text{ m/s})(0.75 \text{ m}) = 2.124 \text{ m}^2/\text{s}$$

Across section 2

$$V_{u2} = \frac{2.124 \text{ m}^2/\text{s}}{r \text{ m}}$$

At the hub $V_u = 2.124/0.225 = 9.44$ m/s, and at the outer edge $V_u = 2.124/0.6 = 3.54$ m/s.

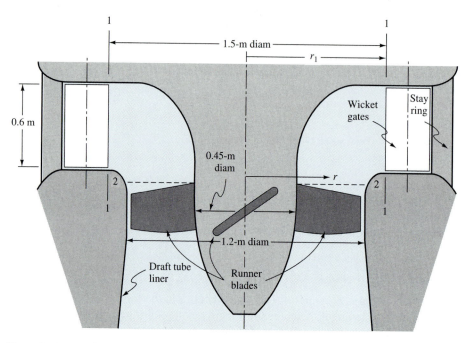

Figure 11.8   Schematic view of propeller turbine.

Head and Energy Relations

Multiplying Eq. (11.3.2) by the rotational speed ω of the runner in radians per second gives

$$T\omega = \rho Q(\omega r_2 V_{u2} - \omega r_1 V_{u1}) = \rho Q(u_2 V_{u2} - u_1 V_{u1}) \qquad \text{(11.3.3)}$$

For no losses the power available from a turbine is $Q \, \Delta p = Q\gamma H$, in which H is the head on the runner since $Q\gamma$ is the weight per unit time and H is the potential energy per unit weight. Similarly, a pump runner produces work, $Q\gamma H$, in which H is the pump head. The power exchange is

$$T\omega = Q\gamma H \qquad \text{(11.3.4)}$$

Solving for H, using Eq. (11.3.3) to eliminate T, gives

$$H = \frac{u_2 V_{u2} - u_1 V_{u1}}{g} \qquad \text{(11.3.5)}$$

For turbines the sign in Eq. (11.3.5) is reversed.

For pumps the *actual* head H_{a_p} produced is

$$H_{a_p} = e_h H = H - H_L \qquad \text{(11.3.6)}$$

and for turbines the actual head H_{a_t} is

$$H_{a_t} = \frac{H}{e_h} = H + H_L \qquad \text{(11.3.7)}$$

in which e_h is the hydraulic efficiency of the machine and H_L represents all the internal fluid losses in the machine. The overall efficiency of the machines is further reduced by bearing friction, by friction caused by fluid between the runner and housing, and by leakage or flow that passes around the runner without going through it. These losses do not affect the head relations.

Pumps are generally so designed that the angular momentum of fluid entering the runner (impeller) is zero. Then

$$H = \frac{u_2 V_2 \cos \alpha_2}{g} \tag{11.3.8}$$

Turbines are so designed that the angular momentum is zero at the exit section of the runner for conditions at best efficiency; hence,

$$H = \frac{u_1 V_1 \cos \alpha_1}{g} \tag{11.3.9}$$

In writing the energy equation for a pump, with Eqs. (11.3.5) and (11.3.6),

$$\begin{aligned}
H_p &= \left(\frac{V_2^2}{2g} + \frac{p_2}{\gamma} + z_2\right) - \left(\frac{V_1^2}{2g} + \frac{p_1}{\gamma} + z_1\right) \\
&= \frac{u_2 V_2 \cos \alpha_2 - u_1 V_1 \cos \alpha_1}{g} - H_L
\end{aligned} \tag{11.3.10}$$

for which it is assumed that all streamlines through the pump have the same total energy. With the relations among the absolute velocity V, the velocity v relative to the runner, and the velocity u of the runner, and from the vector diagrams (Fig. 11.7) by the law of cosines,

$$u_1^2 + V_1^2 - 2u_1 V_1 \cos \alpha_1 = v_1^2 \qquad u_2^2 + V_2^2 - 2u_2 V_2 \cos \alpha_2 = v_2^2$$

Eliminating the absolute velocities V_1 and V_2 in these relations and in Eq. (11.3.10) gives

$$H_L = \frac{u_2^2 - u_1^2}{2g} - \frac{v_2^2 - v_1^2}{2g} - \frac{p_2 - p_1}{\gamma} - (z_2 - z_1) \tag{11.3.11}$$

or

$$H_L = \frac{u_2^2 - u_1^2}{2g} - \left[\left(\frac{v_2^2}{2g} + \frac{p_2}{\gamma} + z_2\right) - \left(\frac{v_1^2}{2g} + \frac{p_1}{\gamma} + z_1\right)\right] \tag{11.3.12}$$

The losses are the difference in centrifugal head, $(u_2^2 - u_1^2)/2g$, and in the head change in the relative flow. For no loss, the increase in pressure head, from Eq. (11.3.11), is

$$\frac{p_2 - p_1}{\gamma} + z_2 - z_1 = \frac{u_2^2 - u_1^2}{2g} - \frac{v_2^2 - v_1^2}{2g} \tag{11.3.13}$$

With no flow through the runner, v_1 and v_2 are zero, and the head rise is as expressed in the relative equilibrium relation, Eq. (2.9.6). When flow occurs, the head rise is equal to the centrifugal head minus the difference in relative velocity heads. For the case of a turbine, exactly the same equations result.

Example 11.4

A centrifugal pump with a 700-mm-diameter impeller runs at 1800 rpm. The water enters without whirl, and $\alpha_2 = 60°$. The actual head produced by the pump is 17 m. Find its hydraulic efficiency when $V_2 = 6$ m/s.

Solution

From Eq. (11.3.8) the theoretical head is

$$H = \frac{u_2 V_2 \cos \alpha_2}{g} = \frac{1800(2\pi)(0.35)(6)(0.50)}{60(9.806)} = 20.18 \text{ m}$$

The actual head is 17 m; hence, the hydraulic efficiency is

$$e_h = \frac{17}{20.18} = 84.2 \text{ percent}$$

EXERCISES

11.3.1 A shaft transmits 150 kW at 600 rpm. The torque in newton-meters is (a) 26.2; (b) 250; (c) 2390; (d) 4780; (e) none of these answers.

11.3.2 What torque is required to give 100-cfs water a moment of momentum so that it has a tangential velocity of 10 ft/s at a distance of 6 ft from the axis? (a) 116 lb·ft; (b) 1935 lb·ft; (c) 6000 lb·ft; (d) 11,610 lb·ft; (e) none of these answers.

11.3.3 The moment of momentum of water is reduced by 27,100 N·m in flowing through vanes on a shaft turning 400 rpm. The power developed on the shaft is, in kilowatts, (a) 181.5; (b) 1134; (c) 10,800; (d) not determinable, insufficient data; (e) none of these answers.

11.3.4 Liquid moving with constant angular momentum has a tangential velocity of 4.0 ft/s at 10 ft from the axis of rotation. The tangential velocity 5 ft from the axis, in feet per second, is (a) 2; (b) 4; (c) 8; (d) 16; (e) none of these answers.

11.4   REACTION TURBINES

In the *reaction* turbine a portion of the energy of the fluid is converted into kinetic energy by the fluids passing through adjustable wicket gates (Fig. 11.9) before entering the runner, and the remainder of the conversion takes place through the runner. All passages are filled with liquid, including the passage (draft tube) from the runner to the downstream liquid surface. The static fluid pressure occurs on both sides of the vanes and, hence, does no work. The work done is entirely due to the conversion to kinetic energy.

The reaction turbine is quite different from the impulse turbine, discussed in Sec. 11.6. In an impulse turbine all the available energy of the fluid is converted into kinetic energy by a nozzle that forms a free jet. The energy is then taken from the jet by suitable flow through moving vanes. The vanes are partly filled, with the jet open to the atmosphere throughout its travel through the runner.

In contrast, in the reaction turbine the kinetic energy is appreciable as the fluid leaves the runner and enters the draft tube. The function of the draft tube is to reconvert the kinetic energy into flow energy by a gradual expansion of the flow cross

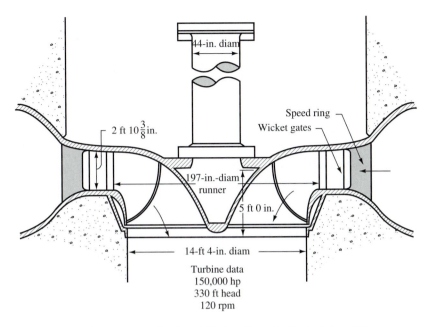

44-in. diam

2 ft 10⅜in.

Speed ring
Wicket gates

197-in.-diam
runner

5 ft 0 in.

14-ft 4-in. diam

Turbine data
150,000 hp
330 ft head
120 rpm

Figure 11.9 Francis turbine for Grand Coulee, Columbia Basin Project. *(Newport News Shipbuilding and Dry Dock Co.)*

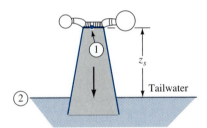

1

z_s

Tailwater

2

Figure 11.10 Draft tube.

section. Application of the energy equation between the two ends of the draft tube shows that the action of the tube is to reduce the pressure at its upstream end to less than atmospheric pressure, thus increasing the effective head across the runner to the difference in elevation between head water and tail water, less losses.

Referring to Fig. 11.10, the energy equation from 1 to 2 yields

$$z_s + \frac{V_1^2}{2g} + \frac{p_1}{\gamma} = 0 + 0 + 0 + \text{losses}$$

The losses include the expansion loss, friction, and velocity head loss at the exit from the draft tube, all of which are quite small; hence,

$$\frac{p_1}{\gamma} = -z_s - \frac{V_1^2}{2g} + \text{losses} \qquad (11.4.1)$$

shows that considerable vacuum is produced at section 1, which effectively increases the head across the turbine runner. The turbine setting must not be too high or cavitation occurs in the runner and draft tube (see Sec. 11.7).

Example 11.5

A turbine has a velocity of 6 m/s at the entrance to the draft tube and a velocity of 1.2 m/s at the exit. For friction losses of 0.1 m and a tail water 5 m below the entrance to the draft tube, find the pressure head at the entrance.

Solution

From Eq. (11.4.1)

$$\frac{p_1}{\gamma} = -5 - \frac{6^2}{2(9.806)} + \frac{1.2^2}{2(9.806)} + 0.1 = -6.66 \text{ m}$$

as the kinetic energy at the exit from the draft tube is lost. Hence, a suction head of 6.66 m is produced by the presence of the draft tube.

There are two forms of the reaction turbine in common use, the *Francis* turbine (Fig. 11.9) and the *propeller* (axial-flow) turbine (Fig. 11.8). In both, all passages flow full, and energy is converted to useful work entirely by changing the moment of momentum of the liquid. The flow passes first through the wicket gates, which impart a tangential and a radially inward velocity to the fluid. A space between the wicket gates and the runner permits the flow to close behind the gates and move as a free vortex, without external torque being applied.

In the Francis turbine (Fig. 11.9) the fluid enters the runner so that the relative velocity is tangent to the leading edge of the vanes. The radial component is gradually changed into an axial component, and the tangential component is reduced as the fluid traverses the vane, so that at the runner exit the flow is axial with very little whirl (tangential component) remaining. The pressure has been reduced to less than atmospheric, and most of the remaining kinetic energy is reconverted into flow energy by the time it discharges from the draft tube. The Francis turbine is best suited for medium-head installations from 80 to 600 ft (25 to 180 m) and has an efficiency between 90 and 95 percent for the larger installations. Francis turbines are designed in the specific speed range of 10 to 110 (ft, hp, rpm) or 40 to 420 (m, kW, rpm) with best efficiency in the range 40 to 60 (ft, hp, rpm) or 150 to 230 (m, kW, rpm).

In the propeller turbine (Fig. 11.8), after passing through the wicket gates, the flow moves as a free vortex and has its radial component changed into an axial component by guidance from the fixed housing. The moment of momentum is constant, and the tangential component of velocity is increased through the reduction in radius. The blades are few in number, relatively flat, and with very little curvature; they are so placed that the relative flow entering the runner is tangential to the leading edge

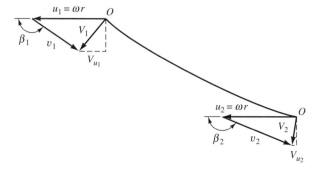

Figure 11.11   Vector diagrams for entrance and exit of a propeller turbine blade at fixed radial distance.

of the blade. The relative velocity is high, as with the Pelton wheel, and changes slightly in traversing the blade. The velocity diagrams in Fig. 11.11 show how the tangential component of velocity is reduced. Propeller turbines are made with blades that pivot around the hub, thus permitting the blade angle to be adjusted for different gate openings and for changes in head. They are particularly suited for low-head installations, up to 30 m, and have top efficiencies around 94 percent. Axial-flow turbines are designed in the specific speed range of 100 to 210 (ft, hp, rpm) or 380 to 800 (m, kW, rpm) with best efficiency from 120 to 160 (ft, hp, rpm) or 460 to 610 (m, kW, rpm).

The windmill is a form of axial-flow turbine. Since it has no fixed vanes to give an initial tangential component to the airstream, it must impart the tangential component to the air with the moving vanes. The airstream expands in passing through the vanes with a reduction in its axial velocity.

Example 11.6

Assuming uniform axial velocity over section 2 of Fig. 11.8 and using the data of Ex. 11.3, determine the angle of the leading edge of the propeller at $r = 0.225$, 0.45, and 0.6 m for a propeller speed of 240 rpm.

Solution

At $r = 0.225$ m

$$u = \frac{240}{60}(2\pi)(0.225) = 5.66 \text{ m/s} \qquad V_u = 9.44 \text{ m/s}$$

At $r = 0.45$ m,

$$u = \frac{240}{60}(2\pi)(0.45) = 11.3 \text{ m/s} \qquad V_u = 4.72 \text{ m/s}$$

At $r = 0.6$ m,

$$u = \frac{240}{60}(2\pi)(0.6) = 15.06 \text{ m/s} \qquad V_u = 3.54 \text{ m/s}$$

The discharge through the turbine is, from section 1,

$$Q = (0.6 \text{ m})(1.5 \text{ m})(\pi)(4.005 \text{ m/s})(\cos 45°) = 8 \text{ m}^3/\text{s}$$

Hence, the axial velocity at section 2 is

$$V_a = \frac{8}{\pi(0.6^2 - 0.225^2)} = 8.24 \text{ m/s}$$

Figure 11.12 shows the initial vane angle for the three positions.

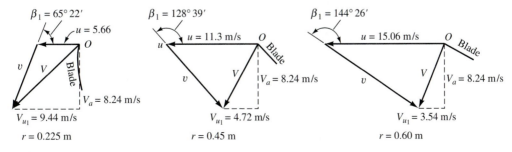

Figure 11.12   Velocity diagrams for the angle of the leading edge of a propeller turbine blade.

Moody [1]† has developed a formula to estimate the efficiency of a unit of a homologous series of turbines when the efficiency of one of the series is known:

$$\eta = 1 - (1 - \eta_1)\left(\frac{D_1}{D}\right)^{1/4}$$

(11.4.2)

in which η_1 and D_1 are usually the efficiency and the diameter of a model, respectively. Stepanoff [2] has used the same equations to relate efficiencies in pumps.

EXERCISE

11.4.1   A reaction-type turbine discharges 34 m³/s under a head of 7.5 m and with an overall efficiency of 91 percent. The power developed is, in kilowatts, (a) 2750; (b) 2500; (c) 2275; (d) 70.7; (e) none of these answers.

11.5   PUMPS AND BLOWERS

Pumps add energy to liquids and blowers to gases. The procedure for designing them is the same for both, except when the density is appreciably increased. Turbopumps and turboblowers use *radial flow, axial flow*, or a combination of the two, called *mixed flow*. For high heads the radial (centrifugal) pump, frequently with two or more stages (two or more impellers in series), is best suited. For large flows under small heads the axial-flow pump or blower (Fig. 11.13a) is best suited. The mixed-flow pump (Fig. 11.13b) is used for medium head and medium discharge.

The equations developed in Sec. 11.3 apply just as well to pumps and blowers as to turbines. The usual centrifugal pump has a suction, or inlet, pipe leading to the center of the impeller, a radial outward-flow runner, as in Fig. 11.6, and a collection pipe or spiral casing that guides the fluid to the discharge pipe. Ordinarily, no fixed

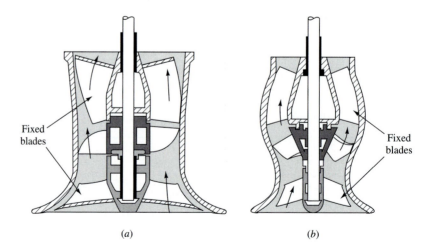

Figure 11.13   Well-type pumps: (a) axial flow and (b) mixed flow. *(Ingersoll-Rand Co.)*

| †Numbered references will be found at the end of this chapter.

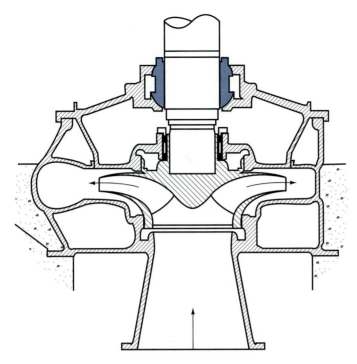

Figure 11.14 Sectional elevation of Eagle Mountain and Hayfield pumps, Colorado River Aqueduct. *(Worthington Corp)*

vanes are used, except for multistage units in which the flow is relatively small and the additional fluid friction is less than the additional gain in conversion of kinetic energy to pressure energy upon leaving the impeller.

Figure 11.14 shows a sectional elevation of a large centrifugal pump. For lower heads and greater discharges (relatively) the impellers vary as shown in Fig. 11.15, from high head at left to low head at right with the axial-flow impeller. The specific speed increases from left to right. A chart for determining the types of pump for best efficiency is given in Fig. 11.16 for water.

Theoretical Head-Discharge Curve

A theoretical head-discharge curve can be obtained by using Eq. (11.3.8) and the vector diagrams of Fig. 11.7. From the exit diagram of Fig. 11.7

$$V_2 \cos \alpha_a = V_{u2} = u_2 - V_{r2} \cot \beta_2$$

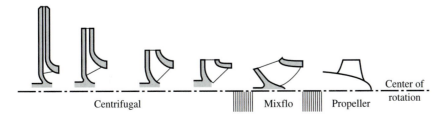

Figure 11.15 Impeller types used in pumps and blowers. *(Worthington Corp)*

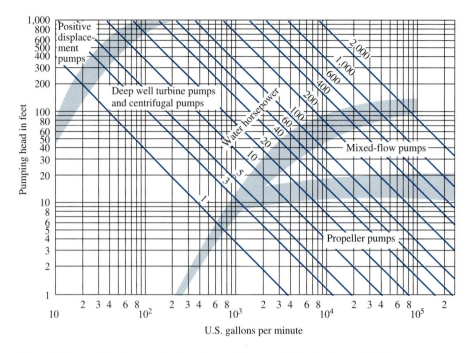

Figure 11.16   Chart for the selection of type of pump. *(Fairbanks, Morse & Co.)*

From the discharge, if b_2 is the width of the impeller at r_2 and vane thickness is neglected,

$$Q = 2\pi r_2 b_2 V_{r2}$$

Eliminating V_{r2} and substituting the preceding two equations into Eq. (11.3.8) give

$$H = \frac{u_2^2}{g} - \frac{u_2 Q \cot \beta_2}{2\pi r_2 b_2 g} \tag{11.5.1}$$

For a given pump and speed, H varies linearly with Q, as shown in Fig. 11.17. The usual design of centrifugal pump has $\beta_2 < 90°$, which gives decreasing head with increasing discharge. For blades radial at the exit, $\beta_2 = 90°$ and the theoretical head is independent of the discharge. For blades curved forward, $\beta_2 > 90°$ and the head rises with the discharge.

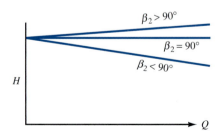

Figure 11.17   Theoretical head-discharge curves.

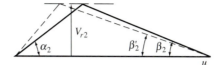

Figure 11.18   Effect of circulatory flow.

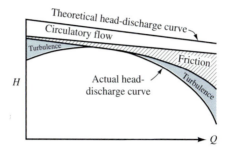

Figure 11.19   Head-discharge relations.

Actual Head-Discharge Curve

By subtracting head losses from the theoretical head-discharge curve the actual head-discharge curve is obtained. The most important subtraction is not an actual loss; it is a failure of the finite number of blades to impart the relative velocity with angle β_2 of the blades. Without perfect guidance (infinite number of blades) the fluid actually is discharged as if the blades had an angle β_2' which is less than β_2 (Fig. 11.18) for the same discharge. This inability of the blades to impart proper guidance reduces V_{u2} and, hence, decreases the actual head produced. This is called *circulatory flow* and is shown in Fig. 11.19. Fluid friction in flow through the fixed and moving passages causes losses that are proportional to the square of the discharge. They are shown in Fig. 11.19. The final head loss to consider is that of turbulence, the loss due to improper relative-velocity angle at the blade inlet. The pump can be designed for one discharge (at a given speed) at which the relative velocity is tangent to the blade at the inlet. This is the point of best efficiency, and shock or turbulence losses are negligible. For other discharges the loss varies about as the square of the discrepancy in approach angle, as shown in Fig. 11.19. The final lower line then represents the actual head-discharge curve. The shut-off head is usually about $u_2^2/2g$, or half of the theoretical shut-off head.

In addition to the head losses and reductions, pumps and blowers have torque losses due to bearing- and packing-friction and disk-friction losses from the fluid between the moving impeller and housing. Internal leakage is also an important power loss, in that fluid which has passed through the impeller with its energy increased escapes through clearances and flows back to the suction side of the impeller.

Characteristic curves showing head, efficiency, and brake horsepower as a function of discharge for a typical centrifugal pump with backward-curved vanes are given in Fig. 11.20. Pumps are not as efficient as turbines, in general, owing to the inherently high losses that result from conversion of kinetic energy into flow energy.

Typical performance curves for centrifugal, mixed-flow, and axial-flow pumps are shown in Figs. 11.21, 11.22, and 11.23, respectively.

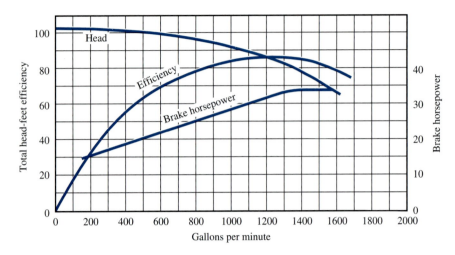

Figure 11.20   Characteristic curves for typical centrifugal pump, 10-in. impeller and 1750 rpm. *(Ingersoll-Rand Co.)*

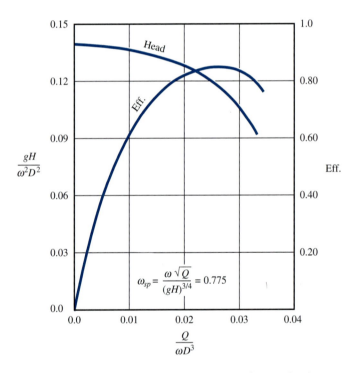

Figure 11.21   Dimensionless homologous curves for centrifugal pump, $D = 10$ in. (consistent units, ω in radians per second).

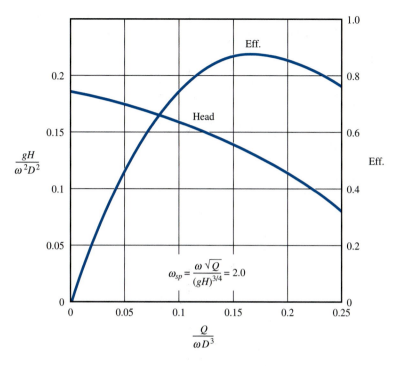

Figure 11.22   Dimensionless homologous curves for mixed-flow pump (consistent units, ω in radians per second).

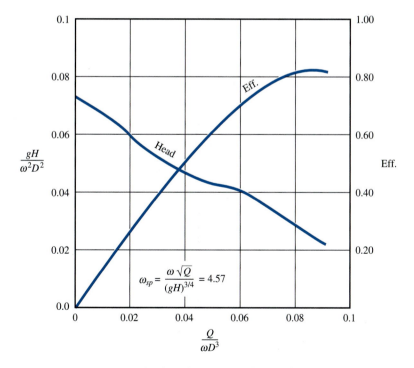

Figure 11.23   Dimensionless homologous curves for axial-flow pump, $D = 336$ mm (consistent units, ω in radians per second).

Example 11.7

A centrifugal water pump has an impeller (Fig. 11.6) with $r_2 = 12$ in., $r_1 = 4$ in., $\beta_1 = 20°$, and $\beta_2 = 10°$. The impeller is 2 in. wide at $r = r_1$ and $\frac{3}{4}$ in. wide at $r = r_2$. For 1800 rpm, neglecting losses and vane thickness, determine (a) the discharge for shockless entrance when $\alpha_1 = 90°$, (b) α_2 and the theoretical head H, (c) the horsepower required, and (d) the pressure rise through the impeller.

Solution

(a) The peripheral speeds are

$$u_1 = \frac{1800}{60}(2\pi)\left(\frac{1}{3}\right) = 62.8 \text{ ft/s} \qquad u_2 = 3u_1 = 188.5 \text{ ft/s}$$

The vector diagrams are shown in Fig. 11.24. With u_1 and the angles α_1 and β_1 known, the entrance velocity may be determined, $V_1 = u_1 \tan 20° = 22.85$ ft/s. Hence,

$$Q = 22.85(\pi)\left(\frac{2}{3}\right)\left(\frac{2}{12}\right) = 7.97 \text{ cfs}$$

(b) At the exit the radial velocity, V_{r2}, is

$$V_{r2} = \frac{7.97(12)}{2\pi(0.75)} = 20.3 \text{ ft/s}$$

By drawing u_2 (Fig. 11.24) and a parallel line distance V_{r2} from it, the vector triangle is determined when β_2 is laid off. Thus,

$$v_{u2} = 20.3 \cot 10° = 115 \text{ ft/s} \qquad V_{u2} = 188.15 - 115 = 73.5 \text{ ft/s}$$

$$\alpha_2 = \tan^{-1}\frac{20.3}{73.5} = 15°26' \qquad V_2 = 20.3 \csc 15°26' = 76.2 \text{ ft/s}$$

From Eq. (11.3.8)

$$H = \frac{u_2 V_2 \cos \alpha_2}{g} = \frac{u_2 \mathbf{V}_{u2}}{g} = \frac{188.5(73.5)}{32.2} = 430 \text{ ft}$$

(c) The fluid power is $Q\gamma H$ and, for this ideal pump, the horsepower required is

$$\text{Power} = \frac{Q\gamma H}{550} = \frac{7.97(62.4)(430)}{550} = 388 \text{ hp}$$

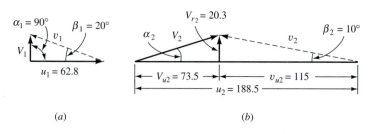

(a)	(b)

Figure 11.24   Vector diagrams for the (a) entrance and (b) exit of a pump impeller.

(*d*) By applying the energy equation from the entrance to exit of the impeller, including the energy H added (elevation change across the impeller is neglected),

$$H + \frac{V_1^2}{2g} + \frac{p_1}{\gamma} = \frac{V_2^2}{2g} + \frac{p_2}{\gamma}$$

and

$$\frac{p_2 - p_1}{\gamma} = 430 + \frac{22.85^2}{64.4} - \frac{76.2^2}{64.4} = 348 \text{ ft}$$

or

$$p_2 - p_1 = 348(0.433) = 151 \text{ psi}$$

EXERCISES

11.5.1   The head developed by a pump with hydraulic efficiency of 80 percent, for $u_2 = 100$ ft/s, $V_2 = 60$ ft/s, $\alpha_2 = 45°$, and $\alpha_1 = 90°$, is (*a*) 52.6; (*b*) 105.3; (*c*) 132; (*d*) 165; (*e*) none of these answers.

11.5.2   The correct relation for pump vector diagrams is (*a*) $\alpha_1 = 90°$ and $v_1 = u_1 \cot \beta_1$; (*b*) $V_{u2} = u_2 - V_{r2} \cot \beta_2$; (*c*) $\omega_2 = r_2/u_2$; (*d*) $r_1 V_1 = r_2 V_2$; (*e*) none of these answers.

11.6   IMPULSE TURBINES

In the impulse turbine all the available energy of the flow is converted by a nozzle into kinetic energy at atmospheric pressure before the fluid contacts the moving blades. Losses occur in the flow from the reservoir through the pressure pipe (penstock) to the base of the nozzle, which can be computed from pipe friction data. At the base of the nozzle (Fig. 11.25) the available energy, or total head, is

$$H_a = \frac{p_1}{\gamma} + \frac{V_1^2}{2g} \tag{11.6.1}$$

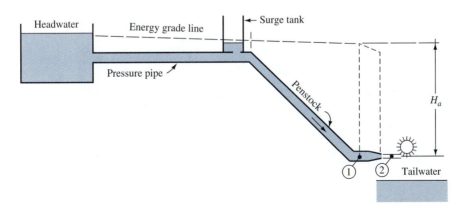

Figure 11.25   Impulse turbine system.

With the nozzle coefficient, C_v, the jet velocity, V_2, is

$$V_2 = C_v\sqrt{2gH_a} = C_v\sqrt{2g\left(\frac{p_1}{\gamma} + \frac{V_1^2}{2g}\right)} \qquad (11.6.2)$$

The head lost in the nozzle is

$$H_a - \frac{V_2^2}{2g} = H_a - C_v^2 H_a = H_a(1 - C_v^2) \qquad (11.6.3)$$

and the efficiency of the nozzle is

$$\frac{V_2^2/2g}{H_a} = \frac{C_v^2 H_a}{H_a} = C_v^2 \qquad (11.6.4)$$

The jet, with velocity V_2, strikes double-cupped buckets (Figs. 11.26 and 11.27), which split the flow and turn the relative velocity through angle θ (Fig. 11.27).

The x component of momentum (Fig. 11.27) is changed by

$$F = \rho Q(v_r - v_r \cos\theta)$$

and the power exerted on the vanes is

$$Fu = \rho Q u v_r (1 - \cos\theta) \qquad (11.6.5)$$

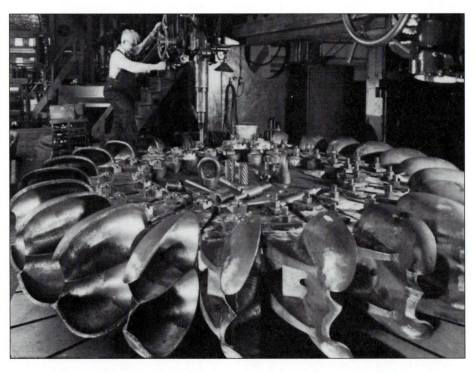

Figure 11.26   Southern California Edison, Big Creek 2A, 8.5-in.-diameter jet impluse buckets and disk in process of being reamed: 56,000 hp, 2200-ft head, and 300 rpm. *(Allis-Chalmer Mfg. Co.)*

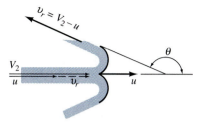

Figure 11.27   Flow-through bucket.

To maximize the power theoretically, two conditions must be met: $\theta = 180°$ and uv_r must be a maximum, that is, $u(V_2 - u)$ must be a maximum. To determine when $(uv_r)_{max}$ occurs, differentiate with respect to u and equate to zero

$$(V_2 - u) + u(-1) = 0$$

The condition is met when $u = V_2/2$. After making these substitutions into Eq. (11.6.5),

$$Fu = \rho Q \frac{V_2}{2}\left(V_2 - \frac{V_2}{2}\right)[1 - (-1)] = \gamma Q \frac{V_2^2}{2g} \tag{11.6.6}$$

which accounts for the total kinetic energy of the jet. The velocity diagram for these values shows that the absolute velocity leaving the vanes is zero.

Practically, when vanes are arranged on the periphery of a wheel (Fig. 11.26), the fluid must retain enough velocity to move out of the way of the following bucket. Most of the practical impulse turbines are Pelton wheels. The jet is split in two and turned in a horizontal plane, and half is discharged from each side to avoid any unbalanced thrust on the shaft. There are losses due to the splitter and to friction between jet and bucket surface, which make the most economical speed somewhat less than $V_2/2$. It is expressed in terms of the *speed factor*, ϕ, as

$$\phi = \frac{u}{\sqrt{2gH_a}} \tag{11.6.7}$$

For most efficient turbine operation ϕ has been found to depend upon specific speed as shown in Table 11.1. The angle θ of the bucket is usually 173 to 176°. If the diameter of the jet is d and the diameter of the wheel is D at the centerline of the buckets, it has been found in practice that the diameter ratio D/d should be about $54/N_s$ (ft, hp, rpm) or $206/N_s$ (m, kW, rpm) for maximum efficiency.

Table 11.1   Dependence of ϕ on specific speed*

N_s (m, kW, rpm)	N_s (ft, hp, rpm)	ϕ
7.62	2	0.47
11.42	3	0.46
15.24	4	0.45
19.05	5	0.44
22.86	6	0.433
26.65	7	0.425

* Modifed from Ref. [3].

In most installations only one jet is used; it discharges horizontally against the lower periphery of the wheel as shown in Fig. 11.25. The wheel speed is carefully regulated for the generation of electric power. A governor operates a needle valve that controls the jet discharge by changing its area. In so doing V_2 remains practically constant for a wide range of positions of the needle valve.

The efficiency of the power conversion drops off rapidly with a change in head (which changes V_2), as is evident when power is plotted against V_2 for constant u in Eq. (11.6.5). The wheel operates in atmospheric air although it is enclosed by a housing. It is, therefore, essential that the wheel be placed above the maximum floodwater level of the river into which it discharges. The head from nozzle to tail-water is wasted. Because of their inefficiency at other than the design head and because of the wasted head, Pelton wheels usually are employed for high heads, for example, from 200 m to more than 1 km. For high heads, the efficiency of the complete installation, from headwater to tailwater, may be in the high 80s.

Example 11.8

A Pelton wheel is to be selected to drive a generator at 600 rpm. The water jet is 75 mm in diameter and has a velocity of 100 m/s. With the blade angle at 170°, the ratio of vane speed to initial jet speed is 0.47. Neglecting losses, determine the (*a*) diameter of the wheel to the centerline of the buckets (vanes), (*b*) power developed, and (*c*) kinetic energy per newton remaining in the fluid.

Solution

(*a*) The peripheral speed of the wheel is

$$u = 0.47(100) = 47 \text{ m/s}$$

Then

$$\frac{600}{60}\left(2\pi\frac{D}{2}\right) = 47 \text{ m/s}$$

or

$$D = 1.495 \text{ m}$$

(*b*) From Eq. (11.6.5) the power, in kilowatts, is computed to be

$$(1000 \text{ kg/m}^3)\frac{\pi}{4}(0.075 \text{ m})^2 100 \text{ m/s}(47 \text{ m/s})(100 - 47) \text{ m/s} \cdot$$

$$[1 - (-0.9848)]\frac{1 \text{ kW}}{1000 \text{ W}} = 2184 \text{ kW}$$

(*c*) From Fig. 3.28, the absolute-velocity components leaving the vane are

$$V_x = (100 - 47)(-0.9848) + 47 = -5.2 \text{ m/s}$$
$$V_y = (100 - 47)(0.1736) = 9.2 \text{ m/s}$$

The kinetic energy remaining in the jet is

$$\frac{5.2^2 + 9.2^2}{2(9.806)} = 5.69 \text{ m·N/N}$$

A small impulse wheel is to be used to drive a generator for 60-Hz power. The head is 100 m, and the discharge is 40 L/s. Determine the diameter of the wheel at the centerline of the buckets and the speed of the wheel. $C_v = 0.98$. Assume an efficiency of 80 percent.

Example 11.9

Solution

The power is

$$P = \gamma Q H_a e = 9806(0.040)(100)(0.80) = 31.38 \text{ kW}$$

By taking a trial value of N_s of 15,

$$N = \frac{N_s H_a^{5/4}}{\sqrt{P}} = \frac{15(100^{5/4})}{\sqrt{31.38}} = 847 \text{ rpm}$$

For 60-Hz power the speed must be 3600 divided by the number of pairs of poles in the generator. For five pairs of poles the speed would be 3600/5 = 720 rpm, and for four pairs of poles it would be 3600/4 = 900 rpm. The closer speed of 900 is selected. Then

$$N_s = \frac{N\sqrt{P}}{H_a^{5/4}} = \frac{900\sqrt{31.38}}{100^{5/4}} = 15.94$$

For $N_s = 15.94$, take $\phi = 0.448$ from Table 11.1,

$$u = \phi\sqrt{2gH_a} = 0.448\sqrt{2(9.806)(100)} = 19.84 \text{ m/s}$$

and

$$\omega = \frac{900}{60} 2\pi = 94.25 \text{ rad/s}$$

The peripheral speed u and D and ω are related as

$$u = \frac{\omega D}{2} \qquad D = \frac{2u}{\omega} = \frac{2(19.84)}{94.25} = 421 \text{ mm}$$

The diameter d of the jet is obtained from the jet velocity V_2; thus,

$$V_2 = C_v\sqrt{2gH_a} = 0.98\sqrt{2(9.806)(100)} = 43.4 \text{ m/s}$$

$$a = \frac{Q}{V_2} = \frac{0.040}{43.4} = 9.22 \text{ cm}^2 \qquad d = \sqrt{\frac{4a}{\pi}} = \sqrt{\frac{0.000922}{0.7854}} = 34.3 \text{ mm}$$

where a is the area of the jet. Hence, the diameter ratio D/d is

$$\frac{D}{d} = \frac{421}{34.3} = 12.27$$

The desired diameter ratio for best efficiency is

$$\frac{D}{d} = \frac{206}{N_s} = \frac{206}{15.94} = 12.92$$

so the ratio D/d is satisfactory. The wheel diameter is 421 mm, and the speed is 900 rpm.

EXERCISES

11.6.1 An impulse turbine (*a*) always operates submerged; (*b*) makes use of a draft tube; (*c*) is most suited for low-head installations; (*d*) converts pressure head into velocity head throughout the vanes; (*e*) operates by an initial complete conversion to kinetic energy.

11.6.2 A 24-in.-diameter Pelton wheel turns at 400 rpm. Select the head, in feet, best suited for this wheel: (*a*) 7; (*b*) 30; (*c*) 120; (*d*) 170; (*e*) 480.

11.7   CAVITATION

When a liquid flows into a region where its pressure is reduced to vapor pressure, the liquid boils and vapor pockets develop. The vapor bubbles are carried along with the liquid until a region of higher pressure is reached, where they suddenly collapse. This process is called *cavitation*. If the vapor bubbles are near or in contact with a solid boundary when they collapse, the forces exerted by the liquid rushing into the cavities create very high localized pressures that cause pitting of the solid surface. The phenomenon is accompanied by noise and vibrations that resemble those of gravel going through a centrifugal pump.

In a flowing liquid, the *cavitation parameter*, σ, is useful in characterizing the susceptibility of the system to cavitate. It is defined by

$$\sigma = \frac{p - p_v}{\rho V^2 / 2} \tag{11.7.1}$$

in which p is the absolute pressure at the point of interest, p_v is the vapor pressure of the liquid, ρ is the density of the liquid, and V is the undisturbed, or reference, velocity. The cavitation parameter is a form of pressure coefficient. Two geometrically similar systems would be equally likely to cavitate or would have the same degree of cavitation for the same value of σ. When $\sigma = 0$, the pressure is reduced to vapor pressure and boiling should occur.

Tests made on chemically pure liquids show that they will sustain high tensile stresses, of the order of megapascals, which is in contradiction to the concept of cavities forming when pressure is reduced to vapor pressure. Since there is generally spontaneous boiling when vapor pressure is reached with commercial or technical liquids, it is generally accepted that nuclei must be present around which the vapor bubbles form and grow. The nature of the nuclei is not thoroughly understood, but they may be microscopic dust particles or other contaminants, which are widely dispersed through technical liquids.

Cavitation bubbles can form on nuclei, grow, then move into an area of higher pressure and collapse, all in a few thousandths of a second in flow inside a turbomachine. In aerated water the bubbles have been photographed as they move through several oscillations, but this phenomenon does not seem to occur in nonaerated liquids. Surface tension of the vapor bubbles appears to be an important property accounting for the high-pressure pulses resulting from collapse of a vapor bubble. Experiments indicate pressures of the order of a gigapascal based on the analysis of strain waves in a photoelastic specimen exposed to cavitation [4]. Pressures of this magnitude appear to be reasonable, in line with the observed mechanical damage caused by cavitation.

Table 11.2   Mass loss in materials used in hydraulic machines

Alloy	Mass loss after 2h, mg
Rolled stellite †	0.6
Welded aluminum bronze ‡	3.2
Cast aluminum bronze §	5.8
Welded stainless steel (two layers, 17 percent Cr and 7 percent Ni)	6.0
Hot-rolled stainless steel (26 percent Cr and 13 percent Ni)	8.0
Tempered, rolled stainless steel (12 percent Cr)	9.0
Cast stainless steel (18 percent Cr and 8 percent Ni)	13.0
Cast stainless steel (12 percent Cr)	20.0
Cast manganese bronze	80.0
Welded mild steel	97.0
Plate steel	98.0
Cast steel	105.0
Aluminum	124.0
Brass	156.0
Cast Iron	224.0

† This material is not suitable for ordinary use, despite its high resistance, because of its high cost and difficulty in machining.
‡ Ampco-Trode 200: 83 percent Cu, 11.3 percent Al, and 5.8 percent Fe.
§ Ampco 20: 83.1 percent Cu, 12.4 percent Al, and 4.1 percent Fe.

The formation and collapse of great numbers of bubbles on a surface subject that surface to intense local stressing, which appears to damage the surface by fatigue. Some ductile materials withstand battering for a period, called the *incubation period*, before damage is noticeable, whereas brittle materials may lose mass immediately. There may be certain electrochemical, corrosive, and thermal effects which hasten the deterioration of exposed surfaces. Rheingans [5] has collected a series of measurements made by magnetostriction-oscillator tests, showing mass losses of various metals used in hydraulic machines (see Table 11.2).

Protection against cavitation should start with the hydraulic design of the system in order to avoid the low pressures if practicable. Otherwise, use of special cavitation-resistant materials or coatings may be effective. Small amounts of air entrained into water systems have markedly reduced cavitation damage, and studies indicate that cathodic protection is helpful.

The formation of vapor cavities decreases the useful channel space for liquid and thus decreases the efficiency of a fluid machine. Cavitation causes three undesirable conditions: lowered efficiency, damage to flow passages, and noise and vibrations. Curved vanes are particularly susceptible to cavitation on their convex sides and may have localized areas where cavitation causes pitting or failure. Since all turbomachinery and ship propellers and many hydraulic structures are subject to cavitation, special attention must be given to it in their design.

A *cavitation index*, σ', is useful in the proper selection of turbomachinery and in its location with respect to suction or tailwater elevation. The minimum pressure in a pump or turbine generally occurs along the convex side of vanes near the low-pressure side of the impeller. In Fig. 11.28, if e is the point of minimum pressure, the energy equation applied between e and the downstream liquid surface can be written as

$$\frac{p_e}{\gamma} + \frac{V_e^2}{2g} + z_s = \frac{p_a}{\gamma} + 0 + 0 + h_l$$

in which p_a is the atmospheric pressure and p_e is the absolute pressure. For cavitation to occur at e, the pressure must be equal to or less than p_v, the vapor pressure. If $p_e = p_v$, then

$$\sigma' = \frac{V_e^2}{2gH} = \frac{p_a - p_v - \gamma z_s + \gamma h_l}{\gamma H} \tag{11.7.2}$$

is the ratio of energy available at e to total energy H across the unit, since the only energy is kinetic energy. The ratio σ' is a cavitation index or number. The critical value σ_c can be determined by a test on a model of the homologous series. For cavitationless performance, the low-pressure setting z_s for an impeller installation must be so fixed that the resulting value of σ' is greater than that of σ_c. When the flow is reversed in Fig. 11.28, as for a pump, the sign of h_l changes.

Example 11.10

Tests of a pump model indicate a σ_c of 0.10. A homologous unit, installed at a location where $p_a = 90$ kPa and $p_v = 3.5$ kPa, has to pump water against a head of 25 m. The head loss from the suction reservoir to the pump impeller is 0.35 N·m/N. What is the maximum permissible suction head?

Solution

Solving Eq. (11.7.2) for z_s and substituting the values of σ_c, H, p_a, and p_v give

$$z_s = \frac{p_a - p_v}{\gamma} - \sigma' H - h_l = \frac{90{,}000 - 3500}{9806} - 0.10(25) - 0.35 = 5.97 \text{ m}$$

The smaller the value of z_s, the greater the value of the plant σ' and the greater the assurance against cavitation.

The *net positive suction head* (NPSH) is frequently used in the specification of minimum suction conditions for a turbomachine. It is defined as

$$\text{NPSH} = \frac{V_e^2}{2g} = \frac{p_a - p_v - \gamma z_s}{\gamma} \pm h_l \tag{11.7.3}$$

where the plus sign on h_l is for turbines and the minus sign for pumps. A test is run on the machine to determine the maximum value of z_s for operation of the machine with no impairment of efficiency and without objectionable noise or damage. Then, from this test NPSH is calculated from Eq. (11.7.3). Any setting of this machine where the suction lift is less than z_s, as found from Eq. (11.7.3), is then acceptable. Note that z_s is positive when the suction reservoir is below the turbomachine, as in Fig. 11.28. A *suction specific speed*, S, for homologous units can be formulated.

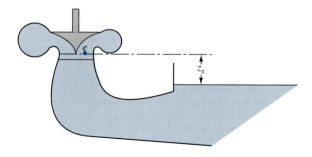

Figure 11.28   Turbine or pump setting.

Elimination of D_e in

$$\text{NPSH} = \frac{V_e^2}{2g} \sim \frac{Q^2}{D_e^4} \qquad \text{and} \qquad \frac{Q}{ND_e^3} = \text{constant}$$

leads to

$$S = \frac{N\sqrt{Q}}{(\text{NPSH})^{3/4}} \qquad (11.7.4)$$

When different units of a series are operating under cavitating conditions, equal values of S indicate a similar degree of cavitation. When cavitation is not present, the equation is not valid.

EXERCISES

11.7.1 The cavitation parameter is defined by

(a) $\dfrac{p_v - p}{\rho V^2/2}$ (b) $\dfrac{p_{\text{atm}} - p_v}{\rho V^2/2}$ (c) $\dfrac{p - p_v}{\gamma V^2/2}$ (d) $\dfrac{p - p_v}{\rho V^2/2}$ (e) none of these answers.

11.7.2 Cavitation is caused by (a) high velocity; (b) low barometric pressure; (c) high pressure; (d) low pressure; (e) low velocity.

PROBLEMS

11.1 Use Eqs. (11.1.1) and (11.1.3), together with $P = \gamma QH$ for power, to develop the homologous relation for P in terms of speed and diameter.

11.2 A centrifugal pump is driven by an induction motor that reduces in speed as the pump load increases. A test determines several sets of values of N, Q, and H for the pump. How is a characteristic curve of the pump for a constant speed determined from these data?

11.3 What is the specific speed of the pump of Ex. 11.1 at its point of best efficiency?

11.4 Plot the dimensionless characteristic curve of the pump of Example 11.1. On this same curve plot several points from the characteristics of the new (1.444-m) pump. Why are they not exactly on the same curve?

11.5 Determine the size and synchronous speed of a pump homologous to the 2-m pump of Ex. 11.1 that will produce 3 m³/s at 100-m head at its point of best efficiency.

11.6 Develop the characteristic curve for a homologous pump of the series of Ex. 11.1 for a 42-cm-diameter discharge and 1800 rpm.

11.7 A pump with a 200-mm-diameter impeller discharges 100 L/s at 1140 rpm and 10-m head at its point of best efficiency. What is its specific speed?

11.8 A hydroelectric site has a head of 100 m and an average discharge of 10 m³/s. For a generator speed of 200 rpm, what specific speed turbine is needed? Assume an efficiency of 92 percent.

11.9 A model turbine, $N_s = 36$, with a 14-in.-diameter impeller develops 27 hp at a head of 44 ft and an efficiency of 86 percent. What are the discharge and speed of the model?

11.10   What size and synchronous speed would be needed by a homologous unit to the one in Prob. 11.9 to discharge 600 cfs at 260 ft of head?

11.11   A pump is needed to deliver 3-m³/s water while producing 25-m head. If it operates at 600 rpm, what type of pump would be best suited for this installation?

11.12   What type of pump should be selected in an application where the pump speed is 1800 rpm and the pressure rise is 900 kPa? Gasoline ($\rho = 680$ kg/m³) is to be pumped at a flow rate of 0.2 m³/s.

11.13   If 22-m³/s water flowing through the fixed vanes of a turbine has a tangential component of 2 m/s at a radius of 1.25 m and the impeller, turning at 180 rpm, discharges in an axial direction, what torque is exerted on the impeller?

11.14   In Prob. 11.13, neglecting losses, what is the head on the turbine?

11.15   A generator with speed $N = 240$ rpm is to be used with a turbine at a site where $H = 120$ m and $Q = 8$ m³/s. Neglecting losses, what tangential component must be given to the water at $r = 1$ m by the fixed vanes? What torque is exerted on the impeller? How much horsepower is produced?

11.16   At what angle should the wicket gates of a turbine be set to extract 9 MW from a flow of 25 m³/s? The diameter of the opening just inside the wicket gates is 3.5 m, and the height is 1 m. The turbine runs at 200 rpm, and flow leaves the runner in an axial direction.

11.17   For a given setting of wicket gates how does the moment of momentum vary with the discharge?

11.18   Assuming constant axial velocity just above the runner of the propeller turbine of Prob. 11.16, calculate the tangential-velocity components if the hub radius is 300 mm and the outer radius is 900 mm.

11.19   Determine the vane angles β_1 and β_2 for entrance and exit from the propeller turbine of Prob. 11.18 so that no angular momentum remains in the flow. (Compute the angles for the inner radius, outer radius, and midpoint).

11.20   Neglecting losses, what is the head on the turbine of Prob. 11.16?

11.21   The hydraulic efficiency of a turbine is 95 percent, and its theoretical head is 80 m. What is the actual head required?

11.22   A turbine test model with 260-mm-diameter impeller showed an efficiency of 90 percent. What efficiency could be expected from a 1.2-m-diameter impeller?

11.23   Construct a theoretical head-discharge curve for the following specifications of a centrifugal pump: $r_1 = 50$ mm, $r_2 = 100$ mm, $b_1 = 25$ mm, $b_2 = 20$ mm, $N = 1200$ rpm, and $\beta_2 = 30°$.

11.24   A centrifugal water pump (Fig. 11.6) has an impeller $r_1 = 2.75$ in., $b_1 = 1\frac{3}{8}$ in., $r_2 = 4.5$ in., $b_2 = \frac{3}{4}$ in., $\beta_1 = 30°$, and $\beta_2 = 45°$ (b_1 and b_2 are impeller width at r_1 and r_2, respectively). Neglect the thickness of the vanes. For 1800 rpm, calculate (*a*) the design discharge for no prerotation of entering fluid, (*b*) α_2 and the theoretical head at the point of best efficiency, and (*c*) for hydraulic efficiency of 85 percent and overall efficiency of 78 percent, the actual head produced, losses in foot-pounds per pound, and brake horsepower.

11.25   A centrifugal pump has an impeller with dimensions $r_1 = 75$ mm, $r_2 = 160$ mm, $b_1 = 50$ mm, $b_2 = 30$ mm, and $\beta_1 = \beta_2 = 30°$. For a discharge of 55 L/s and shockless entry to vanes compute the (*a*) speed, (*b*) head, (*c*) torque, (*d*) power, and (*e*) pressure rise across the impeller. Neglect losses. $\alpha_1 = 90°$.

11.26 A centrifugal water pump with impeller dimensions $r_1 = 2$ in., $r_2 = 5$ in., $b_1 = 3$ in., $b_2 = 1.5$ in., and $\beta_2 = 60°$ is to pump 5 cfs at 64-ft head. Determine (*a*) β_1, (*b*) the speed, (*c*) the horsepower, and (*d*) the pressure rise across the impeller. Neglect losses, and assume no shock at the entrance. $\alpha_1 = 90°$.

11.27 Select values of r_1, r_2, β_1, β_2, b_1, and b_2 of a centrifugal impeller to take 30-L/s water from a 100-mm-diameter suction line and increase its energy by 15 m·N/N. $N = 1200$ rpm; $\alpha_1 = 90°$. Neglect losses.

11.28 A pump has blade angles $\beta_1 = \beta_2$, $b_1 = 2b_2 = 25$ mm, and $r_1 = r_2/3 = 50$ mm. For a theoretical head of 30 m at a discharge at best efficiency of 30 L/s, determine the blade angles and speed of the pump. Neglect the thickness of the vanes and assume perfect guidance. *Hint*: Write down every relation you know connecting β_1, β_2, b_1, b_2, r_1, r_2, u_1, u_2, H_{th}, Q, V_{r2}, V_{u2}, V_1, ω, and N from the two velocity-vector diagrams, and by substitution reduce to one unknown.

11.29 A mercury-water differential manometer, $R' = 700$ mm, is connected from the 100-mm-diameter suction pipe to the 80-mm-diameter discharge pipe of a pump. The centerline of the suction pipe is 300 mm below the discharge pipe. For $Q = 60$-L/s water, calculate the head developed by the pump.

11.30 The impeller for a blower (Fig. 11.29) is 18 in. wide. It has straight blades and turns at 1200 rpm. For 10,000-ft³/min air and $\gamma = 0.08$ lb/ft³, calculate (*a*) entrance and exit blade angles ($\alpha_1 = 90°$), (*b*) the head produced, in inches of water, and (*c*) the theoretical horsepower required.

11.31 An air blower is to be designed to produce pressure of 100-mm H₂O when operating at 3600 rpm. $\gamma = 11.5$ N/m³; $r_2 = 1.1r_1$; $\beta_2 = \beta_1$; width of impeller is 100 mm; and $\alpha_1 = 90°$. Find r_1.

11.32 In Prob. 11.31, when $\beta_1 = 30°$, calculate the discharge in cubic meters per minute.

11.33 A site for a Pelton wheel has a steady flow of 55 L/s with a nozzle velocity of 75 m/s. With a blade angle of 174° and $C_v = 0.98$, for 60-Hz power, determine the (*a*) diameter of the wheel, (*b*) speed, (*c*) power, and (*d*) energy remaining in the water. Neglect losses.

11.34 An impulse wheel is to be used to develop 50-Hz power at a site where $H = 120$ m and $Q = 75$ L/s. Determine the diameter of the wheel and its speed. $C_v = 0.97$ and $e = 82$ percent.

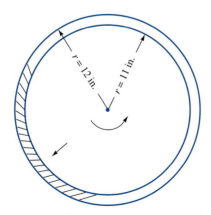

Figure 11.29 Problem 11.30.

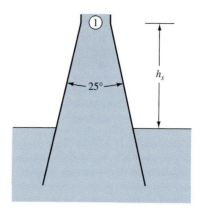

Figure 11.30   Problem 11.35.

11.35   A turbine draft tube (Fig. 11.30) expands from 6- to 18-ft diameter. At section 1 the velocity is 30 ft/s for vapor pressure of 1 ft and barometric pressure of 32 ft of water. Determine h_s for incipient cavitation (pressure equal to vapor pressure at section 1).

11.36   What is the cavitation parameter at a point in flowing water where $t = 20°C$, $p = 14$ kPa, and the velocity is 12 m/s?

11.37   A turbine with $\sigma_c = 0.08$ is to be installed at a site where $H = 60$ m and a water barometer stands at 8.3 m. What is the maximum permissible impeller setting above tailwater?

11.38   The allowable NPSH provided by the pump manufacturer at a flow of 0.06 m³/s is 3.5 m. Determine z_s, the height of the pump above the suction reservoir. The water temperature is 25°C, atmospheric pressure is 101-kPa absolute, and the head loss from the reservoir to pump is 0.3 m·N/N. How would the elevation change if the local barometric pressure were 82-kPa absolute?

REFERENCES

1.   L. F. Moody, "The Propeller Type Turbine," *Trans. ASCE*, vol. 89, pp. 625–647, 1926.

2.   A. J. Stepanoff, *Centrifugal and Axial Flow Pumps,* 2nd ed., Wiley, New York, 1957.

3.   J. W. Daily, "Hydraulic Machinery," in *Engineering Hydraulics*, H. Rouse (ed.), Wiley, New York, 1950.

4.   R. T. Knapp, J. W. Daily, and F. G. Hammett, *Cavitation*, McGraw Hill, New York, 1970.

5.   W. J. Rheingans, "Selecting Materials to Avoid Cavitation Damage," *Mater. Des. Eng.*, pp. 102–106, 1958.

chapter
12

Closed-Conduit Flow

This chapter enlarges the scope of steady pipeline flows that may be studied (Secs. 12.1–12.6) and provides an introduction to unsteady flow in pipelines (Secs. 12.7–12.11). The basic procedures for solving problems in incompressible steady flow in closed conduits are presented in Secs. 6.7 and 6.8, where simple pipe-flow situations are discussed, including losses due to change in cross section or direction of flow. The great majority of practical problems deal with turbulent flow, and velocity distributions in turbulent pipe flow are discussed in Sec. 6.4. The Darcy-Weisbach equation is introduced in Chap. 6 to relate frictional losses to flow rate in pipes, with the friction factor determined from the Moody diagram.

In this chapter exponential friction formulas commonly used in commercial and industrial applications are discussed. The use of the hydraulic and energy grade lines in solving problems is reiterated before particular applications are developed. Complex flow problems are investigated, including hydraulic systems that incorporate various different elements such as pumps and piping networks. The use of spreadsheets and numerical methods in analysis and design becomes particularly relevant when multielement systems are being investigated.

Unsteady flows are a common occurrence in pipelines. Hydraulic-transient analysis deals with the calculation of pressures and velocities during an unsteady-state mode of operation of a system. This may be caused by adjustment of a valve in a piping system, stopping a pump, or innumerable other possible changes in system operation. The analysis of unsteady flow is much more complex than that of steady flow. Another independent variable, time, enters the equations, and equations may be partial differential equations rather than ordinary differential equations. The oscillation of a U tube is first studied with application of the concepts to pipelines, followed by the establishment of flow in a single pipeline system. Each of these unsteady flows is analyzed as a lumped mass, which results in an ordinary differential equation. Equations are next developed for cases with more severe changes in velocity that require consideration of liquid compressibility and pipe-wall elasticity (commonly called waterhammer or liquid hammer). The partial differential equations are solved using numerical methods.

12.1   STEADY FLOW: EXPONENTIAL PIPE-FRICTION FORMULAS

Industrial pipe-friction formulas are usually empirical, of the form

$$\frac{h_f}{L} = \frac{RQ^n}{D^m} \tag{12.1.1}$$

in which h_f/L is the head loss per unit length of pipe (i.e., the slope of the energy grade line), Q is the discharge, and D is the inside pipe diameter. The resistance coefficient R is a function of pipe roughness only. An equation with specified exponents and coefficient R is valid only for the fluid viscosity for which it is developed, and it is normally limited to a range of Reynolds numbers and diameters. In its range of applicability such an equation is convenient, and nomographs are often used to aid problem solution.

The Hazen-Williams formula [1]† for flow of water at ordinary temperatures through pipes is of this form with R given by

$$R = \begin{cases} \dfrac{4.727}{C^n} & \text{USC units} \tag{12.1.2} \\[2ex] \dfrac{10.675}{C^n} & \text{SI units} \tag{12.1.3} \end{cases}$$

with $n = 1.852$, $m = 4.8704$, and C is dependent upon the roughness as follows:

C	Condition
140	Extremely smooth, straight pipes; asbestos-cement
130	Very smooth pipes; concrete; new cast iron
120	Wood stave; new welded steel
110	Vitrified clay; new riveted steel
100	Cast iron after years of use
95	Riveted steel after years of use
60–80	Old pipes in bad condition

One can develop a special-purpose formula for a particular application by using the Darcy-Weisbach equation and friction factors from the Moody diagram or, alternatively, by using experimental data if they are available [2]. Exponential formulas developed from experimental results are generally very useful and handy in the region over which the data were gathered. Extrapolations and applications to other situations must be carried out with caution.

A comparison between the Hazen-Williams equation and the Darcy-Weisbach equation with friction factors from the Moody diagram is presented in Fig. 12.1. It shows equivalent values of f versus Reynolds number for three typical Hazen-Williams roughness values 70, 100, and 140. The fluid is water at 15°C.

By equating the slope of the hydraulic grade line in the Darcy-Weisbach equation, $h_f/L = fQ^2/2gDA^2$, to Eq. (12.1.1), solving for f, and introducing the Reynolds number to eliminate Q (in SI),

$$f = \frac{1014.2}{C^{1.852}D^{0.0184}} \mathbf{R}^{-0.148} \tag{12.1.4}$$

---

†Numbered references will be found at the end of the chapter.

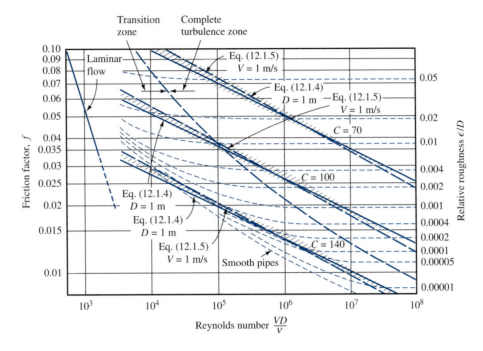

Figure 12.1 Comparison of Hazen-Williams and Darcy-Weisbach equations on the Moody diagram.

For a given Hazen-Williams coefficient C and diameter D, the friction factor reduces with increasing Reynolds number. A similar solution for f in terms of C, Reynolds number, and V can be developed by combining the same equations and eliminating D,

$$f = \frac{1304.56\,V^{0.0184}}{C^{1.852}}\mathbf{R}^{-0.1664} \qquad \textbf{(12.1.5)}$$

Note that f is not strongly dependent upon pipe diameter in Eq. (12.1.4). Similarly, the friction factor is not strongly dependent upon the velocity in Eq. (12.1.5).

In Fig. 12.1, at the three selected values of C, Eq. (12.1.4) is shown for a particular diameter of 1 m and Eq. (12.1.5) is shown for a specific velocity of 1 m/s. The shaded region around each of these lines shows the range of practical variation of the variables (0.025 m < D < 6 m and 0.030 m/s < V < 30 m/s).

The two formulations, Darcy-Weisbach and Hazen-Williams, for calculation of losses in a pipeline can be seen to be significantly different. The Darcy-Weisbach equation is probably more rationally based than other empirical exponential formulations and has received wide acceptance. However, when specific experimental data are available and an exponential formula based on the data is developed, it is preferable to the more general Moody diagram approach. The data must be reliable, and the equation can be considered valid only over the range of gathered data.

12.2 STEADY FLOW: HYDRAULIC AND ENERGY GRADE LINES

The concepts of *hydraulic* and *energy grade lines* are useful in analyzing more complex flow problems. If, at each point along a pipe system, the term p/γ is determined and plotted as a vertical distance above the center of the pipe, the locus of points is

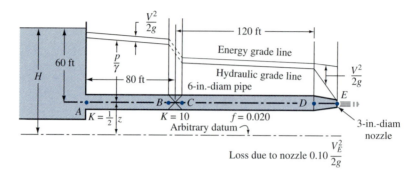

Figure 12.2 Hydraulic and energy grade lines.

the hydraulic grade line. More generally, the plot of the two terms

$$\frac{p}{\gamma} + z$$

as ordinate, against length along the pipe, as abscissa, produces the hydraulic grade line. The hydraulic grade line, or piezometric head line, is the locus of heights to which liquid would rise in vertical glass tubes connected to piezometer openings in the line. When the pressure in the line is less than atmospheric, p/γ is negative and the hydraulic grade line is below the pipeline.

The energy grade line is a line joining a series of points marking the available energy in meter-newtons per newton for each point along the pipe as ordinate, plotted against distance along the pipe as the abscissa. It consists of the plot of

$$\frac{V^2}{2g} + \frac{p}{\gamma} + z$$

for each point along the line. By definition, the energy grade line is always vertically above the hydraulic grade line a distance of $V^2/2g$, neglecting the kinetic-energy correction factor.

The hydraulic and energy grade lines are shown in Fig. 12.2 for a simple pipeline containing a square-edged entrance, a valve, and a nozzle at the end of the line. To construct these lines when the reservoir surface is given, it is necessary first to apply the energy equation from the reservoir to the exit, including all minor losses as well as pipe friction, and to solve for the velocity head $V^2/2g$. Then, to find the elevation of hydraulic grade line at any point, the energy equation is applied from the reservoir to that point, including all losses between the two points. The equation is solved for $p/\gamma + z$, which is plotted above the arbitrary datum. To find the energy grade line at the same point, the equation is solved for $V^2/2g + p/\gamma + z$, which is plotted above the arbitrary datum.

The reservoir surface is the hydraulic grade line and is also the energy grade line. At the square-edged entrance the energy grade line drops by $0.5 \cdot V^2/2g$ because of the loss there, and the hydraulic grade line drops by $1.5 \cdot V^2/2g$. This is made obvious by applying the energy equation between the reservoir surface and a point just downstream from the pipe entrance, that is,

$$H + 0 + 0 = \frac{V^2}{2g} + z + \frac{p}{\gamma} + 0.5\frac{V^2}{2g}$$

Solving for $z + p/\gamma$,

$$z + \frac{p}{\gamma} = H - 1.5\frac{V^2}{2g}$$

shows the drop of $1.5 \cdot V^2/2g$. The head loss due to the sudden entrance does not actually occur at the entrance itself, but over a distance of 10 or more diameters of pipe downstream. It is customary to show it at the fitting.

Example 12.1

Determine the elevation of hydraulic and energy grade lines at points A, B, C, D, and E of Fig. 12.2. $z = 10$ ft.

Solution

Solving for the velocity head is accomplished by applying the energy equation from the reservoir to E,

$$10 + 60 + 0 + 0 = \frac{V_E^2}{2g} + 10 + 0 + \frac{1}{2}\frac{V^2}{2g} + 0.020\frac{200}{0.50}\frac{V^2}{2g} + 10\frac{V^2}{2g} + 0.10\frac{V_E^2}{2g}$$

From the continuity equation, $V_E = 4V$. After simplifying,

$$60 = \frac{V^2}{2g}\left[16 + \frac{1}{2} + 8 + 10 + 16(0.1)\right] = 36.1\frac{V^2}{2g}$$

and $V^2/2g = 1.66$ ft. Applying the energy equation for the portion from the reservoir to A gives

$$70 + 0 + 0 = \frac{V^2}{2g} + \frac{p}{\gamma} + z + 0.5\frac{V^2}{2g}$$

Hence, the hydraulic grade line at A is

$$\frac{p}{\gamma} + z_A = 70 - 1.5\frac{V^2}{2g} = 70 - 1.5(1.66) = 67.51 \text{ ft}$$

The energy grade line for A is

$$\frac{V^2}{2g} + z + \frac{p}{\gamma} = 67.51 + 1.66 = 69.17 \text{ ft}$$

For B,

$$70 + 0 + 0 = \frac{V^2}{2g} + \frac{p}{\gamma} + z + 0.5\frac{V^2}{2g} + 0.02\frac{80}{0.5}\frac{V^2}{2g}$$

and

$$\frac{p}{\gamma} + z_B = 70 - (1.5 + 3.2)(1.66) = 62.19 \text{ ft}$$

the energy grade line is at $62.19 + 1.66 = 63.85$ ft.

Across the valve the hydraulic grade line drops by $10V^2/2g$, or 16.6 ft. Hence, at C the energy and hydraulic grade lines are at 47.25 ft and 45.59 ft, respectively.

At point D,

$$70 = \frac{V^2}{2g} + \frac{p}{\gamma} + z + \left(10.5 + 0.02\frac{200}{0.50}\right)\frac{V^2}{2g}$$

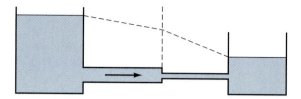

Figure 12.3   Hydraulic grade line for long pipeline where minor losses are neglected or included as equivalent lengths of pipe.

and

$$\frac{p}{\gamma} + z_D = 70 - 19.5(1.66) = 37.6 \text{ ft}$$

with the energy grade line at $37.6 + 1.66 = 39.26$ ft.

At point E the hydraulic grade line is 10 ft, and the energy grade line is

$$z + \frac{V_E^2}{2g} = 10 + 16\frac{V^2}{2g} = 10 + 16(1.66) = 36.6 \text{ ft}$$

The *hydraulic gradient* is the slope of the hydraulic grade line if the conduit is horizontal; otherwise, it is

$$\frac{d(z + p/\gamma)}{dL}$$

The *energy gradient* is the slope of the energy grade line if the conduit is horizontal; otherwise, it is

$$\frac{d(z + p/\gamma + V^2/2g)}{dL}$$

In many situations involving long pipelines the minor losses can be neglected when they are less than 5 percent of the pipe-friction losses or they can be included as equivalent lengths of pipe which are added to the actual length in solving the problem. For these situations the value of the velocity head $V^2/2g$ is small compared with $f(L/D)V^2/2g$ and is neglected.

In this special but very common case, when minor effects are neglected, the energy and hydraulic grade lines are superposed. The single grade line, shown in Fig. 12.3, is commonly referred to as the *hydraulic grade line*. No change in hydraulic grade line is shown for minor losses. For these situations with long pipelines the hydraulic gradient becomes h_f/L, with h_f given by the Darcy-Weisbach equation

$$h_f = f\frac{L}{D}\frac{V^2}{2g} \qquad\qquad \text{(12.2.1)}$$

or by Eq. (12.1.1). Flow (except through a pump) is always in the direction of decreasing energy grade line.

Pumps add energy to the flow, a fact which can be expressed in the energy equation either by including a *negative loss* or by stating the energy per unit weight added as a positive term on the upstream side of the equation. The hydraulic grade line rises

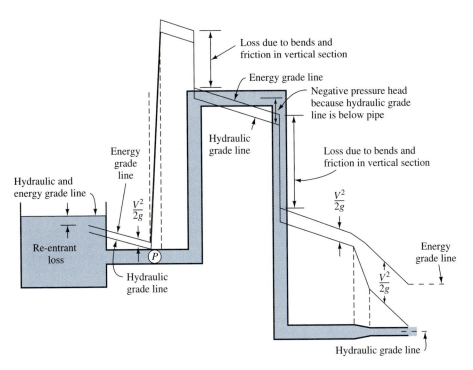

Figure 12.4 Hydraulic and energy grade lines for a system with pump and siphon.

sharply at a pump. Figure 12.4 shows the hydraulic and energy grade lines for a system with a pump and a siphon. The true slope of the grade lines can be shown only for horizontal lines.

Example 12.2

A pump with a shaft input of 7.5 kW and an efficiency of 70 percent is connected in a waterline carrying 0.1 m³/s. The pump has a 150-mm-diameter suction line and a 120-mm-diameter discharge line. The suction line enters the pump 1 m below the discharge line. For a suction pressure of 70 kN/m², calculate the pressure at the discharge flange and the rise in the hydraulic grade line across the pump.

Solution

If the energy added in meter-newtons per newton is symbolized by E, the fluid power added is

$$Q\gamma E = 7500(0.70) \qquad \text{or} \qquad E = \frac{7500(0.7)}{0.1(9806)} = 5.354 \text{ m}$$

Applying the energy equation from suction flange to discharge flange gives

$$\frac{V_s^2}{2g} + \frac{p_s}{\gamma} + 0 + 5.354 = \frac{V_d^2}{2g} + \frac{p_d}{\gamma} + 1$$

in which the subscripts s and d refer to the suction and discharge conditions, respectively.

From the continuity equation

$$V_s = \frac{0.1(4)}{0.15^2 \pi} = 5.66 \text{ m/s} \qquad V_d = \frac{0.1(4)}{0.12^2 \pi} = 8.84 \text{ m/s}$$

Solving for p_d gives

$$\frac{p_d}{\gamma} = \frac{5.66^2}{2(9.806)} + \frac{70,000}{9806} + 5.354 - \frac{8.84^2}{2(9.806)} - 1 = 9.141 \text{ m}$$

and $p_d = 89.6 \text{ kN/m}^2$. The rise in hydraulic grade line is

$$\left(\frac{p_d}{\gamma} + 1\right) - \frac{p_s}{\gamma} = 9.141 + 1 - \frac{70,000}{9806} = 3.002 \text{ m}$$

In this example much of the energy was added in the form of kinetic energy, and the hydraulic grade line rises only 3.002 m for a rise of energy grade line of 5.354 m.

A turbine takes energy from the flow and causes a sharp drop in both the energy and the hydraulic grade lines. The energy removed per unit weight of fluid can be treated as a loss in computing grade lines.

A closed conduit, arranged as in Fig. 12.5, which lifts the liquid to an elevation higher than its free surface and then discharges it at a lower elevation is a *siphon*. It has certain limitations in performance due to the low pressures that occur near the summit s.

Assuming that the siphon flows full, with a continuous liquid column throughout, the application of the energy equation for the portion from 1 to 2 produces the equation

$$H = \frac{V^2}{2g} + K\frac{V^2}{2g} + f\frac{L}{D}\frac{V^2}{2g}$$

in which K is the sum of all the minor-loss coefficients. Factoring out the velocity head gives

$$H = \frac{V^2}{2g}\left(1 + K + \frac{fL}{D}\right) \tag{12.2.2}$$

which is solved in the same fashion as the simple pipe problems of the first or second type. With the discharge known, the solution for H is straightforward, but the solution for velocity with H given is a trial solution started by assuming a value for f.

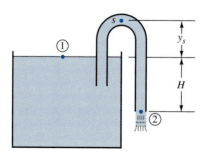

Figure 12.5   Siphon.

The pressure at the summit s is found by applying the energy equation for the portion between 1 and s after Eq. (12.2.2) is solved. It is

$$0 = \frac{V^2}{2g} + \frac{p_s}{\gamma} + y_s + K'\frac{V^2}{2g} + f\frac{L'}{D}\frac{V^2}{2g}$$

in which K' is the sum of the minor-loss coefficients between the two points and L' is the length of conduit upstream from s. Solving for the pressure gives

$$\frac{p_s}{\gamma} = -y_s - \frac{V^2}{2g}\left(1 + K' + \frac{fL'}{D}\right) \qquad (12.2.3)$$

which shows that the pressure is negative and that it decreases with y_s and $V^2/2g$. If the solution of the equation should be a value of p_s/γ equal to or less than the vapor pressure† of the liquid, Eq. (12.2.2) is not valid because the vaporization of portions of the fluid column invalidates the incompressibility assumption used in deriving the energy equation.

Although Eq. (12.2.2) is not valid for this case, theoretically there will be a discharge so long as y_s plus the vapor pressure is less than local atmospheric pressure expressed in length of the fluid column. When Eq. (12.2.3) yields a pressure less than vapor pressure at s, pressure at s can be taken as vapor pressure. Then, with this pressure known, Eq. (12.2.3) is solved for $V^2/2g$, and the discharge is obtained therefrom. It is assumed that air does not enter the siphon at 2 and break, at s, the vacuum that produces the flow.

Practically, a siphon does not work satisfactorily when the pressure intensity at the summit is close to vapor pressure. Air and other gases come out of solution at the low pressures and collect at the summit, thus reducing the length of the right-hand column of liquid that produces the low pressure at the summit. Large siphons that operate continuously have vacuum pumps to remove the gases at the summits.

The lowest pressure may not occur at the summit but somewhere downstream from that point, because friction and minor losses may reduce the pressure more than the decrease in elevation increases pressure.

Example 12.3

Neglecting minor losses and considering the length of pipe equal to its horizontal distance, determine the point of minimum pressure in the siphon of Fig. 12.6.

Solution

When minor losses are neglected, the kinetic-energy term $V^2/2g$ is usually neglected also. Then the hydraulic grade line is a straight line connecting the two liquid surfaces.

Coordinates of two points on the line are

$$x = -40 \text{ m}, \, y = 4 \text{ m} \quad \text{and} \quad x = 56.57 \text{ m}, \, y = 8 \text{ m}$$

The equation of the line is, by substitution into $y = mx + b$,

$$y = 0.0414x + 5.656 \text{ m}$$

---

†A liquid boils when its pressure is reduced to its vapor pressure. The vapor pressure is a function of temperature for a particular liquid. Water has a vapor pressure of 0.0619-m H_2O (0.203-ft H_2O) abs at 0°C (32°F) and 10.33-m H_2O (33.91-ft H_2O) abs at 100°C (212°F). See Appendix C.

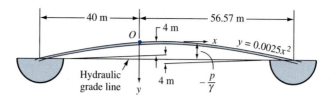

Figure 12.6   Siphon connecting two reservoirs.

The minimum pressure occurs where the distance between the hydraulic grade line and pipe is a maximum, that is,

$$\frac{p}{\gamma} = 0.0025x^2 - 0.0414x - 5.656$$

To find the minimum p/γ, set $d(p/\gamma)/dx = 0$, which yields $x = 8.28$ and $p/\gamma = -5.827$ m of fluid flowing. The minimum point occurs where the slopes of the pipe and of the hydraulic grade line are equal.

12.3   STEADY FLOW: PIPELINE SYSTEMS

Pipes in Series

When two pipes of different sizes or roughnesses are so connected that fluid flows through one pipe and then through the other, they are said to be connected in series. A typical series-pipe problem, in which the head H may be wanted for a given discharge or the discharge wanted for a given H, is illustrated in Fig. 12.7. Applying the energy equation from A to B, including all losses, gives

$$H + 0 + 0 = 0 + 0 + 0 + K_e\frac{V_1^2}{2g} + f_1\frac{L_1}{D_1}\frac{V_1^2}{2g} + \frac{(V_1 - V_2)^2}{2g} + f_2\frac{L_2}{D_2}\frac{V_2^2}{2g} + \frac{V_2^2}{2g}$$

in which the subscripts refer to the two pipes. The last item is the head loss at exit from pipe 2. With the continuity equation

$$V_1D_1^2 = V_2D_2^2$$

V_2 is eliminated from the equations, so that

$$H = \frac{V_1^2}{2g}\left\{K_e + \frac{f_1L_1}{D_1} + \left[1 - \left(\frac{D_1}{D_2}\right)^2\right]^2 + \frac{f_2L_2}{D_2}\left(\frac{D_1}{D_2}\right)^4 + \left(\frac{D_1}{D_2}\right)^4\right\}$$

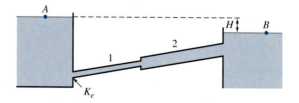

Figure 12.7   Pipes connected in series.

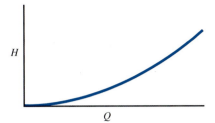

Figure 12.8   Plot of calculated H for selected values of Q.

For known lengths and sizes of pipes this reduces to

$$H = \frac{V_1^2}{2g}(C_1 + C_2 f_1 + C_3 f_2) \tag{12.3.1}$$

in which C_1, C_2, and C_3 are known. With the discharge given, the Reynolds number is readily computed, and the f's can be looked up in the Moody diagram. Then H is found by direct substitution. With H given, V_1, f_1, and f_2 are unknowns in Eq. (12.3.1). By assuming values of f_1 and f_2 (they may be assumed equal), a trial V_1 is found from which trial Reynolds numbers are determined and values of f_1 and f_2 are looked up. With these new values, a better V_1 is computed from Eq. (12.3.1). Since f varies so slightly with the Reynolds number, the trial solution converges very rapidly. The same procedures apply for more than two pipes in series.

   In place of the assumption of f_1 and f_2 when H is given, a graphical solution can be used in which several values of Q are assumed in turn, and the corresponding values of H are calculated and plotted against Q, as in Fig. 12.8. By connecting the points with a smooth curve, it is easy to determine the proper Q for the given value of H.

Example 12.4

In Fig. 12.7, $K_e = 0.5$, $L_1 = 300$ m, $D_1 = 600$ mm, $\epsilon_1 = 2$ mm, $L_2 = 240$ m, $D_2 = 1$ m, $\epsilon_2 = 0.3$ mm, $\nu = 3 \times 10^{-6}$ m²/s, and $H = 6$ m. Determine the discharge through the system.

Solution

From the energy equation

$$6 = \frac{V_1^2}{2g}\left[0.5 + f_1 \frac{300}{0.6} + (1 - 0.6^2)^2 + f_2 \frac{240}{1.0}0.6^4 + 0.6^4\right]$$

After simplifying,

$$6 = \frac{V_1^2}{2g}(1.0392 + 500 f_1 + 31.104 f_2)$$

From $\epsilon_1/D_1 = 0.0033$, $\epsilon_2/D_2 = 0.0003$, and Fig. 6.20, values of f are assumed for the fully turbulent range as

$$f_1 = 0.026 \qquad f_2 = 0.015$$

By solving for V_1 with these values, $V_1 = 2.848$ m/s, $V_2 = 1.025$ m/s, and

$$\mathbf{R}_1 = \frac{2.848(0.6)}{3 \times 10^{-6}} = 569{,}600 \qquad \mathbf{R}_2 = \frac{1.025(1.0)}{3 \times 10^{-6}} = 341{,}667$$

From Fig. 6.20, $f_1 = 0.0265$ and $f_2 = 0.0168$. By solving again for V_1, $V_1 = 2.819$ m/s, and $Q = 0.797$ m³/s.

Equivalent Pipes

Series pipes can be solved by the method of equivalent lengths. Two pipe systems are said to be equivalent when the same head loss produces the same discharge in both systems. From Eq. (12.2.1)

$$h_{f_1} = f_1 \frac{L_1}{D_1} \frac{Q_1^2}{(D_1^2 \pi/4)^2 2g} = \frac{f_1 L_1}{D_1^5} \frac{8 Q_1^2}{\pi^2 g}$$

and for a second pipe

$$h_{f_2} = \frac{f_2 L_2}{D_2^5} \frac{8 Q_2^2}{\pi^2 g}$$

For the two pipes to be equivalent,

$$h_{f_1} = h_{f_2} \qquad Q_1 = Q_2$$

After equating $h_{f_1} = h_{f_2}$ and simplifying,

$$\frac{f_1 L_1}{D_1^5} = \frac{f_2 L_2}{D_2^5}$$

Solving for L_2 gives

$$L_2 = L_1 \frac{f_1}{f_2} \left(\frac{D_2}{D_1} \right)^5 \tag{12.3.2}$$

which determines the length of a second pipe to provide an equivalent friction loss to that of the first pipe. For example, to replace 300 m of 250-mm pipe with an equivalent length of 150-mm pipe, the values of f_1 and f_2 must be approximated by selecting a discharge within the range intended for the pipes. Say $f_1 = 0.020$ and $f_2 = 0.018$, then

$$L_2 = 300 \frac{0.020}{0.018} \left(\frac{150}{250} \right)^5 = 25.9 \text{ m}$$

For these assumed conditions 25.9 m of 150-mm pipe is equivalent to 300 m of 250-mm pipe. Hypothetically, two or more pipes composing a system can also be replaced by a pipe which has the same discharge for the same overall head loss.

Example 12.5

Solve Ex. 12.4 by means of equivalent pipes.

Solution

Expressing the minor losses in terms of equivalent lengths gives, for pipe 1,

$$K_1 = 0.5 + (1 - 0.6^2)^2 = 0.91 \qquad L_{e_1} = \frac{K_1 D_1}{f_1} = \frac{0.91(0.6)}{0.026} = 21 \text{ m}$$

and for pipe 2,

$$K_2 = 1 \qquad L_{e_2} = \frac{K_2 D_2}{f_2} = \frac{1(1)}{0.015} = 66.7 \text{ m}$$

The values of f_1 and f_2 are selected for the fully turbulent range as an approximation. The problem is now reduced to 321 m of 600-mm pipe and 306.7 m of 1-m pipe. Expressing the 1-m pipe in terms of an equivalent length of 600-mm pipe gives, by Eq. (12.4.2),

$$L_e = \frac{f_2}{f_1} L_2 \left(\frac{D_1}{D_2}\right)^5 = 306.7 \frac{0.015}{0.026} \left(\frac{0.6}{1.0}\right)^5 = 13.76 \text{ m}$$

By adding to the 600-mm pipe the problem is reduced to finding the discharge through 334.76 m of 600-mm pipe, $\epsilon_1 = 2$ mm, $H = 6$ m, and

$$6 = f \frac{334.76}{0.6} \frac{V^2}{2g}$$

With $f = 0.026$, $V = 2.848$ m/s, and $\mathbf{R} = 2.848 \times 0.6/(3 \times 10^{-6}) = 569{,}600$.
For $\epsilon/D = 0.0033$, $f = 0.0265$, $V = 2.821$, and $Q = \pi(0.3^2)(2.821) = 0.798$ m³/s. By use of Eq. (6.7.15), $Q = 0.781$ m³/s.

Pipes in Parallel

A combination of two or more pipes connected as in Fig. 12.9 so that the flow is divided among the pipes and then is joined again, is a *parallel-pipe* system. In series pipes the same fluid flows through all the pipes and the head losses are cumulative, but in parallel pipes the head losses are the same in each of the lines and the discharges are cumulative.

In analyzing parallel-pipe systems, it is assumed that the minor losses are added into the lengths of each pipe as equivalent lengths. From Fig. 12.9 the conditions to be satisfied are

$$h_{f_1} = h_{f_2} = h_{f_3} = \frac{p_A}{\gamma} + z_A - \left(\frac{p_B}{\gamma} + z_B\right)$$

$$Q = Q_1 + Q_2 + Q_3 \qquad\qquad \text{(12.3.3)}$$

in which z_A and z_B are elevations of points A and B, respectively, and Q is the discharge through the approach pipe or the exit pipe.

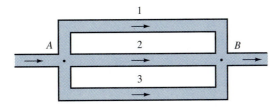

Figure 12.9   Parallel-pipe system.

Two types of problems occur: (1) With known elevations of the hydraulic grade line at A and B, the discharge Q needs to be determined, and (2) with Q known, the distribution of flow and the head loss need to be determined. Sizes of pipe, fluid properties, and roughnesses are assumed to be known.

The first type of problem is, in effect, the solution of simple pipe problems for discharge, since the head loss is the drop in hydraulic grade line. These discharges are added to determine the total discharge.

The second type of problem is more complex, as neither the head loss nor the discharge for any one pipe is known. The recommended procedure is as follows:

1. Assume a discharge Q_1' through pipe 1.

2. Solve for h_{f_1}', using the assumed discharge.

3. Using h_{f_1}', find Q_2' and Q_3'.

4. With the three discharges for a common head loss, now assume that the given Q is split up among the pipes in the same proportion as Q_1', Q_2', and Q_3'. Thus,

$$Q_1 = \frac{Q_1'}{\Sigma Q'}Q \qquad Q_2 = \frac{Q_2'}{\Sigma Q'}Q \qquad Q_3 = \frac{Q_3'}{\Sigma Q'}Q \qquad (12.3.4)$$

5. Check the correctness of these discharges by computing h_{f_1}, h_{f_2}, and h_{f_3} for the computed Q_1, Q_2, and Q_3.

This procedure works for any number of pipes. By judicious choice of Q_1', obtained by estimating the percent of the total flow through the system that should pass through pipe 1 (based on diameter, length, and roughness), Eqs. (12.3.4) produce values that check within a few percent, which is well within the range of accuracy of the friction factors.

If a spreadsheet is available, the second type of problem is easily handled by the optimizer, solver, or equivalent function. For Fig. 12.9, the following three equations are written in terms of the unknown flows, Q_1, Q_2, and Q_3:

$$\frac{f_1 L_1 Q_1^2}{2g(\pi/4)^2 D_1^5} - \frac{f_2 L_2 Q_2^2}{2g(\pi/4)^2 D_2^5} = 0 \qquad (12.3.5)$$

$$\frac{f_2 L_2 Q_2^2}{2g(\pi/4)^2 D_2^5} - \frac{f_3 L_3 Q_3^2}{2g(\pi/4)^2 D_3^5} = 0 \qquad (12.3.6)$$

$$Q - Q_1 - Q_2 - Q_3 = 0 \qquad (12.3.7)$$

The spreadsheet function may be used to set Eq. (12.3.5) to zero, by varying the unknown flows, subject to the constraints that Eq. (12.3.6) and Eq. (12.3.7) must be equal to zero. This is an iterative solution so the friction factors may be corrected at each iteration.

| **Example 12.6** | In Fig. 12.9, $L_1 = 3000$ ft, $D_1 = 1$ ft, and $\epsilon_1 = 0.001$ ft; $L_2 = 2000$ ft, $D_2 = 8$ in., and $\epsilon_2 = 0.0001$ ft; $L_3 = 4000$ ft, $D_3 = 16$ in., and $\epsilon_3 = 0.0008$ ft; and $\rho = 2.00$ slugs/ft^3, $\nu = 0.00003$ ft^2/s, $p_A = 80$ psi, $z_A = 100$ ft, and $z_B = 80$ ft. For a total flow of 12 cfs, determine flow through each pipe and the pressure at B. Solve the problem using the trial and error hand calculation, and check the results using a spreadsheet. |

Solution

Assume $Q_1' = 3$ cfs, then $V_1' = 3.82$, $\mathbf{R}_1' = 3.82(1/0.00003) = 127,000$, $\epsilon_1/D_1 = 0.001$, $f_1' = 0.022$, and

$$h_{f_1}' = 0.022\frac{3000}{1.0}\frac{3.82^2}{64.4} = 14.97 \text{ ft}$$

For pipe 2,

$$14.97 = f_2'\frac{2000}{0.667}\frac{V_2'^2}{2g}$$

Then $\epsilon_2/D_2 = 0.00015$. Assume $f_2' = 0.020$, then $V_2' = 4.01$ ft/s, $\mathbf{R}_2' = 4.01(\frac{2}{3})(1/0.00003) = 89,000$, $f_2' = 0.019$, $V_2' = 4.11$ ft/s, and $Q_2' = 1.44$ cfs. For pipe 3,

$$14.97 = f_3'\frac{4000}{1.333}\frac{V_3'^2}{2g}$$

Then $\epsilon_3/D_3 = 0.0006$. Assume $f_3' = 0.020$, then $V_3' = 4.01$ ft/s, $\mathbf{R}_3' = 4.01(1.333/0.00003) = 178,000$, $f_3' = 0.020$, and $Q_3' = 5.60$ cfs.
The total discharge for the assumed conditions is

$$\Sigma Q' = 3.00 + 1.44 + 5.60 = 10.04 \text{ cfs}$$

Hence

$$Q_1 = \frac{3.00}{10.04}12 = 3.58 \text{ cfs} \qquad Q_2 = \frac{1.44}{10.04}12 = 1.72 \text{ cfs}$$

$$Q_3 = \frac{5.60}{10.04}12 = 6.70 \text{ cfs}$$

Check the values of h_1, h_2, and h_3 as follows

$$V_1 = \frac{3.58}{\pi/4} = 4.56 \qquad \mathbf{R}_1 = 152,000 \qquad f_1 = 0.021 \qquad h_{f_1} = 20.4 \text{ ft}$$

$$V_2 = \frac{1.72}{\pi/9} = 4.93 \qquad \mathbf{R}_2 = 109,200 \qquad f_2 = 0.019 \qquad h_{f_2} = 21.6 \text{ ft}$$

$$V_3 = \frac{6.70}{4\pi/9} = 4.80 \qquad \mathbf{R}_3 = 213,000 \qquad f_3 = 0.019 \qquad h_{f_3} = 20.4 \text{ ft}$$

f_2 is about midway between 0.018 and 0.019. If 0.018 had been selected, h_2 would be 20.4 ft.
To find p_B

$$\frac{p_A}{\gamma} + z_A = \frac{p_B}{\gamma} + z_B + h_f$$

or

$$\frac{p_B}{\gamma} = \frac{80(144)}{64.4} + 100 - 80 - 20.8 = 178.1$$

in which the average head loss was taken. Then

$$p_B = \frac{178.1(64.4)}{144} = 79.6 \text{ psi}$$

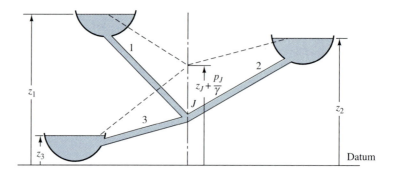

Figure 12.10   Three interconnected reservoirs.

The spreadsheet calculation, using Eqs. (12.3.5) to (12.3.7), yields $Q_1 = 3.573$ cfs, $Q_2 = 1.718$ cfs, $Q_3 = 6.709$ cfs, and $h_f = 20.80$ ft.

Branching Pipes

A simple *branching-pipe* system is shown in Fig. 12.10. In this situation the flow through each pipe is to be determined when the reservoir elevations are given. The sizes and types of pipes and fluid properties are assumed known. The Darcy-Weisbach equation must be satisfied for each pipe, and the continuity equation must also be satisfied. It takes the form that the flow into the junction J must just equal the flow out of the junction. Flow must be out of the highest reservoir and into the lowest; hence, the continuity equation may be either

$$Q_1 = Q_2 + Q_3 \qquad \text{or} \qquad Q_1 + Q_2 = Q_3$$

If the elevation of hydraulic grade line at the junction is above the elevation of the intermediate reservoir, flow is *into* it, but if the elevation of hydraulic grade line at J is below the intermediate reservoir, the flow is *out* of it. Minor losses can be expressed as equivalent lengths and added to the actual lengths of pipe.

The solution is accomplished by first assuming an elevation of hydraulic grade line at the junction and then computing Q_1, Q_2, and Q_3 and substituting into the continuity equation. If the flow into the junction is too great, a higher grade-line elevation, which will reduce the inflow and increase the outflow, is assumed.

Example 12.7	In Fig. 12.10 find the discharges for water at 20°C with the following pipe data and reservoir elevations: $L_1 = 3000$ m, $D_1 = 1$ m, and $\epsilon_1/D_1 = 0.0002$; $L_2 = 600$ m, $D_2 = 0.45$ m, and $\epsilon_2/D_2 = 0.002$; $L_3 = 1000$ m, $D_3 = 0.6$ m, and $\epsilon_3/D_3 = 0.001$; and $z_1 = 30$ m, $z_2 = 18$ m, and $z_3 = 9$ m.

Solution

Assume $z_J + p_J/\gamma = 23$ m. Then

$$7 = f_1 \frac{3000}{1} \frac{V_1^2}{2g} \qquad f_1 = 0.014 \qquad V_1 = 1.75 \text{ m/s} \qquad Q_1 = 1.380 \text{ m}^3/\text{s}$$

$$5 = f_2 \frac{600}{0.45} \frac{V_2^2}{2g} \qquad f_2 = 0.024 \qquad V_2 = 1.75 \text{ m/s} \qquad Q_2 = 0.278 \text{ m}^3/\text{s}$$

$$14 = f_3 \frac{1000}{0.60} \frac{V_3^2}{2g} \qquad f_3 = 0.020 \qquad V_3 = 2.87 \text{ m/s} \qquad Q_3 = 0.811 \text{ m}^3/\text{s}$$

so that the inflow is greater than the outflow by

$$1.380 - 0.278 - 0.811 = 0.291 \text{ m}^3/\text{s}$$

Assume $z_J + p_J/\gamma = 24.6$ m. Then

$$5.4 = f_1 \frac{3000}{1} \frac{V_1^2}{2g} \qquad f_1 = 0.015 \qquad V_1 = 1.534 \text{ m/s} \qquad Q_1 = 1.205 \text{ m}^3/\text{s}$$

$$6.6 = f_2 \frac{600}{0.45} \frac{V_2^2}{2g} \qquad f_2 = 0.024 \qquad V_2 = 2.011 \text{ m/s} \qquad Q_2 = 0.320 \text{ m}^3/\text{s}$$

$$15.6 = f_3 \frac{1000}{0.60} \frac{V_3^2}{2g} \qquad f_3 = 0.020 \qquad V_3 = 3.029 \text{ m/s} \qquad Q_3 = 0.856 \text{ m}^3/\text{s}$$

The inflow is still greater by 0.029 m³/s. By extrapolating linearly, $z_J + p_J/\gamma = 24.8$ m, $Q_1 = 1.183$, $Q_2 = 0.325$, and $Q_3 = 0.862$ m³/s.

Use a spreadsheet to balance the flows in the system in Ex. 12.7. | **Example 12.8**

Solution

The optimizer, solver, or equivalent function, depending on the particular spreadsheet, may be used. When balanced, the sum of flows at the junction must be zero. The elevation of the hydraulic grade line at the junction is varied until the flows are balanced, subject to the constraints that this elevation is limited within the range between the maximum and minimum reservoir levels. It is helpful to adopt the convention that inflow to the junction is positive and the head loss in adjoining pipes is positive when carrying flow to the junction. Equation (6.7.15) may be used to calculate the flow in each pipe, using the absolute value of the head loss. The flow direction is assigned by multiplying Eq. (6.7.15) by the sign of the head loss.

In summary, $h_J = z_J + p_J/\gamma$, Fig. 12.10, is varied until the function

$$\sum Q_{in} = Q_1 + Q_2 + Q_3 = 0$$

is satisfied. The only constraints needed are that $h_J < z_1$ and $h_J > z_3$. The results are

$$h_J = 24.88 \text{ m} \qquad Q_1 = 1.1980 \text{ m}^3/\text{s}$$

$$Q_2 = -0.3293 \text{ m}^3/\text{s} \qquad Q_3 = -0.8687 \text{ m}^3/\text{s}$$

In pumping from one reservoir to two or more other reservoirs, as in Fig. 12.11, the characteristics of the pump must be known. Assuming that the pump runs at constant speed, its head depends upon the discharge. A suitable procedure is as follows:

1. Assume a discharge through the pump.
2. Compute the hydraulic-grade-line elevation at the suction side of the pump.
3. From the pump characteristic curve find the head produced and add it to the suction hydraulic grade line.
4. Compute the drop in the hydraulic grade line to the junction J and determine the elevation of the hydraulic grade line there.
5. For this elevation, compute the flow in pipes connected with the other reservoirs.

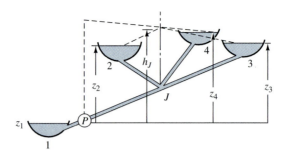

Figure 12.11   Pumping from one reservoir to two other reservoirs.

6. If the flow from the pump into J equals the net flow out of J, the problem is solved. If flow into J is too great, assume less flow through the pump and repeat the procedure.

This procedure is easily plotted on a graph, so that the intersection of two elevations versus the flow curves yields the answer.

More complex branching-pipe problems can be solved with a similar approach by beginning with a trial solution. However, the network-analysis procedures in Secs. 12.4 and 12.5 are recommended for multibranch systems as well as for multiparallel-loop systems.

Example 12.9

A pump is in a pipeline from a suction reservoir to a junction to which three reservoirs are connected with pipes, as in Fig. 12.11. There is a check valve at the pump. Use a spreadsheet to balance the flows in the system. The pump equation is given by

$$H_P = A_0 + A_1 Q + A_2 Q^2 + A_3 Q^3$$

with $A_0 = 100$, $A_1 = -0.2$, $A_2 = -0.03$, and $A_3 = -0.007$; $\nu = 0.000001$ m²/s; $L_{1...4} = 10{,}000, 2000, 2500, 2000$ m, respectively; $D_{1...4} = 4.5, 2.0, 2.5, 2.3$ m, respectively; $\epsilon_{1...4} = 0.00006, 0.00005, 0.00008, 0.00009$ m, respectively; and $z_{1...4} = 0, 12, 18, 25$ m, respectively.

Solution

Alternative approaches are available, including the procedures in the next sections. The method outlined above works well when the optimizer or solver function is used with a spreadsheet. Inflow to the junction is assumed positive, so the goal is to satisfy the relationship,

$$\sum Q_{in} = Q_1 + Q_2 + Q_3 + Q_4 = 0$$

A value of Q_P, the flow through the pump, is used to find the hydraulic grade line elevation at J.

$$h_J = z_1 + H_P - h_{f_1}$$

The value of h_{f_1} is determined with the Darcy-Weisbach equation with $Q_P = Q_1$, and Eq. (6.7.13) is used to evaluate f. With this h_J, flows Q_2, Q_3, and Q_4 are determined with Eq. (6.7.15), by using the absolute values of $(z_i - h_J)$ as h_{f_i}, and using the sign of h_{f_i} to determine the flow direction. The procedure is repeated by varying Q_P until $\sum Q_{in} = 0$, subject to the constraint $h_J > z_2$.

The results are

$$h_J = 22.75 \text{ m} \qquad Q_P = Q_1 = 20.257 \text{ m}^3/\text{s}$$
$$Q_2 = -14.495 \text{ m}^3/\text{s} \qquad Q_3 = -14.761 \text{ m}^3/\text{s} \qquad Q_4 = 8.999 \text{ m}^3/\text{s}$$

12.4 STEADY FLOW: NETWORKS OF PIPES

Interconnected pipes through which the flow to a given outlet may come from several circuits are called a *network of pipes,* in many ways analogous to flow through electric networks. Problems in these networks, in general, are complicated and require trial solutions in which the elementary circuits are balanced in turn until all conditions for the flow are satisfied.

The following conditions must be satisfied in a network of pipes:

1. The algebraic sum of the pressure drops around each circuit must be zero.
2. Flow into each junction must equal flow out of the junction.
3. The Darcy-Weisbach equation, or equivalent exponential friction formula, must be satisfied for each pipe, that is, the proper relation between the head loss and discharge must be maintained for each pipe.

The first condition states that the pressure drop between any two points in the circuit, for example, A and G (Fig. 12.12), must be the same whether through the pipe AG or through $AFEDG$. The second condition is the continuity equation.

Since it is impractical to solve network problems analytically, methods of successive approximations are used. The Hardy Cross method [3] is one in which flows are assumed for each pipe so that continuity is satisfied at every junction. A correction to the flow in each circuit is then computed in turn and applied to bring the circuits into closer balance.

Minor losses are included as equivalent lengths in each pipe. Exponential equations are commonly used, in the form $h_f = rQ^n$, where $r = RL/D^m$ in Eq. (12.1.1). The value of r is a constant in each pipeline (unless the Darcy-Weisbach equation is used) and is determined in advance of the loop-balancing procedure. The corrective term is obtained as follows.

For any pipe in which Q_0 is an assumed initial discharge

$$Q = Q_0 + \Delta Q \qquad (12.4.1)$$

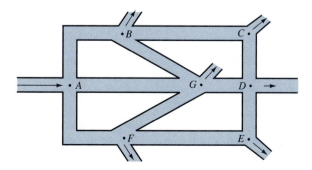

Figure 12.12 Pipe network.

where Q is the correct discharge and ΔQ is the correction. Then for each pipe,

$$h_f = rQ^n = r(Q_0 + \Delta Q)^n = r(Q_0^n + nQ_0^{n-1}\Delta Q + \cdots)$$

If ΔQ is small compared with Q_0, all terms of the series after the second can be dropped. Now for a circuit,

$$\Sigma h_f = \Sigma rQ|Q|^{n-1} = \Sigma rQ_0|Q_0|^{n-1} + \Delta Q \Sigma rn|Q_0|^{n-1} = 0$$

in which ΔQ has been taken out of the summation because it is the same for all pipes in the circuit, and absolute-value signs have been added to account for the direction of summation around the circuit. The last equation is solved for ΔQ in each circuit in the network as

$$\Delta Q = -\frac{\Sigma rQ_0|Q_0|^{n-1}}{\Sigma rn|Q_0|^{n-1}} \tag{12.4.2}$$

When ΔQ is applied to each pipe in a circuit in accordance with Eq. (12.4.1), the directional sense is important, that is, it adds to flows in the clockwise direction and subtracts from flows in the counterclockwise direction.

Steps in an arithmetic procedure can be itemized as follows:

1. Assume the best distribution of flows that satisfies continuity by careful examination of the network.

2. For each pipe in an elementary circuit, calculate and sum the net head loss $\Sigma h_f = \Sigma rQ^n$. Also calculate $\Sigma rn|Q|^{n-1}$ for the circuit. The negative ratio, by Eq. (12.4.2) yields the correction, which is then added algebraically to each flow in the circuit to correct it.

3. Proceed to another elementary circuit and repeat the correction process of step 2. Continue for all elementary circuits.

4. Repeat steps 2 and 3 as many times as needed until the corrections (ΔQ's) are arbitrarily small.

The values of r occur in both numerator and denominator; hence, values proportional to the actual r can be used to find the distribution. Similarly, the apportionment of flows can be expressed as a percent of the actual flows. To find a particular head loss, the actual value of r and Q must be used after the distribution has been determined.

Example 12.10

The distribution of flow through the network of Fig. 12.13 is desired for the inflows and outflows as given. For simplicity n has been given the value 2.0.

Solution

The assumed distribution is shown in Fig. 12.13a. At the upper left the term $\Sigma rQ_0|Q_0|^{n-1}$ is computed for the lower circuit number 1. Next to the diagram on the left is the computation of $\Sigma nr|Q_0|^{n-1}$ for the same circuit. The same format is used for the second circuit in the upper right of the figure. The corrected flow after the first step for the top horizontal pipe is determined as $15 + 11.06 = 26.06$ and for the diagonal as $35 + (-21.17) + (-11.06) = 2.77$. Figure 12.13$b$ shows the flows after one correction and Fig. 12.13c the values after four corrections.

$$70^2 \times 6 = 29{,}400 \quad 2 \times 70 \times 6 = 840$$
$$35^2 \times 3 = 3{,}675 \quad 2 \times 35 \times 3 = 210$$
$$-30^2 \times 5 = -4{,}500 \quad 2 \times 30 \times 5 = 300$$
$$\overline{28{,}575} \qquad \overline{1{,}350}$$
$$\Delta Q_1 = -\frac{28{,}575}{1{,}350} = -21.17$$

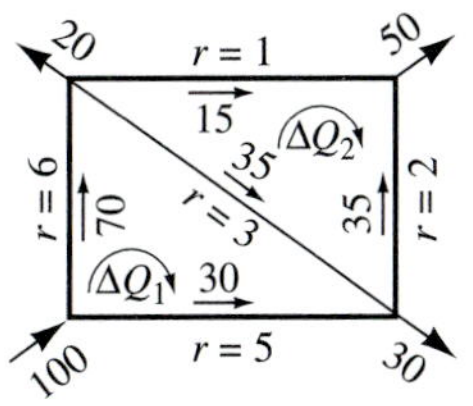

$$15^2 \times 1 = 225 \quad 2 \times \quad 15 \times 1 = 30$$
$$-35^2 \times 2 = -2{,}450 \quad 2 \times \quad 35 \times 2 = 140$$
$$-13.83^2 \times 3 = - 574 \quad 2 \times 13.83 \times 3 = 83$$
$$\overline{-2{,}799} \qquad \overline{253}$$
$$\Delta Q_2 = \frac{2{,}799}{253} = 11.06$$

(a)

$$48.83^2 \times 6 = 14{,}308 \quad 2 \times 48.83 \times 6 = 586$$
$$2.77^2 \times 3 = 23 \quad 2 \times 2.77 \times 3 = 17$$
$$-51.17^2 \times 5 = -13{,}090 \quad 2 \times 51.17 \times 5 = 511$$
$$\overline{1{,}241} \qquad \overline{1{,}114}$$
$$\Delta Q_1 = -\frac{1{,}241}{1{,}114} = -1.114$$

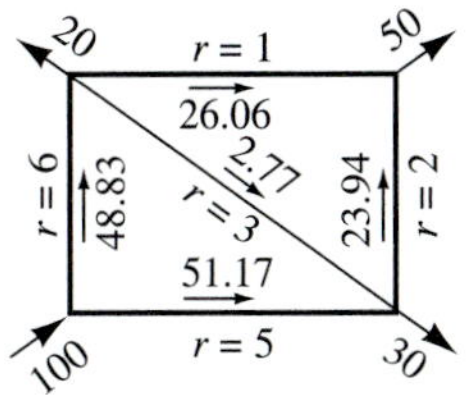

$$26.06^2 \times 1 = 679 \quad 2 \times 26.06 \times 1 = 52$$
$$-23.94^2 \times 2 = -1{,}146 \quad 2 \times 23.94 \times 2 = 96$$
$$-1.656^2 \times 3 = - 8 \quad 2 \times 1.656 \times 3 = 10$$
$$\overline{-475} \qquad \overline{158}$$
$$\Delta Q_2 = \frac{475}{158} = 3.006$$

(b)

$$\Delta Q_1 = 0.0079$$
$$\Delta Q_1 = 0.0013$$

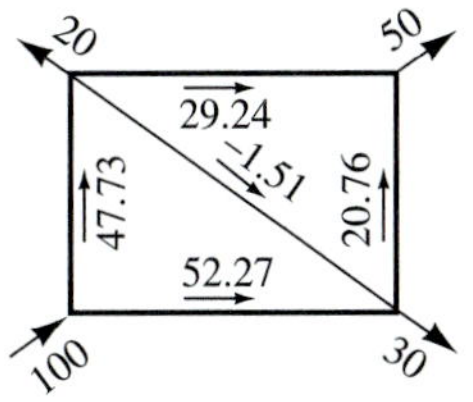

$$\Delta Q_2 = 0.169$$
$$\Delta Q_2 = 0.0003$$

(c)

Figure 12.13 Solution for flow in a simple network.

Very simple networks, such as the one shown in Fig. 12.13, can be solved with a hand-held programmable calculator or a spreadsheet. For networks larger than the previous example or for networks that contain multiple reservoirs, supply pumps, or booster pumps, a spreadsheet may be used as discussed in the next section. For large networks with many circuits and many elements a numerical solution is recommended based upon the principles originally presented by Hardy Cross, and outlined above.

A number of more general methods [4–7] are available, primarily based upon the Hardy Cross loop-balancing or node-balancing schemes. In the more general methods the system is normally modeled with a set of simultaneous equations which are solved by the Newton-Raphson method. Some programmed solutions [5,6] are very useful as design tools, since pipe sizes or roughnesses may be treated as unknowns in addition to junction pressures and flows.

12.5 STEADY FLOW: METHODOLOGIES FOR COMPLEX HYDRAULIC NETWORKS

Hydraulic systems that contain components different from pipelines can be handled by replacing the component with an equivalent length of pipeline. When the additional component is a pump, special consideration is needed. Also, in systems that contain more than one fixed hydraulic-grade-line elevation, a special artifice may be introduced.

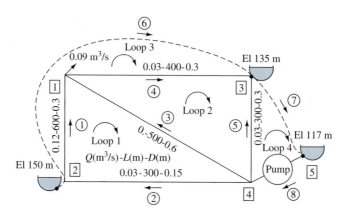

Figure 12.14   Sample network.

Hardy Cross's Loop-Balancing Method

For systems with multiple fixed-pressure-head elevations (Fig. 12.14) *pseudoelements* are introduced to account for the unknown outflows and inflows at the reservoirs and to satisfy continuity conditions during balancing. An imaginary or artificial loop is created by using a pseudoelement that interconnects each pair of fixed pressure levels. These pseudoelements carry no flow but maintain a fixed drop in the hydraulic grade level equal to the difference in elevation of the reservoirs. If the head drop is considered positive in an assumed positive direction in the pseudoelement, the correction in loop 3 (Fig. 12.14) is

$$\Delta Q_3 = -\frac{150 - 135 - r_4 Q_4 |Q_4|^{n-1} - r_1 Q_1 |Q_1|^{n-1}}{n r_4 |Q_4|^{n-1} + n r_1 |Q_1|^{n-1}} \qquad \text{(12.5.1)}$$

This correction is applied to pipes 1 and 4 only. If additional real pipelines exist in an imaginary loop, each would be adjusted accordingly during each loop-balancing iteration. The terms in Eq. (12.5.1) can easily be identified by relating to Eq. (12.4.2). Alternatively, the same equation can be generated by application of Newton's method (see the Web page).

A pump in a system can be considered as a flow element with a negative head loss equal to the head rise that corresponds to the flow through the unit. The pump-head-discharge curve, element 8 in Fig. 12.14, can be expressed by a cubic equation

$$H = A_0 + A_1 Q_8 + A_2 Q_8^2 + A_3 Q_8^3$$

where A_0 is the shut-off head of the pump. The correction in loop 4 is

$$\Delta Q_4 = \frac{135 - 117 - (A_0 + A_1 Q_8 + A_2 Q_8^2 + A_3 Q_8^3) + r_5 Q_5 |Q_5|^{n-1}}{n r_5 |Q_5|^{n-1} - (A_1 + 2 A_2 Q_8 + 3 A_3 Q_8^2)} \qquad \text{(12.5.2)}$$

This correction is applied to pipe 5 and to pump 8 in the loop. Equation (12.5.2) is developed by application of Newton's method to the loop. For satisfactory balancing of networks with pumping stations, the slope of the head-discharge curve should always be less than or equal to zero.

By use of these principles a computer code may be generated to analyze a wide variety of liquid steady-state pipe-flow problems. Pipeline flows described by the Hazen-Williams equation, laminar or turbulent flows analyzed with the Darcy-Weisbach equation, multiple reservoirs or fixed pressure levels, as in a sprinkler system, and systems with booster pumps or supply pumps can be treated.

A network is visualized as a combination of elements that are interconnected at junctions. The elements may include pipelines, pumps, and pseudoelements. All minor losses are handled by estimating equivalent lengths and adding them onto the actual pipe lengths. Each element in the system is uniquely identified. A positive flow direction is assigned to each element, and, as in the arithmetic solution, an estimated flow is assigned to each element such that continuity is satisfied at each junction. The assigned positive flow direction in a pump must be in the intended direction of normal pump operation. Any solution with backward flow through a pump is invalid. The flow direction in the pseudoelement that creates an imaginary loop indicates only the direction of fixed positive head drop, since the flow must be zero in this element. Each junction, which may represent the termination of a single element or the intersection of many elements, is also uniquely identified. An outflow or inflow at a junction is defined during the assignment of initial element flows. A relationship exists between the number of elements, e, in a system that interconnects the junctions or nodes, j, and forms a set of ℓ loops. When properly formulated

$$e = \ell + j - 1 \tag{12.5.3}$$

For example, in Fig. 12.14, $e = 8$, $j = 5$, $\ell = 4$, which satisfies Eq. (12.5.3).

The necessary steps are best visualized in two major parts: the first performs the balancing of each loop in the system successively and then repeats iteratively until the sum of all loop flow corrections is less than a specified tolerance. At the end of this balancing process the element flows are computed. The second part of an analysis involves the computation of the hydraulic grade line elevations at junctions in the system.

Spreadsheet Solution

Spreadsheets have an optimizer, solver, or equivalent function which may be used to find the flow distribution in networks of modest complexity. The two fundamental principles of networks are (1) the algebraic sum of the pressure drops around each circuit must be zero, and (2) continuity of flows must be satisfied at each junction.

For example, in Fig. 12.13, once the assumed flows and flow directions are assigned in each pipeline, these flows are adjusted by the spreadsheet function to force the algebraic sum of the pressure drop around each of the two circuits to zero. At three junctions the constraint that the algebraic sum of inflow to the junctions (including the external flows) must be zero is imposed. It may be noted that if continuity is satisfied at three of the four junctions, overall continuity on the entire system assures that continuity will be satisfied at the fourth node. It may also be noted that Eq. (12.5.3) identifies five elements in the system. Thus, there are five unknown element flows and, therefore, only five constraints can be imposed, that is, two circuits and three nodal conditions.

There are many benefits in the use of a generalized computer code to solve for the steady-state flow distribution in networks. In using a spreadsheet solution all of the generalization benefits of a robust solution procedure are lost. In the spreadsheet procedure each network represents a unique problem and the network must be described in detail within the spreadsheet data. Perhaps the foremost advantage in the spreadsheet solution is the ease with which it may be applied, inasmuch as it uses the fundamental principles directly. Its biggest limitation is that it is only practical for relatively small noncomplex networks. The following example offers flow distribution solutions in networks of modest complexity.

		Spreadsheet solution in water distribution system				
		Solves for Q in each element to balance heads in loop				
uses Hazen Williams Eqn.		and to satisfy continuity at junctions				
HWC =	100	en =		0.852		
A0, A1, A2, A3 =	30	−11.111	−555.55	−6172.84		
				rr1 = 10.675 * B8/(HWC^1.852 * C8^4.8704)		
element	length	diameter		rr		Flow, m^3/s
1	600	0.3	rr1	445.81	qq1	0.1434
2	300	0.15	rr2	6520.10	qq2	−0.0337
3	500	0.6	rr3	12.70	qq3	0.0267
4	400	0.3	rr4	297.21	qq4	0.0801
5	300	0.3	rr5	222.90	qq5	0.0939
					qq8	0.0869
shf1	−6.8E−07	= rr1 * qq1 * ABS(qq1)^en − rr3 * qq3 * ABS(qq3)^en + rr2 * qq2 * ABS(qq2)^en				
shf2	8.82E−07	= rr4 * qq4 * ABS(qq4)^en − rr5 * qq5 * ABS(qq5)^en + rr3 * qq3 * ABS(qq3)^en				
shf3	−1.4E−05	= 150 − 135 − rr4 * qq4 * ABS(qq4)^en − rr1 * qq1 * ABS(qq1)^en				
shf4	2.03E−06	= 135 − 117 − (B5 + C5 * qq8 + D5 * qq8^2 + E5 * qq8^3) + rr5 * qq5 * ABS(qq5)^en				
sumq1	2.78E−17	= qq1 + qq3 − qq4 − 0.09				
sumq4	0	= qq6 − qq2 − qq3 − qq5				
HGL1 =	140.83	= 150 − rr1 * qq1 * ABS(qq1)				

Figure 12.15 Ex. 12.11, Spreadsheet solution for the network in Fig. 12.14.

Example 12.11

Figure 12.14 displays a network of five pipelines, one pump, and three reservoirs. Use a spreadsheet to balance the flows in the system, and to find the hydraulic grade-line elevation at node 1. The Hazen-Williams pipeline coefficient for all the pipes is 100. The pump data, with the coefficients for a cubic equation provided, are as follows:

Q, m³/s	0	0.03	0.06	0.09
H, m	30	29	26	20
$A_{0...3}$	30	−11.11	−555.5	−6172.8

Solution

Nodes, elements, and loops are identified in Fig. 12.14. Figure 12.15 shows a possible setup of a spreadsheet, with the solution. In this example the sum of the head drops around loop 1 are set to zero by changing the flows in each of the six actual elements, subject to the constraints that the sum of the head loss around loops 2, 3, and 4 is zero. Initial guesses for flow in pipelines 1, . . . , 5, and in pump 8 are provided. The procedure usually converges faster if the guesses are reasonable, and if they satisfy continuity at the nodes.

Figure 12.17 provides a spreadsheet solution to the network shown in Fig. 12.16, and Fig. 12.19 provides the solution to the network shown in Fig. 12.18.

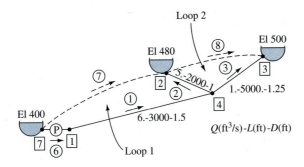

Figure 12.16 Branching system with data.

uses Hazen Williams Eqn.		Solves for Q in each element to balance heads in loop					
HWC =	120	en =		0.852			
A0,A1,A2,A3 =	120	−0.405		−0.0918	−0.00097		
				rr1 = 4.727*B6/(HWC^1.852*C6^4.8704)			
element	length, ft	diameter, ft		rr		**Flow, m^3/s**	
1	3000	1.5	rr1	0.2776	**qq1**	**6.7260**	
2	2000	1	rr2	1.3334	**qq2**	**4.7561**	
3	5000	1.25	rr3	1.1244	**qq3**	**1.9699**	
					qq6	**6.7260**	
shf1	2.8407E−11	=400−480−rr1*qq1*ABS(qq1)^en−rr2*qq2*ABS(qq2)^en+(B3+C3*qq6+D3*qq6^2+E3*qq6^3)					
shf2	8.2569E−11	=480−500−rr3*qq3*ABS(qq3)^en−rr2*qq2*ABS(qq2)^en					
sumq1	0	=−qq1+qq6					
sumq4	6.8834E−15	=qq1−qq2−qq3					

Figure 12.17 Spreadsheet solution for the network in Fig. 12.16.

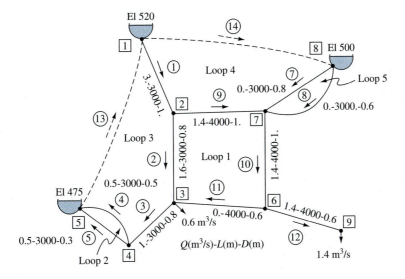

Figure 12.18 Three-reservoir network system with data.

			Loop balancing for Network Solution;						
uses Darcy-Weisbach Eqn.						f = 1.325/(LN(D5 + 5.74/(G5*E5)^0.9))^2			
vnu =	1.2E−06	eps =	0.0005	grav =	9.806				
pipe	length	diameter	eps/(3.7*D)	CRey	QQ	Discharge	f	rr	
1	3000	1	0.00014	1061030	qq1=	**1.9556**	0.01701	rr1 =	4.218754
2	3000	0.8	0.00017	1326288	qq2	**1.0287**	0.01796	rr2	13.5934
3	3000	0.8	0.00017	1326288	qq3	**0.3545**	0.01856	rr3	14.04464
4	3000	0.5	0.00027	2122061	qq4	**0.2813**	0.02025	rr4	160.6562
5	3000	0.3	0.00045	3536768	qq5	**0.0732**	0.02326	rr5	2373.842
7	3000	0.8	0.00017	1326288	qq7	**0.2717**	0.0188	rr7	14.2308
8	3000	0.6	0.00023	1768384	qq8	**0.1272**	0.02034	rr8	64.87928
9	4000	1	0.00014	1061030	qq9	**0.9270**	0.0173	rr9	5.721029
10	4000	1	0.00014	1061030	qq10	**1.3259**	0.01714	rr10	5.667561
11	4000	0.6	0.00023	1768384	qq11	**−0.0741**	0.02118	rr11	90.0722
12	4000	0.6	0.00023	1768384	qq12	**1.4000**	0.01898	rr12	80.71854

shf1	**1E−08**	= rr9*qq9*ABS(qq9) + rr10*qq10*ABS(qq10) + rr11*qq11*ABS(qq11) − rr2*qq2*ABS(qq2)						
shf2	**−4.5E−10**	= rr5*qq5*ABS(qq5) − rr4*qq4*ABS(qq4)						
shf3	**7.8E−08**	= rr1*qq1*ABS(qq1) + rr2*qq2*ABS(qq2) + rr3*qq3*ABS(qq3) + rr4*qq4*ABS(qq4) + 475 − 520						
shf4	**1.4E−09**	= rr7*qq7*ABS(qq7) − rr9*qq9*ABS(qq9) − rr1*qq1*ABS(qq1) + 520 − 500						
shf5	**−9.4E−12**	= rr8*qq8*ABS(qq8) − rr7*qq7*ABS(qq7)						
sumq2	**4.4E−16**	= qq1 − qq9 − qq2						
sumq3	**−5.6E−16**	= qq2 − qq3 + qq11 − 0.6						
sumq4	**−1E−15**	= qq3 − qq5 − qq4						
sumq6	**3.3E−16**	= qq10 − 1.4 − qq11						
sumq7	**−6.7E−16**	= qq9 + qq7 + qq8 − qq10						

Figure 12.19   Spreadsheet solution for the network in Fig. 12.18.

12.6   STEADY FLOW: NONCIRCULAR CONDUITS, AGING OF PIPES, AND ADDITIVES

Noncircular Conduits

In this chapter only circular pipes have been considered so far. For cross sections that are noncircular, the Darcy-Weisbach equation can be applied if the term D can be interpreted in terms of the section. The concept of the *hydraulic radius R* permits circular and noncircular sections to be treated in the same manner. The hydraulic radius is defined as the cross-sectional area divided by the *wetted perimeter*. Hence, for a circular section,

$$R = \frac{\text{area}}{\text{perimeter}} = \frac{\pi D^2/4}{\pi D} = \frac{D}{4} \tag{12.6.1}$$

and the diameter is equivalent to $4R$. If the diameter can be replaced by $4R$ in the Darcy-Weisbach equation, the Reynolds number, and the relative roughness, then

$$h_f = f \frac{L}{4R} \frac{V^2}{2g} \qquad \mathbf{R} = \frac{V4R\rho}{\mu} \qquad \frac{\epsilon}{D} = \frac{\epsilon}{4R} \tag{12.6.2}$$

Noncircular sections can be handled in a similar manner. The Moody diagram applies as before. The assumptions in Eqs. (12.6.2) cannot be expected to hold for odd-shaped sections but should give reasonable values for square, oval, triangular, and similar types of sections.

Determine the head loss, in millimeters of water, required for flow of 300 m³/min of air at 20°C and 100 kPa through a rectangular galvanized-iron section 700 mm wide, 350 mm high, and 70 m long.

Example 12.12

Solution

$$R = \frac{A}{P} = \frac{0.7(0.35)}{2(0.7 + 0.35)} = 0.117 \text{ m} \qquad \frac{\epsilon}{4R} = \frac{0.00015}{4(0.117)} = 0.00032$$

$$V = \frac{300}{60(0.7)(0.35)} = 20.41 \text{ m/s} \qquad \mu = 2.2 \times 10^{-5} \text{ N·s/m}^2$$

$$\rho = \frac{p}{R'T} = \frac{100,000}{287(273 + 20)} = 1.189 \text{ kg/m}^3$$

$$\mathbf{R} = \frac{VD\rho}{\mu} = \frac{V4R\rho}{\mu} = \frac{20.41(4)(0.117)(1.1189)}{2.2 \times 10^{-5}} = 516,200$$

From Fig. 6.20, $f = 0.0165$

$$h_f = f\frac{L}{4R}\frac{V^2}{2g} = 0.0165\frac{70}{4(0.117)}\frac{20.41^2}{2(9.806)} = 52.42\text{-m air}$$

The specific weight of air is $\rho g = 1.189(9.806) = 11.66 \text{ N/m}^3$. In millimeters of water,

$$\frac{52.42(11.66)(1000)}{9806} = 62.33\text{-mm H}_2\text{O}$$

Aging of Pipes

The Moody diagram, with the values of absolute roughness shown there, is for new, clean pipe. With use, pipes become rougher, owing to corrosion, incrustations, and deposition of material on the pipe walls. The speed with which the friction factor changes with time depends greatly on the fluid being handled. Colebrook and White [8] found that the absolute roughness ϵ increases linearly with time,

$$\epsilon = \epsilon_0 + \alpha t \qquad\qquad\text{(12.6.3)}$$

in which ϵ_0 is the absolute roughness of the new surface. Tests on a pipe are required to determine α.

The time variation of the Hazen-Williams coefficient has been summarized graphically [9] for water-distribution systems in seven major cities in the United States. Although it is not a linear variation, the range of values for the average rate of decline in C may typically be between 0.5 and 2 per year, with the larger values generally applicable in the first years following installation. The only sure way of obtaining accurate coefficients for older water mains is through field tests.

Reduction of Head Loss with Additives

Soluble polymers are sometimes added to liquids to reduce the head loss in pipelines. The solutions are referred to as *drag-reducing agents*. Low concentrations of the polymer solution may reduce the head loss by over 50 percent in turbulent flow. There is little, if any, effect in laminar flow. Experiments indicate that the drag reduction is associated with flow near the wall of the pipe, in the viscous sublayer or in the buffer zone. Berman [10], Beaty et al. [11], and Lester [12] provide an indication of the experimental information on the subject. The Alaska oil pipeline has been using drag-reducing agents for a number of years with significant economic implications.

12.7   UNSTEADY FLOW: OSCILLATION OF LIQUID IN A U TUBE

Three cases of oscillations of liquid in a simple U tube are of interest: (1) frictionless liquid, (2) laminar resistance, and (3) turbulent resistance. They are discussed in this section.

Frictionless Liquid

For the frictionless case, Euler's equation of motion in unsteady form [Eq. (3.5.6)] is

$$\frac{1}{\rho}\frac{\partial p}{\partial s} + g\frac{\partial z}{\partial s} + v\frac{\partial v}{\partial s} + \frac{\partial v}{\partial t} = 0$$

can be applied. When sections 1 and 2 are designated (Fig. 12.20) and the incompressible flow equation is integrated from 1 to 2,

$$\frac{p_2 - p_1}{\rho} + g(z_2 - z_1) + \frac{v_2^2 - v_1^2}{2} + \int_1^2 \frac{\partial v}{\partial t}\, ds = 0 \qquad \textbf{(12.7.1)}$$

However, $p_1 = p_2$, $v_1 = v_2$, and $\partial v/\partial t$ is independent of s; hence,

$$g(z_2 - z_1) = -L\frac{\partial v}{\partial t} \qquad \textbf{(12.7.2)}$$

in which L is the length of the liquid column. By changing the elevation datum to the equilibrium position through the menisci, $g(z_2 - z_1) = 2gz$, and since v is a function

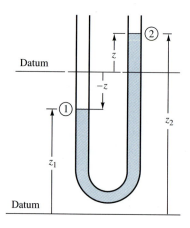

Figure 12.20   Oscillation of liquid in a U tube.

of t only, $\partial v/\partial t$ can be written as dv/dt or d^2z/dt^2. Hence,

$$\frac{d^2z}{dt^2} = \frac{dv}{dt} = -\frac{2g}{L}z \qquad (12.7.3)$$

The general solution of this equation is

$$z = C_1 \cos\sqrt{\frac{2g}{L}}t + C_2 \sin\sqrt{\frac{2g}{L}}t$$

in which C_1 and C_2 are arbitrary constants of integration. The solution is readily checked by differentiating twice and substituting into the differential equation. To evaluate the constants, if $z = Z$ and $dz/dt = 0$ when $t = 0$, then $C_1 = Z$ and $C_2 = 0$, or

$$z = Z\cos\sqrt{\frac{2g}{L}}t \qquad (12.7.4)$$

This equation defines a simple harmonic motion of a meniscus, with a period for a complete oscillation of $2\pi\sqrt{L/2g}$. The velocity of the column can be obtained by differentiating z with respect to t.

A frictionless fluid column 2.18 m long has a speed of 2 m/s when $z = 0.5$ m. Find (*a*) the maximum value of z, (*b*) the maximum speed, and (*c*) the period.

Example 12.13

Solution

(*a*) Differentiating Eq. (12.7.4) after substituting for L gives

$$\frac{dz}{dt} = -3Z\sin 3t$$

If t_1 is the time when $z = 0.5$ and $dz/dt = 2$, then

$$0.5 = Z\cos 3t_1 \qquad 2 = -3Z\sin 3t_1$$

Dividing the second equation by the first equation gives

$$\tan 3t_1 = -\frac{4}{3}$$

or $3t_1 = 5.356$ rad, $t_1 = 1.785$ s, $\sin 3t_1 = 0.8$, and $\cos 3t_1 = 0.6$. Then $Z = 0.5/\cos 3t_1 = 0.5/0.6 = 0.833$ m, the maximum value of z.
(*b*) The maximum speed occurs when $\sin 3t = 1$ or

$$3Z = 3(0.833) = 2.499 \text{ m/s}$$

(*c*) The period is

$$2\pi\sqrt{\frac{L}{2g}} = 2.094 \text{ s}$$

Laminar Resistance

When a shear stress τ_0 at the wall of the tube resists motion of the liquid column, it can be introduced into the Euler equation of motion along a streamline (Fig. 4.7).

The resistance in length δs is $\tau_0 \pi D \delta s$, which after dividing through by the mass of the particle, $\rho A \delta s$, is equal to $4\tau_0/\rho D$. Equation (12.7.1) becomes

$$\frac{1}{\rho}\frac{\partial p}{\partial s} + g\frac{\partial z}{\partial s} + v\frac{\partial v}{\partial s} + \frac{\partial v}{\partial t} + \frac{4\tau_0}{\rho D} = 0 \qquad \text{(12.7.5)}$$

This equation is good for either laminar or turbulent resistance. The assumption is made that the frictional resistance in unsteady flow is the same as for steady flow at the same velocity. From the Poiseuille equation the shear stress at the wall of a tube is

$$\tau_0 = \frac{8\mu v}{D} \qquad \text{(12.7.6)}$$

After making the substitution for τ_0 in Eq. (12.7.5) and integrating with respect to s as before,

$$g(z_2 - z_1) + L\frac{\partial v}{\partial t} + \frac{32 v v L}{D^2} = 0$$

Setting $2gz = g(z_2 - z_1)$, changing to total derivatives, and replacing v by dz/dt give

$$\frac{d^2 z}{dt^2} + \frac{32v}{D^2}\frac{dz}{dt} + \frac{2g}{L}z = 0 \qquad \text{(12.7.7)}$$

In effect, the column is assumed to have the average velocity dz/dt at any cross section.

By substitution

$$z = C_1 e^{at} + C_2 e^{bt}$$

can be shown to be the general solution of Eq. (12.7.7), provided that

$$a^2 + \frac{32v}{D^2}a + \frac{2g}{L} = 0 \qquad \text{and} \qquad b^2 + \frac{32v}{D^2}b + \frac{2g}{L} = 0$$

C_1 and C_2 are arbitrary constants of integration that are determined by given values of z and dz/dt at a given time. To keep a and b distinct, since the equations defining them are identical, they are assigned opposite signs before the radical term in the solution of the quadratic equation. Thus,

$$a = -\frac{16v}{D^2} + \sqrt{\left(\frac{16v}{D^2}\right)^2 - \frac{2g}{L}} \qquad b = -\frac{16v}{D^2} - \sqrt{\left(\frac{16v}{D^2}\right)^2 - \frac{2g}{L}}$$

To simplify the formulas, if

$$m = \frac{16v}{D^2} \qquad n = \sqrt{\left(\frac{16v}{D^2}\right)^2 - \frac{2g}{L}}$$

then

$$z = C_1 e^{-mt+nt} + C_2 e^{-mt-nt}$$

When the initial condition is taken as $t = 0$, $z = 0$, and $dz/dt = V_0$, then by substitution $C_1 = -C_2$ and

$$z = C_1 e^{-mt}(e^{nt} - e^{-nt}) \qquad \text{(12.7.8)}$$

Since

$$\frac{e^{nt} - e^{-nt}}{2} = \sinh nt$$

Eq. (12.7.8) becomes

$$z = 2C_1 e^{-mt} \sinh nt$$

By differentiating with respect to t,

$$\frac{dz}{dt} = 2C_1(-me^{-mt} \sinh nt + ne^{-mt} \cosh nt)$$

and setting $dz/dt = V_0$ for $t = 0$ gives

$$V_0 = 2C_1 n$$

since $\sinh 0 = 0$ and $\cosh 0 = 1$. Then

$$z = \frac{V_0}{n} e^{-mt} \sinh nt \qquad (12.7.9)$$

This equation gives the displacement z of one meniscus of the column as a function of time, starting with the meniscus at $z = 0$ when $t = 0$, and rising with velocity V_0.

Two principal cases† are to be considered. The first case, when

$$\frac{16\nu}{D^2} > \sqrt{\frac{2g}{L}}$$

n is a real number and the viscosity is so great that the motion is damped out in a partial cycle with z never becoming negative, Fig. 12.21 $(m/n = 2)$. The time t_0 for maximum z to occur is found by differentiating z [Eq. (12.7.9)] with respect to t and equating to zero,

$$\frac{dz}{dt} = 0 = \frac{V_0}{n}(-me^{-mt} \sinh nt + ne^{-mt} \cosh nt)$$

or

$$\tanh nt_0 = \frac{n}{m} \qquad (12.7.10)$$

Substitution of this value of t into Eq. (12.7.9) yields the maximum displacement Z

$$Z = \frac{V_0}{\sqrt{m^2 - n^2}} \left(\frac{m - n}{m + n}\right)^{m/2n} = V_0 \sqrt{\frac{L}{2g}} \left(\frac{m - n}{m + n}\right)^{m/2n} \qquad (12.7.11)$$

The second case, when

$$\frac{16\nu}{D^2} < \sqrt{\frac{2g}{L}}$$

---

†A third case, $16\nu/D^2 = \sqrt{2g/L}$, must be treated separately, yielding $z = V_0 t e^{-mt}$. The resulting oscillation is for a partial cycle only and is a limiting case of $16\nu/D^2 > \sqrt{2g/L}$.

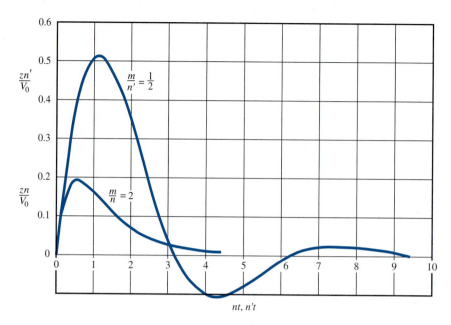

Figure 12.21   Position of meniscus as a function of time for oscillation of liquid in a U tube with laminar resistance.

results in a negative term within the radical,

$$n = \sqrt{-1\left[\frac{2g}{L} - \left(\frac{16\nu}{D^2}\right)^2\right]} = i\sqrt{\frac{2g}{L} - \left(\frac{16\nu}{D^2}\right)^2} = in'$$

in which $i = \sqrt{-1}$ and n' is a real number. Replacing n by in' in Eq. (12.7.9) produces the real function

$$z = \frac{V_0}{in'}e^{-mt}\sinh in't = \frac{V_0}{n'}e^{-mt}\sin n't \qquad \textbf{(12.7.12)}$$

since

$$\sin n't = \frac{1}{i}\sinh in't$$

The resulting motion of z is an oscillation about $z = 0$ with decreasing amplitude, as shown in Fig. 12.21 for the case $m/n' = \frac{1}{2}$. The time t_0 of maximum or minimum displacement is obtained from Eq. (12.7.12) by equating $dz/dt = 0$, producing

$$\tan n't_0 = \frac{n'}{m} \qquad \textbf{(12.7.13)}$$

There are an indefinite number of values of t_0 satisfying this expression, corresponding with all the maximum and minimum positions of a meniscus. By substitution of t_0 into Eq. (12.7.12)

$$Z = \frac{V_0}{\sqrt{n'^2 + m^2}}\exp\left(-\frac{m}{n'}\tan^{-1}\frac{n'}{m}\right) = V_0\sqrt{\frac{L}{2g}}\exp\left(-\frac{m}{n'}\tan^{-1}\frac{n'}{m}\right) \qquad \textbf{(12.7.14)}$$

Example 12.14

A 1.0-in.-diameter U tube contains oil, $\nu = 1 \times 10^{-4}$ ft²/s, with a total column length of 120 in. Applying air pressure to one of the tubes makes the gage difference 16 in. By quickly releasing the air pressure, the oil column is free to oscillate. Find the maximum velocity, the maximum Reynolds number, and the equation for position of one meniscus z, in terms of time.

Solution

The assumption is made that the flow is laminar, and the Reynolds number will be computed on this basis. The constants m and n are

$$m = \frac{16\nu}{D^2} = \frac{16 \times 10^{-4}}{(1/12)^2} = 0.2302$$

$$n = \sqrt{\left(\frac{16\nu}{D^2}\right)^2 - \frac{2g}{L}} = \sqrt{0.2302^2 - \frac{2(32.2)}{10}} = \sqrt{-6.387} = i2.527$$

or

$$n' = 2.527$$

Equations (12.7.12) to (12.7.14) apply to this case, as the liquid will oscillate above and below $z = 0$. The oscillation starts from the maximum position, that is, $Z = 0.667$ ft. By use of Eq. (12.7.14) the velocity (fictitious) when $z = 0$ at time t_0 before the maximum is determined to be

$$V_0 = Z\sqrt{\frac{2g}{L}} \exp\left(\frac{m}{n'} \tan^{-1} \frac{n'}{m}\right) = 0.667\sqrt{\frac{64.4}{10}} \exp\left(\frac{0.2302}{2.527} \tan^{-1} \frac{2.527}{0.2302}\right)$$

$$= 1.935 \text{ ft/s}$$

and

$$\tan n't_0 = \frac{n'}{m} \qquad t_0 = \frac{1}{2.527} \tan^{-1} \frac{2.527}{0.2302} = 0.586 \text{ s}$$

Hence, by substitution into Eq. (12.7.12),

$$z = 0.766 \exp[-0.2302(t + 0.586)] \sin 2.527(t + 0.586)$$

in which $z = Z$ at $t = 0$. The maximum velocity (actual) occurs for $t > 0$. Differentiating with respect to t to obtain the expression for velocity,

$$V = \frac{dz}{dt} = -0.1763 \exp[-0.2302(t + 0.586)] \sin 2.527(t + 0.586)$$

$$+ 1.935 \exp[-0.2302(t + 0.586)] \cos 2.527(t + 0.586)$$

Differentiating again with respect to t and equating to zero to obtain the maximum V produces

$$\tan 2.527(t + 0.586) = -0.1837$$

The solution in the second quadrant should produce the desired maximum, $t = 0.584$ s. Substituting this time into the expression for V produces $V = -1.48$ ft/s.

The corresponding Reynolds number is

$$\mathbf{R} = \frac{VD}{\nu} = 1.48\left(\frac{1}{12} \times 10^4\right) = 1234$$

Hence, the assumption of laminar resistance is justified.

Turbulent Resistance

In most practical cases of oscillation, or surge, in pipe systems there is turbulent resistance. With large pipes and tunnels the Reynolds number is large except for those time periods when the velocity is very near to zero. The assumption of fluid resistance proportional to the square of the average velocity is made (constant f). It closely approximates true conditions, although it yields a resistance that is too small for slow motions, in which case resistance is almost negligible. The equations will be developed for f = constant for oscillation within a simple U tube. The assumption is again made that resistance in unsteady flow is given by steady-flow resistance at the same velocity.

Using Eq. (6.7.2) to substitute for τ_0 in Eq. (12.7.5) leads to

$$\frac{1}{\rho}\frac{\partial p}{\partial s} + g\frac{\partial z}{\partial s} + v\frac{\partial v}{\partial s} + \frac{\partial v}{\partial t} + \frac{fv^2}{2D} = 0 \tag{12.7.15}$$

When this equation is integrated from section 1 to section 2 (Fig. 12.20), the first term drops out as the limits are $p = 0$ in each case, the third term drops out as $\partial v/\partial s \equiv 0$, and the fourth and fifth terms are independent of s. Hence,

$$g(z_2 - z_1) + \frac{\partial v}{\partial t}L + \frac{fv^2}{2D}L = 0$$

Since v is a function of t only, the partial can be replaced with the total derivative

$$\frac{dv}{dt} + \frac{f}{2D}v|v| + \frac{2g}{L}z = 0 \tag{12.7.16}$$

The absolute-value sign on the velocity term is needed so that the resistance opposes the velocity, whether positive or negative. By expressing $v = dz/dt$,

$$\frac{d^2z}{dt^2} + \frac{f}{2D}\frac{dz}{dt}\left|\frac{dz}{dt}\right| + \frac{2g}{L}z = 0 \tag{12.7.17}$$

This is a nonlinear differential equation because of the v-squared term. It can be integrated once with respect to t, but no closed solution is known for the second integration. It is easily handled numerically by the Runge-Kutta methods (see the Web page) when initial conditions are known, that is, $t = t_0$, $z = z_0$, and $dz/dt = 0$. Much can be learned from Eq. (12.7.17), however, by restricting the motion to the $-z$ direction; thus,

$$\frac{d^2z}{dt^2} - \frac{f}{2D}\left(\frac{dz}{dt}\right)^2 + \frac{2g}{L}z = 0 \tag{12.7.18}$$

The equation can be integrated once,† producing

$$\left(\frac{dz}{dt}\right)^2 = \frac{4gD^2}{f^2L}\left(1 + \frac{fz}{D}\right) + Ce^{fz/D} \qquad \text{(12.7.19)}$$

in which C is the constant of integration. To evaluate the constant, if $z = z_m$ for $dz/dt = 0$, then

$$C = -\frac{4gD^2}{f^2L}\left(1 + \frac{fz_m}{D}\right)e^{-fz_m/D}$$

and

$$\left(\frac{dz}{dt}\right)^2 = \frac{4gD^2}{f^2L}\left[1 + \frac{fz}{D} - \left(1 + \frac{fz_m}{D}\right)e^{f(z-z_m)/D}\right] \qquad \text{(12.7.20)}$$

Although this equation cannot be integrated again, numerical integration of particular situations yields z as a function of t. The equation, however, can be used to determine the magnitude of successive oscillations. At the instants of maximum or minimum z, say z_m and z_{m+1}, respectively, $dz/dt = 0$, and Eq. (12.7.20) simplifies to

$$\left(1 + \frac{fz_m}{D}\right)e^{-fz_m/D} = \left(1 + \frac{fz_{m+1}}{D}\right)e^{-fz_{m+1}/D} \qquad \text{(12.7.21)}$$

Since Eq. (12.7.18), the original equation, holds only for decreasing z, z_m must be positive and z_{m+1} negative. To find z_{m+2}, the other meniscus could be considered and z_{m+1} as a positive number substituted into the left-hand side of the equation to determine a minus z_{m+2} in place of z_{m+1} on the right-hand side of the equation.

Example 12.15

A U tube consisting of a 500-mm-diameter pipe with $f = 0.03$ has a maximum oscillation (Fig. 12.20) of $z_m = 6$ m. Find the minimum position of the surface and the following maximum.

Solution

From Eq. (12.7.21)

$$\left[1 + \frac{0.03(6)}{0.5}\right]e^{-0.03(6)/0.5} = (1 + 0.06z_{m+1})e^{-0.06z_{m+1}}$$

or

$$(1 + 0.06z_{m+1})e^{-0.06z_{m+1}} = 0.9488$$

---

†By substitution of $p = dz/dt$, we get

$$\frac{d^2z}{dt^2} = \frac{dp}{dt} = \frac{dp}{dz}\frac{dz}{dt} = p\frac{dp}{dz}$$

Then

$$p\frac{dp}{dz} - \frac{f}{2D}p^2 + \frac{2gz}{L} = 0$$

This equation can be made exact by multiplying by the integrating factor $e^{-fz/D}$. For the detailed method see [13].

which is satisfied by $z_{m+1} = -4.84$ m. Using $z_m = 4.84$ m in Eq. (12.7.21),

$$(1 + 0.06z_{m+1})e^{-0.06z_{m+1}} = [1 + 0.06(4.84)]e^{-0.06(4.84)} = 0.9651$$

which is satisfied by $z_{m+1} = -4.05$ m. Hence, the minimum water surface is $z = -4.84$ m and the next maximum is $z = 4.05$ m.

Example 12.16

The pipe of Example 12.15 is 1000 m long. By numerical integration of Eq. (12.7.17), using the third-order Runge-Kutta procedure (see the Web page), determine the time to the first minimum z and the next maximum z and check the results of Ex. 12.15.

Solution

Let $V = dz/dt$, then Eq. (12.7.17) may be written as

$$\frac{dV}{dt} = F(V, z) = -\frac{f}{2D}V|V| - \frac{2g}{L}z$$

Using the third-order Runge-Kutta with a time step $\Delta t = 0.75$ s, the following statements provide a solution

$$V_1 = \Delta t F(V_n, z_n) \qquad\qquad z_1 = \Delta t V_n$$

$$V_2 = \Delta t F\left(V_n + \frac{V_1}{3}, z_n + \frac{z_1}{3}\right) \qquad z_2 = \Delta t\left(V_n + \frac{V_1}{3}\right)$$

$$V_3 = \Delta t F\left(V_n + \frac{2}{3}V_2, z_n + \frac{2}{3}z_2\right) \qquad z_3 = \Delta t\left(V_n + \frac{2}{3}V_2\right)$$

$$V_{n+1} = V_n + \frac{V_1}{4} + \frac{3}{4}V_3 \qquad\qquad z_{n+1} = Z_n + \frac{z_1}{4} + \frac{3}{4}z_3$$

The minimum and maximum displacements are -4.834 m at 22.5 s and 4.048 m at 45.0 s.

12.8   UNSTEADY FLOW: ESTABLISHMENT OF FLOW

Determining the time needed by a flow in a pipeline to become established after a valve is suddenly opened is easily handled when friction and minor losses are taken into account. After a valve is opened (Fig. 12.22), the head H is available to accelerate the flow in the first instants, but as the velocity increases, the accelerating head is reduced by friction and minor losses. If L_e is the equivalent length of the pipe system, the final velocity V_0 is given by application of the energy equation as

$$H = f\frac{L_e}{D}\frac{V_0^2}{2g} \tag{12.8.1}$$

The equation of motion is

$$\gamma A\left(H - f\frac{L_e}{D}\frac{V^2}{2g}\right) = \frac{\gamma AL}{g}\frac{dV}{dt}$$

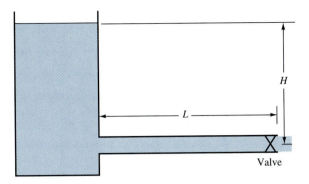

Figure 12.22 Notation for establishment of flow.

Solving for dt and rearranging, with Eq. (12.8.1), gives

$$\int_0^t dt = \frac{LV_0^2}{gH} \int_0^V \frac{dV}{V_0^2 - V^2}$$

After integration,

$$t = \frac{LV_0}{2gH} \ln \frac{V_0 + V}{V_0 - V} \qquad \text{(12.8.2)}$$

The velocity V approaches V_0 asymptotically, that is, mathematically it takes infinite time for V to attain the value V_0. Practically, for V to reach $0.99V_0$ takes

$$t = \frac{LV_0}{gH} \frac{1}{2} \ln \frac{1.99}{0.01} = 2.646 \frac{LV_0}{gH}$$

V_0 must be determined by taking minor losses into account, but Eq. (12.8.2) does not contain L_e.

In Fig. 12.22 the minor losses are $16V^2/2g$, $f = 0.030$, $L = 3000$ m, $D = 2.4$ m, and $H = 20$ m. Determine the time, after the sudden opening of a valve, for velocity to attain nine-tenths the final velocity.

Example 12.17

Solution

$$L_e = 3000 + \frac{16(2.4)}{0.03} = 4280 \text{ m}$$

From Eq. (12.8.1)

$$V_0 = \sqrt{\frac{2gHD}{fL_e}} = \sqrt{\frac{19.612(20)(2.4)}{0.030(4280)}} = 2.708 \text{ m/s}$$

Substituting $V = 0.9V_0$ into Eq. (12.8.2) gives

$$t = \frac{3000(2.708)}{19.612(20)} \ln \frac{1.90}{0.10} = 60.98 \text{ s}$$

12.9 UNSTEADY FLOW: DESCRIPTION OF THE WATERHAMMER PHENOMENON

Waterhammer may occur in a closed conduit flowing full when there is either a retardation or an acceleration of flow, such as with the change in opening of a valve in the line. If the changes are gradual, the calculations can be carried out by lumped methods, considering the liquid incompressible and the conduit rigid. When a valve is rapidly closed in a pipeline during flow, the flow through the valve is reduced. This increases the head on the upstream side of the valve and causes a pulse of high pressure to be propagated upstream at the sonic wave speed a. The action of this pressure pulse is to decrease the velocity of flow. On the downstream side of the valve the pressure is reduced, and a wave of lowered pressure travels downstream at wave speed a, which also reduces the velocity. If the closure is rapid enough and the steady pressure low enough, a vapor pocket may be formed downstream from the valve. When this occurs, the cavity will eventually collapse and produce a high-pressure wave downstream.

Before undertaking the derivation of equations for solution of waterhammer, a description of the sequence of events following sudden closure of a valve at the downstream end of a pipe leading from a reservoir (Fig. 12.23) is given. Friction is neglected in this case. At the instant of valve closure ($t = 0$) the fluid nearest the valve is compressed and brought to rest, and the pipe wall is stretched (Fig. 12.23a). As soon as the first layer is compressed, the process is repeated for the next layer. The fluid upstream from the valve continues to move downstream

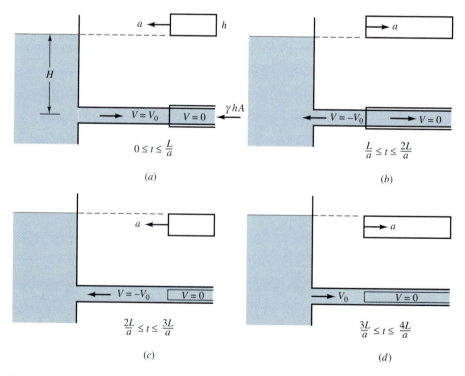

Figure 12.23   Sequence of events for one cycle of sudden closure of a valve.

with undiminished speed until successive layers have been compressed back to the source. The high pressure moves upstream as a wave, bringing the fluid to rest as it passes, compressing it, and expanding the pipe. When the wave reaches the upstream end of the pipe ($t = L/a$ s), all the fluid is under the extra head h, all the momentum has been lost, and all the kinetic energy has been converted into elastic energy.

There is an unbalanced condition at the upstream (reservoir) end at the instant of arrival of the pressure wave, as the reservoir pressure is unchanged. The fluid starts to flow backward, beginning at the upstream end. This flow returns the pressure to the value which was normal before closure, the pipe wall returns to normal, and the fluid has a velocity V_0 in the backward sense. This process of conversion travels downstream toward the valve at the speed of sound a in this pipe. At the instant $2L/a$ the wave arrives at the valve, pressures are back to normal along the pipe, and the velocity is everywhere V_0 in the backward direction.

Since the valve is closed, no fluid is available to maintain the flow at the valve and a low pressure develops ($-h$) such that the fluid is brought to rest. This low-pressure wave travels upstream at speed a and everywhere brings the fluid to rest, causes it to expand because of the lower pressure, and allows the pipe walls to contract. (If the static pressure in the pipe is not sufficiently high to sustain head $-h$ above vapor pressure, the liquid vaporizes in part and continues to move backward over a longer period of time.)

At the instant the negative pressure wave arrives at the upstream end of the pipe, $3L/a$ s after closure, the fluid is at rest but uniformly at head $-h$ less than before closure. This leaves an unbalanced condition at the reservoir, and fluid flows into the pipe, acquiring a velocity V_0 forward and returning the pipe and fluid to normal conditions as the wave progresses downstream at speed a. At the instant this wave reaches the valve, conditions are exactly the same as at the instant of closure, $4L/a$ s earlier.

This process is then repeated every $4L/a$ s. The action of fluid friction and imperfect elasticity of fluid and pipe wall, neglected heretofore, is to dampen the vibration and eventually cause the fluid to come permanently to rest. Closure of a valve in less than $2L/a$ is called *rapid closure; slow closure* refers to times of closure greater than $2L/a$.

The sequence of events taking place in a pipe can be compared with the sudden stopping of a freight train when the engine hits an immovable object. The car behind the engine compresses the spring in its forward coupling and stops as it exerts a force against the engine, and each car in turn keeps moving at its original speed until the preceding one suddenly comes to rest. When the caboose is at rest, all the energy is stored in compressing the coupling springs (neglecting losses). The caboose has an unbalanced force exerted on it, and starts to move backward, which in turn causes an unbalanced force on the next car, setting it in backward motion. This action proceeds as a wave toward the engine, causing each car to move at its original speed in a backward direction. If the engine is immovable, the car next to it is stopped by a tensile force in the coupling between it and the engine, analogous to the low-pressure wave in waterhammer. The process repeats itself car by car until the train is again at rest, with all the couplings in tension. The caboose is then acted upon by the unbalanced tensile force in its coupling and is set into forward motion, followed in turn by the rest of the cars. When this wave reaches the engine, all the cars are in motion as before the original impact. Then the whole cycle is repeated again. Friction acts to reduce the energy to zero in a very few cycles.

12.10   UNSTEADY FLOW: DIFFERENTIAL EQUATIONS FOR CALCULATION OF WATERHAMMER

Two basic mechanics equations are applied to a short segment of fluid in a pipe to obtain the differential equations for transient flow: Newton's second law of motion and the continuity equation. The dependent variables are pressure p and the average velocity V at a cross section. The independent variables are distance x along the pipe measured from the upstream end and time t; hence, $p = p(x, t)$ and $V = V(x, t)$. Poisson's ratio effect is not taken into account in this derivation. For pipelines with expansion joints it does not enter into the derivation. Friction is considered to be proportional to the square of the velocity.

Equation of Motion

The fluid element between two parallel planes δx apart, normal to the pipe axis, is taken as a free body in the application of Newton's second law of motion in the axial direction (Fig. 12.24). In equation form

$$pA - \left[pA + \frac{\partial}{\partial x}(pA)\,\delta x \right] + p\frac{\partial A}{\partial x}\,\delta x - \gamma A\,\delta x \sin\theta - \tau_0 \pi D\,\delta x = \rho A\,\delta x\,\frac{dV}{dt}$$

Dividing through by the mass of the element $\rho A\,\delta x$ and simplifying give

$$-\frac{1}{\rho}\frac{\partial p}{\partial x} - g \sin\theta - \frac{4\tau_0}{\rho D} = \frac{dV}{dt} \qquad (12.10.1)$$

For steady turbulent flow, $\tau_o = \rho f V^2/8$ [Eq. (6.7.2)]. The assumption is made that the friction factor in unsteady flow is the same as in steady flow. Hence, the equation of motion becomes

$$\frac{dV}{dt} + \frac{1}{\rho}\frac{\partial p}{\partial x} + g\sin\theta + \frac{fV|V|}{2D} = 0 \qquad (12.10.2)$$

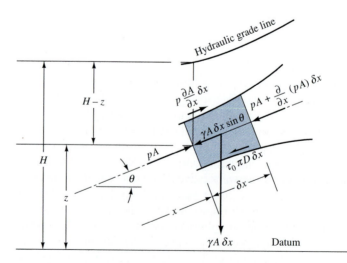

Figure 12.24   Free-body diagram for the derivation of the equation of motion.

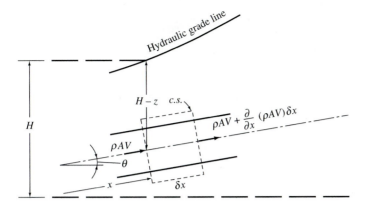

Figure 12.25 Control volume for the derivation of the continuity equation.

Since friction must oppose the motion, V^2 has been written as $V|V|$. By expanding the acceleration term,

$$\frac{dV}{dt} = V\frac{\partial V}{\partial x} + \frac{\partial V}{\partial t}$$

In waterhammer applications the term $V\,\partial V/\partial x$ is generally much smaller than $\partial V/\partial t$; hence, it will be omitted, leaving

$$L_1 = \frac{\partial V}{\partial t} + \frac{1}{\rho}\frac{\partial p}{\partial x} + g\sin\theta + \frac{fV|V|}{2D} = 0 \qquad (12.10.3)$$

The equation is indicated by L_1 to distinguish it from the equation of continuity L_2, which is derived next.

Equation of Continuity

The unsteady continuity equation (3.3.1) is applied to the control volume of Fig. 12.25,

$$-\frac{\partial}{\partial x}(\rho A V)\,\delta x = \frac{\partial}{\partial t}(\rho A\,\delta x) \qquad (12.10.4)$$

in which δx is not a function of t. Expanding the equation and dividing through by the mass $\rho A\,\delta x$ gives

$$\frac{V}{A}\frac{\partial A}{\partial x} + \frac{1}{A}\frac{\partial A}{\partial t} + \frac{V}{\rho}\frac{\partial \rho}{\partial x} + \frac{1}{\rho}\frac{\partial \rho}{\partial t} + \frac{\partial V}{\partial x} = 0 \qquad (12.10.5)$$

The first two terms are the total derivative $(1/A)\,dA/dt$, and the next two terms are the total derivative $(1/\rho)\,d\rho/dt$, yielding

$$\frac{1}{A}\frac{dA}{dt} + \frac{1}{\rho}\frac{d\rho}{dt} + \frac{\partial V}{\partial x} = 0 \qquad (12.10.6)$$

The first term deals with the elasticity of the pipe wall and its rate of deformation with pressure; the second term takes into account the compressibility of the liquid. For the wall elasticity the rate of change of tensile force per unit length (Fig. 12.26)

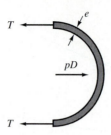

Figure 12.26   Tensile force in a pipe wall.

is $(D/2)\,dp/dt$; when divided by the wall thickness e, it is the rate of change of unit stress $(D/2e)\,dp/dt$; when this is divided by Young's modulus of elasticity for the wall material, the rate of increase of unit strain is obtained, $(D/2eE)\,dp/dt$. After multiplying this by the radius $D/2$, the rate of radial extension is obtained. Finally, by multiplying by the perimeter πD, the rate of area increase is obtained:

$$\frac{dA}{dt} = \frac{D}{2eE}\frac{dp}{dt}\frac{D}{2}\pi D$$

and hence

$$\frac{1}{A}\frac{dA}{dt} = \frac{D}{eE}\frac{dp}{dt} \tag{12.10.7}$$

From the definition of bulk modulus of elasticity of fluid (Chap. 1),

$$K = -\frac{dp}{d\forall/\forall} = \frac{dp}{d\rho/\rho}$$

and the rate of change of density divided by density yields

$$\frac{1}{\rho}\frac{d\rho}{dt} = \frac{1}{K}\frac{dp}{dt} \tag{12.10.8}$$

By Eqs. (12.10.7) and (12.10.8), Eq. (12.10.6) becomes

$$\frac{1}{K}\frac{dp}{dt}\left(1 + \frac{K}{E}\frac{D}{e}\right) + \frac{\partial V}{\partial x} = 0 \tag{12.10.9}$$

It is convenient to express the constants in this equation in the form

$$a^2 = \frac{K/\rho}{1 + (K/E)(D/e)} \tag{12.10.10}$$

Equation (12.10.9) now becomes

$$\frac{1}{\rho}\frac{dp}{dt} + a^2\frac{\partial V}{\partial x} = 0 \tag{12.10.11}$$

When expanded, dp/dt is

$$\frac{dp}{dt} = V\frac{\partial p}{\partial x} + \frac{\partial p}{\partial t}$$

Again, for waterhammer applications the term $V \, \partial p / \partial x$ is usually much smaller than $\partial p / \partial t$ and is neglected, yielding

$$L_2 = \frac{\partial p}{\partial t} + \rho a^2 \frac{\partial V}{\partial x} = 0 \qquad \text{(12.10.12)}$$

which is the continuity equation for a compressible liquid in an elastic pipe. L_1 and L_2 provide two nonlinear partial differential equations in V and p in terms of the independent variables x and t. No general solution to these equations is known, but they can be transformed by the method of characteristics for a convenient finite-difference numerical solution.

12.11 UNSTEADY FLOW: THE METHOD OF CHARACTERISTICS SOLUTION

Equations L_1 and L_2 in the preceding section contain two unknowns. These equations can be combined with an unknown multiplier as $L = L_1 + \lambda L_2$. Any two real, distinct values of λ yield two equations in V and p that contain all the physics of the original two equations L_1 and L_2 and can replace them in any solution. It may happen that great simplification will result if two particular values of λ are found. L_1 and L_2 are substituted into the equation for L, with some rearrangement.

$$L = \left(\frac{\partial V}{\partial x} \lambda \rho a^2 + \frac{\partial V}{\partial t} \right) + \lambda \left(\frac{\partial p}{\partial x} \frac{1}{\rho \lambda} + \frac{\partial p}{\partial t} \right) + g \sin \theta + \frac{f V |V|}{2D} = 0$$

The first term in parentheses is the total derivative dV/dt if $\lambda \rho a^2 = dx/dt$. From the calculus

$$\frac{dV}{dt} = \frac{\partial V}{\partial x} \frac{dx}{dt} + \frac{\partial V}{\partial t} = \frac{\partial V}{\partial x} \lambda \rho a^2 + \frac{\partial V}{\partial t}$$

Similarly, the second term in parentheses is the total derivative dp/dt if $1/\rho \lambda = dx/dt$. For both statements to be correct, dx/dt must have the same value

$$\frac{dx}{dt} = \lambda \rho a^2 = \frac{1}{\lambda \rho} \qquad \text{or} \qquad \lambda = \pm \frac{1}{\rho a} \qquad \text{(12.11.1)}$$

Then

$$\frac{dx}{dt} = \pm a \qquad \text{(12.11.2)}$$

The equation for L now becomes

$$L = \frac{dV}{dt} \pm \frac{1}{\rho a} \frac{dp}{dt} + g \, \sin \theta + \frac{f V |V|}{2D} = 0 \qquad \text{(12.11.3)}$$

subject to the conditions of Eq. (12.11.2). Therefore, two real distinct values of λ have been found that convert the two partial differential equations into the pair of total differential equations (12.11.3), subject to Eqs. (12.11.2).

Since Eq. (12.11.3) is valid only when Eqs. (12.11.2) are satisfied, it is convenient to visualize the solution on a plot of x against t, as in Fig. 12.27. Position x locates a point in the pipeline measured from the upstream end, and t is the time at which the dependent variables V and p are to be determined. First consider the conditions at A are known (that is, V_A, p_A, x_A, and t_A). Then Eq. (12.11.3) with the $+$ sign,

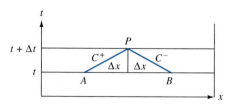

Figure 12.27 *xt* plot of characteristics along which solution is obtained.

called the C^+ compatibility equation, is valid along the line AP or an extension of the line. The slope of line AP is $dt/dx = 1/a$, in which a is the speed of an acoustical wave through the pipe. By multiplying Eq. (12.11.3) through by $\rho a\, dt$, and integrating from A to P,

$$\rho a \int_A^P dV + \int_A^P dp + \int_A^P \rho ag \sin\theta\, dt + \int_A^P \rho a\frac{fV|V|}{2D}\, dt = 0$$

Since $a\, dt = dx$, the equation can be written in finite-difference form as

$$\rho a(V_P - V_A) + p_P - p_A + \rho g \sin\theta\, \Delta x + \frac{\rho \Delta x f|V_A|V_P}{2D} = 0 \qquad \textbf{(12.11.4)}$$

This form of the equation assumes θ constant from A to P along the pipe. The integration of the friction term must be an approximation, since V is not known as a function of x from A to P. The one selected

$$\int_A^P V|V|\, dx = \Delta x|V_A|V_P$$

is a good approximation and becomes exact for steady state, which is a special case of unsteady flow. The corresponding C^- compatibility equation, in a similar manner, becomes

$$\rho a(V_P - V_B) - p_P + p_B + \rho g \sin\theta\, \Delta x + \frac{\rho \Delta x f|V_B|V_P}{2D} = 0 \qquad \textbf{(12.11.5)}$$

since $a\, dt = -dx$. Δx is a positive number, the reach length. These two equations can be solved simultaneously to determine p_P and V_P.

In solving piping problems it is convenient to work with hydraulic grade line H and discharge Q in place of p and V. From Fig. 12.24

$$p_P = \rho g(H_P - z_P) \qquad p_A = \rho g(H_A - z_A)$$

and

$$\begin{aligned}
p_P - p_A &= \rho g(H_P - H_A) - \rho g(z_P - z_A) \\
&= \rho g(H_P - H_A) - \rho g\, \Delta x \sin\theta
\end{aligned} \qquad \textbf{(12.11.6)}$$

Substitution in Eq. (12.11.4), with $V = Q/A$, yields for C^+

$$H_P = H_A - \frac{a}{gA}(Q_P - Q_A) - \frac{\Delta x f Q_P|Q_A|}{2gDA^2} \qquad \textbf{(12.11.7)}$$

and for C^-

$$H_P = H_B + \frac{a}{gA}(Q_P - Q_B) + \frac{\Delta x f Q_P|Q_B|}{2gDA^2} \qquad \textbf{(12.11.8)}$$

The C^- compatibility equation from B to P is obtained in a manner similar to the C^+ compatibility equation. To simplify the equations, let

$$B = \frac{a}{gA} \qquad \text{and} \qquad R = \frac{f\,\Delta x}{2gDA^2}$$

Then for C^+

$$H_P = H_A - B(Q_P - Q_A) - RQ_P|Q_A| \qquad \textbf{(12.11.9)}$$

and for C^-

$$H_P = H_B + B(Q_P - Q_B) + RQ_P|Q_B| \qquad \textbf{(12.11.10)}$$

For the numerical solution of the transient equations (12.11.9) and (12.11.10), or Eqs. (12.11.4) and (12.11.5), a single pipeline is divided into an even number of reaches, N, each Δx in length as shown in Fig. 12.28, $\Delta x = L/N$ and $\Delta t = \Delta x/a$. The C^+ and C^- characteristic lines are then the diagonals of the rectangular grid. A subscripted notation is convenient, as shown in the figure. In applying the equations to solve for an internal section, where H_i and Q_i are desired, conditions at the earlier time are known, that is, Q_{i-1}, H_{i-1}, Q_{i+1}, and H_{i+1}. Collecting the known terms of Eq. (12.11.9) into the constants C_P and B_P gives

$$C_P = H_{i-1} + BQ_{i-1} \qquad B_P = B + R|Q_{i-1}| \qquad \textbf{(12.11.11)}$$

while for Eq. (12.11.10) the constants C_M, B_M are

$$C_M = H_{i+1} - BQ_{i+1} \qquad B_M = B + R|Q_{i+1}| \qquad \textbf{(12.11.12)}$$

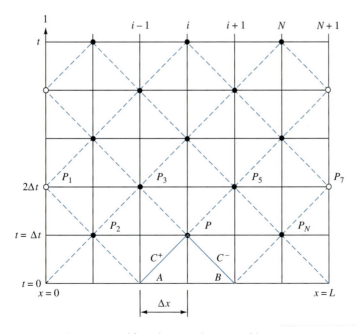

Figure 12.28 x–t grid for solving single-pipe problems.

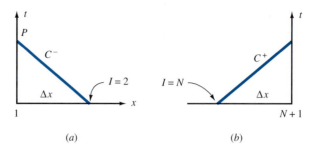

Figure 12.29   Boundary conditions. (a) Upstream end.
(b) Downstream end.

Now Eqs. (12.11.9) and (12.11.10) become for C^+

$$H_i = C_P - B_P Q_i \qquad (12.11.13)$$

and for C^-

$$H_i = C_M + B_M Q_i \qquad (12.11.14)$$

With C_P, B_P, C_M, and B_M known, the solution of Eqs. (12.11.13) and (12.11.14) yields

$$Q_i = \frac{C_P - C_M}{B_P + B_M} \qquad H_i = C_P - B_P Q_i$$

The solution consists of finding H and Q for alternate grid points along $t = \Delta t$, then proceeding to $t = 2\Delta t$, and so on, until the desired time duration has been covered. End points of the pipe are introduced every other time step after the initial conditions. The term *boundary condition* refers to the end condition on each pipeline.

At each end of a single pipe only one of the compatibility equations is available in the two variables. At the upstream end, Fig. 12.29a, Eq. (12.11.14) holds along the C^- characteristic, and for the downstream end, Fig. 12.29b, Eq. (12.11.13) is valid along the C^+ characteristic. These are linear equations in Q_i and H_i. Each conveys to its respective boundary the complete behavior and response of the fluid in the pipeline during the transient. An auxiliary equation is needed in each case, that specifies Q, H, or some relation between them. That is, the auxiliary equation must convey information on the behavior of the boundary to the pipeline. This may be just the end condition of the pipeline, or it may be a different element or facility attached to the end of the pipe. Each boundary condition is solved independently of the other boundary and independently of interior point calculations.

Complex systems can be visualized as a combination of single pipelines that are handled as described above, with boundary conditions at the pipe ends to transfer the transient response from one pipeline to another and to provide interaction with the system terminal conditions. Thus, a complicated system can be treated by a combination of a common solution procedure for the interior of each pipeline, together with a systematic coverage of each terminal and interconnection point in the system. The primary focus in the treatment of a variety of transient liquid-flow problems is on the handling of boundary conditions [14].

Reservoir at the Upstream End

At a large pressure tank or reservoir the elevation of the hydraulic grade line normally may be assumed constant during a short duration transient. This end condition is described by $H_1 = H_R$, in which H_R is the elevation of the reservoir surface above the reference datum. If the reservoir level changes in a known manner, say as a sine wave, the boundary condition is

$$H_1 = H_R + \Delta H \sin \omega t \qquad \text{(12.11.15)}$$

in which ω is the circular frequency and ΔH is the amplitude of the wave. At each instant of calculation of the boundary condition, Q_1 is determined by a direct solution of Eq. (12.11.14) as

$$Q_1 = \frac{H_1 - C_M}{B_M} \qquad \text{(12.11.16)}$$

The subscript 1 refers to the upstream section, at point P in Fig. 12.29a. The values of C_M and B_M, Eq. (12.11.12), are dependent only on known values from the previous time step, in this case point $I = 2$ in Fig. 12.29a.

Valve at the Downstream End

For steady-state flow through the valve, considered as an orifice,

$$Q_0 = (C_d A_v)_0 \sqrt{2gH_0}$$

where Q_0 is the steady-state flow, H_0 is the head drop across the valve, and $(C_d A_v)_0$ is the area of the opening times the discharge coefficient. For another opening, in general,

$$Q_{NS} = C_d A_v \sqrt{2gH_{NS}} \qquad \text{(12.11.17)}$$

in which the subscript $NS = N + 1$ refers to the downstream section in the pipe. The solution of Eq. (12.11.17) and Eq. (12.11.13) yields

$$Q_{NS} = -gB_P(C_d A_v)^2 + \sqrt{[gB_P(C_d A_v)^2]^2 + (C_d A_v)^2 2gC_P} \qquad \text{(12.11.18)}$$

and

$$H_{NS} = C_P - B_P Q_{NS} \qquad \text{(12.11.19)}$$

It is often convenient to use a single reach in pipelines rather than multireaches, as indicated in Fig. 12.28. The decision as to which method is appropriate is based on the magnitude of the friction term in the compatibility equations. For high friction systems short reaches are needed to assure reasonable accuracy. However, for low friction systems longer reaches will yield accurate results, in which case the reach length may extend the full pipe length. When a single reach is used in a pipeline, no interior points are calculated, and end conditions are calculated every time step, $\Delta t = L/a$ s (see Fig. 12.30b).

Example 12.18

In Fig. 12.30a the water surface changes according to the equation

$$H_A = H_R + \Delta H \sin \omega t$$

while the right end of the pipe contains a small orifice. The frequency of the waves is set at the natural period of the pipe, $4L/a$, which yields an ω of $2\pi/(4L/a)$. Determine the resulting motion of fluid in the pipe and the head fluctuations.

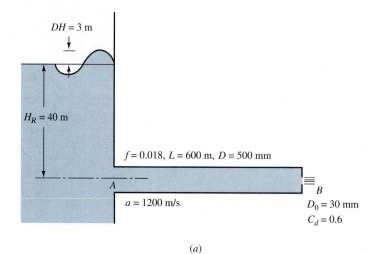

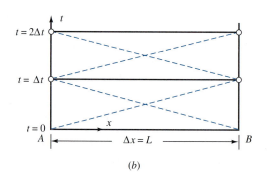

Figure 12.30   Example 12.18.

Solution

A spreadsheet is used for the solution as shown in Fig. 12.31. One reach is used. The upstream section is solved by Eq. (12.11.16) with the head $H_1 = H_A$, a known function of time. The downstream boundary condition solves the orifice equation together with Eq. (12.11.13), as presented in Eq. (12.11.18) with $C_v = g(C_d A)^2$, yielding

$$Q_B = -C_v B_P + \sqrt{(C_v B_P)^2 + 2C_v C_P} \qquad H_B = C_P - B_P Q_B$$

Junction of Two or More Pipes

At a connection of pipelines of different properties, the continuity equation must be satisfied at each instant of time, and a common hydraulic-grade-line elevation can be assumed at the end of each pipe. These statements implicitly assume that there is no storage at the junction and also neglect all minor effects. In multipipe systems it is necessary for each section at the pipe ends at the junction to be separately identified. They have been arbitrarily assigned at node i of the three-pipe junction in Fig. 12.32. If Eqs. (12.11.13) and (12.11.14) are written in the following form, a summation provides a simple solution for the common hydraulic grade line, H_i.

		Transient in single-reach pipe						
		Oscillating reservoir level/pipe/fixed orifice						
		HA = HR+DH*sin(ω*t)						
f =	0.018	AR =	0.19635	=0.7854*D^2				
L =	600.000	B =	623.244	=a/(g*AR)				
D =	0.500	Rf =	28.567	=f*L/(2*g*D*AR^2)				
a =	1200.000	Cv =	1.76E−06	=g*CdA^2				
CdA =	4.24E−04	Q0 =	0.01188	=SQRT(HR/(Rf+.5/Cv))				
HR =	40.000	H0 =	39.996	=HR−Rf*Q0^2				
DH =	3.000	g =	9.806					
ω =	3.1416			QB=(@0.5 s)	= −Cv*G14+SQRT((Cv*G14)^2+2*Cv*F14)			
time	CM	BM	HA	QA	CP	BP	QB	HB
0.0			40.000	0.01188			0.01188	39.996
0.5	32.593	623.584	43.000	0.01669	47.403	623.584	0.01188	39.996
1.0	32.593	623.584	40.000	0.01188	53.402	623.721	0.01267	45.499
1.5	37.603	623.606	37.000	−0.00097	47.403	623.584	0.01188	39.996
2.0	32.593	623.584	40.000	0.01188	36.397	623.272	0.01029	29.987
2.5	23.576	623.538	43.000	0.03115	47.403	623.584	0.01188	39.996
3.0	32.593	623.584	40.000	0.01188	62.415	624.134	0.01378	53.815
3.5	45.228	623.638	37.000	−0.01319	47.403	623.584	0.01188	39.996
4.0	32.593	623.584	40.000	0.01188	28.778	623.621	0.00904	23.143
4.5	17.511	623.503	43.000	0.04088	47.404	623.584	0.01188	39.996
5.0	32.593	623.584	40.000	0.01188	68.478	624.412	0.01448	59.436
5.5	50.412	623.658	37.000	−0.02150	47.403	623.584	0.01188	39.995
6.0	32.592	623.584	40.000	0.01188	23.597	623.859	0.00809	18.550
6.5	13.509	623.476	43.000	0.04730	47.404	623.584	0.01188	39.997
7.0	32.593	623.584	40.000	0.01188	72.480	624.596	0.01493	63.157
7.5	53.854	623.671	37.000	−0.02702	47.402	623.584	0.01188	39.995
8.0	32.592	623.584	40.000	0.01188	20.157	624.017	0.00740	15.537
8.5	10.923	623.456	43.000	0.05145	47.404	623.584	0.01188	39.997
9.0	32.594	623.584	40.000	0.01188	75.066	624.714	0.01521	65.565
9.5	56.086	623.679	37.000	−0.03060	47.402	623.584	0.01188	39.995
10.0	32.592	623.584	40.000	0.01188	17.927	624.119	0.00693	13.603
10.5	9.286	623.442	43.000	0.05408	47.405	623.584	0.01188	39.997
11.0	32.594	623.584	40.000	0.01188	76.703	624.789	0.01538	67.091
11.5	57.503	623.684	37.000	−0.03287	47.401	623.584	0.01188	39.994
12.0	32.591	623.584	40.000	0.01188	16.511	624.184	0.00661	12.385

Figure 12.31 Example 12.18, spreadsheet.

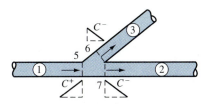

Figure 12.32 Pipeline junction.

$$Q_5 = -\frac{H_i}{B_{P_1}} + \frac{C_{P_1}}{B_{P_1}}$$

$$-Q_7 = \frac{H_i}{B_{M_2}} + \frac{C_{M_2}}{B_{M_2}}$$

$$-Q_6 = \frac{H_i}{B_{M_3}} + \frac{C_{M_3}}{B_{M_3}}$$

$$\Sigma Q = 0 = -H_i \Sigma \frac{1}{B_j} + \frac{C_{P_1}}{B_{P_1}} + \frac{C_{M_2}}{B_{M_2}} + \frac{C_{M_3}}{B_{M_3}}$$

or

$$H_i = \frac{C_{P_1}/B_{P_1} + C_{M_2}/B_{M_2} + C_{M_3}/B_{M_3}}{\Sigma(1/B_j)} \tag{12.11.20}$$

The term $\Sigma(1/B_j)$ is $\Sigma(1/B_{P_j}) + \Sigma(1/B_{M_j})$ where j refers to the pipe numbers. With the common hydraulic grade line computed, the equations above can be used to determine the flow at the end section in each pipe at the junction.

Algebraic or Reach-back Method

In multipipe systems it is necessary to solve for conditions at each junction at the same instant of time, that is, a common time-step size must be used in each pipeline. To accommodate this requirement, while using a single reach in each pipeline, a reach-back method is adopted. This concept is illustrated in Fig. 12.34 for the series system shown in Fig. 12.33. In this system pipe 2 is twice as long as pipe 1, and both pipes have the same wavespeed. Thus, the time step for calculation is $\Delta t = L_1/a_1$ s. Pipe 1 is handled the same as in the single pipe, Ex. 12.18, in that each characteristic reaches back one time step. However, in pipe 2 the characteristic lines must reach back two time steps to retrieve appropriate information from the other end of the pipeline. This is shown in Fig. 12.34 and in Ex. 12.19. To illustrate, calculations at the junction of the two pipes at $t = 0.2$ s use CP and BP in pipe 1, bringing information from the reservoir at $t = 0.1$ s, and CM and BM in pipe 2, bringing information from the valve at $t = 0.0$ s. Similarly, at $t = 0.2$ s at the valve, information at the junction at $t = 0.0$ s is used to calculate CP and BP in pipe 2.

It is emphasized that if friction losses are high, multireaches, as shown in Fig. 12.28, are necessary to assure reasonable accuracy. The reach-back method is most appropriate for low friction systems or short pipelines. In reality this includes many practical cases. The spreadsheet solution is very effective in single pipelines as presented in the previous example, and it can be applied to multipipe systems of limited size. However, it becomes quite cumbersome when there are many computing sections, such as in systems with many elements or in pipelines

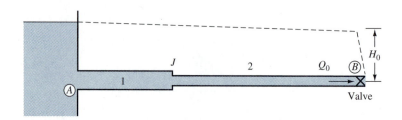

Figure 12.33   Example 12.19, series pipeline.

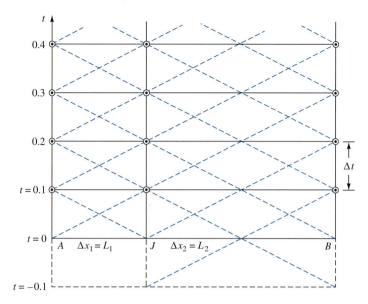

Figure 12.34 Reach-back method in series system.

with multireaches. For these cases the repetitive calculations at computing sections are easily programmed in a formal computer language such as BASIC, FORTRAN, PASCAL, etc., and this procedure is recommended. Several references [14–16] are available for more details.

The system of Fig. 12.33 initially has a valve opening $C_d A = 0.00015$ m^2. The valve is closed by a linear reduction in $C_d A$ from 0.00015 to 0.0 at 0.3 s. Calculate the transients in the system for 2.4 s.

Example 12.19

Solution

The data for the problem are shown in Fig. 12.35, along with the identification of variables, and the calculation of initial constants. Equation (12.11.17) is used at the valve with the data for the hydraulic grade line taken through the centerline of the valve.

$$Q_B = C_d A \sqrt{2gH_B}$$

Hydraulic-grade-line elevations are given in meters and discharges in cubic meters per second. For initial steady flow the energy equation is written from the reservoir to the valve as

$$H_R - f_1 \frac{L_1}{D_1} \frac{Q_0^2}{2gA_{R1}^2} - f_2 \frac{L_2}{D_2} \frac{Q_0^2}{2gA_{R2}^2} = \frac{Q_0^2}{2g(C_d A)^2}$$

from which steady-state flow Q_0 is determined. The initial head at the valve is H_0, and the resistance coefficient R_f in each pipe is

$$R_f = \frac{fL}{2gDA_R^2}$$

Equation (12.11.18) is used to find the flow at the valve until it is closed, when flow at the valve is set to zero. No provision is made for backflow through the valve.

		Transient in series pipe system caused by rapid valve closure								
		Fixed reservoir/pipe/function/pipe/valve								
pipe	**f**	**length**	**diameter**	**a**	**AR**		**B=a/g∗AR**		**Rf**	
1	0.019	100	0.200	1000	0.031	BBB1 =	3246.1	Rf1 =	490.8	
2	0.025	240	0.100	1200	0.008	BBB2 =	15581.1	Rf2 =	49596	
g =	9.806									
CdA =	0.000150	Q0 =	0.00720	= SQRT(HR/(Rf1+Rf2+1/(2∗g∗CdA^2)))						
HR =	120.000	H0 =	117.405	= HR−(Rf1+Rf2)∗Q0^2						
CP & CM to calculate HJ & QJ @.2 s			CP@.2 s =	= HR+BBB1∗B15				CM@.2 s =	= L14−BBB2∗K14	
BP & BM to calculate HJ & QJ @.2 s			BP@.2 s =	= BBB1+Rf1∗ABS(B15)				BM@.2 s =	= BBB2+Rf2∗K14	

	Reservoir	pipe 1		pipe 2		junction		pipe 2		Valve	
time,s	**QA**	**CP**	**BP**	**CM**	**BM**	**HJ**	**QJ**	**CP**	**BP**	**QB**	**HB**
−0.1	0.00720					119.975	0.00720			0.0072	117.405
0.0	0.00720					119.975	0.00720			0.0072	117.405
0.1	0.00720	143.364	3249.6	5.256	15938	119.975	0.00720	232.123	15938	0.0054	146.469
0.2	0.00720	143.364	3249.6	5.256	15938	119.975	0.00720	232.123	15938	0.0030	184.223
0.3	0.00720	143.364	3249.6	63.089	15847	129.704	0.00420	232.123	15938	0.0000	232.123
0.4	0.00121	143.364	3249.6	137.395	15730	142.342	0.00031	232.123	15938	0.0000	232.123
0.5	−0.00657	123.939	3246.7	232.123	15581	142.594	−0.00575	195.202	15790	0.0000	195.202
0.6	−0.01270	98.680	3249.3	232.123	15581	121.706	−0.00709	147.242	15597	0.0000	147.242
0.7	−0.00760	78.790	3252.3	195.202	15581	98.893	−0.00618	53.065	15866	0.0000	53.065
0.8	0.00032	95.317	3249.8	147.242	15581	104.278	−0.00276	11.289	15933	0.0000	11.289
0.9	0.00209	121.042	3246.2	53.065	15581	109.321	0.00361	2.584	15888	0.0000	2.584
1.0	0.00690	126.768	3247.1	11.289	15581	106.853	0.00613	61.314	15718	0.0000	61.314
1.1	0.01017	142.387	3249.4	2.584	15581	118.262	0.00742	165.578	15760	0.0000	165.578
1.2	0.00795	153.026	3251.1	61.314	15581	137.193	0.00487	202.417	15885	0.0000	202.417
1.3	−0.00043	145.809	3250.0	165.578	15581	149.221	−0.00105	233.940	15949	0.0000	233.940
1.4	−0.01005	118.616	3246.3	202.417	15581	133.065	−0.00445	213.073	15823	0.0000	213.073
1.5	−0.00847	87.377	3251.0	233.940	15581	112.678	−0.00778	132.863	15633	0.0000	132.863
1.6	−0.00552	92.505	3250.2	213.073	15581	113.315	−0.00640	63.714	15802	0.0000	63.714
1.7	−0.00434	102.080	3248.8	132.863	15581	107.391	−0.00163	−8.584	15967	0.0000	−8.584
1.8	0.00225	105.916	3248.2	63.714	15581	98.636	0.00224	13.556	15899	0.0000	13.556
1.9	0.00882	127.300	3247.2	−8.584	15581	103.865	0.00722	81.919	15662	0.0000	81.919
2.0	0.01217	148.630	3250.4	13.556	15581	125.316	0.00717	133.558	15692	0.0000	133.558
2.1	0.00553	159.518	3252.0	81.919	15581	146.119	0.00412	216.315	15939	0.0000	216.315
2.2	−0.00393	137.948	3248.8	133.558	15581	137.191	0.00023	237.075	15937	0.0000	237.075
2.3	−0.00506	107.264	3248.0	216.315	15581	126.075	−0.00579	210.318	15785	0.0000	210.318
2.4	−0.00766	103.567	3248.5	237.075	15581	126.600	−0.00709	140.823	15593	0.0000	140.823

Figure 12.35   Example 12.19, series system transient results.

$$Q_B = -gB_P(C_dA)^2 + \sqrt{[gB_P(C_dA)^2]^2 + (C_dA)^2 2gC_P}$$

Transient results are shown in Fig. 12.35. Note that initial conditions are specified at $t = 0.0$ s, and they are also included at $t = -0.1$ s. The latter are needed in pipe 2 when making computations at $t = +0.1$ s, since it is necessary to reach back 0.2 s to $t = -0.1$ s.

PROBLEMS†

12.1 Determine the discrepancy between the head loss calculated with the Darcy-Weisbach equation and with the Hazen-Williams equation for 15°C water flowing in a 1-m-diameter welded steel pipeline. $C = 120$ and $\epsilon = 0.2$ mm. Graph the discrepancy as a function of Reynolds number, $10^4 < \mathbf{R} < 10^7$.

12.2 Sketch the hydraulic and energy grade lines for Fig. 12.36. $H = 10$ m.

12.3 Calculate the value of K for the valve of Fig. 12.36 so that the discharge of Prob. 12.2 is reduced by one-half. Sketch the hydraulic and energy grade lines.

12.4 Compute the discharge of the system in Fig. 12.37. Draw the hydraulic and energy grade lines.

12.5 What head is needed in Fig. 12.37 to produce a discharge of 0.3 m³/s?

12.6 What diameter smooth pipe is required to convey 8-L/s kerosene at 32°C 150 m with a head of 5 m? There is a valve and other minor losses with total K of 7.6.

12.7 Water (15°C) flows through a 12-cm-diameter cast iron horizontal pipe, 300 m long. It is connected to a reservoir with a re-entrant pipe entrance upstream and discharges to the atmosphere downstream. The line contains an open globe valve and three standard elbows. Find the flow rate if the reservoir elevation is (*a*) 8 m and (*b*) 15 m above the exit.

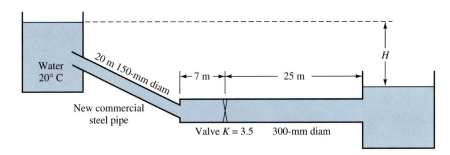

Figure 12.36   Problems 12.2 and 12.3.

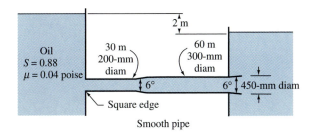

Figure 12.37   Problems 12.4 and 12.5.

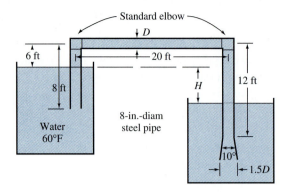

Figure 12.38   Problems 12.8 and 12.9.

12.8   Calculate the discharge through the siphon of Fig. 12.38 with the conical diffuser removed. $H = 4$ ft.

12.9   Calculate the discharge in the siphon of Fig. 12.38 for $H = 8$ ft. What is the minimum pressure in the system?

12.10   Find the discharge through the siphon of Fig. 12.39. What is the pressure at A, which is 150 mm above the outlet? Estimate the minimum pressure in the system.

12.11   Neglecting minor losses other than the valve, sketch the hydraulic grade line for Fig. 12.40. The globe valve has a loss coefficient $K = 4.5$.

12.12   What is the maximum height of point A (Fig. 12.40) for no cavitation? The barometer reading is 29.5-in. Hg.

12.13   Determine the discharge of the system of Fig. 12.41 for $L = 600$ m, $D = 500$ mm, $\epsilon = 0.5$ mm, and $H = 8$ m, with the pump A characteristics given.

12.14   Determine the discharge through the system of Fig. 12.41 for $L = 4000$ ft, $D = 24$-in. smooth pipe, $H = 40$ ft, with pump B characteristics.

12.15   Construct a head-discharge-efficiency table for pumps A and B (Fig. 12.41) connected in series. (SI units).

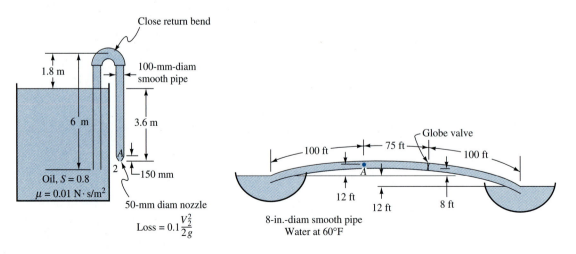

Figure 12.39   Problem 12.10.   **Figure 12.40**   Problems 12.11 and 12.12.

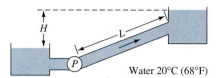

Water 20°C (68°F)

Pump A						Pump B				
H_P		Q_P		e		H_P		Q_P		e
m	ft	l/s	cfs	%		m	ft	l/s	cfs	%
21.3	70	0	0	0		24.4	80	0	0	0
18.3	60	56.6	2.00	59		21.3	70	74	2.60	54
16.8	55	72.5	2.56	70		18.3	60	112	3.94	70
15.2	50	85.8	3.03	76		15.2	50	140	4.96	80
13.7	45	97.7	3.45	78		12.2	40	161	5.70	73
12.2	40	108	3.82	76.3		9.1	30	174	6.14	60
10.7	35	116	4.11	72		6.1	20	177	6.24	40
9.1	30	127	4.48	65						
7.6	25	130	4.59	56.5						
6.1	20	134	4.73	42						

Figure 12.41   Problems 12.13 to 12.20, 12.46, 12.47, and 12.50.

12.16   Construct a head-discharge-efficiency table for pumps A and B (Fig. 12.41) connected in parallel. (USC units.)

12.17   Find the discharge through the system of Fig. 12.41 for pumps A and B connected in series, using 1600 m of 300-mm clean cast-iron pipe and $H = 30$ m.

12.18   Determine the power needed to drive pumps A and B in Prob. 12.17.

12.19   Find the discharge through the system of Fig. 12.41 for pumps A and B connected in parallel, using 2000 m of 500-mm steel pipe and $H = 10$ m.

12.20   Determine the power needed to drive the pumps in Prob. 12.19.

12.21   Water at 60°F is pumped from a pressure tank to a reservoir, 25 ft higher (Fig. 12.42). The commercial steel pipeline length totals 1200 ft, and the diameter is 6 in. Neglect minor effects. The head-discharge curve for the pump is given by $H = 38 - 2Q^2$, with H in feet and Q in cubic feet per second. If the pressure in the tank is atmospheric, what is the flow in the system?

12.22   If the gage pressure in the tank in Prob. 12.21 is 15 psi, what is the flow rate in the system? Sketch the hydraulic grade line for this flow condition.

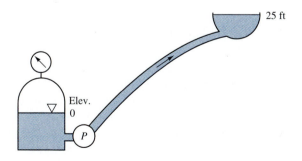

Figure 12.42   Problems 12.21 and 12.22.

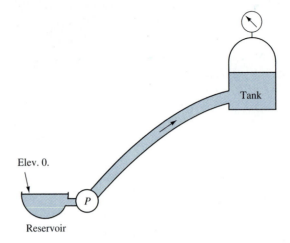

Figure 12.43   Problems 12.23 and 12.24.

12.23   Water is pumped from the large reservoir to the pressure tank at a higher elevation (Fig. 12.43). The total pipeline length is 2000 ft, diameter 8 in., smooth plastic, and $C = 130$. Neglect minor effects. If the pump curve is $H = 48 - 2Q^2$, with H in feet and Q in cubic feet per second, find the flow rate in the system when the pressure in the tank is 12 psi and the water-tank elevation is 10 ft. Sketch the hydraulic grade line.

12.24   If the water-tank elevation in Prob. 12.23 is 35 ft and it is open to the atmosphere, what is the flow rate?

12.25   Two reservoirs are connected by three clean cast-iron pipes in series: $L_1 = 300$ m and $D_1 = 200$ mm; $L_2 = 360$ m and $D_2 = 300$ mm; and $L_3 = 1200$ m and $D_3 = 450$ mm. When $Q = 0.1$-m^3/s water at 20°C, determine the difference in elevation of the reservoirs.

12.26   Solve Prob. 12.25 by the method of equivalent lengths.

12.27   For a difference in elevation of 10 m in Prob. 12.25, find the discharge by use of the Hazen-Williams equation.

12.28   Air at atmospheric pressure and 15°C is carried through two horizontal pipes ($\epsilon = 1.5$ mm) in series. The upstream pipe is 120 m of 720-mm diameter, and the downstream pipe is 30 m of 900-mm diameter. Estimate the equivalent length of 450 mm ($\epsilon = 0.76$ mm) pipe. Neglect minor losses.

12.29   What pressure drop, in millimeters of water, is required for flow of 3 m^3/s in Prob. 12.28? Include losses due to sudden expansion.

12.30   A water company is to be formed to sell water to a city. It will buy water for $(200,000 + 2,000*Q)*Q$, with Q in cubic feet per second, and deliver it 25 miles through a constant wall thickness pipeline. The city will pay $1,900,000 + $265,000*(Q - 5)$ for 5 to 10 cfs. Prices are given on an annual basis. The pipeline costs $1,450,000*D$ (in feet). The pipeline friction factor is $f = 0.02$. The pumping station costs $30,000 + $22,000D_p^3$ (in feet). The pump is one of the homologous series with $H_1 = 200$ ft, $Q_1 = 5$ cfs, $D_{p_1} = 1.167$ ft, $N_1 = 1200$ rpm, and efficiency, $\eta = 0.84$. The cost of money, including sinking fund, and contingencies is 13 percent, and maintenance and operation costs are $100,000 per year. Assume the power cost to be 8 cents per kWh. Select the size of pump, diameter of pipeline, and discharge rate that yield the best return to the water company.

12.31 A pump supplies water through a 14,000-m pipeline to a downstream reservoir 25 m higher in elevation. The pump characteristics are given by $h/\alpha^2 = 1.3 - 0.3(v/\alpha)^2$, and $\eta = -0.27 + 2.32v/\alpha - 1.16(v/\alpha)^2$ near the point of best efficiency, in which $h = H/H_R$, $v = Q/Q_R$, and $\alpha = N/N_R$. Rated conditions for the pump are $H_R = 145$ m, $Q_R = 0.2$ m³/s, $N_R = 400$ rpm, and $f = 0.02$ in the pipeline. If an economical velocity is viewed to be near 1 m/s, find the pipe size to the next larger whole cm, the discharge, the head on the pump, and the synchronous pump speed. The operating efficiency should be greater than 70 percent.

12.32 Two pipes are connected in parallel between two reservoirs: $L_1 = 2500$ m, $D_1 = 1.2$-m-diameter old cast-iron pipe, and $C = 100$; and $L_2 = 2500$ m, $D_2 = 1$ m, and $C = 90$ m. For a difference in elevation of 3.6 m determine the total flow of water at 20°C.

12.33 For 4.5-m³/s flow in the system of Prob. 12.32, determine the difference in elevation of reservoir surfaces.

12.34 Three smooth tubes are connected in parallel: $L_1 = 40$ ft and $D_1 = \frac{1}{2}$ in.; $L_2 = 60$ ft and $D_2 = 1$ in.; and $L_3 = 50$ ft and $D_3 = \frac{3}{4}$ in. For total flow of 30-gpm oil, $\gamma = 55$ lb/ft³, and $\mu = 0.65$ P, what is the drop in hydraulic grade line between junctions?

12.35 Find the head loss and flow distribution for the following system of parallel pipes with water flowing at 15°C.

Pipe	Length, m	Diameter, mm	ϵ, mm
1	600	100	5
2	900	150	0.2
3	1200	200	0.12

The total flow is 75 L/s, the upstream pressure is 0.5 MPa, the upstream junction elevation is 100 m, and the downstream junction elevation is 95 m.

12.36 In Prob. 12.35 the downstream pressure is 0.4 MPa, and the total flow is unknown. Find the flow through each pipe and the total flow using Eq. (6.7.15).

12.37 The 72-in.-diameter pump of Example 11.1 is driven by a variable-speed motor. It discharges into parallel pipe system ($L_1 = 6000$ m, $D_1 = 1.85$ m, and $f_1 = 0.018$; and $L_2 = 7300$ m, $D_2 = 2.1$ m, and $f_2 = 0.02$) leading up to a reservoir at elevation Z. The suction reservoir is at zero elevation. The downstream reservoir varies in elevation from 6 m to 12 m. To operate at best efficiency, the pump speed must change with Z. Find this relationship.

12.38 For $H = 12$ m in Fig. 12.44, find the discharge through each pipe, $\mu = 8$ cP and sp gr $= 0.9$.

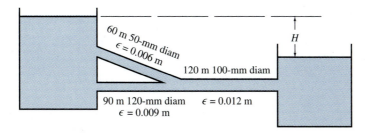

60 m 50-mm diam
$\epsilon = 0.006$ m

120 m 100-mm diam

H

90 m 120-mm diam $\epsilon = 0.012$ m
$\epsilon = 0.009$ m

Figure 12.44   Problems 12.38 and 12.39.

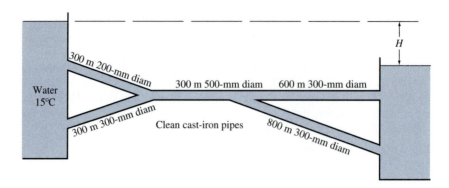

Figure 12.45   Problems 12.40 and 12.41.

12.39   Find H in Fig. 12.44 for 0.03-m³/s flowing, with $\mu = 5$ cP and sp gr $= 0.9$.

12.40   Find the equivalent length of 300-mm-diameter clean cast-iron pipe needed to replace the system of Fig. 12.45. For $H = 10$ m, what is the discharge?

12.41   With velocity of 1 m/s in the 200-mm-diameter pipe of Fig. 12.45 calculate the flow through the system and the head H required.

12.42   In Fig. 12.46 find the flow through the system when the pump is removed.

12.43   If the pump in Fig. 12.46 is delivering 80 L/s toward J, find the flow into A and B and the elevation of the hydraulic grade line at J.

12.44   The pump is adding 7500-W fluid power to the flow (toward J) in Fig. 12.46. Find Q_A and Q_B.

12.45   Use a spreadsheet to balance the water flow (10°C) in the reservoir system given in the table. The pipelines meet at a common junction.

Pipe	Reservoir elevation, m	Length, km	Diameter, mm	ϵ, mm
1	100	12	600	4
2	80	16	800	2
3	70	8	750	0.2
4	50	10	600	0.02
5	0	20	300	0.05

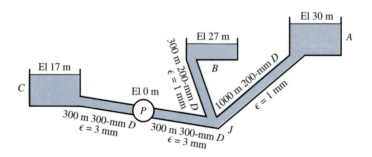

Figure 12.46   Problems 12.42 to 12.44, 12.46 to 12.48, and 12.56.

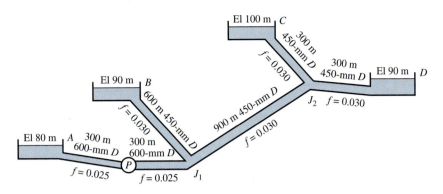

Figure 12.47   Problems 12.49 and 12.50.

12.46   With pump A of Fig. 12.41 in the system of Fig. 12.46, find Q_A, Q_B, and the elevation of the hydraulic grade line at J.

12.47   With pump B of Fig. 12.41 in the system of Fig. 12.46, find the flow into B and the elevation of the hydraulic grade line at J.

12.48   For flow of 30 L/s into B of Fig. 12.46, what head is produced by the pump? For a pump efficiency of 70 percent, how much power is required?

12.49   Find the flow through the system of Fig. 12.47 for no pump in the system.

12.50   (*a*) With pumps A and B of Fig. 12.41 in parallel in the system of Fig. 12.47, find the flow into B, C, and D and the elevation of the hydraulic grade line at J_1 and J_2. (*b*) Assume all the pipes in Fig. 12.47 are cast iron, and allow the friction factor to vary with flow. Use a spreadsheet to balance the flow.

12.51   Water flows at a rate of 1 m³/s from a source at elevation 200 m through a 2600-m pipeline to a branch connection where the flow divides equally to three reservoirs. Three pipes, of lengths $L_A = 2300$ m, $L_B = 3900$ m, and $L_C = 1200$ m, lead to three reservoirs, whose elevations are $H_A = 145$ m, $H_B = 150$ m, and $H_C = 154$ m. Use kinematic viscosity, $\nu = 0.00003$ m²/s, pipe roughness $\epsilon = 0.00005$ m, and $g = 9.806$ m/s². Assume the cost of each pipeline per meter varies as the square of the diameter. Find the diameter of each pipe for the system to be most cost-effective.

12.52   The system shown in Fig. 12.48 is to supply water to the two reservoirs. Flow should be supplied with the ratio Q_B/Q_C within 5 percent of QRATIO $= 2$, that is, $1.9 < Q_B/Q_C < 2.1$. Assume economical velocities to be about 1 m/s for each of the lines. The pump characteristics are given by

$$h/\alpha^2 = 1.3 - 0.3(v/\alpha)^2 \qquad \eta = -0.27 + 2.32v/\alpha - 1.16(v/\alpha)^2$$

near the most efficient point, at which $h = H/H_R$, $v = Q/Q_R$, and $\alpha = N_S/N_R$. Pump rated conditions are $H_R = 145$ m, $Q_R = 0.2$ m³/s, $N_R = 400$ rpm, and pipeline $f = 0.02$. Determine the pipe sizes to the next larger whole centimeter, the discharge through each pipe, the head at the pump, the synchronous pump speed with 60-Hz power supply, and the power output of the electric motor. The pump efficiency should be greater than 0.65.

12.53   The pump in the system shown in Fig. 12.48 has the same pump characteristics as in Prob. 12.52. The pump supplies about 150 L/s to the two reservoirs. Economic velocities are about 1.5 m/s for each of the lines. Pump rated conditions are $H_R = 145$ m, $Q_R = 0.2$ m³/s, $N_R = 450$ rpm, and $f = 0.02$ in each pipeline.

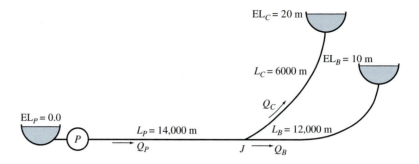

Figure 12.48   Problems 12.52 and 12.53.

Determine the pipe sizes to the next larger whole centimeter, the discharge through each pipe, the head at the pump, the synchronous pump speed with 60-Hz power supply, and the power output of the electric motor. The pump efficiency should be greater than 0.65.

12.54   Use a hand calculation to find the flow through each of the pipes of the network shown in Fig. 12.49. $n = 2$.

12.55   Use a hand calculation to determine the flow through each line of Fig. 12.50. $n = 2$.

12.56   Use a spreadsheet to solve Prob. 12.54.

12.57   Use a spreadsheet to solve Prob. 12.55.

12.58   Determine the slope of the hydraulic grade line for flow of atmospheric air at 80°F through a rectangular 18- by 6-in. galvanized-iron conduit. $V = 30$ ft/s.

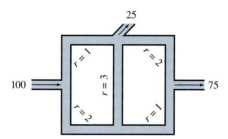

Figure 12.49   Problem 12.54.

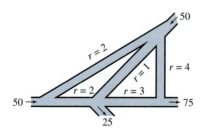

Figure 12.50   Problem 12.55.

12.59 What size square conduit is needed to convey 300-L/s water at 15°C with the slope of the hydraulic grade line = 0.001? ϵ = 1 mm.

12.60 Calculate the discharge of oil, S = 0.85 and μ = 4 cP, through 30 m of 50-by 120-mm sheetmetal conduit when the head loss is 600 mm. ϵ = 0.00015 m.

12.61 A duct whose cross section is an equilateral triangle 1 ft on a side conveys 6-cfs water at 60°F. ϵ = 0.003 ft. Calculate the slope of the hydraulic grade line.

12.62 A clean 700-mm-diameter cast-iron water pipe has its absolute roughness doubled in 5 years of service. Estimate the head loss per 1000 m for a flow of 400 L/s when the pipe is 25 years old.

12.63 An 18-in.-diameter pipe has an f of 0.20 when new for 5-ft/s water flow of 60°F. In 10 years f = 0.029 for V = 3 ft/s. Find f for 4 ft/s at the end of 20 years.

12.64 The hydraulic grade line in a system is (*a*) always above the energy grade line; (*b*) always above the closed conduit; (*c*) always sloping downward in the direction of flow; (*d*) the velocity head below the energy grade line; (*e*) upward in direction of flow when pipe is inclined downward.

12.65 One pipe system is said to be equivalent to another pipe system when the following two quantities are the same: (*a*) h and Q; (*b*) L and Q; (*c*) L and D; (*d*) f and D; (*e*) V and D.

12.66 In parallel-pipe problems (*a*) the head losses through each pipe are added to obtain the total head loss; (*b*) the discharge is the same through all the pipes; (*c*) the head loss is the same through each pipe; (*d*) a direct solution gives the flow through each pipe when the total flow is known; (*e*) a trial solution is not needed.

12.67 Branching-pipe problems usually are solved (*a*) analytically by using as many equations as unknowns; (*b*) by assuming the head loss is the same through each pipe; (*c*) by equivalent lengths; (*d*) by assuming a distribution which satisfies continuity and computing a correction; (*e*) by assuming the elevation of hydraulic grade line at the junction point and trying to satisfy continuity.

12.68 In networks of pipes (*a*) the head loss around each elementary circuit must be zero; (*b*) the (horsepower) loss in all circuits is the same; (*c*) the elevation of hydraulic grade line is assumed for each junction; (*d*) elementary circuits are replaced by equivalent pipes; (*e*) friction factors are assumed for each pipe.

12.69 The following quantities are computed by using $4R$ in place of diameter for noncircular sections: (*a*) velocity and relative roughness; (*b*) velocity and head loss; (*c*) Reynolds number, relative roughness, and head loss; (*d*) velocity, Reynolds number, and friction factor; (*e*) none of these answers.

12.70 Experiments show that in the aging of pipes (*a*) the friction factor increases linearly with time; (*b*) a pipe becomes smoother with use; (*c*) the absolute roughness increases linearly with time; (*d*) no appreciable trends can be found; (*e*) the absolute roughness decreases with time.

12.71 Determine the period of oscillation of a U tube containing 0.5 L of water. The cross-sectional area is 2.4 cm^2. Neglect friction.

12.72 A U tube containing alcohol is oscillating with maximum displacement from an equilibrium position of 120 mm. The total column length is 1 m. Determine the maximum fluid velocity and the period of oscillation. Neglect friction.

12.73 A liquid, ν = 0.0002 m^2/s, is in a 12-mm-diameter U tube. The total liquid column is 1.73 m long. If one meniscus is 40 cm above the other meniscus when the column is at rest, determine the time for one meniscus to move to within 3 cm of its equilibrium position.

12.74   Develop the equations for motion of a liquid in a U tube for laminar resistance when $16\nu/D^2 = \sqrt{2g/L}$. *Suggestion:* Try $z = e^{-mt}(c_1 + c_2 t)$.

12.75   A U tube contains liquid oscillating with a velocity 2 m/s at the instant the menisci are at the same elevation. Find the time to the instant the menisci are next at the same elevation and also the velocity. $\nu = 1 \times 10^{-5}$ m^2/s, $D = 6$ mm, and $L = 750$ mm.

12.76   A 10-ft-diameter horizontal tunnel has 10-ft-diameter vertical shafts spaced 1 mi apart. When valves are closed isolating this reach of tunnel, the water surges to a depth of 50 ft in one shaft when it is 20 ft in the other shaft. For $f = 0.022$, find the height of the next two surges.

12.77   A 10-mm-diameter U tube contains oil, $\nu = 5$ μm^2/s, with a total column length of 2 m. If the initial half-amplitude of displacement is 250 mm, find the first 10 maximum and minimum displacements and the times they occur.

12.78   Equation (12.7.17) may be set up for solution using Runge-Kutta third-order procedures (see the Web page). Use a spreadsheet to carry out this solution and apply it to the following case: $t = 0$, $z = 12$ ft, $V_0 = 0$, $L = 400$ ft, $f = 0.017$, $d = 2$ ft, $dt = 0.1$ s, and $t_{max} = 30$ s.

12.79   A valve is quickly opened in a pipe 1200 m long, $D = 0.6$ m, with a 0.3-m-diameter nozzle on the downstream end. Minor losses are $4V^2/2g$, with V the velocity in the pipe, $f = 0.024$, and $H = 9$ m. Find the time to attain 95 percent of the steady-state discharge.

12.80   A globe valve ($K = 10$) at the end of a pipe 2000 ft long is rapidly opened. $D = 3.0$ ft, $f = 0.018$, minor losses are $2V^2/2g$, and $H = 75$ ft. How long does it take for the discharge to attain 80 percent of its steady-state value?

12.81   A steel pipeline with expansion joints is 1 m in diameter and has a 10-mm wall thickness. When it is carrying water, determine the speed of a pressure wave.

12.82   Benzene ($K = 150,000$ psi and $S = 0.88$) flows through $\frac{3}{4}$-in.-ID steel tubing with $\frac{1}{8}$-in. wall thickness. Determine the speed of a pressure wave.

12.83   Determine the maximum time for rapid valve closure on a pipeline: $L = 1000$ m, $D = 1.3$ m, $e = 12$-mm steel pipe, and $V_0 = 3$-m/s water flowing.

12.84   A valve is closed in 5 s at the downstream end of a 3000-m pipeline carrying water at 2 m/s. $a = 1000$ m/s. What is the peak pressure developed by the closure?

12.85   Determine the length of pipe in Prob. 12.84 subjected to the peak pressure.

12.86   A valve is closed at the downstream end of a pipeline in such a manner that only one-third of the line is subjected to maximum pressure. During what proportion of the time $2L/a$ is it closed?

12.87   A pipe, $L = 2000$ m and $a = 1000$ m/s, has a valve on its downstream end, $V_0 = 2.5$ m/s and $H_0 = 20$ m. It closes in three increments, spaced 1 s apart, each area reduction being one-third of the original opening. Find the pressure at the gate and at the midpoint of the pipeline at 1-s intervals for 5 s after initial closure.

12.88   A pipeline, $L = 600$ m and $a = 1200$ m/s, has a valve at its downstream end, $V_0 = 2$ m/s and $H_0 = 30$ m. Determine the pressure at the valve for closure ($C_d = 0.6$).

A_v/A_{v0}	1	0.75	0.60	0.45	0.30	0.15	0
t, s	0	0.5	1.0	1.5	2.0	2.5	3.0

12.89 In Prob. 12.88 determine the peak pressure at the valve for uniform area reduction in 3.0 s.

12.90 Find the maximum area reduction for $\frac{1}{2}$-s intervals for the pipeline of Prob. 12.88 when the maximum head at the valve is not to exceed 50 m. Increase the head linearly to 50 m in 1 s, then hold constant.

12.91 Derive the characteristics-method solution for waterhammer with the pressure p and the discharge Q as dependent variables.

12.92 Work Ex. 12.18 with one-third of the wave frequency. Do resonance conditions still exist?

12.93 Use a spreadsheet to solve a single-pipeline waterhammer problem with a valve closure at the downstream end of the pipe and a reservoir at the upstream end. The valve closure is given by $C_d A_v/(C_d A_v)_0 = (1 - t/t_c)^m$, where t_c is the time of closure and is 6.2 s, and $m = 3.2$; $L = 5743.5$ ft, $a = 3927$ ft/s, $D = 4$ ft, $f = 0.019$, $V_0 = 3.6$ ft/s, and $H_0 = 300$ ft.

12.94 In Prob. 12.93 place a wave on the reservoir at a period of 1.95 s and obtain a solution with the aid of a spreadsheet.

12.95 In Ex. 12.18 reduce H_0 to 20 m, let $\omega = \pi/2$, and work the problem until $t = 3$ s.

12.96 The head rise at a valve due to sudden closure is (a) $a^2/2g$; (b) $V_0 a/g$; (c) $V_0 a/2g$; (d) $V_0^2/2g$; (e) none of these answers.

12.97 The speed of a pressure wave through a pipe depends upon (a) the length of pipe; (b) the original head at the valve; (c) the viscosity of fluid; (d) the initial velocity; (e) none of these answers.

12.98 When the velocity in a pipe is suddenly reduced from 3 to 2 m/s by downstream valve closure, for $a = 980$ m/s, the head rise, in meters, is (a) 100; (b) 200; (c) 300; (d) 980; (e) none of these answers.

12.99 When $t_c = L/2a$, the proportion of pipe length subjected to maximum heads is, in percent, (a) 25; (b) 50; (c) 75; (d) 100; (e) none of these answers.

12.100 When the steady-state value of head at a valve is 120 ft, the valve is given a sudden partial closure such that $\Delta h = 80$ ft. The head at the valve at the instant this reflected wave returns is (a) -80; (b) 40; (c) 80; (d) 200; (e) none of these answers.

REFERENCES†

1. E. F. Brater, H. W. King, J. E. Lindell, and C. Y. Wei, *Handbook of Hydraulics,* 7th ed., McGraw-Hill, New York, 1996, pp. 6–28.

2. V. L. Streeter and E. B. Wylie, *Fluid Mechanics,* 6th ed., McGraw-Hill, New York, 1975, pp. 545–547.

3. H. Cross, "Analysis of Flow in Networks of Conduits or Conductors," *Univ. Ill. Bull.* 286, November 1936.

4. R. Epp and A. G. Fowler, "Efficient Code for Steady-State Flows in Networks," *J. Hydraul. Div., ASCE,* vol. 96, no. HY1, pp. 43–56, January 1970.

† †References 1–12 are for steady flow and 13–16 are for unsteady flow.

5.   U. Shamir and C. D. D. Howard, "Water Distribution Systems Analysis," *J. Hydraul. Div., ASCE,* vol. 94, no. HY1, pp. 219–234, January 1968.

6.   L. E. Ormsbee and D. J. Wood, "Hydraulic Design Algorithms for Pipe Networks," *J. Hydraul. Div., ASCE,* vol. 112, no. 12, pp. 1195–1207, December 1986.

7.   D. J. Wood and A. G. Rayes, "Reliability of Algorithms for Pipe Network Analysis," *J. Hydraul. Div., ASCE,* vol. 107, no. HY10, pp. 1145–1161, October 1981.

8.   C. F. Colebrook and C. M. White, "The Reduction of Carrying Capacity of Pipes with Age," *J. Inst. Civ. Eng.,* Lond., 1937.

9.   W. D. Hudson, "Computerized Pipeline Design," *Transp. Eng. J., ASCE,* vol. 99, no. TE1, pp. 73–82, 1973.

10.   N. S. Berman, "Drag Reduction by Polymers," *Annual Rev. of Fluid Mechanics,* vol. 10, pp. 47–64, 1978.

11.   W. R. Beaty, W. Carradine, G. Hass, G. Husen, R. Johnston, and M. Mack, "New High-Performance Flow Improver Offers Alternatives to Pipelines," *Oil and Gas J.,* vol. 80, no. 32, pp. 96–102, August 9, 1982.

12.   C. D. Lester, "Four-Part Series on Drag Reducing Agents," *Oil and Gas J.,* February 4 and 18, and March 4 and 11, 1985.

13.   E. D. Rainville, *Elementary Differential Equations,* 3d ed., Macmillan, New York, 1964.

14.   E. B. Wylie and V. L. Streeter, *Fluid Transients in Systems,* Prentice Hall, Englewood Cliffs, NJ, 1993.

15.   M. H. Chaudhry, *Applied Hydraulic Transients,* 2nd ed., Van Nostrand Reinhold, New York, 1987.

16.   G. Z. Watters, *Modern Analysis and Control of Unsteady Flow in Pipelines,* 2nd ed., Butterworth, Woburn, MA, 1984.

chapter

13

Flow in Open Channels

A broad coverage of topics in open-channel flow has been selected for this chapter. Steady uniform flow was discussed in Sec. 6.6, and the application of the momentum equation to hydraulic jump was discussed in Sec. 3.7. Weirs were introduced in Sec. 10.8. In this chapter open-channel flow is first classified, and then the *shape* of optimum canal cross sections is discussed, followed by a section on flow through a floodway. The hydraulic jump and its application to stilling basins is then treated, followed by a discussion of specific energy and critical depth which leads into depth transitions and then gradually varied flow. Water surface profiles are classified and related to channel control sections. In conclusion positive and negative surge waves in a rectangular channel are analyzed, neglecting the effects of friction.

The presence of a free surface makes the mechanics of flow in open channels more complicated than closed-conduit flow. The hydraulic grade line coincides with the free surface, and, in general, its position is unknown. Gravitational forces cause free surface flow, and viscous shear forces along the channel wetted perimeter resist flow. Both the Reynolds number and Froude number are important in characterizing the flow.

For laminar flow to occur, the cross section must be extremely small, the velocity very small, or the kinematic viscosity extremely high. One example of laminar flow is given by a thin film of liquid flowing down an inclined or vertical plane. This case is treated by the methods developed in Chap. 6 (see Prob. 6.10). Pipe flow has a lower critical Reynolds number of 2000. This same value can be applied to an open channel when the diameter D is replaced by $4R$, in which R is the hydraulic radius, defined as the cross-sectional flow area of the channel divided by the wetted perimeter. In the range of Reynolds number, based on R in place of D, if $\mathbf{R} = VR/\nu < 500$, the flow is laminar; if $500 < \mathbf{R} < 2000$, the flow is *transitional* and may be either laminar or turbulent; and if $\mathbf{R} > 2000$, the flow is generally turbulent.

Most open-channel flows are turbulent, usually with water as the liquid. The methods for analyzing open-channel flow are not as developed as those for closed conduits. The equations in use assume complete turbulence, with the head loss proportional to the square of the velocity. Although practically all data on open-channel flow have been obtained from experiments on the flow of water, the equations should yield reasonable values for other liquids of low viscosity. The material in this chapter applies to turbulent flow only.

13.1   CLASSIFICATION OF FLOW

Definitions

Open-channel flow occurs in a large variety of forms, from flow of water over the surface of a plowed field during a hard rain to the flow at constant depth through a large prismatic channel. It can be classified as steady or unsteady, and as uniform or nonuniform. *Steady uniform flow* occurs in very long inclined channels of constant cross section in those regions where *terminal velocity* has been reached, that is, where the head loss due to turbulent flow is exactly supplied by the reduction in potential energy due to the uniform decrease in elevation of the bottom of the channel. The depth for steady uniform flow is called the *normal depth*. In steady uniform flow the discharge is constant and the depth is everywhere constant along the length of the channel. Several equations are in common use for determining the relations between the average velocity, the shape of the cross section, its size and roughness, and the slope, or inclination, of the channel bottom (Sec. 6.6).

Steady nonuniform flow occurs in any irregular channel in which the discharge does not change with time; it also occurs in regular channels when the flow depth and hence the average velocity change from one cross section to another. For gradual changes in depth or section, called *gradually varied flow,* methods are available, by numerical integration or step-by-step means, for computing flow depths for known discharge, channel dimensions and roughness, and given conditions at one cross section. For those reaches of a channel where pronounced changes in velocity and depth occur in a short distance, as in a transition from one cross section to another, model studies are frequently made. The *hydraulic jump* is one example of steady nonuniform flow; it is discussed in Secs. 3.7 and 13.4.

Unsteady uniform flow rarely occurs in open-channel flow. *Unsteady nonuniform flow* is common but difficult to analyze. Wave motion is an example of this type of flow, and its analysis is complex when friction is taken into account.

Flow is also classified as *tranquil* or *rapid*. When flow occurs at low velocities so that a small disturbance can travel upstream and thus change upstream conditions, it is said to be tranquil flow ($\mathbf{F} < 1$; the Froude number $\mathbf{F}$ was defined and discussed in Sec. 5.3). Conditions upstream are affected by downstream conditions, and the flow is controlled by the downstream conditions. When flow occurs at such high velocities that a small disturbance such as an elementary wave is swept downstream, the flow is described as *shooting* or *rapid* ($\mathbf{F} > 1$). Small changes in downstream conditions do not effect any change in upstream conditions; hence, the flow is controlled by upstream conditions. When flow is such that its velocity is just equal to the velocity of an elementary wave, the flow is said to be critical ($\mathbf{F} = 1$).

The terms "subcritical" and "supercritical" are also used to classify flow velocities. *Subcritical* refers to tranquil flow at velocities less than critical, and *supercritical* corresponds to rapid flows when velocities are greater than critical.

Velocity Distribution

The velocity at a solid boundary must be zero, and in open-channel flow it generally increases with distance from the boundaries. The maximum velocity does not occur at the free surface but is usually below the free surface a distance of 0.05 to 0.25 of the depth. The average velocity along a vertical line is sometimes determined by measuring the velocity at 0.6 of the depth, but a more reliable method is to take the average of the velocities at 0.2 and 0.8 of the depth, according to measurements of the U.S. Geological Survey.

EXERCISES

13.1.1 In open-channel flow (*a*) the hydraulic grade line is always parallel to the energy grade line; (*b*) the energy grade line coincides with the free surface; (*c*) the energy and hydraulic grade lines coincide; (*d*) the hydraulic grade line can never rise; (*e*) the hydraulic grade line and free surface coincide.

13.1.2 Tranquil flow must always occur (*a*) above normal depth; (*b*) below normal depth; (*c*) above critical depth; (*d*) below critical depth; (*e*) on adverse slopes.

13.2 BEST HYDRAULIC CHANNEL CROSS SECTIONS

Some channel cross sections are more efficient than others in that they provide more area for a given wetted perimeter. In general, when a channel is constructed, the excavation, and possibly the lining, must be paid for. From the Manning formula (Chap. 6) it is shown that when the area of cross section is a minimum, the wetted perimeter is also a minimum, and so both lining and excavation approach their minimum value for the same dimensions of channel. The *best hydraulic section* is one that has the least wetted perimeter or its equivalent, the least area for the type of section. The Manning formula is

$$Q = \frac{C_m}{n} A R^{2/3} S^{1/2} \tag{13.2.1}$$

in which Q is the discharge (L^3/T), A is the cross-sectional flow area, R (the area divided by the wetted perimeter P) is the hydraulic radius, S is the slope of energy grade line, n is the Manning roughness factor (Table 6.1), and C_m is an empirical constant ($L^{1/3}/T$) equal to 1.49 in USC units and to 1.0 in SI units. With Q, n, and S known, Eq. (13.2.1) can be written as

$$A = c P^{2/5} \tag{13.2.2}$$

in which c is known. This equation shows that P is a minimum when A is a minimum. To find the best hydraulic section for a *rectangular* channel (Fig. 13.1), $P = b + 2y$

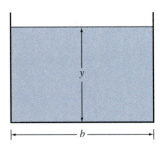

Figure 13.1 Rectangular cross section.

and $A = by$. Then

$$A = (P - 2y)y = cP^{2/5}$$

by elimination of b. The value of y is sought for which P is a minimum. Differentiating with respect to y gives

$$\left(\frac{dP}{dy} - 2\right)y + P - 2y = \frac{2}{5}cP^{-3/5}\frac{dP}{dy}$$

Setting $dP/dy = 0$ gives $P = 4y$, or since $P = b + 2y$, then

$$b = 2y \tag{13.2.3}$$

Therefore, the depth is one-half the bottom width, independent of the size of the rectangular section.

To find the best hydraulic *trapezoidal* section (Fig. 13.2), $A = by + my^2$ and $P = b + 2y\sqrt{1 + m^2}$. After eliminating b and A in these equations and Eq. (13.2.2),

$$A = by + my^2 = (P - 2y\sqrt{1 + m^2})y + my^2 = cP^{2/5} \tag{13.2.4}$$

By holding m constant and by differentiating with respect to y, $\partial P/\partial y$ is set equal to zero; thus,

$$P = 4y\sqrt{1 + m^2} - 2my \tag{13.2.5}$$

Again, by holding y constant, Eq. (13.2.4) is differentiated with respect to m and $\partial P/\partial m$ is set equal to zero, producing

$$\frac{2m}{\sqrt{1 + m^2}} = 1$$

After solving for m,

$$m = \frac{\sqrt{3}}{3}$$

and after substituting for m in Eq. (13.2.5),

$$P = 2\sqrt{3}\, y \qquad b = 2\frac{\sqrt{3}}{3}\, y \qquad A = \sqrt{3}\, y^2 \tag{13.2.6}$$

which shows that $b = P/3$ and hence the sloping sides have the same length as the bottom. As $\tan^{-1}m = 30°$, the best hydraulic section is one-half a hexagon. For trapezoidal sections with m specified (maximum slope at which wet earth will stand) Eq. (13.2.5) is used to find the best bottom-width-to-depth ratio.

The semicircle is the best hydraulic section of all possible open-channel cross sections. The proof of this is left to the readers.

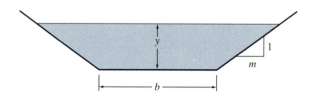

Figure 13.2   Trapezoidal cross section.

Determine the dimensions of the most economical trapezoidal brick-lined channel to carry 200 m³/s with a slope of 0.0004.

Example 13.1

Solution

With Eq. (13.2.6),

$$R = \frac{A}{P} = \frac{y}{2}$$

and by substituting into Eq. (13.2.1),

$$200 = \frac{1.00}{0.016} \sqrt{3}\, y^2 \left(\frac{y}{2}\right)^{2/3} \sqrt{0.0004}$$

or

$$y^{8/3} = 146.64 \qquad y = 6.492 \text{ m}$$

and from Eq. (13.2.6) $b = 7.5$ m.

EXERCISES

13.2.1 The best hydraulic rectangular cross section occurs when ($b = $ bottom width and $y = $ depth) (a) $y = 2b$; (b) $y = b$; (c) $y = b/2$; (d) $y = b^2$; (e) $y = b/5$.

13.2.2 The best hydraulic canal cross section is defined as (a) the least expensive canal cross section; (b) the section with minimum roughness coefficient; (c) the section that has a maximum area for a given flow; (d) the one that has a minimum perimeter; (e) none of these answers.

13.3   STEADY UNIFORM FLOW IN A FLOODWAY

A practical open-channel problem of importance is the computation of discharge through a floodway (Fig. 13.3). In general, the floodway is much rougher than the river channel and its depth (and hydraulic radius) is much lesser. The slope of energy grade line must be the same for both portions. The discharge for each portion is determined separately, using the dashed line of Fig. 13.3 as the separation line for the two sections (but not as a solid boundary), and then the discharges are added to determine the total capacity of the system.

Since both portions have the same slope, the discharge can be expressed as

$$Q_1 = K_1 \sqrt{S} \qquad Q_2 = K_2 \sqrt{S}$$

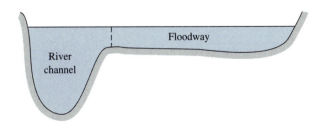

Figure 13.3   Floodway cross section.

or

$$Q = (K_1 + K_2)\sqrt{S} \qquad \text{(13.3.1)}$$

in which the value of K is

$$K = \frac{C_m}{n} A R^{2/3}$$

from Manning's formula and is a function of depth only for a given channel with fixed roughness. By computing K_1 and K_2 for different elevations of water surface, their sum can be taken and plotted against elevation. From this plot it is easy to determine the slope of energy grade line for a given depth and discharge from Eq. (13.3.1).

13.4   HYDRAULIC JUMP AND STILLING BASINS

Hydraulic Jump

The relations between the variables V_1, y_1, V_2, and y_2 for a hydraulic jump in a horizontal rectangular channel were developed in Sec. 3.7. Another way of determining the conjugate depths for a given discharge is the following $F + M$ method. The momentum equation applied to the free body of liquid between y_1 and y_2 (Fig. 13.4) is, for unit width ($V_1 y_1 = V_2 y_2 = q$),

$$\frac{\gamma y_1^2}{2} - \frac{\gamma y_2^2}{2} = \rho q (V_2 - V_1) = \rho V_2^2 y_2 - \rho V_1^2 y_1$$

Rearranging gives

$$\frac{\gamma y_1^2}{2} + \rho V_1^2 y_1 = \frac{\gamma y_2^2}{2} + \rho V_2^2 y_2 \qquad \text{(13.4.1)}$$

or

$$F_1 + M_1 = F_2 + M_2 \qquad \text{(13.4.2)}$$

in which F is the hydrostatic force at the section and M is the momentum per second passing the section. By writing $F + M$ for a given discharge q per unit width,

$$F + M = \frac{\gamma y^2}{2} + \frac{\rho q^2}{y} \qquad \text{(13.4.3)}$$

a plot is made of $F + M$ as abscissa against y as ordinate (Fig. 13.5) for $q = 10$ cfs/ft. Any vertical line intersecting the curve cuts it at two points having the same value of $F + M$; hence, they are conjugate depths. The value of y for a minimum $F + M$, obtained by differentiation of Eq. (13.4.3) with respect to y and setting $d(F + M)/dy$

Figure 13.4   Hydraulic jump in a horizontal rectangular channel.

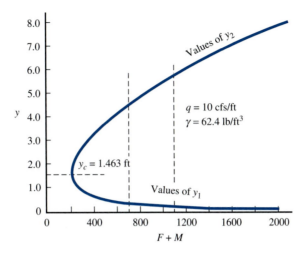

Figure 13.5 $F + M$ curve for a hydraulic jump.

equal to zero, is

$$y_c = \left(\frac{q^2}{g}\right)^{1/3} \tag{13.4.4}$$

The jump must always occur from a depth less than this value to a depth greater than this value. This depth is the *critical depth,* which is shown in the following section to be the depth of minimum energy. Therefore, the jump always occurs from rapid flow to tranquil flow. The fact that available energy is lost in the jump prevents any possibility of its suddenly changing from the higher conjugate depth to the lower conjugate depth.

The conjugate depths are directly related to the Froude numbers before and after the jump as

$$\mathbf{F}_1 = \frac{V_1}{\sqrt{gy_1}} \qquad \mathbf{F}_2 = \frac{V_2}{\sqrt{gy_2}} \tag{13.4.5}$$

From the continuity equation

$$V_1^2 y_1^2 = gy_1^3\mathbf{F}_1^2 = V_2^2 y_2^2 = gy_2^3\mathbf{F}_2^2$$

or

$$\mathbf{F}_1^2 y_1^3 = \mathbf{F}_2^2 y_2^3 \tag{13.4.6}$$

From Eq. (13.4.1)

$$y_1^2\left(1 + 2\frac{V_1^2}{gy_1}\right) = y_2^2\left(1 + 2\frac{V_2^2}{gy_2}\right)$$

Substituting from Eqs. (13.4.5) and (13.4.6) gives

$$(1 + 2\mathbf{F}_1^2)\mathbf{F}_1^{-4/3} = (1 - 2\mathbf{F}_2^2)\mathbf{F}_2^{-4/3} \tag{13.4.7}$$

The value of $\mathbf{F}_2$ in terms of $\mathbf{F}_1$ is obtained from the hydraulic-jump equation (3.7.11) as

$$y_2 = \frac{-y_1}{2} + \sqrt{\left(\frac{y_1}{2}\right)^2 + 2\frac{V_1^2 y_1}{g}} \quad \text{or} \quad 2\frac{y_2}{y_1} = -1 + \sqrt{1 + 8\frac{V_1^2}{g y_1}}$$

By Eqs. (13.4.5) and (13.4.6)

$$\mathbf{F}_2 = \frac{2\sqrt{2}\mathbf{F}_1}{(\sqrt{1 + 8\mathbf{F}_1^2} - 1)^{3/2}} \tag{13.4.8}$$

These equations apply only to a rectangular section. The Froude number is always greater than unity before the jump and less than unity after the jump.

Stilling Basins

A stilling basin is a structure for dissipating available energy of flow below a spill-way, outlet works, chute, or canal structure. In most existing installations a hydraulic jump is housed within the stilling basin and used as the energy dissipator. This discussion is limited to rectangular basins with horizontal floors although sloping floors are used in some cases to save excavation (Table 13.1).

Table 13.1   Classification of the hydraulic jump as an effective energy dissipator

$F_1 = V_1/\sqrt{g y_1}$	Classification	Description
1–1.7	Standing wave	Only a slight difference in conjugate depths; near $F_1 = 1.7$ a series of small rollers develops.
1.7–2.5	Prejump	Water surface quite smooth, velocity fairly uniform, and head loss low; no baffles required if proper length of pool is provided.
2.5–4.5	Transition	Oscillating action of entering jet, from bottom of basin to surface; each oscillation produces a large wave of irregular period that can travel downstream for miles and damage earth banks and riprap; if possible, avoid this F_1 range in stilling-basin design.
4.5–9	Range of good jumps	Jump well balanced and the action at its best; energy absorption (irreversibilities) ranges from 45 to 70 percent; baffles and sills can be used to reduce length of basin.
9 and over	Effective but rough	Energy dissipation up to 85 percent; other types of stilling basins may be more economical.

From Ref. [1].[†]

Baffle blocks are frequently used at the entrance to a basin to corrugate the flow. They are usually regularly spaced with gaps that are about equal to the block widths. Sills, either triangular or dentated, are frequently employed at the downstream end of a basin to aid in holding the jump within the basin and to permit some shortening of the basin.

| [†] Numbered references will be found at the end of this chapter.

The basin should be paved with high-quality concrete to prevent erosion and cavitation damage. No irregularities in floor or training walls should be permitted. The length of the jump, about $6y_2$, should be within the paved basin, with good riprap downstream if the material is easily eroded.

Example 13.2

A hydraulic jump occurs downstream from a 15-m-wide sluice gate. The depth is 1.5 m, and the velocity is 20 m/s. Determine (*a*) the Froude number and the Froude number corresponding to the conjugate depth, (*b*) the depth and velocity after the jump, and (*c*) the power dissipated by the jump.

Solution

(*a*)

$$\mathbf{F}_1 = \frac{V_1}{\sqrt{gy}} = \frac{20}{\sqrt{9.806(1.5)}} = 5.215$$

From Eq. (13.4.8)

$$\mathbf{F}_2 = \frac{2\sqrt{2}(5.215)}{[\sqrt{1 + 8(5.215^2)} - 1]^{3/2}} = 0.2882$$

(*b*)

$$\mathbf{F}_2 = \frac{V_2}{\sqrt{gy_2}} \qquad V_2 y_2 = V_1 y_1 = 1.5(20) = 30 \text{ m}^2/\text{s}$$

Then

$$V_2^2 = \mathbf{F}_2^2 g y_2 = \mathbf{F}_2^2 g \frac{30}{V_2}$$

and

$$V_2 = [0.2882^2(9.806)(30)]^{1/3} = 2.90 \text{ m/s}$$
$$y_2 = 10.34m$$

(*c*) From Eq. (3.11.24) the head loss h_j in the jump is

$$h_j = \frac{(y_2 - y_1)^3}{4y_1 y_2} = \frac{(10.34 - 1.50)^3}{4(1.5)(10.34)} = 11.13 \text{ m·N/N}$$

The power dissipated is

$$\text{Power} = \gamma Q h_j = 9806(15)(30)(11.13) = 49.1 \text{ MW}$$

EXERCISE

13.4.1 Supercritical flow can never occur (*a*) directly after a hydraulic jump; (*b*) in a mild channel; (*c*) in an adverse channel; (*d*) in a horizontal channel; (*e*) in a steep channel.

13.5   SPECIFIC ENERGY AND CRITICAL DEPTH

The energy per unit weight E_s, with reference elevation datum taken at the bottom of the channel, is called the *specific energy*. It is a convenient quantity to use in studying open-channel flow and was introduced by Bakhmeteff [2] in 1912. It is plotted vertically above the channel floor as

$$E_s = y + \frac{V^2}{2g} \tag{13.5.1}$$

A plot of specific energy for a particular case is shown in Fig. 13.6. In a rectangular channel, in which q is the discharge per unit width, with $Vy = q$,

$$E_s = y + \frac{q^2}{2gy^2} \tag{13.5.2}$$

It is of interest to note how the specific energy varies with the depth for a constant discharge (Fig. 13.7). For small values of y the curve goes to infinity along the E_s axis, while for large values of y the velocity head term is negligible and the curve approaches the 45° line, $E_s = y$, asymptotically. The specific energy has a minimum value below which the given q cannot occur. The value of y for minimum E_s is obtained by setting dE_s/dy equal to zero, from Eq. (13.5.2), holding q constant

$$\frac{dE_s}{dy} = 0 = 1 - \frac{q^2}{gy^3}$$

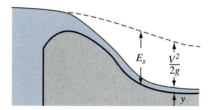

Figure 13.6   Example of specific energy.

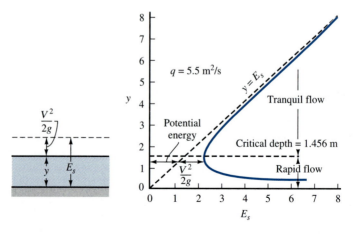

Figure 13.7   Specific energy required for flow of a given discharge at various depths.

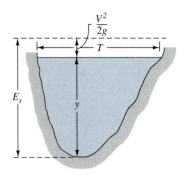

Figure 13.8 Specific energy for a nonrectangular section.

or

$$y_c = \left(\frac{q^2}{g}\right)^{1/3} \tag{13.5.3}$$

The depth for minimum energy y_c is called *critical depth*. Eliminating q^2 in Eqs. (13.5.2) and (13.5.3) gives

$$E_{s_{min}} = \frac{3}{2} y_c \tag{13.5.4}$$

showing that the critical depth is two-thirds of the specific energy. Eliminating E_s in Eqs. (13.5.1) and (13.5.4) gives

$$V_c = \sqrt{g y_c} \tag{13.5.5}$$

The velocity of flow at critical condition V_c is $\sqrt{g y_c}$, which was used in Sec. 10.8 in connection with the broad-crested weir. Another method of arriving at the critical condition is to determine the maximum discharge q that could occur for a given specific energy. The resulting equations are the same as Eqs. (13.5.3) to (13.5.5).

For nonrectangular cross sections (Fig. 13.8), the specific-energy equation takes the form

$$E_s = y + \frac{Q^2}{2gA^2} \tag{13.5.6}$$

in which A is the cross-sectional area. To find the critical depth

$$\frac{dE_s}{dy} = 0 = 1 - \frac{Q^2}{gA^3}\frac{dA}{dy}$$

From Fig. 13.8, the relation between dA and dy is expressed by

$$dA = T\, dy$$

in which T is the width of the cross section at the liquid surface. With this relation

$$\frac{Q^2}{gA_c^3} T_c = 1 \tag{13.5.7}$$

The critical depth must satisfy this equation. Eliminating Q in Eqs. (13.5.6) and (13.5.7) gives

$$E_s = y_c + \frac{A_c}{2T_c} \tag{13.5.8}$$

This equation shows that the minimum energy occurs when the velocity head is one-half the average depth A/T. Equation (13.5.7) can be solved by trial and error for irregular sections by plotting

$$f(y) = \frac{Q^2 T}{g A^3}$$

Critical depth occurs for that value of y which makes $f(y) = 1$.

Example 13.3

Determine the critical depth for 10 m³/s flowing in a trapezoidal channel with bottom width 3 m and side slopes 1 horizontal to 2 vertical (1 on 2).

Solution

$$A = 3y + \frac{y^2}{2} \qquad T = 3 + y$$

Hence,

$$f(y) = \frac{10^2(3 + y)}{9.806(3y + y^2/2)^3} = \frac{10.198(3 + y)}{(3y + 0.5y^2)^3} = 1.0$$

By trial and error

y	2.0	1.2	0.8	1.0	0.99	0.98	0.985	0.984
$f(y)$	0.1	0.53	1.92	0.95	0.982	1.014	0.998	1.0014

The critical depth is 0.984 m. This trial solution is easily carried out with the solver function in a spreadsheet or by use of a programmable calculator.

In uniform flow in an open channel, the energy grade line slopes downward parallel to the bottom of the channel, thus showing a steady decrease in available energy. The specific energy, however, remains constant along the channel, since $y + V^2/2g$ does not change. In nonuniform steady flow the energy grade line always slopes downward, or the available energy is decreased. The specific energy may either increase or decrease, depending upon the slope of the channel bottom, the discharge, the depth of flow, properties of the cross section, and channel roughness. In Fig. 13.6 the specific energy increases during flow down the steep portion of the channel and decreases along the horizontal channel floor.

The specific-energy and critical-depth relations are essential in studying gradually varied flow and in determining control sections in open-channel flow.

The head loss in a hydraulic jump is easily displayed by drawing the $F + M$ curve (Fig. 13.5) and the specific-energy curve (Fig. 13.7) to the same vertical scale for the same discharge. Conjugate depths exist where any given vertical line intersects the $F + M$ curve. The specific energy at the upper depth can be observed to be always less than the specific energy at the corresponding lower conjugate depth.

Water flows at a rate of 16 m³/s at half critical depth in a trapezoidal channel, $b =$ 4 m and $m = 0.4$, before a hydraulic jump occurs. Find the height after the jump and the energy loss in kilowatts.

Example 13.4

Solution

Solve Eq. (13.5.7) for y_c by the bisection method. Then take half the critical depth, y_1, and substitute into the $F + M$ relation,

$$\frac{F + M}{\gamma} = 0.5\, by^2 + \frac{my^3}{3} + \frac{q^2}{gy(b + my)}$$

This equation is now solved again for the root above y_c having the same $(F + M)/\gamma$, again using the bisection method.

$$\text{Loss} = \frac{V_1^2}{2g} - \frac{V_2^2}{2g} + y_1 - y_2 \qquad \text{m·N/N}$$

and

$$\text{Power} = \frac{\gamma Q\, \text{loss}}{1000} \text{ kW}$$

By using the solver function in a spreadsheet, first to find the critical depth and then to find the depth after the jump, the following results are obtained:

$$y_c = 1.132 \text{ m} \qquad\qquad y_2 = 1.974 \text{ m}$$
$$\text{Loss} = 0.727 \text{ m} \cdot \text{N/N} \qquad \text{Power loss} = 114 \text{ kW}$$

EXERCISES

13.5.1 Flow at critical depth occurs when (*a*) changes in upstream resistance alter downstream conditions; (*b*) the specific energy is a maximum for a given discharge; (*c*) any change in depth requires more specific energy; (*d*) the normal depth and critical depth coincide for a channel; (*e*) the velocity is given by $\sqrt{2gy}$.

13.5.2 Critical depth in a rectangular channel is expressed by (*a*) $\sqrt{Vy}$; (*b*) $\sqrt{2gy}$; (*c*) $\sqrt{gy}$; (*d*) $\sqrt{q/g}$; (*e*) $(q^2/g)^{1/3}$.

13.5.3 Critical depth in a nonrectangular channel is expressed by (*a*) $Q^2T/gA^3 = 1$; (*b*) $QT^2/gA^2 = 1$; (*c*) $Q^2A^3/gT^2 = 1$; (*d*) $Q^2/gA^3 = 1$; (*e*) none of these answers.

13.5.4 The specific energy for the flow expressed by $V = 4.43$ m/s and $y = 1$ m, in meter-newtons per newton, is (*a*) 2; (*b*) 3; (*c*) 5.43; (*d*) 9.86; (*e*) none of these answers.

13.5.5 The minimum possible specific energy for a flow is 2.475 ft·lb/lb. The discharge per foot of width, in cubic feet per second, is (*a*) 4.26; (*b*) 12.02; (*c*) 17; (*d*) 22.15; (*e*) none of these answers.

13.6 TRANSITIONS

At entrances to channels and at changes in cross section and bottom slope, the structure that conducts the liquid from the upstream section to the new section is called

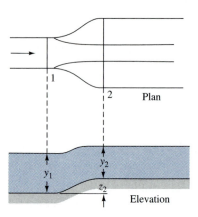

Figure 13.9   Transition from rectangular channel to trapezoidal channel
for tranquil flow.

a *transition.* Its purpose is to change the shape of flow and surface profile in such a manner that minimum losses result. A transition for tranquil flow from a rectangular channel to a trapezoidal channel is illustrated in Fig. 13.9. Applying the energy equation from section 1 to section 2 gives

$$\frac{V_1^2}{2g} + y_1 = \frac{V_2^2}{2g} + y_2 + z_2 + E_\ell \qquad \textbf{(13.6.1)}$$

In general, the sections and depths are determined by other considerations, and z must be determined for the expected available energy loss E_ℓ. By good design, that is, with slowly tapering walls and flooring with no sudden changes in cross-sectional area, the losses can be held to about one-tenth the difference between velocity heads for accelerated flow and to about three-tenths the difference between velocity heads for retarded flow. For rapid flow, wave mechanics is required in designing the transitions [3].

Example 13.5   In Fig. 13.9, 400 cfs flows through the transition; the rectangular section is 8 ft wide, and $y_1 = 8$ ft. The trapezoidal section is 6 ft wide at the bottom with side slopes 1:1, and $y_2 = 7.5$ ft. Determine the rise z in the bottom through the transition.

Solution

$$V_1 = \frac{400}{64} = 6.25 \qquad \frac{V_1^2}{2g} = 0.61 \qquad A_2 = 101.25 \text{ ft}^2$$

$$V_2 = \frac{400}{101.25} = 3.95 \qquad \frac{V_2^2}{2g} = 0.24 \qquad E_\ell = 0.3\left(\frac{V_1^2}{2g} - \frac{V_2^2}{2g}\right) = 0.11$$

Substituting into Eq. (13.6.1) gives

$$z = 0.61 + 8 - 0.24 - 7.5 - 0.11 = 0.76 \text{ ft}$$

The *critical-depth meter* [4] is an excellent device for measuring discharge in an open channel. The relations for determination of discharge are worked out for a

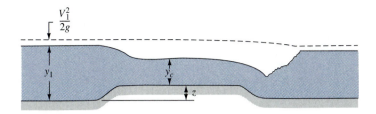

Figure 13.10 Critical-depth meter.

rectangular channel of constant width (Fig. 13.10) with a raised floor over a reach of channel about $3y_c$ long. The raised floor is of such height that the restricted section becomes a control section with critical velocity occurring over it. By measuring only the upstream depth y_1, the discharge per foot of width is accurately determined. Applying the energy equation from section 1 to the critical section (the exact location is unimportant), including the transition-loss term, gives

$$\frac{V_1^2}{2g} + y_1 = z + y_c + \frac{V_c^2}{2g} + \frac{1}{10}\left(\frac{V_c^2}{2g} - \frac{V_1^2}{2g}\right)$$

Since

$$y_c + \frac{V_c^2}{2g} = E_c \qquad \frac{V_c^2}{2g} = \frac{E_c}{3}$$

in which E_c is the specific energy at critical depth,

$$y_1 + 1.1\frac{V_1^2}{2g} = z + 1.033E_c \tag{13.6.2}$$

From Eq. (13.5.3)

$$y_c = \frac{2}{3}E_c = \left(\frac{q^2}{g}\right)^{1/3} \tag{13.6.3}$$

In Eqs. (13.6.2) and (13.6.3) E_c is eliminated and the resulting equation solved for q,

$$q = 0.517g^{1/2}\left(y_1 - z + 1.1\frac{V_1^2}{2g}\right)^{3/2}$$

Since $q = V_1y_1$, V_1 can be eliminated and

$$q = 0.517g^{1/2}\left(y_1 - z + \frac{0.55}{g}\frac{q^2}{y_1^2}\right)^{3/2} \tag{13.6.4}$$

The equation is solved by trial and error. As y_1 and z are known and the right-hand term containing q is small, it can first be neglected for an approximate q. A value a little larger than the approximate q can be substituted on the right-hand side. When the two q's are the same, the equation is solved. Alternatively, the solver function in a spreadsheet may be used to solve Eq. (13.6.4). Once z and the width of the channel are known, a chart or table can be prepared yielding Q for any y_1. Experiments indicate that accuracy within 2 to 3 percent can be expected.

With tranquil flow a jump occurs downstream from the meter, and with rapid flow a jump occurs upstream from the meter.

Example 13.6

In a critical-depth meter 2 m wide with $z = 0.3$ m the depth y_1 is measured to be 0.75 m. Find the discharge.

Solution

Using Eq. (13.6.4)

$$q = 0.517(9.806^{1/2})(0.45^{3/2}) = 0.489 \text{ m}^2/\text{s}$$

As a second approximation, let q be 0.50,

$$q = 0.517(9.806^{1/2})\left[0.45 + \frac{0.55}{9.806}\left(\frac{0.5}{0.75}\right)^2\right]^{3/2} = 0.530 \text{ m}^2/\text{s}$$

and as a third approximation, let q be 0.535,

$$q = 0.517(9.806^{1/2})\left[0.45 + \frac{0.55}{9.806}\left(\frac{0.535}{0.75}\right)^2\right]^{3/2} = 0.536 \text{ m}^2/\text{s}$$

Then

$$Q = 2(0.536) = 1.072 \text{ m}^3/\text{s}.$$

EXERCISES

13.6.1   The loss through a diverging transition is about

(a) $0.1\dfrac{(V_1 - V_2)^2}{2g}$ (b) $0.1\dfrac{(V_1^2 - V_2^2)}{2g}$ (c) $0.3\dfrac{(V_1 - V_2)^2}{2g}$ (d) $0.3\dfrac{(V_1^2 - V_2^2)}{2g}$

(e) none of these answers

13.6.2   A critical-depth meter (a) measures the depth at the critical section; (b) is always preceded by a hydraulic jump; (c) must have a tranquil flow immediately upstream; (d) always has a hydraulic jump downstream; (e) always has a hydraulic jump associated with it.

13.7   GRADUALLY VARIED FLOW

Gradually varied flow is steady nonuniform flow of a special class. The depth, area, roughness, bottom slope, and hydraulic radius change very slowly (if at all) along the channel. The basic assumption required is that the head-loss rate at a given section is given by the Manning formula for the same depth and discharge, regardless of trends in the depth. Solving Eq. (13.2.1) for the head loss per unit length of channel produces

$$S = -\frac{\Delta E}{\Delta L} = \left(\frac{nQ}{C_m A R^{2/3}}\right)^2 \qquad \text{(13.7.1)}$$

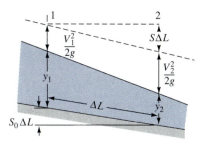

Figure 13.11 Gradually varied flow.

in which S is now the slope of the energy grade line or, more specifically, the sine of the angle the energy grade line makes with the horizontal. In gradually varied flow the slopes of energy grade line, hydraulic grade line, and bottom are all different. Computations of gradually varied flow can be carried out either by the *standard-step method* or by *numerical integration*. Horizontal channels of great width are treated as a special case that can be integrated.

Standard-Step Method

Applying the energy equation between two sections a finite distance ΔL apart (Fig. 13.11), including the loss term, gives

$$\frac{V_1^2}{2g} + S_0\,\Delta L + y_1 = \frac{V_2^2}{2g} + y_2 + S\,\Delta L \tag{13.7.2}$$

Solving for the length of reach gives

$$\Delta L = \frac{(V_1^2 - V_2^2)/2g + y_1 - y_2}{S - S_0} \tag{13.7.3}$$

If conditions are known at one section, for example, section 1, and the depth y_2 is wanted a distance ΔL away, a trial solution is required. The procedure is as follows:

1. Assume a depth y_2, then compute A_2 and V_2.
2. For the assumed y_2 find an average y, P, and A for the reach and compute S. (For prismatic channels $y = (y_1 + y_2)/2$ with A and R computed for this depth.)
3. Substitute in Eq. (13.7.3) to compute ΔL.
4. If ΔL is not correct, assume a new y_2 and repeat the procedure.

At section 1 of a canal the cross section is trapezoidal, $b_1 = 10$ m, $m_1 = 2$, and $y_1 = 7$ m, and at section 2, downstream 200 m, the bottom is 0.08 m higher than at section 1, $b_2 = 15$ m, and $m_2 = 3$. $Q = 200$ m³/s and $n = 0.035$. Determine the depth of water at section 2.

Example 13.7

Solution

$$A_1 = b_1 y_1 + m_1 y_1^2 = 10(7) + 2(7^2) = 168 \text{ m}^2 \qquad V_1 = \frac{200}{168} = 1.19 \text{ m/s}$$

$$P_1 = b_1 + 2y_1\sqrt{m_1^2 + 1} = 10 + 2(7)\sqrt{2^2 + 1} = 41.3 \text{ m}$$

$$S_0 = -\frac{0.08}{200} = -0.0004$$

Since the bottom has an adverse slope, that is, it is rising in the downstream direction, and since section 2 is larger than section 1, y_2 is probably less than y_1. Assume $y_2 = 6.9$ m, then

$$A_2 = 15(6.9) + 3(6.9^2) = 246 \text{ m}^2 \qquad V_2 = \frac{200}{246} = 0.813 \text{ m/s}$$

and

$$P_2 = 15 + 2(6.9\sqrt{10}) = 58.6 \text{ m}$$

The average $A = 207$ and average wetted perimeter $P = 50.0$ are used to find an average hydraulic radius for the reach, $R = 4.14$ m. Then

$$S = \left(\frac{nQ}{C_m A R^{2/3}}\right)^2 = \left[\frac{0.035(200)}{1.0(207)(4.14^{2/3})}\right]^2 = 0.000172$$

Substituting into Eq. (13.7.3) gives

$$\Delta L = \frac{(1.19^2 - 0.813^2)/[2(9.806)] + 7 - 6.9}{0.000172 + 0.0004} = 242 \text{ m}$$

A larger y_2, for example, 6.92 m, would bring the computed value of length closer to the actual length.

The standard-step method is easily implemented with a spreadsheet or with a programmable calculator. In the first trial y_2 is used to evaluate ΔL_{new}. Then a linear proportion yields a new trial $y_{2_{new}}$ for the next step; thus,

$$\frac{y_1 - y_2}{\Delta L_{new}} = \frac{y_1 - y_{2_{new}}}{\Delta L_{given}}$$

or

$$y_{2_{new}} = y_1 + (y_2 - y_1)\frac{\Delta L_{given}}{\Delta L_{new}}$$

A few iterations yield complete information on section 2.

Numerical Integration Method

A more satisfactory procedure, particularly for flow through channels having a constant shape of cross section and constant bottom slope, is to obtain a differential equation in terms of y and L and then perform the integration numerically. When ΔL is considered as an infinitesimal in Fig. 13.11, rate of change of available energy equals rate of head loss $-\Delta E/\Delta L$ given by Eq. (13.7.1), or

$$\frac{d}{dL}\left(\frac{V^2}{2g} + z_0 - S_0 L + y\right) = -\left(\frac{nQ}{C_m A R^{2/3}}\right)^2 \qquad \textbf{(13.7.4)}$$

in which $z_0 - S_0 L$ is the elevation of bottom of channel at L, z_0 is the elevation of bottom at $L = 0$, and L is measured positive in the downstream direction. After performing the differentiation,

$$-\frac{V}{g}\frac{dV}{dL} + S_0 - \frac{dy}{dL} = \left(\frac{nQ}{C_m A R^{2/3}}\right)^2 \qquad (13.7.5)$$

Using the continuity equation $VA = Q$ leads to

$$\frac{dV}{dL}A + V\frac{dA}{dL} = 0$$

and expressing $dA = T\,dy$, in which T is the liquid-surface width of the cross section, gives

$$\frac{dV}{dL} = -\frac{VT}{A}\frac{dy}{dL} = -\frac{QT}{A^2}\frac{dy}{dL}$$

Substituting for V in Eq. (13.7.5) yields

$$\frac{Q^2}{gA^3}T\frac{dy}{dL} + S_0 - \frac{dy}{dL} = \left(\frac{nQ}{C_m A R^{2/3}}\right)^2$$

and solving for dL gives

$$dL = \frac{1 - Q^2 T/gA^3}{S_0 - (nQ/C_m A R^{2/3})^2}\,dy \qquad (13.7.6)$$

After integrating,

$$L = \int_{y_1}^{y_2} \frac{1 - Q^2 T/gA^3}{S_0 - (nQ/C_m A R^{2/3})^2}\,dy \qquad (13.7.7)$$

in which L is the distance between the two sections having depths y_1 and y_2.

When the numerator of the integrand is zero, critical flow prevails; there is no change in L for a change in y (neglecting curvature of the flow and nonhydrostatic pressure distribution at this section). Since this is not a case of gradual change in depth, the equations are not accurate near critical depth. When the denominator of the integrand is zero, uniform flow prevails and there is no change in depth along the channel. The flow is at *normal depth.*

For a channel of prismatic cross section, constant n and S_0, the integrand becomes a function of y only, that is,

$$F(y) = \frac{1 - Q^2 T/gA^3}{S_0 - (nQ/C_m A R^{2/3})^2}$$

and the equation can be integrated numerically by plotting $F(y)$ as ordinate against y as abscissa. The area under the curve (Fig. 13.12) between two values of y is the length L between the sections, since

$$L = \int_{y_1}^{y_2} F(y)\,dy$$

A trapezoidal channel, $b = 3$ m, $m = 1$, $n = 0.014$, and $S_0 = 0.001$, carries 28 m³/s. If the depth is 3 m at section 1, determine the water-surface profile for the next 700 m downstream.

Example 13.8

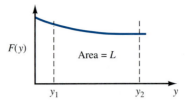

Figure 13.12   Numerical integration of equation for gradually varied flow.

Solution

To determine whether the depth increases or decreases, the slope of the energy grade line at section 1 is computed using Eq. (13.7.1)

$$A = by + my^2 = 3(3) + 1(3^2) = 18 \text{ m}^2$$

$$P = b + 2y\sqrt{m^2 + 1} = 11.485 \text{ m}$$

and

$$R = \frac{18}{11.485} = 1.567 \text{ m}$$

Then

$$S = \left[\frac{0.014(28)}{18(1.567^{2/3})}\right]^2 = 0.00026$$

Substituting into Eq. (13.5.7) the values for A, Q, and $T = 9$ m gives $Q^2 T/gA^3 = 0.12$, showing that the depth is above critical. With the depth greater than critical and the energy grade line less steep than the bottom of the channel, the specific energy is increasing. When the specific energy increases above critical, the depth of flow increases. Δy is then positive. Substituting into Eq. (13.7.7) yields

$$L = \int_3^y \frac{1 - 79.95T/A^3}{0.001 - 0.1537/(A^2 R^{4/3})} \, dy$$

The following table evaluates the terms of the integrand.

y	A	P	R	T	Numerator	Denominator $\times 10^6$	$F(y)$	L
3	18	11.48	1.57	9	0.8766	739	1185	0
3.2	19.84	12.05	1.65	9.4	0.9038	799	1131	231.6
3.4	21.76	12.62	1.72	9.8	0.9240	843	1096	454.3
3.6	23.76	13.18	1.80	10.2	0.9392	876	1072	671.1
3.8	25.84	13.75	1.88	10.6	0.9509	901	1056	883.9

The integral $\int F(y) \, dy$ can be evaluated by plotting the curve and taking the area under it between $y = 3$ and the following values of y. As $F(y)$ does not vary greatly in this example, the average of $F(y)$ can be used for each reach (the trapezoidal rule); and when it is multiplied by Δy, the length of reach is obtained.

Between $y = 3$ and $y = 3.2$

$$\frac{1185 + 1131}{2} \, 0.2 = 231.6$$

Between $y = 3.2$ and $y = 3.4$

$$\frac{1131 + 1096}{2} \, 0.2 = 222.7$$

and so on. Five points on it are known, so the water surface can be plotted. A more accurate way of summing $F(y)$ to obtain L is by use of Simpson's rule (see the Web page). The procedure used is equivalent to a Runge-Kutta second-order solution of a differential equation (see the Web page). A programmable calculator was used to carry out this solution. By taking $\Delta y = 0.1$ m in place of 0.2 m, the length to $y = 3.6$ m is 0.6 m less.

Horizontal Channels of Great Width

For channels of great width the hydraulic radius equals the depth, and for horizontal channel floors, $S_0 = 0$. Hence, Eq. (13.7.7) can be simplified. The width can be considered as unity, that is, $T = 1$, $Q = q$, $A = y$, and $R = y$. Thus,

$$L = -\int_{y_1}^{y} \frac{1 - q^2/gy^3}{n^2 q^2 / C_m^2 y^{10/3}} \, dy \qquad \text{(13.7.8)}$$

or, after performing the integration, as

$$L = -\frac{3}{13}\left(\frac{C_m}{nq}\right)^2 (y^{13/3} - y_1^{13/3}) + \frac{3}{4g}\left(\frac{C_m}{n}\right)^2 (y^{4/3} - y_1^{4/3}) \qquad \text{(13.7.9)}$$

After contracting below a sluice gate water flows onto a wide horizontal floor with a velocity of 15 m/s and a depth of 0.7 m. Find the equation for the water-surface profile, $n = 0.015$.

Example 13.9

Solution

From Eq. (13.7.9), with x replacing L as distance from section 1, where $y_1 = 0.7$, and with $q = 0.7(15) = 10.5$ m²/s,

$$x = -\frac{3}{13}\left[\frac{1}{0.015(10.5)}\right]^2 (y^{13/3} - 0.7^{13/3}) + \frac{3}{4(9.806)}\left(\frac{1}{0.015}\right)^2 (y^{4/3} - 0.7^{4/3})$$

$$= -209.3 - 9.30y^{13/3} + 340y^{4/3}$$

Critical depth occurs [Eq. (13.5.3)] at

$$y_c = \left(\frac{q^2}{g}\right)^{1/3} = \left(\frac{10.5^2}{9.806}\right)^{1/3} = 2.24 \text{ m}$$

The depth must increase downstream, since the specific energy decreases, and the depth must move toward the critical value for less specific energy. The equation does not hold near the critical depth because of vertical accelerations that have been neglected in the derivation of gradually varied flow. If the channel is long enough for critical depth to be attained before the end of the channel, the high-velocity flow downstream from the gate may be drowned or a jump may occur. The water-surface calculation for the subcritical flow must begin with critical depth at the downstream end of the channel.

The numerical computation of water-surface profiles is discussed after the various types of gradually varied flow profiles are classified.

EXERCISE

13.7.1   Gradually varied flow is (*a*) steady uniform flow; (*b*) steady nonuniform flow; (*c*) unsteady uniform flow; (*d*) unsteady nonuniform flow; (*e*) none of these answers.

13.8   CLASSIFICATION OF SURFACE PROFILES

A study of Eq. (13.7.7) reveals many types of surface profiles, each with definite characteristics. The bottom slope is classified as *adverse, horizontal, mild, critical,* and *steep.* In general, the flow can be above the normal depth or below the normal depth, and it can be above critical depth or below critical depth.

The various profiles are plotted in Fig. 13.13; the procedures used are discussed for the various classifications in the following paragraphs. A very wide channel is assumed in the reduced equations which follow, with $R = y$.

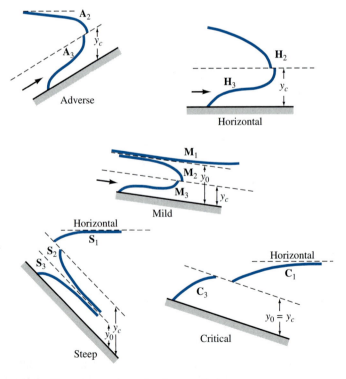

Figure 13.13   Typical liquid-surface profiles.

Adverse Slope Profiles

When the channel bottom rises in the direction of flow (S_0 is negative), the resulting surface profiles are said to be adverse. There is no normal depth, but the flow may be either below or above critical depth. Below critical depth the numerator is negative, and Eq. (13.7.6) has the form

$$dL = \frac{1 - C_1/y^3}{S_0 - C_2/y^{10/3}}\, dy$$

where C_1 and C_2 are positive constants. Here $F(y)$ is positive and the depth increases downstream. This curve is labeled $\mathbf{A}_3$ as shown in Fig. 13.13. For depths greater than critical depth, the numerator is positive and $F(y)$ is negative, that is, the depth decreases in the downstream direction. For y very large, $dL/dY = 1/S_0$, which is a horizontal asymptote for the curve. At $y = y_c$, dL/dy is 0, and the curve is perpendicular to the critical-depth line. This curve is labeled $\mathbf{A}_2$.

Horizontal Slope Profiles

For a horizontal channel $S_0 = 0$, the normal depth is infinite and flow may be either below or above critical depth. The equation has the form

$$dL = -Cy^{1/3}(y^3 - C_1)\, dy$$

For y less than critical, dL/dy is positive, and the depth increases downstream. It is labeled $\mathbf{H}_3$. For y greater than critical ($\mathbf{H}_2$ curve), dL/dy is negative, and the depth decreases downstream. These equations are integrable analytically for very wide channels.

Mild Slope Profiles

A mild slope is one on which the normal flow is tranquil, that is, where the normal depth y_0 is greater than the critical depth. Three profiles may occur, $\mathbf{M}_1$, $\mathbf{M}_2$, and $\mathbf{M}_3$, for depth above normal, below normal and above critical, and below critical, respectively. For the $\mathbf{M}_1$ curve, dL/dy is positive and approaches $1/S_0$ for very large y; hence, the $\mathbf{M}_1$ curve has a horizontal asymptote downstream. As the denominator approaches zero as y approaches y_0, the normal depth is an asymptote at the upstream end of the curve. Thus, dL/dy is negative for the $\mathbf{M}_2$ curve, with the upstream asymptote the normal depth, and $dL/dy = 0$ at critical. The $\mathbf{M}_3$ curve has an increasing depth downstream, as shown.

Critical Slope Profiles

When the normal depth and the critical depth are equal, the resulting profiles are labeled $\mathbf{C}_1$ and $\mathbf{C}_3$ for depth above and below critical, respectively. The equation has the form

$$dL = \frac{1}{S_0} \frac{1 - b/y^3}{1 - b_1/y^{10/3}}\, dy$$

with both numerator and denominator positive for $\mathbf{C}_1$ and negative for $\mathbf{C}_3$. Therefore, the depth increases downstream for both. For large y, dL/dy approaches $1/S_0$; hence, a horizontal line is an asymptote. The value of dL/dy at critical depth is $0.9/S_0$; hence, curve C_1 is convex upward. Curve C_3 also is convex upward, as shown.

Steep Slope Profiles

When the normal flow is rapid in a channel (the normal depth less than the critical depth), the resulting profiles S_1, S_2, and S_3 are referred to as steep profiles. S_1 is above the normal and critical, S_2 between critical and normal, and S_3 below normal depth. For curve S_1 both numerator and denominator are positive, and the depth increases downstream approaching a horizontal asymptote. For curve S_2 the numerator is negative and the denominator positive but approaching zero at $y = y_0$. The curve approaches the normal depth asymptotically. The S_3 curve has a positive dL/dy as both numerator and denominator are negative. It plots as shown on Fig. 13.13.

It should be noted that a given channel may be classified as mild for one discharge, critical for another discharge, and steep for a third discharge, since normal depth and critical depth depend upon different functions of the discharge. The use of the various surface profiles is discussed in the next section.

13.9   CONTROL SECTIONS

A small change in downstream conditions cannot be relayed upstream when the depth is critical or less than critical; hence, downstream conditions do not control the flow. All rapid flows are controlled by upstream conditions, and computations of surface profiles must be started at the upstream end of a channel.

Tranquil flows are affected by small changes in downstream conditions and therefore are controlled by them. Tranquil-flow computations must start at the downstream end of a reach and be carried upstream.

Control sections occur at entrances and exits to channels and at changes in channel slopes, under certain conditions. A gate in a channel can be a control for both the upstream and downstream reaches. Three control sections are illustrated. In Fig. 13.14a the flow passes through critical at the entrance to a channel, and depth

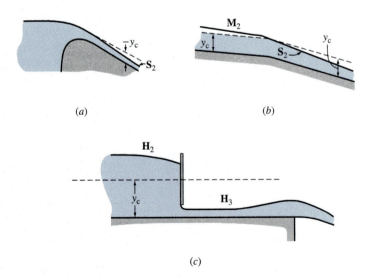

(a)   (b)

(c)

Figure 13.14   Channel control sections.

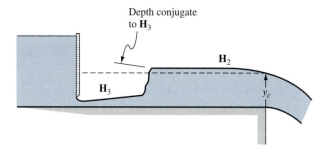

Figure 13.15   Hydraulic jump between two control sections.

can be computed there for a given discharge. The channel is steep; therefore, computations proceed downstream. In Fig. 13.14*b* a change in channel slope from mild to steep causes the flow to pass through critical at the break in grade. Computations proceed both upstream and downstream from the control section at the break in grade. In Fig. 13.14*c* a gate in a horizontal channel provides control both upstream and downstream from it. The various curves are labeled according to the classification in Fig. 13.13.

The hydraulic jump occurs whenever the conditions required by the momentum equation are satisfied. In Fig. 13.15, liquid issues from under a gate in rapid flow along a horizontal channel. If the channel were short enough, the flow could discharge over the end of the channel as an $\mathbf{H}_3$ curve. With a longer channel, however, the jump occurs, and the resulting profile consists of pieces of $\mathbf{H}_3$ and $\mathbf{H}_2$ curves with the jump in between. In computing these profiles for a known discharge, the $\mathbf{H}_3$ curve is computed, starting at the gate (contraction coefficient must be known) and proceeding downstream until it is clear that the depth will reach critical before the end of the channel is reached. Then the $\mathbf{H}_2$ curve is computed, starting with critical depth at the end of the channel and proceeding upstream. The depth conjugate to those along $\mathbf{H}_3$ are computed and plotted as shown. The intersection of the conjugate-depth curve and the $\mathbf{H}_2$ curve locates the position of the jump. The channel may be so long that the $\mathbf{H}_2$ curve is everywhere greater than the depth conjugate to $\mathbf{H}_3$. A *drowned jump* then occurs, with $\mathbf{H}_2$ extending to the gate.

All sketches are drawn to a greatly exaggerated vertical scale, since usual channels have small bottom slopes.

EXERCISE

13.9.1 The hydraulic jump aways occurs from (*a*) an $\mathbf{M}_3$ curve to an $\mathbf{M}_1$ curve; (*b*) an $\mathbf{H}_3$ curve to an $\mathbf{H}_2$ curve; (*d*) an $\mathbf{S}_3$ curve to an $\mathbf{S}_1$ curve; (*d*) below normal depth to above normal depth; (*e*) below critical depth to above critical depth.

13.10   COMPUTER CALCULATION OF GRADUALLY VARIED FLOW

In Sec. 13.7 the standard-step and numerical-integration methods of computing water-surface profiles were introduced. The repetitious calculation in the latter method is easily handled by digital computer using a standard programming language or a spreadsheet. Figure 13.16 shows a spreadsheet to calculate the steady

Water surface profile calculation

Q =	2.5 m^3/s	Gamma =	9806 N/m^3
b =	2.5 m	Cm =	1
m =	0.8	g =	9.806 m/s^2
S0 =	0.0002	n =	0.012
L =	600 m	Ycont =	0.907 m

Calculate critical & normal depths

yc =   1.7802 m   **1st Solver:**   F = 0.0, by varying yc

F = $-1.6571E - 06$ = Q^2*(b + 2*m*yc)/(g*(yc*(b + m*yc))^3) − 1

yn =   3.1899 m   **2nd Solver:** Fn = 0.0, by varying yn

Fn = $1.5921E - 10$ = S0 − (n*Q*(b + 2*yn*SQRT(m^2 + 1))^0.66667/

                                                (Cm*(yn*(b + m*yn))^(5/3)))^2

Calculate water surface profile

dy =   0.04365768 = (yc − Ycont)/20

y	yave	dL	distance	energy	(F+M)/Gamma
0.907		0.000	0.000	4.630	23.013
0.951	0.929	19.401	19.401	4.268	21.921
0.994	0.972	19.166	38.566	3.962	20.949
1.038	1.016	18.879	57.445	3.705	20.083
1.082	1.060	18.537	75.981	3.487	19.310
1.125	1.103	18.136	94.117	3.302	18.621
1.169	1.147	17.672	111.789	3.145	18.007
1.213	1.191	17.141	128.930	3.012	17.461
1.256	1.234	16.538	145.468	2.900	16.976
1.300	1.278	15.858	161.326	2.805	16.549
1.344	1.322	15.095	176.420	2.725	16.173
1.387	1.365	14.242	190.662	2.658	15.845
1.431	1.409	13.292	203.954	2.603	15.562
1.475	1.453	12.239	216.193	2.557	15.320
1.518	1.496	11.073	227.266	2.520	15.116
1.562	1.540	9.785	237.050	2.491	14.949
1.606	1.584	8.364	245.415	2.469	14.816
1.649	1.627	6.800	252.215	2.452	14.715
1.693	1.671	5.080	257.295	2.441	14.644
1.736	1.715	3.190	260.485	2.435	14.603
1.780	1.758	1.113	261.599	2.433	14.590

Figure 13.16   Spreadsheet for water-surface profiles.

gradually varied water-surface profile in any prismatic rectangular, symmetric trapezoidal, or triangular channel. The concepts of physical control sections in a channel must be understood in order to use the spreadsheet successfully.

In the spreadsheet variable names are defined to identify the input data which include channel dimensions, discharge, and water-surface control depth. Critical depth and normal depth (if it exists) are calculated using the solver function. This is followed by the profile calculation, beginning at the control section and calculated upstream for subcritical flow, or downstream for supercritical flow. In the dL column, an average depth is used in calculating $F(y)$ in Eq. (13.7.6). Energy and $(F + M)/\gamma$

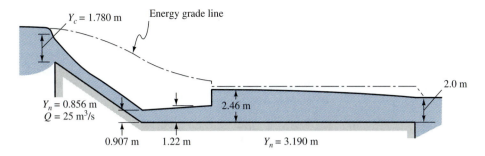

Figure 13.17   Solution to Ex. 13.10 that was obtained from the spreadsheet.

may also be calculated if needed. The data used pertain to the example which follows.

Example 13.10

A trapezoidal channel, $b = 2.5$ m and side slope $= 0.8$, has two bottom slopes. The upstream portion is 200 m long and $S_0 = 0.025$. The downstream portion is 600 m long and $S_0 = 0.0002$. $n = 0.012$. A discharge of 25 m³/s enters at critical depth from a reservoir at the upstream end, and at the downstream end of the system the water depth is 2 m. Determine the water-surface profiles throughout the system, including the jump location.

Solution

The spreadsheet in Fig. 13.16 is used three times to assemble the results used to plot the solution in Fig. 13.17. For the steep upstream channel critical and normal depths are first calculated, followed by the water-surface profile which begins at the upstream critical depth. An interpolation may be necessary to find the end depth at the position 200-m downstream. This depth of 0.907 m is needed for the control depth in the next calculation, which is the upstream (supercritical) flow in the mild channel. These data are shown in Fig. 13.16 along with results. The final calculation uses the 2-m-downstream control depth and computes the water-surface profile in the upstream direction. In Fig. 13.17 the jump is located by finding the position of equal $F + M$ from the output of the last two data sets.

13.11   FRICTIONLESS POSITIVE SURGE WAVE IN A RECTANGULAR CHANNEL

In this section the surge wave resulting from a sudden change in flow (due to a gate or other mechanism) that increases the depth is studied. A rectangular channel is assumed, and friction is neglected. Such a situation is shown in Fig. 13.18 shortly after a sudden, partial closure of a gate. The problem is analyzed by reducing it to a steady-state problem, as in Fig. 13.19. The continuity equation yields, per unit width,

$$(V_1 + c)y_1 = (V_2 + c)y_2 \tag{13.11.1}$$

and the momentum equation for the control volume 1–2, neglecting shear stress on the floor, per unit width, is

$$\frac{\gamma}{2}(y_1^2 - y_2^2) = \frac{\gamma}{g}y_1(V_1 + c)(V_2 + c - V_1 - c) \tag{13.11.2}$$

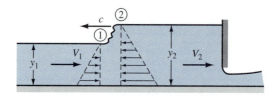

Figure 13.18   Positive surge wave in a rectangular channel.

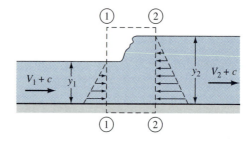

Figure 13.19   Surge problem reduced to a steady-state problem by superposition of surge velocity.

By eliminating V_2 in the last two equations,

$$V_1 + c = \sqrt{gy_1}\left[\frac{y_2}{2y_1}\left(1 + \frac{y_2}{y_1}\right)\right]^{1/2} \tag{13.11.3}$$

In this form the speed of an elementary wave is obtained by letting y_2 approach y_1, yielding

$$V_1 + c = \sqrt{gy} \tag{13.11.4}$$

For propagation through still liquid $V_1 \rightarrow 0$, and the wave speed is $c = \sqrt{gy}$ when the problem is converted back to the unsteady form by superposition of $V = -c$.

In general, Eqs. (13.11.1) and (13.11.2) have to be solved by trial and error. The hydraulic-jump formula results from setting $c = 0$ in the two equations [see Eq. (3.7.11)].

Example 13.11   A rectangular channel 3 m wide and 2 m deep, discharging 18 m³/s, suddenly has the discharge reduced to 12 m³/s at the downstream end. Compute the height and speed of the surge wave.

Solution

$V_1 = 3$, $y_1 = 2$, and $V_2y_2 = 4$. With Eqs. (13.11.1) and (13.11.2),

$$6 = 4 + c(y_2 - 2) \qquad \text{and} \qquad y_2^2 - 4 = \frac{2(2)}{9.806}(c + 3)(3 - V_2)$$

Eliminating c and V_2 gives

$$y_2^2 - 4 = \frac{4}{9.806}\left(\frac{2}{y_2 - 2} + 3\right)\left(3 - \frac{4}{y_2}\right)$$

or

$$\left(\frac{y_2 - 2}{3y_2 - 4}\right)^2 (y_2 + 2)y_2 = \frac{4}{9.806} = 0.407$$

After solving for y_2 by trial and error, $y_2 = 2.75$ m. Hence, $V_2 = 4/2.75 = 1.455$ m/s. The height of the surge wave is 0.75 m, and the speed of the wave is

$$c = \frac{2}{y_2 - 2} = \frac{2}{0.75} = 2.667 \text{ m/s}$$

EXERCISE

13.11.1 An elementary wave can travel upstream in a channel, $y = 4$ ft and $V = 8$ ft/s, with a velocity of (*a*) 3.35 ft/s; (*b*) 11.35 ft/s; (*c*) 16.04 ft/s; (*d*) 19.35 ft/s; (*e*) none of these answers.

13.12 FRICTIONLESS NEGATIVE SURGE WAVE IN A RECTANGULAR CHANNEL

Basic Equations

The negative surge wave appears as a gradual flattening and lowering of a liquid surface. It occurs, for example, in a channel downstream from a gate that is being closed or upstream from a gate that is being opened. Its propagation is accomplished by a series of elementary negative waves superposed on the existing velocity, each wave traveling at less speed than the one at the next greater depth. Application of the momentum equation and the continuity equation to a small depth change produces simple differential expressions relating wave speed, c; velocity, V; and depth, y. Integration of the equations yields the liquid-surface profile as a function of time, and velocity as a function of depth or as a function of position along the channel, x, and time, t. The fluid is assumed to be frictionless, and vertical accelerations are neglected.

In Fig. 13.20*a* an elementary disturbance is indicated in which the flow upstream has been slightly reduced. In order to apply the momentum and continuity equations it is convenient to reduce the motion to a steady one, as in Fig. 13.20*b*, by imposing a uniform velocity c to the left. The continuity equation is

$$(V - \delta V - c)(y - \delta y) = (V - c)y$$

or, by neglecting the product of small quantities,

$$(c - V)\delta y = y\,\delta V \qquad\qquad \textbf{(13.12.1)}$$

The momentum equation produces

$$\frac{\gamma}{2}(y - \delta y)^2 - \frac{\gamma}{2}y^2 = \frac{\gamma}{g}(V - c)y[V - c - (V - \delta V - c)]$$

After simplifying,

$$\delta y = \frac{c - V}{g}\delta V \qquad\qquad \textbf{(13.12.2)}$$

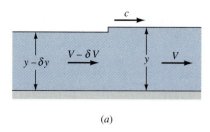

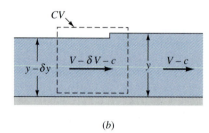

Figure 13.20   Elementary wave.

Equating $\delta V/\delta y$ in Eqs. (13.12.1) and (13.12.2) gives

$$c - V = \pm\sqrt{gy} \qquad\qquad (13.12.3)$$

or

$$c = V \pm \sqrt{gy}$$

The speed of an elementary wave in still liquid at depth y is $\sqrt{gy}$ and with flow the wave travels at the speed $\sqrt{gy}$ *relative* to the flowing liquid.

Eliminating c from Eqs. (13.12.1) and (13.12.2) gives

$$\frac{dV}{dy} = \pm\sqrt{\frac{g}{y}}$$

and integrating leads to

$$V = \pm 2\sqrt{gy} + \text{constant}$$

For a negative wave forming downstream from a gate (Fig. 13.21) after an instantaneous partial closure and using the plus sign, $V = V_0$ when $y = y_0$ and

$$V_0 = 2\sqrt{gy_0} + \text{constant}$$

After eliminating the constant,

$$V = V_0 - 2\sqrt{g}\left(\sqrt{y_0} - \sqrt{y}\right) \qquad\qquad (13.12.4)$$

The wave travels in the $+x$ direction, so that

$$c = V + \sqrt{gy} = V_0 - 2\sqrt{gy_0} + 3\sqrt{gy} \qquad\qquad (13.12.5)$$

If the gate motion occurs at $t = 0$, the liquid-surface position is expressed by $x = ct$, or

$$x = \left(V_0 - 2\sqrt{gy_0} + 3\sqrt{gy}\right)t \qquad\qquad (13.12.6)$$

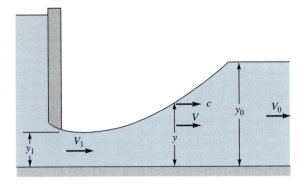

Figure 13.21   Negative wave after gate closure.

Eliminating y from Eqs. (13.12.5) and (13.12.6) gives

$$V = \frac{V_0}{3} + \frac{2}{3}\frac{x}{t} - \frac{2}{3}\sqrt{g y_0}$$ (13.12.7)

which is the velocity in terms of x and t.

In Fig. 13.21 find the Froude number of the undisturbed flow such that the depth y_1 at the gate is just zero when the gate is suddenly closed. For $V_0 = 20$ ft/s, find the liquid-surface equation.

Example 13.12

Solution

It is required that $V_1 = 0$ when $y_1 = 0$ at $x = 0$ for any time after $t = 0$. In Eq. (13.12.4), with $V = 0$ and $y = 0$,

$$V_0 = 2\sqrt{g y_0} \qquad \text{or} \qquad \mathbf{F}_0 = \frac{V_0}{\sqrt{g y_0}} = 2$$

For $V_0 = 20$,

$$y_0 = \frac{V_0^2}{4g} = \frac{20^2}{4g} = 3.11 \text{ ft}$$

By Eq. (13.12.6)

$$x = \left[20 - 2\sqrt{32.2(3.11)} + 3\sqrt{32.2y}\right]t = 17.02\sqrt{y}\,t$$

The liquid surface is a parabola with vertex at the origin and surface concave upward.

In Fig. 13.21 the gate is partially closed at the instant $t = 0$ so that the discharge is reduced by 50 percent. $V_0 = 6$ m/s and $y_0 = 3$ m. Find V_1, y_1, and the surface profile.

Example 13.13

Solution

The new discharge is

$$q = \frac{6(3)}{2} = 9 = V_1 y_1$$

By Eq. (13.12.4)

$$V_1 = 6 - 2\sqrt{9.806}\left(\sqrt{3} - \sqrt{y_1}\right)$$

Then V_1 and y_1 are found by trial and error from the last two equations, that is, $V_1 = 4.24$ m/s and $y_1 = 2.11$ m. The liquid-surface equation, from Eq. (13.12.6), is

$$x = \left(6 - 2\sqrt{3g} + 3\sqrt{gy}\right)t \qquad \text{or} \qquad x = \left(9.39\sqrt{y} - 4.85\right)t$$

which holds for the range of values of y between 2.11 and 3 m.

Dam Break

An idealization of a dam-break water-surface profile (Fig. 13.22) can be obtained from Eqs. (13.12.4) to (13.12.7). From a frictionless, horizontal channel with depth of water y_0 on one side of a gate and no water on the other side of the gate, the gate is suddenly removed. Vertical accelerations are neglected. $V_0 = 0$ in the equations, and y varies from y_0 to 0. The velocity at any section, from Eq. (13.12.4), is

$$V = -2\sqrt{g}\left(\sqrt{y_0} - \sqrt{y}\right) \qquad \text{(13.12.8)}$$

always in the downstream direction. The water-surface profile is, from Eq. (13.12.6),

$$x = \left(3\sqrt{gy} - 2\sqrt{gy_0}\right)t \qquad \text{(13.12.9)}$$

At $x = 0$, $y = 4y_0/9$, the depth remains constant and the velocity past the section $x = 0$ is, from Eq. (13.12.8),

$$V = -\frac{2}{3}\sqrt{gy_0}$$

also independent of time. The leading edge of the wave feathers out to zero height and moves downstream at $V = c = -2\sqrt{gy_0}$. The water surface is a parabola with vertex at the leading edge, concave upward.

With an actual dam break, ground roughness causes a positive surge, or wall of water, to move downstream, that is, the feathered edge is retarded by friction.

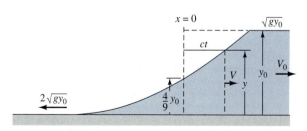

Figure 13.22   Dam-break profile.

EXERCISES

13.12.1   The speed of an elementary wave in a still liquid is given by (*a*) $(gy^2)^{1/3}$; (*b*) $2y/3$; (*c*) $\sqrt{2gy}$; (*d*) $\sqrt{gy}$; (*e*) none of these answers.

13.12.2   A negative surge wave (*a*) is a positive surge wave moving backward; (*b*) is an inverted positive surge wave; (*c*) can never travel upstream; (*d*) can never travel downstream; (*e*) is none of the above.

PROBLEMS

13.1   Show that, for laminar flow to be ensured down an inclined surface, the discharge per unit width cannot be greater than 500ν. (See Prob. 6.10.)

13.2   Calculate the depth of laminar flow of water at 20°C down a plane surface making an angle of 30° with the horizontal for the lower critical Reynolds number. (See Prob. 6.10.)

13.3   Calculate the depth of turbulent flow at $\mathbf{R} = VR/\nu = 500$ for flow of water at 20°C down a plane surface making an angle θ of 30° with the horizontal. Use Manning's formula. $n = 0.01$ and $S = \sin\theta$.

13.4   Water flows in a trapezoidal channel at a depth of 1.2 m. The bottom width is 6 m, side slopes 2 horizontal to 1 vertical (2:1), bottom slope of 0.008, and $n = 0.016$. Find (*a*) the flow rate, (*b*) the Froude number, and (*c*) the critical depth.

13.5   A semicircular channel flowing full carries water at uniform flow depth. Compare this slope with a rectangular channel with the same surface material, the same width, and the same cross-sectional area carrying the same flow rate at a uniform flow depth.

13.6   A rectangular channel is to carry 1.2 m³/s at a slope of 0.009. If the channel is lined with galvanized iron, $n = 0.011$, what is the minimum number of square meters of metal needed for each 100 m of channel? Neglect freeboard.

13.7   A trapezoidal channel, with side slope 2:1, is to carry 20 m³/s with a bottom slope of 0.0009. Determine the bottom width, depth, and velocity for the best hydraulic section. $n = 0.025$.

13.8   A trapezoidal channel made out of brick, with bottom width 6 ft and with bottom slope 0.001, is to carry 600 cfs. What should be the side slopes and depth of channel for the least number of bricks?

13.9   What radius semicircular corrugated-metal channel is needed to convey 2.5 m³/s a distance of 1 km with a head loss of 2 m? Can you find another cross section that requires less perimeter?

13.10   Determine the best hydraulic trapezoidal section to convey 85 m³/s with a bottom slope of 0.002. The lining is finished concrete.

13.11   Calculate the discharge through the channel and floodway of Fig. 13.23 for steady uniform flow, with $S = 0.0009$ and $y = 8$ ft.

13.12   For 200-m³/s flow in the section of Fig. 13.23 when the depth over the floodway is 1.2 m, calculate the energy gradient.

13.13   For 25,000-cfs flow through the section of Fig. 13.23, find the depth of flow in the floodway when the slope of the energy grade line is 0.0004.

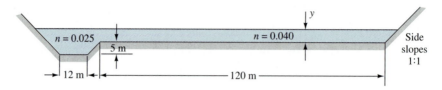

Figure 13.23   Problems 13.11 to 13.13.

13.14   Draw an $F + M$ curve for 2.5 m³/s per meter of width.

13.15   Draw the specific-energy curve for 2.5 m³/s per meter of width on the same chart as Prob. 13.14. What is the energy loss in a jump whose upstream depth is 0.5 m?

13.16   Prepare a plot of Eq. (13.4.7).

13.17   With $q = 100$ cfs/ft and $\mathbf{F}_1 = 3.5$, determine V_1, y_1, and the conjugate depth y_2.

13.18   Determine the two depths having a specific energy of 2 m for 1 m³/s per meter of width.

13.19   A flow rate of 15.0-m³/s water exists at a depth of 1.4 m in a rectangular channel 5.6 m wide. Find the critical depth, specific energy, Froude number, $F + M$, and conjugate depth.

13.20   In Prob. 13.19 if $n = 0.013$, what slope would be required to maintain uniform flow?

13.21   What is the critical depth for flow of 1.5 m³/s per meter of width?

13.22   What is the critical depth for flow of 0.3 m³/s through a triangular channel with a 60°-apex angle?

13.23   Determine the critical depth for flow of 8.5 m³/s through a trapezoidal channel with a bottom width of 2.5 m and side slopes of 1:1.

13.24   Find critical depth for a flow of 0.4 m³/s in a trapezoidal channel with a bottom width of 1.5 m and side slopes at 45°.

13.25   Water flows at uniform flow depth in a wide unfinished concrete channel ($n = 0.014$) of slope 0.0002. A 9-cm bump across the channel creates a slight depression in the water surface, Fig. 13.24. If the depth over the bump is 0.5 m, calculate the flow rate per meter of width and the velocity over the bump. Neglect friction losses at the bump.

13.26   In Prob. 13.25 find the height of the bump to create critical depth over the bump.

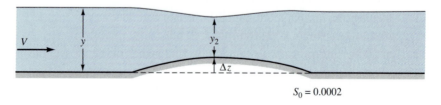

Figure 13.24   Problems 13.25 and 13.26.

13.27 The average water velocity is 2.4 m/s in a rectangular channel when the depth is 1.8 m. What is the change in depth (*a*) for a smooth drop of 15 cm in the channel bottom, and (*b*) for a smooth rise of 15 cm in the channel bottom?

13.28 Uniform flow of water at 1.5 m³/s occurs in a rectangular channel 3 m wide. The bottom slope is 0.00011 and $n = 0.016$. For a width contraction to 1.6 m what would the depth be in the contraction and just upstream of the contraction? Neglect local losses.

13.29 In Prob. 13.28, if the width contraction is to 0.8 m find the same two depths.

13.30 Design a transition from a trapezoidal section, 8 ft bottom width, side slopes 1:1, and depth 4 ft, to a rectangular section, 6 ft wide and 6 ft deep, for a flow of 250 cfs. The transition is to be 20 ft long, and the loss is one-tenth the difference between velocity heads. Show the bottom profile, and do not make any sudden changes in cross-sectional area.

13.31 A transition from a rectangular channel, 2.6 m wide and 2 m deep, to a trapezoidal channel, bottom width 4 m, side slopes 2:1, and depth 1.3 m has a loss four-tenths the difference between velocity heads. The discharge is 5.6 m³/s. Determine the difference between elevations of the channel bottoms.

13.32 A critical-depth meter 16 ft wide has a rise in bottom of 2.0 ft. For an upstream depth of 3.52 ft determine the flow through the meter.

13.33 With flow approaching a critical-depth meter site at 6 m/s and a Froude number of 3, what is the minimum amount the floor must be raised?

13.34 An unfinished concrete rectangular channel 12 ft wide has a slope of 0.0009. It carries 480 cfs and has a depth of 7 ft at one section. By using the step method and taking one step only, compute the depth 1000 ft downstream.

13.35 Solve Prob. 13.34 by taking two equal steps. What is the classification of this water-surface profile?

13.36 A very wide gate (Fig. 13.25) admits water to a horizontal channel. Considering the pressure distribution hydrostatic at section 0, compute the depth at section 0 and the discharge per meter of width when $y = 1.0$ m.

13.37 If the depth at section 0 of Fig. 13.25 is 600 mm and the discharge per meter of width is 6 m²/s, compute the water-surface curve downstream from the gate.

13.38 Draw the curve of conjugate depths for the surface profile of Prob. 13.37.

13.39 If the very wide channel in Fig. 13.25 extends downstream 700 m and then has a sudden dropoff, compute the flow profile upstream from the end of the channel for $q = 6$ m²/s by integrating the equation for gradually varied flow.

13.40 Using the results of Probs. 13.38 and 13.39, determine the position of a hydraulic jump in the channel.

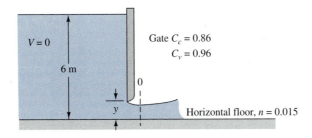

Figure 13.25 Problems 13.36 to 13.40.

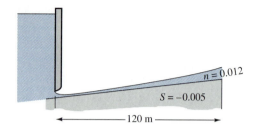

Figure 13.26   Problem 13.41.

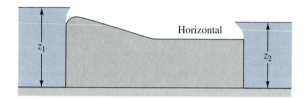

Figure 13.27   Problems 13.42 to 13.44.

13.41   (*a*) In Fig. 13.26 the depth downstream from the gate is 0.6 m and the velocity is 12 m/s. For a very wide channel, compute the depth at the downstream end of the adverse slope. (*b*) Solve part (*a*) by use of a spreadsheet like Fig. 13.16.

13.42   Sketch (without computation) and label all the liquid-surface profiles that can be obtained from Fig. 13.27 by varying z_1, z_2, and the lengths of the channels for $z_2 < z_1$, with a steep, inclined channel.

13.43   In Fig. 13.27 determine the possible combination of control sections for various values of z_1, z_2, and various channel lengths for $z_1 > z_2$, with the inclined channel always steep.

13.44   Sketch the various liquid-surface profiles and control sections for Fig. 13.27 obtained by varying channel length for $z_2 > z_1$.

13.45   Show an example of a channel that is mild for one discharge and steep for another discharge. What discharge is required for it to be critical?

13.46   Use a spreadsheet like Fig. 13.16, or prepare a program, to locate the hydraulic jump in a 90° triangular channel, 0.5 km long, that carries a flow of 1 m³/s, with $n = 0.015$ and $S_0 = 0.001$. The upstream depth is 0.2 m, and the downstream depth is 0.8 m.

13.47   Figure 13.28 shows a profile of a rectangular channel, 4.5 m wide, with a change in slope. The downstream channel has a slope of 0.0011, $n = 0.018$, and $Q = 20$ m³/s. (*a*) Find the depth before the jump for the hydraulic jump to end at uniform flow conditions. (*b*) Calculate the distance to the jump if the uniform flow depth in the upstream channel is 0.62 m. (*c*) Calculate and draw the water surface profile and energy grade line.

13.48   A weir produces a depth of 3.2 m in a horizontal rectangular channel 5 m wide, Fig. 13.29. At 180-m upstream the slope changes to 0.011. Calculate the depth 80-m upstream of the slope change if the flow is 15 m³/s; $n = 0.029$.

13.49   A rectangular channel is discharging 50 cfs per foot of width at a depth of 10 ft when the discharge upstream is suddenly increased to 70 cfs/ft. Determine the speed and height of the surge wave.

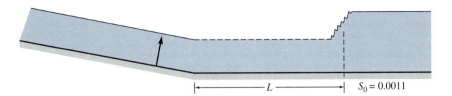

Figure 13.28   Problem 13.47.

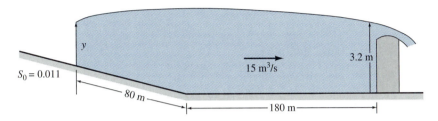

Figure 13.29   Problem 13.48.

13.50   In a rectangular channel with velocity 2 m/s flowing at a depth of 2 m, a surge wave 0.3 m high travels upstream. What is the speed of the wave, and how much is the discharge reduced per meter of width?

13.51   A rectangular channel 3 m wide and 2 m deep discharges 28 m³/s when the flow is completely stopped downstream by closure of a gate. Compute the height and speed of the resulting positive surge wave.

13.52   Determine the depth downstream from the gate of Prob. 13.51 after it closes.

13.53   Find the downstream water surface of Prob. 13.51 at 3 s after closure.

13.54   Determine the water surface 2 s after an ideal dam breaks. The original depth is 30 m.

13.55   Prepare a spreadsheet or write a program to design a transition from a rectangular or trapezoidal channel to a trapezoidal channel. The rate of change of area, bottom width, and side slope are to be zero at each end. Prepare results in tabular form for each one-tenth of the distance, and then solve Prob. 13.31 with the program.

REFERENCES

1.   U.S. Bureau of Reclamation, "Research Study on Settling Basins, Energy Dissipators, and Associated Appurtenances, Progress Report II," *U.S. Bur. Reclam. Hydraul. Lab. Rep.,* Hyd-399, Denver, June 1, 1955.

2.   B. A. Bakhmeteff, *O Neravnomernom Dvizhenii Zhidkosti v Otkrytum Rusle* (Varied Flow in Open Channel), St. Petersburg, Russia, 1912.

3.   A. T. Ippen, "Channel Transitions and Controls," in H. Rouse, (ed.), *Engineering Hydraulics,* Wiley, New York, 1950.

4.   E. F. Brater, H. W. King, J. E. Lindell, and C. Y. Wei, *Handbook of Hydraulics,* 7th ed., pp. 12.22–12.25, McGraw-Hill, New York, 1996.

ADDITIONAL READING

Bakhmeteff, B. A.: *Hydraulics of Open Channels,* McGraw-Hill, New York, 1932.

Chow, V. T.: *Open-Channel Hydraulics,* McGraw-Hill, New York, 1959.

French, R. H., *Open Channel Hydraulics,* McGraw-Hill, New York, 1985.

Henderson, F. M.: *Open Channel Flow,* Macmillan, New York, 1966.

14

Applications of Transport Phenomena

The objective of this chapter is to elaborate on the basic principles of transport phenomena by describing various application areas. The applications selected here represent a mix of both engineered and natural flow-field problems and concentrate on elements of multiphase and interfacial transport, process reactors, and stirring and mixing.

14.1   ENGINEERED VERSUS GEOENVIRONMENTAL TRANSPORT

It is important to reemphasize the difference between (1) engineered flow and transport fields that are designed to elicit a specific behavior, and (2) naturally occurring flow and transport which result from a variety of random processes including, for example, weather, and the associated wind and rain. The methods developed in this text can be applied to both classes of problems but the application viewpoint is quite different for each. Transport resulting from and constrained by natural processes shall be called *geoenvironmental transport* while *engineered transport* will refer to processes designed to have predictable outcomes.

A variety of disciplines exist within which geoenvironmental flow and transport are studied. These include *oceanography, meteorology,* and *hydrogeology.* Even more esoteric fields such as *volcanology* (the study of volcanoes), *plate tectonics* (earth crustal motion), and *astrophysics* (the motion and physics of the stars, etc.) use many of the fundamental laws of fluid mechanics and transport phenomena that are applied to the geoenvironmental class of problems.

While perhaps an oversimplification, one of the primary differences between the geoenvironmental and engineered flows is that the natural flow fields are not yet fully understood. Therefore, the techniques presented in this text are used to understand and describe the various types of naturally occurring flow fields. The scientific elaboration and understanding of these flows must be completed. The chief obstacle to scientific understanding is the complexity of the flow fields. There are both external and internal sources of natural flow complexity. The external sources are the highly variable flow geometry and the highly variable imposed boundary conditions or couplings.

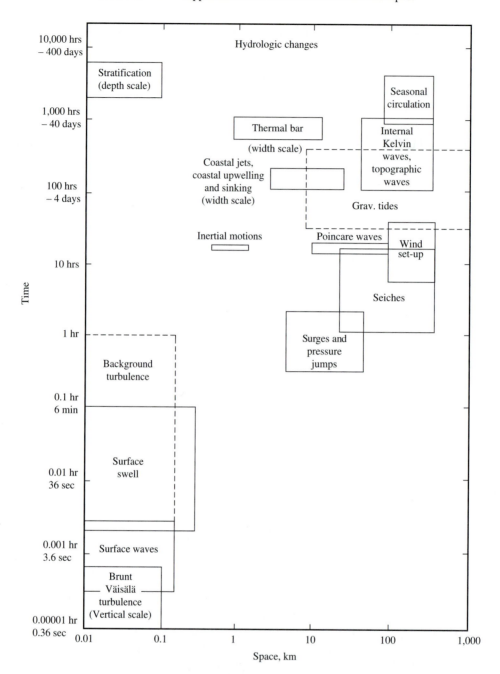

Figure 14.1   Space and time scales in naturally occurring flows, in this case a very large lake.

Characterization of the internal flow-field complexity can first be approached by simply defining a control volume and listing all the fluid mechanics and transport processes in the flow field and their associated length and time scales. Figure 14.1 contains a simple space-time diagram of the processes. It can readily be seen that there are a wide variety of processes each possibly occurring simultaneously and each covering a wide range of length and time scales. The occurrence of these internal processes directly results from the coupling with other complex external systems through boundary loadings (and their mathematical counterpart boundary conditions). For instance, a lake is set into motion by wind shear on the surface resulting from random weather systems, heat flux from the sun which is mitigated by weather-derived clouds, and tributary input and outflow. All these boundary loadings are highly variable in that large gradients in time or space occur, occasionally resulting in extreme events such as storms. The maximum number of observed internal transport processes occurs when the boundary loadings are the most extreme and intense. The fewest internal processes occur between events.

Another external complexity, flow geometry, further confuses the natural flow and transport pattern. The geometry of natural flow fields is also quite variable, ranging from small-scale roughness variability all the way to large-scale changes, for example, river flow in meandering channels or atmospheric flow over mountains. While a generality, the flow of the earth's atmosphere over the ground is the least controlled by the irregularity of the geometry while the flow and transport of solutes in the groundwater system are completely and totally dictated by the geometry. In analogy to the boundary loading impact, it can also be said that the more rapid and extreme the geometry changes become, the more complex the suite of resulting flow and transport processes becomes.

The analytical approach for natural flows is elusive as traditional methods of making a variety of simplifying assumptions and creating an exact or *deterministic* solution is almost impossible. As noted earlier such exact solutions are possible only for reasonably well-controlled conditions with few processes. In lieu of deterministic analyses *computer modeling* or *simulation* of the natural flow and transport is the preferred approach. Here the turbulent Navier-Stokes [Eqs. (6.4.6a-c)] and the corresponding heat [Eq. (9.4.7a)] and mass transport [Eq. (9.4.7b)] equations are approximated with *finite difference* or *finite element* [1]† techniques to achieve a computer solution for the appropriate boundary loadings and geometry. The resulting computed variables reproduce or simulate the processes in the flow field for the imposed conditions.

In direct contrast the *engineered flow field* is one motivated by a desire on the part of the engineer to have almost complete control over the flow and transport process. The minimum control requirements are complete and predictable behavior of the boundary loadings, geometry, and internal flow and transport response. With such dependable conditions the designed flow transport fields can achieve desired outcomes which can be incorporated in hydraulic, heat exchange, or process designs. Such designs can then be scaled in capacity and boundary loadings for use in any number of similar situations. These features are the heart of the chemical engineering profession where *unit processes* are designed and scaled for industrial use. Each process typically depends upon one dominant predictable transport process to achieve the desired result; examples include drying by evaporation, heat exchangers, fermentation, settling ponds and separators, and filtration. The concept of engineered flow and transport or process design has now been extended to the creation of equipment for environmental control. Examples include settling tanks and cooling ponds, filtration and flocculation units, incinerators, and scrubbers. Whether municipal or industrial in application the complete process is typically achieved by connecting a series of functionally different unit processes together to achieve the desired outcome.

Engineered transport systems are consciously designed to be stable and dependable. The boundary loadings are typically in steady state with their constituents and concentrations known for the lifetime of the process, and the geometry

†Numbered references are found at the end of this chapter.

of the flow field is kept quite simple, that is, circular pipes, flat plates, spheres, or cylinders. As seen in the earlier portion of this text a wide variety of exact solutions for distributions, flux rates, and phase changes of the design variables are known for these geometries. Process designs are therefore based upon the geometry and boundary conditions necessary to elicit these exact solution behaviors. This is quite unlike the natural flow case where the observed outcome is rarely if ever known in advance.

What follows are a series of applications which attempt to extend some of the transport concepts presented earlier, as well as to introduce the reader to the use of transport phenomena concepts to solve problems of practical importance. Several applications at the beginning of the chapter are based upon multiphase flow and interfacial transfer. These include particle erosion and transport at a bed-particle interface, and evaporation at an air-water interface. Both processes can be described with boundary layer solutions and both are mixtures with the density of the mixture causing *stratification* effects in the flow field. The rest of the examples concentrate on tanks and reactors. All applications have counterparts in natural and engineered flows.

14.2   MULTIPHASE FLOWS: PARTICLE TRANSPORT

Multiphase flow and transport involve predicting the distribution of one phase of a material being transported by another phase of the same or different material. Examples include the transport of particles or *aerosols* by water or air, or the transport of air bubbles in liquids. Often interface exchange between two media can lead to the origin of the material in the multiphase transport. Two of the most ubiquitous examples include (1) the *erosion* and subsequent *entrainment* of sediment particles off the bottom of a river, lake, or channel, and (2) the formation of atmospheric dust storms near desert areas. In either example the bottom or ground is comprised of particles which are highly consolidated and, therefore, in most respects behave as a solid surface. The flow of the fluid over the bottom creates sufficient shear stress on the surface particles to dislodge them, and the turbulent stresses in the overlying boundary layer carry the particles away from the bottom. The particles are not neutrally buoyant. Therefore, contrasted with the backdrop of turbulent transport away from the bottom is the constant settling of the particles to the bottom.

Being able to parameterize the exchange, transport, and flux intensities especially near the ground bottom or solid surface is of extreme pragmatic importance. Dredging, foundation stability, and construction activities are all affected by erosion. Recent environmental attention on interfacial particle exchange results from the presence of toxic substances in the sediments to be dredged from the harbors. Similar problems can be identified from airborne particulates. For example, recent scientific articles indicate that erosion from desert areas creates such an intense aerosol cloud that global warming climate estimates are in error if they are not accounted for.

The parameters of interest in this multiphase problem include: (1) the distribution of particle concentration with height above bottom, (2) the settling, entrainment, and net vertical flux of particles from the solid phase or ground, (3) the mass, weight, or thickness of material eroded from the solid phase in a unit of time, and (4) the horizontal mass flux of material at the site. A wide variety of boundary layer solution procedures have been used to estimate the magnitude of these variables, and in this section one of the more enduringly useful models will be presented.

Mixture Forms

The combined water column-bottom (or air-ground) site can be thought of as a continuum distinguished only by the relative intensity of particles per unit volume compared to the transporting fluid volume. A detailed description of each state can be found in Refs. [2–7]. Included in these references are extensive discussions ranging from the most basic procedures for formulating continuum descriptions of the motion of discrete particles to approximate engineering solutions for a number of industrial or natural flows. Before proceeding to the equations for the problem solution, the range of possible particle-fluid systems is first summarized. Throughout this entire section the assumption is made that the particles are *cohesionless* as opposed to *cohesive*. The particle surface is assumed to be electrochemically neutral and unable therefore to "stick" to neighboring particles through any agent other than gravity. Clay particles are cohesive but sand and silt particles are not. Therefore the discussion is limited to these and other nonreactive larger particles.

Consolidated Bed   With reference to Fig. 14.2*a*, the consolidated state is the configuration where all the particles are as tightly packed together as their individual geometry and gravity permit. The weight of the individual particles plus the integrated weight of the material above a particular bed elevation are sufficient to overcome whatever internal fluid pressure exists. *Interstitial* spaces exist between the particles, and fluid can be transmitted through these potentially irregular paths. The *porosity* of the packed bed is the ratio of the void volume to the total bed volume. The tortuous nature of the path a fluid parcel must take to travel a straight line distance between two points will be longer (in some cases considerably so) than the straight line distance. *Tortuosity,* a measure of the irregularity, can be defined for a packed bed but is the province of geotechnical engineering and is beyond the scope of this text. In general the particles are tightly packed at the interface and a shear or normal force in excess of the particle weight must be applied to dislodge the particle.

Fluidized Bed   Whether through imposition of a designed pressure gradient such as for a commercial reactor or through the occurrence of a naturally occurring pressure gradient, the internal pore pressure can become sufficient to overcome the particle and burden weight (Fig. 14.2*b*). In this case the particles may be quite near if not close to touching their neighboring particles, but the interparticle shear and normal forces offer quite reduced resistance to the motion of the particles either singly or in a bulk fashion. A hallmark of the fluidization process is the transition to a particulate composition where each individual particle is possibly subject to and responds to an external flow. Due to the particle separation an expansion of the fluidized bed occurs in contrast to the consolidated or packed state. Fluidized beds are designed to achieve specified heat and mass transfer and/or chemical reactions in an industrial or commercial process. The excellent text by Gidaspow [2] expands on these concepts.

Mobile Fluidized Beds   This category refers to the circumstances when the whole mass of the fluidized bed particles is moving in response to an imposed pressure gradient and shear as in Fig. 14.2*c*. Each particle can move randomly relative to this *bulk* motion. Unlike the previous two states the subsequent bulk motion can be described by derivation of fluid mechanics–based continuum equations which are, however, decidedly not based upon Newtonian viscosity representations. The solids volume to total volume ratio or volume concentration of the mixture is quite high, and therefore, the flows are termed *hyperconcentrations,* or *slurries.* Because the particulates are suspended in the flow and have individual flow and force fields

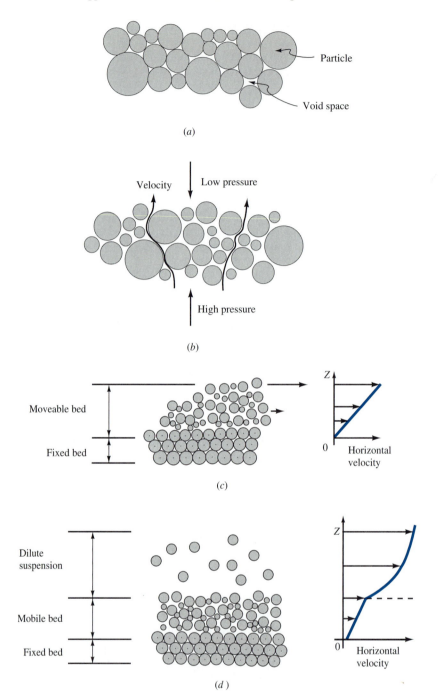

Figure 14.2   Combined mixture forms: (*a*) Consolidated bed, no particle motion. (*b*) Fluidized bed, no particle motion. (*c*) Mobile fluidized bed. (*d*) Dilute suspension.

acting upon them, these flows are also called *suspensions* or in this case hyperconcentrated suspensions.

As noted in Sec. 8.7 water waves are the deformation of a density interface between two moving fluids, air and water. In surface waters (or atmospheric flows over the ground) both the water (or air) and the mobile fluidized sediments are

Figure 14.3   Desert sand dune field.

considered fluids. This would also represent a very large density interface, and waves would be expected to form in the moving bed layer. In the atmospheric case sand dunes are "sand waves" that have been set into motion by the wind shear (see Fig. 14.3). In surface waters, particularly in swiftly flowing streams, sand waves are formed at the interface (Fig. 14.4). The moving bed sediments are referred to as the *bedload* and many analyses have been presented to predict the mass transport rate of the bed load sediments (see Refs. [8,9]).

This layer is a particularly important layer as it represents the transition between a bed description described from the Lagrangian (solid) and Eulerian (fluid) viewpoints (Sec. 3.2). It is a challenging region to analyze as the continuum equations are not yet fully accepted into practical use from the science community.

Dilute Suspensions   This state occurs when the interparticle distances are sufficiently large that the full flow field over each particle or sphere is allowed to develop as in Sec. 7.3 (Fig. 14.2d). If the particle is allowed to move in an unhindered fashion such as this, then the particle velocity differs from the fluid transport velocity only by the fact that the particle is settling relative to the fluid velocity in a direction parallel to gravity [10]. Otherwise the horizontal velocity of the particles is identical to the fluid velocity. A number of simplifications accrue to the analysis of dilute suspensions, and fortunately a wide variety of engineered and natural

Figure 14.4   River channel sand waves.

flows can be so analyzed. Our analysis therefore starts with a dilute suspension boundary layer approach.

Equation Formulation

With regard to Fig. 14.5a a coordinate system is placed at the solid-fluid interface, with the z axis directed positive away from the bottom. In order to achieve a two-dimensional (x,z) boundary layer solution, a streamline following coordinate system is used where the horizontal velocity in the equations is in fact the vector sum of the two horizontal velocities from the field or experimental situation (Fig. 14.5b). In the field, or natural situation, this can be especially troublesome as the direction of the total horizontal velocity vector ($u_H^2 = u^2 + v^2$) will change over time.

The basic equations of particle transport have evolved over time from a single species passive scalar formulation to the more robust mixture theory approach. The earliest use of mixture theory was by Hunt [11]. By using a mixture formulation, the equation validity can be extended to quite high concentrations, thereby encompassing the expected range of concentrations down to the bed. For purposes of this discussion the bed will be a source of material for the suspension but will not move. Therefore, the effect of the bed will be imposed (primitively) as a boundary condition.

As noted in Lumley [10] two formulation possibilities exist: the first requires formulation of a conservation equation for each major grain size class in the particle mixture, and the second forms one mass balance equation for the mixture average concentration and corresponding settling velocity at each point in the domain. In the former each grain size class will have a constant settling velocity. In the latter the mixture average settling velocity will vary from point to point and over time in the domain as the grain size distribution is variable. While the latter only solves "one" transport equation, the number of additional equations required to specify the mixture settling velocity will essentially result in both formulations requiring the same equation solution effort.

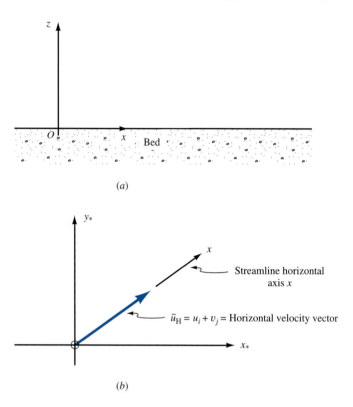

Figure 14.5 Coordinate system at the particle-fluid interface in the (a) vertical plane and (b) horizontal plane (perspective).

Density Equation If c is the mixture *volume-fraction particle concentration,* the mass density of the mixture is given by

$$\rho = \rho_w(1 - c) + \rho_s c \qquad (14.2.1)$$

Here ρ_w and ρ_s are the mass densities $[M/L^3]$ of the water and particles, respectively. The water, sediment, and total mixture mass balances are given by the following equations.

Mass Balance Equations For the *water phase*

$$\frac{\partial}{\partial t}\left[\rho_w(1 - c)\right] + \nabla \cdot \left[\rho_w \mathbf{v}_w(1 - c)\right] = 0 \qquad (14.2.2)$$

For the *particle phase*

$$\frac{\partial}{\partial t}(\rho_s c) + \nabla \cdot (\rho_s \mathbf{v}_s c) = \nabla \cdot \left(\mathcal{D}_{sw}\nabla(\rho_s c)\right) \qquad (14.2.3)$$

For the *total mixture* (see Eq. 14.2.1)

$$\frac{\partial \rho}{\partial t} + \nabla \cdot (\rho \mathbf{v}) = 0 \qquad (14.2.4)$$

In these equations $\mathbf{v}_w$, $\mathbf{v}_s$, and $\mathbf{v}$ are the water phase, particle phase, and mixture velocity vectors, respectively. For high concentrations or transport with large particles it is not a forgone conclusion that these velocity vectors are equal. The sum of Eqs. (14.2.2) and (14.2.3) is Eq. (14.2.4).

Mixture Momentum Field Equation   Given the validity of the Newton viscosity relationship, the mixture momentum balance equation is [Eqs. (4.4.8) and (4.4.12)]

$$\frac{\partial}{\partial t}(\rho\mathbf{v}) + \nabla \cdot (\rho\mathbf{v}\mathbf{v}) = -\rho g\nabla h + \left(\nabla \cdot \overline{\overline{\tau^*}}\right) \tag{14.2.5}$$

Dilute Suspension Equations

If the volume fraction of particles is low ($\sim c < 0.001$), then $1 - c \approx 1.0$ and Eq. (14.2.4) becomes

$$\frac{\partial \rho_w}{\partial t} + \nabla \cdot (\rho_w\mathbf{v}_w) = \frac{\partial \rho}{\partial t} + \nabla \cdot (\rho\mathbf{v}_w) = 0 \tag{14.2.6}$$

and Eq. (14.2.3) remains as does Eq. (14.2.5).

Particle Velocity Field and Volume Concentration Form

For the indicated coordinate system the particle and water phase velocities, $\mathbf{v}_s$ and $\mathbf{v}_w$, respectively, are unified by assuming that the horizontal components are identical, that is, $u_w = u_s = u$ and $v_w = v_s = v$. Only the vertical velocities differ as the particle will settle with velocity w_t [Eq. (7.3.7)] relative to the fluid velocity. Therefore, w for the dilute form of the equations [Eq. (14.2.6)] becomes

$$w_s = w = w_w - w_t \tag{14.2.7}$$

The particle mass conservation equation [Eq. (14.2.3)] becomes

$$\frac{\partial}{\partial t}(\rho_s c) + \nabla \cdot (\rho_s\mathbf{v}c) - \frac{\partial}{\partial z}(\rho_s w_t c) = \nabla \cdot \left[\mathcal{D}_{sw}\nabla(\rho_s c)\right] \tag{14.2.8}$$

and the momentum equation remains as Eq. (14.2.5). It should be noted that for local coordinate systems (x^*, z^*) (such as river channels) with very mild slopes, S_o, w_s will decompose into a component in each local coordinate direction, that is, $w_{tx^*} = w_t S_o$ and $w_{tz^*} = w_t(1 - S_o)$. For very weak slopes approaching zero, $S_o \sim 0$ and the local component in the x direction is negligible, and $w_{tz^*} \approx w_t$. This assumption is used in most river modeling activities to date. Steep sloped channels in mountain watersheds are not so conveniently analyzed.

Mass Concentration Form

As discussed in Chaps. 1, 3, and 4, mass concentration is often used in the governing equation. This is particularly so as sediment measurement instruments yield data in mass concentration form. Hence $C = \rho_s c$, and Eq. (14.2.8) becomes

$$\frac{\partial C}{\partial t} + \nabla \cdot (\mathbf{v}C) - \frac{\partial}{\partial z}(w_t C) = \nabla \cdot (\mathcal{D}_{sw}\nabla C) \tag{14.2.9}$$

Turbulent Form

Equation (14.2.9) as well as the continuity and momentum equations can be Reynolds averaged, and a turbulent form can be derived. Equation (14.2.9) becomes

$$\frac{\partial \overline{C}}{\partial t} + \nabla \cdot (\overline{\mathbf{v}}\overline{C}) - \frac{\partial}{\partial z}(w_t\overline{C}) = \nabla \cdot \left[(\mathcal{D}_{sw} + E_s)\nabla\overline{C}\right] \tag{14.2.10}$$

where E_s is the eddy diffusivity defined from

$$\overline{u'C'} = -E_{sx}\frac{\partial\overline{C}}{\partial x} \qquad \overline{v'C'} = -E_{sy}\frac{\partial\overline{C}}{\partial y} \qquad \overline{w'C'} = -E_{sz}\frac{\partial\overline{C}}{\partial z} \qquad \textbf{(14.2.11)}$$

Flux Forms

Equations (14.2.10) and (14.2.11) can be written in flux form as

$$\frac{\partial\overline{C}}{\partial t} + \nabla \cdot \overline{N} = 0 \qquad \textbf{(14.2.12)}$$

in which

$$\overline{N}_x = \overline{u}\overline{C} + \overline{u'C'} - \mathcal{D}_{sw}\frac{\partial\overline{C}}{\partial x} = \overline{u}\overline{C} - (\mathcal{D}_{sw} + E_{sx})\frac{\partial\overline{C}}{\partial x} \qquad \textbf{(14.2.13a)}$$

and in the vertical

$$\overline{N}_z = \overline{w}\overline{C} + \overline{w'C'} - w_t\overline{C} - \mathcal{D}_{sw}\frac{\partial\overline{C}}{\partial z} = (\overline{w} - w_t)\overline{C} - (\mathcal{D}_{sw} + E_{sz})\frac{\partial\overline{C}}{\partial z} \qquad \textbf{(14.2.13b)}$$

The component fluxes are designated as follows: $\overline{w}\overline{C}$ is the (vertical) advective flux, $w_t\overline{C}$ is the settling flux, $\overline{w'C'}$ is the (vertical) turbulent flux, and $\mathcal{D}_{sw}\partial\overline{C}/\partial z$ is the molecular diffusion flux. At the bottom $N_z(z = 0) = N_{zo}$ and the advective flux is zero since $\overline{w} = 0$ on the bottom, that is,

$$N_z(z = 0) = N_{zo} = -w_t\overline{C} - \mathcal{D}_{sw}\frac{\partial\overline{C}}{\partial z} + E_o \qquad \textbf{(14.2.14)}$$

At the bottom the previously defined fluxes take on new names due to the interfacial exchange. Here $-w_t\overline{C}$ is called the *deposition flux* indicating a loss of material completely out of the domain, while E_o is called the *entrainment flux* indicating a transfer of new material across the interface into the domain from the bed as resulting from shear stress.

When $N_{zo} > 0$, there is a net input of mass into the fluid domain from the bed and (ignoring Fickian diffusion) entrainment is greater than settling; $N_{zo} > 0$ is therefore termed *erosion* or *resuspension*. The flux is in *equilibrium* if $N_{zo} = 0$ and $w_t\overline{C} = E_o$. The flux is termed *depositional* if $N_{zo} < 0$, that is, settling is greater than entrainment. If the flux is erosive, there would be a loss of bed material over time such that for an area, A, the bed height, h_B, lost during time period T would equal

$$\rho_s h_B A = A\int_t^{t+T} N_{zo}\, dt \qquad \Rightarrow \qquad h_B = \frac{1}{\rho_s}\int_t^{t+T} N_{zo}\, dt \qquad \textbf{(14.2.15)}$$

A similar integration during deposition conditions would give the increase in bed height due to net deposition.

Profiles from a Current with Uniform Turbulence

The earliest solution to Eq. (14.2.10) was found by assuming that the eddy diffusivity is constant and the vertical flux is in equilibrium, that is,

$$w_t\overline{C} + E_{sz}\frac{\partial\overline{C}}{\partial z} = 0 \qquad \textbf{(14.2.16)}$$

The integration of Eq. (14.2.16) reveals that

$$\frac{C}{C_r} = \exp\left\{-\frac{w_t}{E_{sz}}(z - z_r)\right\}$$ (14.2.17)

Here C_r is the reference particle concentration which must be known at the reference height $z = z_r$. The resulting form of the concentration profile is a simple exponential decay with distance from the bottom. Even though the turbulence field is overly simplified for boundary layers, the basic profile shape resulting from the relative strengths of settling versus turbulent flux is clear. Figure 14.6 is a schematic of the profile for various values of the nondimensional flux strength ratio ($w_t d/E_{sz}$), a form of the Reynolds number. Here it is noticed that as the nondimensional argument increases settling dominates and that very high concentrations are confined to the near bottom region. Turbulent flux is not strong enough to mix the particles throughout the fluid column. As mixing dominates, $w_t d/E_{sz}$ becomes small and particles are mixed throughout the column, leaving a more uniform profile.

Profile from a Current-Driven Boundary Layer

As discussed in Chap. 6, the turbulence field near a wall is not uniformly distributed with distance away from the wall as assumed above. Mixing length formulations are therefore used to more accurately calculate the profile. One of the more enduring of these boundary layer solutions is due to Smith [12] and is based upon a shear stress distribution originating from the steady flow of a current over a rough bottom. The particle mass balance equation is written for each grain size class, n. Therefore, in *volume concentration* form $\sum c_n = c_s = c$ which is the total volume concentration of the particles. The turbulent eddy diffusivity is independent of grain size but very

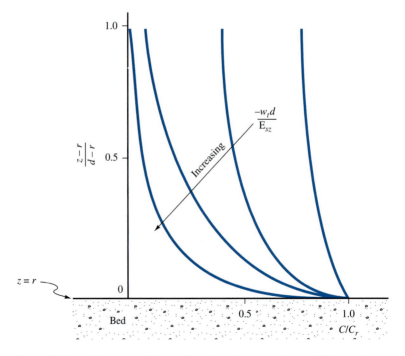

Figure 14.6 Schematic of the nondimensional concentration profile for various values of the nondimensional flux strength ratio.

much varies as a function of z. Smith and MacLean [13] assume a finite water column depth, d. In this case the authors used the data of Klebanoff [14] and Townsend [15] to derive a universal function for the variation of the eddy viscosity in a channel flow. The data, presented in the form of a *similarity function* $f(\xi)$, are of the form

$$\frac{\eta(\xi)}{ku_*d} \approx \frac{E_{sz}}{ku_*d} = \frac{E_{sz}}{E_{szo}} = f(\xi) \qquad (14.2.18)$$

where k is the von Kármán's coefficient ($k = 0.4$), $\xi = z/d$ is a nondimensional height above the bottom, and η is the eddy viscosity. As is typically assumed, $\eta \approx E_{sz}$, and $E_{szo} = ku_*d$ is a reference eddy diffusivity based on the overall parameters of the flow field. Two functions for $f(\xi)$ are identified

For $0 \leq \xi \leq 0.3$,

$$f(\xi) = \left(\xi + 1.32892\xi^2 - 16.86321\xi^3 + 25.22663\xi^4\right) \qquad (14.2.19a)$$

For $0.3 \leq \xi \leq 1$,

$$f(\xi) = \left(0.160552 + 0.075605\xi - 0.1305618\xi^2 - 0.1055945\xi^3\right) \qquad (14.2.19b)$$

Figure 14.7 contains a schematic of the profile from Eqs. (14.19a) and (14.19b). It is immediately seen that for $\xi < 0.2$ the eddy viscosity has a linear increase with height. This is a direct result of the presence of the constant stress boundary layer derived from Prandtl's mixing length hypothesis (see Sec. 6.4). Therefore, in this layer

$$\eta \approx E_{sz} = ku_*z = ku_*\xi d = E_{szo}\xi \qquad (14.2.20)$$

For a single or average grain size with constant settling velocity, w_t, the turbulent form of the particle mass balance equation [Eq. (14.2.8)] for steady horizontally uniform flow reduces to a statement that the vertical flux gradient is zero or that the vertical flux itself is constant. Constant can mean constant resuspension flux,

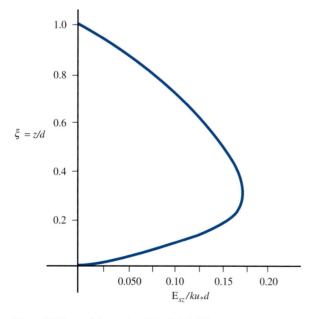

Figure 14.7   Schematic of Eq. (14.2.19).

constant depositional flux, or no flux, an equilibrium condition. For the equilibrium condition and noting that for one grain size $c_n = c_s = c$, the vertical flux is

$$w_t \, \overline{c} \left(1 - \overline{c}\right) + \mathrm{E}_{sz} \frac{\partial \overline{c}}{\partial z} = 0 \tag{14.2.21}$$

By substitution of the expression for E_{sz} [Eq. (14.2.21)] and noting that the validity ceases above $\xi > 0.2$, Eq. (14.2.21) can be integrated to give

$$\frac{\overline{c}(z)}{1 - \overline{c}(z)} = \frac{\overline{c}(z_r)}{1 - \overline{c}(z_r)} \left(\frac{z_r}{z}\right)^{(w_t/ku_*)} \tag{14.2.22}$$

It should be noted that as $z \to z_r$, then $c(z) \to c(z_r)$. The discussion of what z_r is will be pursued in an ensuing subsection as well as the determination of the associated reference concentration at $z = z_r$.

This profile solution is quite straightforward to program, operate, and plot and is so principally because the eddy viscosity-diffusivity variation is linear. Furthermore, the velocity and particle concentration profiles are independent of each other or *uncoupled* in that one does not need knowledge of the fluid velocity to calculate the concentration profile and vice versa. To calculate the profile above $\xi > 0.2$ or to calculate profiles where excessive near-bottom sediment concentration results in a change in the velocity field through density *stratification* requires a more complex numerical solution, which is detailed in the next subsection.

To finish this calculation requires specification of z_r, $c(z_r) = c_r$, and w_t. Therefore, these data are summarized.

Settling Velocity

In Eq. (7.3.7) a formula for the terminal settling velocity was derived from assuming steady flow and a particle Reynolds number ($\mathbf{R}_D$) less than one. A general formula for the settling velocity as a fraction of drag coefficient C_D is repeated here from Eq. (7.3.8) as

$$w_t^2 = \frac{4}{3} \frac{D}{\rho \, C_D} \left(\gamma_s - \gamma_f\right) \tag{14.2.23}$$

For the *Stokes* range $\mathbf{R}_D < 1$ and $C_D = 24/\mathbf{R}_D$. Many particle-laden flows have $\mathbf{R}_D > 1$ and the determination of C_D is complicated by the fact that C_D is a function of the Reynolds number. A considerable body of literature exists parameterizing just how C_D varies with Reynolds number and reviews can be found in Soo [5, 6]. H. Rouse (see Ref. [8], Fig. 2.1) first presented a compilation of laboratory data on this relationship which has become the de facto standard for spherical particles. Soo presented more comprehensive data along with Rouse's data in order to show the variability introduced by nonstandard conditions such as pipe flow, fluidization, and nonspherical particles. With current emphasis on computational forms, however, it is desirable to have a form amenable to computer calculation. A variety of incremental improvements on the Stokes form have been achieved by perturbation theory. However the most comprehensive form is found in Soo [6] which applies up to $\mathbf{R}_D \leq 100$ and is

$$C_D = \frac{24}{\mathbf{R}_D} \left[1 + 0.0975\mathbf{R}_D - 0.636(10^{-3})\mathbf{R}_D^2\right] \tag{14.2.24}$$

For $700 < \mathbf{R}_D < 2(10^5)$ the drag coefficient has a constant value of $C_D = 0.44$. The region between $100 < \mathbf{R}_D < 700$ is complicated by vortex shedding phenomena, and the reader must consult the data in the references for quite specific $\mathbf{R}_D$ values. Fortunately, most of the sediment and aerosol settling problems considered here fall under the region governed by Eq. (14.2.24).

The calculation for w_t proceeds by assuming a value for the Reynolds number, estimating C_D from Eq. (14.2.24), estimating w_t from Eq. (14.2.23), and checking to see if $\mathbf{R}_D$ with the value of w_t is equal to the assumed value of $\mathbf{R}_D$. Iteration proceeds until they are equal, within a specified tolerance.

Finally, it should be noted that there are several mitigating circumstances to this reasonably straightforward picture. The presence of vertical walls can inhibit settling through the introduction of shear-induced spin and transverse motion. The proximity of the bottom and the creation of high concentrations at the "bottom" can result in inhibited settling through increased collisions with particles. Also, irregularly shaped particles can rotate and tumble thereby reducing the net w_t. These factors are reviewed in Soo [6].

Reference Concentration and Critical Shear Stress

The boundary conditions for Eqs. (14.2.17) and (14.2.22) require specification of the reference concentration $[c(z = z_r) = c_r]$ at the reference height z_r. This seemingly simple piece of information is extremely difficult to determine because the value for c_r represents the integrated activity of the thin but complex bed load layer at the bottom. As mentioned earlier the most direct, though complex approach would be to write the full continuum equations for the suspended load, bed load, and consolidated load regions and solve for their subsequent deformation and motion. Such multiphase models are just now being attempted.

Before proceeding to a discussion of c_r and its point of application (z_r), it is necessary to reaffirm the results of Ex. 6.8 which depicted the condition for the *initiation of motion*. The particles sitting in the bottom are subjected to shear stress due to drag, a lifting force, and gravity. Therefore, the angle of repose of the particles in the bed, ϕ, and the channel slope, θ, are key variables as are particle diameter, D, and the specific weight of the fluid particle. Figure 14.8 depicts the situation for spherical

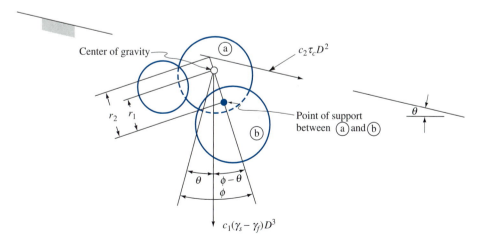

Figure 14.8   Force balance of a particle at rest.

particles. The weight force is given by $c_1(\gamma_s - \gamma_f)D^3$ where c_1 is a shape factor that equals $\pi/6$ for a sphere. When the particle is at the critical condition where the particle is just about to rotate out of its setting, the critical shear stress force to be exceeded is given by $c_2\tau_cD^2$, where τ_c is the critical shear stress and c_2 is a coefficient accounting for irregular particle surface area. If a moment balance is performed at critical conditions [8], then the critical shear stress that must be equaled or exceeded for motion to occur is

$$\tau_c = \frac{c_1 r_1}{c_2 r_2}\left(\gamma_s - \gamma_f\right) D \cos\theta \left(\tan\phi - \tan\theta\right) \tag{14.2.25}$$

which for a flat bed ($\theta = 0$) reduces to

$$\tau_c = \frac{c_1 r_1}{c_2 r_2}\left(\gamma_s - \gamma_f\right)D \tan\phi \tag{14.2.26}$$

Therefore, it is seen in Ex. 6.8 that

$$\frac{c_1 r_1}{c_2 r_2}\tan\phi = 0.040 \tag{14.2.27}$$

The original experimental data from White [16] shows this value to be at the low end of the observed range with the upper range being 0.2.

The criteria for the initiation of motion is that the fluid shear stress, τ_{f_b}, at the bed be equal to or greater than τ_c. One may see this in terms of the size of the particles eroded from the bed. Since $\tau_{f_b} = \tau_c = \rho u_*^2$, then the relation in Eq. (14.2.26) indicates that

$$u_*^2 \sim D_c$$

Since the volume of the particle and therefore its weight is proportional to D^3, then

$$u_{*_c}^6 \to D_c^3$$

That the particle volume eroded at the critical condition is proportional to the sixth power of the friction velocity (or the near-bed velocity) provides empirical support for the observation that geometrically increasingly severe erosion occurs for incremental increases in discharge or velocity.

The most extensive compilation of laboratory data on the values for τ_c was initially collected by A. Shields (see Ref. [8], Fig. 2.43) and has been added to or extended over the years. Figure 14.9 contains a plot of the modified Shields diagram from Glenn and Grant [17]. With the data for τ_c from the plot, S_τ, the normalized shear stress is defined as

$$S_\tau = \frac{\overline{\tau}_{f_b} - \overline{\tau}_c}{\overline{\tau}_c} \tag{14.2.28}$$

The overbar indicates the temporal average and its associated Reynolds average time period.

Once in suspension various particle motions may occur; principally the particles are ejected from the moving bed and carried away by the turbulence field in the *suspended load*. Particles may be ejected from the bottom but reach only a very small height before falling back to the bed. This motion is called *saltation* and this type of particle motion is the basic physical picture from which c_r and z_r are parameterized. Garcia and Parker [18] reviewed and compared a variety of empirical

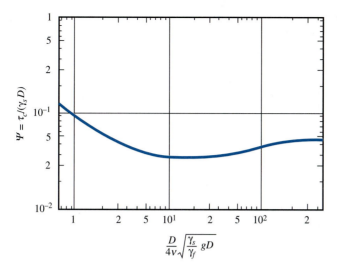

Figure 14.9   Modified Shields diagram (redrawn from Glenn and Grant [17]).

parameterizations for c_r, and the model due to Smith [12] appears to perform most satisfactorily.

The reference concentration c_r is estimated to be a fraction of the concentration of particles (c_b) in the bed layer

$$c_r = \frac{\gamma_o \, c_b \, S_\tau}{1 + \gamma_o S_\tau} \qquad \textbf{(14.2.29)}$$

where c_b is defined as $1 - p$, p is the porosity of the bed material, and γ_o is an empirical coefficient defined from laboratory experiments as $2.4(10^{-3})$. In practice $c_b \sim 0.65$ is used for quartz sand grains. Clearly if $S_\tau < 1$, then $c_r = 0$, while as $S_\tau \to \infty$, $c_r \to c_b \sim 0.65$.

The final question to resolve is the determination of the point of application of c_r, that is, what is z_r? Were the bed fixed with nonmoving sand grains (the limiting case for very slow flow), then the effective height of c_r would be the Nikuradse roughness height [ϵ', Eq. (6.4.22)]. However, the moving bed results in larger scale deformations of the bottom, particularly ripples, and z_r will be effectively higher than ϵ'.

Owen [19] hypothesized in an *aeolian* (airborne sediment) bed-load transport analysis that the effective roughness, z_o, is essentially proportional to the bed-load thickness. Smith and McClean [13, 20] then located z_r at z_o. With reference to the velocity boundary layer, z_o is also the height where the no-slip condition effectively applies. The effective height z_o is based upon the amount of work required to lift the particle to its maximum saltation height and after some algebra

$$z_o = \begin{cases} \alpha_o \dfrac{(\tau_{f_b} - \tau_c)}{(\gamma_s - \gamma_f)} + \epsilon' & \text{for } \tau_{f_b} > \tau_c \\[2ex] \epsilon' & \text{for } \tau_{f_b} < \tau_c \end{cases} \qquad \textbf{(14.2.30)}$$

The empirical correlation coefficient, α_o, is determined to be on the order of 26.0. Columbia River data yielded a value of 26.3 while aeolian equivalent data result in a value of 22.4.

EXERCISES

14.2.1 A fluidized bed (*a*) is a particle condition marked by low internal pore water pressure and high burden weight; (*b*) serves as the first particle assemblage model which allows each particle to independently respond to the flow; (*c*) is affected by density stratification; (*d*) is developed from a Lagrangian formulation as described in Chap. 3; (*e*) none of the above.

14.2.2 In sediment-laden river flows with a very mild bottom slope (*a*) the full mixture theory equation must be used; (*b*) the volume fraction concentrations must be low; (*c*) the settling velocity must be accounted for only in the vertical coordinate; (*d*) horizontal turbulent sediment flux is weak; (*e*) all of the above.

14.2.3 Simple boundary layer exact solutions for concentration profiles (*a*) exist for vertical equilibrium flux conditions; (*b*) exist for uniform vertical turbulence distributions; (*c*) decay exponentially or in power law form with the height; (*d*) are a function of the bed reference concentration; (*e*) all of the above.

14.2.4 The eddy diffusivity for the Smith particle concentration boundary layer model (*a*) decays exponentially with height; (*b*) is inversely proportional to the friction velocity; (*c*) is approximately equal to the eddy viscosity; (*d*) linearly increases with depth ($z \leq 0.2d$); (*e*) *c* and *d*.

14.2.5 The settling velocity formula (*a*) is nonlinear due to the drag coefficient being itself a function of w_t; (*b*) is insensitive to particle specific weight; (*c*) is valid for Reynolds numbers up to vortex shedding Reynolds numbers; (*d*) determines Stokes' settling velocity for $\mathbf{R}_D < 1$; (*e*) *a*, *c*, and *d*.

14.2.6 The criteria for the initiation of motion (*a*) infers that the particle volume eroded increases as the sixth power of the friction velocity; (*b*) is met when fluid shear stress at the bottom is less than the critical erosion shear stress; (*c*) yields a Shields diagram; (*d*) all of the above.

14.3 COUPLED FLOW AND TRANSPORT: STRATIFIED BOUNDARY LAYER

The presence of high concentrations of heat and/or mass can have a profound effect on the flow and transport field due to the change in the density field. Under these conditions the density field can now vary from point to point in the flow field and result in a rearrangement of the velocity distribution and accompanying turbulence. Up to this point the density field has been assumed constant everywhere, the implicit assumption being that the added heat and mass were quite low and resulted in very little change in density. The effect of this assumption then is to *uncouple* the velocity calculation from the temperature or mass species calculation.

Many engineered or natural flows do possess added mass or heat in sufficient quantities to cause quite different density values to exist in different regions of the flow field. For example, in fluidized coal slurries settling coal particles can collect in large quantities just near the bottom resulting in a 3 to 4 order of magnitude increase in concentration at the bottom and a substantial increase in density (e.g., 15 percent) for the mixture. In this case the more dense fluid is laying near the bottom and is therefore *stable*. Here we intuitively suggest that more dense material would naturally lay under lighter density fluid and would tend to return to that position if a parcel were momentarily moved to a higher position with less density. With this simple definition it is seen that the density distribution in Ex. 3.3 is also stable.

In contrast an *unstable* density distribution occurs when a fluid parcel of a specified density is moved upwards into a density which is greater than the parcel. In this case the parcel attains a vertical velocity as buoyancy accelerates the particle continuously upwards. The fluid motion introduced by this density *instability* mechanism is called *convection,* and many examples of convection occur in our everyday existence. The most common form of convection is the type introduced by boiling water on a stove. Here a pan of water is heated from below to a high temperature. The heated layer of fluid at the pan bottom is less dense than the fluid at the top of the pan. During the initial heating stages the upper portion of the fluid remains cooler and more dense. Hot water convects (or floats) to the surface with a speed determined by the temperature difference between the bottom, T_B, and top, T_T, of the pan. After a certain amount of *convective mixing* the temperature difference $(T_B - T_T)$ approaches zero. Convection episodes are frequent in meteorology and are a core process in the formation of thunderstorms, hurricanes, tornadoes, and assorted types of clouds.

It is necessary to incorporate the spatially varying density distribution into the equations of motion, which therefore implies that the solution of them is *coupled* with the solution of the heat and mass transport equation. Therefore, in this section the criteria for the onset of density instability will be summarized as will the most frequently used procedure for coupling the equations. A stably stratified particle concentration boundary layer solution will be used as an example. The humidity and evaporation problem reviewed in the next section will address both stable and unstable coupling.

Stability Criteria

From Chap. 2 (see also Refs. [21, 22]) the net force acting on a parcel is the weight of the fluid displaced by the parcel, that is

$$F = (m - m_p)g \qquad (14.3.1)$$

Here m is the mass of the ambient fluid and m_p is the parcel mass. Newton's second law (in Lagrangian form) says that the particle acceleration is

$$m_p \frac{d^2z}{dt^2} = (m - m_p)g \qquad (14.3.2)$$

or that

$$a_z = \left(\frac{\rho - \rho_p}{\rho_p}\right)g = \frac{\Delta\rho}{\rho_p}g \qquad (14.3.3)$$

Thus, if $\Delta\rho = (\rho - \rho_p) > 0$, then vertical acceleration is positive indicating increasing velocity and unstable density. If $\Delta\rho = (\rho - \rho_p) < 0$, then the parcel will have negative acceleration or declining velocity and stable density.

More specific displacement data can be obtained by using a Taylor series expansion between density and height (ρ, z) and their initial condition $(\rho_p = \rho_o, z_o)$

$$\rho = \rho_o + \left.\frac{\partial\rho}{\partial z}\right|_o (z - z_0) + \frac{1}{2}\left.\frac{\partial^2\rho}{\partial z^2}\right|_o (z - z_0)^2 + \ldots$$

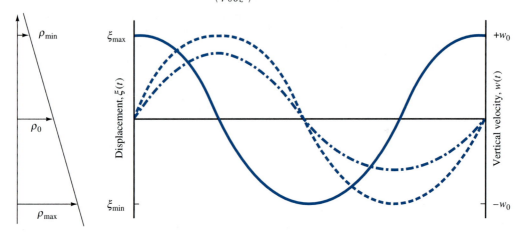

Figure 14.10   Particle acceleration versus time in a stably stratified density field.

From Eq. (14.3.3) then

$$\frac{d^2}{dt^2}(z - z_o) = \frac{d^2\xi}{dt^2} \approx \frac{g}{\rho_o}\frac{\partial \rho}{\partial z}\bigg|_o (z - z_o) + \ldots \approx g\left(\frac{1}{\rho}\frac{\partial \rho}{\partial z}\right)_o (z - z_o) \quad \text{(14.3.4)}$$

If a solution of the form $\xi = c_2 e^{st}$ is assumed, then the equation becomes

$$\frac{d^2\xi}{dt^2} = -\left(\frac{1}{\rho}\frac{\partial \rho}{\partial z}\right)_o \xi = 0 \qquad \text{(14.3.5)}$$

and the characteristic equation becomes

$$s^2 - \left(\frac{1}{\rho}\frac{\partial \rho}{\partial z}\right)_o = 0$$

The term $\left(-\frac{g}{\rho}\frac{\partial \rho}{\partial z}\right)^{1/2}$ is known as the *Brunt-Väisälä frequency,* N_B, with dimension t^{-1}.

Two solutions exist for Eq. (14.3.5). If $\partial \rho/\partial z < 0$, then the solution for the particle motion ($\xi_o = z_o = 0$ and $d\xi/dt = w_o$, a small vertical parcel velocity) is

$$\xi = \frac{w_o}{N_B} \sin N_B t \qquad \text{(14.3.6)}$$

that is, in the absence of friction a displaced parcel in a stable density field will attempt to return to fluid of its own density but overshoot that position (Figure 14.10). The continuation of the overshoot is stopped as the buoyancy force overcomes the acceleration and attempts to return it to its equilibrium position. Friction would, over time, dampen the overshoot.

For the case of the unstable density gradient $\partial \rho / \partial z > 0$ the displacement solution [21] indicates a displacement that grows exponentially with time.

The Effect of Shear on Stability

The above construct for stability is limited by the restriction of a motionless friction-free fluid. Most flows are marked by moving fluids with friction which introduces shear as an additional variable to consider in the parameterization of stable or unstable flows. A robust derivation for this effect is found in the *perturbation analysis* methodology found in Dutton [21]. We have already seen that $\mathbf{N}_B$ has dimensions of [1/t] or frequency. Shear stress, and therefore friction, is proportional to the vertical mean velocity gradient $\partial u / \partial z$, which also has units of [1/t]. Therefore, the relative strength of the density field to the shear stress is assessed by the dimensionless ratio

$$\mathbf{Ri} = \frac{\mathbf{N}_B^2}{\left(\frac{\partial u}{\partial z}\right)^2} = \frac{-\frac{g}{\rho}\frac{\partial \rho}{\partial z}}{\left(\frac{\partial u}{\partial z}\right)^2} \tag{14.3.7}$$

which is called the *gradient Richardson number*. The velocity gradient in Eq. (14.3.7) is found from the streamline average horizontal velocity, that is, $(u^2 + v^2)^{1/2}$.

The original stability analysis by Boussinesq in 1903 established that the flow-density field was unstable for $\mathbf{Ri} > \frac{1}{4}$. Again when the flow is unstable a small particle dislodged from its surroundings will have its displacement grow without bound. Small particle displacements will dampen out and disappear when $\mathbf{Ri} < \frac{1}{4}$. Unstable parcel motion typically results in considerable overturning of the fluid and significant mixing occurs.

Governing Equations

The governing equations in either laminar or turbulent forms are already coupled in a one-way sense in that the velocity field is required to be known for accurate heat and mass transport. Two-way coupling with the density changes arising from high concentrations of heat and mass can occur in two ways. First, the presence of a density gradient will absorb energy and momentum from the turbulent field in order to maintain the progress of a fluid parcel. Essentially then a stable density gradient will suppress or dampen turbulence, and the closure formulations for the eddy viscosity and diffusivity must reflect this effect. Second, the density field felt by the momentum field will result in adjustments of the velocity distribution which could be quite different than the velocity field in the nonstratified case.

From the discussion originating with Boussinesq, and repeated many times [22], it can be argued on dimensional grounds that all the density variations in the momentum equation are small enough to be ignored except those in the gravity term. Therefore, from Eqs. (4.4.12), the x and z horizontal momentum equations can be rewritten, respectively, as

$$\rho_o \frac{Du}{Dt} = -\frac{\partial p}{\partial x} + \mu \nabla^2 u \tag{14.3.8a}$$

$$\rho_o \frac{Dw}{Dt} = -\frac{\partial p}{\partial z} - \rho g + \mu \nabla^2 w \tag{14.3.8b}$$

Here $\rho(\mathbf{x}, t) = \rho_o + \rho'(\mathbf{x}, t)$ where ρ_o is a reference density and ρ' is a departure from the reference density introduced by the heat and mass. From Eqs. (1.5.10) to (1.5.12) a density equation can be written as

$$\rho = \rho_o(1 - \beta_T \Delta T + \beta_C \Delta C) \tag{14.3.9}$$

Inserting this formula into Eqs. (14.3.8a) and (14.3.8b) and dividing by ρ_o gives for x

$$\frac{Du}{Dt} = -\frac{1}{\rho_o}\frac{\partial p}{\partial x} + \nu\nabla^2 u \tag{14.3.10a}$$

and for z

$$\frac{Dw}{Dt} = -\frac{1}{\rho_o}\frac{\partial p}{\partial z} - (1 - \beta_T \Delta T + \beta_C \Delta C)g + \nu\nabla^2 w \tag{14.3.10b}$$

Stratification Due to Particles and the Required Model

This section extends the homogeneous particle boundary layer solution from the previous section to account for the effects of stratification. The volume concentration of particles, c, affects total density, ρ, as expressed by

$$\rho = \rho_w + (\rho_s - \rho_w)c. \tag{14.3.11}$$

This coupling of the particle concentration and density is incorporated into the current-driven boundary model presented in the previous section via the methods presented in Smith and McClean [13, 20]. Here as in the previous section the principal assumption is that the vertical fluxes are in equilibrium and that the principal coupling from the density field to the momentum field is on the turbulent flux suppression as imposed through the eddy viscosity and diffusion coefficients, that is, the first coupling effect mentioned in the previous subsection. As noted before the chief source of density stratification is the excessive particle accumulation at the bottom of the fluid column, a zone already noted for its weak boundary layer velocity field. The turbulent intensity available for mixing and diffusion of near bottom particles and the effect of the increasingly dense particle mixture therefore must be reflected in the eddy viscosity and diffusivity. One of the more enduring approaches used to parameterize this trade-off is a Richardson number formulation. The earliest forms of these parameterizations were used by, for example, Rossby and Montgomery [23], who proposed that the vertical viscosity in stratified conditions, $\eta(z)$, be related to the eddy viscosity in *neutral* (nonstratified) conditions, η_o, as

$$\frac{\eta(z)}{\eta_o} = (1 + c_1 \mathbf{Ri})^{-1} \tag{14.3.12}$$

Here $\mathbf{Ri}$ is the gradient Richardson number as defined in Eq. (14.3.7). Unfortunately a variety of forms for this relationship have been developed over the years, each developed from quite different laboratory and field conditions but each result, unfortunately, lumped with the others in practical application. Munk and Anderson [24] posited a purely empirical form as

$$\frac{\eta(z)}{\eta_o} = \frac{1}{(1 + \beta\mathbf{Ri})^\alpha} \tag{14.3.13}$$

Therefore, in the Rossby and Montgomery form $\alpha = 1$. Munk and Anderson recommended $\beta = 10.0$ and $\alpha = 0.5$. French ([25], Table 10.7) presented a summary of values of α and β from eleven different experiments.

In the particle problem the mixture density ρ is defined from Eq. (14.3.11), and therefore the gradient Richardson number equals

$$\mathbf{Ri} = \frac{-\dfrac{g}{\rho}\dfrac{\partial \rho}{\partial z}}{\left(\dfrac{\partial \overline{u}}{\partial z}\right)^2} = \frac{-g\left[\dfrac{\rho_s - \rho_w}{\rho}\right]\dfrac{\partial c}{\partial z}}{\left(\dfrac{\partial \overline{u}}{\partial z}\right)^2} \qquad (14.3.14)$$

Clearly then the sediment concentration gradient will play a central role in determining the resulting concentration and velocity profiles. From the meteorological community (e.g., Refs. [26, 27]) a formal extension of this approach that is based upon *similarity* procedures is in current use for analyzing boundary layer flows, and the Smith and McClean [13, 20] model adopted this approach.

A nondimensional shear stress, ϕ_m, is defined in the boundary layer as

$$\frac{\tau}{\rho u_*^2} = \phi_m = \frac{\eta}{u_*^2}\frac{\partial \overline{u}}{\partial z} = \frac{kz}{u_*}\frac{\partial \overline{u}}{\partial z} \qquad (14.3.15)$$

or that

$$\phi_m = \frac{kz}{u_*}\left(\frac{\partial \overline{u}}{\partial z}\right) \qquad (14.3.16)$$

In the case of stably stratified boundary layers [28] the nondimensional shear stress is found to be

$$\phi_m = \frac{kz}{u_*}\frac{\partial \overline{u}}{\partial z} = (1 + \beta \mathbf{Pr}_t \mathbf{Ri})^{-1} \qquad (14.3.17)$$

where $\mathbf{Pr}_t$ is the turbulent Prandtl number, E_{sz}/η, and β is selected by Businger et al. [28] to be 4.7 ± 0.5.

Equation (14.2.12) is still the basic governing equation even for stratified conditions. Once again this is a statement of vertical flux being in equilibrium. However unlike Sec. 14.2, $\eta(z)$ and $E_{sz}(z)$ now vary with elevation and the integration of Eq. (14.2.12) and its momentum counterpart cannot be done in exact fashion as in Chap. 6 or Sec. 14.2. Equation (14.2.12) and its momentum counterpart must now be numerically integrated to each level where $c(\xi)$ is defined, that is,

$$\frac{\overline{c}(\xi)}{1 - \overline{c}(\xi)} = \frac{c_r}{1 - c_r}\exp\left\{-p\int_{\xi_o}^{\xi}\left[1 + \beta_*\left(\frac{\partial \overline{c}}{\partial \xi}\right)\left(\frac{\partial \overline{u}}{\partial \xi}\right)^{-2}\right]^{-1}\frac{d\xi}{f(\xi)}\right\} \qquad (14.3.18)$$

and

$$u(\xi) = \frac{u_*}{k}\int_{\xi_o}^{\xi}\frac{(1 - \xi)}{f(\xi)}\left[1 + \beta_*\left(\frac{\partial \overline{c}}{\partial \xi}\right)\left(\frac{\partial \overline{u}}{\partial \xi}\right)^{-2}\right]^{-1}d\xi \qquad (14.3.19)$$

Here $p = (w_t/ku_*)$ and ξ_o is z_o/d in nondimensional coordinates. β_* is defined as

$$\beta_* = gd\beta\alpha\left(\frac{\rho_s - \rho_w}{\rho_w}\right) \qquad (14.3.20)$$

The calculation then becomes a nonlinear iteration and a simple midpoint numerical integration of the integrals is all that is required to complete the calculation.

Table 14.1   Input parameters and model parameters

Variable	Value for variable	NCLASS	FRAC	D_n (cm)
u_*	4.52 cm/s	1	1.0000	0.03077
h	1620 cm			
ρ_w	1.0 g/cm^3	3	0.1550	0.05050
ν	0.0131 cm^2/s		0.6850	0.02775
g	980 cm/s^2		0.1600	0.01503
SG$_{quartz}$	2.65			
SCF	1.0	10	0.0250	0.05950
NSTEP	20		0.0450	0.05000
α_o	26.3		0.0850	0.04200
β	4.7		0.1300	0.03540
α	1.0		0.2100	0.02970
κ	0.4		0.1950	0.02500
γ_o	0.0024		0.1500	0.02100
c_b	0.6		0.0950	0.01770
TOLERANCE	0.01		0.0500	0.01490
ITSTEP	100		0.0150	0.01250

The Computer Program

A computer program has been prepared and will solve the neutral or stratified particle and velocity boundary layer data as described in Secs. 14.2 and 14.3. The source code is available in the book's Web page.

In general the model is configured to run with any consistent set of units; the output data units will be whatever the input data units are. In the program concentration is always in volume concentration units, and multiplication by the particle density will yield mass concentration. Table 14.1 contains a list of the variables in the program. Three classes of variables exist, flow-field variables, particle variables, and calculation variables, and Table 14.2 is annotated to show which are required as input to the program. While straightforward, several limits do exist. The number of particle classes (NCLASS) that a mixture can be broken into is 10 and the minimum recommended number is three. When entering the fraction of the volume concentration in each grain size it is necessary to make sure that they sum to 1. Should the analyst desire, it is possible (as has been done for over 100 years) to analyze the mixture as if it were a single average grain size with a solitary settling velocity. It would be instructive to compare the calculations from each viewpoint. The values of γ_o and c_b are fixed at $2(10^{-3})$ and 0.60, respectively, based upon the original derivation for c_r.

The calculation, though straightforward, does require iteration. The iteration proceeds by estimating the value of β_* [Eq. (14.3.18)] followed by the calculation of the Richardson number [Eq. (14.3.7)] and the nondimensional eddy viscosity. Next $\partial \overline{u}/\partial z$ and $\partial \overline{c}/\partial z$ are calculated by iteration and Eqs. (14.3.16) and (14.3.17) are finally numerically integrated to form the estimated profile. The results are presented in both tabular and graphical form.

Example 14.1   Table 14.1 contains the input parameters for the calculation of particle concentration and velocity over a 16.2-m distance above the bottom. Ten size fractions were included as denoted by the first two columns in Table 14.3, the rest of the calculated input parameters are included in Table 14.3. Table 14.4 contains the calculated

Table 14.2 Variable list for the particle boundary layer model

Variable	Code Variable	Description
D_n	D(NCLASS)	Particle grain diameter, n^{th} class*
w_{tn}	WF(NCLASS)	Terminal particle fall velocity, n^{th} class; [Eq. (14.2.23)]
s_n	S(NCLASS)	Particle specific gravity*
τ_{cn}	TAUCRIT(NCLASS)	Critical shear stress for initiation of particle motion; [Eq. (14.2.28)]
S_*	SSTAR(NCLASS)	$\frac{D_n}{4\nu}[(s_n - 1)gD_n]^{1/2}$; nondimensional ordinates on Fig. (14.10)
ν	NU	Kinematic viscosity of fluid*
	SCF(NCLASS)	Shield's diagram correction factor*
Q_c	SHLDC(NCLASS)	Critical Shield's parameter
	WSGD(NCLASS)	$w_{tn}/[(s_n - 1)gD_n]^{1/2}$
S_n	SN(NCLASS)	Normalized excess skin friction; [Eq. (14.2.28)]
c_{bn}	CBED(NCLASS)	Particle bed concentration; [Eq. (14.2.29)]*
z_{on}	Z0(NCLASS)	Roughness length due to particle class n
c_{rn}	CREF(NCLASS)	Reference concentration at z_o; [Eq. (14.2.29)]
c_m	CMT(NSTEP)	Mean concentration
u	U(NSTEP)	Mean velocity
z	Z(NSTEP)	Height above bottom
g	G	Acceleration due to gravity*
ρ	FDENSITY	Fluid density*
π	PI	Ratio of the circumference of a circle to its diameter
γ_o	GAMMA	Variable in c_r calculation parameter; [Eq. (14.2.29)]*
τ_b	TAUBED	Bed shear stress
u_*	USTAR	Shear velocity*
h	DEPTH	Flow-field depth*
	DEL	Numerically evaluated integral
	NCLASS	Number of particle classes*
	NSTEP	Number of vertical integration steps*
	ITSTEP	Number of iteration steps*
i_n	FRAC(NCLASS)	Fraction of each particle class*
α_o	ALPHA0	26.3; roughness length calculation parameter
β	BETA	4.7; Richardson number calculation parameter
α	ALPHA	1.0; ratio of diffusion coefficient of particle to diffusion coefficient of momentum
z_o	AVEZ0	Weighted average roughness length; [Eq. (14.2.30)]
s	AVES	Weighted average particle specific gravity
p_n	PN(NCLASS)	$w_{tn}/\kappa u_*$
ξ_1	E(NSTEP)	z/h; normalized height above bottom
$f_2(\xi)$	F2(NSTEP)	Normalized eddy viscosity
$\sigma/2\beta$	SIBE	Coefficient ratio
	TOLERANCE	Allowable error in iteration procedure*
	DKUH	$\rho\kappa u_* h/\tau_b$
β_*	BETASTAR	$\alpha\beta(s - \rho)gh/\rho$
$du/d\xi_1$	DUDE(NSTEP)	Velocity gradient
$dc_m/d\xi_1$	DCDE(NSTEP)	Concentration gradient
	DUGUESS	Estimation of $du/d\xi$
	DCGUESS	Estimation of $dc_m/d\xi$
	ERROR1	Relative error in $dc_m/d\xi_1$ iteration
	ERROR2	Relative error in $du/d\xi_1$ iteration
$\mathbf{Ri}_n$	RICH(NSTEP)	Richardson number

* Indicates input data.

Table 14.3   Particle parameters by class (NCLASS = 10)

D_n, cm	i_n	z_o (cm) Eq. (14.2.30)	c_{rn}, Eq. (14.2.29)	w_{tn} (cm/s), Eq. (14.2.23)	τ_{cn} (dyne/cm^2), Eq. (14.2.28)
0.0595	0.0250	0.3402	0.1931E−3	8.164	3.175
0.0500	0.0450	0.3368	0.4020E−3	6.614	2.799
0.0420	0.0850	0.3328	0.8430E−3	5.252	2.553
0.0354	0.1300	0.3298	0.1426E−3	4.190	2.332
0.0297	0.2100	0.3273	0.2543E−3	3.322	2.132
0.0250	0.1950	0.3243	0.2494E−3	2.639	2.028
0.0210	0.1500	0.3219	0.2026E−3	2.087	1.928
0.0177	0.0950	0.3202	0.1354E−3	1.658	1.835
0.0149	0.0500	0.3183	0.7386E−3	1.315	1.774
0.0125	0.0150	0.3156	0.2191E−3	1.038	1.793

Table 14.4   Calculated parameters by height

n	z (cm)	C_m (mg/L)	u (cm/s)	E (cm^2/s)	$-w_t C_m$ (mg/cm^2/s)	$-\overline{w'C'}$ (mg/cm^2/s)
1	0.33	32431.85	0.00	0.59	−105.63	136.24
2	0.50	18772.66	4.95	0.90	−61.14	41.27
3	0.76	11431.74	9.90	1.38	−37.23	17.46
4	1.17	7235.73	14.85	2.12	−23.57	8.56
5	1.79	4722.25	19.80	3.24	−15.38	4.57
6	2.74	3160.22	24.74	4.96	−10.29	2.59
7	4.19	2160.05	29.67	7.60	−7.04	1.54
8	6.41	1503.48	34.59	11.65	−4.90	0.94
9	9.81	1063.25	39.49	17.88	−3.46	0.59
10	15.02	762.63	44.36	27.45	−2.48	0.38
11	22.99	554.03	49.20	42.20	−1.80	0.25
12	35.18	407.17	53.99	64.94	−1.33	0.17
13	53.84	302.30	58.70	99.91	−0.98	0.12
14	82.39	226.22	63.33	153.02	−0.74	0.08
15	126.09	169.78	67.89	230.96	−0.55	0.06
16	192.97	126.33	72.41	335.49	−0.41	0.04
17	295.32	90.57	77.07	445.64	−0.30	0.02
18	451.96	57.96	82.34	495.17	−0.19	0.02
19	691.68	28.97	88.88	471.00	−0.09	0.01
20	1058.55	7.90	96.99	365.38	−0.03	0.00

values for 20 increments within the water column and Figure 14.10 presents a plot of the resulting profiles.

Solution

In Figure 14.11 the concentrations from each of the ten grain size classes have been summed to get the plotted concentrations, C_m, at each level in the plot (see also Table 14.4). In reviewing the plot it is noticed that the concentration profile contains quite large concentrations near the bottom. A three order of magnitude drop in concentration occurs as the upper portion of the profile is reached.

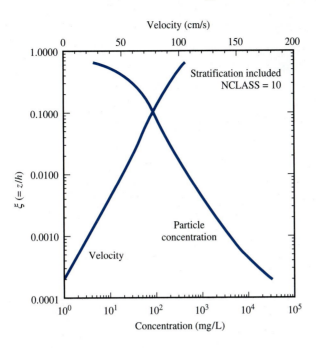

Figure 14.11 Calculated vertical profiles of velocity and particle concentration as determined by the conditions in Tables 14.3 and 14.4.

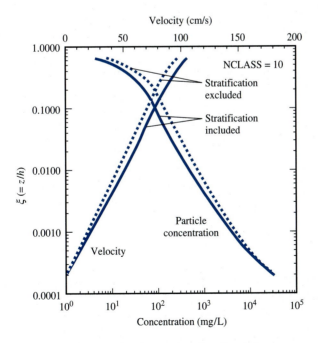

Figure 14.12 Calculated vertical profiles of particle concentration and velocity comparing the effect of including corrections for stratification.

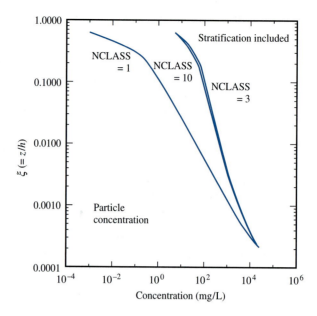

Figure 14.13   Calculated profiles of particle concentration demonstrating the sensitivity of the calculation to the number of grain size classes (NCLASS) in the calculation.

Results Sensitivity

Several questions about the behavior of these profiles and the impact of various components of the model can be addressed by a *sensitivity study*. Figures 14.12 and 14.13 present calculated profiles, each resulting from changing one variable and contrasting the results to the previous calculation. Of chief interest here is the impact of including the stratification effects versus not including them (Fig. 14.12), and the impact of the number of size classes used in the calculation.

Figure 14.12 (see page 669) shows that including the effect of stratification on eddy diffusivities lowers the turbulent transport of particles to the upper reaches of the column. A water column with less total concentration in the water column results in increasing discrepancies between the neutral ($\mathbf{Ri} = 0$) and stratified formulations. With more mass to carry in the upper water column for the neutral case the velocity profile is reduced in comparison to the profile for the stratified case.

Finally with the importance of grain size suggested in the previous discussions the last item of interest is how many grain size ranges should comprise the distribution. Figure 14.13 presents profiles for NCLASS equal to 1, 3, and 10 size classes. Clearly the ensemble average representation, NCLASS = 1, severely underestimates the particle concentration profiles in the upper portion of the column. However both NCLASS = 3 and NCLASS = 10 give nearly identical profiles. NCLASS = 1 suffers from its inability to account for the finer-scale particles being propagated higher in the column. Three class sizes is a minimum but satisfactory representation in that one size range is above and below the size class encompassing the mixture average.

EXERCISES

14.3.1 In contrast to parcel motion in a stable density gradient, parcel motion in an unstable density gradient (*a*) leads to a sinusoidal oscillation of a parcel of fixed density fluid; (*b*) is controlled by the Brunt-Väisäila frequency; (*c*) can be parameterized by the gradient Richardson number; (*d*) occurs for $\mathbf{Ri} < \frac{1}{4}$; (*e*) *c* and *d*.

14.3.2 For the multigrain size particle stratified boundary layer model (*a*) three grain size classes are minimally acceptable for accurate calculation; (*b*) the impact of including particle stratification is to suppress turbulent transport into the upper reaches of the column; (*c*) the concentration decreases as the distance from the bottom increases; (*d*) all of the above; (*e*) *a* and *c*.

14.4 INTERFACIAL TRANSFER: EVAPORATION

The phase change from liquid to vapor is called *evaporation* while its counterpart, a phase change from vapor to liquid, is called *condensation*. Examples of both are readily found in geoenvironmental transport and in engineered transport. In the chemical and food processing industries unit processes for *drying* (evaporation) are an important portion of a manufacturing process or form the basis for *cooling towers* used to reduce waste heat load from power-utility plants. Naturally occurring evaporation and condensation are a fundamental aspect of the *hydrologic cycle* and a strong determinant of weather. As an example consider Lake Erie in the North American Laurentian Great Lakes. Because water is slow to conduct heat relative to the atmosphere, the Lakes retain temperatures above freezing until January. The land is much quicker in conducting heat and consequently cools to a lower temperature more quickly than the Lake in early winter. A storm passing over the Lakes during December and January consequently absorbs considerable quantities of evaporated moisture from the Lakes. As the wind carries the moist air over the colder shore, the air temperature drops and the moisture condenses and precipitates as snow. Nearshore communities such as Cleveland, Ohio, and Erie, Pennsylvania, have received as much as 8 inches of snow per hour during so-called lake effect snows. The objective of this section is to introduce transport phenomena–based parameterizations for evaporation and flux rates. The corresponding unit processes encompassing this transport phenomenon are fully explained in the chemical engineering literature (e.g., Geankopolis [29]). Finally, only water is considered here.

Water Phases

The various phases of water are well described in meteorology [30], hydrology [31], climatology [26], thermodynamics [32], or chemical engineering [33] textbooks and the reader is referred to these texts for detailed descriptions.

To change from one state to another requires either the use or release of energy. The amount of heat used or given up per unit mass is called the *latent heat* which has units of joules/kg or calories/g. Table 14.5, amalgamated from Eagleson [34] and Bras [31], summarizes the definitions for the latent heats required to proceed from state 1 to state 2.

Table 14.5   Definition and computation of latent heats for water

State 1	State 2	Definition and energy absorption (+) or release (−)	Temperature dependance, L (cal/g) or T (°C)
Liquid	Vapor	L_e = latent heat of evaporation (+)	$L_e = 597.3 - 0.57T$
Vapor	Liquid	L_c = latent heat of condensation (−)	$L_c = -L_e$
Ice	Vapor	L_s = latent heat of sublimation (+)	$L_s = 677 - 0.07T$
Ice	Liquid	L_m = latent heat of melting (+)	$L_m = 79.7$
Liquid	Ice	L_f = latent heat of freezing (−)	$L_f = -L_m$

The transformation between liquid, vapor, and ice is a function of temperature, pressure, and density. In oceanographic studies it must be remembered that density is a function of not only temperature but salinity as well. Diagrams describing the relationship between the phases are numerous. Figure 14.14 based upon a schematic in Iribarne and Godson [30] is typical. The contour lines are lines of constant temperature. Two points labeled C^* and T^* mark interesting features. The critical point, C^*, marks the only point where the liquid, gas, and vapor coexist. Above the line of constant temperature, T_c, passing through point C^*, the phase is a gas. On temperature contours below C^*, the water may exist as a pure vapor, a mixture of water and vapor, or water. The bell shaped, dashed curve denotes the region where liquid and vapor phases coexist. Point T^* is the triple point where ice, water, and vapor coexist. The pressure, temperature, and specific volume values at points C^* and T^* are noted in the figure.

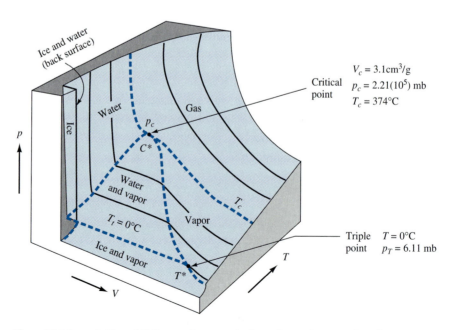

Figure 14.14   P, V, and T diagram for water (redrawn from Iribarne and Godson [30]).

Thermodynamic and Humidity Variables

From chemistry it is well known from Dalton's law for an ideal gas that the total pressure of the dry air and the vapor resulting from evaporation is the sum of partial pressures of the gases, that is,

$$p = p_{dry} + p_v \qquad (14.4.1)$$

Here p_v is the vapor pressure and p_{dry} is the dry air pressure. The density of the vapor ρ_v is found from the ideal gas law as

$$\rho_v = \frac{p_v}{R_v T} \qquad (14.4.2)$$

The vapor gas constant is related to the ideal gas constant R as

$$R_v = \frac{R}{M_v} = \frac{M_{dry}}{M_v} R_{dry} \qquad (14.4.3)$$

Here M_v and M_{dry} are the molecular weights of the water vapor and dry air, respectively. Using values from Appendix C, $M_{dry}/M_v = 1.61$; therefore,

$$\rho_v = \frac{p_v}{\left[\dfrac{M_{dry}}{M_v} R_{dry} T\right]} = \frac{p_v}{1.61 R_{dry} T} = 0.622 \frac{p_v}{RT} \qquad (14.4.4)$$

The vapor density is defined as the *absolute humidity*.

In order to define other types of moisture concentrations or humidities, it is necessary to define the mixture density ρ by use of Eq. (14.4.1)

$$\rho = \rho_{dry} + \rho_v s = \frac{p - p_v}{R_{dry} T} + 0.622 \frac{p_v}{R_{dry} T} \qquad (14.4.5)$$

$$= \frac{p}{R_{dry} T} \left(1 - 0.378 \frac{p_v}{p}\right)$$

The salutory feature of this relationship is that it shows that moist air (ρ) is less dense than dry air (ρ_{dry}).

The *relative humidity* is defined as the ratio of the vapor density to the saturation vapor density. *Saturation* is defined as the condition of the moist air where the maximum mass of water vapor is held in the mixture for the given temperature (*not* pressure). Therefore, relative humidity is

$$r(\%) = 100 \frac{\rho_v}{\rho_s} = 100 \frac{p_v}{p_s} \qquad (14.4.6)$$

Specific humidity is defined, in volume fraction units, as

$$c_q = \frac{\rho_v}{\rho} = \frac{0.622 p_v}{p - 0.378 p_v} \approx 0.622 \frac{p_v}{p} \qquad (14.4.7)$$

The *mixing ratio* is defined as

$$\Gamma = \frac{\rho_v}{\rho_{dry}} = \frac{0.622 p_v}{p - p_v} \qquad (14.4.8)$$

Estimation of Humidity Variables

Based upon a variety of field measurements the saturation vapor pressure over water can be estimated by empirical formulae from Bosen [35] as

$$p_s \approx 3363.9 \left[(0.00738\,T + 0.8072)^8 - 0.000019\,(1.8\,T + 48)\,1.8\,T + 0.001316 \right]$$

(14.4.9)

where p_s is in units of N/m^2 and T is the dewpoint-temperature in degrees Centigrade. From Bras ([31], Eq. 3.33) the dewpoint temperature can be approximated to within 0.3°C by

$$T - T_d \approx (14.55 + 0.114T)x + \left[(2.5 + 0.007T)x \right]^3 + (15.9 + 0.117T)x^{14}$$

(14.4.10)

where $x = 1 - r/100$. The range of applicability is -40 to $50°C$. Finally relative humidity can be estimated via Bosen's [36] formula

$$r = 100 \left(\frac{112 - 0.1T + T_d}{112 + 0.9T} \right)$$

(14.4.11)

where temperatures are in degrees Centigrade.

What follows is a review of the procedures for predicting humidity profiles and evaporation fluxes.

Humidity Profiles and Boundary Layer

Once again as in the momentum, heat, and particle boundary layer problem the interfacial flux between the ground or water and the atmosphere is to be parameterized. As before we shall follow the procedure of first using boundary layer methods to create the profiles of humidity (as well as temperature and velocity) and then attempt to summarize this information into bulk representations parameterized by the appropriate nondimensional numbers. Evaporation of other liquids into either the atmosphere or other gases is straightforward.

As already noticed the problems we approached in this fashion are increasingly plagued by geomtric complexity and more nonlinear coupling between the flow and transport components. Figure 14.15 (adapted from Brutsaert [26]) presents a schematic of the structure of the atmospheric boundary layer flowing over a solid or fluid boundary. Essentially this is the same conceptual picture as was drawn for the particle boundary layer. The physical differences are more pronounced. The *outer* or *defect* region is once again the layer whose physics are unaffected by the nature and character of the bottom. The *inner* or *wall layer* is dominated by the bottom geometry. At the very bottom is the *interfacial* or *viscous* sublayer. The inner layer contains two subregions, an *overlap* region between the inner and defect layers and the *constant shear layer* containing logarithmic profiles of the variables. The constant stress layer is marked by constant fluxes of momentum, mass, and heat transport. As Brutsaert [26] and Stull [27] noted the outer layer in the atmospheric boundary layer is the region where natural convection effects are strong, frequently noted, and affect the various predicted profiles. This is unlike the nearly always stable particle bottom boundary layer in the previous section.

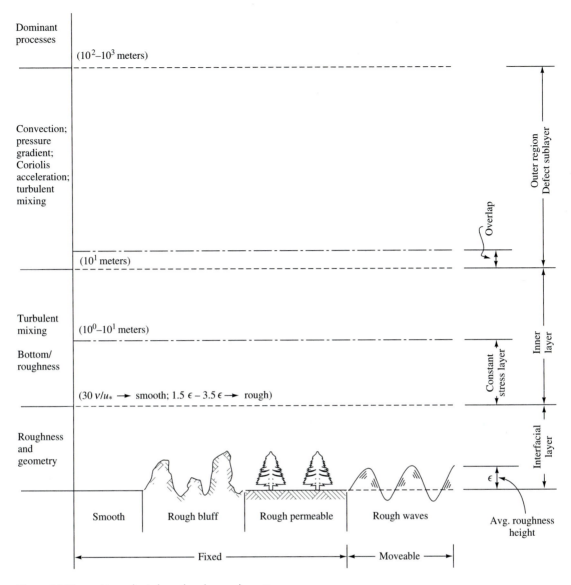

Figure 14.15   Atmospheric boundary layer schematic.

The interfacial layer is referred to (as noted in Chap. 6) as the viscous sublayer if the bottom is hydrodynamically smooth. However, two types of hydrodynamically rough bottom conditions can occur which constitute the *roughness sublayer*. First, as discussed in Chap. 6 and in the preceeding Secs. 14.2 and 14.3, hydrodynamically rough conditions can exist which result in the disappearance of the viscous sublayer and the extension of the constant stress layer to the bottom roughness height z_o. Of importance is that in this type of roughness the flow between or through the elements causing the roughness, for example, sand grains, is irrelevent. For example, in the atmospheric boundary layer the elements forming the roughness can consist of dispersed but tall elements such that the flow through the elements causes a significant reduction in momentum. Examples include corn, trees, shrubs, wind waves on water surfaces, and buildings.

As in the previous sections we will first concentrate on the constant stress and constant flux layer, particularly as it lends itself to adopting answers from the momentum boundary layer results through *similarity*. It also provides a simple method (the two-point method) for making measurements of the flux. The development in this section largely follows the comprehensive work of Brutsaert [26].

Profiles in Neutral Conditions

Assuming that c_q is the water vapor volume concentration yields the following turbulent conservation equation

$$\frac{\partial \overline{c}_q}{\partial t} + \nabla \cdot (\overline{\mathbf{v}}\overline{c}_q) = \mathcal{D}_v \nabla^2 \overline{c}_q - \left[\frac{\partial}{\partial x}(\overline{u'c'}_q) + \frac{\partial}{\partial y}(\overline{v'c'}_q) + \frac{\partial}{\partial z}(\overline{w'c'}_q) \right] \quad \textbf{(14.4.12)}$$

Here $\mathcal{D}_v$ is the molecular diffusivity coefficient for water vapor in air. By use of the boundary layer assumptions this equation quickly becomes

$$\mathcal{D}_v \frac{\partial^2 \overline{c}_q}{\partial z^2} - \frac{\partial}{\partial z}(\overline{w'c'}_q) = 0 \quad \textbf{(14.4.13)}$$

If the molecular diffusion term is quite small relative to the turbulent flux (which is typically the case), then

$$\frac{\partial}{\partial z}(\overline{w'c'}_q) = \frac{d}{dz}(\overline{w'c'}_q) = 0 \qquad \Longrightarrow \qquad \overline{w'c'}_q = constant \quad \textbf{(14.4.14)}$$

The same line of reasoning is applied to the heat flux equation; that is, assuming no radiation effects or other sources or sinks, then

$$\frac{\partial}{\partial z}(\rho C_p \overline{w'T'}) = \frac{d}{dz}(\rho C_p \overline{w'T'}) = 0 \qquad \Longrightarrow \qquad \overline{w'T'} = constant \quad \textbf{(14.4.15)}$$

In order to account for the coupling between the heat content as affected by the coupling to pressure and moisture content [Eq. (14.4.5)], it is useful to define the *potential temperature, θ.* This is defined as the temperature that would result if the moist air (including the special case of dry air) were taken adiabatically to a standard reference pressure level typically $p_o = 1000$ mbars or 10^5 N/m^2. Therefore, the potential temperature, θ, is defined as

$$\theta = T \left(\frac{p}{p_o} \right)^{\kappa} \quad \textbf{(14.4.16)}$$

Here κ equals

$$\kappa = \frac{R_{\text{dry}}}{C_p}(1 - 0.23c_q) \quad \textbf{(14.4.17)}$$

which is essentially the specific heat difference ratio

$$\kappa = \frac{C_p - C_v}{C_p}$$

For dry air $\kappa = 1/1.41$, but accommodation for humidity effects has been accounted for in Eq. (14.4.16).

From Eq. (14.4.15) the *sensible heat flux* is then defined as constant

$$\frac{\partial}{\partial z}(\rho C_p \overline{w'\theta'}) = \frac{d}{dz}(\rho C_p \overline{w'\theta'}) = 0 \qquad \Longrightarrow \qquad \overline{w'\theta'} = constant \quad \textbf{(14.4.18)}$$

The solution for the velocity, potential temperature, and humidity profiles is developed from *similarity* methods. Similarity here refers as before to the fact that the turbulent processes giving rise to all three profiles result from similar turbulent processes. Under this hypothesis the velocity profile is given by the familiar logarithm function

$$\overline{u}_2(z = z_2) - \overline{u}_1(z = z_1) = \frac{u_*}{k} \ln\left(\frac{z_2}{z_1}\right) \tag{14.4.19}$$

or

$$\overline{u}(z) = \frac{u_*}{k} \ln\left(\frac{z}{z_{om}}\right) \tag{14.4.20}$$

where z_{om} is the momentum roughness height. The uncertainty in parameterizing and placing z_{om} has been discussed in Chap. 6 and Secs. 14.2 and 14.3, and the parameterizations fully apply to this problem. The displacement distance formulation, d_o, is often invoked to reduce these uncertainties for larger rough obstacles

$$\overline{u}(z) = \frac{u_*}{k} \ln\left(\frac{z - d_o}{z_{om}}\right) \tag{14.4.21}$$

Therefore, d_o just shifts the profile to better include the effects of roughness.

For the humidity or vapor profile the profile equation [Eq. (14.4.14)] integrates to

$$\overline{c}_{q1}(z = z_1) - \overline{c}_{q2}(z = z_2) = \frac{E}{a_v \rho k u_*} \ln\left(\frac{z_2 - d_o}{z_1 - d_o}\right) \tag{14.4.22}$$

or in displacement form referenced to the surface, $c_{q0}(z = z_{ov})$,

$$c_{qo} - \overline{c}_q(z) = \frac{E}{a_v \rho k u_*} \ln\left(\frac{z - d_o}{z_{ov}}\right) \tag{14.4.23}$$

In these two equations $k_v = a_v k =$ the von Kármán coefficient for water vapor, which for most practical purposes is quite close to 1 (i.e., 1.0 ± 0.1). Unless quite stable or unstable conditions exist, $a_v = 1.0$ is fine. The vapor roughness height, z_{ov}, is not equal to z_{om} principally because in heat or moisture content the molecular diffusion will be important at the interface. Such is not the case for momentum. The displacement height will be equal for momentum, moisture, and heat. Finally, E is the evaporation rate which is the flux of moisture $[M/L^2/t]$ at the interface between the ground (or water) and the air. Essentially it is the constant of integration in the flux equation.

For potential temperature the relation becomes

$$\overline{\theta}_1 - \overline{\theta}_2 = \frac{H}{a_h \rho C_p k u_*} \ln\left(\frac{z_2 - d_o}{z_1 - d_o}\right) \tag{14.4.24}$$

or in displacement form referenced to the potential temperature at the surface, θ_o,

$$\theta_o - \overline{\theta}(z) = \frac{H}{a_h \rho C_p k u_*} \ln\left(\frac{z_2 - d_o}{z_{oh}}\right) \tag{14.4.25}$$

Here H is the heat transfer flux at the interface.

The Effect of Stratification and Buoyancy

In the constant stress or flux range of the inner layer, buoyancy is not an important process. In the inner layer above the constant stress and flux layer it is. Much of the approach used in the particle boundary layer methodology was adapted from the more complex case of the atmospheric boundary layer. Therefore, much of the material in Sec. 14.3 will be familiar. The boundary layer in Sec. 14.3 was stable. Here the physics may be stable or unstable.

A nondimensional boundary layer coordinate is defined as

$$\xi = \frac{z - d_o}{L} \tag{14.4.26}$$

where d_o is the displacement height and L is called the *Monin-Obukov* length. This length is another measure of the stability of the air column which has gained more recent widespread use in the atmospheric community than the Richardson number [Eq. (14.3.12)]. For a moist air column, Eqs. (14.4.5) and (14.4.7) may be combined with Eq. (14.3.3) to give the vertical acceleration of a buoyant parcel of moist air

$$a_z = \frac{g}{T}\left(\frac{\partial \theta}{\partial z} + 0.61 T_A \frac{\partial c_q}{\partial z}\right) \tag{14.4.27a}$$

which in the inner layer with relatively constant fluxes can be approximated by surface conditions as

$$a_z = \frac{g}{\rho T}\left(\frac{H}{C_p} + 0.61 T E\right) \tag{14.4.27b}$$

This buoyancy-induced acceleration may be compared to the work done by the turbulent shear $(\rho u_*^2)u_*$ and the ratio has units of length

$$L = \frac{-\rho u_*^3}{gk\left[\dfrac{H}{TC_p} + 0.61E\right]} \tag{14.4.28}$$

For stable density fields L is positive, for unstable fields L is negative, and for neutral fields the denominator approaches 0 and $L \rightarrow \infty$.

From Eq. (14.4.21) for neutral conditions (differentiating to find $d\bar{u}/dz$)

$$\frac{k(z - d_o)}{u_*}\frac{d\bar{u}}{dz} = 1$$

and similar expressions can be derived from Eqs. (14.4.23) and (14.4.25) for vapor and potential temperature.

For buoyancy effects

$$\frac{k(z - d_o)}{u_*}\frac{d\bar{u}}{dz} = \Phi_m(\xi) \tag{14.4.29a}$$

$$-\frac{ku_*(z - d_o)}{E}\frac{d\bar{c}_q}{dz} = \Phi_v(\xi) \tag{14.4.29b}$$

$$-\frac{ku_*(z - d_o)\rho c_p}{H}\frac{d\bar{\theta}}{dz} = \Phi_h(\xi) \tag{14.4.29c}$$

Here the stability functions for momentum, vapor, and potential temperature Φ_m, Φ_v, and Φ_h, respectively, now account for stratification effects. For neutral conditions $\Phi_m = 1, \Phi_v = a_v^{-1}$, and $\Phi_h = a_h^{-1}$.

From Brutsaert [26] the integrated forms of these equations can be expressed as

$$\bar{u}_2 - \bar{u}_1 = \frac{u_*}{k}\left[\ln\left(\frac{\xi_2}{\xi_1}\right) - \Psi_m(\xi_2) - \Psi_m(\xi_1)\right] \tag{14.4.30a}$$

$$\bar{c}_{q1} - \bar{c}_{q2} = \frac{E}{a_v k u_* \rho}\left[\ln\left(\frac{\xi_2}{\xi_1}\right) - \Psi_v(\xi_2) - \Psi_v(\xi_1)\right] \tag{14.4.30b}$$

$$\bar{\theta}_1 - \bar{\theta}_2 = \frac{H}{a_h k u_* \rho C_p}\left[\ln\left(\frac{\xi_2}{\xi_1}\right) - \Psi_h(\xi_2) - \Psi_h(\xi_1)\right] \tag{14.4.30c}$$

where

$$\Psi_m(\xi) = \int_{(z_{om}/L)}^{\xi}\left[1 - \Phi_m(\zeta)\right]\frac{d\zeta}{\zeta} \tag{14.4.31a}$$

$$\Psi_v(\xi) = \int_{(z_{ov}/L)}^{\xi}\left[1 - \Phi_v(\zeta)\right]\frac{d\zeta}{\zeta} \tag{14.4.31b}$$

$$\Psi_h(\xi) = \int_{(z_{oh}/L)}^{\xi}\left[1 - \Phi_h(\zeta)\right]\frac{d\zeta}{\zeta} \tag{14.4.31c}$$

In these formulas ζ is a dummy variable of integration. Using interfacial conditions as in the neutral case, Eqs. (14.4.31a) through (14.4.31c) can be written as

$$\bar{u}(z) = \frac{u_*}{k}\left[\ln\left(\frac{z - d_o}{z_{om}}\right) - \Psi_m(\xi)\right] \tag{14.4.32a}$$

$$c_{qo} - \bar{c}_q(z) = \frac{E}{a_v k u_* \rho}\left[\ln\left(\frac{z - d_o}{z_{ov}}\right) - \Psi_v(\xi)\right] \tag{14.4.32b}$$

$$\theta_o - \bar{\theta}(z) = \frac{H}{a_h k u_* \rho C_p}\left[\ln\left(\frac{z - d_o}{z_{oh}}\right) - \Psi_h(\xi)\right] \tag{14.4.32c}$$

The remaining question becomes what do the flux profile functions $\Psi(\xi)$ look like for stable and unstable density fields?

Various forms for the Φ and Ψ functions have been derived and have been formulated in terms of both ξ and **Ri** variables. At present $\Phi_v(\xi)$ and $\Phi_h(\xi)$ are assumed equal. From Brutsaert [26] the following forms for the unstable and stable forms are used.

For *unstable* conditions ($\xi < 0$)

$$\Phi_v = \Phi_m^2 = \Phi_h = (1 - 16\xi)^{-1/2} \tag{14.4.33}$$

$$\Psi_m(\xi) = 2\ln\left[\frac{1 + \chi}{2}\right] + \ln\left[\frac{1 + \chi^2}{2}\right] - 2\arctan(\chi) + \frac{\pi}{2} \tag{14.4.34a}$$

$$\Psi_v = \Psi_h = 2\ln\left[\frac{1 + \chi^2}{2}\right] \tag{14.4.34b}$$

$$\chi = (1 - 16\xi)^{1/4}$$

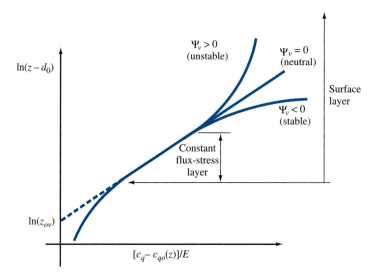

Figure 14.16   Schematic of specific humidity profiles for neutral, stable, and unstable conditions (redrawn from Brutsaert [26]).

For *stable* conditions ($\xi > 0$)

$$\Phi_v = \Phi_m = \Phi_h = \begin{cases} (1 + 5\xi) & 0 < \xi < 1 \\ 6 & \xi > 1 \end{cases} \tag{14.4.35}$$

for weakly stable conditions ($0 < \xi < 1$)

$$\Psi_m = \Psi_v = \Psi_h(\xi) = -5\xi \tag{14.4.36a}$$

while for strongly stable conditions ($\xi > 1$)

$$\Psi_m = \Psi_v = \Psi_h(\xi) = -5(\ln \xi + 1) \tag{14.4.36b}$$

Figure 14.16 is a schematic of the specific humidity profiles for neutral, stable, and unstable conditions.

Bulk Interfacial Flux Parameterizations

As developed in Sec. 9.2 the overall heat or mass flux can be expressed as the difference between two readily measured temperatures or concentrations and an overall heat or mass transfer coefficient. The overall bulk transfer coefficient includes the aggregate effect of both molecular diffusion and turbulent transport. Example 9.4 in Sec 9.2 presented an empirical mass transfer correlation for evaporation as developed from boundary layer experiments. We finish this section by presenting a bit more detail on this approach.

Two critical dimensionless numbers appear in many of these formulations, the *Stanton number*

$$\mathbf{S}_t = \frac{H}{\rho C_p u_{zh}(\theta_o - \overline{\theta}_{zh})} \tag{14.4.37}$$

and the *Dalton number*

$$\mathbf{D}_a = \frac{E}{\rho u_{zh}(c_{qo} - \overline{c}_{qzh})} \tag{14.4.38}$$

In these equations u_{zh}, $\bar{\theta}_{zh}$, and $\bar{c}_{qzh}$ are the velocity, potential temperature, and moisture concentration at height $z = h$ above the interface. A corresponding interfacial drag is defined as

$$\mathbf{C}d_h = \frac{u_*^2}{\bar{u}_{zh}^2} \tag{14.4.39}$$

The *bulk transfer equation* for evaporation and sensible heat exchange is found from the rearrangement of Eqs. (14.4.37) and (14.4.38), that is,

$$E = \rho u_{zh} \mathbf{D}a(c_{qo} - \bar{c}_{qzh}) \tag{14.4.40}$$

and

$$H = \rho C_p u_{zh} \mathbf{St}(\theta_o - \bar{\theta}_{zh}) \tag{14.4.41}$$

Equation (14.4.40) is called the *Dalton equation* for evaporation and has been the subject of intense laboratory and field experiment scrutiny. For the presentation of results the *Dalton number* is rewritten as

$$\mathbf{D}_a = \frac{\mathbf{C}d_h^{1/2}}{\left[(a_v B)^{-1} + a_v^{-1} \mathbf{C}d_h^{-1/2}\right]} \tag{14.4.42}$$

The value for B can be presented in dimensionless combinations of the Schmidt number $\mathbf{S}_c$ and the roughness Reynolds number, $\mathbf{R}_{zo}$ $(u_* z_o/\nu)$. For smooth surfaces: $(\mathbf{R}_{zo} < 0.13)$

$$(a_v B)^{-1} = 13.6\mathbf{S}_c^{2/3} - 13.5 \tag{14.4.43}$$

and for bluff roughness elements $(\mathbf{R}_{zo} > 2.0; 0.6 < \mathbf{S}_c < 6)$

$$(a_v B)^{-1} = 7.3\mathbf{R}_{zo}^{1/4}\mathbf{S}_c^{1/2} - 5 \tag{14.4.44}$$

The above two formulas were prepared by Brutsaert [26]. Helfrich et al. [37] reviewed *over 100* empirical evaporation equations which have been suggested for use (see Table 5.3 in Ref. [31] for a partial summary). Because it is easier to measure vapor pressure, the Dalton equation forms are often expressed in terms of p_v instead of c_q. It is remembered that the vapor pressure and specific humidity c_q are related by $c_q = 0.622 p_v/p$ [Eq. (14.4.7)].

Finally, Table 14.6 from Eagleson [34] summarizes the roughness heights (z_o) for various types of smooth, bluff, and permeable surfaces. These data are drawn from field experiments with these surfaces supplemented by estimates of z_o made with neutral boundary layer theory [Eq. (14.4.19)] with velocity measurements made at 2 m above the surface.

TABLE 14.6   Values for roughness height

Type	Roughness, z_o (cm)	Type	Roughness, z_o (cm)
Open water	0.001	Alfalfa (20–30 cm)	1.4
Smooth mud	0.001	Maize (170 cm)	9.5
Smooth snow	0.005	Brush (135 cm)	14.0
Desert	0.03	Orange grove (350 cm)	50.0
Wet soil	0.02	Deciduous forest (1700 cm)	270
Grass (1.5 cm)	0.2	Pine forest (2700 cm)	300

Adapted from Ref. [34].

Example 14.2

A temporary field station in the sandy plains of North Dakota is measuring conditions over a recently irrigated piece of land. At 1 and 10 m above the surface the wind speeds are measured as 1.2 m/sec and 2.5 m/sec, respectively, along with temperature readings of 25°C ($z_1 = 1$ m) and 20°C ($z_2 = 10$ m). A *hygrometer* has measured the specific humidity at 1 m above the surface to be 0.002, the irrigation preparation has left rigid bluffs in the form of furrows. We are asked to calculate the evaporation rate, E, in both mass flux units and the equivalent loss rate of water from the surface of the irrigation channel. The weather service reports a typical day with a local barometric pressure of 101.325 kPa.

Solution

The solution to this problem depends upon use of the Dalton equation, Eq. (14.4.40), which requires knowledge of the mixture density, ρ; the wind velocity at the reference height, $u_1 = u_{zh}(z = z_1 = h = 1$ m); the Dalton number, $\mathbf{D}_a$; the humidity at the surface, c_{qo}; and the humidity at the reference height, $\bar{c}_{q1} = \bar{c}_{qzh}(z = z_1 = h = 1$ m). The Dalton number is examined first:

1. *Dalton number.* From Eq. (14.4.42)

$$\mathbf{D}_a = \frac{Cd_h^{1/2}}{\left[(a_v B)^{-1} + a_v^{-1} \mathbf{C}d_h^{-1/2} \right]}$$

and the first order of business is to calculate the friction velocity, u_*. Assuming neutral conditions, the velocities at two heights in the constant stress layer are related as

$$u_2(z_2) - u_1(z_1) = \frac{u_*}{k} \ln\left(\frac{z_2}{z_1}\right)$$

or

$$u_* = \frac{k(u_2 - u_1)}{\ln(z_2/z_1)} = \frac{0.4(2.5 - 1.2) \text{ m/s}}{\ln(10/1)}$$

$$= 0.23 \text{ m/s}$$

From Table 14.6 the sandy but rough land will correspond to a roughness $z_o = 0.1$ cm, and therefore, from Eq. (14.4.4) the roughness Reynolds number, $\mathbf{R}_{zo}$, is

$$\mathbf{R}_{zo} = \frac{u_* z_o}{\nu} = \frac{(0.23 \text{ m/s})(0.1 \text{ cm})(0.01 \text{ m/cm})}{1.5(10^{-5}) \text{ m}^2/\text{s}}$$

$$= 18.6$$

and

$$(a_v B)^{-1} = 7.3\mathbf{R}_{zo}^{1/4}\mathbf{S}_c^{1/2} - 5.0 = 7.3(18.6)^{1/4}\left(\frac{\nu}{\mathcal{D}_v}\right)^{1/2} - 5.0$$

$$= 7.3(2.08)\left(\frac{1.5(10^{-5})}{2.42(10^{-5})}\right)^{1/2} - 5.0$$

$$= 12.0 - 5.0$$

$$= 7.0$$

The Dalton number also requires the value of Cd_h; therefore, from Eq. (14.4.39)

$$Cd_h = \frac{u_*^2}{u_{zh}^2(z = z_1)} = \frac{(0.28 \text{ m/s})^2}{(1.2 \text{ m/s})^2} = 0.54$$

The Dalton number can now be calculated

$$\mathbf{D}_a = \frac{Cd_h^{1/2}}{(a_v B)^{-1} + a_v^{-1} Cd_h^{1/2}} = \frac{(0.054)^{-1/2}}{[(7.0) + (1)(0.054)^{-1/2}]} = 0.021$$

2. *Mixture density.* From Eq. (14.4.5) the mixture density is found from

$$\rho = \rho_m = \frac{p}{RT}\left(1 - 0.378\frac{p_v}{p}\right)$$

Here R is the ideal gas constant for dry air ($R = 287$ m·N/kg·K). Therefore,

$$\rho_m = \frac{101.3(10^3) \text{ N/m}^2}{(287 \text{ m·N/kg·K})(298 \text{ K})}\left[1 - 0.378\frac{3.17(10^3) \text{ N/m}^2}{101.3(10^3) \text{ N/m}^2}\right]$$

The vapor pressure is taken for the 25°C which is now assumed to be constant from $z = 1$ m to the surface. After the calculation the mixture density is found to be

$$\rho_m = 1.17 \text{ kg/m}^3 = 1.17(10^{-3}) \text{ g/cm}^3$$

3. *Humidities.* If the mixture density is known, then the vapor density, ρ_v, is readily calculated from Eq. (14.4.2) as

$$\rho_v = \frac{p_v}{1.61 R_{dry} T} = \frac{3.17(10^3) \text{ N/m}^2}{(1.61)(287 \text{ m·N/kg·K})(298 \text{ K})}$$
$$= 0.023 \text{ kg/m}^3 = 0.023(10^{-3}) \text{ g/cm}^3$$

Then

$$\rho_m = \rho_{dry \ air} + \rho_v$$
$$\rho_{dry \ air} = \rho_m - \rho_v = 1.17 - 0.023 = 1.14 \text{ kg/m}^3 = 1.14(10^{-3}) \text{ g/cm}^3$$

The *specific humidity* at the surface is found by

$$c_{qo} = \rho_v/\rho_m = \frac{0.023}{1.17} = 0.02$$

We can check the *relative humidity* by finding the saturation vapor pressure, ρ_s, that is, Eq. (14.4.6)

$$r(\%) = \frac{\rho_v}{\rho_s}(100) = 1(100) = 100\%$$

Since we have assumed the saturation value for p_v at the surface (3.17 kPa), it is anticipated that $r = 100$. However, an empirical formula for p_s, the saturation vapor pressure, can be adapted from Bras ([31], Eq. (3.31)) where T is in Centigrade units and p_s is in millibar units

$$p_s \approx 33.8639\left[(0.00738 + 0.8072)^8 - 0.000019\right] - 0.000019(1.8T + 48) + 0.00136$$

$$\text{(14.4.45)}$$

Using the data from this problem, Eq. (14.4.45) gives a value of $p_s = 31.66$ millibars or 3.16 kPa. This value would be anticipated as we chose the saturation vapor pressure. Since the values are within 0.01kPa some measure of the error is also gained.

4. *Evaporation estimate.* Returning to the evaporation equation

$$E(\text{kg/m}^2\text{s}) = \rho u_{zh}(z = z_1)\mathbf{D}_a\left(c_{qo} - \bar{c}_{qzh}(z = z_1)\right)$$
$$= (1.17\text{kg/m}^3)(1.2\text{ m/s})(0.021)(0.020 - 0.018)$$
$$= 5.3(10^{-5})\text{ kg/m}^2\text{s}$$

Over a 1-square-meter surface area the total mass evaporated over 1 day (86,400 seconds) is

$$M_{\text{day}} = \left(5.3(10^{-5})\text{ kg/m}^2\text{s}\right)(1\text{ m}^2)(86,400\text{ sec/day}) = 4.58\text{ kg}$$

As in Ex. 9.4 the equivalent height of water surface, h_w, lost per day per square meter of surface area is

$$M_{\text{day}} = 4.58\text{ kg} = \rho_w h_w A \qquad \Longrightarrow \qquad h_w = \frac{M_{\text{day}}}{\rho_w A}$$

so

$$h_w = \frac{4.58\text{ kg}}{(997.1\text{ kg/m}^3)(1\text{ m}^2)} = 0.00459\text{ m/day} = 0.46\text{ cm/day}$$

Therefore, $h_w = 0.46$ cm of water is lost per day per square meter of surface area.

5. *Sensible heat.* From Table 14.5 the latent heat of evaporation L_e is found from

$$L_e = 597.3 - 0.57(25°\text{C}) = 583.05\text{ cal/g}$$

This can be converted to joules per second

$$L_e(\text{J/g}) = L_e(\text{cal/g})4.186\text{ J/g} = 2440.6\text{ J/g}$$

During one day 4.6 kg of mass is evaporated per square meter; therefore, the total heat input into the equivalent water mass is

$$Q_e = (4.6\text{ kg})(2440.6\text{ J/g})(1000\text{ g/kg}) = 11.2(10^6)\text{ J}$$

EXERCISES

14.4.1 While the latent heat of condensation is the heat given up in a gram mass when water changes phase from vapor to a liquid, the latent heat of sublimation (*a*) is temperature dependent; (*b*) is constant; (*c*) accounts for heat absorbed as vapor changes to ice; (*d*) accounts for heat absorbed as ice changes to vapor; (*e*) all of the above.

14.4.2 The triple point is the point on the pressure, specific volume, temperature diagram (*a*) denoting where water may exist as a pure vapor or a mixture of vapor and water; (*b*) is the only point in the *p-V-T* diagram where ice, water, and vapor coexist; (*c*) occurs at a value of 0°C; (*d*) occurs at a value of 6.11 mbar; (*e*) *b*, *c*, and *d*.

14.4.3 Relative humidity (*a*) is always less than or equal to the saturation vapor density; (*b*) is inversely proportional to the dew point temperature; (*c*) equals $(\rho_v/\rho_s) \times 100$ percent; (*d*) is the ratio of ρ_v multiplied by the absolute humidity and 100; (*e*) all of the above.

14.4.4 In comparison to the boundary layer explored in Chap. 6, the atmospheric boundary layer (*a*) is marked by an outer layer where natural convection effects are important; (*b*) also has a constant stress layer; (*c*) contains potentially quite dispersed but large roughness elements; (*d*) contains no hydrodynamically smooth viscous sublayer; (*e*) all but *d*.

14.4.5 Similarity solution methods result in neutral, nonstratified boundary layer solutions for velocity, humidity, and temperature (*a*) that all vary logarithmically with distance from the wall; (*b*) that all employ the same roughness and displacement heights; (*c*) that are functions of the Richardson number; (*d*) that have linearly increasing turbulent flux profiles; (*e*) none of the above.

14.5 PROCESS REACTORS AND TANKS

As discussed in the opening section of this chapter engineered flows and transport fields are rationally conceived by the analyst to achieve a specific purpose or outcome. The outcome should be predictable and when used operationally should be cost efficient, dependable, and durable. Often the intended outcome may take a succession of processes or transformations to achieve the outcome. For example, the combining of water, grains, hops, and yeast to create beer requires more than 20 processes to be applied to these inputs. Of relevance to the civil or environmental engineer the transformation of municipal wastes into clear water that can be reinserted into the natural stream or ground water requires waste water treatment plants containing as many as 70 *unit process operations* [38].

Some of these processes are quite simple, such as the land application of sludge, and as such there is little except drying and evaporation that fluid mechanics and transport phenomena methods can help quantify. Most of the processes, however, are dominated by fluid mechanics, transport phenomena, and associated biological and chemical transformations. The physical units containing the processes are geometrically designed to elicit precise transport fields and are called *process reactor tanks, vessels,* or *ponds.* Tanks and vessels are completely closed reservoirs with access provided solely by well-placed conduits. Ponds are distinguished from tanks only by virtue of a large expanse of the fluid surface which is exposed to the atmosphere and therefore subject to the wind, radiation, convection, and evaporation effects of the atmosphere.

The purpose of the next two sections is to review two areas of reaction vessel fluid mechanics which affect their design and selection: The first is the relationship between the type of flow through the reactor, the reactor process, and the required vessel size; and the second is on vessel mixing and agitation.

Reactor Process Types and Analysis

As seen in the many textbooks on wastewater treatment plant design or industrial unit processes (e.g., [29, 38–40]) there are six types of process reactors. Figure 14.17 contains schematics of the various types. Four types (A–D) are fairly general in

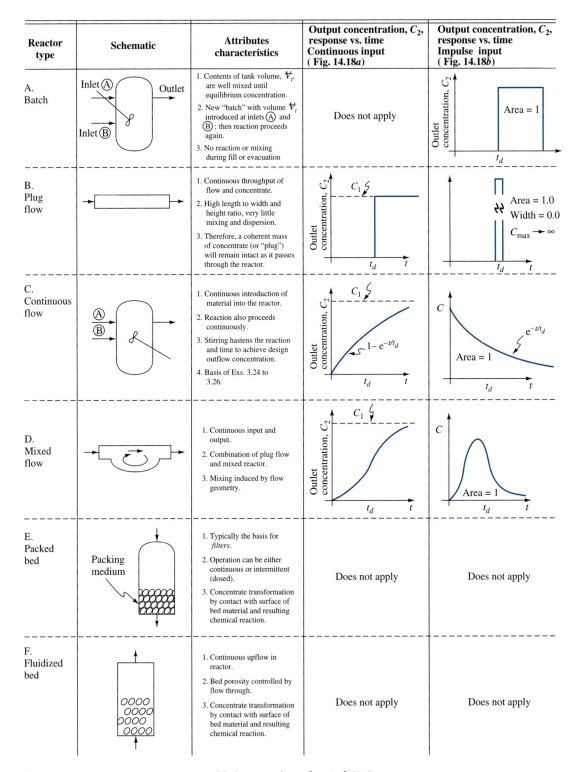

Figure 14.17   Reactor process types and behavior (redrawn from Ref. [38]).

application and can be used for a number of processes. The *packed and fluidized reactors* (E and F) serve quite specialized functions, and although they are quite fascinating in their complex fluid mechanics and transport, their analysis is typically left to more advanced coursework.

The following is a brief description of the response of reactors A–D. The basic mass balance approach for the reactor was introduced in Sec. 3.9 from a control volume perspective. Examples 3.24 to 3.26 elaborated on the basic principles for a continuously stirred reactor undergoing a first-order decay rate. Equation (3.9.12) is the basic control volume equation for a one-inlet, one-outlet system and is repeated here for the case of one inflow (1) and outflow (2) with cross-section average concentration and velocity values. V is the tank volume and S is the source-sink term.

$$V\frac{dC}{dt} = V_1 A_1 C_1 - V_2 A_2 C_2 + S \qquad (14.5.1)$$

For simpler cases, the source-sink term for various reactions can usually be selected from the three forms listed in Table 14.7. The zero-, first-, and second-order reaction labels essentially refer to the exponent on the concentration variable on the right-hand side of the ordinary differential equation. The parameters k_o, k_1, and k_2 are the rate coefficients or k-rate coefficients whose dimensions are distinct, that is, $[M/L^3 t]$, $[t^{-1}]$, and $[L^3/Mt]$, respectively. Once again reference should be made to Exs. 3.24 to 3.26 for a detailed calculation based upon first-order reaction kinetics for both decay (uptake) and generation.

Figure 14.17 contains a summary of the various features of each device. Particular attention is drawn to the last two columns in Fig. 14.17 which describes the time histories of the concentration at the *outlet* of the system. With regard to Fig. 14.18 two possible inputs are created to show these reactor differences, the continuous and impulse loading. The continuous loading essentially starts at time zero and proceeds to input material into the tank at predictable (or known) flows and concentrations. One or more inputs may be permitted. The impulse condition essentially puts a known mass of material (which translates into a fixed area in the diagram) into the system instantaneously. The functional description for this behavior is a Kronecker delta function. The two figures say nothing about how the concentration varies within the tank such that these outlet behaviors result.

The Batch Reactor

The batch reactor is the simplest to analyze from a transport perspective as there is no system throughput or transport. Rather new material is introduced into the tank, the tank is sealed, and the concentrate stirred for the time necessary to achieve a desired concentration in the fluid. The material in the tank is mixed to this uniform concentration throughout the tank. The *detention time,* or residence time, t_d (Fig. 14.17), is

TABLE 14.7 Forms of reaction terms

Reaction order	Rate equation form	Integrated form	Reference time (half life)
Zero	$V\frac{dC}{dt} = S = -Vk_o$	$C = C_o(t = 0) - k_o t$	$t_r = \frac{C_o}{2k_o}$
First	$V\frac{dC}{dt} = S = -Vk_1 C$	$\frac{C}{C_o} = \exp(-k_1 t)$	$t_r = \frac{1}{k_1}\ln 2$
Second	$V\frac{dC}{dt} = S = -Vk_2 C^2$	$\frac{1}{C} - \frac{1}{C_o} = k_2 t$	$t_r = \frac{1}{k_2 C_o}$

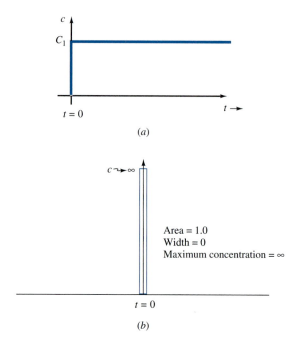

Figure 14.18   Reactor input concentration condition: (a) continuous loading and (b) impulse loading.

the time the fluid is retained in the tank, and this time is the amount required for the mixing or chemical and/or biological reaction to achieve the desired concentration.

Equation (14.5.1) written for the batch reactor is quite simple, that is,

$$V\frac{dC}{dt} = S$$

Therefore, the batch reactor is governed solely by the volume of the tank, the type of reaction, and the reaction rates. The results in Table 14.7 are then directly applicable to the batch reactor. Mechanical agitation by stirring is used to ensure that the entire contents of the tank are allowed to achieve the maximum chemical or biological transformation yielded by the reaction rate. Figure 14.17, column 5, contains a schematic of the concentration behavior at the outlet. Essentially a known volume (V_t) of material is released with a uniformly distributed concentration. The outlet behavior is therefore a well mixed concentration enduring over the time required to evacuate the tank volume.

The Continuous Flow Reactor

This situation has been dealt with at some length in Exs. 3.24 to 3.26. The outlet response curves in Fig. 14.17, column 4, can be derived as a special case of Eq. (3.9.16) by assuming no chemical or biological reaction ($k = k_1 = 0$) and redoing the calculation for the coefficients using the initial condition that the outlet concentration $C_2(t = 0) = 0$. With this initial condition the outlet response over time is

$$C_2(t) = C_1\left[1 - e^{-t(Q/V_t)}\right] = C_1\left(1 - e^{-t/t_d}\right) \qquad \textbf{(14.5.2)}$$

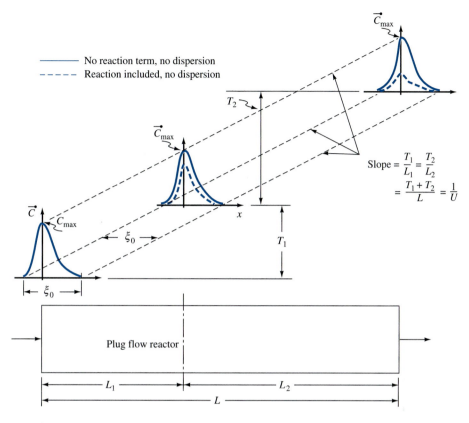

Figure 14.19 Concentration distribution advance for a plug flow reactor.

Here the hydraulic *retention time*, t_d, has been used and defined as Q/Ψ_t where Q is the net through flow and Ψ_t is the tank volume. The time variation is portrayed in the schematic in column 4. If the process is reversed and $C_2(t = 0) = C_1$, then a *purge* condition exists and the solution of Eq (3.9.16) gives

$$C_2(t) = C_1 e^{-t/t_d} \tag{14.5.3}$$

The Plug Flow Reactor

The ideal plug flow reactor is essentially the pipe or channel version of the area average transport equation [Eq. (9.4.15)] with the dispersion coefficient assumed to be so small that it can be ignored. The plug flow reactor is quite long in the direction of the hydraulic throughput and has a constant cross-sectional area. If x is the long channel direction, then from Eq. (9.4.15)

$$\frac{\partial \overline{C}^\bullet}{\partial t} + U \frac{\partial \overline{C}^\bullet}{\partial x} = K \frac{\partial^2 \overline{C}^\bullet}{\partial x^2} \tag{14.5.4}$$

Again the —● symbol refers to the cross-section average and U is defined as Q/A. The plug flow reactor is designed to have no dispersion effects. Therefore, the dispersion term is zero. Further, in analyzing this equation it is useful to review Eq. (4.1.6) which shows the relationship between the total derivative dC/dt and its Eulerian counterpart. Therefore, the concentration equation can be written as

$$\frac{\partial \overline{C}^{\bullet}}{\partial t} + U\frac{\partial \overline{C}^{\bullet}}{\partial x} = \frac{d\overline{C}^{\bullet}}{dt} = 0 \qquad (14.5.5)$$

The physical implication of this statement is that $\overline{C}^{\bullet}(x, t)$ is a nonvarying function in that whatever shape or function the initial concentration distribution takes at time $t = 0$, it will retain that same shape. The additional implication is that mass is completely conserved. Figure 14.7 depicts this circumstance in columns 4 and 5. The only motion allowed is the simple advection of the initial concentration distribution down the channel. Figure 14.19 on page 689 additionally illustrates this point for an asymmetric concentration loading being advected down the reactor at speed U. The distribution at three different points along the reactor is noted as are a series of dashed lines with constant slopes (slope $= 1/U$) showing the evolution of the distribution in time as well. In general the differential equation, $d\overline{C}^{\bullet}/dt$, admits shape-preserving solutions of the form $G(x - Ut) = 0$ such that if a distribution anywhere in a flow where Eq. (14.5.5) governs along with the velocity can be measured, then the solution can be reconstructed backward or forward in time. Such techniques form the basis for the *method of characteristics* and have been used to provide methodologies for tracking a river pollution spill backward in time and space to determine its origin.

The primary assumption for plug flow reactor theory is that there is no dispersion. Most actual plug flow reactors have dispersion and consequently initial distributions with sharp gradients will have them smoothed over the length of the reactor due to dispersion. Figure 14.20 depicts this behavior for a hypothetical rectangular initial distribution. The shape changes considerably but it is remembered that mass will be conserved over the reactor length. The diffusion solutions developed in Chap. 9 directly apply to the nonideal plug flow conditions.

Continuously Stirred Tanks in Series

Following the development in Ref. [38], it is often quite efficient to replace one very large reactor with a series of smaller reactor tanks. Figure 14.21 contains a schematic of the setup and the variable index definitions. In this example all the tank volumes will be equal, $V_{i-1} = V_i = V_{i+1}$, and the flow through the system will be constant,

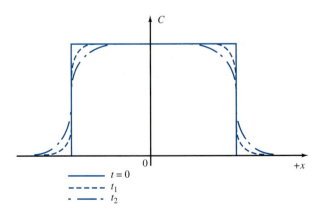

Figure 14.20   Nonideal plug flow concentration variation with time.

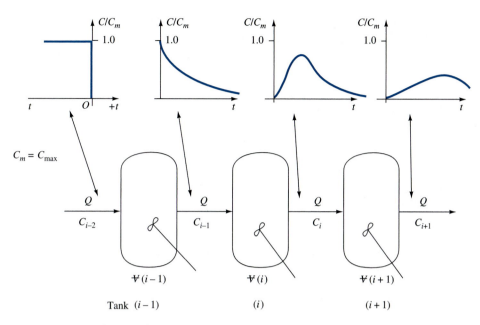

Figure 14.21 Schematic of reactors in series.

$Q = Q_1 = Q_2 = Q_3$. It is anticipated that as the input concentration is processed by each tank that the maximum concentration will decrease at each stage, and for the case of a chemical or biological reaction the overall mass will decrease. Continuing with the case of no reaction term, Eq. (3.9.12) can be rewritten for the ith tank with volume V/n, where n is the total number of reactors as

$$\left(\frac{V}{n}\right)\frac{dC_i}{dt} = V_i\frac{dC_i}{dt} = V_{i-1}A_{i-1}C_{i-1} - V_iA_iC_i \qquad \text{(14.5.6a)}$$

For $Q = A_iV_i = $ constant, the control volume equation becomes

$$\frac{dC_i}{dt} + \left(\frac{nQ}{V}\right)C_i = \left(\frac{nQ}{V}\right)C_{i-1} \qquad \text{(14.5.6b)}$$

From Eq. (14.5.3) for a purging system, the inlet concentration C_{i-1} can be specified as

$$C_{i-1}(t) = C_{i-2}e^{-n(Q/V)t} = C_{i-2}e^{-nt/t_d} \qquad \text{(14.5.7)}$$

It should be noted that the detention time, t_d, still refers to the detention time of the entire system, that is, over $V = \sum_{i=1}^{n} V_i$. It does not refer to the individual tank detention time, although it is easy to redefine the equations from this perspective. Equation (14.5.6b) then becomes

$$\frac{dC_i}{dt} + \left(\frac{nQ}{V}\right)C_i = \left(\frac{nQ}{V}\right)C_{i-2}e^{-n(Q/V)t}$$

which is of the form that the general solution, Eq. (3.9.16), applies. After applying boundary conditions,

$$C_i(t) = C_{i-2}\left(n\frac{Q}{V}t\right)e^{-n(Q/V)t} = C_{i-2}\left(\frac{nt}{t_d}\right)e^{-nt/t_d} \qquad \text{(14.5.7)}$$

If $C_o(t)$ denotes the initial inlet concentration history (i.e., $i = 1$ and $i - 1 = 0$), then the outlet concentration at tank i is related to the inlet concentration by

$$C_i(t) = \frac{C_o(t)}{(n-1)!} \left(\frac{nt}{t_d}\right)^{i-1} e^{-nt/t_d} \qquad \textbf{(14.5.8a)}$$

or in primitive variables

$$C_i(t) = \frac{C_o(t)}{(n-1)!} \left(\frac{ntQ}{V}\right)^{i-1} e^{-nt/t_d} \qquad \textbf{(14.5.8b)}$$

Figure 14.21 also contains schematics of the concentration response versus time at the various intertank locations. As noted the inlet condition is a purge, that is, $C_o(t < 0) = C_{max}$ and $C(t \geq 0) = 0$.

The Effect of Reaction Kinetics on Plug Flow Performance

A plug flow reactor with a source-sink term is analyzed as follows. Equation (14.5.5) now includes a source-sink term; therefore,

$$\frac{\partial \overline{C}^{\bullet}}{\partial t} + U \frac{\partial \overline{C}^{\bullet}}{\partial x} = -k\overline{C}^{\bullet} \qquad \textbf{(14.5.9)}$$

This can again be written in total derivative form as

$$\frac{d\overline{C}^{\bullet}}{dt} = -k\overline{C}^{\bullet} \qquad \textbf{(14.5.10)}$$

The initial condition is $\overline{C}_o^{\bullet}(x, t \geq 0) = C_o(x)$, and Eq. (14.5.10) has the formal solution

$$\overline{C}^{\bullet}(x - Ut) = C_o(x)e^{-kt}$$

The implication of this finding is that the initial condition is being advected downstream with velocity $U = U(t)$ (possibly nonsteady) and subject to continuous exponential degradation due to the first-order reaction term. Figure 14.19 has been annotated to show the effect of the exponential degradation with time. For the steady state case the problem is simpler in that $\partial \overline{C}^{\bullet}/\partial t = 0$, and after some manipulation, the resulting equation becomes

$$\frac{d\overline{C}^{\bullet}}{dx} = -k\frac{A}{Q}\overline{C}^{\bullet} \qquad \Longrightarrow \qquad \frac{d\overline{C}^{\bullet}}{\overline{C}^{\bullet}} = -k\frac{A}{Q}dx$$

Integrating, the solution becomes

$$\int \frac{d\overline{C}^{\bullet}}{dx} = -k\frac{A}{Q}\int_0^L dx \qquad \Longrightarrow \qquad \frac{\overline{C}^{\bullet}}{C_o} = e^{-kt_d} \qquad \textbf{(14.5.11)}$$

Clearly then in the steady-state case the reactor volume, throughflow, and reaction rate fix the outflow concentration at a steady value.

Continuously Stirred Tanks with Reaction

While single, continuously stirred reactors with reaction have been dealt with earlier, the case for a series of them with reaction has not. As before [Eq. (14.5.6)] the previous approach leads to the following control volume equation for the ith of n reactors

$$\frac{\Psi}{n}\frac{dC_i}{dt} = V_{i-1}A_{i-1}C_{n-1} - V_iA_iC_i - k\frac{\Psi}{n}C_i$$

$$\frac{dC_i}{dt} + \left[\frac{nQ}{\Psi} + \frac{\Psi}{n}k\right]C_i = \left(\frac{nQ}{\Psi}\right)C_{i-1} \qquad \text{(14.5.12)}$$

The solution form can be found by the general solution in Eq. (3.9.17b) for the purge plus reaction case. Therefore, Eq. (3.9.17b) is inserted into Eq. (14.5.12) and the result integrated. This integration is extremely tedious with the resulting differential equation written in the form

$$\frac{dC_i}{dt} + [D]C_i = [R]C_{i-2} \qquad \text{(14.5.13)}$$

where

$$D = \left[\frac{nQ}{\Psi} + \frac{\Psi}{n}k\right]$$

$$R = [E]\left(F + Ge^{-\alpha t}\right)$$

$$E = \left[\frac{nQ}{\Psi}\right] \quad F = \left(\frac{\beta}{\alpha}\right) \quad G = \left(\frac{\beta}{\alpha} + 1\right)$$

Here β and α are defined as in Eq. (3.9.16). The solution also takes the form of Eqs. (3.9.16) and (3.9.17a), that is,

$$C_i(t) = \left(\frac{R}{D}\right)C_{i-2}\left(1 - e^{-Dt}\right) + C_{i-2}e^{-Dt} \qquad \text{(14.5.14)}$$

A slightly rewritten version becomes

$$C_i(t) = \left[\frac{R}{D} + \left(\frac{R}{D} + 1\right)e^{-Dt}\right]C_{i-2} \qquad \text{(14.5.14)}$$

This progression continues on for n tanks in succession.

A steady-state version is useful as well. Equation (3.9.5) can be written for any successive pair of stages to give

$$C_i = C_i\left[\frac{1}{1 + \left(\frac{k\Psi}{nQ}\right)}\right] \qquad \text{(14.5.13)}$$

Repeated application of Eq. (14.5.13) between the input (C_o, $i = 0$) and output stage (C_n) gives

$$C_n = C_o\left[\frac{1}{1 + \left(\frac{k\Psi}{nQ}\right)^n}\right] \qquad \text{(14.5.14)}$$

The required total tank volume for the process to achieve the $C_o \rightarrow C_n$ change via the solution of Eq. (14.5.14) can be calculated as

$$\Psi = \frac{nQ}{k}\left[\left(\frac{C_o}{C_n}\right)^{1/n} - 1\right] \qquad \text{(14.5.15)}$$

TABLE 14.8   Comparison of tank reactor attributes

Removal efficiency, percent		Nondimensional Tank Volume V_*†				
85	5.67	3.16	2.42	2.23	2.14	2.09
95	19.0	6.94	4.46	3.88	3.63	3.50

† For 1, 2, 4, 6, 8, and 10 stirred tanks in series, n.

Example 14.3

In addition to Exs. 3.24 to 3.26 the following calculations illustrate the concepts presented here. Bacteria naturally die over time, a process which can be described by first-order reaction kinetics. Develop a table relating the number of tanks, their volume, the flow through, and the reaction rates as a function of *removal efficiency*. Removal efficiency, r_t, is defined as $[1 - C_n/C_o](100)$.

Solution

Equation (14.5.15) can be written in nondimensional volume as

$$V_* = \frac{V}{V_{\text{ref}}} = n \left[\left(\frac{1}{1 - \frac{r_t}{100}} \right)^{1/n} - 1 \right] \tag{14.5.16}$$

Here the *reference volume* is defined as (Q/k). Table 14.8 contains the comparison between r_t, n, and V_*.

It is interesting to note that the volumes (V_*) required for a plug flow reactor to achieve the same removal efficiencies as in Table 14.8 are 1.90 and 3.00 for the 85- and 95-percent removal efficiency, respectively. Further, it is noticed that as the number of tanks increases for a given removal efficiency, the total volume required approaches that required for the theoretical plug flow reactor. Cost savings might accrue since a number of smaller tanks might have a total cost of construction less than the one monolithic plug flow system. Additionally, for a given number of re-actors large incremental increases in tank volume are required for small increases in removal efficiency when the efficiencies required are above 90 percent. One can compare the 98-percent efficiency volume to the 95-percent efficiency volume and compare the incremental volume increase to that between 85 percent and 90 percent.

Example 14.4

Four reactor tanks are connected in series to achieve a 90-percent reduction in a pathogen concentration. If the reaction rate is 1.125 per day and the flow rate is 1 million gallons per day (mgd), compute the removal efficiency for an increase of Q to 1.45 mgd.

Solution

From Eq. (14.5.15) the volume of the four-tank series is computed from

$$V = \frac{nQ}{k} \left[\left(\frac{C_o}{C_n} \right)^{1/n} - 1 \right] = \frac{nQ}{k} \left[\left(\frac{1}{1 - \frac{r_t}{100}} \right)^{1/n} - 1 \right]$$

Since $Q = 1$ mgd $= 1.55$ ft^3/sec, $k = 1.125$ day$^{-1} = 1.3(10^{-5})$ sec^{-1} and $n = 4$. Then

$$V = \frac{4(1.55 \text{ ft}^3/\text{s})}{1.35(10^{-5}) \text{ sec}^{-1}}(1.77 - 1.0) = 3.53(10^5) \text{ ft}^3$$

This means that each tank has a volume of $0.88(10^5)$ ft^3.

For an increase in throughflow to 1.45 mgd, Eq. (14.5.14) is used to calculate the new efficiency as

$$\frac{C_n}{C_o} = \left\{ \frac{1}{1 + \left(\frac{k\Psi}{nQ} \right)^n} \right\}$$

$$= \left\{ \frac{1}{1 + \left[\frac{1.3(10^{-5})3.53(10^5)}{4(1.55)} \right]^4} \right\} = 1.77$$

Therefore, the new removal efficiency is

$$r_t = \left(1 - \frac{C_n}{C_o} \right) 100 = (0.23)100 = 23 \text{ percent}$$

Clearly a more rapid flowthrough results in far less time for a volume to remain in each tank to degrade by the reaction.

EXERCISES

14.5.1 In a continuous flow tank (a) the material is introduced into the tank in one operation; (b) will approach the final process concentration as $1 - \exp(-t/t_d)$; (c) is less efficient when performed in a series of small tanks as opposed to one large tank; (d) has a smaller detention time than a batch reactor of equal volume; (e) none of the above.

14.5.2 A plug flow reactor (a) can be quantitatively analyzed by the channel advection-dispersion equation and solutions presented in Chap. 9; (b) can be analyzed by the method of characteristics discussed in Chap. 12; (c) will achieve exponential decay for a first-order reaction rate in steady state; (d) will contain highly stratified flow; (e) all but c and d.

14.5.3 A second-order reaction rate process for a tank refers to (a) the highest degree of the temporal derivative in the governing equation for the tank; (b) the highest degree of spatial derivative in the governing equation; (c) the number of tanks in a series of reactor vessels; (d) the exponent or power on the dependent variable in the source-sink term for the tank equation; (e) all of the above.

14.6 MECHANICAL MIXING AND AGITATION

Reactor vessels and engineered ponds all serve to hasten the occurrence of a desired outcome. In the previous section the analysis of simple reactor responses employed a control volume analysis which contained a temporal term dC/dt. To evade the complications caused by complex internal tank flow and transport patterns, two simplifications were applied to this term, either steady state was assumed such that it equaled zero, or, C, the *volume average* concentration, was allowed to equal the outlet concentration. This second assumption was discussed in Sec. 3.9 and clearly

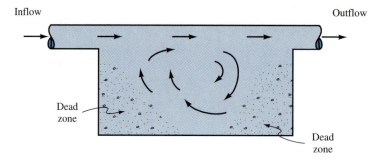

Figure 14.22   Schematic of short circuiting and dead zones in a tank.

this assumption, though analytically expedient, would be in considerable jeopardy especially if the inlet and outlet flows were time varying. Even under steady inlet and outlet conditions this assumption is in jeopardy especially if the momentum exchange between the inlet and outlet is the sole source of energy required for the mixing in the tank. As seen in Fig. 14.22 it is quite possible to develop *dead zones* or *short circuiting* in the tank which results in only part of the tank contents being mixed to the uniform concentration required to invoke this assumption.

Mechanical *stirrers* and *agitators* are frequently employed to create uniform conditions in the tank. Processes requiring such mixing include mixing of two fluids, dispersing gases in the liquid, dissolving solids (such as salt) in liquids, distributing particles throughout the tank (flocculation), or hastening heat transfer.

Agitator Types

Mechanical devices may be either *active* devices which mix through the external input of energy into a bladed propellor or stirrer or *passive* devices by which irregular flow paths and hence enhanced mixing are designed into the vessel geometry. Figure 14.23a contains schematics of several blade type stirrers, while Fig. 14.23b contains a passive device consisting of internal *baffles*. Natural or baffle-based devices serve to increase detention times and thus serve a dual function.

In Fig. 14.23a the *paddle* agitators (1 and 2) are used for low-speed installations and typically have even-numbered increments of blades (2, 4, etc.). At higher speeds rotational imbalances develop and an odd number of blades (3, 5, etc.) are required. These blades are designed to occupy 60 to 70 percent of the vessel width with each blade width being $\frac{1}{8}$ to $\frac{1}{10}$ the tank length [29]. Wall baffles are used at slightly higher speeds (Figs. 14.24 and 14.25). The paddle agitator does not yield any vertical flow and hence it is a poor mixer, but it does provide for the high shear necessary to keep particles in suspension. Typical examples of a nonindustrial origin include household paint mixers attached to electric drills or household electric mixers for mixing cake dough and making pastries. The *propellor* or *turbine* agitator turns at higher rates of speed. The turbine agitator usually refers to bladed systems operating at higher revolutions per minute (rpm). A typical household example is the ubiquitous ceiling fan which encourages even heat flow in the winter by operating in a downflow mode to mix ceiling-trapped hot air or uniform cool air distribution in the summer by operating in updraft mode to mix floor-trapped cool air.

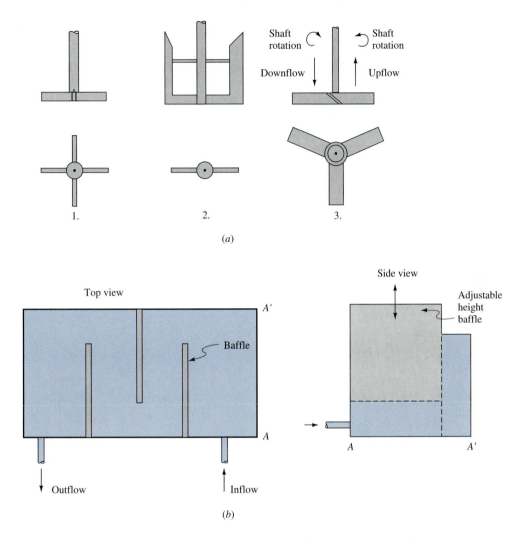

Figure 14.23 Stirring devices. (a) Active devices—stirrers. (b) Passive devices—baffles.

Circulation Patterns

From Brodkey and Hershey [41] and Geankopolis [29] schematics of the circulation fields in stirred tanks are compiled in Fig. 14.25. Figure 14.24a contains a schematic of the fluid behavior in an unbaffled tank. The simplest cylindrical tank driven by a paddle wheel agitator placed down the centerline of the tank will give rise over time to an ideal flow which could be described by simple potential flow theory cast in cylindrical coordinates. The centrifugal acceleration is the dominant feature of the system and no mixing or agitation is achieved (Fig. 14.24b). To alleviate this problem baffles are placed in the tank as shown in Figs. 14.24b and 14.25. Without baffles one could achieve successful mixing by placing the mixer-agitator off the tank center.

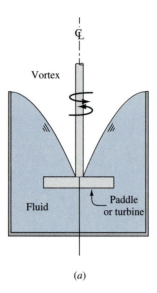

(a)

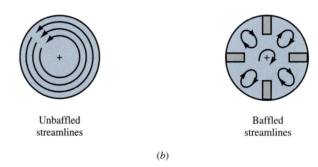

(b)

Figure 14.24   (a) An ideal stirred cylindrical tank. (b) Baffled and unbaffled flow.

Sizing and Scale Ratios

In the design of agitation tank systems the following parameters are used to create dimensionless ratios for scale up and sizing (Fig. 14.26): the tank internal diameter, D_i, and height, H_i; the diameter, D_p, and thickness, t_p, of the propellor; the height of the propellor above the tank bottom, H_p; and the baffle width, B_w. It is assumed that there are four baffles per tank. Table 14.9 summarizes design ratios for sizing a single tank.

To preserve similarity between a bench scale test of a process and the field scale prototype, these ratios can also be used as initial scaling factors between model, m, and prototype, p, as

$$\left.\frac{D_p}{D_i}\right|_m = \left.\frac{D_p}{D_i}\right|_p \qquad \left.\frac{B_w}{D_i}\right|_m = \left.\frac{B_w}{D_i}\right|_p$$

$$\left.\frac{H_p}{D_i}\right|_m = \left.\frac{H_p}{D_i}\right|_p \qquad \left.\frac{H_i}{D_i}\right|_m = \left.\frac{H_i}{D_i}\right|_p \qquad \left.\frac{t_p}{D_p}\right|_m = \left.\frac{t_p}{D_p}\right|_p \qquad \textbf{(14.6.1)}$$

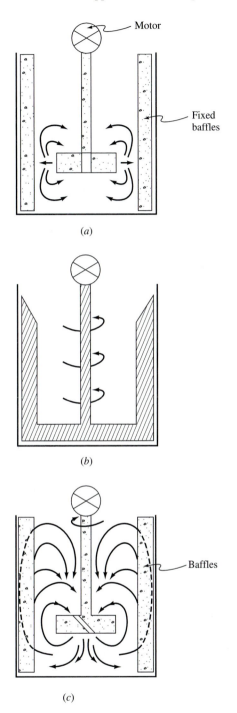

(a)

(b)

(c)

Figure 14.25 Circulation fields for various agitators. (a) Flat blade paddle or turbine. (b) Anchor, tangential flow circulation. (c) Pitched blade, axial flow.

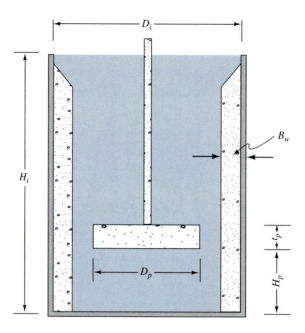

Figure 14.26   Definitions for tank scaling dimensions.

TABLE 14.9   Design ratios for sizing a single tank

$0.2 < D_p/D_i < 0.5$	(turbulent flow)
$0.7 < D_p/D_i < 1.0$	(laminar flow)
$B_w/D_i = \frac{1}{12}$	
$H_p/D_i = \frac{1}{3}$	
$H_i/D_i = 1$	
$t_p/D_p = \frac{1}{5}$	

Adapted from Ref. [29, 41, 42].

Power Requirements and Mixing Times

A number of dimensionless groups are derived to prepare design correlations. Chief amongst these are the tank Reynolds number, $\mathbf{Re}_T$, tank power number, and mixing time. The tank or *impellor* Reynolds number and power number are defined as

$$\mathbf{R}_T = \frac{D_p^2 N \rho}{\mu} \qquad \mathbf{N}_p = \mathbf{P}_o = \frac{P_o}{\rho N^3 D_p^5} \qquad \text{(14.6.2)}$$

where N is the rotational speed in revolutions per second and P_o is the power in joules/sec or watts. The relationship between these two numbers has been explored for a variety of blade types [42], and Fig. 14.27 is a reproduction of the correlation.

The mixing time or *blend time*, θ_T, has been correlated with the impellor Reynolds number [43] via the nondimensional mixing number, $\mathbf{M}_t$, as

$$\mathbf{M}_t = \frac{\theta_T \left(N D_p^2\right)^{2/3} g^{1/6} D_p}{H_i^{1/2} D_i^{3/2}} \qquad \text{(14.6.3)}$$

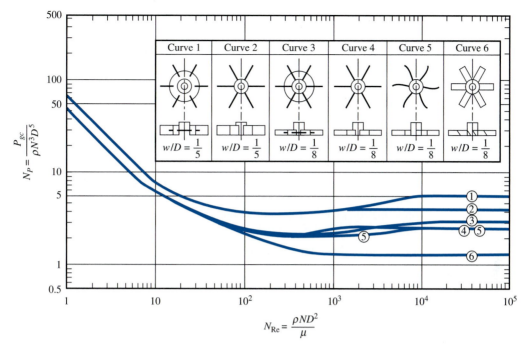

Figure 14.27 Power number–Reynolds number correlation in Newtonian fluids for various turbine impellor designs (from Bates, Fondy, and Corpstein [42]).

The correlation data indicate that $\mathbf{M}_t$ is constant for $\mathbf{R}_T > 1000$. By some algebra it is possible to show that mixing times between a model and prototype with the same power per unit volume are related by

$$\left.\frac{\theta_T\big|_p}{\theta_T\big|_m}\right. = \left(\frac{D_p\big|_p}{D_p\big|_m}\right)^{11/18}$$ (14.6.4)

To keep the same mixing times the ratio becomes

$$\left.\frac{P/\forall\big|_p}{P/\forall\big|_m}\right. = \left(\frac{D_p\big|_p}{D_p\big|_m}\right)^{11/4}$$ (14.6.5)

Hence, in order to keep θ_T constant, the power per unit volume increases substantially for the larger prototype tank.

More in-depth treatment is available in the references by Tatterson [44, 45].

EXERCISES

14.6.1 The purpose of tank baffles is to (*a*) increase mixing efficiency; (*b*) prevent dead zones; (*c*) prevent short circuiting; (*d*) hasten the chemical reaction; (*e*) all of the above.

14.6.2 The mixing number (*a*) is a linear function of the size of the tank; (*b*) is inversely proportional to the impellor Reynolds number; (*c*) is constant for a tank Reynolds number greater than 1000; (*d*) is linearly dependent upon impellor diameter; (*e*) *c* and *d*.

PROBLEMS

14.1 Calculate the settling velocity of 500-μm sand particles in water which has a temperature of 10°C.

14.2 Using Eq. (14.2.23) for the terminal velocity of a settling particle, derive an equation which expresses the settling velocity of a particle in the Stokes range. Express the equation in terms of the diameter of the particle, the density of the particle, and the density and dynamic viscosity of the fluid medium.

14.3 An experiment is performed in a channel with water flowing such that the turbulence can be considered uniform. The water temperature is 10°C and the bottom is made up of well-sorted sand grains with $D_{50} = 0.5$ mm. The concentration was measured at 5 cm above the bottom to be 23.5 mg/L and at 50 cm above the bottom to be 21.6 mg/L. Determine the eddy diffusivity of the sediment in water.

14.4 Derive the turbulent form of the mass transport equation [Eq. (14.2.10)] using the eddy diffusivity defined in Eq. (14.2.11) and the appropriate Reynolds averaging rules.

14.5 Calculate the eddy diffusivity for 300-μm sand in a 1.5-m-deep channel filled with water at 20°C. The channel is tilted slightly to generate a current which imparts a 1000-N/m^2 shear stress on the bottom. The diffusivities are required at 10, 15, 20, 40, 60, 100, and 120 cm.

14.6 A 3-m tank with water at 15°C is set up so that a current imparts a 2500-N/m^2 shear stress on the bottom. The bottom of the tank is made up of well-sorted sand which has a measured $D_{50} = 300$ μm. The concentration of sand was measured at 5 cm above the bottom to be 964 g/L. Calculate the concentration at 10 and 50 cm.

14.7 A simple rectangular settling tank must be designed so that all particles larger than 0.01 in. will collect at the bottom before the liquid flows through the tank. The flow rate is expected to reach 14 million gallons per day to the weir overflow tank. Using a length to width ratio of 4, what is the length of the tank? Assume that the liquid has $T = 70$°F.

14.8 Estimate the reference concentration (in g/L) above a sand bed with uniformly sized sand grains of $D = 0.6$ mm for a range of shear velocities (due to a uniform current) from 0.01 to 1 m/s. Also calculate the effective height, z_o. Use $\alpha_o = 26$, $\epsilon' = 2D$, $\gamma_o = 0.0024$, and $c_b = 0.65$.

14.9 A uniformly turbulent, current-induced flow field carries a dilute suspension of sand with a median diameter of 300 μm. Using an eddy diffusivity of 1.0 m^2/s, calculate the sand concentration 60 cm above the bottom if it is known that the concentration is 131 mg/L at a point 5 cm above the bottom. The mixture temperature is 20°C.

14.10 For the single grain size conditions given in Prob. 14.8, use the computer program to calculate the same variables as in that problem as well as the concentration profile. Plot the results.

14.11 Using the computer program, calculate the concentration profiles and various parameters using the conditions given in Ex. 14.1. However, in addition the shear velocity (or shear stress) should be halved and doubled, and the three results compared. The different profiles should be plotted on the same graph.

14.12 At the top of Long's Peak in the Rocky Mountains (*a*) what is the vapor pressure if boiling a container of eggs (without a lid) occurs at 190°F? (*b*) What is the ideal gas constant for the vapor? (*c*) What is the vapor density corresponding to the above condition?

14.13 What is the total pressure at the surface of the boiling water in Prob. 14.12? (*Hint:* Use the constant temperature pressure change equation derived for an ideal gas in Chap. 2 to find the dry pressure, *p*. Assume that Long's Peak is 12,000 ft high and that $T = 80°F$ and $\rho = 1.24$ kg/m^3 at sea level.)

14.14 For the conditions given in Probs. 14.12 and 14.13, compute the total density.

14.15 In Vicksburg, Mississippi, summertime conditions are marked by quite high relative humidities approaching 85 percent or higher. For afternoon temperatures of 35°C what are the saturation density and dewpoint temperature corresponding to $r = 85$ percent.

14.16 Explain why, when fog forms directly over a field of snow, the snow will melt faster than if no fog forms at all.

14.17 What are the potential temperature equivalents of the air temperature in Probs. 14.12 and 14.15.

14.18 For Ex. 14.2 how would the Dalton, roughness Reynolds, and Schmidt numbers change if the roughness, z_o, were typical of brush (Table 14.6).

14.19 Continuing Prob. 14.18, what is the evaporation rate for flow over brush. What is the total mass of moisture evaporated from one square meter for one day.

14.20 Referring to the reference velocity, $u_r(z = 10 \text{ m})$, specified in Ex. 14.2 what would be the displacement height, d_o, for the furrows ($z_o = 0.1$ cm) and for the brush in Prob. 14.18.

14.21 Using the temperature data in Ex. 14.2, estimate the heat flux H at the surface.

14.22 By calculation of the vertical acceleration of the moisture and the Monin-Obukov length for the conditions in Ex. 14.2, determine if the flow is buoyant, neutral, or stable.

14.23 Polluted water is to be treated in a series of completely mixed contact chambers. The concentration of the bacteria in the water is to be reduced from 10,000 organisms/mL to 10 organisms/mL. The detention time in the chambers is 42.5 min. Assuming first-order kinetics with a rate coefficient of 5.85 hr^{-1}, find the number of chambers required to reduce the bacteria to 10 organisms/mL.

14.24 Assume that the flocculation of suspended sediment particles is defined by a first-order reaction and the reference time is 2 h. At a time $t = 43$ min, the concentration of the suspended sediment is 98 particles/mL. Determine the initial concentration of the suspended particles and the corresponding rate coefficient. What is the time required for the concentration to be reduced to 10 percent of that of the initial concentration?

14.25 A number of reactor tanks are connected in series to achieve a reduction in pathogen concentration. The volume of each tank is 2000 m^3. If the flow rate is 1.32 m^3/s and the reaction rate is 0.12 hr^{-1}, determine the number of reactors required for a 73-percent removal efficiency.

14.26 Find the time required for a 90-percent removal of a bacteria population assuming second-order reaction.

14.27   The steady state, defined at 90-percent removal, in a continuous flow reactor is reached at time 1.85 hours after the process started. If the volume of the reactor is 16,500 m^3, determine the flow rate in m^3/s.

14.28   Instead of the continuous flow reactor in Prob. 14.25, a plug flow reactor is used. Assuming in this case first-order reaction and the conditions described in the problem, determine the value of the rate coefficient.

14.29   In a given steady-state filtration operation it has been found that the spatial concentration gradient can be evaluated as $dC/dx = -kC$. If the concentration at $x = 0$ is C_o, determine the distance x, where the concentration is $0.7C_o$.

14.30   The rate of generation of a bacteria population in a closed vessel is very well described by second-order kinetics. Derive an expression for the concentration of the bacteria population as a function of time and the rate coefficient. What is the time for the bacteria population to be doubled?

14.31   The contents of a circular tank of diameter 30 ft are to be mixed with an 8-ft-diameter turbine impeller that has 6 flat blades. The impeller is installed 8 ft above the bottom of the 30-ft-tall tank and is rotated at 50 rev/min. The mixture temperature is 40°C. By using the properties of water for the mixture, determine the tank Reynolds number. If the tank power number is 4.0, what is the power in watts?

14.32   For the conditions described in Prob. 14.31 find the flow rate through the tank in m^3/s if the head loss between the inlet and the outlet is 12 ft.

REFERENCES

1.   E. Oran and J. Boris, *Numerical Simulation of Reactive Flow,* Elsevier, New York, 1987.

2.   D. Gidaspow, *Multiphase Flow and Fluidization,* Academic Press, Boston, MA, 1994.

3.   R. Meyer, *Theory of Dispersed Multiphase Flow,* Academic Press, New York, 1983.

4.   G. Papanicolaou, *Hydrodynamic Behavior of Interacting Particle Systems,* Springer-Verlag, New York, 1987.

5.   S. Soo, *Particulates and Continuum,* Hemisphere Pub. Co., New York, 1989.

6.   S. Soo, *Multiphase Fluid Dynamics,* Glower Technical, Alsershot-Brookfield, 1990.

7.   T. Tadros, *Solid/Liquid Dispersions,* Academic Press, New York, 1987.

8.   V. Vanoni, *Sedimentation Engineering,* Amer. Soc. Civil Engrs., New York, 1975.

9.   M. Yalin, *Mechanics of Sediment Transport,* 2nd ed., Pergammon Press, U.K., 1977.

10.   J. Lumley, "Two-Phase and Non-Newtonian Flows," in *Turbulence,* ed. P. Bradshaw, Springer Verlag, Berlin, pp. 290–324, 1976.

11.   J. Hunt, "The Turbulent Transport of Sediment in Open Channels," *Proc. Royal Soc. of London,* 224A, pp. 322–335, 1954.

12.   J. Smith, "Modeling of Sediment Transport on Continental Shelves," in *The Sea,* Vol. 6, ed. E. Goldberg, I. McCave, J. O'Brien and J. Steele, Wiley Interscience, New York, p. 539, 1976.

13. J. Smith, and S. McClean, "Spatially Averaged Flow over a Wavy Surface," *J. Geophys Res.,* 82, pp. 1735–1746, 1977.

14. P. Klebanoff, *Natl. Adv. Comm. Aeronautical Technical Notes*, Rept. No. 3178.

15. A. Townsend, *The Structure of Turbulent Shear Flow,* 2nd ed., Cambridge University Press, New York, 1976.

16. C. White, "The Equilibrium of Grains on the Bed of a Stream," *Proc. Royal Soc. London*, Ser. A, 174, pp. 322–338, 1940.

17. S. Glenn and W. Grant, "A Suspended Sediment Stratification Correction for Combined Wave and Current Flows," *J. Geophysical Research,* 92, pp. 8244–8264, 1986.

18. M. Garcia and G. Parker, "Entrainment of Bed Sediment into Suspension," *J. Hydraul. Eng.*, 117, pp. 414–435, 1991.

19. P. Owen, "Saltation of Uniform Grains in Air," *J. Fluid Mech.,* 20, pp. 225–242, 1964.

20. J. Smith and S. McClean, "Boundary Layer Adjustments to Bottom Topography and Suspended Sediments," in *Bottom Turbulence,* ed. J. Nihoul, Elsevier, New York, pp. 123–151, 1977.

21. J. Dutton, *The Ceaseless Wind, An Introduction to Meteorology,* McGraw-Hill, New York, 1976.

22. J. Turner, *Buoyancy Effects in Fluids,* Cambridge Univ. Press, U.K., 1973.

23. C. Rossby and R. Montgomery, "The Layers of Frictional Influence in Wind and Ocean Currents," Mass. Inst. Tech./Woods Hole Ocean. Inst., *Papers in Physical Oceanography and Meteorology,* 3, No. 3, 1935.

24. W. Munk and E. Anderson, "Notes on the Theory of the Thermocline," *J. Marine Res.,* 1, pp. 276–295, 1948.

25. R. French, *Open Channel Hydraulics,* McGraw-Hill, New York, 1985.

26. W. Brutsaert, *Evaporation into the Atmosphere,* D. Reidel Pub. Co., Holland, 1982.

27. R. Stull, *An Introduction to Boundary Layer Meteorology,* Kluwer Pub. Co., Boston, MA, 1988.

28. J. Businger, J. Wyngaard, Y. Izumi, and E. Bradley, "Flux Profile Relationships in the Atmospheric Surface Layer," *J. Atmos. Science,* 28, pp. 181–189, 1971.

29. C. Geankopolis, *Transport Operations and Unit Operations,* 3rd ed., Prentice Hall, New Jersey, 1993.

30. J. Iribarne and W. Godson, *Atmospheric Thermodynamics,* J. Wiley and Sons, New York, 1981.

31. R. Bras, *Hydrology,* Addison-Wesley, New York, 1990.

32. M. Moran and H. Shapiro, *Fundamentals of Engineering Thermodynamics,* D. Reidel Pub. Co., Holland, 1992.

33. C. Bennet and J. Meyers, *Momentum, Heat and Mass Transfer,* 2nd ed., McGraw-Hill, New York, 1974.

34. P. Eagleson, *Dynamic Hydrology,* McGraw-Hill, New York, 1970.

35. J. Bosen, "A Formula for Approximation of the Saturation Vapor Pressure over Water," *Monthly Weather Review,* 88, p. 275, 1960.

36. J. Bosen, "An Approximation Formula to Compute Relative Humidity from Dry Bulk and Dew Point Temperatures," *Monthly Weather Review,* 86, p. 486, 1958.

37.   K. Helfrich, E. Adams, A. Godbey, and D. Harleman, *Evaluation of Models for Predicting Evaporative Water Loss in Cooling Ponds,* Electric Power Resources Inst., Palo Alto, CA, 1982.

38.   Metcalf and Eddy, Inc., revised by G. Tchobanoglous and F. Burton, *Wastewater Engineering,* 3rd ed., McGraw-Hill, New York, 1991.

39.   J. Clark, W. Viessman Jr., and M. Hammer, *Water Supply and Pollution Control,* 3rd ed., Harper and Row Publ., New York, 1977.

40.   G. Fair, J. Geyer, and D. Okun, *Elements of Water Supply and Wastewater Disposal,* 2nd ed., J. Wiley and Sons, New York, 1971.

41.   R. Brodkey and H. Hershey, *Transport Phenomena: A Unified Approach,* McGraw-Hill, New York, 1988.

42.   R. Bates, P. Fondy, and R. Corpstein, *I and EC,* Design Development, 2, p. 310, 1963.

43.   K. Norwood, and A. Metzner, *AIChEJ*, 6, p. 432, 1960.

44.   G. Tatterson, *Scaleup and Design of Industrial Mixing Processes,* McGraw-Hill, New York, 1994.

45.   G. Tatterson, *Fluid Mixing and Gas Dispersion in Agitated Tanks,* McGraw-Hill, New York, 1991.

A

FORCE SYSTEMS, MOMENTS, AND CENTROIDS

The material in this appendix has been assembled to aid in working with force systems. Simple force systems are briefly reviewed, and first and second moments, including the product of inertia, are discussed. Centroids and centroidal axes are defined.

A.1   SIMPLE FORCE SYSTEMS

A free-body diagram for an object or portion of an object shows the action of all other bodies on it. The action of the earth on the object, called a *body force,* is proportional to the mass of the object. In addition, forces and couples may act on the object by contact with its surface. When the free body is at rest or is moving in a straight line with uniform speed, it is said to be in *equilibrium.* By Newton's second law of motion, since there is no acceleration of the free body, the summation of all force components in any direction must be zero and the summation of all moments about any axis must be zero.

Two force systems are equivalent if they have the same value for the summation of forces in every direction and the same value for the summation of moments about every axis. The simplest equivalent force system is called the *resultant* of the force system. Equivalent force systems always cause the same motion (or lack of motion) of a free body.

In coplanar force systems the resultant is either a force or a couple. In noncoplanar parallel force systems the resultant is either a force or a couple. In general noncoplanar systems the resultant may be a force, a couple, or a force and a couple.

The action of a fluid on any surface may be replaced by the resultant force system that causes the same external motion or reaction as the distributed fluid-force system. In this situation the fluid may be considered to be completely removed, the resultant acting in its place.

A.2   FIRST AND SECOND MOMENTS: CENTROIDS

The moment of an area, volume, weight, or mass may be determined in a manner analogous to that of determining the moments of a force about an axis.

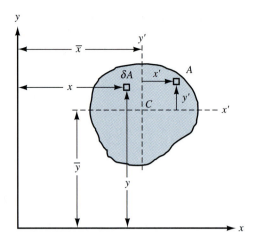

Figure A.1   Notation for first and second moments.

First Moments

The moment of an area A about the y axis (Fig. A.1) is expressed by

$$\int_A x \, dA$$

in which the integration is carried out over the area. To determine the moment about a parallel axis, for example, $x = k$, the moment becomes

$$\int_A (x - k) \, dA = \int_A x \, dA - kA \tag{A.1}$$

which shows that there will always be a parallel axis $x = k = \bar{x}$ about which the moment is zero. This axis, called a *centroidal axis,* is obtained from Eq. (A.1) by setting it equal to zero and solving for $\bar{x}$,

$$\bar{x} = \frac{1}{A} \int_A x \, dA \tag{A.2}$$

Another centroidal axis may be determined parallel to the x axis as

$$\bar{y} = \frac{1}{A} \int_A y \, dA \tag{A.3}$$

The point of intersection of centroidal axes is called the *centroid* of the area. It may easily be shown, by rotation of axes, that the first moment of the area is zero about any axis through the centroid. When an area has an axis of symmetry, it is a centroidal axis because the moments of corresponding area elements on each side of the axis are equal in magnitude and opposite in sign. When location of the centroid is known, the first moment for any axis may be obtained without integration by taking the product of area and distance from centroid to the axis, that is,

$$\int_A z \, dA = \bar{z}A \tag{A.4}$$

The centroidal axis of a triangle, parallel to one side, is one-third the altitude from that side; the centroid of a semicircle of radius a is $4a/3\pi$ from the diameter.

By taking the first moment of a volume V about a plane, say the yz plane, the distance to its centroid is similarly determined as

$$\bar{x} = \frac{1}{V} \int_V x \, dV \tag{A.5}$$

The mass center of a body is determined by the same procedure,

$$x_m = \frac{1}{M} \int_M x \, dm \tag{A.6}$$

in which dm is an element of mass and M is the total mass of the body. For practical engineering purposes the *center of gravity* of a body is at its mass center.

Second Moments

The second moment of an area A (Fig. A.1) about the y axis is

$$I_y = \int_A x^2 \, dA \tag{A.7}$$

It is called the *moment of inertia* of the area, and it is always positive, since dA is always considered positive. After transferring the axis to a parallel axis through the centroid C of the area,

$$I_c = \int_A (x - \bar{x})^2 \, dA = \int_A x^2 \, dA - 2\bar{x} \int_A x \, dA + \bar{x}^2 \int_A dA$$

Since

$$\int_A x \, dA = \bar{x}A \qquad \int_A x^2 \, dA = I_y \qquad \int_A dA = A$$

therefore,

$$I_c = I_y - \bar{x}^2 A \qquad \text{or} \qquad I_y = I_c + \bar{x}^2 A \tag{A.8}$$

In words, the moment of inertia of an area about any axis is the sum of the moment of inertia about a parallel axis through the centroid and the product of the area and square of the distance between the axes. Figure A.2 shows moments of inertia for four simple areas.

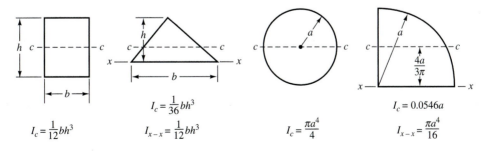

$$I_c = \frac{1}{12}bh^3$$

$$I_c = \frac{1}{36}bh^3$$
$$I_{x-x} = \frac{1}{12}bh^3$$

$$I_c = \frac{\pi a^4}{4}$$

$$I_c = 0.0546a$$
$$I_{x-x} = \frac{\pi a^4}{16}$$

Figure A.2   Moments of inertia of simple areas about centroidal axes.

The *product of inertia I_{xy}* of an area is expressed by

$$I_{xy} = \int_A xy \, dA \tag{A.9}$$

with the notation of Fig. A.1. It may be positive or negative. By writing the expression for product of inertia about the xy axes, in terms of $\bar{x}$ and $\bar{y}$, Fig. A.1,

$$I_{xy} = \int_A (\bar{x} + x')(\bar{y} + y') \, dA = \bar{x}\bar{y}A + \int_A x'y' \, dA$$

$$+ \bar{x} \int_A y' \, dA + \bar{y} \int_A x' \, dA = \bar{x}\bar{y}A + \bar{I}_{xy} \tag{A.10}$$

$\bar{I}_{xy}$ is the product of inertia about centroidal axes parallel to the xy axes. Whenever either axis is an axis of symmetry of the area, the product of inertia is zero.

The product of inertia I_{xy} of a triangle having sides b and h along the positive coordinate axes is $b^2 h^2 / 24$.

APPENDIX

B

COMPUTER PROGRAMMING AIDS

The earlier versions of this text were amongst the first textbooks in either solid or fluid mechanics to fully incorporate numerical and computer methods into the solution of problems. When first introduced in earlier editions only mainframe computers existed. They executed batch jobs and BASIC and FORTRAN were the only widely used languages. There was little agreement on what mathematical concepts should lie at the heart of a particular numerical solution. With the widespread availability of powerful, inexpensive PCs, numerical solution procedures are so accepted and routine they have become a button on hand-held calculators (e.g., Runge-Kutta solutions for ordinary differential equations) or an icon on mathematical software packages. Of more importance, however, is that unlike the time period of the prior edition, numerical analysis courses are widely available, if not required, in engineering and science curricula, and therefore, numerical procedures do not have to be a subject of instruction in these texts. However, the materials that were in this appendix in the prior edition are still considered to be important because they form the basis of the example calculations performed in this edition of the text. For reference they have been placed in the Web pages for the book, and in the unlikely event they have not been learned in prior coursework they can be reviewed there if necessary.

C

PHYSICAL PROPERTIES OF FLUIDS

TABLE C.1   Physical properties of water in SI units

Temp, °C	Specific weight γ, N/m³	Density ρ, kg/m³	Viscosity $\mu \times 10^3$, N·s/m²	Kine-matic viscosity $\nu \times 10^6$, m²/s	Surface tension $\sigma \times 10^2$, N/m	Vapor-pressure head absolute p_v/γ,† m	Bulk modulus of elasticity $K \times 10^{-7}$, N/m²	Thermal conduc-tivity k, W/m·K
0	9806	999.9	1.792	1.792	7.62	0.06	204	0.561
5	9807	1000.0	1.519	1.519	7.54	0.09	206	0.571
10	9804	999.7	1.308	1.308	7.48	0.12	211	0.580
15	9798	999.1	1.140	1.141	7.41	0.17	214	0.589
20	9789	998.2	1.005	1.007	7.36	0.25	220	0.598
25	9778	997.1	0.894	0.897	7.26	0.33	222	0.607
30	9764	995.7	0.801	0.804	7.18	0.44	223	0.615
35	9749	994.1	0.723	0.727	7.10	0.58	224	0.623
40	9730	992.2	0.656	0.661	7.01	0.76	227	0.630
45	9711	990.2	0.599	0.605	6.92	0.98	229	0.637
50	9690	988.1	0.549	0.556	6.82	1.26	230	0.643
55	9666	985.7	0.506	0.513	6.74	1.61	231	0.649
60	9642	983.2	0.469	0.477	6.68	2.03	228	0.654
65	9616	980.6	0.436	0.444	6.58	2.56	226	0.659
70	9589	977.8	0.406	0.415	6.50	3.20	225	0.663
75	9560	974.9	0.380	0.390	6.40	3.96	223	0.667
80	9530	971.8	0.357	0.367	6.30	4.86	221	0.670
85	9499	968.6	0.336	0.347	6.20	5.93	217	0.673
90	9466	965.3	0.317	0.328	6.12	7.18	216	0.675
95	9433	961.9	0.299	0.311	6.02	8.62	211	0.677
100	9399	958.4	0.284	0.296	5.94	10.33	207	0.679

† $\gamma = 9806$ N/m³.

TABLE C.2 Physical properties of water in USC units

Temp, °F	Specific weight γ, lb/ft³	Density ρ, slugs³	Viscosity $\mu \times 10^5$, lb·s/ft²	Kinematic viscosity $\nu \times 10^5$, ft²/s	Surface tension $\sigma \times 10^2$, lb/ft	Vapor-pressure head absolute p_v/γ,† ft	Bulk modulus of elasticity $K \times 10^{-3}$, lb/in²	Thermal conductivity k, Btu/(hr·°F·ft)
32	62.42	1.940	3.746	1.931	0.518	0.20	293	1.006
40	62.42	1.940	3.229	1.664	0.514	0.28	294	1.021
50	62.42	1.940	2.735	1.410	0.509	0.41	305	1.039
60	62.35	1.938	2.359	1.217	0.504	0.59	311	1.058
70	62.29	1.936	2.050	1.059	0.500	0.84	320	1.075
80	62.22	1.934	1.799	0.930	0.492	1.17	322	1.093
90	62.13	1.931	1.595	0.826	0.486	1.61	323	1.109
100	62.00	1.927	1.424	0.739	0.480	2.19	327	1.124
110	61.87	1.923	1.284	0.667	0.473	2.95	331	1.138
120	61.71	1.918	1.168	0.609	0.465	3.91	333	1.151
130	61.55	1.913	1.069	0.558	0.460	5.13	334	1.162
140	61.39	1.908	0.981	0.514	0.454	6.67	330	1.172
150	61.19	1.902	0.905	0.476	0.447	8.58	328	1.182
160	61.00	1.896	0.838	0.442	0.441	10.95	326	1.190
170	60.81	1.890	0.780	0.413	0.433	13.83	322	1.197
180	60.58	1.883	0.726	0.385	0.426	17.33	313	1.203
190	60.36	1.876	0.678	0.362	0.419	21.55	313	1.208
200	60.10	1.868	0.637	0.341	0.412	26.59	308	1.213
212	59.84	1.860	0.593	0.319	0.404	33.90	300	1.217

† $\gamma = 62.4$ lb/ft³.
Adapted from American Society of Civil Engineers, "Hydraulic Models," Manual of Engineering Practice, No. 25, ASCE, N.Y., 1942.

TABLE C.3 Physical properties of gases in SI units at low pressures and 26.67°C (80°F)

Gas	Molecular weight	Thermal conductivity $k \times 10^3$, W/m·K	Vapor pressure temperature at 1 atm, °C	Gas constant R, m·N/kg·K	Specific heat, kJ/kg·K c_p	Specific heat, kJ/kg·K c_v	Specific heat ratio κ
Acetylene, C_2H_2	26.0	—	—	—	1.703	1.377	1.24
Air	29.0	26.2	—	287	1.004	0.716	1.40
Ammonia, NH_3	16.0	24.4	−33.6	519.5	2.061	—	1.31
Carbon monoxide, CO	28.0	25.0	−191.3	297	1.043	0.745	1.40
Carbon dioxide, CO_2	46.0	—	—	—	0.850	0.661	1.29
Ethanol, C_2H_5OH	46.07	14.4	78.4	180.7	1.419	—	1.13
Helium, He	4.00	156.7	−268.6	2077	5.233	3.153	1.66
Hydrogen, H_2	2.02	186.9	−252.5	4121	14.361	10.216	1.40
Methane, CH_4	16.04	34.1	−161.5	519.5	2.238	1.381	1.62
Nitrogen, N_2	28.0	26.0	−195.8	297	1.038	0.741	1.40
Oxygen, O_2	32.0	26.3	−182.9	260	0.917	0.657	1.40
Water vapor, H_2O	18.0	18.7	100.0	462	1.863	1.403	1.33

TABLE C.4   Physical properties of gases in USC units at low pressures and 80°F (26.67°C)

Gas	Molecular weight	Thermal conductivity k, Btu/ft·s·°R	Vapor pressure temperature at 1 atm, °F	Gas constant R, ft·lb/lb$_m$·°R	Specific heat, Btu/lb$_m$·°R		Specific heat ratio κ
					c_p	c_v	
Acetylene, C_2H_2	26.0	—	—	—	—	—	1.24
Air	29.0	13.6	—	53.3	0.240	0.171	1.40
Ammonia, NH_3	16.0	12.7	−28.5	96.6	0.493	—	1.31
Carbon monoxide, CO	28.0	13.0	−312.3	55.2	0.249	0.178	1.40
Carbon dioxide, CO_2	46.0	—	—	—	—	—	1.29
Ethanol, C_2H_5OH	46.07	7.5	173.1	33.6	0.339	—	1.13
Helium, He	4.0	81.5	−451.5	386.0	1.25	0.753	1.66
Hydrogen, H_2	2.02	97.2	−422.5	766.0	3.43	2.44	1.40
Methane, CH_4	16.04	17.7	−258.7	96.6	0.535	0.330	1.62
Nitrogen, N_2	28.0	13.5	−320.4	55.2	0.248	0.177	1.40
Oxygen, O_2	32.0	13.7	−297.2	48.3	0.219	0.157	1.40
Water vapor, H_2O	18.0	9.7	212.0	85.8	0.445	0.335	1.33

TABLE C.5   Thermal conductivities, k, of liquids and solids

k (W/m·K) for liquids (25°C)			
Water	0.607	Acetone	0.161
Mercury	8.25	Benzene	0.141
Methanol	0.200	Toluene	0.131
Ethanol	0.169	Furan	0.126
Ethylene glycol	0.256	Glycerine	0.292

k (W/m·K) for solids (≤ 27°C)			
Calcium	$2.00(10^2)$	Asbestos (0°C)	0.09
Aluminum	$2.37(10^2)$	Asbestos (100°C)	1.0
Copper	$4.01(10^2)$	Brick, dry	0.04
Iron	$0.802(10^2)$	Brick, fireclay (400°C)	1.0
Lead	$0.353(10^2)$	Cement mortar (90°C)	0.55
Mercury	$0.0834(10^2)$	Concrete	0.8
Sodium	$1.41(10^2)$	Bakelite	1.4
Rock, limestone	1.0	Urethane foam	0.06
Rock, sandstone	1.3	Glass, pyrex (100°C)	1.25
Snow	0.16	Glass, silica (50°C)	1.40
Sawdust	0.06		
Wood, plywood	0.11		
Wood, oak	0.16		
Zinc	$1.16(10^2)$		

TABLE C.6   Diffusion coefficients of binary mixtures in air or water

Mixture	Diffusion coefficient, $\mathcal{D}$ (cm²/s)	Mixture	Diffusion coefficient, $\mathcal{D}$ (cm²/s)
CO in air	0.208	Acetone in water	$1.28(10^{-5})$
CO_2 in air	0.160	Benzene in water	$1.02(10^{-5})$
H_2O in air	0.242	Ethanol in water	$1.24(10^{-5})$
H_2 in air	0.627	Glycene in water	$1.05(10^{-5})$
		Methane in water	$1.49(10^{-5})$
		Toluene in water	$0.85(10^{-5})$

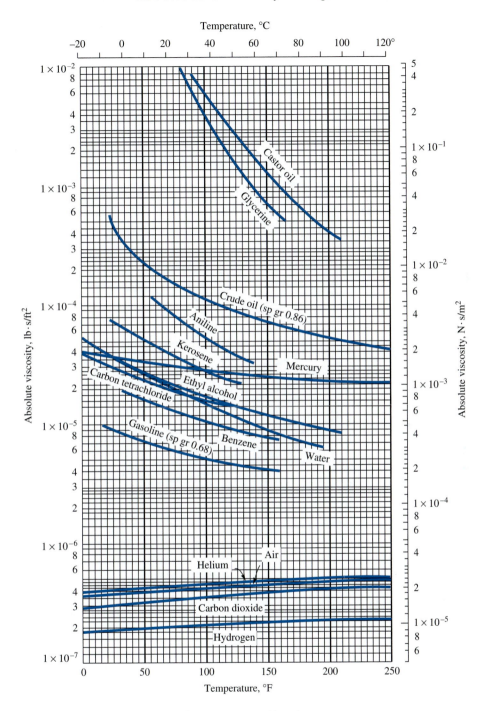

Figure C.1   Absolute viscosities of certain gases and liquids.

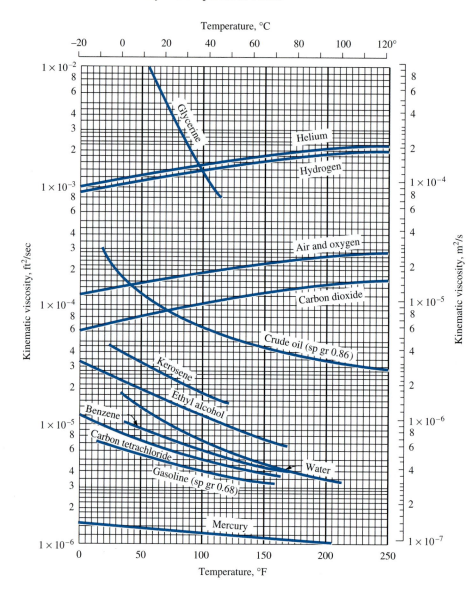

Figure C.2 Kinematic viscosities of certain gases and liquids. The gases are at standard pressure.

APPENDIX

D

VARIABLE NOTATION

Symbol	Quantity	Unit SI	Unit USC	Dimensions
a	Constant, pulse wave speed	m/s	ft/s	LT^{-1}
a	Acceleration	m/s^2	ft/s^2	LT^{-2}
$\mathbf{a}$	Acceleration vector	m/s^2	ft/s^2	LT^{-2}
a_x, a_y, a_z	Acceleration components	m/s^2	ft/s^2	LT^{-2}
a^*	Velocity	m/s	ft/s	LT^{-1}
A	Area	m^2	ft^2	L^2
$\mathbf{A}$	Adverse slope	none	none	
b	Distance	m	ft	L
b	Constant			
B	Constant			
B_w	Baffle width	m	ft	L
c	Speed of surge wave	m/s	ft/s	LT^{-1}
c	Speed of sound	m/s	ft/s	LT^{-1}
c_p	Specific heat, constant pressure	J/kg·k	ft·lb/slug·°R	
c_q	Specific humidity	none	none	
c_v	Specific heat, constant volume	J/kg·k	ft·lb/slug·°R	
C	Concentration (mass)	kg/m^3	slug/ft^3	ML^{-3}
C	Wave speed, celerity	m/s	ft/s	LT^{-1}
C	Coefficient	none	none	
C_b	Bed concentration	kg/m^3	slugs/ft^3	ML^{-3}
C_D	Drag coefficient	none	none	
C_i	Mass concentration, ith component	kg/m^3	slugs/ft^3	ML^{-3}
C_i	Inertia coefficient	none	none	
C_L	Lift coefficient	none	none	
C_m	Empirical constant	m$^{1/3}$/s	ft$^{1/3}$/s	$L^{1/3}T^{-1}$
C_r	Reference concentration	kg/m^3	slugs/ft^3	ML^{-3}
C^*	Critical point	K	°R	Φ
$\mathbf{C}$	Critical slope	none	none	
d	Water depth	m	ft	L
D'	Volumetric displacement	m^3	ft^3	L^3
D	Diameter	m	ft	L
Dd	Dispersion coefficient	m^2/s	ft^2/s	L^2T^{-1}
D_i	Tank internal diameter	m	ft	L
$\mathcal{D}$	Mass diffusion coefficient	m^2/s	ft^2/s	L^2T^{-1}
$\mathbf{D}_N$	Damkholer number	none	none	
$\mathbf{D}_a$	Dalton number	none	none	
e	Efficiency	none	none	
e	Internal energy per unit mass	J/kg	ft·lb/slug	L^2T^{-2}
e	Absolute error			
E	Internal energy	J	ft·lb	ML^2T^{-2}
E	Specific energy	m·N/N	ft·lb/lb	L
E	Losses per unit weight	m·N/N	ft·lb/lb	L

Symbol	Quantity	Unit SI	Unit USC	Dimensions
E	Modulus of elasticity	Pa	lb/ft^2	$ML^{-1}T^{-2}$
E	Eddy diffusivity	m^2/s	ft^2/s	$ML^{-2}T^{-1}$
E_o	Bottom sediment entrainment flux	kg/m^2·s	slug/ft^2·s	$ML^{-2}T^{-1}$
f	Friction factor	none	none	
F	Force	N	lb	MLT^{-2}
$\mathbf{F}$	Force vector	N	lb	MLT^{-2}
$\mathbf{F}$	Froude number	none	none	
F_B	Buoyant force	N	lb	MLT^{-2}
g	Acceleration of gravity	m/s^2	ft/s^2	LT^{-2}
g_0	Gravitation constant	kg·m/N·s^2	lb$_m$·ft/lb·s^2	
G	Mass flow rate per unit area	kg/s·m^2	slug/s·ft^2	$ML^{-2}T^{-1}$
$\mathbf{G}_r$	Grashof number	none	none	
h	Head, vertical distance	m	ft	L
h	Enthalpy per unit mass	J/kg	ft·lb/slug	L^2T^{-2}
h	Convective heat transfer coefficient	J/m^2·s/K	Btu/ft^2·s/°R	$MT^{-3}\Theta^{-1}$
h_m	Mass transfer coefficient	m·s	ft·s	LT
H	Head, elevation of hydraulic grade line	m	ft	L
H	Wave height	m	ft	L
H_i	Tank height	m	ft	L
H_p	Pump head	m	ft	L
H_t	Turbine head	m	ft	L
H_T	Overall heat transfer coefficient	J/m^2·s/K	Btu/ft^2·s/°R	$MT^{-3}\Theta^{-1}$
$\mathbf{H}$	Horizontal slope	none	none	
HP	Horsepower			
I	Moment of inertia	m^4	ft^4	L^4
J	Junction point	none	none	
k, k_1	Decay rate coefficients	1/s	1/s	T^{-1}
k	Wave number	1/m	1/ft	l^{-1}
k	Thermal conductivity	J/m·s/K	Btu/ft·s/°R	$MLT^{-3}\Theta^{-1}$
K	Bulk modulus of elasticity	Pa	lb/ft^2	$ML^{-1}T^{-2}$
K	Minor loss coefficient	none	none	
K	Dispersion coefficient	m^2/s	ft^2/s	L^2T^{-1}
l	Length, mixing length	m	ft	L
l_L	Lagrangian length scale	m	ft	L
L	Length	m	ft	L
L	Liter	m^3	none	L^3
L	Lift	N	lb	MLT^{-2}
L	Monin-Obukhov length	m	ft	L
L	Wavelength	m	ft	L
L	Latent heat	J/kg	Btu/slug	L^2T^2
ln	Natural logarithm	none	none	
m	Mass	kg	slug	M
m	Form factor, constant	none	none	
m	Strength of source	m^3/s	ft^3/s	L^3T^{-1}
$\dot{m}$	Mass flow rate or mass per unit time	kg/s	slug/s	MT^{-1}
M	Molecular weight			
M	Total mass	kg	slug	M
M_t	Mixing number	none	none	
M_ϕ	Mean diameter, ϕ-units	m	ft	L
M	Momentum flux	N	lb	MLT^{-2}
M	Quantity of markers			
$\mathbf{M}$	Mild slope	none	none	
$\overline{\mathbf{M}}$	Mach number	none	none	
$\overline{MG}$	Metacentric height	m	ft	L

Symbol	Quantity	Unit SI	Unit USC	Dimensions
n	Exponent, constant	none	none	
n	Normal direction			
n	Manning roughness factor			
n	Number of moles			
$\mathbf{n}$	Unit normal at surface	none	none	
$\mathbf{n}_1$	Normal unit vector			
N	Rotation speed	1/s	1/s	T^{-1}
N	Control volume property			
$\mathbf{N}$	Flux vector	none	none	
$\mathbf{N}$	Unit normal at bottom	none	none	
$\mathbf{N}_b$	Brunt-Väisälä frequency	1/s	1/s	T^{-1}
$\mathbf{N}_p$	Power number	none	none	
$\mathbf{N}_u$	Nusselt number	none	none	
$NPSH$	Net positive suction head	m	ft	L
p	Pressure	Pa	lb/ft^2	$ML^{-1}T^{-2}$
p	Force	N	lb	MLT^{-2}
P	Height of weir	m	ft	L
P	Transfer rate per unit area	1/(s·m^2)	1/(s·ft^2)	$T^{-1}L^{-2}$
P	Wetted perimeter	m	ft	L
P	Probability distribution			
$\mathbf{P}_e$	Peclet number	none	none	
$\mathbf{P}_{em}$	Mass Peclet number	none	none	
P_o	Power number	none	none	
P_o	Pipe circumference	m	ft	L
$\mathbf{P}_r$	Prandtl number	none	none	
q	Discharge per unit width	m^2/s	ft^2/s	L^2T^{-1}
q	Velocity	m/s	ft/s	LT^{-1}
$\mathbf{q}$	Velocity vector	m/s	ft/s	LT^{-1}
q_H	Heat transfer per unit mass	J/kg	ft·lb/slug	L^2T^{-2}
Q	Discharge	m^3/s	ft^3/s	L^3T^{-1}
Q_H	Heat content	J	ft·lb	ML^2T^{-2}
r	Coefficient			
r	Radial distance	m	ft	L
$\mathbf{r}$	Position vector	m	ft	L
R	Position vector	m	ft	L
R	Hydraulic radius	m	ft	L
R	Gas constant	J/kg·K	ft·lb/slug·°R	
R	Thermal resistance	s·K/J	s·°R/Btu	$ML^{-2}T^2\Theta$
R, R'	Gage difference	m	ft	L
R_L	Lagrangian correlation coefficient	none	none	
$\mathbf{R}$	Reynolds number	none	none	
$\mathbf{R}_i$	Richardson number	none	none	
s	Slip	none	none	
s	Salt concentration	kg/m^3	slugs/ft^3	ML^{-3}
s	Streamline coordinate	m	ft	L
S	Entropy per unit mass	J/kg·K	ft·lb/slug·°R	
S	Entropy	J/K	ft·lb/°R	
S	Specific gravity, slope	none	none	
S	Source-sink term	kg/s	slug/s	ML^{-1}
$\mathbf{S}$	Distance	m	ft	L
$\mathbf{S}$	Steep slope	none	none	
S_o	Bottom slope	none	none	
S_ϕ	Skewness, ϕ-units	none	none	
S_τ	Normalized shear stress	none	none	
$\mathbf{S}_c$	Schmidt number	none	none	

Symbol	Quantity	Unit SI	Unit USC	Dimensions
S_h	Sherwood number	none	none	
S_t	Strouhal number	none	none	
S_t	Stanton number	none	none	
t	Time	s	s	T
t, t'	Distance, thickness	m	ft	L
t_d	Detention time	s	s	T
t_p	Propeller thickness	m	ft	L
T	Averaging time period	s	s	T
T	Temperature	K	°R	
T	Torque	N·m	lb·ft	ML^2T^{-2}
T	Tensile force/unit length	N/m	lb/ft	MT^{-2}
T	Top width	m	ft	L
T_L	Lagrangian time scale	s	s	T
T_o	Averaging period	s	s	T
T^*	Triple point	K	°R	Θ
u	Velocity, velocity component	m/s	ft/s	LT^{-1}
u	Peripheral speed	m/s	ft/s	LT^{-1}
u_*	Shear stress velocity	m/s	ft/s	LT^{-1}
u^{**}	Intrinsic energy	J/kg	ft·lb/slug	L^2T^{-2}
U	Velocity	m/s	ft/s	LT^{-1}
v	Velocity, velocity component	m/s	ft/s	LT^{-1}
v_s	Specific volume	m³/kg	ft³/slug	$M^{-1}L^3$
$\forall$	Volume	m³	ft³	L^3
$\mathcal{V}$	Volume	m³	ft³	L^3
$\mathbf{v}$	Velocity vector	m/s	ft/s	LT^{-1}
V	Velocity	m/s	ft/s	LT^{-1}
w	Velocity component	m/s	ft/s	LT^{-1}
w	Work per unit mass	J/kg	ft·lb/slug	L^2T^{-2}
w_t	Terminal settling velocity	m/s	ft/s	LT^{-1}
W	Work per unit time	J/s	ft·lb/s	ML^2T^{-3}
W	Work of expansion	m·N	ft·lb	ML^2T^{-2}
W	Width of channel	m	ft	L
W_c	Plume width	m	ft	L
W_s	Shaft work	m·N	ft·lb	ML^2T^{-2}
W	Weight	N	lb	MLT^{-2}
$\mathbf{W}$	Weight vector			
$\mathbf{W}$	Weber number	none	none	
x	Distance	m	ft	L
$\mathbf{x}$	Coordinate, position vector	m	ft	L
x_p	Distance to pressure center	m	ft	L
X	Body-force component per unit mass	N/kg	lb/slug	LT^{-2}
y	Distance, depth	m	ft	L
y_p	Distance to pressure center	m	ft	L
Y	Expansion factor	none	none	
Y	Body-force component per unit mass	N/kg	lb/slug	LT^{-2}
z	Vertical distance	m	ft	L
z_{om}	Momentum roughness length	m	ft	L
z_{ov}	Vapor roughness length	m	ft	L
z_{oh}	Humidity roughness length	m	ft	L
Z	Vertical distance	m	ft	L
Z	Body-force component per unit mass	N/kg	lb/slug	LT^{-2}
α	Kinetic-energy correction factor	none	none	
α	Angle, coefficient	none	none	
α	Thermal diffusivity	m²/s	ft²/s	L^2T^{-1}
β	Momentum correction factor	none	none	
β	Blade angle	none	none	
Γ	Circulation	m²/s	ft²/s	L^2T^{-1}

Symbol	Quantity	Unit		Dimensions
		SI	USC	
Γ	Vorticity	1/s	1/s	T^{-1}
Γ	Relative humidity	none	none	
Γ	Mixing ratio	none	none	
∇	Vector operator	1/m	1/ft	L^{-1}
γ	Specific weight	N/m^3	lb/ft^3	$ML^{-2}T^{-2}$
δ	Boundary-layer thickness	m	ft	L
δ'	Laminar-sublayer thickness	m	ft	L
ϵ	Kinematic eddy viscosity	m^2/s	ft^2/s	L^2T^{-1}
ϵ	Roughness height	m	ft	L
ϵ	Strain rate	1/s	1/s	T^{-1}
η	Control volume property (N) per unit mass			
η	Eddy viscosity	N·s/m^2	lb·s/ft^2	$ML^{-1}T^{-1}$
η	Head ratio	none	none	
η	Efficiency			
η	Free surface elevation	m	ft	L
θ	Angle	none	none	
θ	Angular displacement	radians	radians	
θ	Potential temperature	K	°R	Θ
κ	Ratio of specific heats	none	none	
κ	Universal constant	none	none	
λ	Scale ratio, undetermined multiplier	none	none	
μ	Viscosity	N·s/m^2	lb·s/ft^2	$ML^{-1}T^{-1}$
μ	Constant			
ν	Kinematic viscosity	m^2/s	ft^2/s	L^2T^{-1}
ϕ	Velocity potential	m^2/s	ft^2/s	L^2T^{-1}
ϕ	Function			
ϕ	Phi, units of particle diameter			
ϕ_m	Nondimensional stress	none	none	
Φ	Stability function	none	none	
π	Constant	none	none	
Π	Dimensionless parameter	none	none	
ρ	Density	kg/m^3	slug/ft^3	ML^{-3}
ρ_w	Water density	kg/m^3	slug/ft^3	ML^{-3}
ρ_v	Vapor density, absolute humidity	kg/m^3	slug/ft^3	ML^{-3}
σ	Surface tension	N/m	lb/ft	MT^{-2}
σ	Cavitation index	none	none	
σ	Normal stress	Pa	lb/ft^2	$ML^{-1}T^{-2}$
σ_ϕ	Standard deviation, ϕ-units	m	ft	L
σ^2	Variance			
τ	Shear stress	Pa	lb/ft^2	$ML^{-1}T^{-2}$
τ_o	Wall shear stress	N/m^2	lb/ft^2	$ML^{-1}T^{-2}$
τ_c	Critical erosion shear stress	N/m^2	lb/ft^2	$ML^{-1}T^{-2}$
ψ	Stream function, two dimensions	m^2/s	ft^2/s	L^2T^{-1}
ψ	Stokes' stream function	m^3/s	ft^3/s	L^3T^{-1}
Ψ	Flux profile function	none	none	
Ω	Angular velocity	1/s	1/s	T^{-1}
ω	Angular velocity	rad/s	rad/s	T^{-1}
ω_i	Mass fraction, ith component	none	none	
ω	Wave frequency	1/s	1/s	T^{-1}
ξ	Nondimensional height			

APPENDIX

E

VECTOR OPERATIONS AND NOTATION

In this appendix, the basic definitions and operations of vectors are reviewed. The material reviewed here is presented in support of the operations used in the text and is not to be taken as an exhaustive treatment of vectors.

E.1 NOTATION AND DEFINITIONS

A *vector* is a quantity having both magnitude and direction. In contrast a *scalar* is a quantity having only magnitude but no direction. Examples of a vector include velocity and force, while examples of a scalar include mass, length, and time. In this text a vector, **A**, is denoted in bold face type; in many other publications a vector is denoted with an overbar, $\overline{A}$, or underbar $\underline{A}$. As an overbar often denotes averaging in fluid mechanics, great care must be taken to ensure the authors' symbols for vector or average are known in advance. A scalar quantity is denoted by a simple letter definition, for example, m.

In three-dimensional Cartesian coordinates all vectors can be described as linear combinations of three *unit vectors* denoted **i**, **j**, and **k**. As in Figure E.1 the unit vectors are mutually orthogonal, meet at the origin of the coordinate system, and are 1 unit in length. The coordinate position of the tip of the unit vectors is denoted in

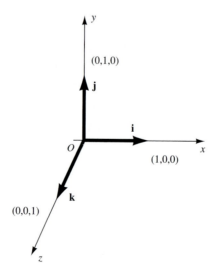

Figure E.1   Unit vector definitions.

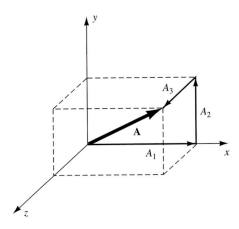

Figure E.2   Component contributions to vector **A**.

the figure. These three unit vectors are called the *basis functions*. Any three-dimensional vector in Cartesian space can be represented by a combination of these basis functions, for example,

$$\mathbf{A} = A_1\mathbf{i} + A_2\mathbf{j} + A_3\mathbf{k} \tag{E.1}$$

$$\mathbf{B} = B_1\mathbf{i} + B_2\mathbf{j} + B_3\mathbf{k} \tag{E.2}$$

Figure E.2 depicts the **A** vector and its components geometrically added to yield the total vector **A**.

The length or magnitude of **A**, denoted as $|\mathbf{A}|$, is found from

$$|\mathbf{A}| = A = \sqrt{A_1^2 + A_2^2 + A_3^2} \tag{E.3}$$

This is an extension of the well-known fact that if a position is specified to be at point (x_1, y_1, z_1) in space, then the radius or distance to this point from the origin is given by

$$r = \sqrt{x_1^2 + y_1^2 + z_1^2} \tag{E.4}$$

Vectors are not only comprised of numbers and surrogates for numbers, they can include mathematical operations as well. A frequently used math operation vector is the vector differential operator defined by the symbol ∇ and equivalent to

$$\nabla(\ ) = \frac{\partial(\ )}{\partial x}\mathbf{i} + \frac{\partial(\ )}{\partial y}\mathbf{j} + \frac{\partial(\ )}{\partial z}\mathbf{k} \tag{E.5}$$

∇ is itself a vector.

E.2   VECTOR ALGEBRA

Two vectors **A** and **B** are equal if they have the same magnitude and direction regardless of the position of their origin. The sum or difference of two vectors is formed geometrically by the "head to tail" approach or the parallelogram approach; $\mathbf{A} + \mathbf{B} = \mathbf{C}$. If $\mathbf{A} - \mathbf{B} = \mathbf{0}$ then the *null vector* is obtained which has zero magnitude and direction.

If m is a scalar then the following laws apply

$$\mathbf{A} + \mathbf{B} = \mathbf{B} + \mathbf{A}$$

$$m\mathbf{A} = \mathbf{A}m$$

$$(m + n)\mathbf{A} = m\mathbf{A} + n\mathbf{A} \tag{E.6}$$

$$m(\mathbf{A} + \mathbf{B}) = m\mathbf{A} + m\mathbf{B}$$

$$\mathbf{A} + (\mathbf{B} + \mathbf{C}) = (\mathbf{A} + \mathbf{B}) + \mathbf{C}$$

E.3   VECTOR OPERATIONS

The *dot product* is the projection of one vector upon another. With reference to Fig. E.3, the dot product of $\mathbf{A}$ and $\mathbf{B}$ is denoted as $\mathbf{A} \cdot \mathbf{B}$ and is defined as

$$\mathbf{A} \cdot \mathbf{B} = |\mathbf{A}| \, |\mathbf{B}| \cos \theta = AB \cos \theta \tag{E.7}$$

The dot product of two vectors always yields a scalar quantity. In components the dot product is evaluated as

$$\mathbf{A} \cdot \mathbf{B} = (A_1\mathbf{i} + A_2\mathbf{j} + A_3\mathbf{k}) \cdot (B_1\mathbf{i} + B_2\mathbf{j} + B_3\mathbf{k})$$

$$= A_1B_1 + A_2B_2 + A_3B_3 \tag{E.8}$$

$$\mathbf{A} \cdot \mathbf{A} = |\mathbf{A}|^2 = A_1^2 + A_2^2 + A_3^2 \tag{E.9}$$

Applying the dot product to unit vectors gives the following: If

$$\mathbf{i} = 1, 0, 0$$

$$\mathbf{j} = 0, 1, 0$$

then

$$\mathbf{i} \cdot \mathbf{i} = (1 \cdot 1) + (0 \cdot 0) + (0 \cdot 0) = 1 \tag{E.10}$$

or

$$\mathbf{i} \cdot \mathbf{i} = |\mathbf{i}| \, |\mathbf{i}| \cos 0° = (1)(1)(1) = 1$$

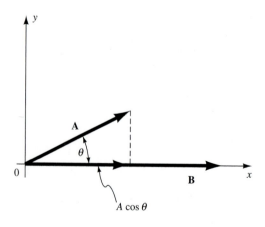

Figure E.3   Geometric representation of **A** and **B**.

Additionally,

$$\mathbf{i} \cdot \mathbf{j} = (1 \cdot 0) + (0 \cdot 1) + (0 \cdot 0) = 0 \qquad \text{(E.11)}$$

or

$$\mathbf{i} \cdot \mathbf{j} = |\mathbf{i}| \, |\mathbf{j}| \cos 90° = (1)(1)(0) = 0$$

It is known from the geometric definition of $\mathbf{i}$ and $\mathbf{j}$ that they are perpendicular to each other, and therefore it is seen that the dot product of these two perpendicular or *orthogonal* vectors is zero. It is the general case that if the dot product of any two vectors is zero they are orthogonal.

The *cross product* of two vectors is computed as

$$\mathbf{A} \times \mathbf{B} = \begin{vmatrix} \mathbf{i} & \mathbf{j} & \mathbf{k} \\ A_1 & A_2 & A_3 \\ B_1 & B_2 & B_3 \end{vmatrix} = \begin{vmatrix} A_2 & A_3 \\ B_2 & B_3 \end{vmatrix} \mathbf{i} - \begin{vmatrix} A_1 & A_3 \\ B_1 & B_3 \end{vmatrix} \mathbf{j} + \begin{vmatrix} A_1 & A_2 \\ B_1 & B_2 \end{vmatrix} \mathbf{k} \qquad \text{(E.12)}$$

or

$$\mathbf{A} \times \mathbf{B} = (A_2 B_3 - A_3 B_2)\mathbf{i} - (A_1 B_3 - A_3 B_1)\mathbf{j} + (A_1 B_2 - A_2 B_1)\mathbf{k} \qquad \text{(E.13)}$$

The unit vectors' cross products are as follows:

$$\begin{aligned} \mathbf{i} \times \mathbf{i} &= 0 & \mathbf{i} \times \mathbf{j} &= \mathbf{k} & \mathbf{j} \times \mathbf{i} &= -\mathbf{k} \\ \mathbf{j} \times \mathbf{j} &= 0 & \mathbf{j} \times \mathbf{k} &= \mathbf{i} & \mathbf{k} \times \mathbf{j} &= -\mathbf{i} \\ \mathbf{k} \times \mathbf{k} &= 0 & \mathbf{k} \times \mathbf{i} &= \mathbf{j} & \mathbf{i} \times \mathbf{k} &= -\mathbf{j} \end{aligned} \qquad \text{(E.14)}$$

If $\mathbf{A}$ and $\mathbf{B}$ are not null vectors, then if $\mathbf{A} \times \mathbf{B} = \mathbf{0}$, the vectors are parallel.

E.4   UNIT NORMALS AND PLANES

The intersection of two vectors (e.g., $\mathbf{A}$ and $\mathbf{B}$) defines a plane surface and a vector perpendicular to the plane can be found by taking the cross product $\mathbf{A} \times \mathbf{B}$. For example, if

$$\mathbf{A} = 2\mathbf{i} - 6\mathbf{j} - 3\mathbf{k} \quad \text{and} \quad \mathbf{B} = 4\mathbf{i} + 3\mathbf{j} - \mathbf{k}$$

then the vector perpendicular to the plane formed by $\mathbf{A}$ and $\mathbf{B}$ is

$$\mathbf{C} = \mathbf{A} \times \mathbf{B} = \begin{vmatrix} \mathbf{i} & \mathbf{j} & \mathbf{k} \\ 2 & -6 & -3 \\ 4 & 3 & -1 \end{vmatrix} = 15\mathbf{i} - 10\mathbf{j} + 30\mathbf{k}$$

As noted in Chap. 3 a unit vector or *unit normal* to the plane defined by $\mathbf{A}$ and $\mathbf{B}$ can be found by normalizing $\mathbf{C}$ with the length or magnitude of $\mathbf{C}$. Therefore,

$$\mathbf{n} = \frac{\mathbf{C}}{|\mathbf{C}|} = \frac{\mathbf{A} \times \mathbf{B}}{|\mathbf{A} \times \mathbf{B}|} = \frac{15\mathbf{i} - 10\mathbf{j} + 30\mathbf{k}}{[15^2 + 10^2 + 30^2]^{1/2}} = \frac{3}{7}\mathbf{i} - \frac{2}{7}\mathbf{j} + \frac{6}{7}\mathbf{k} \qquad \text{(E.15)}$$

A quick check of the magnitude of $\mathbf{n}$ reveals it has unit "length." Therefore, the vector $\mathbf{n}$ serves to give the direction of the unit vector perpendicular to the plane defined by $\mathbf{A}$ and $\mathbf{B}$. The area vector [e.g., Eq. (3.2.2)] $\mathbf{A}$ is formally defined as the scalar area of the surface times the unit normal, that is, $\mathbf{A} = A\mathbf{n}$.

E.5   DIFFERENTIAL OPERATIONS

If **A** and **B** are differentiable vector functions of a scalar s and ϕ is a differentiable function of scalar s, then the following rules apply:

$$\frac{d}{ds}(\mathbf{A} + \mathbf{B}) = \frac{d\mathbf{A}}{ds} + \frac{d\mathbf{B}}{ds}$$

$$\frac{d}{ds}(\mathbf{A} \cdot \mathbf{B}) = \mathbf{A} \cdot \frac{d\mathbf{B}}{ds} + \mathbf{B} \cdot \frac{d\mathbf{A}}{ds}$$

$$\frac{d}{ds}(\mathbf{A} \times \mathbf{B}) = \mathbf{A} \times \frac{d\mathbf{B}}{ds} + \frac{d\mathbf{A}}{ds} \times \mathbf{B} \qquad \text{(E.16)}$$

$$\frac{d}{ds}(\phi\mathbf{A}) = \phi\frac{d\mathbf{A}}{ds} + \frac{d\phi}{ds}\mathbf{A}$$

If **A** is a differentiable vector, then $d\mathbf{A}$ is defined as $d\mathbf{A} = dA_1\mathbf{i} + dA_2\mathbf{j} + dA_3\mathbf{k}$ and the following differential operations apply:

$$d(\mathbf{A} \cdot \mathbf{B}) = \mathbf{A} \cdot d\mathbf{B} + d\mathbf{A} \cdot \mathbf{B}$$

$$d(\mathbf{A} \times \mathbf{B}) = \mathbf{A} \times d\mathbf{B} + d\mathbf{A} \times \mathbf{B}$$

If **A** is a function of x, y, and z, then

$$d\mathbf{A} = d\mathbf{A}(x, y, z) = \frac{\partial\mathbf{A}}{\partial x}dx + \frac{\partial\mathbf{A}}{\partial y}dy + \frac{\partial\mathbf{A}}{\partial z}dz$$

Extension to time gives

$$d\mathbf{A} = d\mathbf{A}(x, y, z, t) = \frac{\partial\mathbf{A}}{\partial x}dx + \frac{\partial\mathbf{A}}{\partial y}dy + \frac{\partial\mathbf{A}}{\partial z}dz + \frac{\partial\mathbf{A}}{\partial t}dt$$

which after division by dt gives the *total derivative* as

$$\frac{d\mathbf{A}}{dt} = \frac{\partial\mathbf{A}}{\partial t} + \frac{\partial\mathbf{A}}{\partial x}\frac{dx}{dt} + \frac{\partial\mathbf{A}}{\partial y}\frac{dy}{dt} + \frac{\partial\mathbf{A}}{\partial z}\frac{dz}{dt}$$

$$= \frac{\partial\mathbf{A}}{\partial t} + u\frac{\partial\mathbf{A}}{\partial x} + v\frac{\partial\mathbf{A}}{\partial y} + w\frac{\partial\mathbf{A}}{\partial z} \qquad \text{(E.17)}$$

The vector ∇ operation in Eq. (E.5) can be used to compress and organize the various calculus-vector operations.

The *gradient* of a scalar differentiable function, ϕ, is given by

$$\nabla\phi = \left(\mathbf{i}\frac{\partial}{\partial x} + \mathbf{j}\frac{\partial}{\partial y} + \mathbf{k}\frac{\partial}{\partial z}\right)\phi = \frac{\partial\phi}{\partial x}\mathbf{i} + \frac{\partial\phi}{\partial y}\mathbf{j} + \frac{\partial\phi}{\partial z}\mathbf{k} \qquad \text{(E.18)}$$

and is itself a vector. The component of this gradient in the direction **n** is given by $\nabla\phi \cdot \mathbf{n}$.

The *divergence* of a differentiable vector field, **A**, is given by

$$\nabla \cdot \mathbf{A} = \left(\mathbf{i}\frac{\partial}{\partial x} + \mathbf{j}\frac{\partial}{\partial y} + \mathbf{k}\frac{\partial}{\partial z}\right) \cdot \left(A_1\mathbf{i} + A_2\mathbf{j} + A_3\mathbf{k}\right)$$

$$= \frac{\partial A_1}{\partial x} + \frac{\partial A_2}{\partial y} + \frac{\partial A_3}{\partial z} \qquad \text{(E.19)}$$

and is a scalar quantity.

The *Laplacian* is defined as the divergence of the gradient of a scalar differentiable function, ϕ, and is given by

$$\nabla \cdot \nabla \phi = \nabla^2 \phi = \left(\frac{\partial^2}{\partial y^2} + \frac{\partial^2}{\partial y^2} + \frac{\partial^2}{\partial z^2} \right) \phi = \frac{\partial^2 \phi}{\partial y^2} + \frac{\partial^2 \phi}{\partial y^2} + \frac{\partial^2 \phi}{\partial z^2} \qquad \text{(E.20)}$$

The *curl* of a differentiable vector field, $\mathbf{A}$, results in a vector field and is defined by

$$\nabla \times \mathbf{A} = \left(\mathbf{i} \frac{\partial}{\partial x} + \mathbf{j} \frac{\partial}{\partial y} + \mathbf{k} \frac{\partial}{\partial z} \right) \times \left(A_1 \mathbf{i} + A_2 \mathbf{j} + A_3 \mathbf{k} \right)$$

$$= \left(\frac{\partial A_3}{\partial y} - \frac{\partial A_2}{\partial z} \right) \mathbf{i} + \left(\frac{\partial A_1}{\partial z} - \frac{\partial A_3}{\partial x} \right) \mathbf{j} + \left(\frac{\partial A_2}{\partial x} - \frac{\partial A_1}{\partial y} \right) \mathbf{k} \qquad \text{(E.21)}$$

With a small amount of algebra you can show that for the scalar, ϕ, and vector, $\mathbf{A}$,

$$\nabla \times (\nabla \phi) = 0 \qquad \text{(E.22}a\text{)}$$

$$\nabla \cdot (\nabla \times \mathbf{A}) = 0 \qquad \text{(E.22}b\text{)}$$

Finally, if Eq. (E. 17) is rewritten as

$$\frac{d\mathbf{A}}{dt} = \left(\frac{\partial \mathbf{A}}{\partial t} + u \frac{\partial}{\partial x} + v \frac{\partial}{\partial y} + w \frac{\partial}{\partial z} \right) \mathbf{A}$$

and attention focused upon the terms with spatial derivatives, it is straightforward to show that they can be written as

$$(\mathbf{v} \cdot \nabla) = (u \mathbf{i} + v \mathbf{j} + w \mathbf{k}) \cdot \left(\frac{\partial}{\partial x} \mathbf{i} + \frac{\partial}{\partial y} \mathbf{j} + \frac{\partial}{\partial z} \mathbf{k} \right)$$

$$= u \frac{\partial}{\partial x} + v \frac{\partial}{\partial y} + w \frac{\partial}{\partial z}$$

Returning to Eq. (E.17) then gives

$$\frac{d\mathbf{A}}{dt} = \frac{\partial \mathbf{A}}{\partial t} + (\mathbf{v} \cdot \nabla) \mathbf{A} \qquad \text{(E.23)}$$

You are encouraged to show yourself that there is not one iota of relationship between $(\nabla \cdot \mathbf{v})$ and $(\mathbf{v} \cdot \nabla)$; the former yields a scalar result while the latter yields a differentiation which must operate upon a scalar or vector function. The general relation in Eq. (E.23) is used in Eqs. (4.3.3b), (4.4.2), (4.4.11), (4.7.1), and (4.8.10a).

APPENDIX

F

ANSWERS TO EVEN-NUMBERED PROBLEMS

Chapter 1

1.2 Ideal plastic, $\tau_y = 15$ kPa

1.4 $\frac{du}{dy} = 0$ at $y = t$; nonlinear variation of u

1.6 95.1 lb

1.8 $m = 450$ kg, $W = 45.89$ N $= 10.32$ lb, $m = 450$ kg, $W = 275.3$ N

1.10 8.5 m/s^2

1.12 2.5 m/s

1.14 0.04 N·s/m^2

1.16 0.02387 lb·s/ft^2

1.18 0.0582 mm

1.20 6.12 mm/s

1.22 8.059(10^{-5}) ft^2/s, 7.488(10^{-2}) Stokes

1.24 2.51(10^{-3}) in.

1.26 2.044 ft^2/s

1.28 $v_s = \frac{g}{\gamma}$

1.30 $\frac{8312}{M}\frac{mN}{kgK}$

1.32 602.4 kW

1.34 1025.243 kg/m^3

1.36 $\rho_m = \rho_s/[\omega_s + (1 - \omega_s)\rho_s/\rho_w]$

1.38 999.975 kg/m^3

1.40 $\lambda = 1$; $\rho_m^{max} = 2\rho_A\rho_B/(\rho_A + \rho_B)$

1.42 $\rho = 1147.3$ kg/m^3; $C_1 = 306.7$ kg/m^3; $C_2 = 427.9$ kg/m^3; $C_3 = 367.27$ kg/m^3

1.44 3.2504 kg/m^3, 0.13 kg

1.46 3.006 kg/m^3

1.48 6.39 MN/m^2

1.50 895.32 kJ

1.52 $p_{mixture} = 95.8$ MPa; $p_{NH_3} = 14.91$ MPa; $p_{CO_2} = 4.84$ MPa; $p_{H_2} = 76.06$ MPa

1.54 $K = \rho\frac{dp}{d\rho}$

1.56 34,722 psi; 300 MPa

1.58 0.0217 slug; 0.697 lb

1.62 0.4 MPa abs

1.64 6000 kPa

1.66 10.27 kPa

1.68 $D = 13$ mm for tap water

1.70 0.00919 N

1.74 $\theta = 67.03°$

Chapter 2

2.2 541.7 psi

2.4 0.3858 ft; 0.376 ft

2.8 60.6 kPa; 0.773 kg/m^3

2.10 374.9 mm; 5.099 m; 1.734 m

2.12 31.14 m

2.14 4.082 m H$_2$O; 4.92 m kerosene; 1.389 m acetylene tetrabromide

2.16 -2.125 m, -0.425 m

2.18 1.372 psi

2.20 32.20 ft

2.22 0.2176 in. of Hg

2.24 -1.8 m

2.26 -2.6 mm

2.28 12,060 Pa; 2942 Pa

2.30 110.53 mm

2.32 36 MN

2.34 -441 N

2.36 (a) 51.08 kN; (b) 58.9 kN

2.38 $\gamma bh^3/3$

2.40 1125 b^2h^2 Nm

2.42 1636.5 lb·ft

2.44 20.36 kN

2.46 35,947 lb·ft

2.48 5.855 ft

2.50 $\frac{3}{4}h$ from A

2.52 2.156 m

2.54 (a) 6.708 ft, (b) 6.5 ft

2.56 $\frac{h}{2}$, $\frac{b}{2}$

2.58 0.3334 m

2.62 $0.433b$

2.64 $y_P = \frac{12.25}{3(5h-7)} + \frac{5h-7}{4}$

2.66 2116 ft

2.68 $h = 5$ ft, $F = 430$ lb

2.70 (a) 11.588 in from A
(b) $\sigma_{min} = 5.12\gamma$, $\sigma_{max} = 69.44\gamma$

2.72 9984 lb-ft

2.74 4.227 ft

2.76 2.357 kN

2.78 (a) 156.9 kN, 4.083 m, 1.083 m;
(b) 179.3 kN, 0.948m; (c) 0; (d) 0

2.80 28.81 kN

2.82 $F_H = 154.4$ kN, $y_p = 0.786$ from 0
$F_V = 167.1$ kN, $x_p = 0.774$ from 0
$F = 167.1$ kN

2.84 549.1 lb

2.86 1.633 ft

2.88 $-40r\rho\pi$ lb, 0 lb

2.90 (a) $\gamma r^2/2$, $y_p = 2r/3$ above BD
(b) $-\gamma\pi r^2/4$, $x_p = 4r/3\pi$ to left of AC

2.92 (a) 23.95 kN, (b) 6.654 kN, (c) 17.3 kN

2.94 30.806 kN

2.96 equivalent

2.98 815.4 lb, 2242 lb

2.100 13.3 m

2.102 1.334 ft

2.104 0.1 m, 980.4 N

2.106 126.96 lb

2.108 18.207 kN

2.110 11.5 mm

2.112 (a) 1083 lb, (b) stable

2.114 No

2.116 (a) $1.348(\sin^2\theta\cos\theta)^{1/3}$ m
(b) $2.19° \leq \theta \leq 54.74°$

2.118 11.71 ft/s^2

2.120 $p_B = 0.347$ psi, $p_C = 0.069$ psi,
$p_D = 1.109$ psi, $p_E = 0.693$ psi

2.122 $p_B = -0.52$ psi, $p_C = -0.26$ psi,
$p_D = 1.30$ psi, $p_E = 1.04$ psi

2.124 $p_A = 0$, $p_B = 22.8$ kPa,
$p_C = 16.42$ kPa

2.126 $a_x = 0$, $a_y = -g$

2.128 $a_x = 2.0394$ m/s^2, $a_y = -1.178$ m/s^2, $\theta = 13.3°$

2.132 31.52 rpm

2.134 0.5 ft left of A, 54.16 rpm

2.136 $2\sqrt{gh_0}/r_0$

2.138 $\frac{2(p-p_0)}{\rho\omega^2} + \frac{g^2}{\omega^4} = x^2 + (y - \frac{g}{\omega^2})^2$

2.142 $p = \left[p_0^{(n-1)/n} + \frac{n-1}{n}\frac{\rho_0\omega^2 r^2}{2p_0^{1/n}} \right]^{n/(n-1)}$

2.144 220.5 rpm

2.146 2.3907 kN

2.148 $132.38(1 + 0.0102\omega^2)$

Chapter 3

3.2 16.33 m/s, 54.04 kg/s

3.4 $\frac{1.273}{(0.7-0.4x/L)^2}$

3.6 Yes

3.8 $n = 1.85$, turbulent

3.10 72.78 kJ

3.12 $F\,ds = pA\,ds - p\,d\forall$

3.14 -250ρ mN/s (J/s or W)

3.16 $y = 0.755$ m or 2.74 m

3.18 8.857 m/s

3.20 $r = \frac{1}{4(1+y/H)^{0.25}}$

3.22 $Q = A_2\left[\frac{2gR(\rho_m/\rho-1)}{(1-(D_2/D_1)^4)}\right]^{1/2}$

3.24 $R = 0$ for all H

3.26 68.1%, 28.14 ft·lb/lb, 88.14 ft·lb/lb

3.28 107.4% greater

3.30 1.037

3.32 0.82 m

3.34 A to B

3.36 3.94 cfs

3.38 6.11

3.40 0.0136 m^3/s, 2.731 m

3.42 91.66 m^3/s, -77.09 kPa

3.44 101 L/s, -24.73 kN/m^2, 9.59 kN/m^2

3.46 8.75 hp

3.48 18,152 hp

3.50 568.6 gpm

3.52 $0.0667 H$

3.54 80.68 kPa

3.56 -0.0033 mN/kg·K

3.58 $v_{max}\frac{2n^2}{(n+1)(2n+1)}$, $\frac{(2n+1)^2(n+1)}{4n^2(n+2)}$

3.60 $F_y = 62.4$ lb

3.62 $F_y = 0$, $F_x = 0.652$ kN

3.64 tension

3.66 No change, reversible

3.68 $F_x = F_y = 29.3$ N

3.70 $F_x = 18,713.9$ N, $F_y = 38,355$ N

3.72 $F_x = -682.7$ lb, $F_y = -1450.4$ lb

3.74 $F_y = -16,859$ N, downwards

3.76 (a) 10,662 lb, (b) 1,251,700 ft · lb/s
(c) 7961 hp, (d) 71.4%, (e) 212.2 psf

3.78 $F = 35,606$ lb, Power $= 3165$ hp

3.82 $\theta = 84.26°$

3.84 $u = -V_0/3$

3.86 $F_x = -185.8$ lb, $F_y = 35.75$ lb

3.88 $F_x = 66.18$ N, $F_y = 20.82$ N

3.92 (a) 985.8 hp, (b) 1577.3 hp

3.94 $\theta = 158.67°$

3.96 $A_J/A_T = \frac{1}{2}$

3.98 21.22-m air, 7.1-cm water

3.102 $y_1^2 + y_1 y_2 + y_2^2 - 3V_1^2 y_1^2 (y_1 + y_2)/y_2^2 = 0$

3.104 $y_2 = 11.04$ ft, loss $= 8.36$ ft · lb/lb,
$V_0^2/2g < y_1$

3.106 32.87 cfs/ft

3.108 0.161 ft/s^2

3.110 242.1 lb

3.112 $T = 25$ mN, Power $= 3141.6$ W, added energy $= 12.82$ mN/N

3.114 463 rpm

3.116 285 rpm

3.118 $V = 11.229$ ft/s, $s = 6.0$ ft

3.120 0.003495 W/cm·°C

3.122 $\frac{1}{2}\alpha_1 T^2 + \alpha_0 T = \frac{1}{2}\alpha_1 T_0^2 + \alpha_0 T_0 - q_H(x/A)$

3.124 1 Btu $= 251.906$ cal

3.126 $c_p^G = c_p^W[m_1 \Delta T_1 (\Delta T_3 - \Delta T_2) + m_2 \Delta T_2 (\Delta T_1 - \Delta T_3)]/[m_3 \Delta T_3 (\Delta T_1 - \Delta T_2)]$

3.128 25.34 °C

3.130 $c_p^{oil} = 486.25$ J/kg·K

3.132 80 s

3.134 2565.4 W/m^2; from hot to cold

3.136 $C_n = C_0\{\beta/\alpha + (1 - \beta/\alpha)e^{-\alpha t}\}^n$

3.138 0.0521 m^3/s; 20.44 min

3.140 $J_A = D_{AB}(p_1 - p_2)/RT\Delta x$

3.142 2.34 hr

Chapter 4

4.2 2.88°

4.4 $-\frac{1}{29}$

4.6 $(2xyz^{1/2})\mathbf{i} + (x^2 z^{1/2})\mathbf{j} + (\frac{1}{2}x^2 yz^{-1/2})\mathbf{k}$

4.8 scalar; $u\frac{\partial}{\partial x} + v\frac{\partial}{\partial y} + w\frac{\partial}{\partial z}$;
rate of change due to motion

4.10 $-38\mathbf{j} - 5520\mathbf{k}$

4.12 $-0.2673\mathbf{i} + 0.5346\mathbf{j} - 0.8019\mathbf{k}$

4.14 $-(\nabla^2 \psi)\mathbf{k}$

4.16 yes

4.18 yes

4.22 $U + \omega x \sin \omega t/(1.5 + \cos \omega t)$

4.24 $u = 1; v = 0$

4.26 $\alpha_0(\frac{1}{2}x^2 - \frac{1}{2}y^2 - \frac{3}{4})$; yes

4.28 $\bar{\sigma} = \frac{16}{3}(x^2 + y^2)$; 106.67 units

4.34 (a) FL^{-4}
(b) $2\alpha\sqrt{x^2 + y^2}$

4.36 $-9.84(10^5)\mathbf{i} - 15.75(10^5)\mathbf{j} - 0.14(10^5)\mathbf{k}$ N/m^3

4.38 0

4.40 62.4 lb

4.42 $\partial(ru_r)/\partial r + \partial u_\theta/\partial \theta = 0$

4.44 $u_r\frac{\partial C}{\partial r} + u_\theta \frac{\partial C}{\partial \theta} = \frac{1}{r}\frac{\partial(r\frac{\partial C}{\partial r})}{\partial r} + \frac{1}{r^2}\frac{\partial^2 C}{\partial \theta^2}$

4.46 $du/dx = 0$
$\rho u\, du/dx = -dp/dx + \mu\, d^2u/dx^2$

4.52 $v(x) = \frac{x}{L}v_w - \frac{\rho g + \frac{\partial p}{\partial y}}{2\mu}(Lx - x^2)$

4.54 $\partial u/\partial x + \partial v/\partial y = 0$
$\rho(u\,\partial u/\partial x + v\,\partial u/\partial y) = -\partial p/\partial x + \mu\nabla^2 u$
$\rho(u\,\partial v/\partial x + v\,\partial v/\partial y) = -\partial p/\partial y + \mu\nabla^2 v$

4.56 $g\sin\alpha(t^2 - x^2)/2\nu$

4.58 $\rho g x \sin \alpha$

4.62 55.9 ft

4.64 508 lb/ft^2

4.66 2

4.68 $52.9(10^6)$ ft-lb

4.70 -250ρ W

4.72 A to B

4.74 12.386 ft

4.76 2528 ft, 3.15 hp, 51.15%

4.78 -39.23 kPa

4.82 0.16 m^3/s

4.84 $T(x) - T_0 = R_0 L^2(1 - b\frac{x}{L} - e^{-b\frac{x}{L}})/\alpha b^2$

4.86 $T(x) = -\frac{1}{2}q_H x^2 + (T_2 + \frac{1}{2}q_H d^2)(x/d) + T_1$

4.88 $\nabla^2 C_b = -\frac{k}{D_b}C_b$

4.90 $\frac{\partial C_M}{\partial t} = D\frac{1}{r^2}\frac{\partial}{\partial r}(r^2\frac{\partial C_M}{\partial r}) + kC_M$

Chapter 5

5.2 (a) $\rho V^2/\Delta p$; (b) $Fg^2/(\rho V^6)$; (c) $t\Delta p/\mu$

5.4 $86.4(10^6)$ lb · s^2/ft

5.6 Dimensionless, T^{-1}, FLT^{-1}, FL, FL, FLT

5.8 $f\left(\frac{\Delta h}{l}, \frac{\mu D}{Q\rho}, \frac{Q^3\rho^5 g}{\mu^5}\right) = 0$

5.10 $\Delta p = c'\gamma\Delta z$

5.12 $F_B = c'\forall\rho g$

5.14 $\mathbf{M} = f\left(\frac{V}{\sqrt{p/\rho}}, k\right)$

5.18 3.162 m/s

5.20 $\gamma H^4 f\left(\frac{\omega H^3}{Q}, e\right)$

5.24 $\alpha t/x^2$; $h_c x/k$; $x^2 q_H/kT$

5.26 $kL_{ref}/\rho\mathcal{D}$; $uL_{ref}/\mathcal{D}$; $\mu/\rho\mathcal{D}$

5.28 $DT_*/Dt_* = \frac{1}{\mathbf{Pe}}\nabla^2 T_* + (L/U_m T_m)S$

5.30 $0.6\,\mathbf{Re_L}^{-1/2}x_*^{-1/2}$; $1.2\,\mathbf{Re_L}^{-1/2}$; $x_* = x/L$

5.34 no; $C_A(z,t)/C_{Ao} = \alpha\,e^{-z^2/D_z t}$

5.36 $\mathbf{Re} = 7.4(10^6)$; $\mathbf{Pr} = 5.98$; $(\mathbf{Re}\cdot\mathbf{Pr})^{1/2} = 6652$

5.38 $\rho V^2 D^2 f'(\mathbf{R}, \mathbf{M})$

5.40 0.195

5.42 Choose model size $\frac{1}{75}$ or less of prototype size; $loss_p = loss_m\left(\frac{D_m \nu_p}{D_p \nu_m}\right)$

5.44 g, d, yes

5.46 $\omega D^{3/2}\rho_f^{1/2}/\sigma^{1/2}$

5.48 36.98 m/s, 18.59 m³/s; losses the same when expressed in velocity heads

Chapter 6

6.2 $2\mu U/a^2$, $Ua/3$

6.4 $1.636(10^{-8})$ m³/s, 0.59 N

6.6 $Q = \frac{(p_2-p_1)}{2\mu l}\frac{2}{3}a^3 - Va + \frac{U+V}{2}a$

6.8 $\alpha = 1.543$, $\beta = 1.20$

6.12 2.0 mm

6.14 $U = -0.064$ m/s (upward)

6.16 $2\rho U^2 a/15$, $0.02857\rho U^3 a$

6.18 $\frac{4}{3}$

6.20 $0.707r_0$

6.24 0.0412 lb/ft²

6.26 $4.21(10^{-3})$ ft³/s, 122.6

6.28 0.00152 ft³/s

6.30 8 m

6.32 93 mm

6.34 $6.5(10^{-4})$ N s/m²

6.36 $\frac{7}{6}\kappa\frac{y}{r_0}$

6.38 0.223

6.42 2.39 m/s

6.44 0.004326

6.46 41.97 m³/s

6.48 0.000482

6.50 234 ft³/s

6.52 10.11 ft/s

6.54 $Q \sim y^{8/3}$

6.56 1.997 ft

6.58 0.538 a

6.60 0.016

6.62 3916 m

6.64 50 ft

6.66 0.005 ft

6.70 0.0146

6.72 0.34 in.

6.74 yes, 1.096 MW

6.76 1.121 MPa

6.78 15.6 hp

6.80 0.054 L/s

6.82 1456 m³/min

6.84 86.2 kW

6.86 26.2 N/s

6.88 0.654 m

6.90 $62,196/km

6.92 0.2117 m

6.100 0.52 m

6.102 0.306 m³/s

6.104 29.73 ft

6.106 0.59 m

6.108 37.91 ft

6.110 12.05 L/s

Chapter 7

7.6 0.332 m

7.8 0.90°

7.10 1238.4 lb, 528.4 hp

7.12 4 days + 11.7 h

7.14 1.292 m/s

7.16 8.92 ft

7.18 22.31 ft

7.20 Power$_{cups}$ = 87.5 W, Power$_{rods}$ = 11.84 W, Total Power = 99.38 W

7.22 $F = 51.1$ kN, $M = 1406$ kNm

7.24 7.59 ft/s, 11.78 ft/s

7.26 1.604 N·s/m²

7.28 8.57 N

7.30 21.25

7.32 3.59 ft

7.34 11,820 rpm

7.36 115 m, 116 m, 113m

Chapter 8

8.2 0, 0, 0

8.4 $\omega_x = 1.5, \omega_y = -2, \omega_z = -0.5$

8.6 $\omega = -2z(x + y)$

8.8 $\phi = -4x + \frac{7}{2}(x^2 - y^2) - 6y + c$

8.12 $\psi = \theta + c$

8.14 $\phi = 18x + c$

8.16 $\frac{\partial \phi}{\partial r}\big|_{r=a} = 0, -\frac{\partial \phi}{\partial r}\big|_{r=\infty} = V_r,$
$-\frac{1}{r}\frac{\partial \phi}{\partial \theta}\big|_{r=\infty} = V_\theta$

8.18 3790 lb

8.22 $\phi = -\frac{\mu}{2}\ln\{[(x - 1)^2 + y^2][(x + 1)^2 + y^2]\}$

8.26 Half body

8.30 80 m³/s

8.38 470 mi/hr, or 695 ft/s

8.40 19.0 ft/s; 76.2 ft; 0.0825 ft^{-1}; 1.57 s^{-1}

8.42 $u_s = \frac{gHT}{L}\frac{\cosh[k(z+d)]}{\cos(kd)}\cos(kx)\cos(\omega t)$

Chapter 9

9.2 415,800 Btu/hr; from the roof to the air

9.4 40.47 W/m²

9.6 12.84 W/m²·K

9.8 (a) $-2.45°C$; (b) 50.37 W/m²

9.10 490.5 kW

9.12 (a) 721.97 W/m; (b) 234.06 W/m; (c) 67.58%

9.14 $q = \dfrac{4\pi(T_1 - T_3)}{\frac{1}{k_s}\left(\frac{1}{R_1} - \frac{1}{R_2}\right) + \frac{1}{k_3}\left(\frac{1}{R_2} - \frac{1}{R_3}\right)}$

9.16 $T(t) = -386.96 + 19t + 428.96e^{-0.0387t}$ °F

9.18 0.0673 kg/s

9.20 0.74

9.22 $2.18(10^{-4})$

9.24 (a) 0.00197; (b) 0.00394; (c) 0.00975 N;
(d) 6.088 W/m²· K; (e) 12.176 W/m²· K;
(f) 153.42 W

9.26 1080.4 m

9.28 $\frac{k_0}{L}[(T_0 - T_L) + \frac{\alpha_1}{2}(T_0^2 - T_L^2) + \frac{\alpha_2}{3}(T_0^3 - T_L^3)]$

9.30 (a) glass fiber; (b) 5.5 Btu/hr·ft²

9.32 3.7 in. or 0.3083 ft

9.34 3.97%; 20.71%; 29.73%

9.38 $0.288\left[\dfrac{\mathbf{Pr}}{1 - \left(\frac{L}{x}\right)^{3/4}}\right]^{1/3}\mathbf{Re_x}^{1/2}$

9.40 $0.327\mathbf{Re_x}^{1/2}\mathbf{Pr}^{1/3}$

9.42 (a) 0.14; 0.05; (b) 0.090;
(c) 950.95 kPa; 172.9 kPa; 61.75 kPa; 49.6 kPa

9.48 (a) 0.52
(b) 2.321 kg/m³; 0.9323 kg/m³; 1.389 kg/m³
(c) 60.7 mole/m³; 24.4 mole/m³; 36.32 mole/m³
(d) 0.1762 m/s; 0.01244 m³/s
(e) 1.313 mole/m²·s; -1.315 mole/m²·s

9.50 2.352 g

9.52 $-4.0523(10^{-5})$ lb·mole/hr

9.54 $\xi^3 = 1/\mathbf{Sc}; \xi = \delta_c/\delta$

9.64 $\xi^3[1.62 - 0.12\xi^2] = 1/\mathbf{Sc}; \xi = \delta_c/\delta$

9.66 no

9.72 140.6 ft²/s, or 13.06 m²/s

9.76 $(C_s - C)/(C_s - C_0) = \text{erf}(z/\sqrt{4Dt})$

9.78 6.8 m

9.80 $\bar{E}_z = 0.0371$ m²/s; $\bar{E}_y = 0.0831$ m²/s;
48.17 hr; 97,110.7 m

9.82 (a) 26.30 mg/L; (b) 91.34 km; (c) 22.74 °C

9.84 273.82 ft; 6,133.57 ft³/s

9.86 3.91 hr; 7.03 km

9.88 $C L_m(E_y/K)^{1/2}$
$C = 0.406$, centerline; $C = 0.203$, side

Chapter 10

10.2 $\bar{u} = 11.06$ cm/s; $\sigma = 0.06$ cm/s;
$e_{\bar{u}} = 0.02$ cm/s; $e_{rms} = 0.31$ cm/s

10.4 42.67 mm

10.6 0.863 m/s

10.10 1.323

10.12 40.54 ft³/s

10.14 2.336 L/min

10.16 28.42 gpm

10.18 $y = 0.0452x^2$

10.20 $y = H\cos^2\alpha$

10.24 $C_d = 0.773, C_v = 0.977, C_c = 0.791$

10.26 0.287 m·N/N, 62.4 N·m/s

10.28 0.273 J/N, 163.3 W

10.30 $C_v = 0.8847, C_c = 0.7762, C_d = 0.687$

10.32 33.246 s

10.34 0.827

10.36 1384 s

10.38 61.44 s

10.42 61.36 kPa

10.44 1.329 kg/s

10.46 1.389 kg/s

10.48 42.93 L/s

10.50 260.3 m³/min

10.52 0.655 m³/s

10.54 1.94 m

10.56 $Q = 1.36H^{2.48}$

10.58 0.687 m

10.60 S_1 : 0.92ϕ(0.53 mm) 0.95ϕ(0.52 mm) $\times$
     0.36ϕ(0.78 mm) 0.08
     S_2 : 0.85ϕ(0.55 mm) 0.92ϕ(0.53 mm) $\times$
     0.34ϕ(0.79 mm)0.21

10.62 0.0069 slug/ft·s = 3.31 poise

Chapter 11

11.2 $Q_c = (Q/N)n, H_c = (H/N^2)n^2,$
     c = corrected, n = const speed

11.4 Synchronization not exact

11.6 $Q = 0.07049Q_1, H = 2.822H_1$

11.8 1900 (based on watts)

11.10 89 in., 300 rpm

11.12 centrifugal pump

11.14 4.81 m

11.16 13.04°

11.18 19.1 m/s, 57.3 m/s

11.20 36.71 m

11.22 93.18%

11.24 4.12 cfs, 33.17°, 93.93 ft,
     79.84 ft, 14.09 ft·lb/lb, 47.85 hp

11.26 43.68°, 1146 rpm, 36.3 hp, 17.15 psi

11.28 30.51°, 1238.2 rpm

11.30 $\beta_1 = 9.51°, \beta_2 = 25.31°$, 5.305-in. H₂O, 8.36 hp

11.32 229.87 m³/min

11.34 553 mm, 750 rpm

11.36 0.1604

11.38 6.18 m, 4.9 m

Chapter 12

12.2 $V_1 = 6.971$ m/s, $V_2 = 1.746$ m/s

12.4 0.1026 m³/s

12.6 0.0822 m

12.8 2.57 cfs

12.10 13.62 L/s, 23.47 kPa, -26.48 kPa

12.12 30.47 ft above left reservoir surface

12.14 5.6 ft³/s

12.16 H: 50 ft, Q: 7.99 cfs, e: 78.3%

12.18 45.68 kW

12.20 44.1 kW

12.22 1.627 cfs

12.24 2.005 cfs

12.26 19.66 m

12.28 13.42 m

12.30 1.685 ft, 2.049 ft, 8.76 cfs

12.32 2.046 m³/s

12.34 56.91 ft

12.36 6.53, 26.4, 51.1, 84.0 L/s

12.38 $Q_1 = 1.018$ L/s, $Q_2 = 8.44$ L/s, $Q_3 = 9.46$ L/s

12.40 357.8 m, 159.3 L/s

12.42 $Q_A = 31.69$ L/s, $Q_B = 46.23$ L/s, $Q_C = 77.84$ L/s

12.44 $Q_A = 9.81$ L/s, $Q_B = 39.12$ L/s

12.46 $Q_A = 15.64$ L/s, $Q_B = 45$ L/s, $h_j = 32.02$ m

12.48 12.53 m, 3.44 kW

12.50 $Q_B = 0.2375$ L/s, $Q_C = 0.368$ L/s,
     $Q_D = 0.336$ L/s, $h_{j1} = 94.55$ m,
     $h_{j2} = 94.555$ m

12.52 $D_P = 47$ cm, $D_B = 36$ cm, $D_C = 25$ cm,
     $H_P = 96.98$ m, 400 rpm

12.54 58.51, 41.49, 2.36, 31.15, 43.85

12.58 0.392

12.60 5.59 L/s

12.62 2.044 m

12.72 0.5314 m/s, 1.419 s

12.74 $z = V_o t e^{-mt}$

12.76 14.682 ft, 14.371 ft

12.80 12.9 s

12.82 3509 ft/s

12.84 2 MPa

12.86 $\frac{2}{3}$

12.88 $h_{max} = 98.44$ at 3.0 s

12.90 time of closure = 7.0 s

Chapter 13

13.2 0.677 mm

13.4 56.35 m³/s, 1.63, 1.707 m

13.6 175.1 m²/100 m

13.8 $m = 0.686, y = 7.775$ ft

13.10 $b = 3.6103$ m, $m = 1/\sqrt{3}$

13.12 0.000173

13.18 1.987, 1.667 m

13.20 0.0000583

13.22 0.56 m

13.24 0.185 m

13.26 0.227 m

13.28 1.0196 m, 1.0512 m

13.32 92.0 cfs

13.34 7.39 ft

13.36 0.86 m, 8.289 m^2/s

13.48 2.421 m

13.50 $c = 2.924$ m/s, 0.877 m^2/s

13.52 $y = 0.01133(\frac{x}{t} + 4.1904)^2$

13.54 $y = 0.0133(\frac{x}{3} + 4.1904)^2$ for 3 seconds

Chapter 14

14.2 $\frac{d^2}{18\mu}(\rho_s - \rho)g$

14.4 $\partial \bar{c}/\partial t + \nabla \cdot (\bar{u}\bar{c}) - \partial(w_t\bar{c})/\partial z = \nabla \cdot [(\mathcal{D}_{sw} + E_s)\nabla\bar{c}]$

14.6 939 g/L; 882 g/L

14.8 u_* (m/s) : 0.01 0.02 0.04 0.08 0.16 0.32 0.64 1.0
z_0 (mm) : 1.2 1.2 1.2 1.2 1.2 1.2001 1.2007 1.2059
C_0 (g/L) : 0 0.73 15.19 76.83 260.66 1240.68
19,351 97,923

14.24 126 particles/mL; 0.347 hr^{-1}; 6.64 hr

14.26 $9t_r$

14.28 0.00178 s^{-1}

14.30 $1/C_0 - 1/C = k_2t; t = \frac{1}{2} k_2 C_0$

14.32 5.56 m^3/s

INDEX